11th Conference on Electrorheological Fluids and Magnetorheological Suspensions 2008

Journal of Physics: Conference Series Volume 149

Dresden, Germany
25-29 August 2008

ISBN: 978-1-60560-995-9

11th Conference on Electrorheological Fluids and Magnetorheological Suspensions 2008

Journal of Physics: Conference Series Volume 149

Dresden, Germany
25-29 August 2008

ISBN: 978-1-60560-995-9

IOP Institute of Physics

Printed from e-media with permission by:

Curran Associates, Inc.
57 Morehouse Lane
Red Hook, NY 12571
www.proceedings.com

Some format issues inherent in the e-media version may also appear in this print version.

Printed by Curran Associates, Inc. (2009)

Institute of Physics
Dirac House, Temple Back
Bristol BS1 6BE UK

Phone: 44 1 17 929 7481
Fax: 44 1 17 920 0979

techtracking@iop.org

TABLE OF CONTENTS

PREFACE

11th International Conference on Electrorheological Fluids and Magnetorheological Suspensions

Stefan Odenbach *et al* 2009 *J. Phys.: Conf. Ser.* **149** 011001 (1p) doi: 10.1088/1742-6596/149/1/011001 Help

Stefan Odenbach and Dmitry Borin

Materials with properties controllable by external fields become more and more important for modern product design. Magnetorheological suspensions, electrorheological fluids and ferrofluids are classical examples for such smart materials exhibiting magnetic or electric field dependent properties. Their development over the past 60 years has shown how complex research and development of fluids with tailored properties can be. The demands of potential applications are the driving force for the synthesis of new fluids. The changes in the synthetic process and the composition of the fluids based on an increasing understanding of the relevant microscopic processes leads to certain macroscopic properties of the material. Only intense cooperation between basic and application oriented research, synthesis, characterisation, theory and application design can finally lead to fluids suitable for the envisaged use.

The International Conferences on Electrorheological Fluids and Magnetorheological Suspensions have provided over more than 20 years a platform for interdisciplinary discussions strengthening the scientific progress in the field. Besides being a forum for scientific exchange about recent developments in electro- and magnetorheological fluids. The 11th International Conference on Electrorheological Fluids and Magnetorheological Suspensions (ERMR08) which has been organized by the chair of Magnetofluiddynamics at the Technische Universität Dresden in August 2008 in Dresden has continued this fruitful tradition.

With more than 180 participants from 24 different countries it has been the largest ERMR meeting during the last decade – a tendency showing the high potential and promising development of the field of electrically and magnetically controllable fluids. A significant proportion of the participants were PhD students, a fact that also highlights the sustainability of the field. In total 85 oral presentations – including 8 plenary talks – and 81 posters were presented at a high scientific level. The scientific program focused primarily on the classical topics in basic and applied research with electrorheological fluids, magnetorheological suspensions and ferrofluids. But among the program highlights were also magnetorheological elastomers, which have been included into the conference program as a special topic in separate sessions for the first time.

The present proceedings include 116 peer reviewed papers covering all conferences topics. In contrast to former ERMR proceedings issues we've decided to subdivide this issue into four thematic sections: Electrorheological Fluids, Magnetorheological Suspensions, Ferrofluids and Magnetorheological Elastomers – and to sort the contributions inside these sections alphabetically by the name of the first author. The reason for this sorting has been the fact that more and more contributions have a clear interdisciplinary character and cannot be assigned to a certain subtopic. The high amount of interdisciplinary papers is indeed a very positive and forward-looking development in this research field, strengthening the future development.

The organization of a large international conference requires the hard effort of many individuals and we would like to take this opportunity to thank everybody involved in this project. We would also like to thank the members of the international advisory and program committees, the plenary speakers, the sponsors and of course all participants for the very successful conference.

Stefan Odenbach and Dmitry Borin
Technische Universität Dresden, Germany
Editors

Journal of Physics: Conference Series **149** (2009) 012001 doi:10.1088/1742-6596/149/1/012001

Particle Shape Dependence of Polar Molecule Dominated Electrorheological Effect

W Bao[1], X F Wu[1], J G Cao[1], L W Zhou[1], Y Gao[1], J P Huang[1], Z J Yang[2], N Ren[2] and Y Tang[2]

[1] *Surface Physics Laboratory (National Key Laboratory) and Department of Physics, Fudan University, Shanghai 200433, China*
[2] *Department of Chemistry, Fudan University, Shanghai 200433, China*

E-mail: lwzhou@fudan.edu.cn

Abstract. We investigated the relations between the yield stress of polar-molecule dominated electrorheological (PM-ER) fluids and the shape of ellipsoidal particles of the fluids. It is found that, when the short axis of an ellipsoidal particle is along an external electric field while the volume of the particle keeps constant, the shorter of the short axis is, the stronger the attraction between the two adjacent particles would be. The equipment used for yield stress determination, the experiment and calculation of the condition under which the sample are short-axis chained, the detailed process of sample preparation and sample characterization will be described and discussed in an extended form of this paper.

1. Introduction

In the PM-ER model, the interaction between charges and polar molecules adsorbed on particle surface is a key to the PM-ER effect.[1] In this work, we will discuss the relations between the shape of ellipsoidal dielectric particles and their attractive force according to the PM-ER theory. The result indicates that the yield stress increases by changing the shape of particle from spherical to ellipsoidal.

2. Theory

We will first discuss the interaction between charges and polar molecules adsorbed on particle surfaces. According to PM-ER theory [1], the polar molecules are relatively free for rotation and potentially to be oriented in the direction of a strong electric field.[2,3] The attractive force density $f_{\mathrm{m-e}}$ of molecule-charge can be expressed as $f_{m-e} = e\mu\left(2\pi\varepsilon_0\varepsilon_r d_{m-e}^3\right)^{-1}$.When two or more polar molecules, which are adsorbed on particle surface, exist in the gap between two adjacent particles, the attractive force $F_{m-e}=Nf_{m-e}$, where N is the number of active polar-molecule attractions in the gap. $N=S\rho_{\mathrm{m}}n(E)$, where ρ_m is the density of the polar molecules adsorbed on the particles, the probability $n(E)$ of the aligning polar molecules being nearly proportional to μE_{loc}[4]. When the dielectric particle is spherical, the effective area S is nearly $\pi r d_{m-e}$,[1] where d_{m-e} is the distance from molecule centre to charge, and the molecule size $d\approx 2\, d_{m-e}$.

When the particle is ellipsoidal, the effective area S can be calculated by following process.

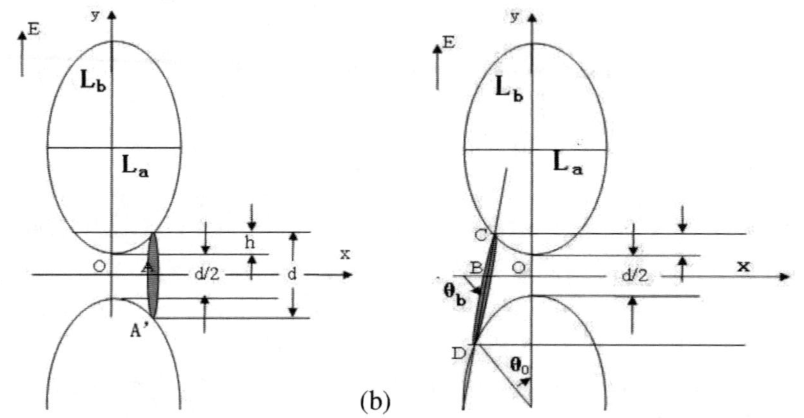

Figure 1 Ellipsoid particles and a polar molecule parallel to E (a) and not parallel to E (b).

Polar molecules rotate under external electric field E and turn to the external field direction, as shown in figure 1(a). The molecule size d=0.4nm, which is approximately a double of the distance $d_{m\text{-}e}$ from molecule center to charge. Suppose the gap between two particles is $d/2$, so $h=d/4$. The length of semi-axis along x is L_a, and that along y is L_b. The Langevin function [4] can be expressed as $L(a)=\overline{\cos\theta}= a/3-a^3/45+2a^5/945-2a^7/9450+\ldots$, where $a=\mu E/kT$. When a is not very large, $L(a)\approx a/3$, or $L(a)=\mu E_{loc}/3kT$.

If polar molecules orientate along external field direction, $\theta=0$, one has $L(a)$=1, $E_{loc}=3kT/\mu\approx$ 186 kV/mm when the dipole moment μ=2D.

In figure 1(b), the angle between the polar molecule and the external field direction is θ. When L_a and L_b are constants, the elliptic equations of two particles are expressed as

$$x_1^2/L_a^2+(y_1-L_b-d/4)^2/L_b^2=1, \tag{1}$$

$$x_2^2/L_a^2+(y_2+L_b+d/4)^2/L_b^2=1. \tag{2}$$

The slope s of line CD can be calculated by coordinates of the point C(x_1, y_1) and D(x_2, y_2), $(y_1-y_2)=s(x_1-x_2)$ The distance d between C and D is polar molecule size, $(x_1-x_2)^2+(y_1-y_2)^2=d^2$. Solving above equations, we can get the value of x_1, x_2, y_1 and y_2. When polar molecules turn to the external field direction and the molecules touch both two adjacent particles as shown in figure 1(a), coordinate of contact point A'($x_{A'}$, $y_{A'}$) on the lower particle can be expressed as $x_{A'}=L_a\left[1-(y_{A'}+L_b+d/4)^2/L_b^2\right]^{1/2}$, and $y_{A'}=-d/2$.

The distribution of local electric field calculated with finite element analysis shows a sharp peak with peak value of about 200 kV/mm when the external electric field E=2kV/mm and L_a=15nm. E_{locA}, the local electric field at centre point A of the polar molecule parallel to the external field can be obtained through local electric field distribution. E_{locA} is close to $E_{locA'}$.The Langevin function $L(a)=\overline{\cos\theta}\approx\mu E_{loc}/3kT$, where k is Boltzmann constant, T=300K, μ=2D. It is supposed that the polar molecule does not align along the external field direction as shown in figure 1(b) and x_B is slightly smaller than x_A .We can also get the value of E_{locB} through the local electric field distribution. The value of slope s is tan(θ_B), so $\theta_B=\arccos(\mu E_{locB}/3kT)$.

Define d_B as the distance between two intersections of two ellipses with a line which is passing through point B. In figure 1(b),

$$d_B=CD=(x_1-x_2)(\cos\theta_B)^{-1}=3kT(x_1-x_2)(\mu E_{locB})^{-1} \tag{3}$$

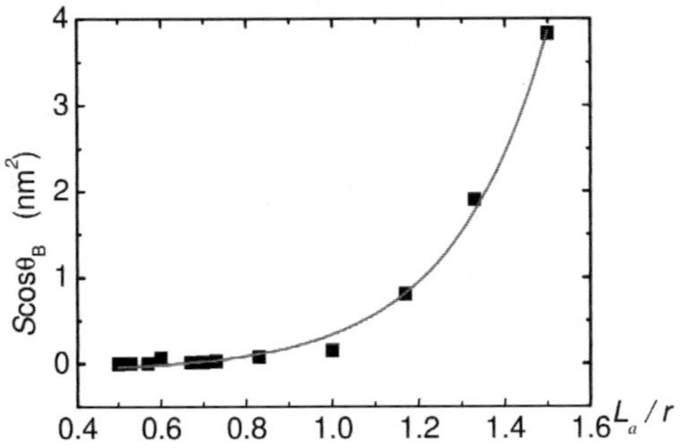

Figure 2 Effective area which is proportional to attractive force vs. axis length perpendicular to E.

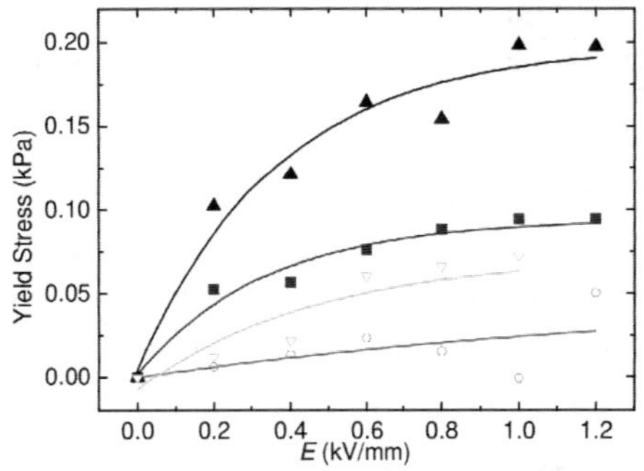

Figure 3 The yield stress of ellipsoidal particles versus electric field (see text for notations).

The value of x_B can be changed slightly and CD=d_B in figure 1b can be closer to the molecule size d. When $|d_B - d|/d \le 10^{-3}$, the polar molecule is thought to be adsorbed on two adjacent particles. The effective contact area S is an area of a particle on which all polar molecules are adsorbed on two adjacent particles respectively under an electric field. The effective contact areas $S_<$ (for $L_a<L_b$), $S_=$ (for $L_a=L_b$) and $S_>$ (for $L_a>L_b$) can be calculated as follows.

$$S_< = \pi\int_0^{\theta_0} L_a \sin\theta\left[(-L_b\sin\theta)^2 + (L_a\cos\theta)^2\right]^{1/2} d\theta$$

$$= \pi L_a L_b\left[(1-\varepsilon^2)^{1/2} + \varepsilon^{-1}\arcsin\varepsilon - \cos\theta_0\left(1-\varepsilon^2\cos^2\theta_0\right)^{1/2} - \varepsilon^{-1}\arcsin(\varepsilon\cos\theta_0)\right]; \quad (4)$$

$$S_= = \pi L_a L_b \int_0^{\theta_0}\sin\theta d\theta = \pi L_a L_b (1-\cos\theta_0); \quad (5)$$

$$S_> = \pi\int_0^{\theta_0} L_a \sin\theta\left[(-L_b\sin\theta)^2 + (L_a\cos\theta)^2\right]^{1/2} d\theta$$

$$= \pi L_a L_b\left\{(\varepsilon^2+1)^{1/2} + \varepsilon^{-1}\ln\left[\varepsilon + (\varepsilon^2+1)^{1/2}\right] - \cos\theta_0\left(\varepsilon^2\cos^2\theta_0+1\right)^{1/2}\right\} \quad (6)$$

$$-\ln\left[\cos\theta_0 + \varepsilon^{-1}\left(1+\varepsilon^2\cos^2\theta_0\right)^{1/2}\right],$$

where $\varepsilon = \left[(L_b^2 - L_a^2)/L_b^2\right]^{1/2}$, and $\theta_0 = \arctan\left[-x_2\left(L_b + d/4 + x_2\tan\theta_B\right)^{-1}\right]$.

Effective area S can be calculated through an integral and expressed analytically. E_{locB} is the local field around the edge of effective area on the plane across the center point of the particle gap. This plane is perpendicular to the external field direction. N is the number of active polar-molecule interactions in gap, and $N=S\rho_m n(E)$. If the density of polar molecule adsorbed on particle surface keeps constant, N is proportional to $Sn(E)$.

3. Results and discussions

The particle-particle attractive force F can be expressed as $F=Nf_{m-e}$, $F\propto Sn(E)\propto S\cos\theta_B$, so we get figure 2, where r is the diameter of spherical particle when the volume of a particle keeps constant.

As shown in figure 2, the effective contact area, which is proportional to the attractive interaction of two ellipsoid particles, is simulated when L_a/r=1.5, and it is about 25 times larger than that of two spherical ones.

If the ellipsoid particles are soft and the contact surface of particles can be flattened, the effective area increases, and so is the yield stress. If we can get the image of particles and the density of polar molecule adsorbed on particle surface through experiment, an exact model is found through a finite element analysis and a definite attractive force will be presented.

This modeling approach has many advantages. Finite element analysis can deal with non-sphere particles such as ellipsoid. The model can be easily integrated into complex shape particle simulation. Multi-field including magnetic field and optical field is allowed in finite element analysis. Finally, the model can be extended to hundreds or more particles simulation.

Ellipsoidal dielectric particles will orient by its longest dimension along the external field direction (z-direction) because z component of dielectric constant is larger than x or y component. If x or y component of dielectric constant is larger than z component, ellipsoidal dielectric particles will align along short axis under external electric field so as to larger yield stress.

4. Experimental Section

We prepared submicrometer-sized mesoporous silica SBA-15 with cuboidlike morphology. The channels of the material run parallel to the short axis of the cuboidlike SBA-15. A ferroelectric material such as potassium dihydrogen phosphate (KH_2PO_4) with large dielectric constant was interpolated in the channels. The channels of the material run parallel to short axis.

The yield stress of several cuboid-like SBA-15ER fluids (solid volume fraction is about 25%) versus E is shown in figure 3. Solid upper triangles (or solid squares) denote the yield stress of the elliptic plate sample doped with KH_2PO_4 and adsorbed with (or without) polar molecules; open lower triangles (or open circles) denote that for samples without KH_2PO_4 but adsorbed with (or without) polar molecules. Solid curves are the fitting for the corresponding data points. The yield stress of ER fluid consisting of the KH_2PO_4-doped material with coating of polar molecules in silicon oil has highest yield stress. This is because that adding ferroelectric materials in the channels parallel to the short axis would make the particles contact to each other with their short axis parallel to the external electric field, and this would increase the effective area, as well as the yield stress. The calculation of the condition under which the sample are short-axis chained, as well as the detailed process of sample preparation and sample characterization will be described and discussed elsewhere.

5. Conclusion

The shape of particles suspended in a liquid has a direct effect on ER effects. Effective area on particle surface plays a crucial role in PM-ER fluids. The ellipsoid particles with short axis parallel to the external field direction have much larger effective area and attractive force than spherical ones when the volume of one particle keeps constant. The yield stress will increase by 25 times when the length of the axis perpendicular to field direction is 1.5 times longer than the sphere radius according to our simulation. Adding ferroelectric materials in the channels parallel to the short axis would make the short axis parallel to the field direction, and the PM-ER fluid consisting of such ellipsoidal particles should have higher yield stress in comparison with spherical ones.

Acknowledgement

This work is supported by the National Natural Science Foundation of China under Grant Nos. 10334020, 10574027 and 10604014. J.P.H. acknowledges the financial supports by CNKBRSF under Grant No. 2006CB921706, the Pujiang Talent Project (No. 06PJ14006) of STCSM and the Shu Guang project. The authors thank Prof. Rongjia Tao for stimulating and fruitful discussions.

References

[1] Lu K Q, Shen R, Wang X Z, Sun G, Wen W J, and Liu J X, *Chin. Phys.* **15** (2006) 2476.
[2] Shih Y T, Chuu D S and Mei W N 1996 *Solid State Commun.* **99** 819.
[3] Shih Y T, Liao Y Y and Chuu D S 2003 *Phys. Rev. B* **68** 075402.
[4] Bottcher C J F 1973 *Theory of Electric Polarization* (Amsterdam:Elsevier) p. 163.
[5] Zhang H, Sun J M, Ma D, Bao X H, R. Schlogl. *J. Am. Chem. Soc.* **126**(2004) 7440-7441.
[6] Zhao DY, Huo QS, Feng JL, Chmelka BF, Stuck GD., *J. Am. Chem. Soc.* **126**(1998) 6024-6036.

Journal of Physics: Conference Series **149** (2009) 012002 doi:10.1088/1742-6596/149/1/012002

Development of 3 DOF Manipulator Using ER Fluid Clutches for Reduction of Collision Force

Kazuhiko Boku and Taro Nakamura

Chuo University, Faculty of Science and Engineering, Department of Precision Mechanics, 1-13-27 Kasuga, Bunkyo-ku, Tokyo 112-8551, Japan

k_boku@bio.mech.chuo-u.ac.jp

Abstract.With robots and users more commonly sharing space such as in the fields of medicine and home automation, the possibility of a physical collision has increased, even though many robots use actuators with high-ratio gear trains to minimize the effects of impact. We developed a 3-DOF manipulator having a smart flexible joint using an ER fluid and a sensor-equipped pneumatic cushion. Results of position control and collision experiments using the manipulator demonstrated its effectiveness.

1. Introduction

Robots currently used in industrial applications are also expected to be used in homes and offices [1] alongside users. This means that robots coexist with users. Conventional safety strategies used for industrial robots cannot be applied to symbiotic robots working with users; therefore, safety strategies in coexisting space must be comprehensively studied. One critical consideration is user safety in an unexpected collision, not to mention damage to the robot. Approaches to collision avoidance contributing to user safety can be briefly summarized as follows: Collision can be suppressed by minimizing the robot's effective arm mass with redundant degrees of freedom [2] or by specifying initial parameters, such as approach speed and force control gain. [3],[4] Sensor-guided control using optical, ultrasonic, and force/torque sensors on a robot's wrists, for example, [5],[6] may pose problems with dead angles and disturbances, and thus, it is difficult to ensure optimum safety using these approaches. Many robot motors use mechanisms with a high reduction ratio and cannot sufficiently ensure safety in a collision. [7],[8],[9] Thus, mechanical flexibility must be added to the robot's functions.

This study describes a manipulator we developed, whose joints and links have lower viscoelasticities than conventional models and whose rigid joints are controlled to a high degree of accuracy so that it can minimize collisions, and when it does collide with an object in the external environment (e.g., a human), the manipulator can ease the impact.

2. Soft Manipulator

We developed a three-link soft manipulator (Figure 1) that consists of three actuator-driven joints with a clutch on each joint, which uses an electrorheological (ER) fluid. [10],[11] The rotation axis is fixed by giving a viscous parameter to the ER clutch, and motor rotation is transmitted to the link. In this case, the motor's rotation is transmitted to each pulley of joints by a timing belt.

A pneumatic cushion sensor on all links, [12] which provides a soft touch, operates while a pneumatic sensor encountering the object in the external environment exerts an appropriate initial pressure against the pneumatic cushion (Figure 2).

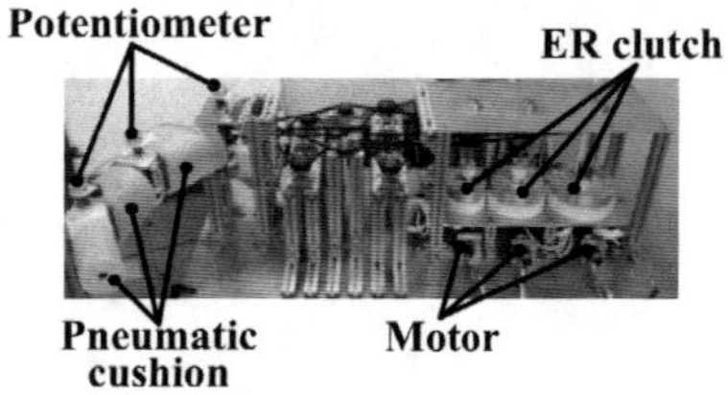

Figure 1. Experimental apparatus of the soft manipulator

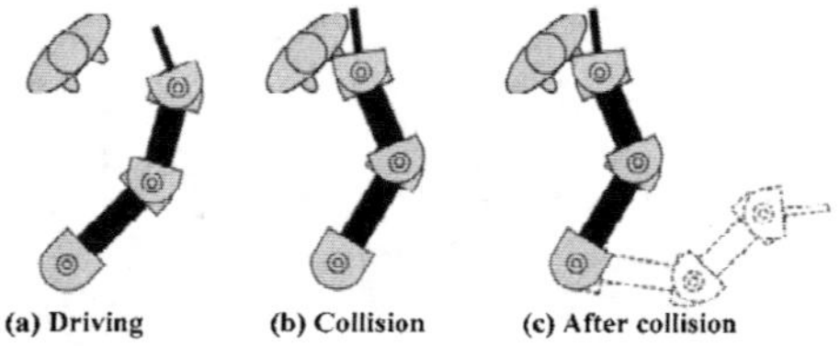

Figure 2. Motion patterns of the soft manipulator

First, when the manipulator is driving, its behavior can achieve a highly accurate position control due to the provision of appropriate viscosity to the ER clutches. Next, when the manipulator collides with the external object (in this case, a human), the contact force can be decreased by reducing the flexibility of the buffer material and the viscous parameter of the joint, which minimizes the effect of contact and eases the collision force on the external object. Finally, it is the behavior of the manipulator that controls the reaction when the manipulator collides, evades the external object, and returns to trajectory after collision.

3. Pulley Layout

Each joint of the manipulator is driven by the timing belt. Figure 3 shows the pulley layout of each joint. Each joint layout consists of three layers. The direction of the belt is controlled by the idler pulley, which is not seen in Figure 3.

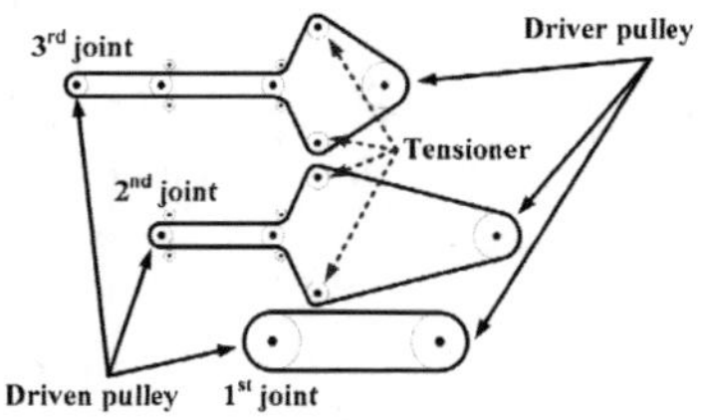

Figure 3. Pulley layout

4. ER Clutch

A decentralized ER fluid is used for the clutch's operational fluid. The viscosity of this fluid changes with the application of an electrical field. The fluid used in the clutch function in the joint has the following two advantages: [11],[13]

- Because fluid viscosity changes instantaneously, the joint responds more quickly to a collision than a conventional electromagnetic clutch.
- The clutch is lightweight and easily miniaturized because it uses only one electrode, suppressing the influence of manipulator inertia.

The torque required for the ER clutches must be calculated. Because the torque of ER clutches must be considered in the manipulator's function, a value that compensates for the torque of each joint by dynamics is preferable. Figure 4 shows a schematic diagram of the ER clutch in this study. As shown in this figure, the sheets in the disk of the ER clutch were assumed to be composed of four multiple-layer structures. Therefore, because the electrode's surface area was increased, miniaturization and high torque output could be achieved. Moreover, both the minus and plus electrodes are integrated within a signal housing installed directly along the I/O axis. In addition, the use of a cover diminishes the chance of an electric shock caused by the high voltage applied to the ER fluid. Figure 5 shows the relationship between torque and the electric field. This figure shows that torque increases when the applied electric field increases. Table 1 shows the specifications of the ER clutch in this study.

5. Pneumatic Cushion Sensor

The pneumatic cushion sensor uses the compressibility (elasticity) of internal air in a midair cylinder device as a buffer. In addition, by installing a pressure sensor in the air ingate of the buffer material

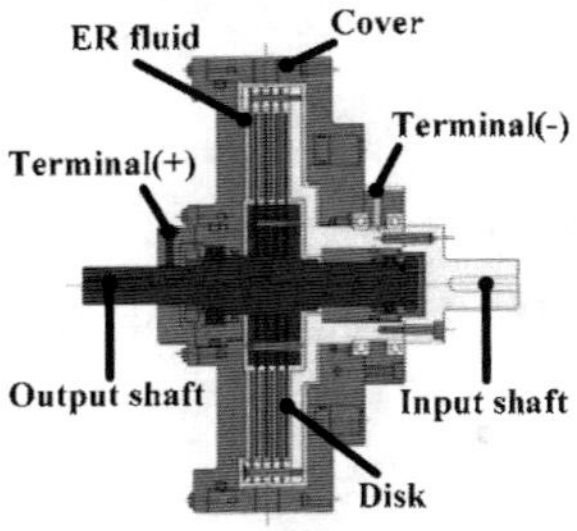

Figure 4. Configuration of the ER clutch

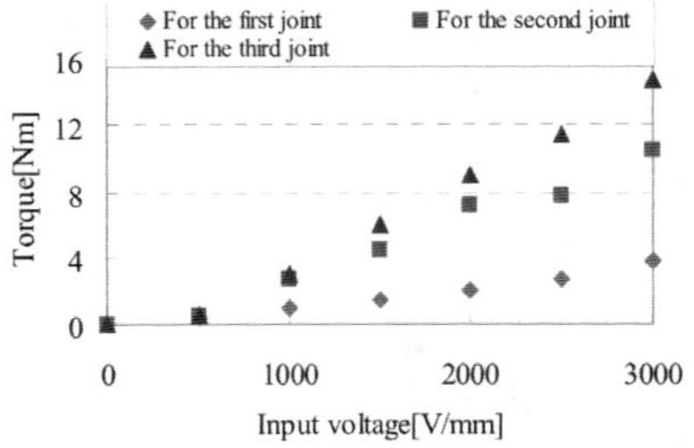

Figure 5. Characteristics of the ER clutch

Table 1. Specification of the ER clutch

Joint number	1st	2nd	3rd
Torque [Nm]	12	7	2
Diameter [mm]	215	185	132
Weight [kg]	8.2	6.1	3
Outside diameter of the disk [mm]	170	140	90
Inside diameter of the disk [mm]	90	70	54
Number of sheets of the disk [disk]	4	4	4
Thickness of disk [mm]	2	2	2

and measuring pressure changes, the pneumatic cushion is used as a sensor that has contact detection capability over a wide range.

It is necessary that the material of the pneumatic cushion has a low degree of elasticity and plasticity, because the buffer material used in this study may also function as a sensor. Thus, this study focuses on a polyurethane pneumatic cushion because of its strong tensile strength.

A shape that would cover the link of the manipulator was designed. Figure 6 shows a schematic diagram of the pneumatic cushion sensor (for the third link). There is a hole for the air enclosing the pneumatic cushion where a tube can be installed.

Figure 7 shows the relationship between the pressure and contact force, when the initial pressure is changed. This figure shows that irrespective of initial pressure values, the curves of the contact force vs. pressure are almost parallel. In addition, the range of detection of the contact force is wide when the initial pressure is higher. Therefore, the pneumatic cushion sensor can trade off the sensor effect and the buffer effect. Table 2 shows the specifications of the pneumatic cushion sensor developed in this study.

6. Experimental Results and Discussion

6.1. Experimental System

In experiments, the angle of each link is measured using a potentiometer (Figure 1). Maximum impact exerted on the link is detected by the pneumatic cushion (force sensor) in each link. The motor and ER clutch are controlled by a PC based on measured data. Human pain tolerance is the limiting value of the mechanical stimulation that causes pain in a human subject [14]. If this is exceeded, each joint is made flexible by ceasing to apply the ER clutch's electrical field.

6.2. Position Control Experiment

We studied the positioning control of the manipulator by comparing when the ER clutch is and is not installed in the arm.

Figure 8 show the experimental results for the third joint, comparing cases with and without the ER clutch, in which the target angle ranges from 0° to 60°. When the angle is detected by the potentiometer, feedback control will be applied. When the ER clutch is installed (soft manipulator),

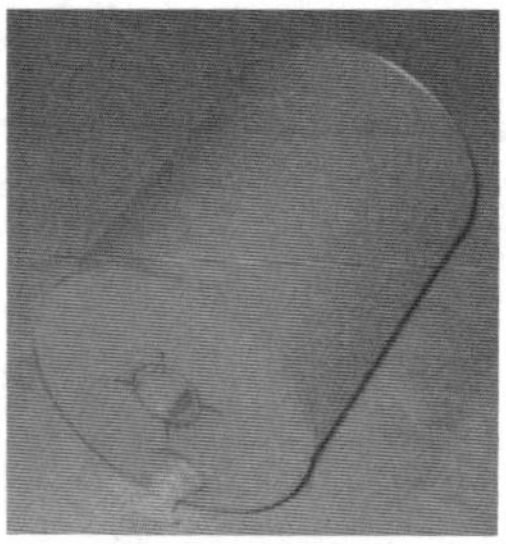

Figure 6. Schematic diagram of the pneumatic cushion sensor for the third link

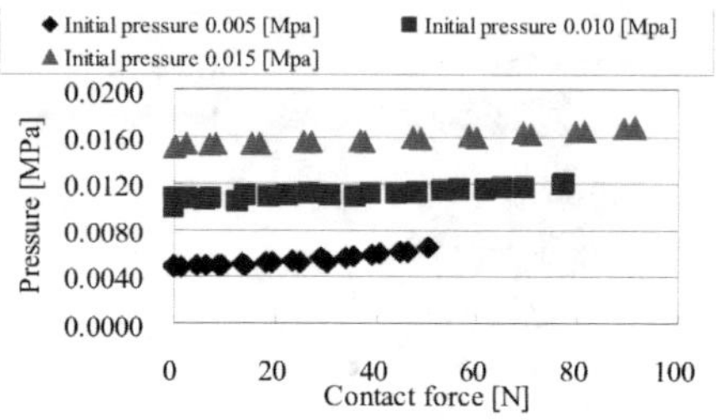

Figure 7. Relationship between contact force and pressure as a parameter of initial pressure for the first link

Table 2. Specification of pneumatic cushion sensor

Pneumatic cushion	1st link	2nd link	3rd link
Outside diameter [mm]	160	160	100
Inside diameter [mm]	70	70	20
Length [mm]	190	140	140
Thickness [mm]	0.05	0.05	0.05

the electric field is applied at 0.0 kV/mm and 3.0 kV/mm, and when the ER clutch is not installed (direct gear drive manipulator), the joint is driven directly by the motor. From this figure, the manipulator can guide the arm to an exact position when an electric field of 3.0 kV/mm is applied to the ER clutch. However, it can be inferred that the raising of the arm is delayed when an electric field of 0.0 kV/mm is applied to ER clutch compared to the case direct gear drive manipulator, and the joint does not reach the target angle. The joint cannot follow the rotation, because the inertial torque of the motor is larger than the retained torque of the ER clutch in this case. The experimental results for the first joint and the second joint is the same as the results for the third joint.

6.3. Collision Experiment

We studied four conditions related to collisions between the manipulator and something in its external environment when (1) the buffer element was not installed (direct gear drive), (2) only the ER clutch was installed, (3) only the pneumatic cushion was installed, and (4) both the ER clutch and the pneumatic cushion were installed (soft manipulator). Manipulator's first joint drives at 10 rpm on the experiment. At the end point of the third link, the manipulator collides with an object in the outside environment that consists of a force sensor. Figure 9 shows the experimental results for the collision force.

In this figure, when the direct gear drive manipulator collides, the maximum impact force is generated and the contact force keeps generating after the collision. When the pneumatic cushion sensor was installed, the maximum impact force decreased compared with the direct gear drive, but the contact force remained. When the ER clutch was installed, the maximum impact force decreased. Moreover, the contact force could decrease to zero because when the third link collided with the external object, it never contacted the external object again by making the joint flexible after the collision. When the soft manipulator is installed in both the pneumatic cushion sensor and the ER clutch, the maximum impact force decreased most, and the contact force could decrease to zero.

7. Conclusion

This study describes a soft manipulator we have developed that ensures user safety in collisions between robots and an external object (e.g., humans), using a smart flexible joint composed of an ER fluid and a pneumatic cushion. We came to the following conclusions:

- It was shown that the pneumatic cushion sensor could enable the manipulator to detect a collision with the external object and has sufficient buffer ability to cushion the maximum impact force.
- To decrease external contact, we installed a clutch using an ER fluid on the manipulator's joint. Control was sufficient in a positioning control experiment to confirm the effectiveness of the manipulator's functionality. The collision experiment confirmed a decrease in contact impact.

In future research, we will examine the collision simulation of the manipulator and manipulator's behavior after collision. Also, collision patterns will be compared to the soft manipulator and this simulation.

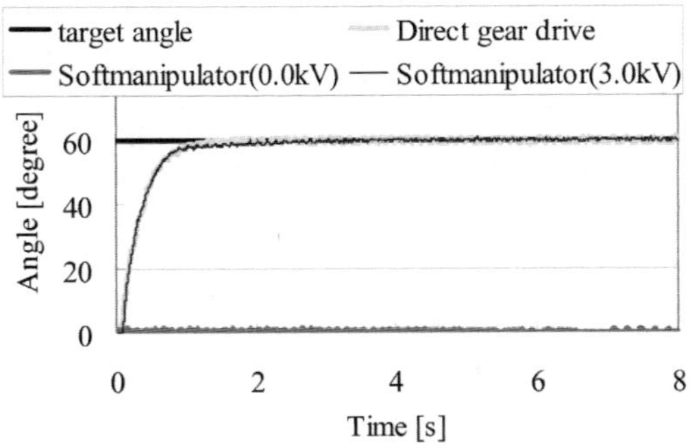

Figure 8. Step responses of position control for the third joint

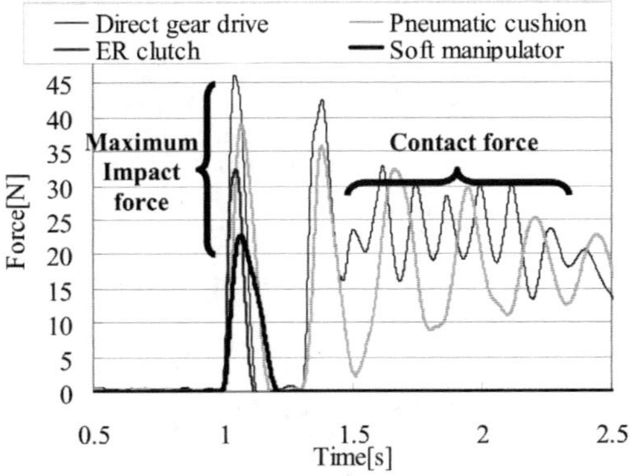

Figure 9. Collision experiment of the soft manipulator

References

[1] Z. Li and A. Ming 2004 *Mobile manipulator collision control with hybrid joints in human-robot symbiotic environments: Proc. of IEEE/RSJ International Conference on Intelligent Robots and Systems* pp 154–161

[2] I. D. Walker 1990 *The use of kinematic redundancy in reducing impact and contact effects in manipulation: Proc. of IEEE/RSJ International Conference on Intelligent Robotics and Automation* pp 433–449

[3] K. Kitagaki and M. Uchiyama 1992 *Optimal approach velocity of end effecter to the environment: Proc. of IEEE/RSJ International Conference on Intelligent Robotics and Automation* pp 1928–1934

[4] R. Volpe and P. Khosla 1991 *Experimental verification of strategy for impact control: Proc. of IEEE/RSJ International Conference on Intelligent Robotics and Automation* pp 1854–1860

[5] K. Youcef-Toumi and D. A. Gutz 1989 *Impact and force control: Proc. of IEEE/RSJ International Conference on Intelligent Robotics and Automation* pp 410–416

[6] J. K. Mills 1990 *Manipulator transition to and from contact task (A discontinuous control approach): Proc. of IEEE/RSJ International Conference on Intelligent Robotics and Automation* pp 440–446

[7] T. Nakamura and N. Saga 2004 *Development of a Soft Manipulator Using a Smart Flexible Joint for Safe Contact with Humans: J. of the Society of Instrument and Control Engineer* vol 40 number 10 pp 1060–1066 (in Japanese)

[8] T. Takahashi et al. 2000 *A control method to reduce the oscillation of a robot manipulator with elastic joints and non-backdrivable gear trains: J. of Robotics Society of Japan* vol 18 number 1 pp 142–149 (in Japanese)

[9] T. Nakamura, Y. Akamatsu, and Y. Kusaka 2008 *Development of soft manipulator with variable rheological joints and pneumatic sensor for collision with environment: J. of Robotics and Mechatronics* vol 20 number 4 pp 634–640

[10] S. Norihiko and T. Nakamura 2001 *Development and fundamental characteristic of one-shaft type er clutch: J. of Robotics and Mechatronics* vol 13 number 4 pp 534–539

[11] M. Sakaguchi, G. Zhang, and J. Furusho 2000 *Modeling and motion control of an actuator unit using er clutches: Proc. of IEEE International Conference on Intelligent Robotics and Automation* vol 2 pp 1347–1353

[12] Y. Eguchi and T. Nakamura 2006 *Development of a pneumatic cushion with an impact sensor: Proc. of 4th IFAC-Symposium on Mechatronics Systems* pp 1002–1005

[13] W. M. Winslow 1949 *J. Applied Physics* vol 20 pp 1137–1140

[14] Y. Yamada, K. Suita, and K. Imai 1997 *Human-robot contact in the safeguarding space: IEEE/ASME Transactions on Mechatronics* vol 2 number 4 pp 230–236

Journal of Physics: Conference Series **149** (2009) 012003 doi:10.1088/1742-6596/149/1/012003

CFD Study of the Flow in a Radial Clutch with a Real Electrorheological Fluid

S M Chen[1], W A Bullough[2] and J Hart[2]

[1]College of Mechanical Engineering, Fuzhou University, Fujian Provice, 35002, China

[2]Department of Mechanical Engineering, The University of Sheffield, Sheffield S1 3JD, UK

E-mail: smchen@fzu.edu.cn

Abstract Results are presented from a computational fluid dynamics (CFD) study into the complex flow of an excited electrorheological fluid (ERF) between a driven rotating and a parallel stationary radial shear plate's form of clutch. The real fluid is modelled using a technique which can be employed for any fluid/field characterisation. It is shown that the induced 'cooling' through flow rate can not be maintained when a substantial field is switched on and the clutches remain slipping.

1. Introduction

The cooling of ER clutches has been investigated previously [1-4]. In the present instance, the thermodynamic problem is not confronted in an exercise to study the interactional flows in radial and circumferential directions in order to establish the degree of through flow (potentially cooling fluid) which occurs and their effect on torque.

A well described ER fluid Bayer 3565 x 60% solid volume fraction at temperature $T = 40^0$C is chosen for completeness of characterisation data and Fluent6 for the adaptability of the sub-routines for the purpose stated.

The representation of these characteristics was developed in references [5-7] – it has been named the Bingham power-law, or Herschel-Bulkley method.

Apparent viscosity $$\mu_a = f(E, \dot{\gamma}, T, k, n) \quad (1)$$

Under a given electric field E and constant T (the flow is taken to be steady and isothermal, and the fluid incompressible) with a local shear rate of $\dot{\gamma}$.

$$\mu_a = f(\dot{\gamma}, k, n) \quad (2)$$

and $$\mu_a = \frac{\tau_y + k\left[\dot{\gamma}^n - \left(\frac{\tau_y}{\mu_0}\right)^n\right]}{\dot{\gamma}} \quad (3)$$

Where τ_y is the yield stress and μ_0 the zero shear viscosity as $\dot{\gamma} \to o$ (the pre-yield zone is not further studied in this work) and n indicates the power law index. k, the consistency factor, and n, are determined from the Couette flow test data of Figs 1 – see also Figs 2 and 3. The form of the cross section of the radial plates clutch is given in Fig 4. When the inter plate gap h is constant with radius and $\dot{\gamma}$ is assumed constant for a Bingham plastic,

torque
$$N = 2\pi R^3\left(\frac{\tau_y}{3} + \frac{\mu_e R\Omega}{4h}\right) \tag{4}$$

which is used in the construction of the results figures. R is the plate outer radius, r an intermediate radius, and Ω is the rotational speed.

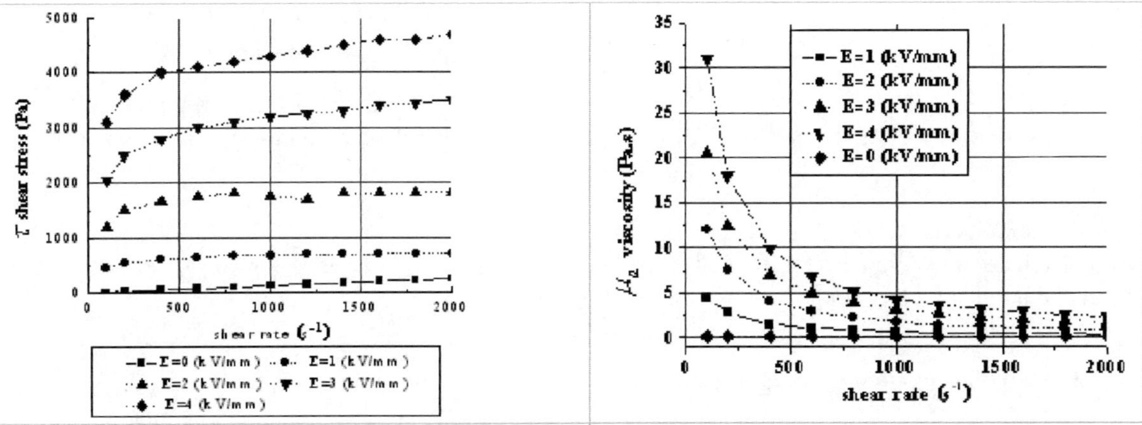

(a) Bayer curves of shear stress vs. shear rate

(b) Bayer curves of apparent viscosity vs. shear rate

Figure 1 Bayer 3565(60%) experimental [8] electricalrheological fluid characteristics

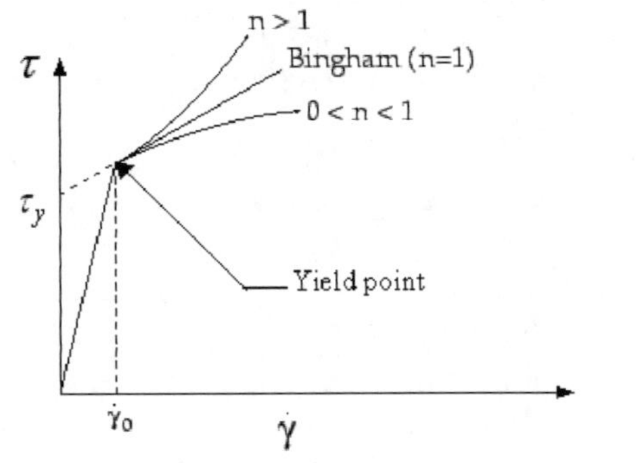

μ_0 the zero shear viscosity as $\dot{\gamma} \to o$

μ_e the shear viscosity as $\dot{\gamma} > \dot{\gamma}_0$ (n=1)

Figure 2 Bingham – power law model

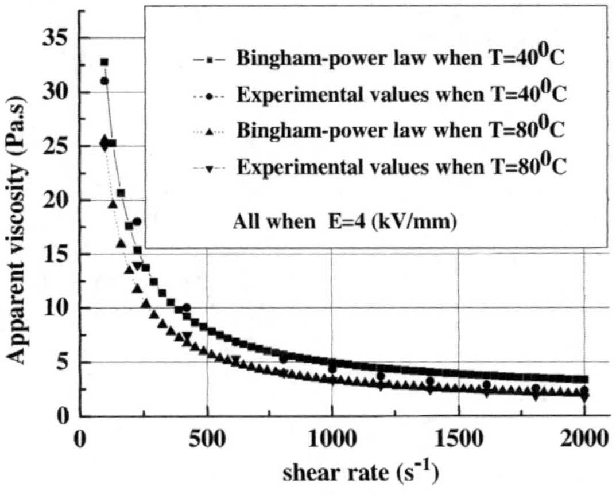

Figure 3 Curves of Bingham-power law mode fit with experimental data from Bayer data

Journal of Physics: Conference Series **149** (2009) 012003 doi:10.1088/1742-6596/149/1/012003

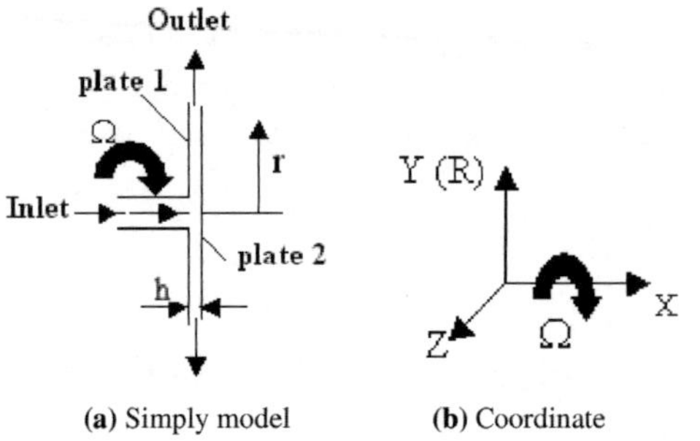

Figure 4 Model of radial clutch

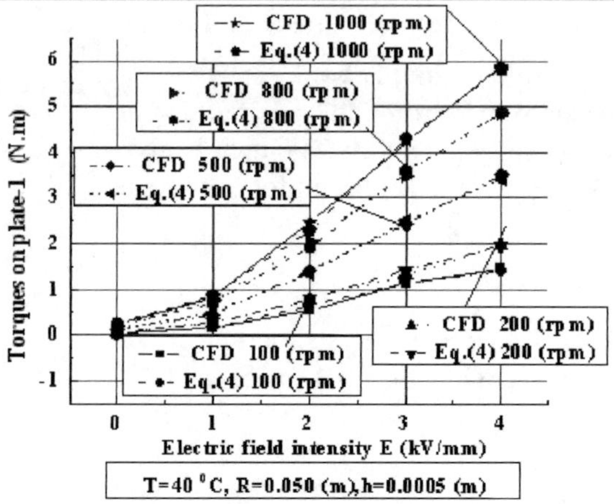

Figure 5(a) Effect of electric field intensity and rotational speed on torque (CFD=power law)

2. Numerical analysis

Fitted coefficients are given by tables 1. The input conditions are given in Table 2 and results are computed for a density of 1000kg/m^3 and a temperature of 40^{0}C.

Table 1 The values of parameters k and n in equation (3) (material: type number 3565 (60%)

E (kV/mm)	k	N	R-Square
E=1	0.23014	0.97353	0.9868
E=2	1.23984	0.96723	0.98972
E=3	2.09916	0.9701	0.98951
E=4	2.25096	0.97138	0.98921

Table 2 Scheme of CFD

Software	Fluent 6.2, finite volume, segregated solver	
Physical model	2D, steady, laminar, incompressible, isothermal flow	
Fluid properties	(1) Bayer ER fluid, viscosity (constant) and density (1000kg/m^3) for zero electric field	
	(2) Bayer ER fluid, viscosity (Herschel-Bulkley model in Fluent 6.2) and density (1000kg/m^3) with active electric field	
Equations: flow swirl velocity	Interpolation scheme	2nd order for all equations
	Pressure-velocity coupling	SIMPLE
	Discretisation pressure	PROESTO
	Grid (x, y) [γ = radial direction]	60 (axial) x 6000 (radial)
Boundary conditions	Plate 1	Rotating
	Plate 2	Stationary
	Inlet	Total pressure inlet with initialisation 0
	Outlet	Static pressure outlet with initialisation 0

3. Results and analysis

Bayer data was available only at constant temperatures and hence the only computations were made at 40^{0}C under isothermal conditions. CFD software package (Fluent 6.2) [9] was used with the data from Table 1 and settings in Table 2. The plates were also set at 40^{0}C. Density was set to 1000 kg/m^3.

Journal of Physics: Conference Series **149** (2009) 012003 doi:10.1088/1742-6596/149/1/012003

Yield stresses were derived from Figure 1(a) by extrapolating the $\tau - \dot{\gamma}$ curve at any given voltage to $\dot{\gamma} = 0$. Likewise the 'solid' viscosity was derived from Figure 1(b). After $\dot{\gamma}$ = 2000 l/s the apparent viscosity is taken to depend only on E.

4. Discussions

Only partial verification of the computed results is available at this time via: a comparison of the CFD generated torque and that approximation from equation 4, and a specimen comparison of Bingham-power law CFD and Bingham CFD (bi-viscous) solution: Figs 5 (a), 5 (b) and 5(c) respectively.
The reason for the closeness of the models is seen from Fig. 5(c): there is little induced flow. Hence the fluid follows circumferential stream liner, as assumed in Equation (4).

Doubling h, Ω or R fails to induce extra flow. Farther computation validates the $\dot{\gamma} = \Omega r / h$ assumption.

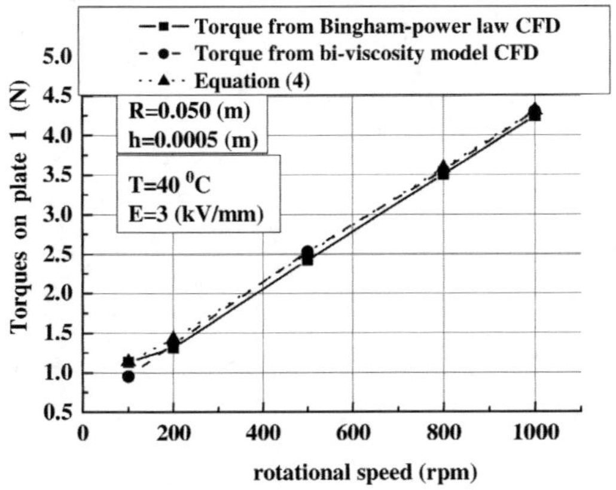

Figure 5(b) Effect of model on torque

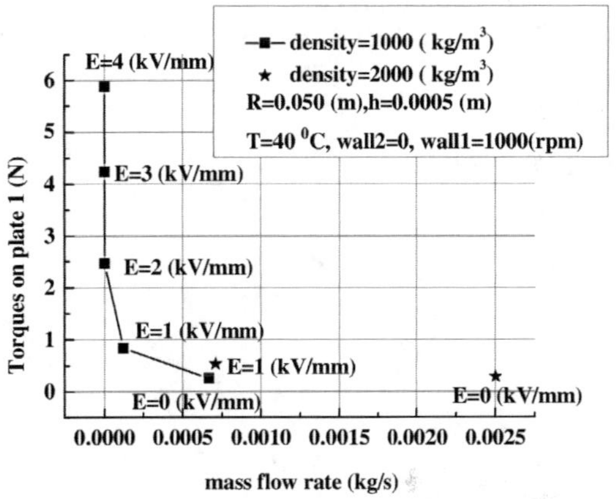

Figure 5(c) Interaction between electric field intensity, mass flow rate, torque and density

5. Conclusions

Based on real ERF material (characteristics by Bayer), this paper investigated the apparent viscosity expression and obtained the coefficients for a Bingham-power law, and showed that this representation is quite consistent with real ERF apparent viscosity. In the main, the ERF behaves as shear-thinning, and its stability is good for a range of temperature and electric field intensity.
A bi-viscous sub-routine in a CFD package has been combined with a power law constitutive form through constants and yield stresses derived from commercial (real fluid) characteristic data. This has been done to provide the basis of a technique for the pre-prototype prediction of machine/device performance. The technique has been tested on a parallel plate radial clutch to relate through flow rates, torque and velocity profiles with geometry and operating parameters etc.

References

[1] S. M. Chen , W. A. Bullough and D J Ellam. Examination of Through Flow in a Radial ESF Clutch. Journal of Intelligent Material Systems and Structures. Dec.2007. 1175-1179.
[2] Makin J et al. Methodology Performances and Problems in ER Clutch Based Positing Mechanisms. Proc. 6th Int. Conf. ERMR fluid, Yonezawa, Japan, July 1997. P623- 630 World Scientific Pub.

[3] D.J.Ellam, W.A. Bullough and V. Oravsky. Heat Transfer from an ESF Radial Plate Clutch Surface, Proc 9th Int. Conf. ER/MR Fluids held Nice July 2001. P 104 – 110. World Scientific Pub.

[4] D.J.Ellam, W A Bullough, and RJ Atkin. Modellinf the flow of an electrostructured fluid in transient operation. Proc. IMechE 2005, vol.219 Part A : J. Power and Energy. P61-76.

[5] Chen Shumei and Wei Chenguan. Experimental Study on Static Yield Stress and Dynamic Shear Stress For Electrorheological Fluid, Ninth International Conference on Electrorheological (ER) Fluids and Magneto-rheological (MR) Suspensions, Aug.29-Sept.3, 2004, Beijing, China. P850-910. World scientific pub.

[6] Shu-Mei Chen and Chen-GuanWei. Experimental study on percolation structure of electrorheological suspensions. International Journal of Applied Mechanics and Engineering. 2007, vol.12, No.1, pp.299-308.

[7] Shu-Mei Chen and Chen-Guan Wei. Experimental Study on Rheological Behaviour for Electrorheological Fluids. Smart Materials and Structures 15 (2006) 371-377.

[8] Bayer. Private communication.

[9] Fluent 6.2 User's Guide, Copyright c 2005 by Fluent Inc.

Journal of Physics: Conference Series **149** (2009) 012004 doi:10.1088/1742-6596/149/1/012004

Effect of Cluster Formation on Dynamic Response of an Electrorheological Fluid in Shear Flow

B H Choi[1], Y J Nam[1], R Yamane[2] and M K Park[1]

[1] Graduated School of Mechanical Engineering, Pusan National University, Busan 609-735, Republic of Korea

[2] Department of Mechanical Engineering, Kokushikan University, 4-28-1, Setagaya-Ku, Tyoko 154-8515, Japan

Abstract. Transient shear stress responses of an electrorheological fluid to step-wise electric fields are investigated experimentally. The characteristic time constants of the electrorheological fluid sheared between two concentric cylinders are obtained under various electric field strengths and shear rates. Especially, two experimental modes are adopted to investigate the effect of the cluster formation on the dynamic response of the fluid; one is that the electric field is induced before shearing, and the other is the electric field is induced after shearing. The cluster formation time is obtained from the difference in the response times between the two modes. The cluster formation time is monotonously increased with the increase of shear rate and converged on a constant value.

1. Introduction

An electrorheological (ER) fluid is a kind of suspensions which consist of polarizable particles dispersed in insulation solvents. The ER particles which are induced by the electrostatic interaction upon applying electric field are aligned to form chains or columns along the field direction. Due to the controllability of the apparent viscosity, ER fluids are increasingly being considered in a large variety of devices such as shock absorbers, vibration insulators, brakes or clutches. In order to effectively design and control devices based on ER fluids, the knowledge of the behaviors and properties of the ER fluids during the phase transition is required inevitably. [1]

The previous researches on the dynamic response of the ER fluid have been achieved through the rheological stress behavior or the microstructural configuration change based on optical visualization or the simulation. [2, 3] However, it is difficult to quantitatively specify many of the parameters employed in the simulation as well as the flow visualization conditions are rather different from the operating conditions in the real situation. Especially, such studies have focused on the ER fluid at quiescent state because of the limited size of the inspection window. Therefore, in this study, the dynamic stress behavior of the ER fluid sheared between two concentric cylinders is investigated under various electric field strengths and shear rates.

In general, the dynamic response of the ER fluid is determined by the microstructural behavior of the ER particles in the competition of the electric-field induced particle attractive interaction (electrostatic

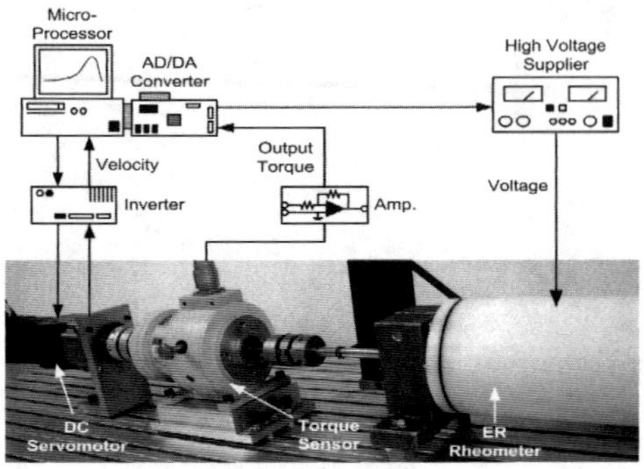

Figure 1. Schematic of the test system of ER fluid sheared between two concentric cylinders

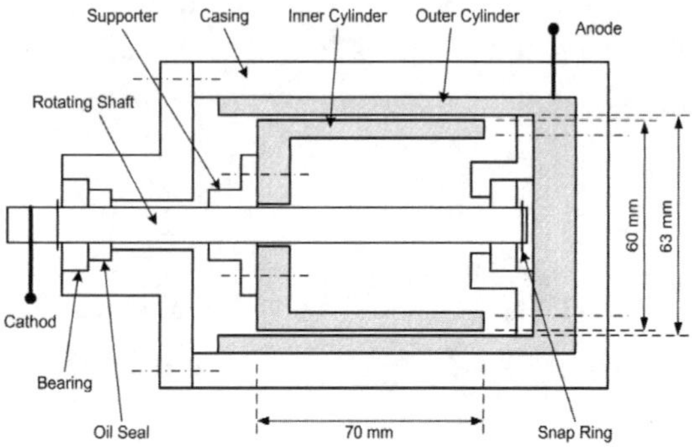

Figure 2. A cross section of the two concentric cylinder type rheometer

force) and the hydrodynamic force. [4, 5] Such a microstructural behavior is mostly separated by the formation, the deformation, the destruction and the reformation processes of the cluster. While the static response of the ER fluid is determined by the circulating close-loop process including the cluster destruction and reformation, the dynamic response is closely related to the cluster formation and deformation processes. Specifically, the cluster formation can be also divided by the particle polarization, the particle aggregation and the cluster densification. [6]

In this study, two experimental modes are adopted; one is that the electric field is induced before shearing (EH mode), and the other is the electric field is induced after shearing (HE mode). Compared to the HE mode, the cluster formation process in the EH mode can be considered to be already completed prior to the cluster deformation due to the hydrodynamic force. Consequently, the cluster formation time of the ER fluid in shear mode can be obtained from the difference in the response times between the two modes.

2. Experimental setup

The ER fluid used for the investigation of the shear stress response is TX-ER8 composed with sulfonated polymer particles dispersed in a mixture of fluorine oil and silicone oil. This fluid sample has a viscosity of 115 mPa s and a density of 1.6 g/cm^3 at room temperature (about 25°C). [6]

Figure 1 depicts the schematic diagram of the experimental apparatus. The switching module installed in a high voltage power supplier is used to turn on or off DC electric fields, whose operation can be controlled by a micro-processor. The switching time of the module is about 2 msec, which is in the negligible order.

As seen in figure 2, a concentric type cylindrical rheometer is fabricated to investigate the shear stress response of ER fluid. The gap size between the internal (cathode) and outer (anode) cylinders which are used for electrodes is designed to be 1.5mm, considering low viscous effect due to the ER fluid. The length of the electrode is designed to be 70mm. The internal cylinder is designed in the form of a cup to reduce its inertial load. A shaft-supporter system is employed to keep the gap size of between the cylinders constant. The internal cylinder is rotated at a certain velocity in one direction by a DC servomotor, where the inverter is used to keep the rotational velocity of the DC servomotor uniform irrespectively to an external load torque. And, the output torque is measured by the rotary-type torque sensor connecting the rheometer with the DC servomotor and recorded using a PC-based data acquisition system with the sampling speed of 1 kHz.

The characteristic time constants of the ER fluid are obtained from the transient stress behaviors at various rotating velocities and electric field strengths.

3. Results and discussions

Figures 3 and 4 show the shear stress response of the EH mode and the HE mode, respectively. The shear rate and the electric field strength are changed in the ranges of 4-125 s^{-1} and 0-3kV/mm,

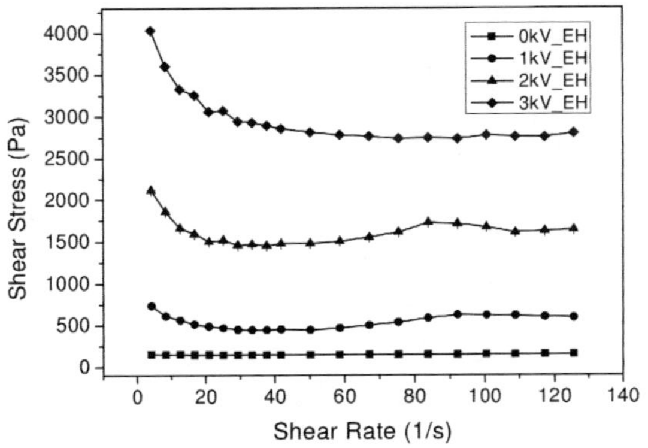

Figure 3. Shear stress of ER fluids where Electric field is induced before shearing

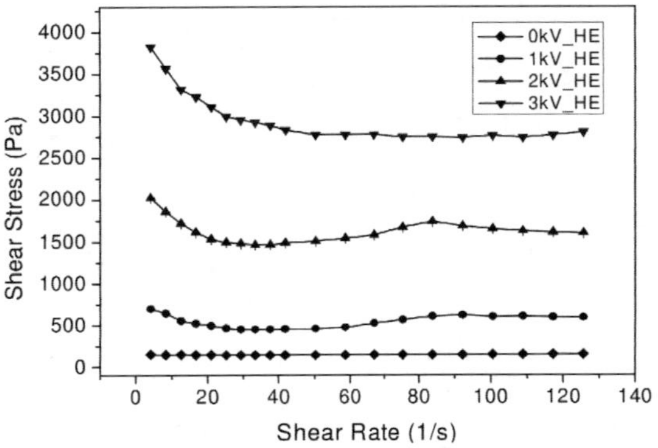

Figure 4. Shear stress of ER fluids where Electric field is induced after shearing

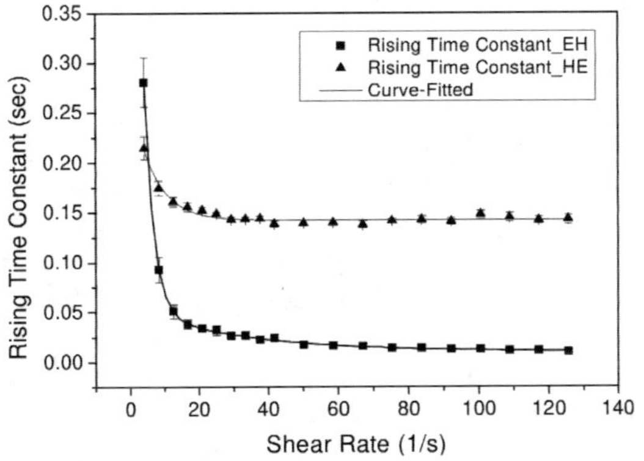

Figure 5. Rising time constant of ER fluids

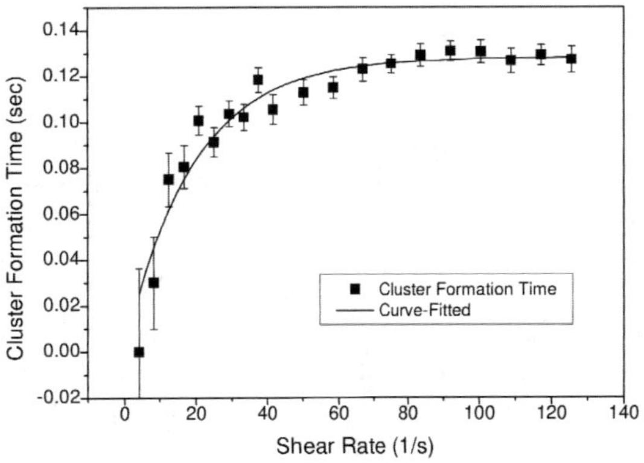

Figure 6. Cluster formation time of ER fluid

respectively. There is no significant difference in the steady state response of the shear stress according to those two experimental modes. As can be seen in these figures, the shear stress is increased with the increase of the electric field strength. However, it is hardly influenced by the shear rate, with the exception of low shear rate region having relatively high stress. At the relatively low shear rate, the cluster formation is easily achieved because of small hydrodynamic force. Therefore, the cluster structure at the low shear rate is similar to that of quiescent state. In addition, the attracting force between cluster structure and two electrodes might be larger than that of ER particles. Therefore, the slipping phenomenon between the clusters and the electrodes hardly take place at the relatively low shear rate. [7] At the moment that cylinder is about to rotate from the quiescent state, a relatively large torque is required to destroy the cluster structure due to the friction between the clusters and the electrodes. Because the friction can be considered as the sticking friction, the yield stress is highly affected by the static yield stress. Instead, as the shear rate is increased, the cluster structure is less dense than those at the relatively low shear rate. Also, the strength of the cluster structure becomes weak gradually due to large hydrodynamic force. Therefore, the slip phenomenon is occurred in the high shear rate. [7] The friction can be considered as the sliding friction. The yield stress of this case is known as dynamic yield stress.

Figure 5 shows the dynamic response time of the two modes. As can be seen in the figure, the response times and their decreasing rates are decreased with the increase of the shear rate and are unchanged with the electric field strength applied. At the shear rates more than about 15 s^{-1}, the response times tend to be kept nearly constant. The inverse relationship between the stress response time and the shear rate is similar to the previous researches.

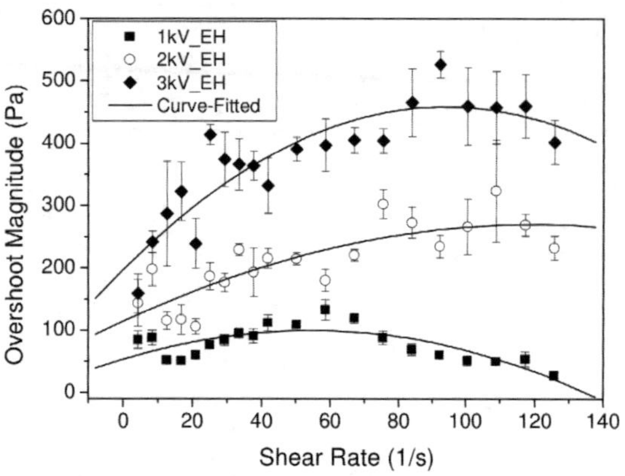

Figure 7. Overshoot magnitude where Electric field is induced before shearing

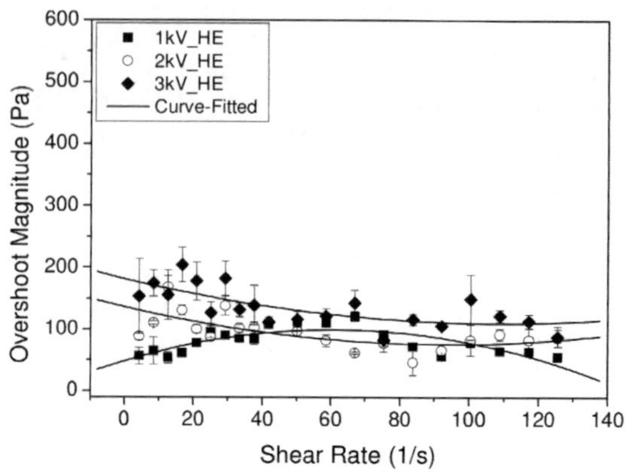

Figure 8. Overshoot magnitude where Electric field is induced after shearing

In addition, it is observed that the EH mode has much shorter response times than the HE mode under the same shear rates. The ER particles under an external electric field are polarized and then chains or columns are formed between the two electrodes due to the dipole interactions (cluster formation). By applying shearing load, the ER chains and columns exhibit the elastic behavior where its shear stress is proportional to the shear strain (cluster deformation). After this yielding region, the clusters are repeatedly destructed and reformed. Here, the cluster formation and deformation processes are closely related to the dynamic response of the ER fluid, while the cluster destruction and reformation processes give an effect on the static response. At the EH mode, the hydrodynamic force is applied after having enough time to form the stable cluster structure. On the contrary, at the HE mode, the cluster begins to form the quasi-stable structure as the electric field is induced on the ER fluid already subjected to the hydrodynamic force. In other words, the dynamic response at the EH mode is determined only by the cluster deformation process, while the cluster formation and deformation are occurred simultaneously at the HE mode. Therefore, it is concluded that the cluster formation time of the ER fluid in shearing can be found from the difference in the dynamic response time between the EH mode and the HE mode, as shown in figure 6. The cluster formation time is monotonously increased with the increase of shear rate and converged on a constant value. This is because the collision chance among the aggregated ER particles and the deformation rate of the cluster structure are increased with the increase of hydrodynamic force. [8] Therefore, It is also verified that shearing force inactivates the cluster formation of the ER fluid and promotes the cluster deformation, but consequently results in the decrease of the ER response time.

On the other hand, it is verified that there is an overshoot phenomenon in the shear stress rising process, as shown in figures 7 and 8. The magnitude of the overshoot at the EH mode is relatively large, compared to the HE mode. In the microstructural viewpoint, the cluster formation can be divided by the particle polarization, the particle aggregation and the cluster densification. The cluster structure of the EH mode is fully stable because of the cluster densification process at the quiescent state, as reported in the previous visualization study. [3] However, the cluster structure of the HE mode is less densely formed because its formation and deformation is occurred simultaneously. In other word, the chains or columns at the HE mode are mainly formed by the particle aggregation, but not the cluster densification. Therefore, the clusters at the EH mode are stronger than the counterpart, leading to the overshoot in the shear stress rising process. Also, it can be seen at the EH mode that the overshoot magnitude tends to be increased with the increase of the shear rate. It is considered that it is closely related to relatively high shear stress at the low shear rate. In other word, because the shear rate at the EH mode is increased from the quiescent state to the constant level, the corresponding shear stress is reached at its steady state via the yielding region at the low shear rate. Therefore, the overshoot at the EH mode can be also considered as the effect of the static yield stress in relatively high level.

4. Conclusions

The dynamic response of the ER fluids in shear flow to step wise electric fields is investigated experimentally. With the characteristic time constants of the ER fluid obtained under various experimental conditions, the relationship between the ER response and the cluster formation is examined. The conclusions drawn from this research can be summarized as follows.

(1) The shear stress at the steady state is increased with the increase of the electric field strength. And, the shear stress at low shear rate is larger than those at relatively high shear rate. This is because the clusters subject to weak hydrodynamic force are relatively densely formed as well as the stick friction between the clusters and the electrodes drives the static yielding stress to the ER fluid.

(2) If the electric field is induced before shearing (EH mode), the response time of the ER fluid is dominated by the deformation process of the cluster structure. When the electric field is induced after shearing (HE mode), the response time is determined by the cluster formation and deformation processes. The cluster formation time can be derived from the difference in the response time between the two modes.

(3) The cluster formation time is increased monotonically and then converged constantly with the increase of the shear rate. Instead, the dynamic response of the ER fluid is decreased with the increase of the shear rate. Under the hydrodynamic field, the increased cluster deformation rate has more effect on the dynamic response time of the ER fluid than the increased collision chance of the ER particle.

5. References

[1] Rhee E J, Park M K, Yamane R and Oshima S 2003 "A study on the relation between flow characteristics and cluster formation of electrorheological fluid using visualization" *Experiments in Fluids* **34** 316-23

[2] Ginder J M 1993 "Diffuse optical probes of motion and structure formation in an electrorheological fluid" *Physical review E* **47(5)** 3418-29

[3] Wen W, Zheng D W and Tu K N 1999 "Chain/column evolution and corresponding electrorheological effect"*Journal of applied physics* **85(1)** 530-33

[4] Tian Y, Zhang M, Meng Y and Wen S 2005 "Transient response of compressed electrorheological fluid" *Journal of Colloid and Interface Science* **290** 289-97

[5] Tian Y, Meng Y and Wen S 2004 "Dynamic responses of zeolite-based ER fluid sheared between two concentric cylinder" *Journal of Intelligent Material Systems and Structures* **15** 621-26

[6] Nam Y J, Park M K and Yamane R 2008 "Dynamic responses of electrorheological fluid in steady pressure flow" *Experiments in Fluids* **44** 915-26

[7] Ryoichi H, Koji H, Hidenobu A, Koji S and Shinichi K 2000 "Internal structure and ER properties in ER suspensions of disperse system under DC electric field" *Electrical Engineering In Japan* **132(4)**, pp. 9-18

[8] Tanaka K, Hashimoto S, Takenouchi T, Sugimoto I, Kubono A and Akiyama R 2001 "Shearing effects on the electrorheological response" *Intelligent. Journal of Modern Physics B* **15** 930-37

Journal of Physics: Conference Series **149** (2009) 012005 doi:10.1088/1742-6596/149/1/012005

Heterogeneous electrorheological fluids with liquid crystalline matrices

M Ciszewska, T Sadoś and J Płocharski

Faculty of Chemistry, Warsaw University of Technology, ul. Noakowskiego 3, 00-664 Warszawa, Poland

E-mail: janusz@ch.pw.edu.pl

Abstract. Hybrid ER fluids comprising powders of ionic, electronic and mixed conductivity dispersed in lyotropic and thermotropic LC matrices were studied. Flow curves of the ER fluids were recorded under electric field and compared to analogous data obtained for dispersions in silicone oil. The best results were obtained for electronically conducting pyrolysed polyacrylonitrile powders dispersed in poly(*n*-hexyl isocyanate) solutions showing a significant enhancement of the ER effect. A qualitative model based on local electric field increase has been proposed for the enhancement phenomenon.

1. Introduction

Classic ER fluids consist of solid particles dispersed in inert liquids. Under electric fields the particles form structures giving rise to the increase of the fluid viscosity while the inert matrix does not contribute to this change. The so called homogeneous ER fluids comprise liquid crystalline (LC) solutions or emulsions whose viscosity increases upon electric field. Combination of these two phenomena leads to a concept of a hybrid ER fluid in which solid particles are dispersed in a LC matrix. Possibly, the two effects – from interactions between solid grains and from viscosity increase of the matrix itself – sum up giving an ER fluid of improved properties. Despite the apparent attractiveness of this concept it found only a limited response in literature. Its realisation was first described by Quist and Filisko [1] who dispersed an ion-exchange resin and a synthetic zeolite in a lyotropic LC solution and reported a strong ER activity of the dispersions. It was considerably higher than the expected sum of the two mechanisms. The aim of the presented study was to compare results obtained for two systems of LC matrices and tree types of dispersed solids in order to reveal the nature of the extraordinary enhancement of the ER effect.

2. Experimental

Poly(*n*-hexyl isocyanate) (PHIC) was prepared by polymerisation of the monomer in a DMF/toluene mixture [2]. Viscosity of a xylene PHIC solution depends on molecular weight of the polymer and its concentration. To obtain the required viscosity of 0.1 Pas, the same as for the silicone oil used as the reference matrix liquid, a 15%wt. solution was prepared. The PHIC solution at this concentration is isotropic but its viscosity does increase with electric field. Application of 1.6kV/mm resulted in increase of shear rate from 20 to about 70Pa@200s^{-1}. This solution was used as an active LC matrix.

LC polysiloxane (LCPSi) was prepared according to Inoue *et al.* [3]. 4-cyanophenyl 4-[2-(propyloxy)etoxy]benzoate mesogenic groups were attached to a polydimethylsiloxane main chain. The mesogen was connected to every second (50%) or to every twenty fifth (4%) carbon atom in the main siloxane chain. The prepared polymers were thoroughly characterized.

Solid electrolyte (SE) samples – polyacrylonitrile complexes of KSCN, NaSCN, $KClO_4$ and Bu_4NI were prepared by dissolving the salts and the polymer in DMF, evaporation of the solvent and mechanical grinding/screening to obtain powders within 1÷22 μm range [4].

Fine powders of poly(*p*-phenylene) (PPP) were obtained from benzene at 75°C [5,6] and doped in a nitromethane/$FeCl_3$ solution. Depending on the doping level a total conductivity was $10^{-8}\div10^{-7}$ S/cm with equal electronic and ionic contributions. The grains were within 1÷8 μm range.

Pyrolysed polyacrylonitrile (PPAN) samples were prepared from a copolymer of acrylonitrile and methyl methacrylate that was oxidized in air (230°C, 30 min), pyrolysed in nitrogen (500°C, 10 min), then ground and screened. PPAN had a pure electronic conductivity of about 10^{-9}S/cm with polar functional groups on a grain surface. The particle size of PPAN was in the range of 1÷36 μm.

The ER fluids were prepared by dispersing weighted amounts of the powders in dimethyl silicone oil (PDMS) of 0.1Pas viscosity and in the two LC matrices. The concentration of solids was 15%wt.

The ER experiments were performed in a concentric cylinders geometry with Bohlin Visco 88 viscometer within a shear rate range of 6÷530s^{-1} and at DC voltages of 0-5kV/mm. Currents were recorded. The measurements were done at ambient temperature of about 23°C (if not indicated otherwise). The yield stress values given were obtained by extrapolation of the measured data to the zero shear rate. The results neglected a behaviour of the systems at shear rates below 6s^{-1}.

Raman spectra of the ER dispersions were studied with the Nicolet Almega spectrometer combined with the confocal microscope Olympus BX-50.

3. Results and Discussion

3.1. ER fluids containing PHIC based matrix

In most cases of heterogeneous ER fluids the dispersed phase has ionic conductivity, so the first experiments were with SE samples. These powders dispersed in PDMS gave stable ER fluids of moderate strength. The yield stress values reached 200Pa@3kV/mm with currents not exceeding 4μA/cm^2. Replacing the inert matrix with the active one based on the PHIC polymer resulted in a higher ER effect, however, only at low field strengths. The maximum recorded yield stress was about 400 Pa@1.6 kV/mm that was accompanied by high currents exceeding 60μA/cm^2.

Dispersions of the doped PPP in PDMS showed moderate ER activity [6]. The flow curves were smooth and of Bingham type. The yield stress reached about 100Pa@3kV/mm that was accompanied by currents in the range of 10μA/cm^2. Replacing the inert matrix with the active one (PHIC/xylene) resulted in poorly reproducible results caused by electrophoresis taking place during the ER experiments.

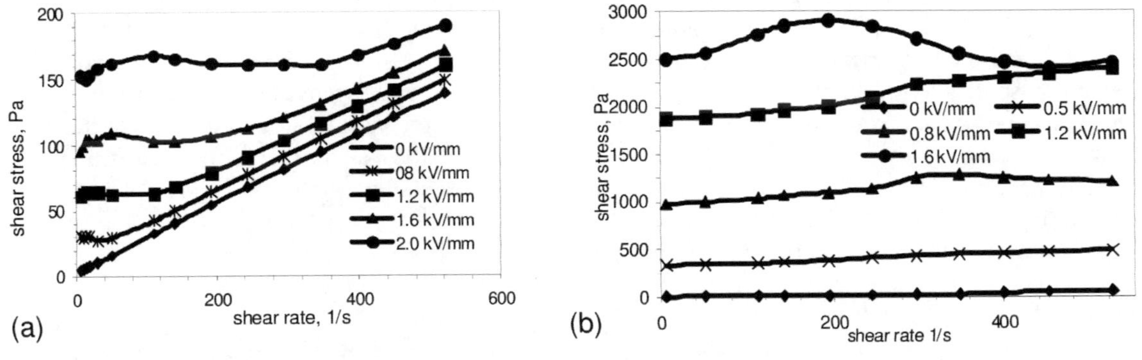

Figure 1. Flow curves of PPAN dispersion in silicone oil (a) and in 15%wt. PHIC in xylene (b).

In contrast to SE and PPP the PPAN samples exhibited stable ER activity both in PDMS and PHIC/xylene as shown in Fig.1. A very strong increase of the ER activity is clearly seen exemplified by the yield stress change from 100Pa@1.6 kV/mm for PPAN/PDMS up to 2500Pa@1.6 kV/mm for PPAN/PHIC. It is worth to note that this high ER effect was obtained at relatively low voltages and low concentration of solid phase. The system, however, had a high current – about 75μA/cm^2.

Journal of Physics: Conference Series **149** (2009) 012005 doi:10.1088/1742-6596/149/1/012005

The observed ER effect in the PPAN/PHIC system is about 18 times higher than the expected straight sum of the two mechanisms – the arrangement of PPAN grains and the increased viscosity of the PHIC matrix. The reasons causing this phenomenon have not been clearly revealed to date. Quist and Filisko suggested that some specific sorption of LC took place on the grain surface [1]. To verify this concept we measured local concentrations of PHIC in ER suspension in various conditions. The experiments were performed in a 0.5 mL quartz cuvette comprising two electrodes. A thin glass cover was used to prevent xylene evaporation. The cuvette was placed in the microscope chamber and Raman spectra were taken point by point to record maps of the PHIC concentration. The spectra of PHIC showed that its concentration was exactly the same within the whole sample – at the ER column formed from PPAN and far from it, for a pristine dispersion, for an electrified dispersion with PPAN columns formed and for samples with the voltage switched off. Considering these results we concluded that the concept of the specific sorption did not found an experimental support.

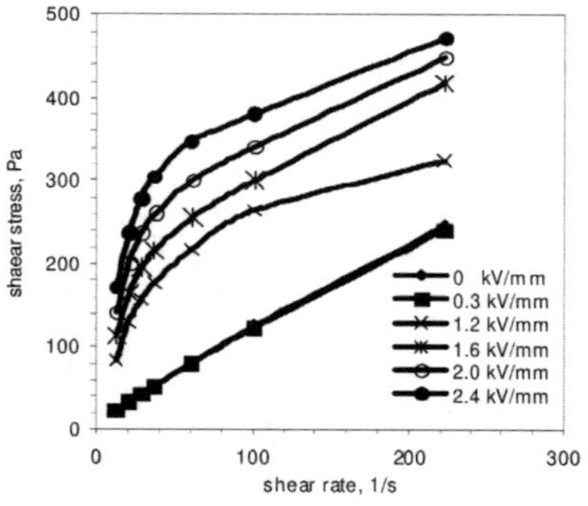

Figure 2. Flow curves of 9:1 mixture of LCPSi's containing 4 and 50% of mesogens, respectively. Measurements at 60ºC.

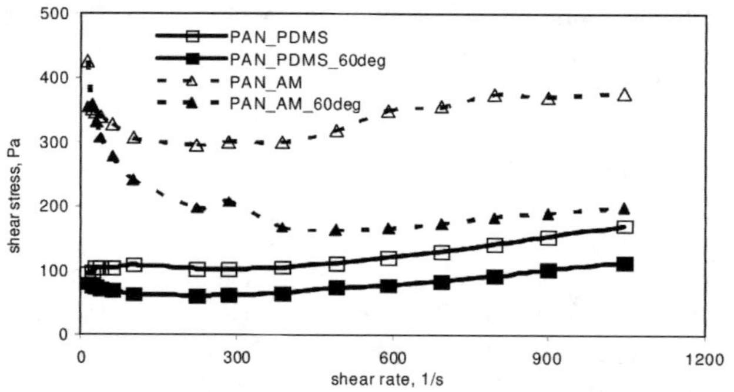

Figure 3. Flow curves of 15%wt. PPAN dispersions in PDMS (solid line) and in LCPSi/PDMS emulsion (dashed line) at 1.6kV/mm. Open and solid symbols show data for ambient temperature and 60ºC, respectively

3.2. ER fluids containing LCPSi based matrix

The LCPSi with 4% of the mesogen was in a liquid form of about 1Pas viscosity that did not change upon electric field. The one with 50% of the mesogen took a form of a semi-solid paste making ER experiments difficult. Therefore, we mixed the two LCPSi's in 9:1 proportion to get a liquid of a reasonable viscosity. Fig.2 shows the flow curves of this mixture at 60°C proving its ER activity.

Further experiments were performed with dispersions of PPAN in a LCPSi based active matrix. Following the suggestions of Inoue *et al.* [3] we prepared emulsions of the "9:1" mixture in PDMS having 10%wt. of LCPSi. The resulting stable emulsion had a negligible ER effect at ambient temperature that increased when heated to 60°C. The shear stress measured at $100s^{-1}$ was about 20Pa@2kV/mm with a considerably high current. Finally, the PPAN powder was added to this emulsion. The flow curves were recorded and shown in Fig.3. Some enhancement of the ER effect could be observed but it was lower than for PPAN/PHIC. In addition, the ER effect was poorly reversible. Switching the voltage off resulted in a very slow decrease of the shear stress and a continuous shearing for a few minutes was necessary to reset the system to the initial state.

In order to determine the distribution of the ER dispersion components the experiments with the microscope/Raman spectrometer were performed. The results obtained for an electrified dispersion are presented in Fig.4. They clearly show that the PPAN particles were surrounded with the LCPSi polymer filling spaces between the grains. The Raman maps taken before application of electric field did not show this phenomenon. The "9:1" LCPSi mixture had a relatively high viscosity what explains the observed limited reversibility of the ER effect in this system.

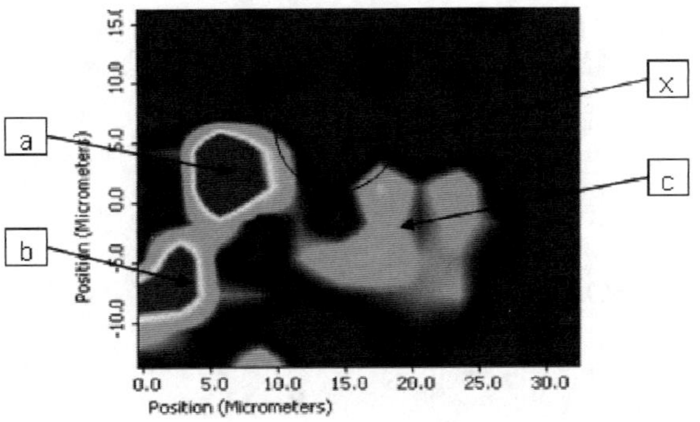

Figure 4. Intensity distribution of peak characteristic of LCPSi polymer; **a**, **b**, and **c** letters indicate areas of highest concentration; **x** indicates area of lowest concentration.

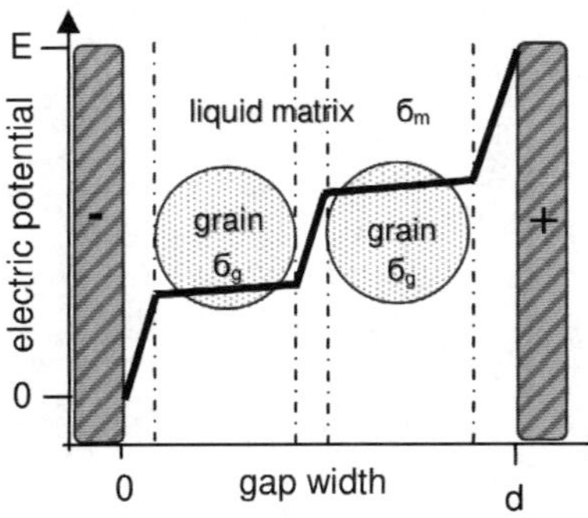

Figure 5. Schematic distribution of electric potential; conductivity of solid phase (σ_g) is higher than conductivity of liquid matrix (σ_m).

3.3. Mechanism of the ER effect enhancement

Any explanation of the extraordinary enhancement of ER effect in LC matrices should account for a few important points evidenced by the presented results. First of all, the strong enhancement took place in homogeneous solutions of lyotropic LC and only when the dispersed phase had relatively high electronic conductivity, higher that the optimal for a classic ER fluid. In addition, the concept of specific sorption of LC on dispersed grains found no experimental support. The enhancement mechanisms for lyotropic LC solutions and LCPSi/PDMS emulsions were apparently different.

Considering that we propose a simple qualitative model of the enhancement. It assumes that in the case when electric conductivity of a solid phase is higher than the conductivity of a liquid carrier the main drop of an electric potential takes place within intergranular spaces. This is because the potential drop amounts to a product of current and resistance ($u_i = j \times R_i$). The current j is constant in steady conditions of an ER experiment and all local potential drops must sum up to the E value between electrodes ($\Sigma u_i = E$). This situation is schematically presented in Fig.5. The active matrix between grains is exposed to this high field and increases its viscosity far more than initially expected. This increase contributes to the total enhancement of the shear stress value. In case of ionically conductive particles DC electric field applied causes transport of ions to grain boundaries resulting in generation of additional potentials reducing the electric field intensity in intergranular spaces.

4. Conclusions

Hybrid ER fluids comprising dispersions of polarisable powders in LC matrices were studied. The experiments showed that powders of pure electronic conductivity dispersed in a homogeneous lyotropic matrix gave fluids of the best ER performance. The observed extraordinary enhancement of the ER effect was attributed to local increase of electric field in itergranular spaces.

5. Acknowledgements

The authors thank dr. G. Żukowska for her invaluable assistance during the Raman experiments. The work was supported by statutory funds of Faculty of Chemistry, WUT.

References

[1] Quist G P, Filisko F E 1999 *Int. J. Mod. Phys. B* **13** 675
[2] Aharoni S M, Walsh E K 1979 *Macromolecules* **12** 271
[3] Inoue A, Maniwa S, Ide Y 1997 *J. Appl. Phys. Sci.* **64** 303
[4] Krztoń-Maziopa A, Ciszewska M, Płocharski J 2005 *J. Electrochim. Acta*, **50** 3838
[5] Hirai H, Toshima N, Tanaka K, 1987 *Chem. Lett.* 1461
[6] Krztoń-Maziopa A, Wyciślik H, Płocharski J 2005 *J. Rheol.* **49** 1177

Journal of Physics: Conference Series **149** (2009) 012006 doi:10.1088/1742-6596/149/1/012006

Electrical characterization of the solid phase (particles) of electrorheological fluids

Q Guegan and J-N Foulc
Grenoble Electrical Engineering Laboratory, UMR 5269, CNRS and University of Grenoble, BP 166, 38042 Grenoble Cedex 9, France

E-mail: jean-numa.foulc@grenoble.cnrs.fr

Abstract. Electrorheological (ER) fluids are suspensions of fine particles dispersed in a dielectric liquid. To understand and model the ER phenomenon, it is necessary to take into account the dielectric properties (conductivity and permittivity) of the solid and liquid phases. For the experimental investigation, the main difficulty is to measure the conductivity and the permittivity of the micro-sized (or nano-sized) particles. In this paper, we propose a method based on the arrangement of the particles in columns between electrodes under high applied field. The measurement on a cellulose/silicone oil suspension of the resistive and capacitive parts of the dc and ac time-varying currents allows us to obtain an apparent conductivity and permittivity of the cellulose particles. Finally, we discuss on the interest of the proposed characterization method from the specific conduction properties of the ER fluids.

1. Introduction

Electrorheological (ER) fluids are suspensions of fine particles dispersed in a dielectric liquid [1]. Upon application of an electric field, the solid phase (particles) fibrillates along the field lines between the electrodes. Thus, the rheological properties of the suspension are drastically modified (strongly dependence of the viscosity of the suspension on the applied field strength).

All the models proposed to describe the ER effect take into account the dielectric properties of the host liquid and the suspended particles [2,3]. There are several methods to evaluate the electrical parameters of randomly dispersed particles in a liquid medium [4]. Here we propose an alternative approach, better suited to the electrorheology, based on the characterization of the fibrillated suspension under high electric field to obtain the dielectric properties of the particles. The measurement of the electrical current under dc and ac voltages allows us to calculate the resistive and capacitive parts of the fibrillated ER fluid and then to deduce the conductivity and the permittivity of the particles.

2. Principle

For the calculation of the conductivity and the permittivity of the powder (fine particles), we assume that the ER fluid subjected to a high electric field is composed of a pure liquid phase (silicone oil) and a solid-like phase (thick cellulose column). Figure 1 describes the system. Thus, we estimate the section areas of the cellulose solid part S_S and of the liquid part S_L as $S_S = S_T\,\Phi$ and $S_L = S_T\,(1-\Phi)$ with $S_T = S_S + S_L$. Φ denotes the volume fraction of the particles and S_T the electrode surface. The thickness e is the same for the solid and the liquid parts. Then each of the two parts of the ER fluid can be represented by a resistance and a capacitance in parallel.

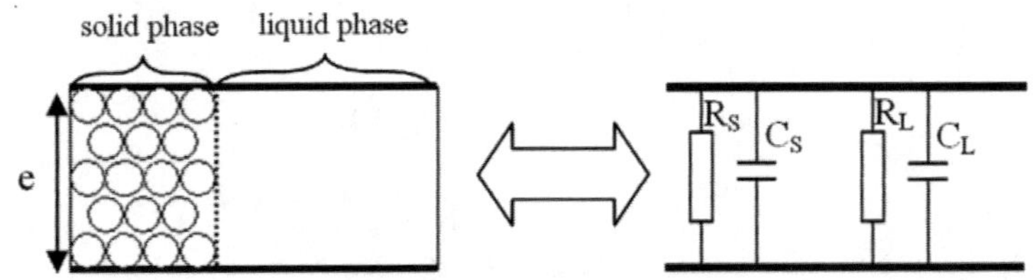

Figure 1. Sketch of the modelled arrangement of the ER fluid in two parts under high field and corresponding electrical circuit.

2.1. Conductivity

When the particles are fibrillated under dc voltage, the equivalent electrical model (Fig. 1) leads to the following expression:

$$\frac{1}{R_T}=\frac{1}{R_S}+\frac{1}{R_L}=\frac{\sigma_S\cdot S_S}{e}+\frac{\sigma_L\cdot S_L}{e}=\frac{I}{U} \tag{1}$$

where σ_S, S_S, σ_L and S_L are the conductivity and the section area of the solid and the liquid parts respectively. I is the measured current, U is the applied voltage and R_T is the total resistance of the ER fluid.

From Eq. 1, we can deduce the conductivity of the particles:

$$\sigma_S=\frac{j/E-\sigma_L\cdot(1-\Phi)}{\Phi} \tag{2}$$

If the conductivity of the liquid and the volume fraction of the solid phase are known, the measurement of the global current density $j = I / S_T$ for a given electric field strength E leads us to obtain the conductivity of the cellulose.

2.2. Permittivity

Under ac voltage and in capacitive regime ($I \propto f$), we can write (Fig. 1):

$$C_T=C_S+C_L=\frac{\varepsilon_S\cdot S_S}{e}+\frac{\varepsilon_S\cdot S_S}{e}=\frac{I}{U\cdot\omega} \tag{3}$$

where C_T is the total capacitance, ε_S and ε_L are the permittivities of the solid and the liquid parts respectively and $\omega = 2\pi f$ (f is the frequency). Thus, we obtain:

$$\varepsilon_S=\frac{\frac{j}{E\cdot\omega}-\varepsilon_L\cdot(1-\Phi)}{\Phi} \tag{4}$$

3. Experimental set-up

The measurement cell is shown in figure 2. The ER fluids consisted of a suspension of cellulose microcrystalline powder (Avicel, by Merck), with a typical size of 30 µm, relative humidity of 5.2% and bulk density of 1.4 g/cm^3, dispersed in silicone oil (Rhodia H47V20), with $\sigma_L \cong 10^{-13}$ S/m and relative permittivity $\varepsilon_L \cong 2.7$. Several samples of ER fluid with different values of particle concentration (in g/ml) were prepared and characterized (m grams of solid particles dispersed in 20 ml of silicone oil).

To apply the dc or ac high voltage in the cell, we used a generator function (*Hameg HM8030*) and a high voltage amplifier (*Trek 5/80*). The current was monitored with a home made pico-ampermeter with an analog output. All measurements were performed at room temperature.

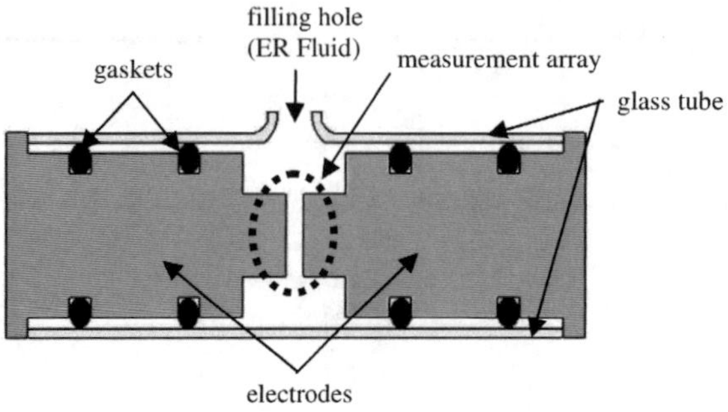

Figure 2. Schematic of the measurement cell.

4. Results and discussion

The *j-E* and *j-f* characteristics are show in figure 3. We find a $j\text{-}E^{3/2}$ law (Fig. 3a) as indicated by the conduction model [3]. It may be noted in figure 3b that the capacitive regime occurs for $f > 80$ Hz. Therefore, to calculate the particle permittivity we used the data obtained at $f = 100$ Hz. Figures 4a and 4b show that the calculated values of the conductivity and the permittivity of the cellulose are weakly dependents on the concentration (except for the high concentration). We also see in figure 5a that the conductivity of the cellulose increases with the electric field ($\sigma_S \propto E^{1/2}$). Figure 5b shows that the permittivity of the cellulose decreases when the frequency increases. We expect that for larger frequencies (in the full capacitive regime) the permittivity tends to a constant value.

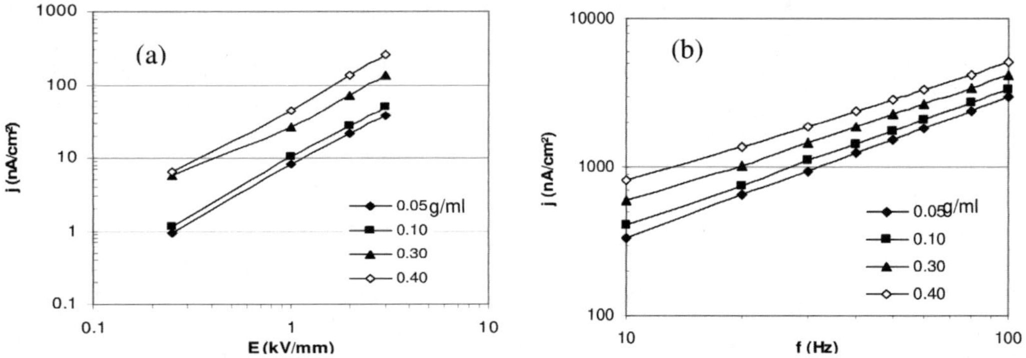

Figure 3. (a) Current density vs electric field in dc voltage and (b) current density vs frequency in ac voltage ($E = 2$ kV/mm) for different particle concentrations.

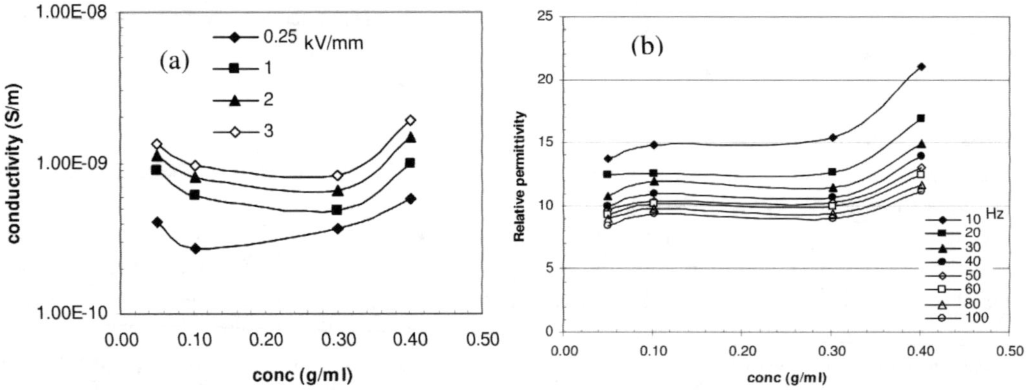

Figure 4. (a) Cellulose conductivity vs particle concentration and (b) cellulose relative permittivity vs particle concentration ($E = 2$ kV/mm), for different frequencies.

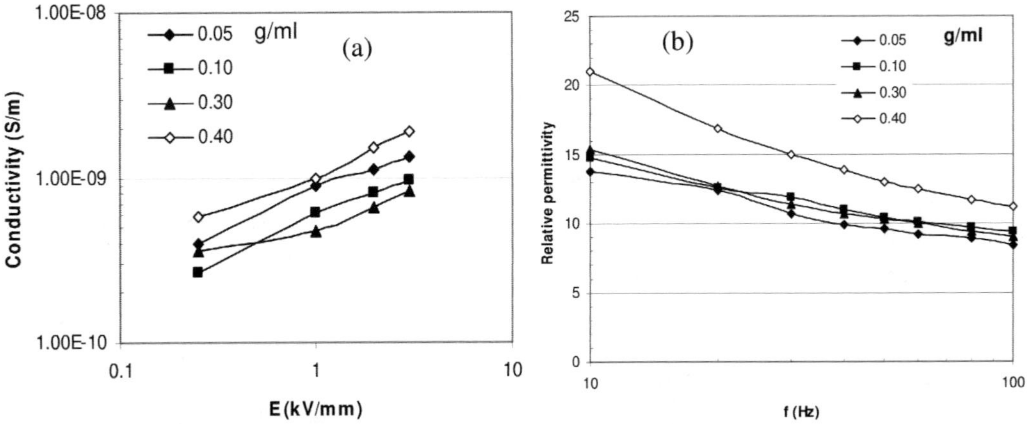

Figure 5. (a) Cellulose conductivity vs electric field and (b) cellulose relative permittivity vs frequency (E = 2 kV/mm), for different particle concentration

The conductivity and the permittivity of the cellulose powder greatly depend on its water content. In our case we obtain $\sigma_S \approx 10^{-9}$ S/m and $\varepsilon_{S\ rel} \approx 9$ which is rather consistent with the expected values [5].
In previous works [3,6] we have shown that in dc condition, the conduction properties of the solid and liquid components of the ER fluids determine both the current density and the yield stress of the fluids. More precisely, in the case of two particles (conductivity σ_S) in contact, immersed in dielectric liquid (conductivity $\sigma_L < \sigma_S$) and subjected to an electric field, the conduction model highlight the existence of an electrical contact zone of radius δ through which the current escapes from the particle and crosses the liquid. The conductance C_S of the electrical contact zone is given by $C_S = 2\ \sigma_S\delta$ with $\delta = f(\sigma_S/\sigma_L)$ and the current flowing the particle chain is $I = C_SU$ (U is a particle-particle voltage). Therefore the current flowing the ER fluids (which is essentially the current flowing the particle chains because $\sigma_S >> \sigma_L$) depends on both the particles and liquid conductivities. Regarding the characterization specificities of the particles used in ER fluids one can deduce the following remarks:
i) for very small particle sizes (< μm) the measurement of the conductivity and the permittivity appears to be very difficult in practice (choice and availability of a specific measurement set-up)
ii) an alternative method to obtain the dielectric properties of the particles is that we propose in this paper: fibrillation under dc voltage (with a large dc and a weak ac signals) of the particles immersed in dielectric liquid, measure of the dc and ac values of the current and then calculate the apparent conductivity and permittivity of the particles for a given field strength and volume fraction.

5. Conclusion

In this paper we proposed a method to obtain the dielectric properties of the particles used in ER fluids. We noted that the electrical characterization of a fine isolated particle is difficult but not essential to determine the efficiency of this material. Our method is based to the characterization of a fibrillated particles immersed in insulating liquid to deduce the apparent conductivity and permittivity of the particles. The results obtained on cellulose powder/silicone oil suspensions are promising and many others ER materials in particular the nanosuspensions need to be tested and a specific measurement procedure detailed before to validate the method.

6. References

[1] Block H and Kelly J P 1988 *J.Phys. D: Appl. Phys.* **21** 1661
[2] Klingenberg D J van Swol F and Zukoski C F 1991 *J. Chem. Phys.* **94** 6170
[3] Atten P Foulc J N and Felici N 1994 *Int. J. Mod. Phys. B* **8** 2731
[4] Van Beek L K H 1967 in *Progress in Dielectrics* **7** 71
[5] Murphy E J 1960 *J. Phys. Chem. Solids* **16** 115
[6] Boissy C Atten P and Foulc J N 1996 *Int. J. Mod. Phys. B* **10** 2991

Characterization and effectiveness of ER fluids: an attempt to unify methods and criteria

Q Guegan and J-N Foulc
Grenoble Electrical Engineering Laboratory, UMR 5269, CNRS and University of Grenoble, BP 166, 38042 Grenoble Cedex 9, France

E-mail: jean-numa.foulc@grenoble.cnrs.fr

Abstract. This paper concerns a proposal to define a way to evaluate the effectiveness of ER fluids. After recalling the basic parameters of the ER fluids: static yield stress τ_y, current density j, and zero-field viscosity η_0, we determine for a lot of ER suspensions the experimental values of two corresponding parameters $E_1 = \tau_y / j$ and $E_2 = \tau_y / \eta_0$, and thereafter of a global effectiveness parameter $E_n = E_1E_2$. We also compare the results obtained in the different cases of micrometer-sized and nanometer-sized particles-based ER fluids.

1. Introduction

The fundamental property of the ER fluids [1] is their capacity to change reversibly from a natural liquid state to a solid-like state in the presence of an electric field E. In its liquid state ($E = 0$) they should have low viscosity in order to limit the viscous losses when the fluid flows. In the solid state ($E \neq 0$) it must withstand a strong yield stress to be able to transmit large mechanical forces. In addition, to increase the energy efficiency, it is necessary to reduce the consumption of electric current passing through the fluid. Finally, taking into account that the practical electrical field ranges from 3 to 5 kV/mm (electric field strength can be regarded as a nearly constant value) it can be considered that the main features of ER fluids are: i) the static yield stress and the current density, and ii) the zero-field viscosity. Since the discovery of the electrorheological (ER) effect more than fifty years, significant advances including the synthesis of new efficient ER fluids have emerged. The first important step forward led to replace the water-activated particles by the intrinsic conductivity particles [2,3]. This has provided more thermally stable fluids. The second significant development has emerged more recently. The use of nanometer-sized particles (having specific physical properties) has significantly increased the value of the static yield stress of the ER fluids under electric field. An increase in excess of 20 times or more of the yield stress appears in this case. This is known as the giant ER effect [4].

2. Characterization and effectiveness of ER fluids

2.1. Experimental procedure

The steady rheological response of ER fluids is usually modelled by the following equation:

$$\tau(E, \dot{\gamma}) = \tau_y(E) + \eta_p \dot{\gamma} = \eta_{app} \cdot \dot{\gamma} \qquad (1)$$

where the measured values are: the shear stress τ, the shear rate $\dot{\gamma}$, the electric field E and the calculated parameters are: the (static or dynamic) yield stress τ_y, the plastic viscosity η_p and the apparent viscosity η_{app}.
Unfortunately, the different parameters defined in Eq. (1) are more or less dependent on the experimental conditions. So, if we want to properly compare the electrical and rheological responses of ER fluids it is necessary to apply at the characterization stage a well established experimental procedure. The temperature of the ER fluids is an important parameter influencing the responses of the fluids (for the yield stress and current density particularly). Other specific experimental conditions are: application way of the electric field and mechanical stress to the fluid prior to the measurements, order, duration and magnitude of the application of the field and the shear rate (or shear stess) during experiment, type of experimental set-up (shear or squeeze flow), geometry of measurement cell (parallel plates, concentric cylinders …).
Thus it seems like a necessity to clearly specify the whole experimental conditions for the electrorheological measurements (ERM) and better still to define standards of ERM. In this case one may expect to compare more effectively the performances of the ER fluids.

2.2. *Proposal of an effectiveness ER parameter E_n*

Previous studies have highlighted a strong correlation between the yield stress and the electric current density for an ER fluid at rest or in solid-like state. This result is based on the theory (the conduction model) and the experience [5,6]. Furthermore it may also be noted that the recent nanoparticles-based ER fluids shows most of the time a relatively high value of the zero-field viscosity.

These remarks lead us to propose:

i) a first efficiency parameter E_1 in connection with the static yield stress τ_y and the corresponding current density j, at a given electric field strength:

$$E_1 = \tau_y / j \tag{2}$$

ii) a second efficiency parameter E_2 including the static yield stress and the zero-field viscosity η_0:

$$E_2 = \tau_y / \eta_0 \tag{3}$$

iii) an ER fluids effectiveness parameter E_n:

$$E_n = E_1 E_2 \tag{4}$$

For the micrometer-sized-based ER fluids the zero-field viscosity is still quite limited (typical values: $\eta_0 < 0.2$ Pa·s for a volume fraction $\phi \sim 25$ %) whereas for the nanometer-sized-based fluids the zero-field viscosity is widely greater ($\eta_0 > 2$ Pa·s). In some cases, for the latter ER fluids, a gel of the suspension at rest appears due to the strong interaction between nanoparticles. In this situation, one can used, a zero-field viscosity for a very low value of shear rate (for example $\dot{\gamma} = 1\ \mathrm{s}^{-1}$).

3. Some experimental results and discussion

We have gathered below some experimental data from the literature and from our previous works. We note that the data used (see legends of Figs. 1 and 2) correspond to a large variety of ER fluids (size and formulation of the dispersed phase, various host insulating liquid and volume fraction of the solid phase …). We note in figure 1, for E ranging from 3 to 5 kV/mm, a widely dispersion of the values of the static yield stress τ_y and current density j (τ_y and j varie over three order of magnitude) whereas a much closer values for the ratio τ_y / j (one order of magnitude). These results confirm the correlation between the static yield stress and the current density for ER fluids subjected to a dc electric field both for the micrometer-sized and the nanometer-sized ER suspensions.

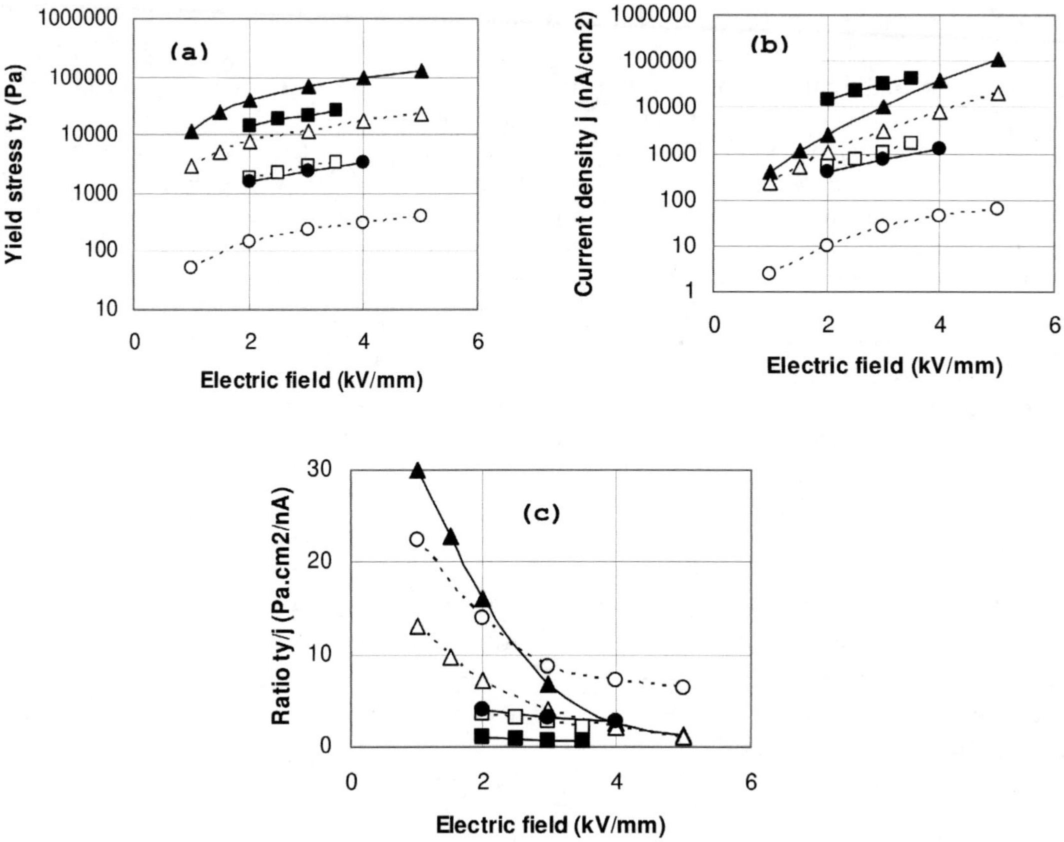

Figure 1. (a) Static yield stress τ_y, (b) current density j and (c) ratio τ_y / j vs applied electric field for various ER fluids. Symbols: nanometer-sized particle-based ERF, ▲△ [4], ■□ [7], ● Q. Quegan *et al.* unpublished, micrometer particles-based ERF, ○ [8].

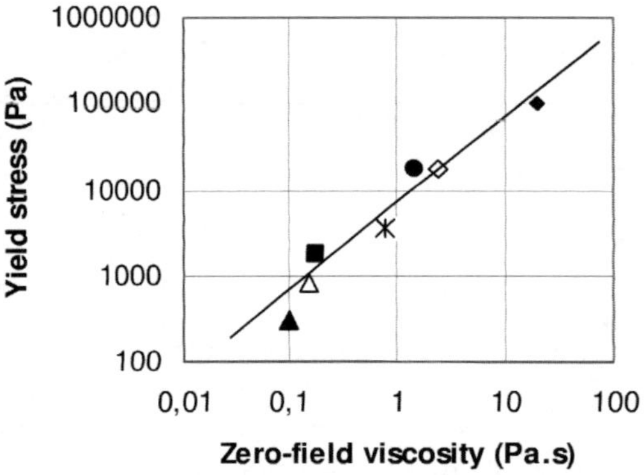

Figure 2. Static yield stress at $E = 4$ kV/mm vs zero-field viscosity for various ER fluids. Symbols: nanometer particles-based ERF, ◆◇ [9], ● [10], ✱ [11], ▲ Q.Guegan *et al.* unpublished, micrometer particles-based ERF ■ [12], △ [7]. Dots: experimental points. Solid line: linear fit.

Figure 2 shows the variation of the static yield stress τ_y as a function of the zero-field viscosity η_0 for a constant field strength E. A rather surprising result appears: all the data seem to follow a master straight line with a linear law (τ_y = k η_0) meaning that E_2 can be roughly considered as a constant value k (Eq. 3). Here we find for E = 4 kV/mm, E_2 = k ≅ 0.6 x10^4 s^{-1}. Noting that above E = 2 kV/mm the static yield stress of the ER fluids varies almost linearly with the applied field, we can consider in the first approximation that k is proportional to E: k = k'E with k' ≅ 0.15x10^4 (s kV/mm)$^{-1}$. Obviously, the experimental data used is not enough to justify a universal law for the dependence of the static yield stress on the zero-field viscosity for the ER fluids but this should be deepened in future work. Taking into account a constant value for E_2 (E_2 = k) for a given electric field strength we then obtain for the effectiveness parameter: $E_n = k\,E_1$.

If further experiments confirm that the zero-field viscosity is more or less proportional to the static yield stress (for a same field), this will imply that the ideal ER fluid concept (i.e. a strong yield stress and a low zero-field viscosity) is illusory. This situation is probably the consequence of the dispersion forces (van der Waals) between manometer-sized particles in "contact" and more generally to the importance of the dielectric interfacial properties at the nanometric scales [13]. However future progress on material science should lead to provide a better stabilization of the ER nanosuspensions and to obtain a satisfactory stress/viscosity compromise.

4. Conclusion

In this paper we proposed a global parameter E_n intended to compare the effectiveness of the ER fluids. Firstly, we have noted the importance to well define experimental conditions for the relevance of this comparison. Then, we have defined two basic efficiency parameters. The first one E_1 specifies the correlation between the ER effect (static yield stress) and the conduction properties of the suspension (current density). The second parameter E_2 shows other correlation based on the rheological responses of the fluid, under electric field (static yield stress) and without field (zero-field viscosity). It would appear that the zero-field viscosity rises with the static yield stress. This behaviour seems to exist for a large variety of ER fluids (among others, for the micrometer-sized and the nanometer-sized particles-based fluids) and may be explained by the size effect of the dispersed phase of the suspensions. Finally, an effectiveness ER parameter E_n is proposed taking into account the main chatacteristics of the ER fluids: static yield stress and current density at a given field strength, zero-field viscosity, and temperature (in this regard E = 4 kV/mm and T = 25 °C may be considered as a typical standard values).

References

[1] Gast A P and Zukoski C F 1989 *Adv. Coll. Int. Sci.* **30** 153
[2] Filisko F E and Radzilowski L H 1990 *J. Rheol.* **34** 539
[3] Block H Kelly J P Qin A and Watson T 1990 *Langmuir* **6** 6
[4] Wen W Huang X Yang S Lu K and Sheng P 2003 *Nature Materials* **2** 727
[5] Atten P Foulc J N and Felici N 1994 *Int. J. Mod. Phys. B* **8** 2731
[6] Wu C W and Conrad H 1996 *J. Phys. D: Appl. Phys.* **29** 3147
[7] Wu Q Zhao B Y Fang C and Hu K A 2005 *Eur. Phys. J. E* **17** 63
[8] Foulc J N and Atten P 1994 in *Proc. Int. Conf. on ER fluids* Tao R and Roy G D ed. World Scientific 358
[9] Wen W Huang X and Sheng P 2004 *Appl. Phys. Lett.* **85** 299
[10] Yin J and Zhao X 2005 *Int. J. Modern Phys. B* **19** 1071
[11] Wang X Shen R Wen W and Lu K 2005 *Int. J. Modern Phys. B* **19** 1110
[12] Ishino Y Maruyama T Ohsaki T Endo S Saito T and Goshima N 1995 in *Progress in electrorheology* Havelka K O and Filisko F E ed. Plenum 137
[13] Lewis T J 2004 *IEEE Trans. DEI* **11** 739

Journal of Physics: Conference Series **149** (2009) 012008 doi:10.1088/1742-6596/149/1/012008

New electrorheological fluids – characteristics and implementation in industrial and mobile applications

M Gurka[1,2], D Adams[1], L Johnston[1] and R Petricevic[2]

[1]Fludicon GmbH, Landwehrstrasse 55, 64293 Darmstadt, Germany

[2]ERF Produktion Würzburg GmbH, Landwehrstrasse 1, 97249 Eisingen, Germany

E-mail: gurka@erfgmbh.de

Abstract. Various applications for controllable dampers in the industrial and automotive sector are demanding an improved ER-Fluid, concerning performance and long time behavior. The new ER-Fluid RheOil®3.0 developed by us and presented here overcomes the main dis-advantages like sedimentation and re-dispersing behavior. Besides this, better ER-performance (control ratio, current-density, step response time) could also be achieved. During tests of the response to changing temperature and long time behavior no significant degradation of the fluid or abrasive wear in the components was found over a wide temperature range.

1. Introduction

The way to control movements of mechanisms or to realize vibration damping e.g. of critical components in large structures has changed since various so-called "smart materials" became available. For mechanisms, solid state actuators like piezo-ceramics or shape memory alloys are widely used, but for shock absorption in automotive or industrial applications the hydraulic principle is preferred. For this, two kinds of smart fluid technologies emerged in the last few years, relying on the magnetorheological (MR-Fluids) and electrorheological effect (ER-Fluids) respectively. Both technologies enable the construction of controllable dampers with a relatively simple design. Different damping characteristics are realized by modifying the fluid properties rather than by adjusting complex mechanical elements. Combining a direct, dynamic (<1 ms fluid reaction time), continuous and reversible manipulation of the apparent viscosity over a large control range with favorable basic hydraulic properties, ER-fluids are well suited for such intelligent hydraulic shock absorbers.

This article describes the development and testing of a new ER-Fluid (RheOil®3.0) with improved properties concerning the ER-behavior (max. shear strength, reduced current density and improved step response time), improved sedimentation and re-dispersing behavior and a wider usable temperature range. Therefore this fluid is particularly suited for use in dampers for the automotive sector and industrial applications. Together with improved system components (amplifiers, control algorithms and an advanced valve-design) the new ER-fluid was successfully implemented in standardized damper systems for industry and health-care [1] as well as a semi-active suspension system for off-road vehicles [2].

2. Experimental Results and discussion

2.1. Synthesis of ER Fluids and testing apparatus

The development was driven by the needs of industrial applications. For that, the ER-fluid of choice has to be based on LiCl or $ZnCl_2$ doped Polyurethane particles (PUR) dispersed in silicone-oil as base fluid. This type of ER-Fluid exhibits some outstanding properties: good ER-Properties, non abrasive behavior, low base-viscosity. It can be synthesized in a scalable process from standard raw-materials. The work, namely the optimization of polymer-composition and processing parameters, resulting in optimal particle morphology was carried out on a lab-scale production plant with the possibility to keep all relevant processing parameters under well-controlled conditions. Progress was checked with standard laboratory equipment as well as custom-made test rigs for measurement of ER-fluid-properties or testing the whole system. Part of the testing procedures is summarized in an industrial test-standard [3], some more data on the ERF-properties can be found in [4].

ER-specific tests were carried out on a custom-made test rig in flow-mode. Measurements were done under constant flow rate of 10000 s^{-1} and defined temperatures in the range between 25°C and +80°C. The ER-behavior was recorded by two pressure sensors before (inlet) and behind (outlet) a cylindrical ER-valve with an electrode diameter of 40 mm, a length of 100 mm and a valve height of 0.5 mm. The electric field was applied via a voltage amplifier RheCon2 from FLUDICON, with a step response time of 0.3 ms, which was quite below the sampling rate of 1kS/s per channel of the recording system. The mean temperature of the fluid was determined from sensors at the in- and outlet of the ER valve.

Other rheological measurements were carried out with a commercially available Rheometer MCR300 from Anton Paar with a double gap cylinder (DG26.7) for base viscosity measurements and a cylindrical high voltage cylinder (CC17) for the shear behavior under electrical field influence. These rheometric measurements, used for characterization of the long term behavior under elevated temperatures were carried out in the lower shear rate range between 0…450 s^{-1} under oscillating shear-mode in contrary to the measurements in flow-mode.

Re-dispersing measurements were carried out on the samples from the sedimentation tests (settling time 8 weeks). The remixing was done by a vortex-shaker (Type VV3, VWR International) which homogenizes the sample by circular motions of the glass tube. The remixing behavior was determined by the time needed for full homogenization while the vortex-shaker was set to a constant value.

2.2. Test results

2.2.1. *Sedimentation / Re-dispersing:* RheOil®3.0 shows improved sedimentation stability compared to the former commercially available ER-Fluid Rhe Oil®2.0 (also known as Rheobay 3566 developed by Bayer AG [5], [6], [7]) (Figure 1). More relevant for most applications is the re-dispersing behavior which could be improved by a factor of 25 for the new ERF. Compared to RheOil®2.0 the expended energy necessary for re-homogenization after settlement is a factor of 25 less for RheOil®3.0 which practically means that settlement becomes less critical for applications where fast access to the full performance of the damper is necessary even after a long term stand-by.

2.2.2. *Improved ER-performance:* Unique properties are constantly high shear stress of about 4000 Pa across a wide temperature range between 20°C and 80°C (Figure 2). Towards higher temperatures the current density increases starting at a very low level (Figure 2). Thus for an average operating temperature of 60°C the mean current density is in the range of 25 μA/cm², a factor of four less than for RheOil®2.0. The control ratio $(\sigma-\sigma_0)/\sigma_0$ is in the range of 18 (60°C, 10000 s^{-1}, 5 kV/mm) and can be increased to 25 (at 6 kV/mm). The field dependence of the shear force and current density can be approximated by a simple polynomial of the third degree (Figure 3). The higher standard deviation for the current density compared to the shear stress is due to the lower signal/noise ratio of the current monitor compared to the pressure signal.

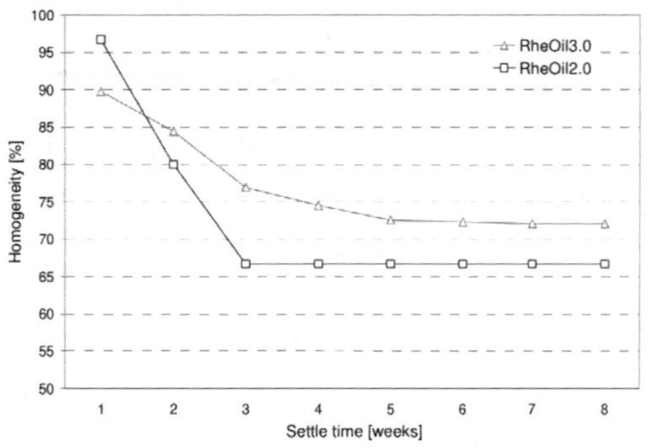

Figure 1. Sedimentation behavior of RheOil® 2.0 and RheOil® 3.0. Starting point was 100%. The new ERF shows a 15 % slower settlement.

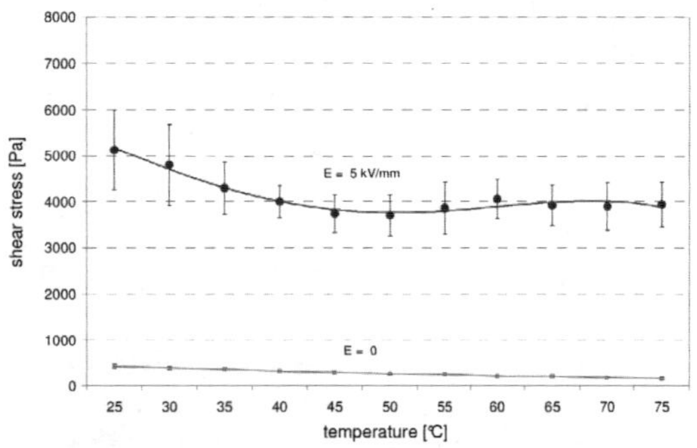

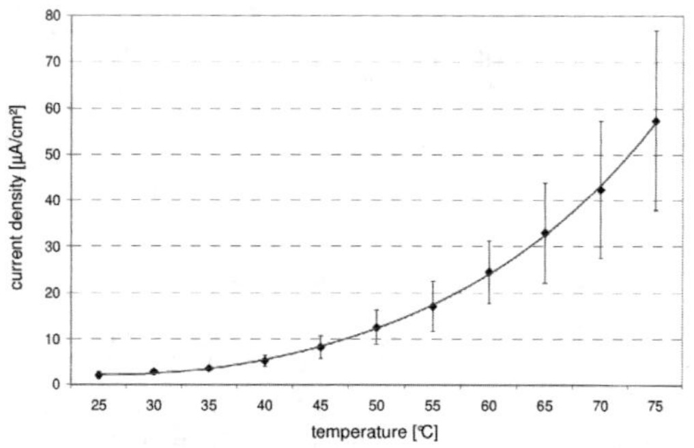

Figure 2. Mean shear stress vs. temperature (left) and current density vs. temperature (right) for RheOil®3.0 in flow-mode under applied high voltage (E = 5 kV/mm) and for voltage turned off (E = 0). Applied shear rate: 10000 s^{-1}.

Figure 4 shows an example of a shear stress step response in flow mode for a shear rate 10000 s^{-1} at 60°C after a field step from zero to 5kV/mm. Within the first millisecond, 90 % of the equilibrium shear stress (3700 Pa) is reached.

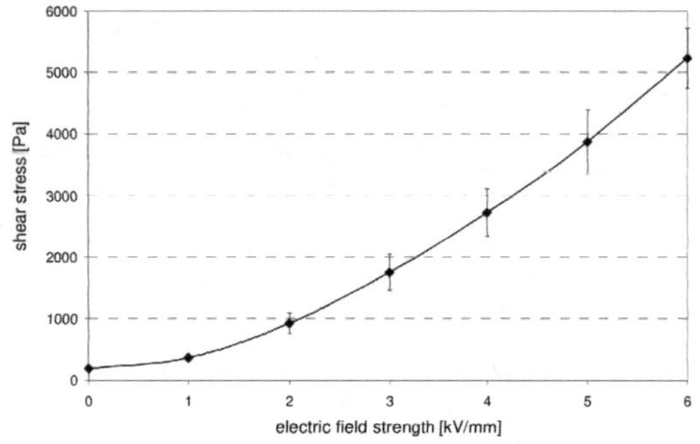

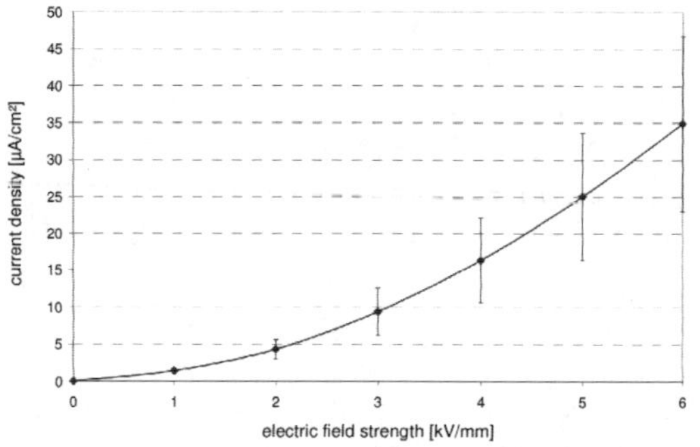

Figure 3. Mean shear stress vs. electric field strength (left) and mean current density vs. electric field strength (right) for RheOil®3.0 in flow-mode at a temperature of 60°C. Applied shear rate: 10000 s^{-1}.

2.2.3. *An extended usable temperature range* of RheOil®3.0 was proved in a range between -40°C and 120°C. Within the whole range the performance with regard to shear forces and current were evaluated.

It has been observed that the ER-effect decreases rapidly between 5°C and -5°C and approaches the base viscosity caused shear stress slope asymptotically at the polymers glassy point of -17°C. The base viscosity at this temperature approaches 200 mPas increasing to 500 mPas towards -40°C.

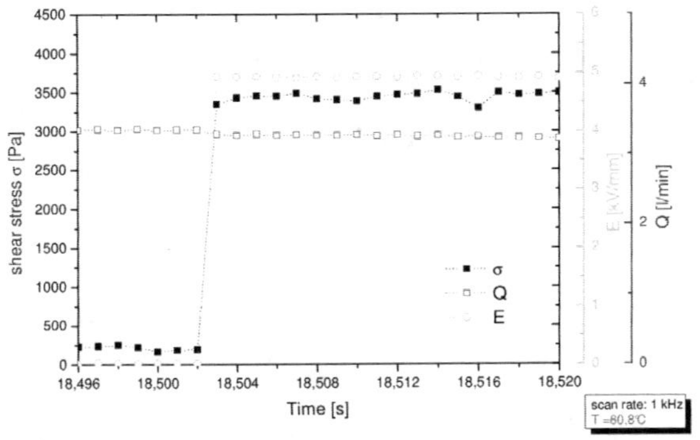

Figure 4. Step response of the shear force after E-field step from 0 to 5 kV/mm. (T= 60°C; shear rate 10000 s^{-1}). The Flow-Rate Q remains constant at about 3.2 l/min due to the large pump of the system.

From DSC/TG measurements fluid decomposition was observed to be relevant above 140°C. The fluid performance does not suffer after a short term (< 2 h) temperature cure in air at 120°C. Regarding the long term stability of the ERF under elevated temperatures the ER-response of the ERF on a cure at various temperatures and time is shown in Figure 5.

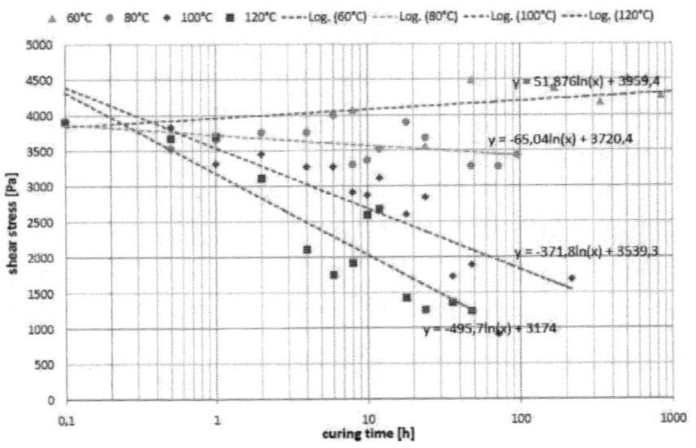

Figure 5. Response of the ER-shear force, measured by a Rheometer in shear-mode after curing the ERF at elevated temperatures (60, 80 100, and 120 °C) for up to 1000 h. The view-graph shows the ER-shear stress at shear rates of 150 s^{-1} and electric field strength of 5 kV/mm at 40°C cell temperature after a long term cure under air at above mentioned temperatures.

After a long term cure at 60°C no degradation was observed at all. The performance improved yet. For a long term temperature stress at 80°C the performance seems to change only slightly with time. Tempering at 100°C, however, leads to a measureable degradation of the shear stress as a function of time. The performance drop after 100 h at 100°C falls below 50 %. After a four hour cure at 120°C the ER-performance reveals a degradation of only 40 %.

In comparison, flow mode data measured under conditions comparable to Figure 2 suggest a degradation of about 10 - 15 % for the equally cured sample (Figure 6). The difference can be attributed to the different measuring conditions. Thus the above shown shear mode results give an upper range estimation for the thermal load capacity rather than an absolute behavior for a special application. Therefore, the MCR300 data might not be representative for the ER fluid under real application conditions, since neither air contact is expected nor a pure shear stress applies in common applications.

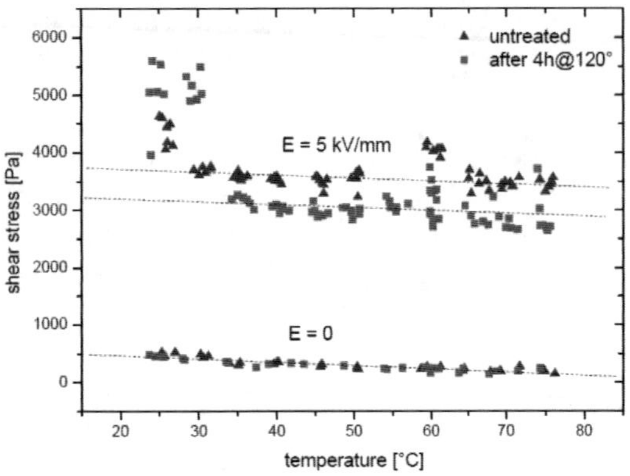

Figure 6. The graph shows shear stress measurements vs. operating temperature measured in the flow mode test rig under application relevant conditions.
The degradation of the ERF sample after a four hour cure at 120°C is in the range of 10-15 %. In contrasts to the MCR 300, the flow mode test rig operates with larger fluid volume (> 2000 ccm) and at larger shear rates (10000 s^{-1}).

2.3. Testing in Applications

The new RheOil®3.0 was successfully tested in various applications. For industrial applications the ER-Fluid succeeded various performance and long-term tests in the now commercially available damper series RheDamp, where a total damping energy of up to $1.7*10^7$ J was dissipated in about 200 ml of ER-fluid. Remarkably, neither the fluid nor the sealing of the damper was affected during the whole test with a total stroke of about 40 km, which can be awarded to the excellent non-abrasive behaviour of the Polyurethane-based ERF. For automotive applications RheOil®3.0 was successfully tested in a semi-active suspension system for a military off-road vehicle. The excellent response time and very low power consumption allowed for an outstanding field-test performance.

3. Conclusion

The new ER-Fluid RheOil®3.0 exhibits major advantages over former existing ER-Fluids or other controllable fluids. In laboratory measurements and practical testing under field-test conditions, advantages for the practical use could be proved: outstanding sedimentation and re-dispersing behaviour, improved ER-performance – short response time (< 1 ms), large control ratio (up to 25), low base viscosity (70 mPas@25°C) and current density, broad working temperature range (-20…+120°C) – and non-abrasive behaviour.

References

[1] Johnston L 2008 Electrorheological Dampers for Industrial and Mobile Applications – an Overview of Design Variations, Product Realisation and Performance *Actuator 2008, 11th International Conference on New Actuators* **Proceedings** 499
[2] Schneider S, Holzmann K, Kemmetmüller W, Johnston L, Lißman D 2008 Development and Testing of a Semi-active Suspension System for Off-Road Trucks using Electrorheological Dampers *Actuator 2008, 11th International Conference on New Actuators* **Proceedings** 485
[3] DIN 51480-1 Elektrorheologischen Suspensionen — Teil 1: Grundlagen und Definitionen *Industrial Test Standard* **DIN 51480-1**
[4] ERF Production Würzburg GmbH (2008) Product Information on RheOil 3.0
[5] Bloodworth R, Wendt E, (1994) Polymer Preprints 35:356
[6] Bayer AG, Leverkusen (1994) Provisional product information, Bayer Silicone
[7] Zaun M, Elektrorheologische Ventile als Stellelemente in Zylinderantrieben, Shaker Verlag 2007, ISBN 978-3-8322-6207-5.

Haptic Device for Vehicular Instrument Controls Using Electrorheological Fluids

Y M Han[1], K W Noh[1], S B Choi[1] and H J Choi[2]

[1] Smart Structures and Systems Laboratory, Department of Mechanical Engineering, Inha University, Incheon 402-751, Korea

[2] Department of Polymer Science and Engineering, Inha University, Incheon 402-751, Korea

E-mail: seungbok@inha.ac.kr

Abstract. This paper presents control performances of an electrorheological (ER) fluid-based multifunctional haptic device which is applicable to realization of in-vehicle comfort functions. By combining the functions into a single device, the proposed haptic device can transmit various reflection forces for each comfort function to a driver without requiring the driver's visual attention. As a multifunctional haptic device, a single ER knob, which is capable of both rotary and push motions, is designed and manufactured. In-vehicle comfort functions are constructed in virtual environment which makes the functions to communicate with the haptic device. Subsequently, a feed-forward controller using torque/force maps is formulated for the force tracking control. Control performances such as reflection force of the haptic device are experimentally evaluated via the torque/force map-based feed-forward controller.

1. Introduction

In recent years, vehicles provide various comfort functions to a driver such as audio, window, road and traffic information, air conditioner, etc. The growing number of comfort functions requires for several different control devices to be installed in dashboard as secondary controls of vehicles. These various secondary control devices can decrease the attention of a vehicle driver, and which is one of the most common causes of traffic accidents [1]. However, by combining the in-vehicle functions, information can be transmitted without requiring a driver's visual attention. Therefore, several research works have been performed on driver-vehicle interaction. Among them, haptic devices with the sense of touch have been actively researched to consolidate the in-vehicle functions into a single device [1-4]. The traditional consolidating control device using motors has difficulty in realizing precise tactile feeling and stopping force [1]. Therefore, some researches have been recently made by adopting smart materials such as electrorheological (ER) fluids whose advantages are fast response, simple mechanism, continuous force control and high stability [2-4]. However, these research works are still limited to rotary motion although the in-vehicle comfort functions need not only rotary motion but also push motion. Furthermore, advanced haptic device should be evaluated in feasible situation such as integrated haptic architecture of driver and vehicle. Consequently, the main contribution of the

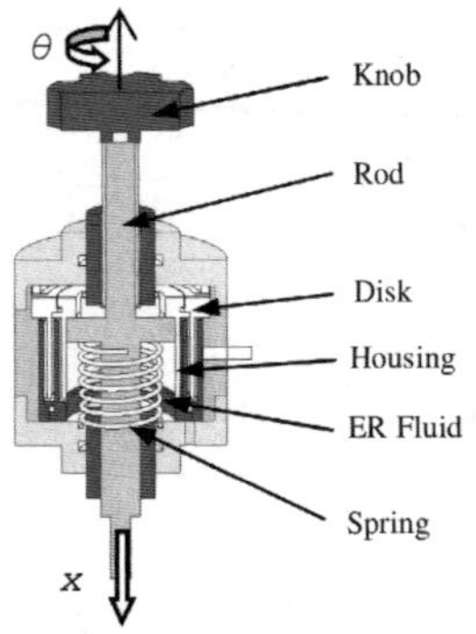

Figure 1. Schematic configuration.

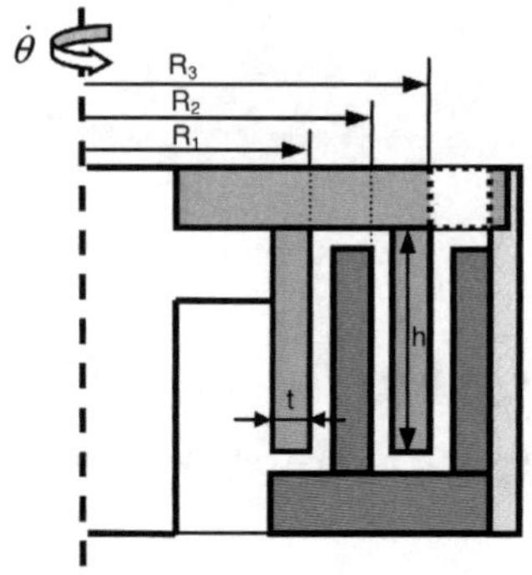

Figure 2. Geometric parameters.

present work is to propose the ER fluid-based haptic knob and to evaluate its control performances within integrated haptic architecture of driver and in-vehicle functions.

2. ER fluid-based haptic knob

In this work, in order to achieve both rotary motion and push motion within a single device, a cylindrical type mechanism is devised. Figure 1 shows a schematic configuration of the proposed multifunctional haptic knob. The knob is composed of upper disk electrode, lower disk electrode and reaction spring. The between upper and lower disk is a specific gap fully filled with ER fluid. The ER fluid has been employed as a controllable fluid. When a electric field of E is applied to the ER fluid, the shear stress (τ) can be expressed by

$$\tau = \tau_y(E) + \eta\dot{\gamma}, \qquad \tau_y(E) = \alpha E^{\beta} \tag{1}$$

where η is the dynamic viscosity, γ is the shear rate, and $\tau_y(E)$ is the dynamic yield stress of the ER fluid. α and β are the intrinsic values of the ER fluid. E is the input voltage which is the control input for the ER haptic knob.

Among several operation modes of an ER fluid-based device, in this work, the shear and flow modes are significantly developed for the rotary and push motions, respectively. Therefore, the torque(T)/force(F) model is mathematically described for each motion as follows:

$$\begin{aligned} T &= \sum_{i=1}^{3}\left\{2\pi R_i^{\,2} h + \pi\left(R_i^{\,2} - (R_i - t)^2\right)\frac{R_i + (R_i - t)}{2}\right\}\alpha(E)^{\beta}\,\mathrm{sgn}(\dot{\theta}) \\ &\quad + \sum_{i=1}^{3}\left\{2\pi R_i^{\,3}\eta\dot{\theta} + \eta\frac{R_i}{h}\dot{\theta}\pi\left(R_i^{\,2} - (R_i - t)^2\right)\frac{R_i + (R_i - t)}{2}\right\} + C_{cf}\,\mathrm{sgn}(\dot{\theta}) + C_{vf}\dot{\theta} \\ F &= 3\frac{c(h+t)}{g}(A_p - A_r)\,\alpha(E)^{\beta}\,\mathrm{sgn}(\dot{x}) + \sum_{i=1}^{3}\frac{12\eta(h+t)}{w_i g^3}(A_p - A_r)^2\dot{x} + kx + D_{cf}\,\mathrm{sgn}(\dot{x}) + D_{vf}\dot{x} \end{aligned} \tag{2}$$

where $\dot{\theta}$ is the rotational velocity of the knob in rotary motion. g is the gap size. c is a coefficient which depends on flow velocity profile. w is the electrode width. x and $\dot{x}$ are the displacement and velocity of the knob in push motion. k is a spring constant. C_{cf} and D_{cf} are the coefficient of Coulomb friction for rotary mode and push mode, respectively. C_{vf} and D_{vf} are the coefficient of viscous friction for rotary mode and push mode, respectively. A_d and A_r are the disk area and the rod area, respectively. $\mathrm{sgn}(\cdot)$ is a signum function. The other geometric parameters are shown in Figure 2 and the proposed ER haptic device is then manufactured and shown in Figure 3.

Journal of Physics: Conference Series **149** (2009) 012009 doi:10.1088/1742-6596/149/1/012009

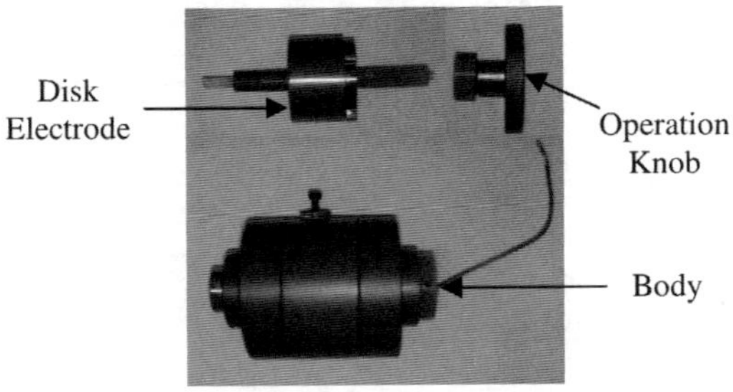

Figure 3. Manufactured haptic knob.

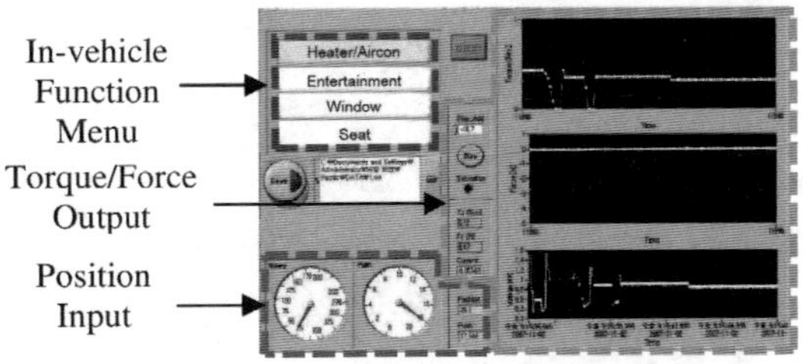

Figure 4. Virtual environment.

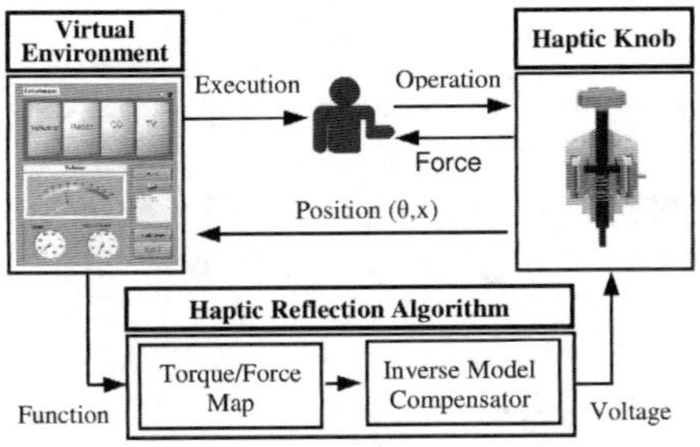

Figure 5. Haptic architecture.

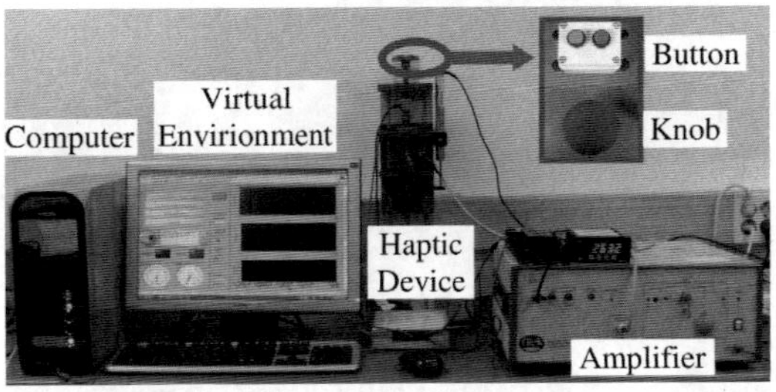

Figure 6. Experimental configuration.

3. Haptic architecture with virtual environment

In this work, a virtual environment is established by considering control operations of various in-vehicle functions within a real passenger vehicle such as menu shifting, window opening, etc. Figure 4 shows the established virtual environment which has the data display windows for input and output measurements, and four in-vehicle function menus: heater/air conditioner, entertainment, window and seat. The proposed haptic architecture is composed of the manufactured ER haptic knob, virtual in-vehicle functions, torque/force reflection algorithm and driver as shown in Figure 5. The haptic knob then interacts with the virtual environment of in-vehicle functions. When the driver operates the ER knob, the position information is transferred to the virtual environment. Then the torque/force map to generate the desired torque and force for the ER knob is selected according to the corresponding event such as menu shift event, window opening event, etc.

In order to achieve the desired torque and force at the ER knob, a feed-forward controller based on inverse model of the torque/force model in Equation (2) is adopted in this work.

$$V_T = g\left(\frac{\tilde{T}}{\sum_{i=1}^{3}\left\{2\pi R_i^{\,2}h + \pi\left(R_i^{\,2} - (R_i - t)^2\right)\frac{R_i + (R_i - t)}{2}\right\}\alpha\,\mathrm{sgn}(\dot{\theta})}\right)^{\frac{1}{\beta}},\ V_F = g\left(\frac{\tilde{F}}{3\frac{c(h+t)}{g}(A_p - A_r)\,\alpha\,\mathrm{sgn}(\dot{x})}\right)^{\frac{1}{\beta}} \quad (3)$$

where V_T and V_F are the input voltages for rotary motion and push motion, respectively, as control inputs. $\tilde{T}$ and $\tilde{F}$ are calculated by eliminating fluid viscous term and friction term from the desired torque and force trajectories which are obtained by the torque/force map upon the measured position information.

4. Results and discussions

Figure 6 shows the experimental configuration to evaluate control performance of the proposed multifunctional haptic device in the virtual environment realized by LabVIEW. The knob connected to the disk is operated by a person who is a vehicle driver. When he or she operates the ER knob, the

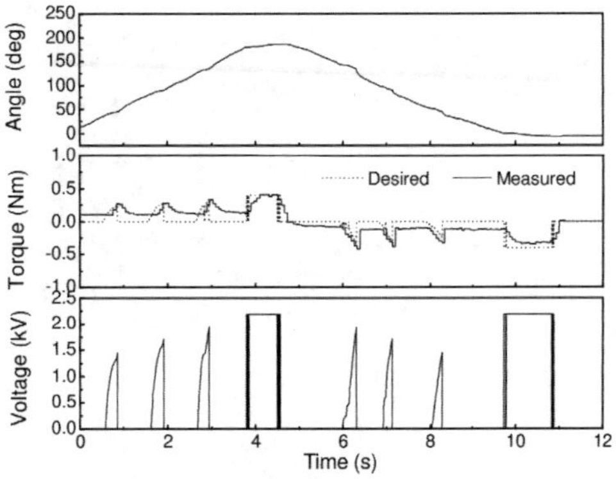

Figure 7. Control results for menu shifting.

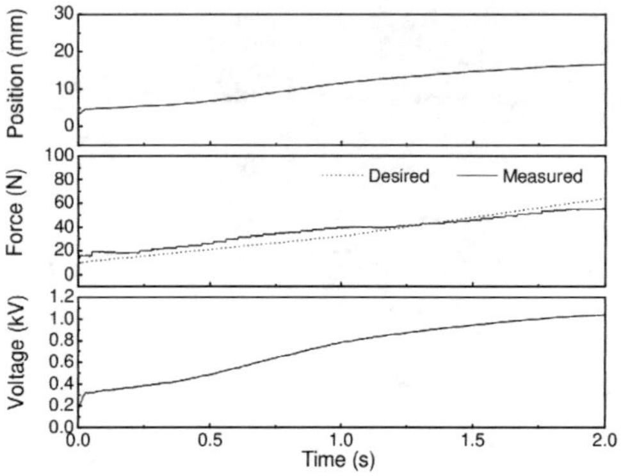

Figure 8. Control results for window opening.

position information is obtained from the encoder and LVDT. A specific in-vehicle function in the virtual environment is then visually operated and generates the desired torque or force by the map upon the position information. The feed-forward controller given by Equation (3) is then activated for the ER knob to reflect the desired torque or force to a driver. In this work, the torque or force is measured by a 6 axes force/torque transducer (ATI, Nano25). Figure 7 shows torque tracking control results for the event of menu shifting in rotary motion. A person rotates the knob from 0 deg to 180 deg, and then rotates back to original position. In this work, the event of menu shifting occurs at 45 deg , 90 deg and 135 deg, which are the shifting positions from heater/air conditioner to entertainment, from entertainment to window, and from window to seat, respectively. And the stopping torque is applied at 0 deg and 180 deg. As clearly observed from the results, the desired torque from virtual environment is favourably followed by the feed-forward controller. Figure 8 shows force tracking control results for the event of window opening in push motion. According to the displacement of knob, the window opening position was adjusted from 0mm to 20mm and the corresponding force from 0N to 80N was reflected to a driver. From the results, it is observed that the desired force is well generated by the virtual in-vehicle function and achieved from the ER knob by activating the controller.

5. Conclusion

In this work, a new type of electrorheological (ER) fluid-based haptic knob was proposed as a multifunctional haptic device and its control performances were evaluated in virtual environment of in-vehicle comfort functions. The ER haptic knob has been innovated with actuation mechanism capable of both rotary and push motions with a single knob. The manufactured knob was interacted with the virtual environment, in which various events of in-vehicle functions including menu shifting and window opening were constructed by LabVIEW. Subsequently, it has been demonstrated that the desired torque and force trajectories were well achieved for both rotary and push motions via the feed-forward controller based on inverse model. The results presented in this work are quite satisfactory justifying that the proposed haptic device is one of potential candidates which can be effectively utilized in vehicular comfort functions.

References

[1] BMW iDrive Controller http://www.bmwworld.com /models/e65.htm
[2] Vitrani M A, Nikitczuk J, Morel G, Mavroidis C and Weinberg B 2006 *J. Dyn. Syst.-T. ASME* **128** 216
[3] Li W H, Du H, Guo N Q and Kosasih P B 2004 *Sensor Review* **24** 68
[4] Han Y M and Choi S B 2006 *Smart Mater. Struct.* **15** 1438

Journal of Physics: Conference Series **149** (2009) 012010 doi:10.1088/1742-6596/149/1/012010

Electrorheological behavior of titanate nanotubes suspensions under oscillatory shear

Y He[1,2], Q Cheng[1], V Pavlínek[1,3], C Li[2] and P Sáha[1]

[1] Polymer Centre, Faculty of Technology, Tomas Bata University in Zlin, TGM 275, 762 72 Zlin, Czech Republic
[2] Key Laboratory for Ultrafine Materials of Ministry of Education, School of Materials Science and Engineering, East China University of Science and Technology, 200237 Shanghai, China

E-mail: pavlinek@ft.utb.cz

Abstract. In this study, titanate nanotubes (TNTs) were synthesized by hydrothermal treatment of TiO_2 powder (P25) in a NaOH solution. The as-synthesized TNTs exhibit high surface area and large aspect ratio. Rheological properties of TNTs suspensions were then investigated under oscillatory shear. The TNTs fluid shows viscoelastic behavior and the dynamic moduli (G',G'') increase significantly by about 4 orders of magnitude as the electric field strength is up to 2.0 kV/mm. The complex modulus of TNTs fluids is sensitive to temperature while that of P25 fluid become insensitive at higher temperature. Dynamic viscoelastic behavior suggests that structure of P25 to TNTs transition merits the enhancement of ER activity of TNTs fluid.

1. Introduction

The past few years have witnessed significant progress in the design and fabrication of versatile nanomaterials with novel structures due to their unique properties. Among such materials, one-dimensional (1D) nanostructures (nanotubes, nanowires and nanofibers) have attracted extraordinary attention recently, mainly because of their novel physical properties as well as potential applications in nanoscale electronic and optoelectronic devices. However, very recently interesting electrorheological (ER) behavior induced by 1D nanomaterials based fluids [1-3] has also become the research focus inspired by the discovery of the giant electrorheological (GER) effect [4].

1D nanotubes have received particular interest in ER systems owing to their high aspect ratios and good suspended stability, which probably favor the enhanced ER activity. In this respect, Zhao et al. recently reported a new ER fluid based on titanate nanotubes (TNTs) in steady flow, which shows better dispersion stability and a notable ER activity [5,6]. In this work, in order to further investigate the changes of the microstructures of 1D nanotubes based ER fluids, the ER behavior of TNT fluids is measured under oscillatory shear. Investigation of ER properties of TNTs by the dynamic oscillation tests will be helpful in understanding the nature of the interactions among particles forming the internal structures which is vital for the optimization of 1D ER materials.

2. Experimental

2.1. Synthesis of TNTs

TNTs were prepared according to the previous report [7] with a minor modification. In a typical synthesis, 1 g of TiO_2 powder (P25, Degussa AG, Germany) was mixed with 60 ml of 10 M NaOH solution, which was followed by hydrothermal treatment of the mixture at 130 °C in a 100 ml Teflon-lined autoclave for 24 h. The white precipitate was separated by filtration and washed with ethanol and distilled water until pH~7 was reached. The samples were then dried in an oven at 60 °C for 12 h.

2.2. Preparation of ER fluids

The apparent density of TNTs was measured to be around 1.95 g/cm^3. Before the preparation of the suspensions, the dried TNTs were calcined in an oven at 250 °C for 2 h to get dehydrated TNTs. Then the TNTs and dried P25 particles were dispersed by ultrasonic in silicone oil (Fluid 200, Dow Corning, UK; viscosity η_c = 108 mPa s, density ρ_c = 0.965 g/cm^3) to form stable suspension with a particle volume fraction of 5 % (v/v) for each material separately.

2.3. Electrorheological measurements

Measurements of rheological properties of the prepared fluids were carried out using a coaxial cylinder viscometer (Bohlin GEMINI, Malvern Instruments, UK). The suspensions were placed in the Couette cell with the rotating inner cylinder of 14 mm diameter and the outer cylinder separated by a 0.7 mm gap. They were connected to a DC power supply producing a field strength E = 0–2 kV/mm. Further, dynamic viscoelastic tests were performed by dynamic strain sweeps and frequency sweeps. The strain sweep was carried out with applied strains of 0.0001 to 1.0 at a frequency of 62.8 rad/s under an electric field to determine the linear viscoelastic region. The rheological parameters were then obtained from the frequency sweep tests (0.1 to 100 rad/s) at a fixed strain amplitude in the linear viscoelastic region. All experiments were performed at 25 ºC.

3. Results and discussion

3.1. Characterization of materials

The morphologies of P25 nanoparticles and TNTs were characterized by TEM. Most particles are nearly globular and the average particle size of P25 is around 20 nm. While after hydrothermal treatment, the P25 particles almost convert into TNTs with diameters in the range of 8-10 nm and lengths of up to several hundreds of nanometers. The results are in good agreement with previous report [8]. Meanwhile, TNTs exhibit obviously larger length-diameter ratios than that of P25 particles.

The structure effects of P25 particles to TNTs transition were further investigated by the N_2 adsorption/desorption isotherm. As P25 is converted into TNTs, there is a remarkable increase for S_{BET} from 49.7 to 381.9 m^2/g and for pore volume from 0.05 to 1.02 cm^3/g. This significant increase in surface area and pore volume mainly results from the formation of tubular nanostructures.

3.2. Rheological properties of ER suspensions

To study the dynamic characteristics of microstructures formed in ER suspension, oscillatory shear tests were performed. Figure 1 depicts storage modulus, G', and loss modulus, G'', versus strain for TNTs fluid under various electric field strengths. Without an electric field, the suspension shows liquid-like behavior, i.e. viscosity is dominant over elasticity and G'' is larger than G'. When an electric field is applied to the TNTs suspension, both G' and G'' increase rapidly with the electric field strength over the whole strain range, but G' becomes significantly higher than G'' in the linear viscoelastic region. Especially in the case of E = 2.0 kV/mm, G' is higher than that at zero electric filed by four orders of magnitude. This dramatic change in rheological properties arises from the formation of chain-or column-like structures in TNTs suspension. Under an external electric field the polarized TNTs particles attract each other due to dipole-dipole interactions to form particle chains along the field direction. These chain-like structures make the suspension in the electric field to behave as solid at small strains. With increasing *E* the particular microstructures become stiffer to sustain the larger strain, and the elasticity dominates viscosity (i.e. $G' > G''$) in the linear viscoelastic region. However, as the strain is increased, the chains experience continuous deformation and finally destruction occurs above some critical value, and the TNTs suspension starts to flow demonstrating viscous behavior (i.e. $G'' > G'$) [9,10]. Note that with increasing *E* linear viscoelastic region is shorter and the critical strain becomes lower, indicating the particle chain-like structures are also fragile at relatively small strain.

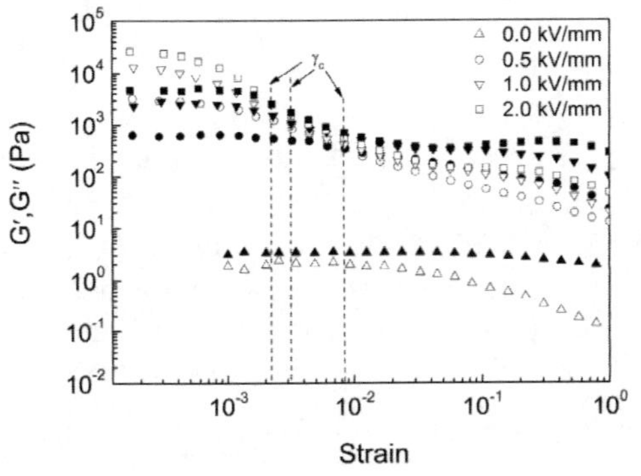

Figure 1. Strain dependence of G' and G'' for 5 vol.% TNTs ER fluid under different electric field strengths: open symbols for G', solid symbols for G''.

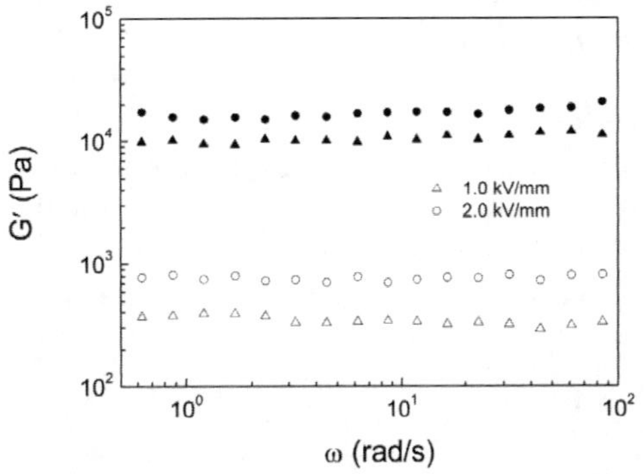

Figure 2. G' as a function of frequency at E = 1.0 and 2.0 kV/mm for both suspensions: open symbols for P25, solid symbols for TNTs.

The structural effect of P25 to TNTs transition on viscoelastic properties of both fluids are also reflected in figure 2. It is evident that G' for TNTs fluid is much higher than that for P25 fluid. As mentioned above, TNTs possess larger aspect ratios and higher surface area than P25, which may bring about large interfacial polarization of TNTs in suspension. Larger interfacial polarization in turn merits the optimal ER effect. Therefore, the stronger polarizability of TNTs particles in ER fluid induced by the electric field results in the stiffer chain structures showing a large G' than that of P25 fluid in the linear viscoelastic region. Similar behavior of TNTs in steady shear was also observed in [6].

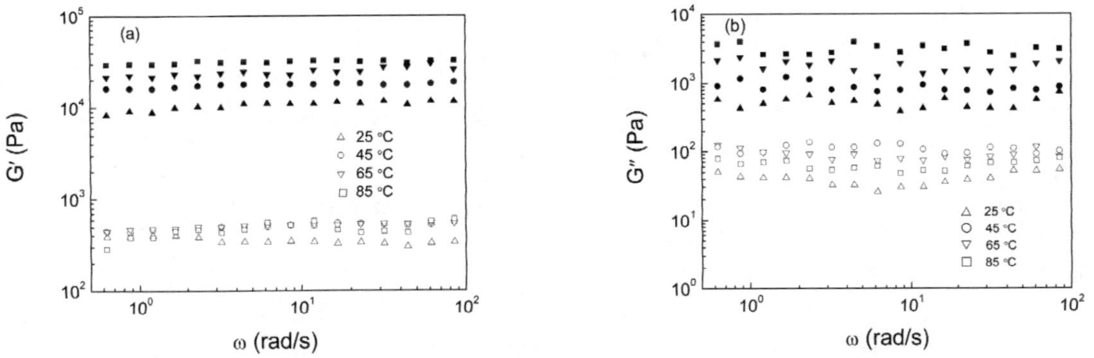

Figure 3. Dependence of storage modulus G' (a) and loss modulus G'' (b) for both suspensions on frequency at various temperatures (E = 1.0 kV/mm): open symbols for P25, solid symbols for TNTs.

Viscoelastic properties are also influenced by the temperature. G' and G'' dependence on frequency for both suspensions at various temperatures, T, is shown in figure 3. As can be seen, G' for TNTs fluid is larger than that for P25 fluid at the same E and T, and it is nearly independent of frequency. With increasing T, G' for TNTs fluid increases whereas that for P25 fluid does not follow this trend; and there is no distinct difference in G' between 45 and 85 °C. Meanwhile, G'' for both fluids shown in figure 3b also shows similar change with T. Usually, the temperature effect on the rheological properties of ER suspension includes the heat-induced change of polarization behavior and the thermal convection of fluid [11,12]. Recent study [6] indicates that the temperature positively affects the dielectric or polarization properties of Na-titanate nanotubes based suspension. Thus, the heat-induced polarization is dominant and the chain structures become more rigid with increasing T, manifested in an increase of G'. As for P25 fluid, with increase T the thermal convection of fluid dominates gradually to weaken the structures formed by polarized P25 particles, so G'' start to decrease at higher temperature (e.g. 85 °C).

Conclusions

Hydrothermal treatment of P25 in a NaOH solution results in the formation of TNTs with large surface area. This TNTs based fluid exhibits typical viscoelastic behavior under applied electric field. In comparison with P25, TNTs fluid shows larger G' in the linear viscoelastic region, which originates from the large aspect rations and high surface area of TNTs. Furthermore, the temperature plays a different impact on the dynamic modulus of both suspensions due to their variously inherent chemical structure. Investigation of ER properties of TNTs fluid will further favor the optimization of 1D ER materials.

Acknowledgements

This work was supported by the Ministry of Education, Youth and Sports of the Czech Republic (MSM 7088352101) and Grant Agency of the Czech Republic (202/06/0419) for financial support. The authors also wish to thank to the National Natural Science Foundation of China (20236020), the National High Technology Research and Development Program of China (2006AA03Z358), and 973 program (2004CB719500).

References

[1] Feng P, Wan Q, Fu X Q, Wang T H and Tian Y 2002 *Appl. Phys. Lett.* **87** 033114
[2] Lozano K, Hernandez C, Petty T W, Sigman M B and Korgel B 2006 *J. Colloid Interface Sci.* **297** 618
[3] Lin C and Shan J W 2007 *Phys. Fluids* **19** 121702
[4] Wen W, Huang X, Yang S, Lu K and Shen P 2003 *Nat. Mater.* **2** 727
[5] Yin J and Zhao X 2006 *Nanotechnology* **17** 192
[6] Yin J and Zhao X 2008 *Colloid Surf. A* **329** 153
[7] Kasuga T, Hiramatsu M, Hoson A, Sekino T and Niihara K 1999 *Adv. Mater.* **11** 1307
[8] Lan Y, Gao X, Zhu H, Zheng Z, Yan T, Wu F, Ringer S P and Song D 2005 *Adv. Mater.* **15** 1310
[9] Cho M S, Choi H J and Ahn W S 2004 *Langmuir* **20** 202
[10] Hiamtup P, Sirivat A and Jamieson A M 2006 *J. Colloid Interface Sci.* **295** 270
[11] Block H and Kelly J P 1988 *J. Phys. D: Appl. Phys.* **21** 1661
[12] Conrad H, Sprecher A F, Choi Y and Chen Y 1991 *J. Rheol.* **35** 1393

Journal of Physics: Conference Series **149** (2009) 012011 doi:10.1088/1742-6596/149/1/012011

Modeling and control of an off-road truck using electrorheological dampers

K Holzmann[1], W Kemmetmüller[1], A Kugi[1], M Stork[2], H Rosenfeldt[2] and S Schneider[3]

[1] Automation and Control Institute (ACIN), Vienna University of Technology, Gusshausstrasse 27-29 / E376, A-1040 Vienna, Austria

[2] FLUDICON GmbH, Landwehrstrasse 55, D-64293 Darmstadt, Germany

[3] Bundeswehr Research Institute for Materials, Explosives, Fuels and Lubricants (WIWEB), Institutsweg 1, D-85435 Erding, Germany

E-mail: kugi@acin.tuwien.ac.at

Abstract. This work deals with the mathematical modeling and control of the semi-active suspension of an MAN off-road truck with a payload of 5 tons which comprises electrorheological dampers. Thereby, a cascaded control structure with four controllers for the control of a quarter-car in the inner control loop and a superimposed control strategy for the overall vehicle is used. The main goal of the control strategy is to reduce the motion of the chassis (especially roll, pitch and vertical movement) while increasing driving stability. The capability of the overall control strategy is demonstrated by means of simulation studies and measurement results.

1. Introduction

Increasing demands on driving comfort and driving stability over the last years require the development of new active and semi-active suspension systems for automotive applications. In general, in semi-active suspension systems as considered in this paper, the damping characteristics of the shock absorber has to be varied over a large range. Nowadays, this change of the damping characteristics is in most cases achieved by using electro-hydraulic adjustable dampers, see, e. g., [1].

In order to circumvent the drawback of an electro-hydraulic suspension system regarding e. g. the dynamics, magnetorheological and electrorheological semi-active suspension systems were introduced in the recent years, see, e. g., [2], [3] and [4]. Such dampers usually exhibit a symmetrical damping characteristics whereas passive shock absorbers used in this field of application possess an asymmetrical damping characteristics, i. e. the damping is considerably higher for rebound than for compression. This is mainly due to the fact that such a characteristics yields a good compromise between driving stability and driving comfort. Therefore, one basic idea in the design of the ER dampers used in this work was to adopt the asymmetrical behavior of a passive automotive shock absorber and to combine it with an active adjustment of the damping characteristics by means of an ER valve.

Based on the mathematical modeling of the ER damper and the vehicle in the next section, a nonlinear control strategy for the damper and the vehicle is developed, which is designed to meet the requirements on driving stability and comfort. The feasibility of the proposed control strategy is shown by means of simulation studies and measurement results.

2. Mathematical Modeling

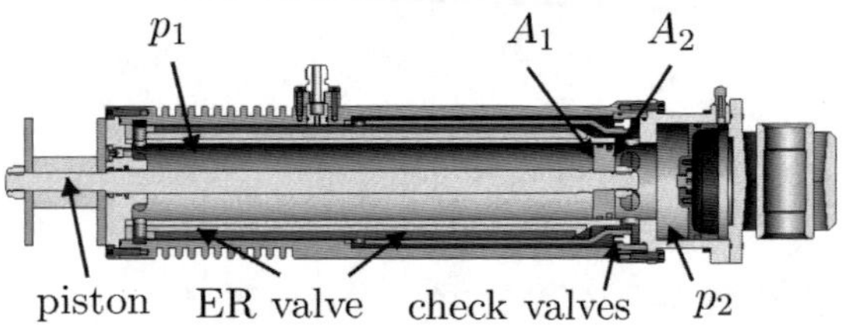

Figure 1. Longitudinal section of the ER damper.

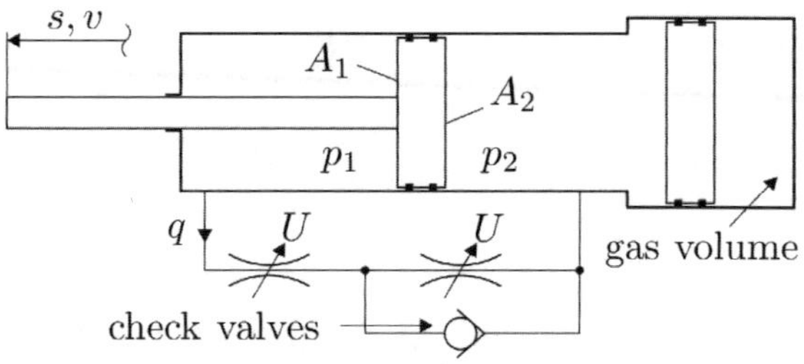

Figure 2. Simplified schematics of the ER damper in figure 1.

A longitudinal section of the ER damper is depicted in figure 1. Therein, A_1 and A_2 are the effective piston areas and p_1 and p_2 denote the pressure in chamber 1 and 2 of the cylinder. Furthermore, s and $v = \frac{\mathrm{d}}{\mathrm{d}t}s$ denote the position and the velocity of the piston. In figure 1, it can be seen that the ER valve forms an annular gap and is together with a number of check valves integrated within the ER damper. The corresponding schematics in figure 2 shows the fluid path between the two chambers composed of the ER valve and the check valves. During compression ($v < 0$), the check valves are open and the left hand part of the ER valve dominates the pressure difference while during rebound ($v > 0$) the check valves are closed and the fluid has to flow through the full length of the ER valve leading to higher pressure differences and thus to higher damping forces. The mathematical modeling of a quite similar ER damper prototype under the assumption of a Bingham-like material model for the ER fluid is show in [5]. Accordingly, the volume flow q through the ER valve and the check valves can be derived in the form $q = q(p_2 - p_1, U)$ with U the voltage applied to the ER valve. For the subsequent mathematical modeling of the truck, one is mainly interested in the total force $F_{ER} = F_d + F_c$ generated by the ER damper acting on the truck, where the damping part $F_d = A_1(p_2 - p_1)$ is responsible for the dissipation of energy and the part $F_c = (A_2 - A_1)p_2$ refers to a spring force resulting from the different areas A_1 and A_2 of the damper.

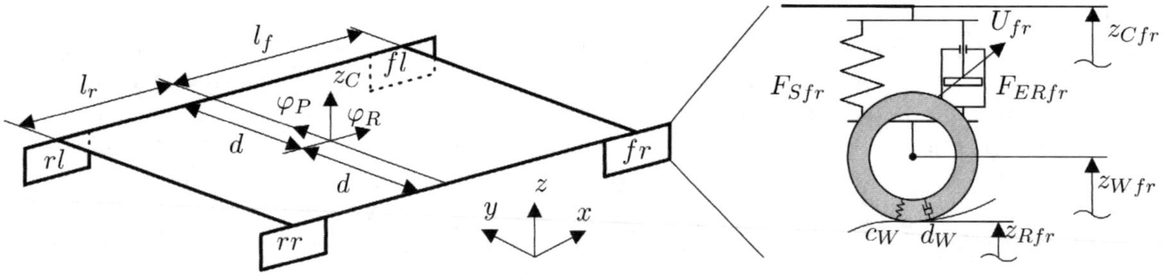

Figure 3. Schematics of the vehicle with the relevant degrees-of-freedom.

The mathematical model of the truck incorporates the vertical degree-of-freedom and the roll and pitch movement of the chassis while effects like breaking, cornering or acceleration are not considered. This leads to the generalized positions of the chassis $\mathbf{p}_C = [\varphi_R, \varphi_P, z_C]^T$ with the roll angle φ_R, the pitch angle φ_P and the vertical position of the center of gravity of the chassis z_C. Furthermore, each of the four wheels has an independent vertical degree-of-freedom, whereas the tires are modeled in form of a linear spring/damper system with the tire stiffness c_W and the tire damping d_W, cf. figure 3. In order to indicate each of the four wheels individually, the index pair $\alpha\gamma$ with $\alpha \in \{f = \text{front}, r = \text{rear}\}$ and $\gamma \in \{l = \text{left}, r = \text{right}\}$ is used. Doing so, the vertical excitation of the wheels due to the road surface is denoted by $\mathbf{z}_R = [z_{Rfl}, z_{Rfr}, z_{Rrl}, z_{Rrr}]^T$ and the vertical position of the wheels and the points of the chassis where the dampers and springs are mounted read as $\mathbf{z}_W = [z_{Wfl},\ z_{Wfr},\ z_{Wrl},\ z_{Wrr}]^T$ and $\mathbf{z}_C = [z_{Cfl},\ z_{Cfr},\ z_{Crl},\ z_{Crr}]^T$, respectively. With the four spring forces $\mathbf{F}_S = [F_{Sfl},\ F_{Sfr},\ F_{Srl},\ F_{Srr}]^T$, which are nonlinear functions of $\mathbf{z}_W - \mathbf{z}_C$, and the mass matrix of the chassis $\mathbf{M}_C = \mathrm{diag}\,(J_R,\ J_P,\ m_C)$ (J_R and J_P are the mass moments of inertia about the φ_R and φ_P axis, respectively and m_C is the total mass of the chassis) the equations of motion of the chassis read as

$$\frac{\mathrm{d}}{\mathrm{d}t}\mathbf{p}_C = \mathbf{v}_C \qquad \frac{\mathrm{d}}{\mathrm{d}t}\mathbf{v}_C = \mathbf{M}_C^{-1}\mathbf{T}_C\,(\mathbf{F}_S + \mathbf{F}_{ER}) - [0,\ 0,\ g]^T \tag{1}$$

with

$$\mathbf{T}_C = \begin{bmatrix} d & -d & d & -d \\ -l_f & -l_f & l_r & l_r \\ 1 & 1 & 1 & 1 \end{bmatrix} \qquad \text{and} \qquad \mathbf{F}_{ER} = [F_{ERfl}, F_{ERfr}, F_{ERrl}, F_{ERrr}]^T\ . \tag{2}$$

Thereby, a small angle approximation is used and g denotes the gravitational constant. Furthermore, the four equations of motion of the wheels take the form

$$\frac{\mathrm{d}}{\mathrm{d}t}\mathbf{z}_W = \mathbf{v}_W \qquad \frac{\mathrm{d}}{\mathrm{d}t}\mathbf{v}_W = \mathbf{M}_W^{-1}\,(\mathbf{F}_W - \mathbf{F}_S - \mathbf{F}_{ER}) - g\,[1,\ 1,\ 1,\ 1]^T \tag{3}$$

with the mass matrix of the wheels $\mathbf{M}_W = \mathrm{diag}\,(m_{Wf},\ m_{Wf},\ m_{Wr},\ m_{Wr})$. Therein, the forces due to the tire deflection $\mathbf{F}_W = [F_{Wfl},\ F_{Wfr},\ F_{Wrl},\ F_{Wrr}]^T$ are of the form $\mathbf{F}_W = \mathbf{C}_W\,(\mathbf{z}_R - \mathbf{z}_W) + \mathbf{D}_W\,(\mathbf{v}_R - \mathbf{v}_W)$ with $\mathbf{v}_R = \frac{\mathrm{d}}{\mathrm{d}t}\mathbf{z}_R$, the matrix of the tire stiffnesses $\mathbf{C}_W = \mathrm{diag}\,(c_W,\ c_W,\ c_W,\ c_W)$ and the matrix of the tire damping coefficients $\mathbf{D}_W = \mathrm{diag}\,(d_W,\ d_W,\ d_W,\ d_W)$.

3. Controller Design

The fundamental task is to track a desired damping force $\tilde{F}_{d\alpha\gamma}$ on each ER damper as accurate as possible within the physical limits. Using the velocities $v_{\alpha\gamma} = v_{C\alpha\gamma} - v_{W\alpha\gamma}$, the required voltage $U_{\alpha\gamma}$ to be applied to the ER valves can be calculated in the form [5]

$$U_{\alpha\gamma} = U_{\alpha\gamma}\left(v_{\alpha\gamma}, \tilde{F}_{d\alpha\gamma}\right), \quad \alpha \in \{f, r\}, \quad \gamma \in \{l, r\} \tag{4}$$

where the desired damping forces $\tilde{F}_{d\alpha\gamma}$ serve as virtual control inputs for the superimposed controller. The superimposed control comprises of two parts. The first one is a combination of sky-hook and ground-hook control, the so-called hybrid-control strategy [6], for each wheel individually to assure good road handling. The second part is concerned with a strategy to reduce the roll, pitch and vertical movement of the chassis. Thereby, the dampers are controlled in such a way that linear damping forces are realized for these three degrees-of-freedom

$$\mathbf{F}_{Cd} = -\mathbf{K}\mathbf{v}_C \tag{5}$$

with $\mathbf{K} = \mathrm{diag}\,(k_R,\ k_P,\ k_Z)$. The resulting desired force $\tilde{F}_{d\alpha\gamma}$ of each damper in (4) is determined by a suitable force allocation algorithm from the three generalized forces (5) and the forces calculated by the hybrid-control strategy.

4. Simulation Studies

The results of the simulation studies are demonstrated by means of a trapezoidal obstacle [7] with a height of 200 mm, a length of 500 mm and a slope of 30°. Since both (pitch and roll) degrees-of-freedom are of interest, two virtual obstacles are used in simulation, one on the left hand side and another on the right hand side, where the contact of the right wheels with the trapezoid is 0.5 s later than on the left hand side. With a velocity of the truck of 3.75 km/h, the resulting roll angle is depicted in figure 4 which shows a significant improvement with the ER damper (ERD) compared to the currently used commercial original damper (OD). The corresponding accelerations of the chassis show similar results. By contrast, the driving stability is amongst others determined by the tire deflection as depicted in figure 5. One can see a reduced tendency for take-off of the wheel also leading to a better driving stability.

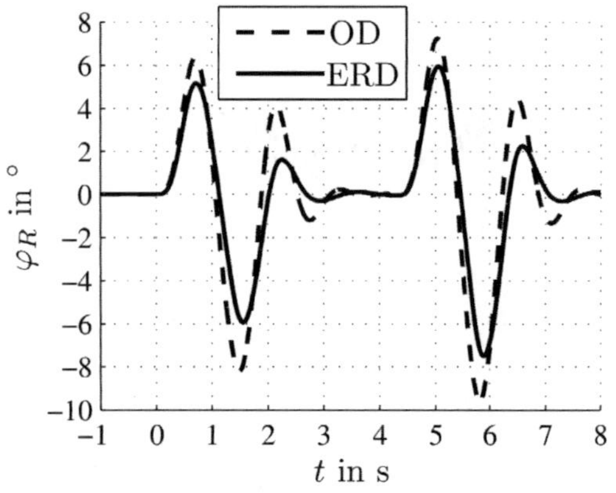

Figure 4. Roll angle.

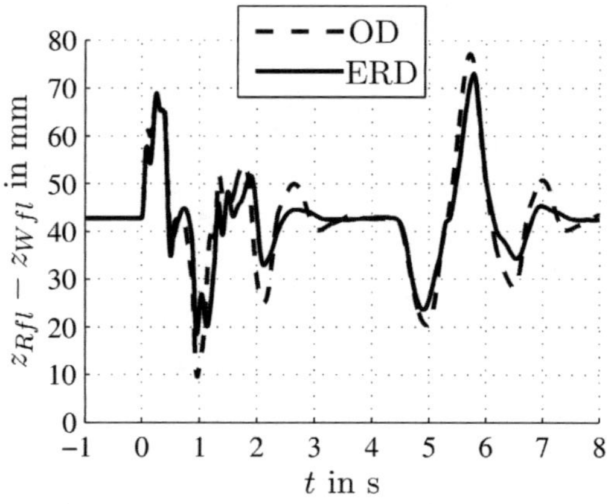

Figure 5. Tire deflection.

5. Measurement Results of the Truck

Measurements of the truck equipped with ER-dampers and of the original truck were performed on different test tracks with various velocities. Since it is almost impossible to perform identical measurements twice for the same track, it is not meaningful to compare and depict the time evolutions of the measurement results in a plot. Therefore, the measurement data is analyzed by means of the rms-value [7]

$$\varepsilon_m = \sqrt{\frac{1}{T}\int_0^T \varepsilon^2(t)\,\mathrm{d}t} \tag{6}$$

for a given signal ε. The improvement for example w. r. t. the acceleration of the chassis on the so-called Schweizer Bahn with 12 km/h and 20 km/h was approximately $11 - 18\%$ in both cases.

6. Conclusion

The proposed control strategy for a truck equipped with ER dampers results in improved driving stability and comfort. This was shown by means of simulation studies and measurement results.

References

[1] Ieluzzi M, Turco P and Montiglio M 2006 *Control Engineering Practice* **14** 305–312
[2] Adams D and Johnston L 2001 *Proc. of the 8th International Conference on Electrorheological Fluids and Magnetorheological Suspensions* (Nice, France) pp 37–42
[3] Choi Y T, Bitman L and Wereley N M 2005 *J. of Intelligent Material Systems and Structures* **16** 383–394
[4] Wang H and Hu H 2004 *Proc. of the 9th International Conference on Electrorheological and Magnetorheological Suspensions* (Beijing, China) pp 595–601
[5] Holzmann K, Kemmetmüller W, Kugi A and Stork M 2006 *Preprints of the 4rd IFAC Symposium on Mechatronic Systems* (Heidelberg, Germany) pp 372–377
[6] Goncalves F D and Ahmadian M 2003 *Shock and Vibration* **10** 59–69
[7] Schneider S, Holzmann K, Kemmetmüller W, Johnston L and Lißmann D 2008 *11th International Conference on New Actuators* (Bremen, Germany) pp 485–488

Journal of Physics: Conference Series **149** (2009) 012012 doi:10.1088/1742-6596/149/1/012012

Unsteady Laminar Flow Analysis of ER Valve Systems: Modeling and Simulation

Y S Jeon[1], Y M Han[2], Q H Nguyen[2] and S B Choi[2]

[1] Department of Tool and Mold Design, Yuhan College, Gyeonggi-Do 442-749, Korea
[2] Smart Structures and Systems Laboratory, Department of Mechanical Engineering, Inha University, Incheon, 402-751, Korea

E-mail: seungbok@inha.ac.kr

Abstract. This paper proposes a new method to model electrorheological valve systems based on the unsteady flow analysis. After describing the mechanism and operational principle of the ER valve, the unsteady flow in the valve is identified. A method to model this unsteady flow is developed based on the assumptions that the ER fluid follows Bingham behavior and the pressure drop of the unsteady flow results from the independent pressure drops due to the yield stress, inertia and fluid viscosity. The dynamic model of the valve is then derived based on the unsteady flow analysis. The time response of ER effect is also considered in the dynamic model. Using the proposed model, the performance of the ER valve is predicted and validated by comparing with experimental results. In addition, the flow rate predicted by the unsteady flow model is compared with that by the steady flow model.

1. Introduction

A hydraulic system is very important in industrial applications where large inertia and torque loads have to be handled [1]. Various types of valves have been developed as the key element of the hydraulic system. Among them, the electrorheological (ER) valve has been successfully employed in many engineering systems such as cylinder for position control, pressure modulator for pressure control, damper for suspension control, etc. When the ER fluid is employed in valve systems, its rheological change causes a controllable pressure drop which can be continuously tuned by the intensity of the imposed electric field. This feature has triggered a large amount of research to develop accurate control model and design mechanism of the ER valve [2,3]. However, in previous studies, the behaviour of the ER valve was predicted based on the assumption of the steady flow of ER fluid through the valve duct although the ER fluid unsteadily flows at real situations. Consequently, the main contribution of the present work is to propose a new modeling methodology for the ER valve based on unsteady laminar flow analysis of the ER fluid. After manufacturing a cylindrical type of ER valve, its pressure drop is analyzed under unsteady flow conditions with randomly varying electric field in which the ER fluid follows Bingham behaviour. The dynamic model of the valve is then derived by considering the pressure drop under unsteady flow conditions and time response of the ER fluid. In order to validate the proposed model, the predicted valve performance is compared with experimental result.

Journal of Physics: Conference Series **149** (2009) 012012 doi:10.1088/1742-6596/149/1/012012

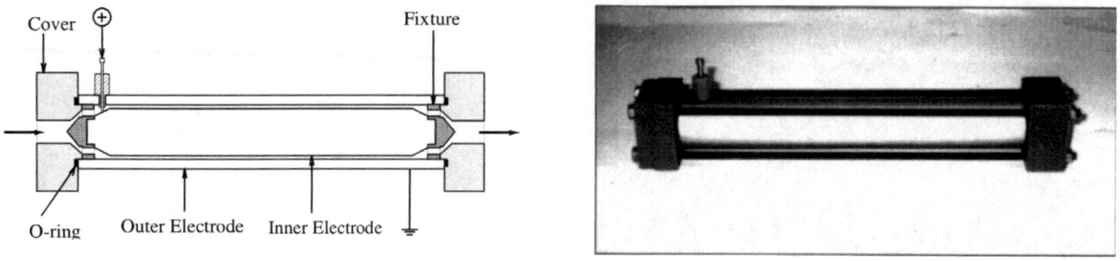

(a) Schematic configuration (b) Photograph

Figure 1. The proposed ER valve

2. Modeling of the ER Valve

The geometric configuration and photograph of a cylindrical ER valve devised in this work is shown in figure 1. As shown in the figure 1(a), it is composed of cylindrical inner and outer electrodes which are fixed by two fixtures at both ends. The ER fluid flows through the gap between the inner and outer cylinders. The pressure drop is then continuously controlled by the intensity of the input electric field. Figure 1(b) shows the manufactured ER valve. The materials for the outer electrode and inner electrode are aluminum and stainless steel, respectively. A nonconductive fixture is used to maintain the electrode gap, and the 'o' ring is adopted to prevent the leakage of the ER fluid.

Figure 2 shows the flow through an annular duct. By neglecting the gravity and assuming the linear axial distribution of pressure, the axial momentum equation of motion for the steady-state laminar flow of an incompressible fluid can be expressed in cylindrical coordinate system as:

$$\frac{1}{r}\frac{\partial}{\partial r}(r\tau) = \frac{dp}{dz} = -\frac{\Delta P}{L}; \quad R_i \le r \le R_o \tag{1}$$

where τ is the shear stress across the flow at radius r, L is the duct length, R_i and R_o are respectively the inner and outer radius of the duct, respectively, and ΔP is the pressure drop of the flow through the duct. Because the radius of the duct is much larger than its gap it is assumed that the flow through the annular duct is equivalent to the flow through the duct between two parallel plates shown in figure 3. Therefore, equation (1) becomes

$$\frac{\partial \tau}{\partial y} = -\frac{\Delta P}{L}; \quad (0 \le y \le d) \tag{2}$$

where d is the duct gap. In this work, the ER fluid is assumed to follow Bingham rheological laws, thus the following rheological equation of state can be obtained.

$$\tau = \tau_y \,\mathrm{sgn}(u) + \mu \frac{\partial u}{\partial y} \tag{3}$$

where u is the velocity of the flow at coordinate y, τ_y is the yield stress of the ER fluid induced by the applied voltage, μ is the field-independent plastic viscosity.

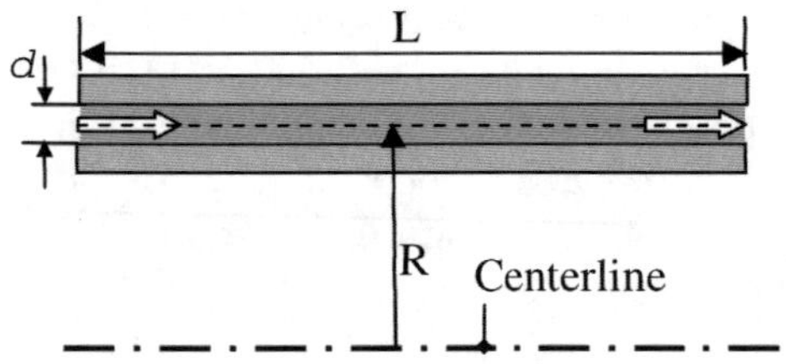

Figure 2. ER fluid flow in the annular valve duct

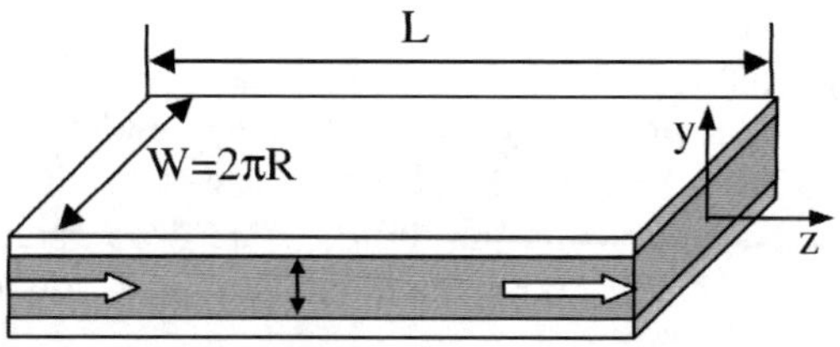

Figure 3. Equivalent ER fluid flow.

From equations (2) and (3), the flow rate of steady state ER fluid flow through the duct is derived as follows [4]:

$$Q = \frac{W\Delta P}{12\mu L}(d-\delta)^2(d+\delta/2) \tag{4}$$

In the above, W is the width of the equivalent duct which is calculated by $W=2\pi R$; R is the average radius of the annular duct, and δ is plug thickness of the ER flow given by

$$\delta = \frac{2L\tau_y}{|\Delta P|} \tag{5}$$

Equation (4) is usually used to calculate steady state flow rate of ER fluid flow due to the given pressure at the inlet and the outlet of the duct. On the other hand, when the flow rate is given, the pressure drop of the steady state ER fluid flow can be calculated by

$$\Delta P = \frac{12\mu L}{Wd^3}Q + c\frac{L}{d}\tau_y \tag{6}$$

where the coefficient c can be approximately estimated as follows:

$$c = 2.07 + \frac{12Q\mu}{12Q\mu + 0.8\pi Rd^2\tau_y} \tag{7}$$

In the case of unsteady state flow, the momentum equation of the ER fluid in the duct is expressed as follows:

$$\rho\frac{\partial u}{\partial t} - \frac{\partial \tau}{\partial y} = \frac{\Delta P_u}{L}, \quad (0 \le y \le d) \tag{8}$$

where ρ is the density of the fluid. By multiplying W in both sides of Eq. (8) and then integrating with y, equation (8) can be rewritten by

$$\Delta P_u = \frac{\rho L}{Wd}\frac{dQ}{dt} - \frac{L}{d}\int_0^d \frac{\partial \tau}{\partial y}dy = \Delta P_1 + \Delta P_2; \quad (0 \le y \le d) \tag{9}$$

where ΔP_1 is pressure drop due to the flow inertia and ΔP_2 is pressure drop due to fluid rheological behavior. The pressure drop due to fluid rheological behavior of the unsteady state flow can be approximately calculated from that of the steady state flow as follows:

$$\Delta P_2 = -\frac{L}{d}\int_0^d \frac{\partial \tau}{\partial y}dy = k\Delta P \tag{10}$$

The coefficient k depends on the variation of unsteady flow rate. Substitution of ΔP obtained from equations (4) and (6) into equation (10) for ΔP_2, then plug ΔP_2 into equation (9) yields

$$\Delta P_u = \frac{\rho L}{Wd}\frac{dQ}{dt} + \frac{12k\mu LQ}{W(d-\delta)^2(d+\delta/2)} \tag{11}$$

$$\Delta P_u = \frac{\rho L}{Wd}\frac{dQ}{dt} + \frac{6k\mu L}{\pi d^3 R}Q + k\frac{cL}{d}\tau_y \tag{12}$$

Noteworthily, equation (11) is suitable for calculating flow rate of the unsteady flow due to the given variation of pressure drop while equation (12) is used for calculating pressure drop of unsteady flow due to the given variation of flow rate.

By neglecting time response of electronic devices of power supply unit, the time response of ER fluid to the applied electric field is equal to the switching time of the ER fluid itself. Thus, the effect of electric field change on the ER yield stress can be described by

Journal of Physics: Conference Series **149** (2009) 012012 doi:10.1088/1742-6596/149/1/012012

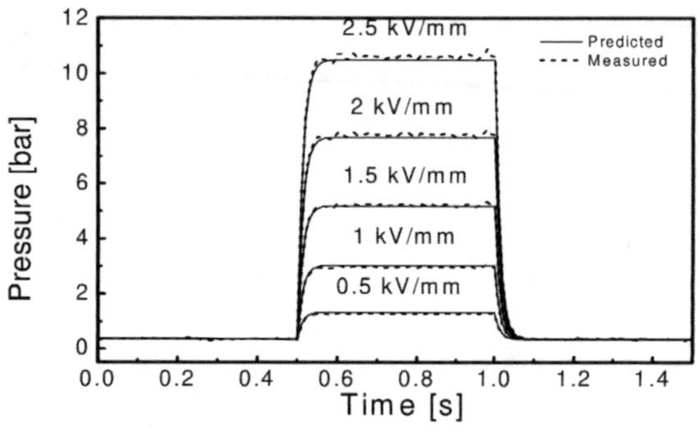

Figure 4. Pressure drop of the valve.

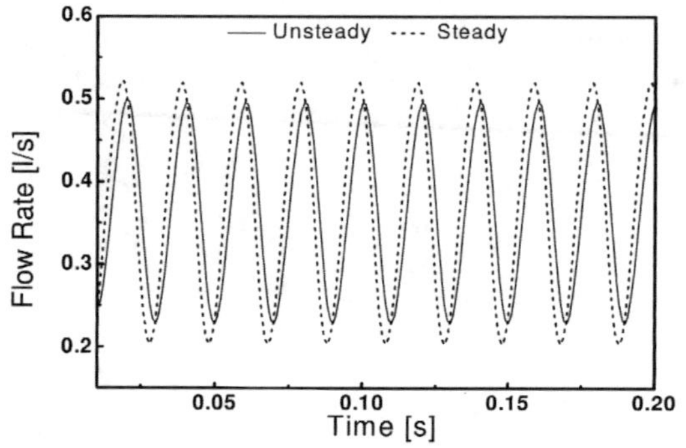

Figure 5. Flow rate of the valve.

$$\tau_y + \lambda \dot{\tau}_y = \alpha E^{\beta} \tag{13}$$

where λ is the switching time of ER fluid that is same as the response time of the ER fluid itself. In this study it is assumed that λ=5*ms*.

3. Results and Discussion

In this study, the predicted results obtained by the proposed model are presented and compared with experimental results to validate the proposed model. The significant geometric dimensions of the ER valve are as follows: L=265mm; d=0.8mm; R_i=12mm. For the ER fluid, chemically-treated starch and silicone oil are chosen as particles and liquid, respectively. The viscosity of the base oil is 0.003 $Pa \cdot s$. The size of the particles ranges from 10μm to 50μm. The weight ratio of the particles to the ER fluid is 40%. At testing temperature of 45□, the coefficients α and β are respectively evaluated by 405.7 and 1.45 using least square curve fitting method [2]. Figure 4 shows the pressure drop of the ER valve calculated from equation (11) due to step input electric fields. In this case, the flow rate through the valve is kept constant at 0.015*l/s*. As clearly observed from the results, the predicted results by the proposed model are well matched with experimental results. Figure 5 shows flow rate response of the ER valve due to sinusoidal input electric fields whose frequency and amplitude are 50Hz and 1kV, respectively. The pressure at the inlet of the valve is kept constant at 15*bar*. It is obvious that the flow rate under steady flow condition is quite different from that under unsteady flow condition. These results indicate that the proposed dynamic model based on the unsteady state flow can accurately describe the behaviour of the ER valve.

4. Conclusion

In this work, a new modeling method based on unsteady flow analysis has been proposed to provide accurate control model for ER valve systems. The pressure drop of the cylindrical ER valve was analytically expressed with unsteady flow condition of the ER fluid and varying electric field. After designing and manufacturing of the ER valve, the predicted pressure drop for the proposed dynamic model has been compared with experimental results in order to validate the proposed model. The difference of the flow rate between steady and unsteady analysis cases has been also demonstrated.

Reference

[1] Watton J 1988 Fluid Power Systems, Prentice Hall, Upper Saddle River, NJ
[2] Choi S B and Cho M S 2005 *International Journal of Vehicle Design* **38** 196
[3] Nakano M and Yonekawa T 1994 *4th International Conference on Electrorheological Fluids* 477
[4] Wereley N M and Li P 1998 *Smart Mater. Struct.* **7** 732

Development of High-performance ER Gel Produced by Electric-field assisted Molding

Y Kakinuma[1], T Aoyama[1] and H Anzai[2]

[1]Department of System Design Engineering, Keio University,
3-14-1 Hiyoshi Kouhoku-ku Yokohama, Japan
[2]Fujikura kasei Co., Ltd. 2-6-15 Shibakouen, Minato-ku, Tokyo, Japan

kakinuma@sd.keio.ac.jp

Abstract. Electro-rheological gel (ERG) is a novel functional elastomer whose surface frictional and adhesive property varies according to the intensity of applied electric field. This peculiar phenomenon is named as Electro-adhesive effect. A generated shear stress of ERG under applied electric field is approximately 30~40 times higher than that of ERF because of high adhesive strength. However, the performances of ERG vary widely due to its surface condition, especially density and distribution of ER particles at the surface. In order to stabilize and improve the performance of ERG, the electric- filed assisted molding process is proposed as the producing method of ERG. In this study, first, the principle of electro-adhesive effect is theoretically investigated. Second, a high-performance ERG produced by the proposed process, in which ER particles are aligned densely at the surface, is developed and its performance is evaluated experimentally. As the experimental result, the high-performance ERG shows twice higher shear stress than the conventional ERG.

1. Introduction

Electro-rheological gel (ERG) is a novel functional elastomer whose surface frictional and adhesive property varies according to the intensity of applied electric field [1]. This peculiar phenomenon is named as Electro-adhesive effect in order to distinguish from ER effect of ER fluids due to cluster formation of ER particles, which induces change of visco-elastic property under electric field [2, 3]. Figure 1 shows the photo of the actual ERG. ERG is composed of ER particles and silicone gel and has a texture like rubber in appearance. When electric field is applied, ERG provides approximately 30~40 times higher shear stress than that of ERF since high adhesive strength is induced under low electric field such as 0.5kV/mm~1.0kV/mm [4].

However, the performances of ERG strongly depend on its surface condition, especially density and distribution of ER particles at the surface. In order to stabilize and improve the performance of ERG, higher density and precise alignment of ER particles are expected to the production process. In this study, the high-performance ERG is produced by the new method called "the *Electric- field assisted Molding Process*" and its performance is evaluated experimentally.

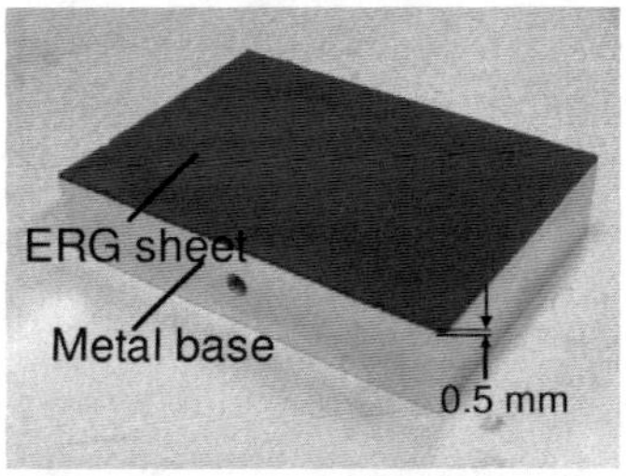

Figure 1. Overview of ERG

2. Estimated principle of Electro-adhesive effect

The ERG sheet was sandwiched between a glass electrode pasted by the conductive ITO film and a metal electrode. A microscope was set above the glass electrode as shown in figure 2(a), and the electro-adhesive phenomenon that occurred on the contact surface between the ERG and glass electrode were carefully monitored. Figure 2(b) shows the electro-adhesive phenomenon of ERG observed in the boundary region under a variable electric field. Upon application of the electric field, the gel adhered to the electrode surface. The higher the applied electric field was, the wider the area of adhesion.

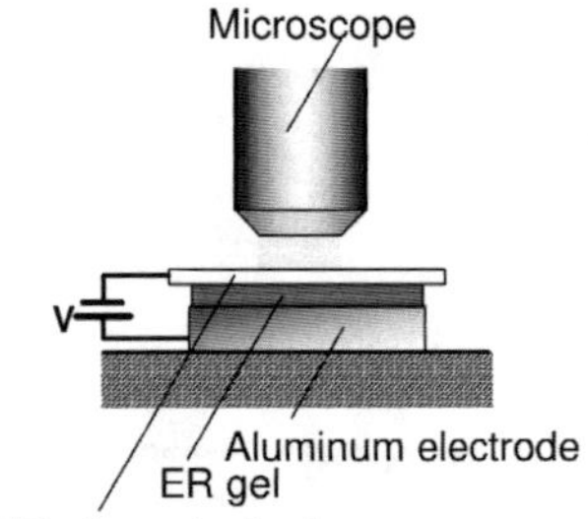

(a) Schematic diagram of surface observation

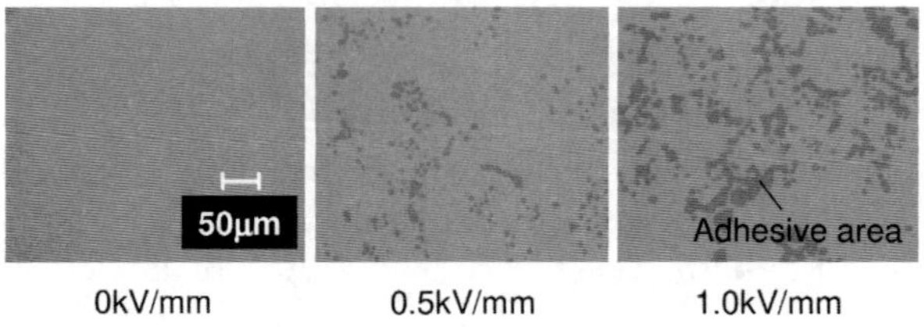

(b) Appearance of adhesion under various electric field

Figure 2. Electro-adhesive effect of ERG

On the basis of electromagnetics, the principle of electro-adhesive effect of ERG is investigated theoretically. In the absence of an electric field, the electrode is in contact with slippery ER particles because ER particle is protruded from the silicone gel. The distribution of electric field at the boundary between ERG and electrode under applied voltage between electrodes is analysed numerically, using ANSYS 11.0 software. Figure 3 represents the electric vector distribution at the boundary area. Four components are microscopically existed at the boundary area; ER particle, electrode, silicone gel and air gap. Electric field in small air gap surrounded between ER particles, silicone gel and electrode becomes high since the permittivity of air is lower than any other material. Moreover, the electric field of air gap increases near the boundary between ER particle and electrode. When the boundary surface between dielectric materials is in parallel and perpendicular with the direction of electric field, the Maxwell stress exerted on its boundary is represented as Eq. (1) and Eq.(2) respectively.

$$F_{n//} = \frac{1}{2}(\varepsilon_1 - \varepsilon_2)E^2\,[Pa] \tag{1}$$

$$F_{n\perp} = \frac{1}{2}\left(\frac{1}{\varepsilon_2} - \frac{1}{\varepsilon_1}\right)D^2\,[Pa] \tag{2}$$

Where ε_1 and ε_2 are the permittivity of dielectric materials, E is electric field intensity, D is dielectric flux density. As shown in these equations, the resultant total stress due to Maxwell stress acts toward from low permittivity to high permittivity. The interaction of Maxwell stress at each boundary is illustrated in figure 4(a). According to these interactions, ER particles protruding the surface slightly sink down inside though their range of motion is restricted by the elastic deformation of the gel. The soft silicon gel comes up to electrode and, in addition, it is deformed toward the area with higher electric field, i.e. toward the boundary between ER particle and electrode. This is because Maxwell stress increases with increase of electric field intensity. As the result, the sliding electrode makes contact with the sticky gel surface as shown in figure 4(b) and shear force increases. This theoretical estimation based on the Maxwell stress agrees with the observed result and could represent the electro-adhesive effect of ERG [3].

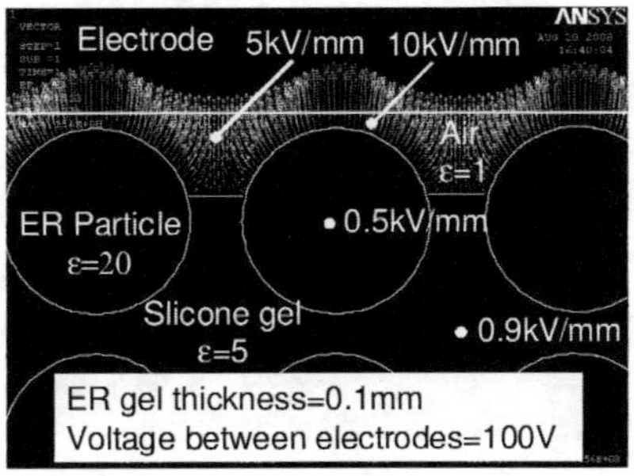

Figure 3. Distribution of electric field at the boundary

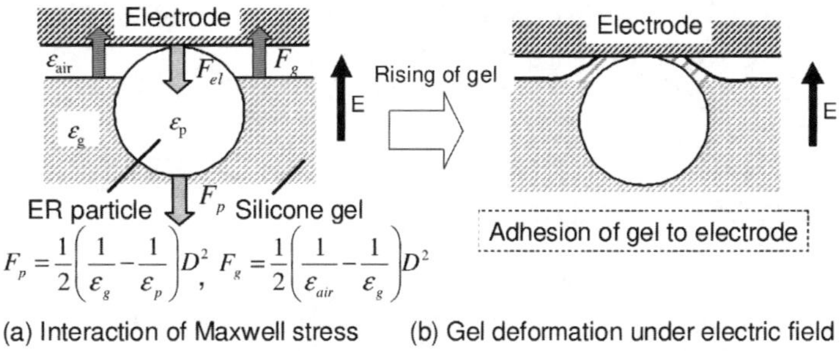

$$F_p = \frac{1}{2}\left(\frac{1}{\varepsilon_g} - \frac{1}{\varepsilon_p}\right)D^2, \quad F_g = \frac{1}{2}\left(\frac{1}{\varepsilon_{air}} - \frac{1}{\varepsilon_g}\right)D^2$$

Figure 4. Estimated principle of electro-adhesive effect

3. Electric-field assisted molding process to produce the high-performance ERG

From the view point of the generation mechanism of ERG effect, it is suggested that the dispersion state of ER particles at the surface significantly influences on the performance of ERG. In order to align ER particles at the surface uniformly and densely, an *Electric-field assisted Molding Process* is proposed as the new production method of high-performance ERG. The proposed new process is explained in Figure 5. A mixed solution of ER fluid and gelling agent is prepared and poured into a mold. When the metal base and the metal upper plate serve as anode and cathode respectively, the electric field is applied to its solution. By gelling the solution under an electric field, high-performance ERG is produced because the ER particles are arranged densely at the surface along electric lines of force. Figure 6 shows the surfaces of conventional ERG and the high-performance ERG. These ERGs have the same ER particle concentration of 50wt%. As shown in figure 6, it is clarified that the particles at the surface of the high performance ERG are arranged uniformly and densely.

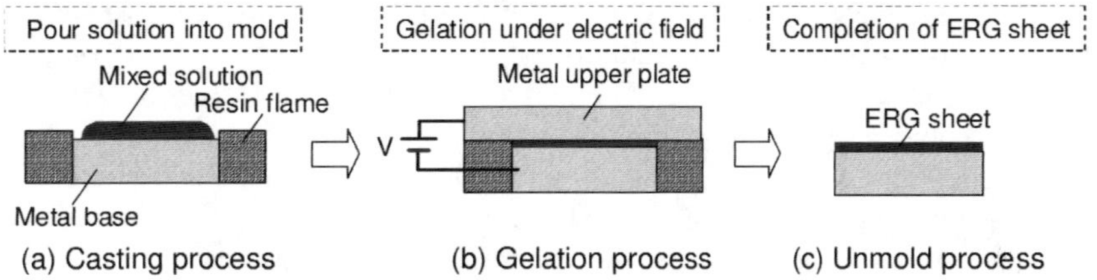

Figure 5. Electric field assisted molding process

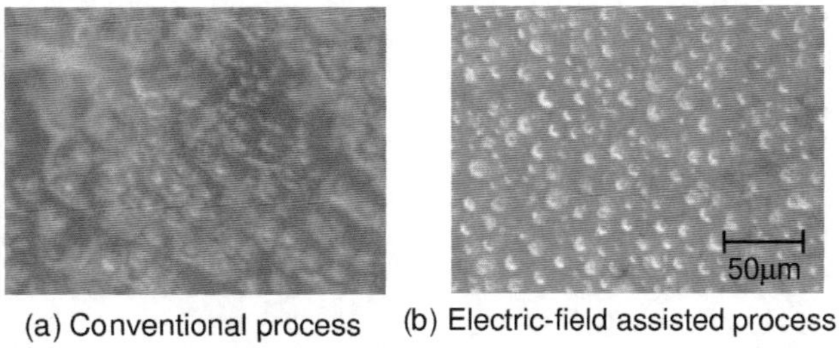

Figure 6. Overview of ERG surface

4. Performance evaluation of the high-performance ERG

4.1. Experimental procedure

The basic performance of the high-performance ERG is evaluated by the simple test schematized in figure 7. Two ERG sheets with 0.5 mm thickness are prepared; one is conventional ERG and the other is high-performance ERG gelled under electric field of 1.5kV/mm. A single sheet of ERG is fixed physically on the lower electrode. The upper electrode only put on the ERG sheet and could slide on its surface. Since ERG sheet is sandwiched between a pair of metal plate electrodes, an electric field can be applied to ERG sheet. In this test, the lower and upper electrodes serve as anode and cathode respectively. When an electric field is applied to ERG sheet and the upper electrode slides at the constant feed rate, the variation of shear stress between the upper electrode and the ERG sheet is measured using a strain gauge load cell.

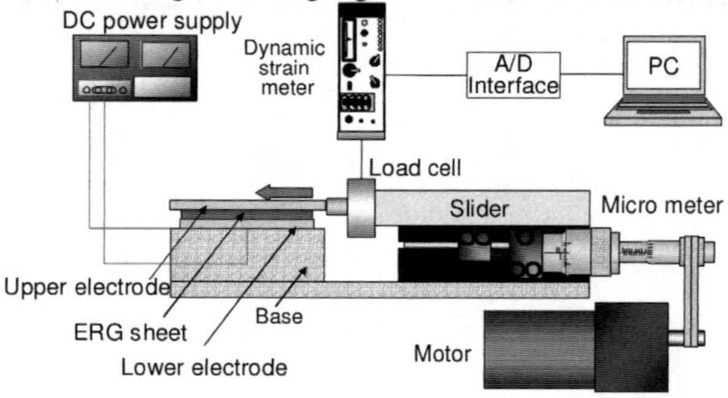

Figure 7. Schematic diagram of experimental setup for shear test

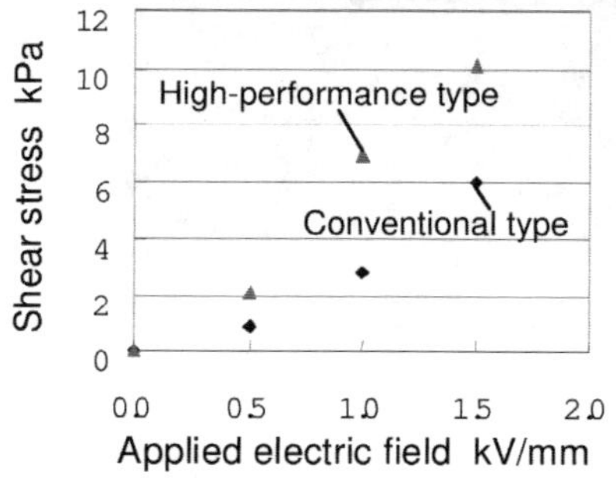

Figure 8. Relation between shear stress and electric field

4.2. Basic performance

The relation between the generated shear stress and applied electric field is shown in figure 8. Each deviation of measured shear stress is less than 0.3kPa. Both conventional and the high-performance ERG exhibit that generated shear stress increases with increase of applied electric field intensity. Compared with two types of ERG, generated shear stress of the high performance ERG is approximately twice as high as that of conventional one. In order to investigate the reason why the high shear stress is generated in high-performance type, the number of ER particles in contact with electrode under electric field is counted by observation of the boundary between ERG and glass electrode. Table 1 shows the comparison of ER particle number per unit area in contact with electrode between conventional and high-performance ERG. The counted particle of high-performance type is twice larger than conventional one. These results indicate that the number of ER particles in contact with electrode corresponds to the generated shear stress. When the ER particles are arranged densely and uniformly at its surface, it is clear that the electro-adhesive effect of ERG can be obtained effectively

Table 1. Average particle number per unit area

	Conventional ERG	High-performance ERG
Average particle number of unit area □ 100μm	25.5	53.5

5. Conclusion

The high-performance ERG, produced by electric-field assisted molding process, is developed and its basic performance is evaluated experimentally. The high performance ERG shows the higher ERG effect. The generated shear stress is twice as high as conventional one. In future work, the dynamic performance and durability will be evaluated.

Acknowledgments

This work was partially supported by Research for Promoting Technological Seeds, Japan Science and Technology Agency.

References

[1] Kakinuma Y, et al. 2005 Basic Properties of Gel-structured Electro-rheological Fluids *Int. J. Modern Phys. B*, **19,** 7-9, 1339-1345

[2] Tomas C. and Halsey 1992 Electrorheological Fluids *Science* **258**, 5083, 761-766

[3] Wen W, et al. 2003 The Giant Electrorheological Effect in Suspensions of Nanoparticles *Nature Materials*, **2**, 727-730

[4] Kakinuma Y, et al. 2006 Application of ER gel with Variable Friction Surface to the Clamp System of Aerostatic slider *Precision Engineering* **30**, 280-287

Electrorheological fluids based on the modified aromatic polyimides

A Karabko[1,3], I Globa[2], Z Novikova[1], E Krutko[2], N Globa[2] and E Korobko[1]
[1] Luikov Heat and Mass Transfer Institute of Academy of Sciences of Belarus, 15 P Brovka Street, 220072, Minsk, Belarus
[2] Belarusian State Technological University, Sverdlova Street 220006, Minsk, Belarus

E-mail: eva@itmo.by

Abstract. Fine-dispersed particles of aromatic polyimides, modified by cobalt acetylacetonate, are synthesised. The electrorheological activity of the dispersed systems, based on mineral oil and such particles, is investigated. It is shown, that polyimides and polyamide acids are perspective to be used as dispersed phase in electrorheological compositions.

1. Introduction

Electrorheological fluids (ERF) are complex fluids consisting of micron-sized or, rarely, nano-sized dielectric particles (so-called filler) dispersed in a nonconducting fluid [1]. As fillers, various oxides, silica modifications, silica gels, aerosil, diatomite, and also calcium titanates, alumosilicates, metallic oxides in the form of a single dispersed powder with a particle size from 1 to 10 μm are widely used [2]. However, many well-known compositions have a number of disadvantages, such as: a rather low stability in time, a limitation in the temperature stability range, a limitation in the chemical stability of the dispersion medium, and a rather low electrostructural response to the electric impact. Recently, a number of new organic compositions [3], natural and synthetic polymers [4], namely, cellulosic materials and polypyrroles [5] have been regarded as potential candidates for ERF fillers. Since such kinds of the organic fillers are of a key interest nowadays, we have developed a new type of the composition for ERF based on the aromatic polyimides [6], and investigated the electrorheological properties of it.

Aromatic polyimides are a class of polymers obtained by the interaction of aromatic diamines with dianhydrides of aromatic tetra carbon acids according to the so-called two-stage scheme. They have a unique complex of properties, including high dielectric and mechanical properties in the temperature range of –200 — +350°, and good chemical and radiation stability. The first stage of the synthesis includes the interaction of diamines and dianhydrides in the medium of aprotic amide solvents [6]:

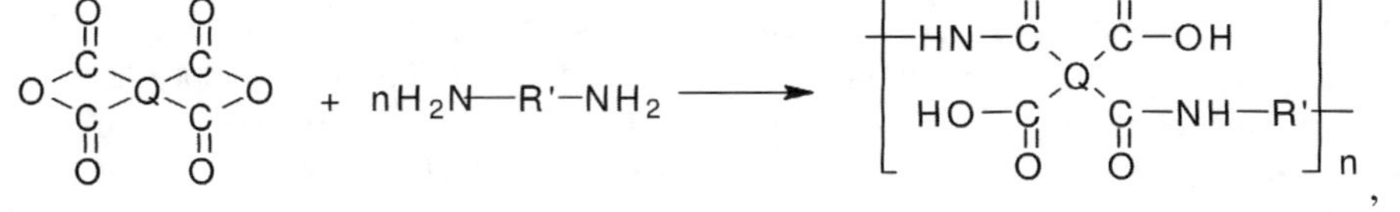

where

R = 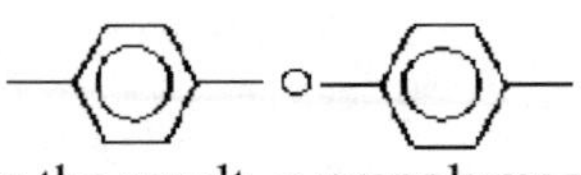 – divalent organic radical; Q = 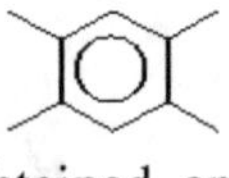 – tetravalent organic radical.

As the result, a prepolymer — polyamide acid (PAA) is obtained, and the final product is received by the reaction of the solid phase thermal dehydrocyclization by means of a step by step heating up to 300–320 °C or by a so-called chemical imidization, consisting in the treatment of PAA by hydro retaining reagents and carried according to the scheme [6]:

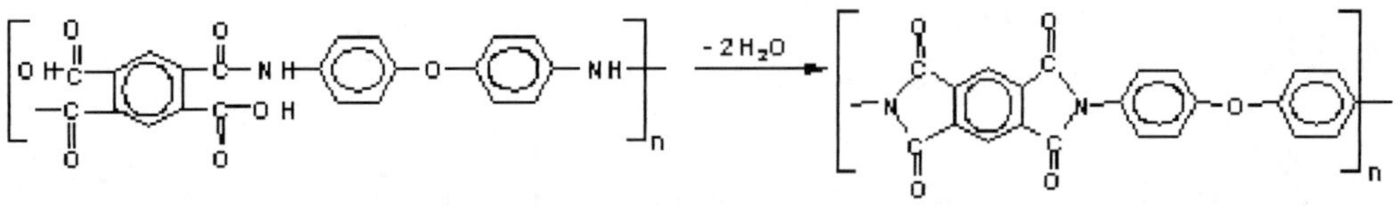

The structure of the polymers of this class allows us to give them peculiar properties, needed for the certain areas of application, while the complex of main properties is retained. Therefore a particular interest as dispersed phase of ERF presents the class of modified aromatic polyimides. The use of aromatic polyimides is perspective in the enhancement of sedimentation stability of composition because of substantially low density of material (1.1 – 1.2 g/cm^3) in comparison with the dispersed phases of an inorganic nature, and the decrease of the particle aggregation will make easier the process of dispersion of the particles up to the needed sizes.

High sorption capacity of polymeric materials allows us to carry out a wide range of possibilities of controlling the electrorheological properties of the dispersed medium by the introduction of activators. The most important is the implementation of the purposed formation of the electrosensitivity for the filler by the aid of different types of modifies due to the creation of optimum quantity of active groups (hydroxyl-carboxyl containing), responsible for effective surface polarization of the particles in the electric field, and also their mobility, that regulates the kinetics of structural processes.

2. Experimental

As dispersed phase we have chosen polyimides, based on the pyromellitic dianhydrides and diaminodiphenyloxide, modified by cobalt acetonate. Aromatic polyimides we obtained by the above described two-step synthesis process, in which at the first stage a prepolymer — polyamidacid (PAA) is formed by the interaction of equimolecular quantity of pyromellitic dianhydride and diaminodiphenyloxide in dimethylformamide. The final polymer — polyimide (PI) is formed at the second stage of the synthesis — by the solid phase high temperature or low temperature chemical dehydrocyclization of a prepolymer [5]. A modifier – cobalt acetylacetonate was introduced by two methods:

- by the introduction of its calculated quantity in PAA solution, mixing during 30-40 min, sedimentation of the solvent in acetone, separation of its solid polymeric precipitaties and its drying, and solid phase dehydrocyclization (imidization) by means of smooth temperature elevation up to 300 °C in the ambient of nitrogen;
- by the introduction of its calculated quantity in solid PAA, obtained by the sedimentation of its solution of dimethylformamide in acetone, separation of its solid polymeric precipitation by filtering, and solid phase dehydrocyclization (imidization) by means of smooth temperature elevation up to 300 °C in the ambient of nitrogen;

Polymeric dispersed phases based on the modified polyimide compositions were prepared by attrition in an agate mortar until a high dispersion powder was obtained. Thermal properties of the samples were investigated by means of thermograph device Paulik-Paulik-Erdei, the degree of the transition of the prepolymer into polyimide — by means of IR spectroscopy, using IR-Furier spectrometer FT-IR «Thermo Nicolet». Suspensions were prepared by accurate mixing of the calculated quantity of the milled polymeric dispersed phase with mineral oil in a ceramic mortar.

Electrorheological characteristics of compositions were tested in a coaxial cylindrical unit of a rotational viscometer.

3. Results and discussion

Thermographic investigation of the PI samples, modified by cobalt acetylacetonate has shown that the losses of 5 % of mass are reached only at approximately 300 °C. It means that the chosen thermal treatment conditions of the modified polyimide provide its rather high thermal stability.

As IR spectra of the modified polyimide have shown (not presented in paper), the process of imidization runs rather deep in the obtained modified polymer under thermal treatment, which is indicated by the intensive absorption bands at 1780, 1720, 1380 and 720 cm^{-1}. They are due to the valence and deformational vibrations of the five termed imide cycle, the formation of which occurs at PAA transition to PI. However, in the IR-spectrum it is impossible to identify absorption bands of modified macromolecules that have fragments of a modifier, which is due to the low concentration of the modifier.

The results of SEM measurements indicate that the polyimide, modified by cobalt acetylacetonate, is composed of uniformed particles of the size about 100 μm (Figure 1). Although it is a bigger particle size than generally used and expected in ERF composition, it did not affect the electrorheological properties of the fluid. Further improvement of the milling technique, concerning this material, should be done.

Figure 1. SEM of polyimide dispersed phase used in electrorheological compositions.

In the experiment we have used 10 wt. % suspensions of PAA in mineral oil. By preparing the sample *1*, PAA was milled in the agate mortar with the periodical cooling in liquid nitrogen, and the sample *2* was prepared in the orbed ultrasound mill. Sample 1 and sample 2 were not modified by cobalt acetylacetonate. The composition of ERF containing polyimide, modified by cobalt acetylacetonate is shown in Table 1. There were prepared suspensions with a concentration of 2.6 % and 5.7% based on the modified 11.5 % and 32.0 % cobalt acetylacetonate, differed by the way of the introduction of the modifier – in sample *4* cobalt acetonate is introduced by the reaction in solution, while in sample *5* – by the solid phase method. The results of investigation of the electrorheological activity of compositions with the use of polyamidacid and modified aromatic polyimides as dispersed phase are presented in Figures 2 and 3.

As it is seen from the Figure 2, for ERF with polyimide the process of imidization and modification by cobalt acetylacetonate does not yield any substantial enhancement of shear stress under the electric field. The current density j is changed simbate to the shear stress τ at increasing the intensity of the electric field for samples *1* and *2* (Figure 3). The values j for the samples *4* and *5* coincide, and for the sample *3* are somewhat bigger. The use of PAA (samples *1* and *2*) as dispersed phase of electrorheological compositions leads to the substantial enhancement of their shear stress in the electric field – up to 2 orders at the strength 2 kV/mm. This can be explained by the enhanced concentration of hydroxyl groups in the polymeric chain of PAA in comparison with their composition in the modified PI.

Table 1. The composition of electrorheological suspensions, based on transformer oil and polyimide.

Indication	Sample №		
	3	4	5
Quantity of an introduced modifier in relation to dry PAS, m. %	11.5	32.0	32.0
Content of cobalt in the modified PI, %	3.32	23.57	23.57
Concentration of PI in suspension, %	5.7	2.6	2.6
The way of introduction of the modifier in PI	Reaction in solution	Reaction in solution	Solid phase reaction

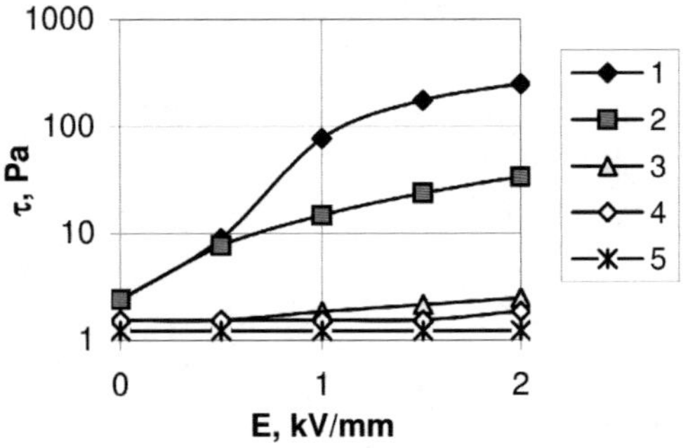

Figure 2. The dependence of shear stress of ERF (samples *1* – *5*) on the electric field strength (the shear rate value is 27 s^{-1}).

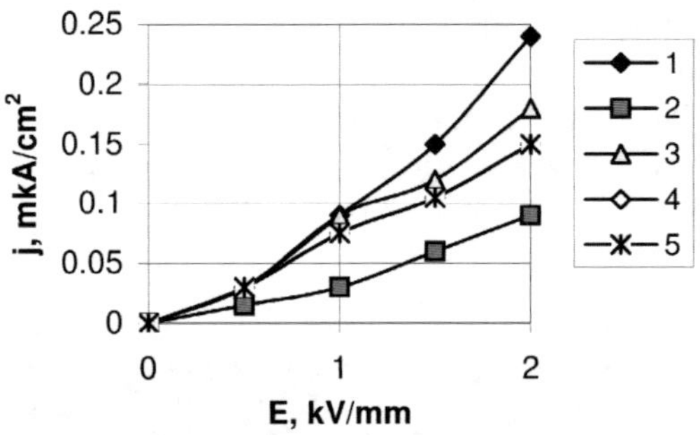

Figure 3. The dependence of current density in ERF (samples *1* – *5*) on the electric field strength (the shear rate value is 27 s^{-1}).

In the course of the study we have noticed that the properties of electrorheological compositions are highly dependent on the way of the preparation. Comparing the properties of the samples *1* and *2* we can conclude, that the preferential way of milling is the milling in the agate mortar with a periodical cooling by liquid nitrogen, which was used for sample *1*. Milling in the orbed grinder, used for the preparation of the sample *2*, results in a slight decrease of the electrosensitivity of the composition, that can be due to the imidization process occurring at such conditions and consequently with the decrease in OH-group concentration.

It is also possible that by the synthesis of PI, modified by cobalt acetylacetonate (samples *3-5*), the content of cobalt was not enough for the formation of the needed quantity of the mobile charge careers, to which electrorheological effect is due to. Nevertheless, the performed investigation has shown that modified polyimides are perspective to be used, although further investigations are needed.

References

[1] Korobko E V 1996 *Electrostructured (Electrorheological) Fluids: the Peculiarities of Hydromechanics and the Possibilities of Application* (Minsk: ITMO NANB)
[2] Hao T 2002 *Advances in Colloid and Interface Science* **97** 1
[3] Endo S *European Patent 0964053* / 1999
[4] Uejima H 1972 *Jpn. J. Appl. Phys.* **11** 319
[5] Bryant C P, Lal K, Pialet J W *US Patent 5435932*/1995
[6] Krutko E T, Prokopchuk N R, Martinkevich A A and Drozdova D A 2002 *Polyimides. Synthesis, Properties, Application* (Minsk: BSTU) p 302

Journal of Physics: Conference Series **149** (2009) 012015 doi:10.1088/1742-6596/149/1/012015

Development of Quasi-3DOF Upper Limb Rehabilitation System Using ER brake: PLEMO-P1

T Kikuchi[1], K Fukushima[1], J Furusho[1] and T Ozawa[1,2]

[1]Graduate School of Engineering, Osaka University, 2-1 Yamadaoka, Suita, Osaka 565-0871, Japan
[2] Kano General Hospital, 7-5-15 Tenjinbashi, Kitaku, Osaka 531-0041, Japan
E-mail: kikuchi@mech.eng.osaka-u.ac.jp

Abstract. In recent years, many researchers have studied the potential of using robotics technology to assist and quantify the motor functions for neuron-rehabilitation. Some kinds of haptic devices have been developed and evaluated its efficiency with clinical tests, for example, upper limb training for patients with spasticity after stroke. However, almost all the devices are active-type (motor-driven) haptic devices and they basically require high-cost safety system compared to passive-type (brake-based) devices. In this study, we developed a new practical haptic device 'PLEMO-P1'; this system adopted ER brakes as its force generators. In this paper, the mechanism of PLEMO-P1 and its software for a reaching rehabilitation are described.

1. Introduction

The increasing numbers of the aged people and their physical deteriorations have become one of the most serious problems in many countries. Many stroke patients suffer from disabilities which restrict Activities of Daily Living (ADL), e.g., reaching actions. Therefore, sufficient rehabilitative training is necessary for such patients.

In general, therapists make rehabilitation programs based on their inspections and measurements of each patient. However, it is difficult to adopt appropriate rehabilitative programs for all patients, because the evaluation method is based on experiences of each therapist. Nowadays, Evidence Based Medicine (EBM) is required strongly in the field of rehabilitation. Therefore, rehabilitation systems that utilize robotics technologies are expected for; 1) quantification of the effect of the rehabilitative training; 2) enhancement of the motivation for patients with creating new training methods (many patients are giving up rehabilitative training because of its boredom); 3) improvement of the efficiency of physical therapist's works.

Krebs H.I., et al. [1] have developed active (motor-driven) haptic device (MIT-MANUS) and conducted many clinical tests for upper limbs rehabilitation. However, the motor-driven robots basically require high-cost safety system compared to passive (brake-based) devices. Book W.J., et al. [2] have developed passive haptic devices. In their system, conventional powder brakes were used as haptic generators. Generally, the response time of the powder brake is more than 100ms and it causes lack in quality of force feedback.

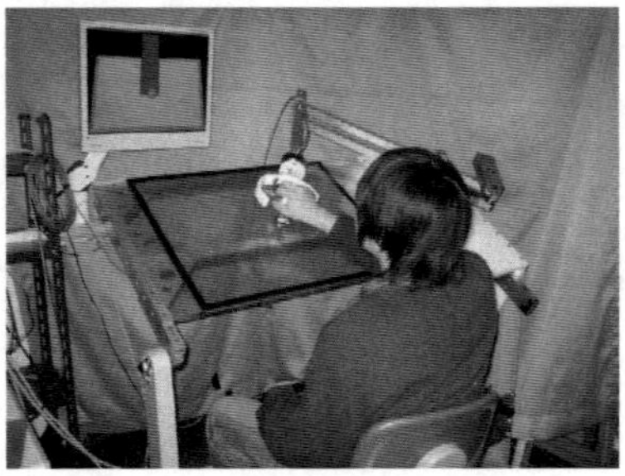

Figure1. Quasi-3-DOF rehabilitation system “PLEMO-P1”

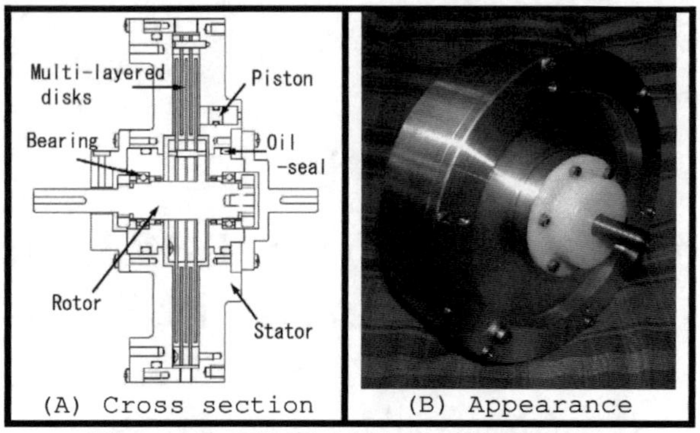

Figures2. ER Fluid Brake (ER Brake)

To solve these problems, we have developed passive haptic devices using ER Fluid Brakes (ER Brake) [3]. Due to the rapid response of the ER Brake (2~3ms), our haptic device presented good performance as a force display. In this study, we improved this system (ER brake-based system) as a practical rehabilitation system for upper limbs. To meet demands for the rehabilitative training in 3-D space, we developed a new haptic device "Quasi-3-DOF Rehabilitation System for Upper Limbs" or "PLEMO-P1" (shown in Fig.1). PLEMO-P1 has 2-DOF force-feedback function in its working plane but the working plane can be adjusted its inclination. In this paper, the mechanism of PLEMO-P1 and its software are described.

Table 1. Specifications of ER brake

Total weight	2.3kg
Diameter	150mm
Height	40mm
Maximum brake torque	4.0Nm (for 3.0kV/mm)
Idling torque	0.1Nm
Num. of rotor disks	3
Disk gap	1mm
Time constant of response	2~3 ms

2. Rehabilitation System "PLEMO-P1"

PLEMO-P1 is a stand-alone robot and one of the passive haptic devices: it can present several kinds of virtual force, for example resistance, viscosity force, vibration etc. This machine has two controllable degrees of freedom (DOF) in a working plane and an uncontrollable DOF for the angle-adjustment of the working plane. Two ER brakes are installed as the torque generators.

Under the training, an operator grasps a handle of PLEMO-P1, watches a display in front of him/her and plays several kinds of application software as rehabilitation trainings or evaluation tests. This system can measure the position of the hand and its manipulating forces with rotary encoders installed on link joints and a force sensor built in its grip, respectively. In this section, we describe both the ER brake and the mechanism of PLEMO-P1.

2.1. ER brake

ER fluid is one of functional fluids, whose rheological properties can be changed by applying an electrical field [4]. To use this fluid as working fluid, we can construct an electrically controllable brake with the ER fluid, which has high-performance (good rapidity and repeatability of brake torque) [5]. We use this brake for the force generators of a new rehabilitation system (force-feedback system) in this paper. Figures 2 show the developed ER brake for PLEMO-P1. This brake consists of multi-layered disks. ER fluid is filled between the rotor-disks and stator-disks. As a result, six layers of ER fluid generate the brake torque. Piston mechanism prevents the fluid from leaking with its expansion. Table 1 shows specifications of the brake. We can control the brake torque from 0.1Nm to 4.0Nm with the electric field from 0.0kV/mm to 3.0kV/mm, respectively.

As shown in the left figure of Figs.3, the time constant of response is 2~3ms [4]. Due to this rapidity of the response, a haptic device using this ER breaks can realize a high frequency response (e.g. an impact force of virtual hockey).

The right figure of Figs.3 shows the brake torque of the ER brake depending on the electric field applied to it. This characteristic has good repeatability and we can formulate this relation as follows;

$$T = 0.39E^2 + 0.12E + 0.10 \quad (1)$$

'T' represents the brake torque [Nm] and 'E' represents the electric field [kV/mm].

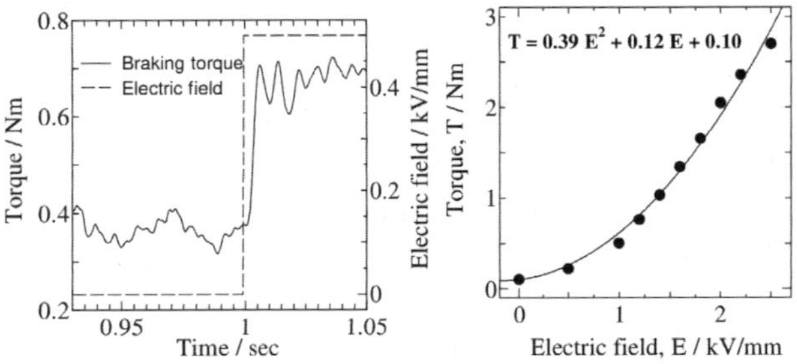

Figures3. Basic characteristics of the ER brake: Step response (left) & braking torque vs. electric field (right)

2.2. Mechanism of Force Control

Force control unit consists of two ER brakes and the brake torques control external force on a handle built on the front edge of the parallel linkage shown in Fig.4. As shown in this figure, T1 or brake torque of brake1 generates force F1 and similarly torque T2 generates force F2. If the movement of the handle is vertical direction like this figure and this system generates force F1 and F2, the user feel resistance against his movement direction. If these forces are constant, users feel constant resistance. In other case, if these forces are proportion with the velocity of the movement, the users feel viscous force. Necessary torques T1 and T2 are calculated based on the inverse kinematics. Lengths of the link 1 and the link 2 are 450mm. This value is designed on the basis of the manipulability-analysis.

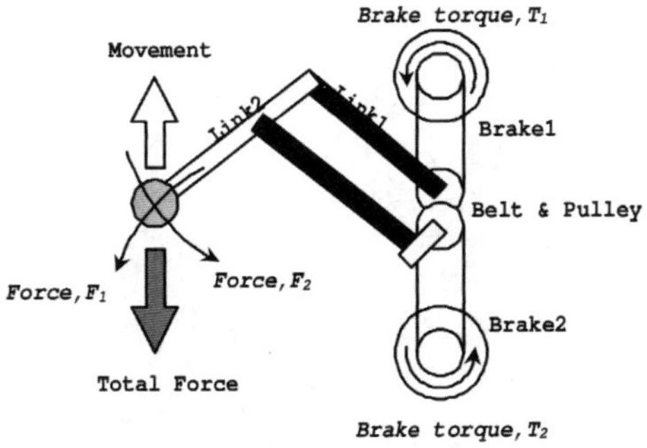

Figure4. Parallel linkage of PLEMO

3. Software of PLEMO-P1

We suggest application software "Virtual Sanding Training" for reaching trainings shown in Figs.5 (A, B, C). This software is developed on the basis of conventional sanding training in the occupational therapy (shown in the picture D of Figs.5 [5]). In the sanding-training, the patient is instructed to actively push the sander higher and higher which provide some resistance against the extensors. This resistance activates patient's intention to move their own arm. Actually, in physical therapy and occupational therapy, reaching training is done without pulling patient's hand in the correct direction, but with giving resistance against correct direction. The

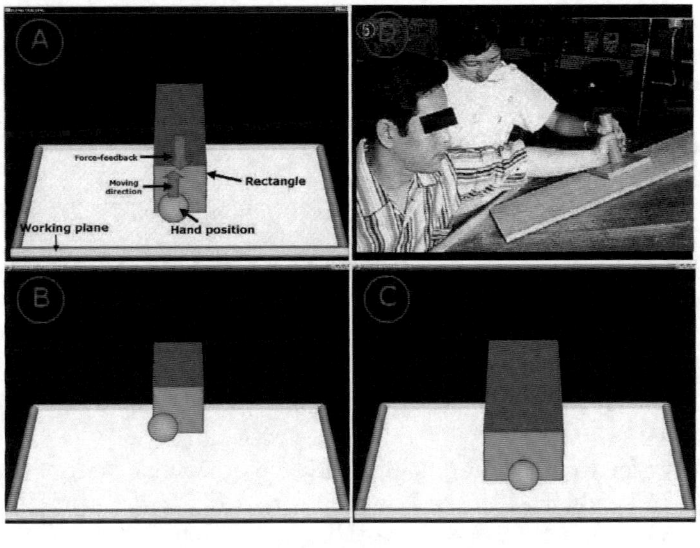

Figures5. Virtual Sanding Training
(Picture D is modified from [6])

inclination of the sanding board is adjusted depending on the recovery level of the patients. So therapists gradually change the inclination as the patient improves.

In the "Virtual Sanding Training", patients are instructed to actively push the rectangle with extending their arm toward front direction like a sanding-training (shown in Fig.5 (A-B)). The position of the hand is feed backed visually, so patients can check physical relationship between the object (a rectangle box) and their own hand (a sphere). In the only case that the patients push the box to the proper direction (straight front), the box move to the front and resistive force was generated on the end-effecter. Moreover, if the system detects their abnormal movements, the system stops presenting the force. Following the hand-position, velocity and operating force, this system can judge the patient's movements. Objective of this training is to help the patients recognize a correct movement and to restructure their motion control function which they lost following stroke (Post stroke patients basically lose independent control of individual joints and coordination among the joints. Especially, it is difficult for them to control movements toward extension). The width of the box can be adjusted depending on the recovery level of the patients.

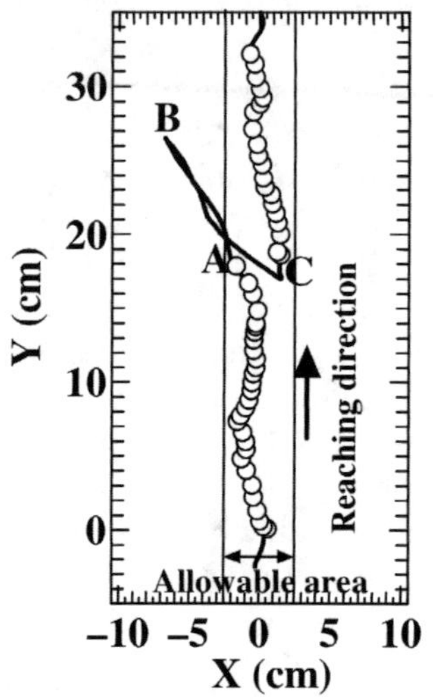

Figure6. Experimental result

Figure 6 shows an experimental result of the Virtual Sanding Training. An "Allowable area" means the width of the box. The solid line is a trajectory of the hand and white circles represent hand's position under the proper training. As shown in this figure, the hand departed from the allowable area (box width) at the point A. Once the operator departed from the allowable area, he cognized disappearance of the resistive force (point B), modified his trajectory to the allowable area and restarted the proper reaching (point C).

4. Conclusion

In this paper, we described the development of "Quasi-3DOF Rehabilitation System for Upper Limb" or "PLEMO-P1" and its software. PLEMO-P1 has two controllable DOF on a working plane and one uncontrollable DOF to adjust the inclination of the working plane. Using ER brakes as a torque generator, PLEMO-P1 has a good responsibility for its force feedback and guarantees the safety for humane because PLEMO-P1 is a passive haptic device; it dose not have any actuators and it cannot move automatically.

References

[1] Krebs H.I., Volpe B.T., Aisen M.L. and Hogan N., 2000, Increasing productivity and quality of care: Robot-aided neuron rehabilitation, Journal of Rehabilitation Research and Development, 37/6, 639-652

[2] Munir S., Tognetti L. and Book W.J., 1999, Experimental Evaluation of a New Braking System for Use in Passive Haptic Displays, Proceedings of the American Control Conference, 4456-4460

[3] Koyanagi K. and Furusho J., 2005, Progress of Passive-type Force Display System with ER Fluid Devices, International Journal of Modern Physics B, 19/7-9, 1641-1647

[4] Bossis I. G., Ed., 2002, Proceedings of the Eighth International Conference on Electrorheological Fluid and Magnetorheological Suspensions, World Scientific.

[5] Kikuchi T., Furusho J. and Oda K., 2003, Development of Isokinetic Exercise Machine Using ER Brake, Proceedings of 2003 IEEE International Conference on Robotics and Automation, 214-219

[6] Ueda S., 1981, Stroke Rehabilitation Illustrated, University of Tokyo Press, in Japanese

Journal of Physics: Conference Series **149** (2009) 012016 doi:10.1088/1742-6596/149/1/012016

Study on Development of Active-Passive Rehabilitation System for Upper Limbs: Hybrid-PLEMO

T Kikuchi, Y Jin, K Fukushima, H Akai and J Furusho

Department of Mechanical Engineering, Graduate School of Engineering, Osaka University, Osaka, Japan

E-mail: kikuchi@mech.eng.osaka-u.ac.jp

Abstract. In recent years, many researchers have studied the potential of using robotics technology to assist and quantify the motor functions for neuron-rehabilitation. Some kinds of haptic devices have been developed and evaluated its efficiency with clinical tests, for example, upper limb training for patients with spasticity after stroke. Active-type (motor-driven) haptic devices can realize a lot of varieties of haptics. But they basically require high-cost safety system. On the other hand, passive-type (brake-based) haptic devices have inherent safety. However, the passive robot system has strong limitation on varieties of haptics. There are not sufficient evidences to clarify how the passive/active haptics effect to the rehabilitation of motor skills. In this paper, we developed an active-passive-switchable rehabilitation system with ER clutch/brake device named "Hybrid-PLEMO" in order to address these problems. In this paper, basic structures and haptic control methods of the Hybrid-PLEMO are described.

1. Introduction

The increasing numbers of the aged people and their physical deteriorations have become one of the most serious problems in many countries. Many stroke patients suffer from disabilities which restrict Activities of Daily Living (ADL), e.g., reaching actions. Therefore, sufficient rehabilitative training is necessary for such patients. Nowadays, Evidence Based Medicine (EBM) is required strongly in the field of rehabilitation. Therefore, rehabilitation systems that utilize robotics technologies are expected for; 1) quantification of the effect of the rehabilitative training; 2) enhancement of the motivation for patients with creating new training methods (many patients are giving up rehabilitative training because of its boredom); 3) improvement of the efficiency of physical therapist's works.

Krebs H.I., et al. [1] have developed active (motor-driven) haptic device (MIT-MANUS) and conducted many clinical tests for upper limbs rehabilitation. However, the motor-driven robots basically require high-cost safety system compared to passive (brake-based) devices. Book W.J., et al. [2] have developed passive haptic devices. In their system, conventional powder brakes were used as haptic generators. Generally, the response time of the powder brake is more than 100ms and it causes lack in quality of force feedback.

To solve these problems, we have developed two-types (active-type [3] and passive-type [4]) of rehabilitation system for upper limbs with ER fluid (Electrorheological fluid) actuators / brakes [3]. Due to the rapid response of ER fluids, our haptic device presented good performance as a force display. ER fluid actuator-type system has high safety and good performance for haptic control [3], but it needs a lot of cost and size for actuator. On the other hand, ER fluid brake-type system is very compact and inherently safe; however, it has strong limitation of haptic varieties. At present, we do not have guidelines to determine which type is suitable for an individual case, because there are not sufficient evidences to clarify how the passive/active haptics effect to rehabilitative trainings for motor skills. In this study, we have developed an active/passive switchable rehabilitation system with ER clutch/brake device named "Hybrid-PLEMO" (see Fig.1) in order to address these problems in common environment. In this paper, basic structures and haptic control methods of the Hybrid-PLEMO are described.

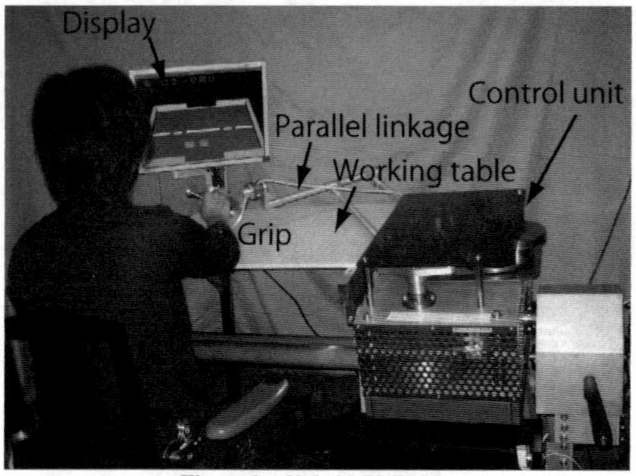

Figure1. Hybrid-PLEMO

2. ER fluid Device for Active / passive switchable mechanism

2.1. Mechanism

ER fluid is one of the functional fluids of which rheological properties can be changed by applying electrical fields [5]. In this paper, particle-dispersed-type ER fluid is used.

Figure 2 shows a basic structure of the ER-fluid-clutch-type actuator (or simply, ER Actuator). As mentioned in the previous report [3], this actuator is safer than other conventional actuators. In this paper, we utilized this actuator as a torque generator. Additionally, by fixing its input rotation, we can switch it as a brake system. In this study, we adopted this actuator (active) / brake (passive) switchable system.

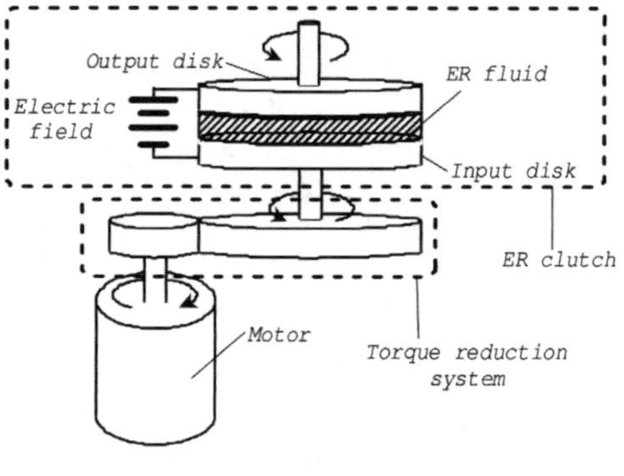

Figure2. ER-clutch-type actuation system for safety

2.2. ER fluid clutch / brake device with double shafts

Figures 3 show a developed ER fluid clutch / brake device (or simply, ER device in this paper). This device has two controllable shafts with a double-shaft structure. An inner shaft and an outer shaft are connected to the multi-layered disks (output disks) for each. Each disk is laminated with case-side disks (input disks) in alternative shift with 1mm gap. The ER fluid is filled in the gap. When the casing is rotated with

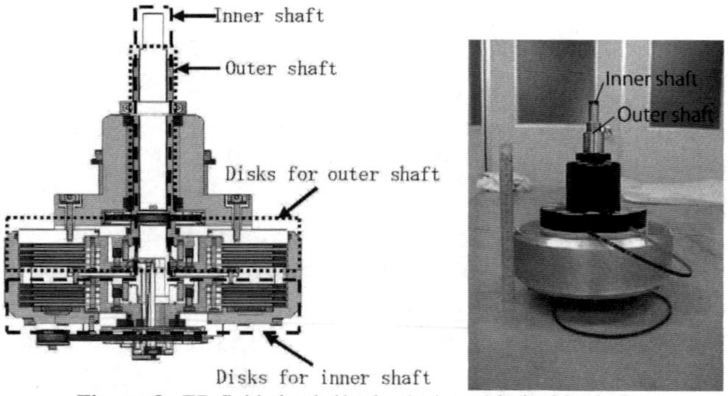

Figures3. ER fluid clutch / brake device with double shaft

TABLE 1. SPECIFICATIONS OF HYBRID-PLEMO

Size	W0.6m*D0.5m*H1.0m
Motion region	W0.6m*D0.5m Adjustable angle of the inclination is -30~90deg
Maximum force	4kfg at end effecter
Num. of 2 DOF ER clutch	2
Motor	40W

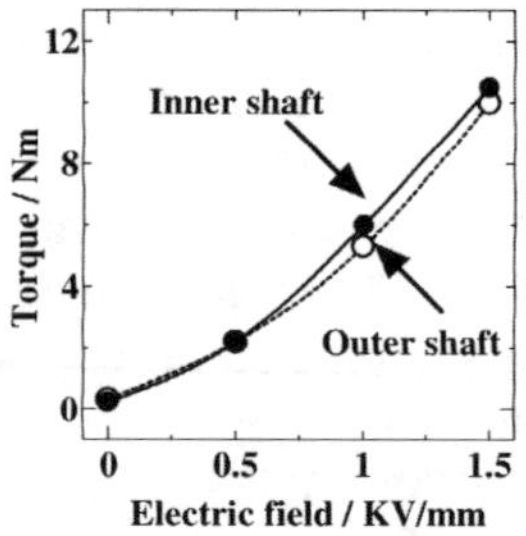

Figure4. Output torque of double-shaft ER fluid clutch

motors, input disks are rotated simultaneously and the device works as a clutch. On the other hand, when the casing is fixed, the device works as a brake.

Diameter and height of this device are 192mm, 225mm, respectively. Diameter, thickness and gap of disks are 155mm, 1.0mm, 1.0mm, respectively. Numbers of disks are 4 for each group. Figure 4 shows output toque of this device.

3. Hybrid-PLEMO

3.1. Concept

The Hybrid PLEMO has 2 controllable DOF on a working table and 1 uncontrollable DOF of the inclination of the working table (see Fig.1). We defined this working space as a "Quasi-3-DOF Workings pace". An operator grasps a handle on the end-effecter of its arm, watches visual information on a display and plays application software as rehabilitative training and evaluation test.

In previous report [4], we used only ER brake for its torque control. Therefore, it was a passive haptic device. In Hybrid PLEMO, we used the active/passive switchable device mentioned above with the same Quasi-3-DOF mechanism. Table 1 shows specifications of the Hybrid-PLEMO.

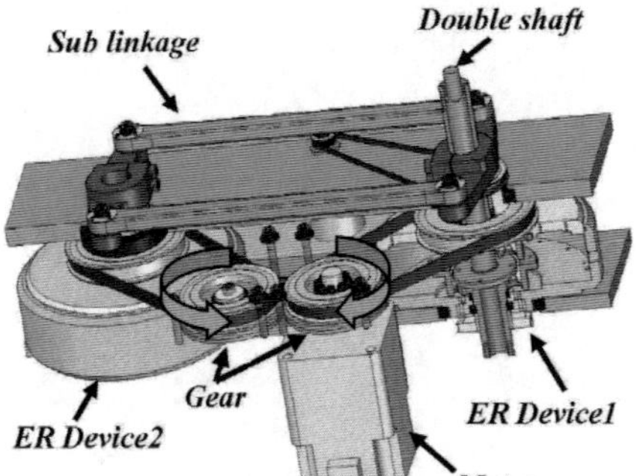

Figure5. Torque control unit for Hybrid-PLEMO

3.2. Force Control Mechanism

Haptic force on the end-effecter of Hybrid PLEMO is controlled with torque control unit including ER devises mentioned above. Figure 5 shows the torque control unit for the Hybrid-PLEMO system. In this structure, a "motor" (AC servo-motor) rotates slowly and generate two directional rotations (clockwise; CW and counter-clockwise; CCW) with "gears". These rotations are transmitted to two ER devices. Therefore, "ER device 1" controls torques of CW direction. At the same time, "ER device 2" controls torques of CCW direction.

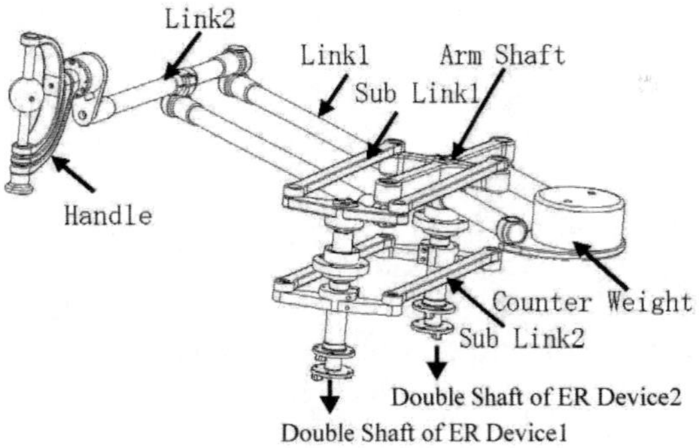

Figure6. Parallel linkage system for Hybrid PLEMO

Figure 6 shows a parallel linkage mechanism for the end-effecter of the Hybrid-PLEMO. Inner shafts of two ER devices are connected with "Sub Link1" and it controls torques of "Link2". In the same manner, outer shafts of two ER devices are also connected with "Sub Link2" and it controls torques of "Link1". Finally, we can control haptic forces on the handle with the parallel linkage that consist of "Link1" and "Link2".

3.3. Control System

Figure 7 shows control system for the Hybrid PLEMO. Absolute encoders (FA Coder, TS566N320, Tamagawa Seiki Inc., Japan, resolution: 17bits) measure the rotational angle of each link. We can calculate the position and the velocity of the handle depending on each angle. Digital Input/ Output (DIO) board (PCI-2154C, Interface Inc., Japan) loads this information to a controller (personal computer). Operating handle includes a force sensor (OPFT-220N, Minebea Co. Ltd., Japan), and operating force is measured by this sensor. A potentiometer (CP-2F, Midori Precision Inc., Japan) measures the inclination of the worktable and the angle is loaded by Analog/Digital (A/D) converter board (PCI-3165, Interface Inc., Japan, resolution: 16bits). The torque of the ER devices is controlled

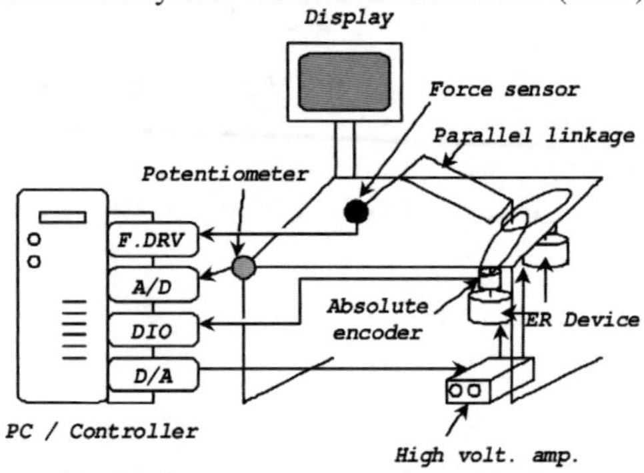

Figure7. Control System for Hybrid PLEMO

by application of the voltage from high voltage

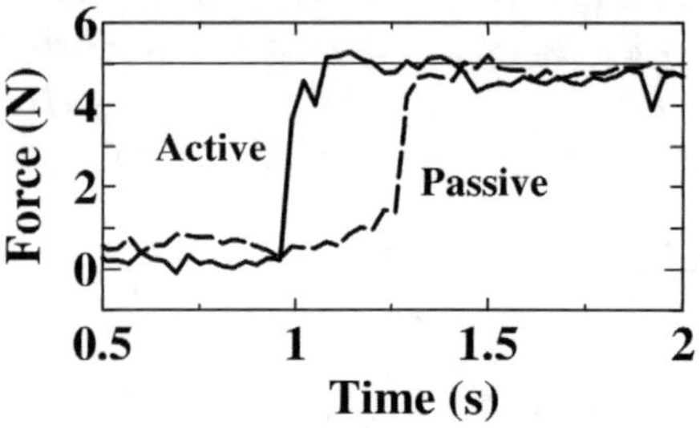

Figure8. Constant resistive force by passive / active mode

amplifiers (High voltage amplifier, MAX-ELECTRONICS, Co. Ltd., Japan). Digital/Analog (D/A) converter board (PCI-3338, Interface Inc., Japan, resolution: 12bits) outputs the reference signal to the amplifiers.

A controller is a personal computer (DOS/V), and an operating system (OS) is Vine Linux 3.2 and ART-Linux (kernel 2.4.31) as a real-time OS. Open-GL and Glut3.7 are used for the graphic library. Graphic process and control process are executed by one PC. Multi-process programming is used to realize it. The control process is repeated by 1 [ms] exactly.

3.4. Experimental result for constant resistive force

To evaluate the performance of the haptic force of the Hybrid-PLEMO, constant resistive forces with active / passive haptic control were measured. Figure 8 shows experimental results. A solid line represents force with the active haptic control, and a dashed line represents force with the passive haptic control. In these experiments, an operator manipulated a handle forward from initial position. Initially, resistive force was zero, but when the hand entered a friction area, resistive force of constant 5 N instantly presented. As shown in this figure, haptic forces with both of active/passive haptic methods were accurately controlled.

4. Conclusion

This paper described the development and evaluation of "Hybrid PLEMO", which is one of the rehabilitation systems for upper limbs with quasi-3DOF mechanism and switchable mechanism of active/passive haptic controls. We expect to use this system to clarify the effects and roles of active / passive force feedback in rehabilitative training.

References

[1] Krebs H.I., Volpe B.T., Aisen M.L. and Hogan N., 2000, Increasing productivity and quality of care: Robot-aided neuron rehabilitation, Journal of Rehabilitation Research and Development, 37/6, 639-652

[2] Munir S., Tognetti L. and Book W.J., 1999, Experimental Evaluation of a New Braking System for Use in Passive Haptic Displays, Proceedings of the American Control Conference, 4456-4460

[3] Furusho J. and Kikuchi T., 2007, A 3-D Rehabilitation System for Upper Limbs "EMUL", and a 6-DOF Rehabilitation System "Robotherapist", and Other Rehabilitation System with High Safety, Rehabilitation Robotics (Edited by Sashi S Kommu), Chapter 8, I-Tech Education and Publishing, 115-136

[4] Kikuchi T. , Xinghao H., Fukushima K. Oda K., Furusho J. and Inoue A., 2007, Quasi-3-DOF Rehabilitation System for Upper Limbs: Its Force-Feedback Mechanism and Software for Rehabilitation, Proceedings of International Conference on Rehabilitation Robotics, pp.24-27.

[5] Winslow, W.M., 1949, Journal of Applied Physics, 20, 1137-1140

Journal of Physics: Conference Series **149** (2009) 012017 doi:10.1088/1742-6596/149/1/012017

Preparation of polystyrene/polyaniline composite particles and their electrorheology

Y H Kim[1], B J Park[1], H J Choi[1] and S B Choi[2]

[1]Department of Polymer Science and Engineering, Inha University, Incheon 402-751, Korea

[2]Department of Mechanical Engineering, Inha University, Incheon 402-751, Korea

E-mail: hjchoi@inha.ac.kr

Abstract. Monodisperse polystyrene (PS) seed latex was initially prepared by a dispersion polymerization, and then polyaniline (PANI)-coated PS composite particles were synthesized by diffusion and polymerization of aniline at the interface. Characteristics of the obtained PS/PANI composite particles such as morphology and thermal property were examined using SEM and TGA, respectively. In addition, we employed a rotational rheometer to investigate rheological behaviour of the ER fluids based on both PANI particles and PS/PANI composites. Yield stress and flow response were investigated under electric field strengths showing typical ER behaviours. Further, the Cole-Cole plot and the dielectric spectra gave relaxation times of the ER systems, confirming the correlation of dielectric properties with ER performance.

1. Introduction

Electrorheological (ER) fluids, suspensions of polarizable particles in non-conducting liquids, exhibit drastic changes in their rheological properties including a large increase in the apparent viscosity and formation of reversible suspension microstructures with an applied electric field [1-4]. When a certain electric field is imposed to the ER fluids, they can flow by an applied shear stress exceeding a minimum limiting value. Therefore, the ER fluids are often considered as ‘Bingham fluids’, and the minimum limiting stress for flowing is designated by a yield stress. In addition, the apparent viscosity and shear stress of the ER fluids increase under an electric field.

These ER fluids also represent a phase transition from a liquid-like state to a solid-like behavior by forming a chain-like structure within milliseconds when an electric field is applied [5]. This can be understood qualitatively in terms of particles polarization induced by the mismatch between the dielectric constants of the particles and suspending fluid. Due to these unique variations of the rheological properties, the ER fluids have great potential applications in designing torque transducer, damper, actuator and other control systems [6, 7].

As a dispersed phase of the ER fluid, the latex particles are coated with thin overlayers of conducting polymers to form core-shell composite particles [8]. The combination of polymer latex and conducting polymers is expected to produce composite materials with new and unique properties intrinsic to dispersed microspheres and the improvement of polymer processability [9]. These are based not only on the electrical parameter, but also they often employ the chemical, electrochemical, optical and other physical properties of conducting polymers [10-12].

Polyaniline (PANI) is known to be an air-stable organic conducting polymer which is relatively easy to polymerize and to control its electrical conductivity [13]. As a potential material for the ER fluids having an environmental stability [14-16], the PANIs are adopted in this study.

In this work, we prepared surface conductive particles with a polystyrene (PS) microbead as a core and PANI as a shell. First, aniline was added to swell PS microbead under vigorous stirring, then aniline swollen PS was transferred into a water phase consisting of the APS, by which way oil in water system was formed. An overlayer of the conducting polymer is formed by a diffusion-interface-polymerization process of the aniline pre-absorbed on PS seed latex without any surface modification [17]. Core-shell structured particles prepared by this process were found to have uniform thickness of the conducting PANI shell, and these particles were adopted as an ER material.

2. Experimental

2.1. Synthesis of PS seed particles

The PS seed latex was prepared by a dispersion polymerization. Initially, the purified styrene was dispersed in methanol containing polyvinylpyrrolidone as a stabilizer at room temperature. The temperature of the mixture was increased up to 60°C and then a radical initiator 2, 2-azobisisobutyronitrile (AIBN) was added into the reaction system. The polymerization continued for 24h at 60°C. After the reaction, products were centrifuged with both methanol and distilled-water, and then dried using a freezing-drier. Final product was obtained in a powder form.

2.2. Preparation of PS/PANI composite particles

PS/PANI composite particles were formed by a diffusion-interface polymerization process of the aniline pre-absorbed on PS seed latex without any surface modification. Aniline (DC chemical, Korea) and ammonium persulfate (APS) (Daejung, Korea) were used as monomer and initiator, respectively. The PS seed latex was dispersed in 500ml of distilled water by vigorous stirring (200rpm) at 0°C, after then aniline was added to the reactor. The PS seed latex/monomer weight ratio was fixed at 3:1. The mixture was stirred for 12h at 0°C to allow sufficient absorption of aniline into the PS seed particles. The aqueous solution of APS was added to the dispersion in reactor and the initiator/aniline molar ratio was fixed at 1:0.6. This was followed by the addition of hydrochloric acid (1mol/L), also in an equi-molar amount relative to aniline, being dropped via syringe. Thus, aniline existed within polystyrene while the APS was dispersed in water phase, and there was no obvious reaction. In order to initiate polymerization of PANI, hydrochloride acid which diffused to polystyrene to initiate the reaction was dropped. The mixture was kept stirring with 200rpm at 0°C for the first 6h of the polymerization, after which the polymerization was carried out for 20h at room temperature. The resulting PS/PANI composite was filtered with distilled-water and acetone sequentially to remove the excess initiator, monomer and oligomer. It was finally dried at 60°C in vacuum oven for 2days. Furthermore, PANI particles were also synthesized using the conventional oxidation polymerization of aniline to produce a fine emeralidine hydrochloride form following our previous method and the details of PANI particles are same as previously reported [17].

2.3. Measurements

The particle size and surface feature were observed by scanning electron microscopy (SEM, S-4300, Hitachi). In order to study thermal degradation of the sample, thermogravimetric analysis (TGA, TA instrument Q50, USA) was carried out in the temperature range of 100-650°C at a heating rate of 20°C /min under an atmospheric condition (air). Electrical conductivity of the composite pellets was measured by a 4-pin probe resistivity meter (LORESTA-GP) and a MCP probe resistivity meter (HIRESTA-UP). To be applied as an ER material, the PS/PANI composites have to possess appropriate value of electrical conductivity. Therefore, we controlled the electrical conductivity of the composite via dedoping process and monitored the pH value to be 10 by adding either 1M NaOH or

1M HCl solution. The product was filtered, dried, and sieved. The conductivity of PANI-coated PS composites, 10^{-5}S/cm was decreased to 10^{-10}S/cm by dedoping.

We then prepared an ER fluid with 10vol% particle concentration of which the PS/PANI composites were dispersed in silicone oil (kinematic viscosity: 50cS, density: 0.96g/cm^3). ER behaviours were investigated by rotational rheometer (Physica MC120, Stuttgart) equipped with a DC high voltage generator. The dielectric spectra of ER fluids were also measured using the HP 4284A Precision LCR meter with HP 16452A Liquid Test Fixture for liquids to investigate their interfacial polarization. Frequency of the AC electric fields varied from 20Hz to 1MHz.

3. Results and discussion

Monodisperse PS latex was prepared initially by a dispersion polymerization, and then polyaniline (PANI)-coated PS composite particles were synthesized by a diffusion-interface polymerization process and used without post-treatment following the similar method we did for PANI-coated PMMA based ER fluid [13]. In order to initiate polymerization of PANI, hydrochloride acid which diffused to polystyrene to initiate the reaction was dropped. At the surface of PS, the hydrochloride acid modified aniline monomer got initiated with the aid of APS. Finally, more and more aniline monomer diffused to the interface and underwent polymerization [18].

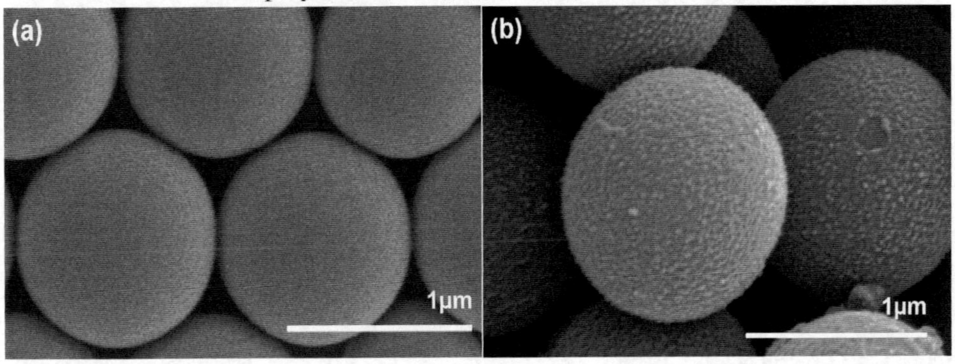

Figure 1. SEM images of the PS seed particles (a), PS/PANI composite particles (b).

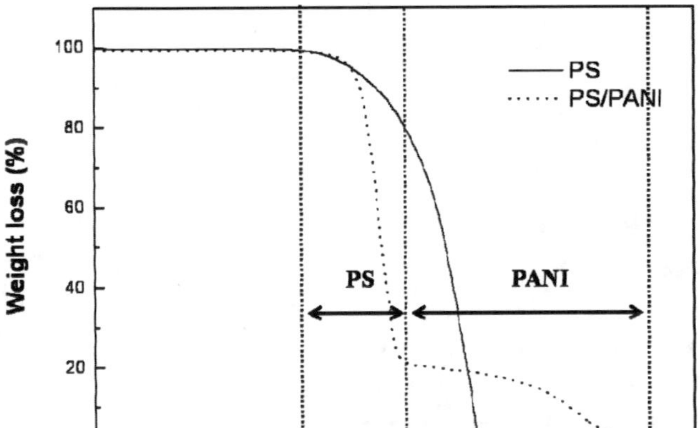

Figure 2. TGA analysis of PS seed particles and PANI-coated PS composite particles.

SEM images of both PS seed particles and PS/PANI composite particles are shown in Fig. 1 (a) and (b), respectively. We can find that the PS seed particles posses a nearly spherical morphology and uniform size distribution. Figure 1(b) represents that the PS/PANI composite particles are also spherical, indicating that the PANI-coated PS composites possess uniformity and well-defined core-shell morphology. As for the composite, we observed that the size of the composites gets bigger than that of the PS seed particles. These images demonstrate that the surface of the PS/PANI composite particles is rough because of the deposition of the PANI component.

Figure 2 represents TGA curves of both PS seed particles and PS/PANI composites exhibiting the residual weight percentage versus the temperature. It was shown that the sharp weight loss beginning at nearly 280ºC and continuing till around 450ºC for PS is attributed to the large scale thermal degradation of PS chains. TGA curves of the PS/PANI composites were found to have 79% weight loss in the range between 280~390ºC which are in agreement with the decomposition of PS seed particles. Furthermore, the 21% weight loss of the composites from 390ºC to 610ºC is regarded to the degradation of PANI particles.

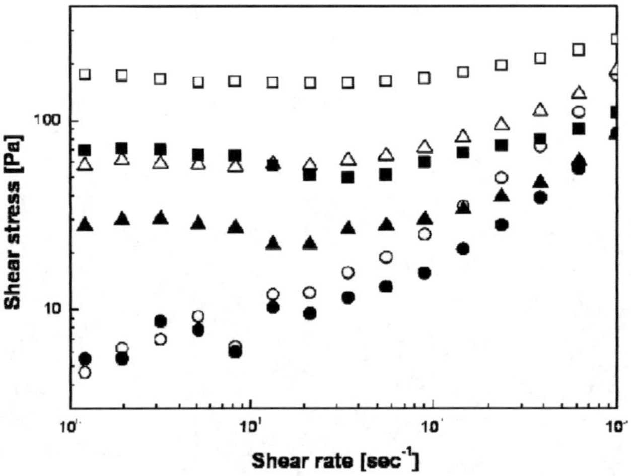

Figure 3. Shear stress *vs.* shear rate for the PANI particles and PS/PANI composites based ER fluids (10 vol%) under various electric fields. (□: 0kV/mm, Δ: 0.7kV/mm, ○: 1.5kV/mm. Open symbol for PANI particles, closed symbol for PS/PANI composites)

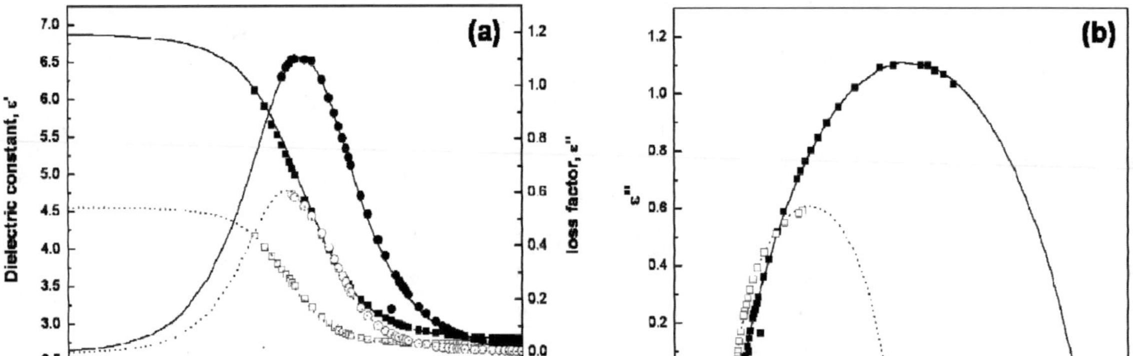

Figure 4. (a) Dielectric spectra of PANI particles and PS/PANI composites based ER fluids, (b) Cole-Cole fitting curves for PANI particles and PS/PANI composites based ER fluids. (closed symbol and solid line for PANI particles, open symbol and dash line for PS/PANI composites)

The ER behaviors of shear stress (τ) as a function of the shear rate under various electric field strengths for the PANI particles and PS/PANI composites based ER fluids are shown in Fig. 3. The dedoped PS/PANI suspensions in silicone oil show typical ER behaviors such that the suspensions exhibit Bingham fluid behavior with a yield stress under an externally applied electric field. By applying electric fields, the shear stress increases within the whole shear rate with yielding behaviors such as a plateau with shear rate in flow curves. This could be explained by the reformation of the broken chain-like structure. In addition, the region of plateau becomes wider with the strength of electric field, indicating that the ER effects can be maintained up to a higher shear rate with applied electric fields.

Figure 4 (a) and (b) shows the dielectric spectra (ε' and ε'') as a function of the frequency and Cole-Cole plots for 10 vol% PANI particles and PS/PANI composites based ER fluids, respectively [19, 20]. These are typical results for the interfacial polarization of suspensions including the ER fluids consisting of polarizable dispersed phases in an insulating media. Even though ε_0s of ER fluids are not shown in experimental results, they can be determined by fitting data with the Cole-Cole plots.

Lines in Figure 4 are fitted from the famous Cole-Cole formula (Eq. 1) [21] with parameters given in Table1.

$$\varepsilon^* = \varepsilon' + i\varepsilon'' = \varepsilon_\infty + \frac{\varepsilon_0 - \varepsilon_\infty}{1 + (i\omega\lambda)^{1-\alpha}} \qquad (1)$$

$\Delta\varepsilon \equiv \varepsilon_0 - \varepsilon_\infty$ is an achievable polarizability in ER fluids. The $\Delta\varepsilon$ dependence on ER properties also has some positive effect on good ER performance. Hence, in Table 1, we can see that the $\Delta\varepsilon$ of PS/PANI composite based ER fluids is lower than that of the PANI particles based ER fluids suggesting a weak ER behaviour.

Table 1. Parameters in Equation 1 for PANI particles and PS/PANI composites (10 vol%) based ER fluids.

Parameter	PANI	PS/PANI
ε_0	6.87	4.55
ε_∞	2.79	2.70
$\Delta\varepsilon$	4.08	1.85
α	0.365	0.260
λ	0.0012	0.0022

In addition, the relation time (λ) for interfacial polarization of ER fluids is known to be related with yield stress and stress enhancement under the applied electric field. When analyzing the dielectric spectra, we can see that the relaxation times are 1.2ms and 2.2ms for the two ER fluids separately. In general, short relaxation time is known to be related with higher shear stress. Obviously, the above analysis on the dielectric spectra coincide with the results obtained from comparing the shear stress.

4. Conclusion

Core-shell structure of PS/PANI composites was synthesized via a diffusion-interface-polymerization process of the aniline to precipitate onto the PS surface without any surface modification. The core-shell structure with rough surface was confirmed by SEM images. TGA data showed that the weight percent of conductive PANI was about 21%, indicating the deposition of the PANI component in the polystyrene core. We also analyzed the flow response as a function of various electric fields to study the different behaviors for the two kinds of ER fluids. PS/PANI composites based on ER fluid indicated typical ER behaviors, showing lower shear stress than that of pure PANI which was attributed to the influence of PS particles on conducting PANI due to its low dielectric performance. At last, investigated dielectric spectra well interpreted the difference in ER performances.

Acknowledgements
This work was supported by Korea Science and Engineering Foundation (Project No. R17-2007-028-01000-0).

References

[1] Parthasarathy M and Klingenberg DJ 1996 *Mater. Sci. Eng.R* **17** 57
[2] Cho M S, Choi H J and Ahn W S 2004 *Langmuir* **20** 202
[3] Fang F F, Choi H J and Joo J 2008 *J. Nanosci. Nanotechnol.* **8** 1559
[4] Lengalova A, Pavlinek V, Cheng Q L and Saha P 2007 *Int. J. Mod. Phys. B* **21** 4883
[5] Lee I S, Cho M S and Choi H J 2005 *Polymer* **46** 1317
[6] Bossis G, Abbo C, Cutillas S and Metayer C 2001 *Int. J. Mod. Phys. B* **15** 564
[7] Kim J W, Kim C A, Choi H J and Choi S B 2006 *Korea-Aust. Rheol. J.* **18** 25
[8] Sung B H, Choi U S, Jang H G and Park Y S 2006 *Colloids Surf. A: Physicochem. Eng. Asp.* **274** 37
[9] Ruckenstein E and Yang S 1993 *Synth. Met.* **53** 283
[10] Barthet C, Armes S P, Lascelles S F, Luk S Y and Stanley H M E 1998 *Langmuir* **14** 2032
[11] Lei T, Aoki K 2000 *J. Electroanal. Chem.* **482** 149
[12] Aoki K, Chen J, Ke Q, Armes S P and Randall D P 2003 *Langmuir* **19** 5511
[13] Cho M S, Cho Y H, Choi H J and Jhon M S 2003 *Langmuir* **19** 5875
[14] Fang F F, Kim J H, Choi H J and Seo Y 2007 *J. Appl. Polym. Sci.* **105** 1853
[15] Choi H J, Kim J W, Joo J and Kim B H 2001 *Synth. Met.* **121** 1325
[16] Lim J Y, Kim S T, Park B J and Choi H J 2007 *Int. J. Mod. Phys. B* **21** 5003
[17] Cho M S, Kim J W, Choi H J, Webber R M and Jhon M S 2000 *Colloid Polym. Sci.* **278** 61
[18] Wu Q, Wang Z and Xue G 2007 *Adv. Funct. Mater.* **17** 1784
[19] Choi H J, Hong C H and Jhon M S 2007 *Int. J. Mod. Phys. B* **21** 4974
[20] Hong C H, Park B J and Choi H J 2006 *J. Colloid Interef. Sci.* **300** 818
[21] Fannin P C and Giannitsis A T 2004 *J. Mol. Liq.* **114** 89

Journal of Physics: Conference Series **149** (2009) 012018 doi:10.1088/1742-6596/149/1/012018

Peculiarities of Particle Polarization in Water Containing Fillers for ERF

E V Korobko[1,3], L S Eshenko[2] and N A Bedik[1]

[1] Luikov Heat and Mass Transfer Institute Academy of Sciences of Belarus, 15 Brovka Street, Minsk 220072 Belarus

[2] Belarusian State Technological University, 13 a Sverdlov Street, Minsk 220006 Belarus

E-mail: eva@itmo.by

Abstract. In the present paper the experimental rheometric information on the investigation of ERF, based on the water containing fillers, is analyzed and generalized. A mechanism of surface polarization for particles of the dispersed medium is suggested.

Introduction

There are plenty of reports on polarization phenomena in ERF, based on the water containing dispersed phase. However, the question about the nature of the processes with interphase transport of protons, which determines surface electroconductivity of the particles, has still not been answered unambiguously. There are several hypotheses on the mechanism of the surface polarization for colloid systems [1]. It is known [2] that different types of hydroxyl groups are feasible to appear on the surface of hydrated hydroxyl containing compounds. However, the nature of the dominating proton containing groups, their role in the polarization processes on the surface of particles of a dispersed phase of ERF is not clear. Therefore, a further discussion and experimental investigations of these problems are needed.

In the present paper we have analyzed and generalized the experimental rheometric material on the investigation of ERF, based on the water containing fillers, and we suggest the mechanism of surface polarization for particles of the dispersed medium.

1. Materials and methods

The experimental investigation of the rheological properties of ERF has been carried out by means of a rotational viscometer of a coaxial-cylindrical type at room temperature. The measurement of the extreme shear stress τ_0 has been carried out by means of a vibrorheometer with coaxial cylinders. As extreme shear stress we have assigned the value of the shear stress, at which the elastic deformations change into the plastic deformations. The measurements have been carried out in the regime of a continuous deformation at a constant rate of change of the deformation in the range of temperatures 20 – 120 °C. As fillers for ERF we have used aluminum and chromium hydrated oxides, obtained by means of sol-gel synthesis. These samples were used for preparation of 40 wt. % suspensions in mineral oil. The total content of H_2O in the samples was determined by means of thermogravimetrically in the polythermical regime by heating up to 850 °C. On the basis of DTA-measurements, the quantity of the different types of water were determined in the following regimes: the structured H_2O (in the form of OH-groups) – in the polythermal regime in the temperature range of 350-520 °C, the unnstructured H_2O – in the isothermal regime in the temperature range of 170-350 °C (the interlayer H_2O) and 80–170 °C (the adsorbed H_2O).

2. Results

A qualitative evaluation of the ER-activity of some fluids (obtained by us and known from literature) on the basis of aqueous dispersed phase of different compositions and preparation process is displayed in table. The studied fillers are nominally divided into two groups: the first one refers to the fillers that are obtained by soaking and keeping of the synthetic and natural compounds in the vapor ambient, namely metal oxides, diatomite and etc., that after drying contain a certain quantity of the adsorbed water (column 1). The second group contains inorganic compounds, synthesized by means of chemical deposition from aqueous solutions of reagents and characterized by a certain chemical, phase, dispersed composition (column 2). The investigated samples represent aluminum and chromium hydrated oxides iffered by the structure, the water content in the form of H_2O molecules and OH groups, the nature of bonding of these forms of water, the place of localization of water molecules (in between layers in the crystalline lattice and on the surface of particles). In correspondence with the classification [3] in the hydrated metal oxides two water categories are distinguished: the structural (in form of OH-groups) and the nonstructural (in form of H_2O molecules) in two varieties – the interlayer and the adsorbed water. The presence of the interlayer water is due to the fact that in the interburst space of the lattice of the layered structures of hydrated metal oxides, the water molecules are incorporated, which form a hydrogen bond with hydroxyls. The molecules of the adsorbed water are known to join OH-groups by a hydrogen bond. In particular, we have investigated aluminum hydrated oxides (rhombic γ-AlOOH (boehmite), pseudo boehmite, monoclinic α-$Al(OH)_3$ (baehrite)), and also $Cr_2O_3 \cdot nH_2O$. In the structure of boehmite, every Al^{3+} ion is surrounded by six ions of O^{2-}, situated at peaks of octahedrons, linked to each other by edges where they are connected with the peaks, forming the double layers, bonded by hydrogen bonds. Pseudo boehmite comparing to boehmite (AlOOH or $Al_2O_3 \cdot nH_2O$), has a content such as $Al_2O_3 \cdot 1.5\text{-}2.0H_2O$. The structure of the latter is similar to the structure of boehmite, but is composed of two-dimensional crystals. Super stoichiometrical water in the quantity of 0.5-1 mol is localized in the interlayer space of the structure and is bonded by strong hydrogen bonds. Such water molecules are somewhat deformed and thermally stable. Baehrite and threehydrate of aluminum oxide α-$Al(OH)_3 \cdot 3H_2O$ have a monoclinic lattice. The content and conditions of obtaining the modifications of hydrated aluminum oxide are shown in table. X-ray amorphous hydrated chromium oxide, depending on the conditions of its preparation contained a variable quantity of nonstructural, namely, adsorbed H_2O (see table). According to the table, the filler, containing only structured H_2O, has a low ER-activity. The presence of nonstructural water in the filler leads to their substantial activating. Moreover, the degree of the activation of the filler is probably defined not by the place of localization of H_2O molecules, but by their quantity and nature of bonding. Thus, the decrease of the content of adsorbed water in the hydrated chromium oxide leads to the vanishing of ER activity (see the table). Therefore, ERE occurs only due to the presence of adsorbed water, consisting in the composition of hydrated chromium oxide in the same way as in the fluids on their basis, for example, diatomite.

In Figures 1, 2 the dependencies of the relative yield stress on the temperature and the yield stress on the electric field intensity for ERF, based on the samples of two groups (diatomite, pseudo boehmite), are shown. As it follows from Figure 1, the ER-effect for ERF, based on the fillers of two types, is comparable at room temperature. By heating higher than 40 °C, ERE in the fluid, based on diatomite, begins to lower, and at $T = 80$ °C is almost vanished. For ERF, based on pseudo boehmite, the growth of τ_{0E}/τ_0 up to 80 °C is seen, than the values of τ_{0E}/τ_0 begin to lower, but even at $T = 120$ °C they are higher than that at room temperature.

From figure 2 it is seen that by heating of ERF, based on diatomite, up to 80 °C, the application of the high electric field (more than 1.5 kV/mm) leads to the rapid decrease of the increment of τ_{0E} (curve 3). Moreover, high values of current are seen – 1.2 mA. For ERF on the basis of pseudo boehmite, the values of τ_{0E} at the increase of temperature under all intensities of the electric field are increased. The values of the conductivity current for this sample are not high, which allows us to perform the following heating of the substance.

Table. Characteristics of aqueous fillers for ERF

Method of preparation of the filler		ER-activity of the fluids based on these investigated fillers **
Saturation of materials by H_2O	sol-gel synthesis	
1	2	3
Al_2O_3 , Cr_2O_3		++
	SiO_2[4]	+++
	$Al_2O_3{\cdot}H_2O$ or AlOOH (boehmite)	+
TiO_2 , NiO, SiO_2		+
	$Al_2O_3{\cdot}1.7\text{-}2.0H_2O$ (pseudo boehmite)	+++*
	$Al_2O_3{\cdot}2.97H_2O$ (baehrite)	++
	$Cr_2O_3{\cdot}5.1H2O$	+++
Diatomite, cellulose [5]		+++
	$Cr_2O_3{\cdot}3.6H2O$	+
Al_2O_3 in the form of fiber [1]		++*

*** - are thermo stable**

** +++ - high (ratio of shear stress under maximum field to that under no field – from 50 up to 100 and higher); ++ - medium (÷ from 10 up to 50); + - low (÷from 1 up to 10)

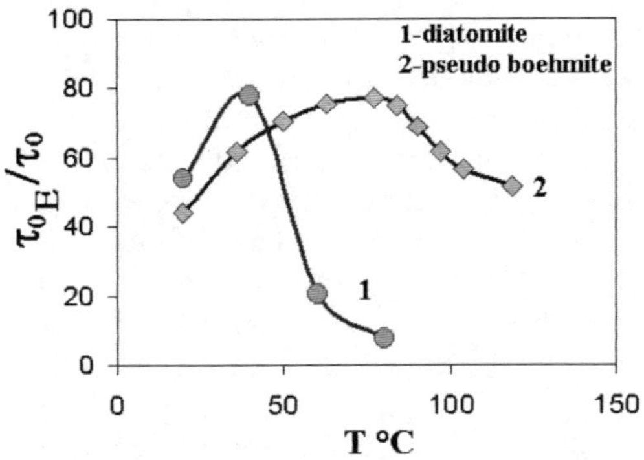

Figures 1. The dependence of the relative yield stress on the temperature at E = 2.5 kV/mm

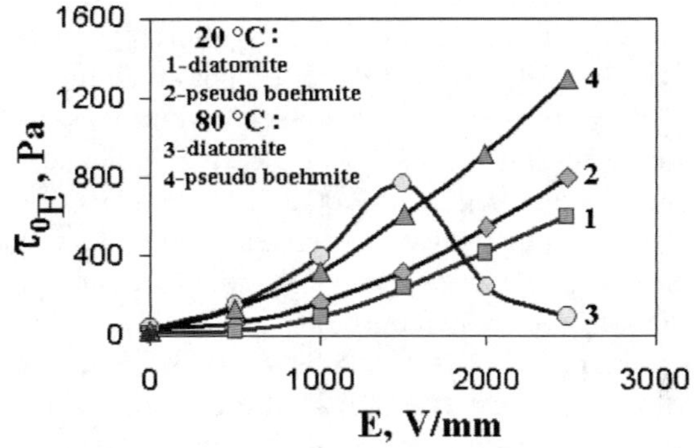

Figures 2. The dependence of the yield stress on the electric field intensity for ERF, based on the samples of two groups

3. Discussion

Analysis of the obtained experimental and published in the literature data allowed us to suggest the following mechanism of the surface polarization of the particles of water containing fillers. As it has been noted, the fillers of the first group are activated by means of surface saturation by water. Such a method of preparation water containing fillers provides a high filling-in of the surface by water molecules and formation of polymolecular layers of H_2O. According to [2], first portions of water are adsorbed by the coordination mechanism. Formed proton ionized molecules of H_2O become active centers for subsequent adsorption. At high filling-in degree of the surface, clusters of molecules grow up in the tangential direction with the formation of polymolecular film. Such behavior of adsorption with the appearance of coordination bonds is typical for a number of metal oxides [2]. Transportation of protons in the H_2O layers provides the surface proton conductivity of the filler particles. The surface polarization in this case can be due to the film polarization, when charge carriers move only along the surface. Under the constant electric field, the polarization of a dielectric with the conductive continuous film is the same as for the fully conductive particle. Evaluations of the concentration of moving protons performed in [2] give the value of $1\text{-}2{\cdot}10^{13}$ cm^{-2}, which is close to the concentration of cooditanionally bonded water on the SiO_2 surface.

The second group is the aqueous fillers for ERF, in which water is not absorbed on the surface, but is localized in the interlayer space of the crystalline structure, as it occurs in the case of the hydrates aluminum oxide in the form of pseudo boehmite. According to [3], super stoichiometric water in pseudo boehmite is bonded by a super strong hydrogen bridges in the lattice space of boehmite. Such water molecules, as it is known, are somewhat deformed, thermally stable and as the results of investigation have shown, are activators for the particles of dispersed phase, creating their enhanced conductivity. The appearance of proton conductivity in this case can be explained by the dissociation of H_2O molecules, having strong hydrogen bonds in the electric field and by the formation of protons and OH groups.

The polarization mechanism is probably due to the thermal proton polarization. It is due to the presence, on the volume or on the surface, of the particles the chains of hydroxyl or other ionogenic groups, bonded by hydrogen bonds, by which a migration of protons occurs. Two mechanisms of proton migration are known [1]. According to first one, after applying the electric field, protons of hydroxyl move to the free place between two neighboring atoms of oxygen in the chain of hydroxyl groups and the combination of O…O is moving opposite to the field that provides the consequent reorientation of all OH-groups in one orientation. The second mechanism is due to the fact, that under applying the electric field, a proton from a neighboring hydroxyl group will move to a free place and it will continue until the full reorientation of all OH-groups takes place and dipole moment of the whole chain is changed. However, the explanation of the migration mechanism of protons both by the first and by the second variant of transitions of protons of hydroxyl groups arises many questions. Authors do not indicate the types of OH-groups, participating in the reorientation, their concentration and the condition on the surface of the dispersed phase. On the assumption of a number of investigations of proton processes in various interphase phenomena, we suggest that in the aluminum oxides investigated by us in the form of pseudo boehmite as the filler for ERF, H^+ migrate, the source of which are protonoionized in the electric field interlayer water molecules. Protolytic dissociation of water molecules in the electric field and the presence of mobile protons due to this, most probably will lead to the transition of H^+ to the oxygen atoms, situated in octahedron groups. Hence, a chain of hydroxyl groups, interacting through a hydrogen bond, is formed, which indicates the sufficient role of the proton mechanism in charge transfer. Therefore, the electroconductivity in this case will be carried out by H^+ transfer between OH-groups. The fact that the suspensions on the basis of the fillers of the second type show ER–effect at higher temperatures (see figures 1, 2) is explained by a higher thermal stability of water molecules in the composition of these fillers. The increase of ER-effect in the range of temperatures 60-80 °C is because of the increase in the proton concentration due to the increase of a dissociation constant for water molecules, which depends on the permittivity of the medium and temperature.

4. Conclusions

- Two groups of aqueous fillers for ERF are emphasized, different in the type of nonstructural water in the form of the interlayer or the adsorbed water.
- It is shown, that the ERF, based on the fillers, that containing interlayer water, namely pseudo boehmite – a hydrated aluminum oxide, have a high ER-activity and are thermally stable.
- Based on the generalization of the analysis of our investigations and literature data, a polarization mechanism for the filler containing both adsorbed and interlayer water has been suggested.

References

[1] Korobko E V, Levin M L 2007 Physica-chemical aspects of structuring in ERF. Heterogeneous ER-compositions, (Luikov Heat and Mass Transfer Institute 2007) (*Preprint № 2)* 61

[2] Golovanova G F, Kvilidze V I, Kiselev V F 1977 Nature of the protonodonor centers on the surface of oxides SiO_2 and Al_2O_3 Digest [in Russian] *Bonded water in dispersed systems* (Moscow: Moscow State University Press), **4** 178-209

[3] Chalyi V P 1972 Hydroxides of Metals [in Russian] (Navukova Dumka Press.) 150

[4] Gehin C, Persello J, Charraut D, Cabane B 2004 Electrorheological properties and microstructure of silica suspensions *J. of Colloid and Interface Science* **273** 658-667

[5] Zhang S, Winter W T, Stipanovic A J 2005 Water-activated cellulose-based electrorheological fluids *Cellulose* **12** 135-144

Journal of Physics: Conference Series **149** (2009) 012019 doi:10.1088/1742-6596/149/1/012019

Time response model of ER fluids for precision control of motors

Ken'ichi Koyanagi
Toyama Prefectural University, 5180 Kurokawa, Imizu, Toyama, Japan
E-mail: koyanagi@pu-toyama.ac.jp

Abstract. For improvement of control performance or new control demands of mechatronics devices using particle type ER fluids, it will be needed to further investigate a response time of the fluids. It is commonly said around 5-mili seconds, however, the formula structure of that delay has not been clear. This study aims to develop a functional damper (attenuators), that can control its viscous characteristics in real time using ER fluids as its working fluid. ER dampers are useful to accomplish high precision positioning not to prevent high speed movement of the motor. To realize the functional damper that can be manipulated according to situations or tasks, the modeling and control of ER fluids are necessary. This paper investigates time delay affects of ER fluids and makes an in-depth dynamic model of the fluid by utilizing simulation and experiment. The mathematical model has a dead-time and first ordered delays of the fluid and the high voltage amplifier for the fluid.

1. Introduction

Particle type ER fluids have been tried to many applications of mechatronics; such as dampers (attenuators), clutches, and brakes to improve control performance or to satisfy new control demands [1]. This study aims to develop a functional damper, that can control its viscous characteristics in real time using ER fluids as its working fluid. ER dampers are useful to accomplish high precision positioning not to prevent high speed movement of the motor. It will concretely be effective in high speed and precision position control area as such industrial robots for assembly, welding, painting, etc.

Fluids of this type commonly have a response time of about 5-mili seconds. The delay needs to be taken into account when electric fields are controlled in real time. Especially, servo systems introduced with ER fluid devices have sampling times for their control loops that are about 1-mili second and so are shorter than the response times of ER fluids. The delay in this case has a strong possibility of having an effect on the performance or stability of these systems.

Though, there is little research in this area mentioning the problem. This may be because the delay is not a serious issue in the case of damping structures or engine mounts [2, 3]. Most of these studies use the damper impressed constant electric fields or constant amplitude ones, and do its viscosity as a damping force. There have been many studies to clarify the mechanism of the ER effect [1], but few have studied the speed of the response itself. Many research groups on materials observe the real behaviors of the particles, or how the response changes when composition of the ER fluid changes. These studies work on static investigation or dynamic observation at micro-level, and do not put emphasis on the modeling of ER fluids and applied control.

In direct-drive (DD) motor systems [4] without reduction gears, disturbances directly influence motor dynamics and control accuracy because of the small amount of damping. Time delays of computers and amplifiers also destabilize the controller [5]. While a time lag cannot be compensated mechanically, damping can be done with an additional mechanical damper [6]. In an earlier study, we introduced a particle type ER fluid damper for a DD motor system for fast and precise positioning [6, 7]. Viscosity of the damper was controlled continuously in real time to improve the performance of the DD system through dynamic changing of the viscosity [7]. In the reference [8], a tentative mathematical model considered the response characteristics of the particle type ER fluid were suggested and confirmed by simulation.

This paper investigates time delay affects of ER fluids and makes an in-depth dynamic model of the fluid by utilizing fine simulation and experiment.

2. Damping models of ER fluid with delay

In the previous study [8], the response of the ER damper was captured macroscopically, and the model was then experientially considered as being approximated with a time-lag of the first order and a dead-time. Here, the basic viscosity is Q_0 $\mathrm{N \cdot m \cdot rad/s}$, the variance of the viscosity is Q_1 $\mathrm{N \cdot m \cdot rad/s}$ specified by the magnitude of the electric field, E V/mm, and the rotational speed of the rotor of the damper, the dead-time is L_Q s, and T_Q s is the time constant. The transient response of the viscosity, $Q(s)$ $\mathrm{N \cdot m \cdot rad/s}$, can then be written as follows:

$$Q(s) = \frac{e^{-L_Q s}}{T_Q s + 1} Q_1 + Q_0 \tag{1}$$

However, more realistic situation was suggested that the time delay effects were inserted after the command of E was outputted to the damper. That is, the electric field that accords to the command value of the viscosity at the sampling time is reflected in the system model after the time lag. In view of the damping property, the new Q is affected by the angular velocity when a new E has been inflicted on the system. Thus Q may differ from the value of the Q desired when calculating the command for the viscosity.

Those has been confirmed by experiment and simulation with a delay model of E that included both dead-time and 1st-ordered delay; $L_Q = 2$ ms, and $T_Q = 3$ ms. However, the control system was so complicated, that it was unclear whether the model of delay was optimal.

3. Experimental System

A DD motor and a ER damper are set on the desk as shown in figure 1. The ER damper is for experimental purpose and simply constructed as shown in figure 2; the ER fluid is between the load inertia of the DD motor and the shallow plate fixed to the torque sensor. The gap between electrodes is 0.5 mm. The torque sensor is to detect fine share stress from ER fluids. Thus, it has been hand-made with strain gages as shown in figure 3 because commercial sensors have high possibility of nonconformity in setting accuracy. A high voltage amplifier is introduced for energizing the ER fluid, that has 100%-response time under 1 millisecond with no overshooting. The controller is constructed in a PC, which has a Pentium4-1.5 GHz CPU, 512 MB memory and ART-Linux for OS. The sampling time is under 1 ms for real time sensing.

While the angular velocity of the DD motor is controlled in regulated speed, the ER fluid energized, and then resistive torques are measured with the torque sensor. The applied electric field changes stepwise or in other way.

4. Experiments and modeling

In this section, time response of an ER fluid was measured in real time. Electric fields were applied as simple step to reconsider the brief model of the ER fluid suggested in the passed study.

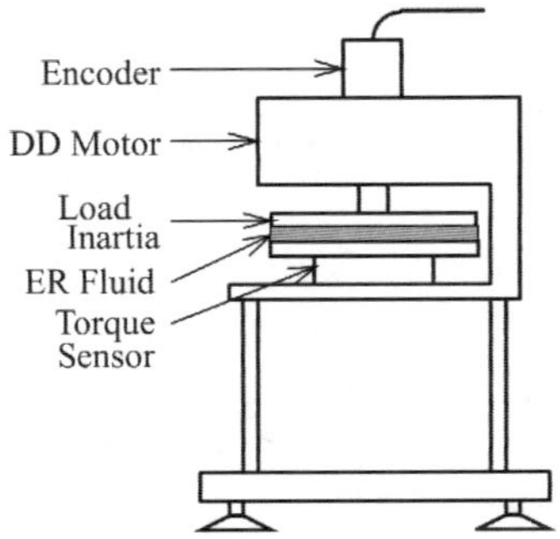

Figure 1. ER damper and DD motor

Figure 2. ER Fluid and electrodes

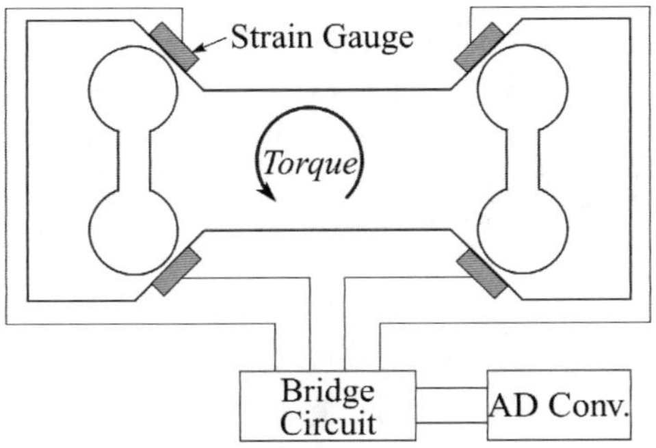

Figure 3. Torque sensor

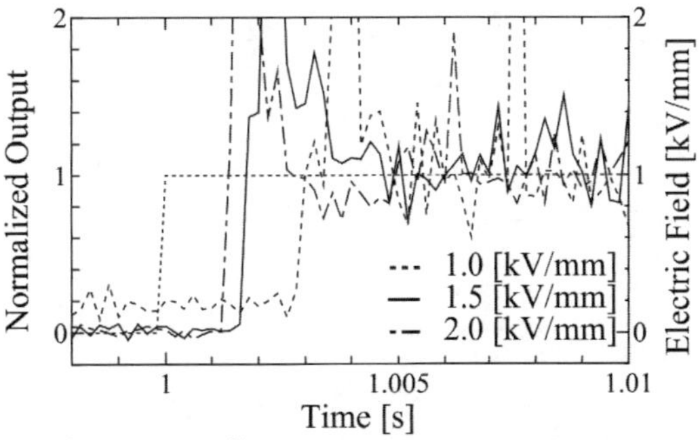

Figure 4. Step response experiments

The motor rotated in about 1/2 round/s (π rad/s), without speed control but applying a constant voltage in this case. The sampling time was 0.2 ms (5 kHz). Step electric fields from 0 kV/mm were applied at 1.0 s, and then removed at 2.0 s. Twenty times experiments of the same condition were carried out. Data were not filtered, but averaged from raw data, normalized by steady state value and analyzed.

Figure 4 shows the normalized data; the dotted line means the command value of the electric field, the short dashed line shows the result applied 1.0 kV/mm electric field, the solid line is the case of 1.5 kV/mm, and the dot-dashed line shows the case of 2.0 kV/mm. All results show big overshoot according to some impact load. The electrodes were always under slipping, so that those were not because of stick-slip. Rather it may be affected by the moment of the load inertia, it will be better to lighten the sensor. Electrical countermeasures against noise will also be required. Further investigation is needed on the difference of the response when 1 kV/mm. Anyway, the dead-time was around 2 ms, and the value settled into the 100%-response in about 4 ms; these were near to the value commonly mentioned.

We focused on the response when 1.5 kV/mm, and tried to identify the delay with simulation. In the simulation, the normalized output τ as follows was implemented in C language;

$$\tau = \frac{e^{-Ls}}{T_{ER}s + 1} \cdot \frac{1}{T_{HV}s + 1} \qquad (2)$$

Where, the sampling time was 0.20 ms, the dead time $L = 1.5$ ms, and the time constant of the ER fluid $T_{ER} = 0.35$ ms, which was adjusted as the 100%-response of the ER fluid was in 2 ms. The time constant $T_{HV} = 0.18$ ms was for the high voltage amplifier; it set the 100%-response of the amplifier as 1 ms. The dead time element e^{-Ls} was not written in the Pade approximation, but in FIFO queue.

The responses in rising of the experiment and simulation are shown as figure 5. The simulation

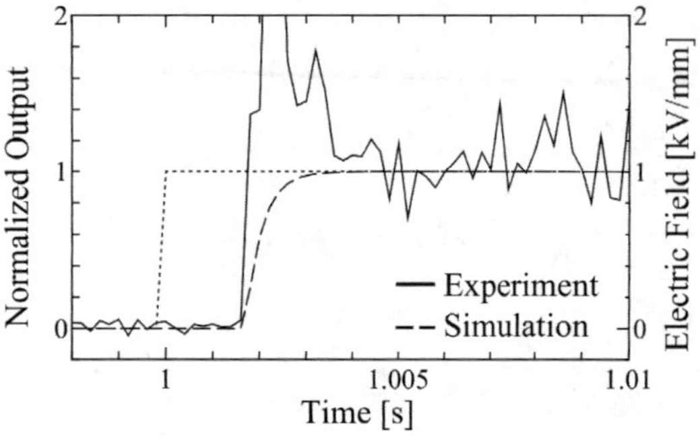

Figure 5. Step responses (rising)

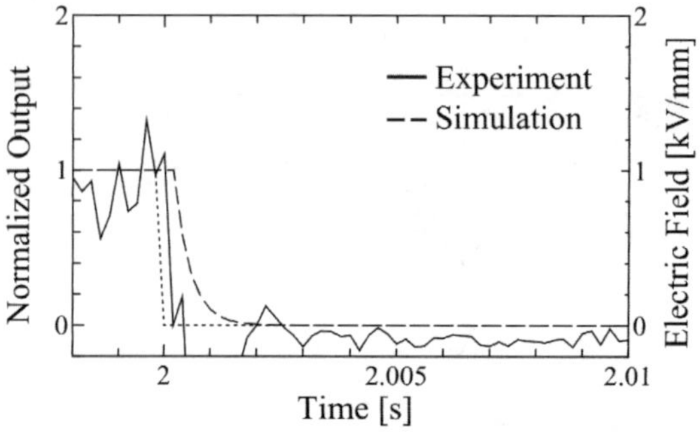

Figure 6. Step responses (falling)

result shows the 100%-response of the ER fluid was about 4 ms. While overshooting made their fitness slightly unclear, they can seem similar. On the other hand, the responses in falling are shown as figure 6. In this case, L was adjusted as zero; it implies different mechanisms of de-energized ER fluids from energized ones. Past experiments say those overshooting wa at most 10%. Those in this paper might be from softness and the moment of inertia of the torque sensor, which were caused by high sensitivity of the sensor.

5. Discussions and Conclusions

In this paper, we investigated the effects of time delays from ER fluids through experiments with very small sampling time. The mathematical models of delay have been updated to more accurate ones reflected in dynamic characteristics of a dead-time and first ordered delays of the fluid and the high voltage amplifier for the fluid. Those were confirmed with simulation.

During step responses, difference of the order of the delay model between the first and higher was not apparent. Then, we used the first ordered one in this paper. Frequency responses from the other experimental device, while it has not been modeled yet, seemed having the characteristics of the first ordered delay model. In future, we will continue to confirm the validity of the order of the model, and also try to make models by more objective method.

After the modeling, we will design a controller that explicitly includes a delay for the ER dampers; it will be an algorithm to use the ER damper suiting for high speed motion clearly involving the delay model.

Acknowledgments

This work has been partially supported by KAKENHI(19700445). We wish to thank Prof. J. Furusho of Osaka University, Prof. N. Takesue of Tokyo Metropolitan University and Dr. A. Inoue of ER-tec Co. for useful advice.

References

[1] Gordaninejad F, Graeve O, Fuchs A and York D (eds) 2008 *Proceedings of the Tenth International Conference on Electrorheological Fluids and Magnetorheological Suspensions* (World Scientific)
[2] Adams D and Johnston L 2001 *Proc. of the Eighth Int. Conf. on Electro-Rheological Fluids and Magneto-Rheological Suspensions* (World Scientific) pp 37–42
[3] Hong S, Choi S, Jung W and Jeong W 2002 *Journal of Intelligent Material Systems and Structures* **13** 421–424
[4] Ono Y 1992 *Advanced Robotics* **6** 243–254
[5] Kempf C J and Kobayashi S 1999 *IEEE Transactions on Control Systems Technology* **7** 513–526
[6] Takesue N, Zhang G, Furusho J and Sakaguchi M 1999 *Journal of Intelligent Material Systems and Structures* **10** 723–727
[7] Koyanagi K and Furusho J 2005 *International Journal of Modern Physics B* **19** 1661–1667
[8] Koyanagi K 2007 *Proceedings of the Tenth International Conference on Electrorheological Fluids and Magnetorheological Suspensions* (World Scientific) pp 519–526

Journal of Physics: Conference Series **149** (2009) 012020 doi:10.1088/1742-6596/149/1/012020

Basic research of electro-rheological gel drum for novel linear actuator

K Koyanagi[1], T Yamaguchi[1], Y Kakinuma[2], H Anzai[3], K Sakurai[3], and T Oshima[1]

[1]Toyama Prefectural University, Toyama, Japan
[2]Keio University, Kanagawa, Japan
[3]Fujikura Kasei, Tochigi, Japan

E-mail: koyanagi@pu-toyama.ac.jp

Abstract. In recent years, robots that coexist with human are under active researches. Such robots need high safety, backdriveability, large generative force, high speed response, good controllability, and so on. We have put emphasis on the backdriveability and created a novel prototype linear actuator with an ER (Electro-Rheological) gel, a new functional material. It is an all-purpose actuator satisfying characteristics for human-coexistence welfare robot.

1. Introduction

Nowadays, robots manipulated by human are actively researched and developed for virtual reality, rehabilitation, and other application [1, 2]. Such robots need high safety, backdriveability, large generative force, high speed response, good controllability, and so on. Among them, safety has been sometimes mentioned as important [3]. Additionally, high backdriveability is also important for Human-coexistence robots, because it helps manipulating these robots from their end point without large force. If above requirements are realized with actuators or structures, they will also be realized easily in a whole of a robot system.

This study uses an ER gel which is a new functional material, creates a linear actuator which has novel structures utilizing characteristics of the gel, and aims to develop an all-purpose linear actuator combining features required in human-coexistence welfare robots. In concrete, through design, construction and basic experiments of the actuator, we aim to develop hardware and software reaching to practical use [4]. This paper proposes such an ER Gel Linear Actuator (ERGLA), shows a improved design, and presents results of basic experiments.

2. Basic Characteristics of ER Gel

ER gels are a gel structured ER fluids and make a breakthrough two shortcomings of ERF [5]. These issues are sealing problem for use in mechanical elements and ER particle sedimentation. ER gel is that ER particles about 20 μm in diameter are dispersed in a insulating oil, and then a gelling agent is added to that. When an electric field is applied to electrodes sandwiching the ER gel, dielectric polarized particles and the gel component increase the shear stress of the surface of the ER gel several tenfold by the mechanism as shown in figure 1. The history of ER gels' studies is not so long [6] and applications with ER gels have scarcely been developed in

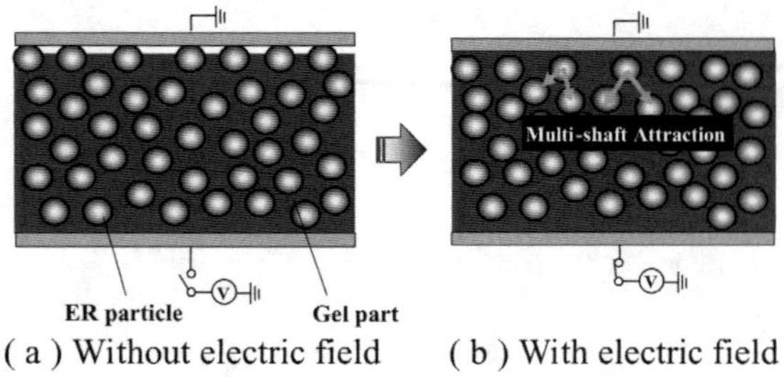

Figure 1. Mechanism of ER Gel Effect

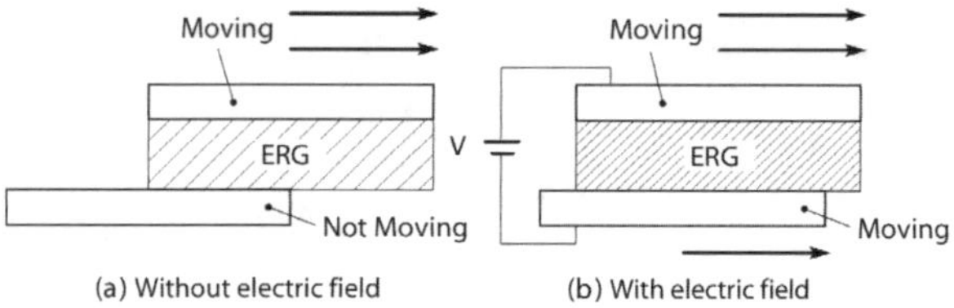

Figure 2. Image of ER Gel Clutch

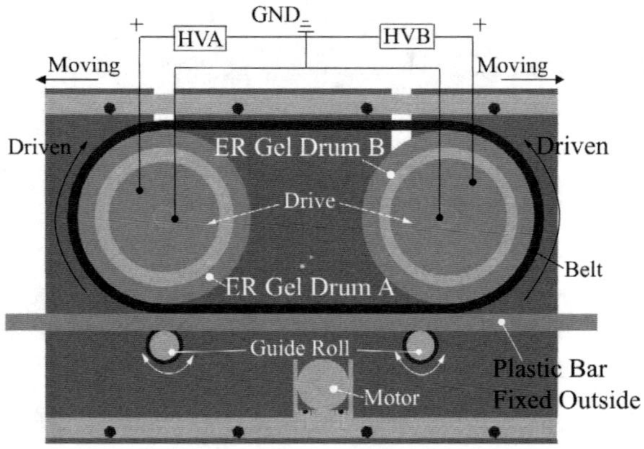

Figure 3. Design of the ERGLA Unit

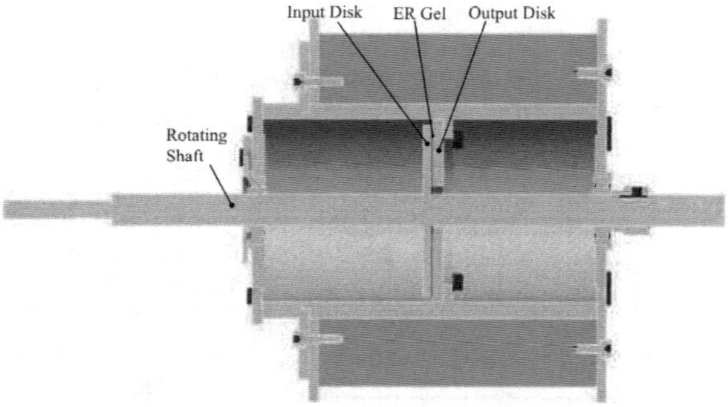

Figure 4. Design of the ER Gel Drum

mechatronics. Recently, Kakinuma etc. proposed a new theory on the mechanism of effects of ER gels and reported samples of ER gels which have large ER effects [7, 8].

For example, an ER gel can be applied to a clutch surface as shown in figure 2. While the ER gel is de-energized, the clutch input force is not transferred because of minimal shear stress. When the ER gel is energized by an electric field, increased shear stress contributes force transfer, that is, the clutch is engaged. The magnitude of the transferred force can be controlled since its shear stress can be controlled by the magnitude of the electric field.

3. The ER Gel Linear Actuator

The designed structures of the ERGLA are shown in figure 3. The ERGLA unit is attached and moves on a fixed bar. The ERGLA unit is connected to loads. Main parts of the ERGLA unit are two ER gel drums (ERGDs), a timing belt, and guide rollers. The fixed bar is pinched by the belt and guide rollers. Each input part of the ERGDs rotates contrarily in the same rotational speed by one driving motor and reversing gears. The ERGLA moves in different ways according to the way of energizing.

Figure 4 shows a sectional view of the ERGD. The gel was fixed on a disk type clutch structure in the drum. An arbitrary torque can be transmitted to the output part of the ER gel drum by control of the electric field for the ER gel. The advantage of this structure is easy increase of the output torque of the drum by multiple input and output gel disks used in the drum.

The produced ERGLA is shown in figure 5. Specifications of the ERGLA and ERGD are described as table 1.

Figure 5. Prototype ERGLA

Table 1. Design Spec of the Prototype

	ERGD		ERGLA
Weight	1.24kg	Weight	5.32kg
Height	0.10m	Height	0.22m
Diameter	0.11m	Width	0.26m
		Depth	0.17m
		Max.force	10N

4. Basic Experiments, Results and Discussions

The basic experiment of ERG drum was evaluated by the simple test schematized in figure 6, which was for gauging the output torque of the ERGD while the motor rotated in a constant speed.

Figure 7 shows the basic characteristic of the ERG on ERGD. The generative torque of ERGD is increased with the magnitude of applied electric field.

Figure 8 and figure 9 show results of basic experiments on step responses of an ERGD. The step inputs of the 2.0 kV/mm electric field energized the ER gel at 3.0 s and de-energized at 7.0 s. The four responses in the figures differ in the pressure on the surface of the ER gel. The pressures were able to adjust with weights on the input disk of the drum's shaft. Each results was averaged from raw data of ten times experiments of the same condition.

The results show that the response time of the ERGD's torque was about 20~30 ms. Response of the ERG may be enough fast for mechatronics or robotics applications as physical supporting in welfare situations. However, the relation of the pressure on the surface of the ER gel and generative force should be invested and the surface pressure ratio should be improved in future study. It seems that the responses occurred at several milliseconds before 3.0 s, but those might be from torque ripples caused by unbalance of the pressure on the gel. Uniformalizing of the pressure is also future issue.

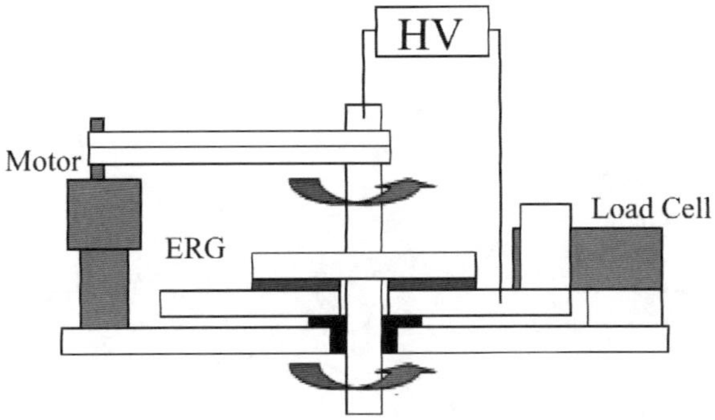

Figure 6. Schematic diagram of ERGD test

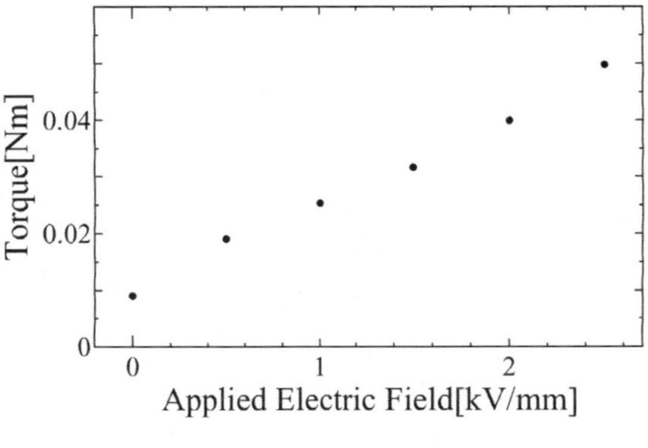

Figure 7. Basic Character of the ERG

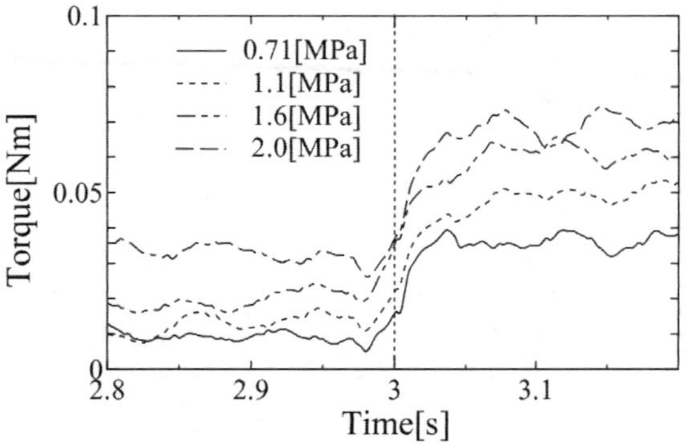

Figure 8. Rise Time of the ERGD Torque

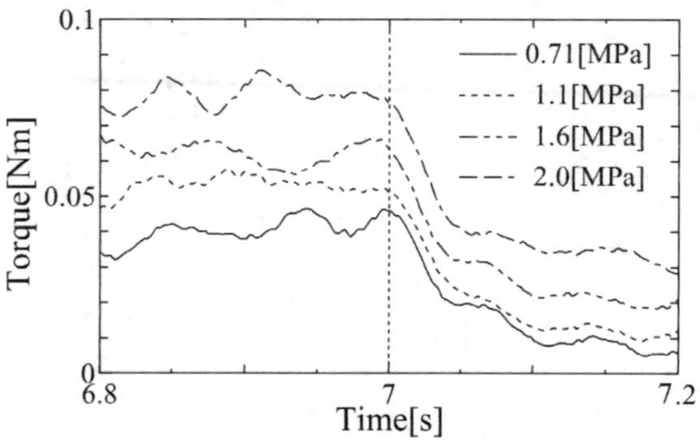

Figure 9. Fall Time of the ERGD Torque

5. Conclusions

This paper showed a novel linear actuator with ER gels, and results of its basic experiments. The generative torque of ERGD was increased with rising ER gel surface pressure. However, the pressure will be needed to set small to keep the backdriveability of ERGLA. In future work, we will improve the surface pressure ratio on the ERG, and then will apply it to robotics applications of welfare situation.

References

[1] G.S. Guthart and K.J. Salisbury Jr. 2000 Intuitive telesurgery system: overview and application *Proc. of the IEEE 2000 Int. Conf. on Robotics and Automation* **1** 618-621

[2] Proc. of the IEEE 9th Int. Conf. on Rehabilitation Robotics(2005), IEEE

[3] D. Engel, J. Raczkowsky, and H. Wörn 2001 A safe robot system for craniofacial surgery *Proceedings of the 2001 IEEE International Conference on Robotics and Automation* **2** 2020-2024

[4] K. Koyanagi, Y. Kakinuma, H. Anzai, K. Sakurai, T. Yamaguchi, T. Oshima, N. Momose and T. Matsuno 2007 Basic Structure and Prototype of Novel Linear Actuator with Electro-Rheological Gel *Mechatronics and Automation, 2007. ICMA 2007. International Conference* 3266-3271

[5] M. Nakano and T. Nagata 2002 ER Properties and Flow-induced Microstructures of an ER Fluid between Two Parallel Disk Electrodes in Squeeze Flow Mode *Int. J. of Modern Physics B* **16** 17,18 2555-2561

[6] K. Matsuura, T. Yakoh, T. Aoyama, H. Anzai, K. Sakurai and K. Isobe 2002 Smooth Tactile Display in Mouse using Electro-rheological Gel *Proceedings of the 2002 International Symposium on Industrial Electronics(IEEE-ISIE 2002) L'Aquila Italy July 8-11 459-462*

[7] Y. Kakinuma, T. Aoyama, H. Anzai, H. Sa-kurai, K. Isobe, and K. Tanaka 2005 Basic proper-ties of gel-structured electro-rheological fluids *Int. J. of Modern Physics B* **19** 7-9 1339-1345

[8] Y. Kakinuma, T Aoyama, H. Anzai, H. Sa-kurai, K. Isobe, and K. Tanaka 2006 Application of ER gel with variable friction surface to the clamp system of aerostatic slider *Precision Engineering* **30** 280-287

Journal of Physics: Conference Series **149** (2009) 012021 doi:10.1088/1742-6596/149/1/012021

Ionically conductive polymers for ER fluid preparation

A Krztoń-Maziopa and J Płocharski

Warsaw University of Technology, ul. Noakowskiego 3, 00-664 Warszawa, Poland

E-mail: anka@ch.pw.edu.pl

Abstract. Two groups of electrorheological (ER) suspensions containing polyelectrolytes and solid polymer electrolytes as dispersed phases of ionic conductivity were synthesized and their ER effect at elevated temperatures was studied. Polymers were carefully characterised and detailed analysis of their ionic conductivity change with temperature was performed. It was found that in suspensions comprising both types of materials ER effect and current density were closely related to cation type presented in resin and in salt added to solid electrolyte. If small monovalent cations of high mobility were present in the solid phase the ER effect of suspensions as well as their current densities increased strongly with temperature both for polyelectrolytes and solid electrolytes. Presence of spatial cations resulted in lower power consumption and reasonable ER effect within wide temperature range.

1. Introduction

Over the last years a lot attention has been paid to materials being a dispersed phase in ER suspensions. Although a lot of work has been done on ER fluid preparation there are still problems with their usage in practical devices. The main drawback here is their susceptibility to overheating during the prolonged work, which results in strong current density increase [1]. Among the materials used in ER fluid synthesis the most promising for ER fluids synthesis are polymers with ionic conductivity like: polyelectrolytes with different counterions and solid polymer electrolytes – complexes of polymers with various salts. In both cases conductivity of the solid phase and its electrorheological behavior can be relatively easy controlled by the change of counterion in polyelectrolyte or the salt type and concentration in polymer electrolyte. In the scientific literature there are publications concerning ER fluids based on polyelectrolytes like sulphonated polystyrene-*co*-divinylbenzene (SPS), phosphorylated cellulose [2], polymer electrolytes based on polyacrylate copolymers [3] and polyurethane polymer electrolytes [1]. ER effect of the reported fluids was related to the humidity of the dispersed material [4], solid phase content [5], grain size [6], electric field strength and frequency [7] but the information about their temperature characteristics are still unsatisfactory. It is necessary therefore to shed more light to this matter.

The aim of our work was a synthesis and investigation of an ionically conductive polymer with good ER response for application in ER devices working at room and elevated temperature without significant current density increase. To this end detailed analysis of rheological and electric properties of ER fluids with different solid phases of ionic conductivity were performed. Measurements were carried out on six suspensions containing: sulphonated polystyrene-co-divinylbenzene with sodium cation (SPS-Na), sulphonated polystyrene-*co*-divinylbenzene with N-trimethoxysilylpropyl-N,N,N-trimethylammonium cation (SPS-NTSi), aminofunctional polystyrene-co-divinylbenzene resin neutralized with sulfuric acid (PS-A) and polycrylonitrile-co-methylmethacrylate (PAN) complexes with lithium trifluoromethanesulfonate (LiTF), zinc chloride ($ZnCl_2$), N-trimethoxysilylpropyl-N,N,N-trimethylammonium chloride (NTSi).

2. Experimental

Polystyrene-*co*-divinylbenzene (PS) was prepared from monomers by the well known suspension polymerization method in a presence of benzoyl peroxide as initiator and aqueous solution of polyvinylalcohol as a protective colloid then washed with water and dried at 80°C over 24h. Product containing not more than 10wt.% of divinylbenzene, was sulphonated at elevated temperature with concentrated H_2SO_4 in the presence of $CoCl_2$. Then obtained acidic resin was filtered under reduced pressure, washed with diluted H_2SO_4 and distilled water. Next protons were replaced to sodium and NTSi cations and two cationites were obtained: SPS-Na and SPS-NTSi. Anionite (PS-A) was prepared by chloromethylation of the initial PS copolymer followed by amination of the chloromethylated product with triethylamine. The preparation method was adopted from [8]. The alkaline resin was then filtered, washed with distilled water and neutralized with diluted H_2SO_4. Final product was filtered and washed with distilled water to neutrality. After synthesis all ionomers were dried in a dryer at 120°C to constant mass. Polymer electrolytes were prepared from polycrylonitrile/methylmethacrylate copolymer (Dralon®, Bayer AG). Polymer was dissolved in dimethylformamide solutions of LiTF, $ZnCl_2$ and NTSi. Then solutions were poured into Petri dishes and the solvent was evacuated in vacuum at 80°C. The obtained thin foils of PAN-LiTF, PAN-ZnCl2 and PAN-NTSi were powdered with impact mill, sieved and dried again in high vacuum about 10^{-5} Tr at 80°C over 24 hours. Salts concentrations in prepared samples were constant and amounted 0.2 mol/kg of polymer. Electrorheological suspensions were prepared by dispersing the weighted amounts of prepared samples in silicone oil (PDMS) of 0.01Pa·s viscosity to obtain 40wt.% suspensions.

Conductivities of samples were calculated from impedance spectra recorded with PARSTAT 2263 system equipped with PowerSuite software at frequencies 100mHz - 100kHz and temperatures from 20 to 80°C. Impedance spectra were analyzed with Equivalent Circuit v.455. Morphology of grains was determined with scanning electron microscope FEI Quanta 200. Quantitative data on grain size distribution of the solid phase were taken from optical images recorded with Nikon Alphaphoto-2 YS-2 microscope with CCD camera. Images were analysed with the MicroScan v.1.3. ER effect of the prepared fluids at temperatures from 20 to 70°C was measured with a rotational rheometer Physica MCR301 (Anton Paar) with ER cell operating in a coaxial cylinders geometry, under dc electric fields up to 3 kV/mm. Leak currents drawn through the studied suspensions were also measured and recorded. Yield stresses were calculated from flow curves using Herschel-Bulkley extrapolation.

3. Results

Ion-exchange resins were prepared from polymer obtained by suspension polymerization and have spherical shape of grains as figure 1a shows.

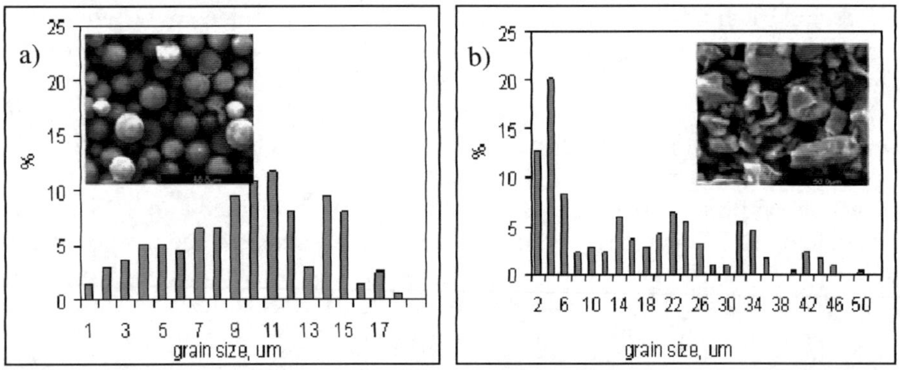

Figure 1. SEM image and grain size distribution of SPS-Na (a) and PAN-LiTF (b).

Grain size distributions of the prepared polyelectrolytes were identical as for the resin presented in figure 1a with the average diameters of particles within 2 to 18 μm with most of them below 12 microns. Solid electrolyte materials were in a form of polyhedral particles with smooth surface as the example SEM image in figure 1b shows. In this case the average particle diameter was taken into account for determination of grain size distribution.

ER effect is related to polarization processes coming from ionic polarization governed by ionic conductivity of dispersed solid phase [9]. Ionic conductivity increases with temperature even a few orders of magnitude that is rather unfavorable because this growth is related to higher power consumption and usually with ER effect drop. In practical applications, especially in shock absorbers and brakes ER suspensions work at elevated temperature and detailed knowledge about temperature characteristic of working fluid is indispensable. ER suspension is a multicomponent system and studying the temperature dependence of solid phase only is not sufficient. Full information can be obtained from direct measurements taken on samples simulating the aggregates formed in ER fluid upon field, where solid phase is densely packed, grains touch each other and small empty spaces between them are filled with carrier fluid. Taking the above into account dense (85%) pastes made out of solid phase and small amount of silicone oil are good objects for carrying such experiments. Results were plotted in the Arrhenius type coordinates shown in figure 2a.

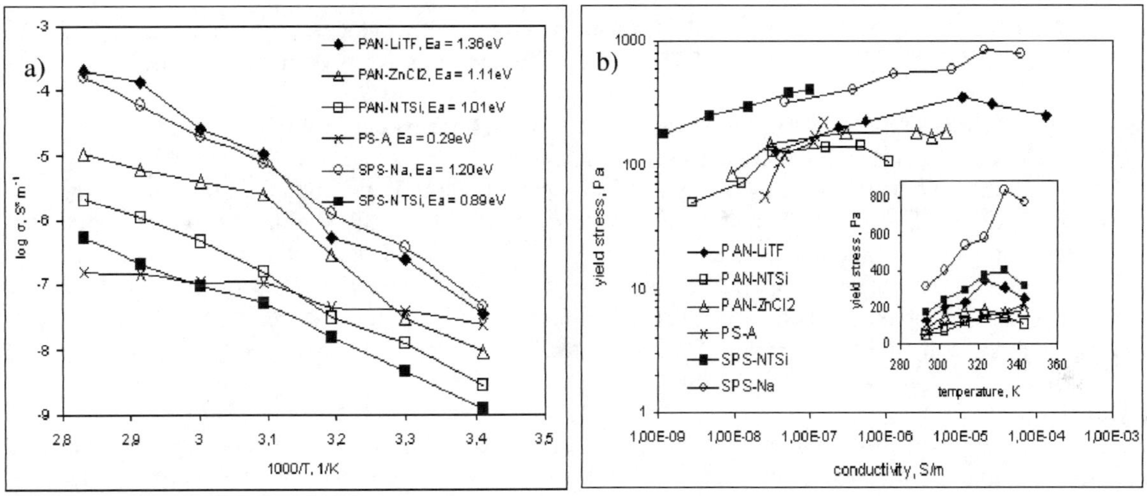

Figure 2. (a) Arrhenius plots of investigated materials, (b) relation between conductivity and yield stress of suspensions for different temperatures

The conductivity values were calculated from total resistance (bulk + intergranular) together with geometry of a particular sample. Activation energies (Ea) of ionic conductivities for all materials were also determined. Ionic conductivities at 20ºC calculated for the investigated samples ranged from 1.20 $x10^{-9}$S/m (SPS-NTSi) to $4.8x10^{-8}$S/m (SPS-Na). After heating to 80ºC the conductivity values increased significantly and were within $1.60x10^{-7}$S/m (for PS-A) to $2.10x10^{-4}$S/m (PAN-LiTF). Comparing the two cationites (SPS-Na and SPS-NTSi) we can see that introduction of spatial organic cation instead of Na^{+} resulted in flattened conductivity-temperature characteristics. Similar relation was observed for polymer electrolytes where the samples can be set up in a series of decreasing conductivity: PAN-LiTF > PAN-ZnCl2 > PAN-NTSi.

Figure 3a illustrates the increase of yield stresses of investigated suspensions with electric field strength. The corresponding change of the yield stress with temperature and conductivity is shown in figure 2b. The changes of current densities at 2kV with temperature are shown in figure 3b.

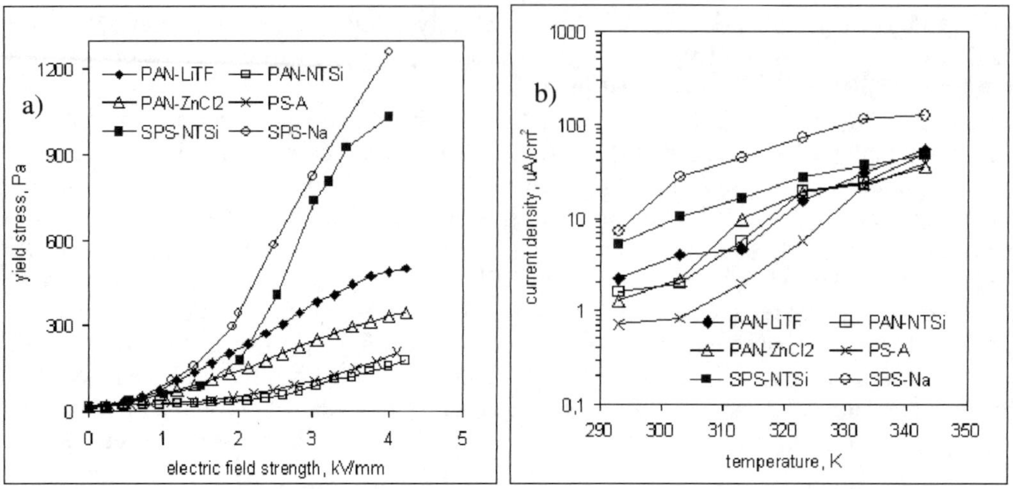

Figure 3. a) Yield stress at 2kV/mm and 20°C on electric field dependences for investigated fluids, b) changes of current density at 2kV/mm with temperature for investigated fluids

Suspensions of SPS-Na and SPS-NTSi exhibited strong ER effect while the PAN complexes as well as PS-A were less active. PAN-NTSi suspension had ER effect comparable to PS-A. In PAN-complexes the ER effect can be related to the cation type of salt applied as an ionic dopant. If the salt comprised big organic cation (NTSi+) the ER effect of suspension was rather low although the conductivity measured for this sample was higher in comparison to cationite-based suspension with the same cation (SPS-NTSi). The strongest temperature dependences were observed for ion exchange-based suspensions where current density increased even three orders of magnitude. In the case of solid electrolyte samples this growth was significantly lower.

4. Conclusions

The above mentioned results proved that all investigated materials of ionic conductivity can be easily modified and applied as solid phases in ER fluids. It means that ER effect in suspensions of solid electrolytes can be tuned by introduction of spatial cations of limited mobility. It was found that ER effect of suspensions of polyelectrolytes and solid electrolytes can be related with the type of cation present in a given system. When small monovalent cations of high mobility were present in the solid phase the ER effect of suspension as well as current density of suspensions increased strongly with temperature both for polyelectrolytes and solid electrolytes. Application of spatial cations resulted in lower power consumption and reasonable ER effect within wide temperature range. Suspension of PAN-complexes with different salts are good candidates for practical application within wide temperature range.

References

[1] Böse H, Trendler A 2001 *Int. J. Mod. Phys.B* **6&7** 626
[2] Kim S, Choi HJ, Ihon MS 2001 *Macromol. Chem. Phys.* **202(4)** 521
[3] Guan JG, Yuan RZ, Gie HQ 1998 **39(22)** 5307
[4] Aizawa R, Vieira S,Nakano M, Asako Y 2001 *Int. J. Mod. Phys. B* **15** 1070
[5] Klingenberg DJ, Parthasarathy M 1996 *Mat. Sci. Eng.* R**17** 57
[6] See H, Kawai A, Ikazaki F 2002 *Rheol. Acta* **41** 55
[7] See H, Kawai A, Ikazaki F 2002 *Colloid Polym. Sci.* **280** 24
[8] Bourier JL, Grammont JE, Herbin E 1980 *US Patent* No. 4225677
[9] Krztoń-Maziopa A, Ciszewska M., Płocharski J 2006 *Polym. Adv. Technol.* **17(1)** 37

Heat Transfer of ERF in a Flat Channel Taking Dissipative Factor Into Account

A A Makhaniok, E V Korobko and N A Bedik

A. V. Luikov Heat and Mass Transfer Institute of the Academy of Sciences of Belarus, P.Brovki 15 Str., 220072 Minsk, Belarus

E-mail: eva@itmo.by

Abstract. The results of an investigation of rheophysical properties of a thermally stable dielectric controllable heat carrier — electrorheological fluid on the basis of hydrated aluminum oxide in the form of pseudo benit, are presented. Considering the obtained experimental data, we have received and discuss the numerical solution of a conjugated problem of hydrodynamics and heat transfer in a flat channel of ERF, the flow curve of which is described by a modified Bulkley-Herschel model. It is shown that the electrorheological effect results in the increase of an average over the channel length Nusselt number up to 17 %.

1. Introduction

Available models of hydrodynamics [1-4] and heat transfer [5-6] in narrow flat and circular channels are regarded to be difficult systems in the rheological aspect, as electrorheological fluids (ERF) are based on substantial assumptions relative to rheological and thermophysical properties of these fluids in respect of the dependence of these properties on the temperature and intensity of electric field. One of the main simplifications applied was to neglect the dissipation of energy into heat due to the internal friction. In a number of devices, operating on the basis of the electrorheological effect, the flux of electrorheological fluid (ERF) along a narrow circular or a flat channel is realized. Moreover, the dissipative heating of the whole system takes place. The change of the temperature results in the change of rheological properties of ERF. At high enough temperatures such devices can become inoperative. This investigation is dedicated to the problems of heat transfer of ERF in such conditions. The main aim of the paper is to evaluate the influence of the intensity of the electric field and the temperature on heat transfer at ERF flowing in the flat channel with and without consideration of the dissipative factor. The second aim of this paper is to analyze the possibilities of the intensification of heat transfer due to the electrorheological effect.

2. Statement of the problem of numerical modeling hydrodynamics and heat transfer of ERF in a narrow channel

In our approximation we consider that the channel length substantial excesses the width of its slot. It allows us to simplify the set of equations of motion due to the neglect of the summands that contain space derivatives of the transversal part of velocity. In practice, the possible supposition is also a supposition on the incompressibility of ERF and the limitation due to the consideration of only the laminar regime of flowing of such fluid. In the result, the equations of motion are reduced to one equation, identical to the boundary-layer equation. In such approximation the pressure depends only on the longitude coordinate and is constant across the channel. In Cartesian reference system the set of nondimensionalized equations of mass and momentum conservation are changed into the following:

$$\frac{\partial u'}{\partial x'}+\frac{\partial \upsilon'}{\partial y'}=0 \tag{1}$$

$$\frac{\mathrm{Re}}{\nu'}\left(u'^{*}\frac{\partial u'}{\partial x'}+\upsilon'^{*}\frac{\partial u'}{\partial y'}\right)\approx-\frac{1}{\mu'}\frac{\partial P'}{\partial x'}+\frac{\partial^2 u'}{\partial y'^2} \tag{2}$$

$$u'^{*}=u'-\frac{2}{\mathrm{Re}}\frac{\partial v'}{\partial x'},\quad \upsilon'^{*}=\upsilon'-\frac{1}{\mathrm{Re}}\frac{\partial v'}{\partial y'} \tag{3}$$

$$u'=\frac{u}{U_0},\quad \upsilon'=\frac{\upsilon}{U_0},\quad x'=\frac{x}{H},\quad y'=\frac{y}{H},\quad P'=\frac{P\,\mathrm{Re}}{\rho U_0^2},\quad \nu'=\frac{\nu}{\nu_0},\quad \rho'=\frac{\rho}{\rho_0},$$

where H is the channel width, P is pressure, ρ is ERF density, U0 is the mean velocity at section of channel. We use only the x component of the Navier-Stokes equation because second equation is $\partial p/\partial y=0$ at made approximation. Rheological properties of a thermally stable ERF were given for the thermal-and-hydraulic calculations. Rheological behavior of such medium is described rather well by the model of Bulkley –Herschel [7-10]:

$$\tau(E,T,\dot{\gamma})=\tau_0(E,T)+K(E)\dot{\gamma}^{n(E)}, \tag{4}$$

where τ_0 is the yield stress, τ is the shear stress, $\dot{\gamma}$ is the shear rate, *K* is the consistency parameter.

According to (4), sheer relative viscosity of ERF is described by the equation

$$\mu'=\frac{\tau_{00}}{\tau_{0ref}}\frac{\tau_0'}{\dot{\gamma}'}+K'\dot{\gamma}'^{n-1}, \tag{5}$$

where

$$\dot{\gamma}'=\frac{H}{U_0}\dot{\gamma}=\left[\left(\frac{\partial u'}{\partial x'}\right)^2+\left(\frac{\partial \upsilon'}{\partial y'}\right)^2+\left(\frac{\partial u'}{\partial y'}+\frac{\partial \upsilon'}{\partial x'}\right)^2\right]^{1/2} \tag{6}$$

$$\tau_0'=\frac{\tau_0}{\tau_{00}},\quad \tau_{00}=\tau_0(E=0,T=T_0),\quad K'=(\tau_{0ref})^{n-1}\delta^n\mu_p',\quad \tau_{0ref}=\frac{\mu_0 U_0}{H},\quad \mu_p'=\frac{\mu_p}{\mu_0}, \tag{7}$$

In (7) the parameter δ=1 $\mathrm{Pa}^{1/n-1}$. It is introduced to comply with dimensionality rule.
When convectional heat transfer along the channel and dissipation of hydraulic energy into the heat are taken into consideration, the energy equation is:

$$\frac{\partial\Theta}{\partial\,\mathrm{Fo}}+\mathrm{Pe}\frac{\partial\Theta}{\partial x'}u'=\frac{\partial}{\partial x'}\left(a\frac{\partial\Theta}{\partial x'}\right)+\frac{\partial}{\partial y'}\left(a\frac{\partial\Theta}{\partial y'}\right)+\Phi. \tag{8}$$

Non-dimensional complexes included in (8) are described by the following equations: $\mathrm{Pe}=U_0H/a_0$, $\Phi=\Pi\nu\dot{\gamma}'^2$, $\Pi=\nu_0U_0^2/[a_0(T_w-T_0)]$, where a_0 is a thermal conductivity at the temperature T_0 of ERF at the entrance of the channel, ν_0 is a reference value of the apparent viscosity of ERF, T_w is an average temperature of the channel wall, $\Theta = (T - T_0)/(T_w - T_0)$ is a non-dimensional temperature. The initial condition is chosen as the condition of constant temperature in the channel. Further we assume that the temperature in the channel is the same as at the entrance of the channel. At the boundary $y = 0$ we use the heat transfer condition of the first kind (the boundary condition of H1):

$$\Theta(\mathrm{Fo}=0, x', y') = 0, \quad \Theta(\mathrm{Fo}>0, x'=0, y') = 0, \quad \Theta(\mathrm{Fo}>0, x'>0, y'=0) = 1,$$

$$\left.\frac{\partial \Theta}{\partial x'}\right|_{x'=L/H} = 0, \quad \left.\frac{\partial \Theta}{\partial y'}\right|_{y'=1} = 0,$$

where L is the channel length. Local and integrated values of Nusselt number are determined according expressions

$$\mathrm{Nu} = -2\left.\frac{\partial \Theta}{\partial y'}\right|_{y'=0}, \quad <\mathrm{Nu}> = -\frac{2H}{L}\int_0^{L/H} \mathrm{Nu}(x')dx'$$

The influence of the intensity of the electric field and temperature on Bulkley-Herschel model parameters is shown in figure 1, in which experimental data are displayed.

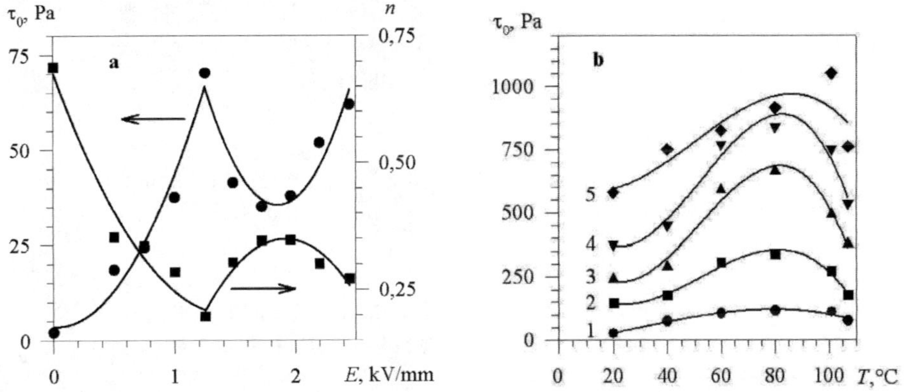

Figure 1. Experimental dependencies of Bulkley-Herschel model parameters on the electric field intensity (left) and temperature (right). In right panel, curve 1 is for E = 0.5, 2 – 1.0, 3 – 1.5, 4 – 2.0, 5 – 2.5 kV/mm

These data were used to perform a computational experiment on quantitative evaluation of the effectiveness of heat transfer for developed ERF in a narrow circular slot (or a flat slot). In the numerics we find also the influence of the intensity of electric field on Nusselt number. The width of the slot was equal to 1 mm. The width of the plate was 100 mm. The length of the channel was varied from 0.25 m up to 1 m. The ERF temperature at the entrance was equal to 25 °C. As the result of the solution of thermal-and-hydraulic problem the drop of the pressure along the channel, providing given mass velocity of ERF flowing, was determined.

3. Numerical modeling and its results

The hydrodynamic and heat problems (1-8) have been solved by the finite difference method. We have used the implicit scheme. The equations (4) – (6) describe interrelations between a non-dimensional shear rate and shear stress, and a non-dimensional apparent viscosity depending on non-dimensional shear rate. The expressions (7) give the procedure for nondimensionalizing. Evaluation of the impact of the electric field intensity on the mean and local (at the exit of the channel) value of Nusselt number and also thermal condition for heat transfer device is carried out in comparison for ERF and water. With this aim thermal-and-hydraulic calculation was performed at the boundary condition H1, mass rate of heat carrier (ERF or water) of 0.01 kg/s and the value of heat stream averted from the channel wall of 10 kWt/m^2. The result of such evaluation is shown in figure 2. The effect of electric field application is 10.4 % for the mean value of Nusselt number and 7.7% for the local value of Nusselt number at the channel exit (to the side of enhancement in comparison with the heat response coefficient without field). For ERF <Nu > is 17.1 % higher in comparison to water at application of electric filed intensity of 2.5 kV/mm, and local Nusselt number is 11.1% higher. Nevertheless, due to the approximately two times higher thermal-capacity and about five times higher thermal conductivity, water shows more effective cooling of the channel wall. We will note only that there was heat stabilization of the temperature profile (the local value of Nusselt number was put constant at the substantial distance from the exit of the channel, the flow was stationary and laminar) for the modeling conditions in both cases. The creeping type of flowing is characteristic for ERF at mass rate of 0.01 kg/s (Re<<1). Moreover the shear rate near the channel wall has been changed substantially depending on the value of electric field intensity (from 600 s^{-1} without electric field, up to 2000 s^{-1} at $E = 2.5$ kV/mm).

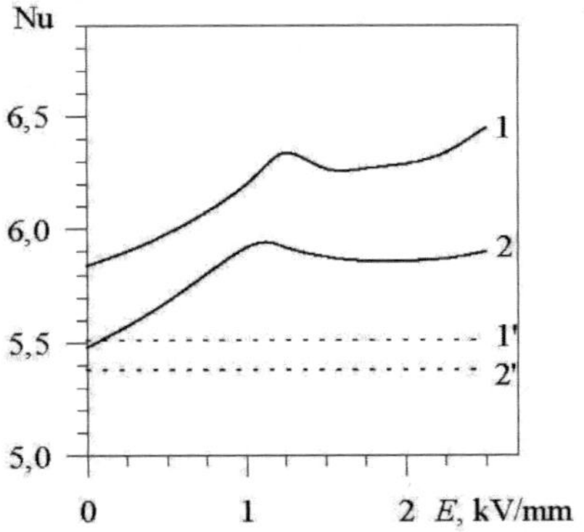

Figure 2. The influence of the electric field intensity on Nusselt number. Curves 1 and 1′ mean the value of Nusselt number along the channel length; curves 2 and 2′, the local value of Nusselt number at the exit of the channel. The solid curves are for ERF, and the dash curves are for water

Dissipative heating of ERF was less than 0.1 °C. Dissipative heating of ERF has started to appear at mass rates more than 0.1 kg/s and was accompanied by the decrease of heat transfer coefficient by 8 %, and also by the increase of the average temperature in the exit cross section by more than 0.6 °C.

4. Conclusions

Electrorheological effect leads to a substantial intensification of heat transfer that should be considered at projecting devices based on it. The increase of nondimensionalized coefficient of heat, due to the increase of shear rate at the walls of the channel as a result of formation of a quasi solid core of flowing is not high enough for ERF to become a more effective heat carrier by an increase of the intensity of the electric field than water. Also by an increase of the intensity if the electric field the increase of hydraulic resistance and dissipative heating of ERF is observed. As the result, the increase of Nusselt number is changed by the decrease of this parameter.

5. References

[1] Dietz P W and Melcher J R 1978 *I&EC Fundamentals.* **17** 28-32
[2] Gavin H P 2001 *J. Rheology.* **45** 983-94
[3] Shin Morishita and Tamaki Ura 1992 *Proc. of the recent advances in adaptive and sensory materials and their applications. Virginia Politechnic Institute and State University Blacksburg, Virginia. April 27-29, 1992* 537-47
[4] Stevens N G et al.1987 *Transaction of the ASME. J. of Applied Mechanics.* **54** 456-58
[5] Giu Z Y, Hu L, Korobko E V et al. 1996 *Proc. at the 5 Int. Conf. On ERF / Ed. N. Bullouph 1996* 486-93
[6] Davis L C 1992 *J. Appl. Phys.* **72** 1334-40
[7] Herschel W H, Bulkley R 1926 *Kolloid Z.* **39** 291-300.
[8] Choi Y T, Cho J U, Choi S B, Wereley N M 2005 *Smart Mater. Struct.***14**1025-1036
[9] Garcia E. J., Steffe J F 2007 *Journal of Food Process Engineering* **8** 117 – 136
[10] Lee D Y, Choi Y T and Wereley N M 2002 *J. of Intel. Material Systems and Struct.* **13** 525-531

Journal of Physics: Conference Series **149** (2009) 012023 doi:10.1088/1742-6596/149/1/012023

DNA as a Component of ER Materials

K Minagawa[1], Y Aoki[1], M R Berber[1], T Mori[2], M Tanaka[3]

[1]Institute of Technology and Science, The University of Tokushima, Tokushima 770-8506, Japan, [2]Department of Applied Chemistry, Faculty of Engineering, Kyushu University, Fukuoka 819-0395, Japan, [3]Faculty of Pharmaceutical Science, Tokushima Bunri University, Tokushima 770-8514, Japan

minagawa@chem.tokushima-u.ac.jp

Abstract. Deoxyribonucleic acid (DNA), which is known as a typical biopolymer, has been utilized for a few types of ER materials. Suspensions were prepared with the particles of DNA, DNA/lipid complexes, and LDH (layered double hydroxide)/DNA composites. The purified DNA showed larger ER effect than the others, but this particle tended to absorb water, which caused less stability. Preliminary experiments of preparing composite with LDH indicated that this inorganic material would be useful for hydrophobic modification of DNA particles, although further optimization of composite preparation is needed. In addition, the LDH/DNA suspensions showed interesting behaviours under some conditions, which indicated possibility for controlling ER property in a wide range.

1. Introduction

Deoxyribonucleic acid (DNA) is not only one of the most important biomolecules but also an interesting functional material because of its unique structure and property. The base pairs inside the double-strand specifically interact with appropriate chemical species, and also adsorb hydrophobic substances. The outside double helix is negatively charged, and binds cationic substances. For example, cationic lipids can form a complex with DNA, and its possible application as a functional material has been reported [1]. It is also known that the charge of DNA is used for separation of them according to size, i.e. electrophoresis, which is based on the movement of the charged polymer chain by an electric field. This electric-field response suggests utilization of DNA as a component of electrorheological (ER) fluids. We have focused on DNA as one of the series of study on polymeric ER materials [2, 3, 4]. The application of DNA as a bio-based functional material can also be effective use of the 'waste natural product'. For example, salmon testes, which are produced in the process of food industry, are typical DNA-rich waste.

The salmon testes, and other commonly available DNA's are combined with proteins, and purification and/or modification are necessary for utilization as a material. Here suspensions of DNA, DNA/lipid complexes, and LDH (layered double hydroxide) /DNA composites were prepared in order to evaluate the effects of chemical modification of DNA. Some cationic lipids are known to bind with DNA, which should be useful for DNA modification. LDH was also selected here because of its unique property to form composites with organic compounds. The resulting composites have the properties of both inorganic and organic components. For example, LDH/drug composites are expected to be useful for biomedical applications, which is another subject of our recent work [5] as the basis of this study.

2. Materials and methods

2.1. Preparation of DNA and complex particles

Salmon sperm DNA was commercially available as powders containing some amount of proteins. The as-received DNA, purified DNA, and DNA/lipid complex were compared here. The purification of DNA was carried out according to the standard methods of phenol/chloroform extraction and ethanol precipitation. DNA/lipid complex was prepared by mixing aqueous solutions of DNA and of n-hexadecylammonium bromide followed by filtration. The complex was freeze-dried and milled to give fine particles. The LDH/DNA composites were prepared according to the co-precipitation method. Aqueous $MgCl_2$ and $AlCl_3$ solutions (Mg/Al = 2, 3, or 4) were mixed in the presence of DNA. The obtained composite particles were filtered and dried in vacuum.

2.2. Characterization of particles

X-ray powder diffraction patterns of LDH samples were recorded on a Rigaku X-ray diffractometer. An optical microscope (Nikon) and a scanning electron microscope (Hitachi FE-SEM S-4700) were used for the observation of appearance of the particles.

2.3. Preparation of ER fluids

The as-received DNA, purified DNA, DNA/lipid complex, and LDH/DNA composite particles were respectively dispersed in silicone oil (DMS; 1000cSt) to examine the influence of proteins, the cationic lipid, and LDH bound to DNA.

2.4. ER measurement

The ER measurements were performed by use of a parallel-disc rotation rheometer (Rheometric ARES) equipped with a high-voltage power supply. The disc diameter and the gap between the electrodes were 40 mm and 1.0 mm, respectively. Steady shear viscosity and dynamic viscoelasticity were measured in the presence and absence of a d.c. electric field.

3. Results and discussion

3.1. DNA and DNA/lipid particles

Figure 1 (a-c) shows the appearance of the DNA particles prepared in this study; i.e. (a) as-received DNA, (b) purified DNA and (c) DNA/lipid complex. It is found that all these particles have similar irregular shape and rather wide size distribution. Figure 2 exemplifies the results of the ER measurements with the suspensions of purified DNA and DNA/lipid complex (particle concentration: 20wt%). Here the shear rate dependence of the shear stress is plotted in the absence and presence of an electric field ranging between 0 and 2kV/mm. The both fluids behaved like a Newtonian and a Bingham fluid in the absence and presence of an electric field, respectively. This ER effect was caused by the formation of chain-like clusters of DNA particles bridging the electrodes under the electric field as shown in Fig. 1(d).

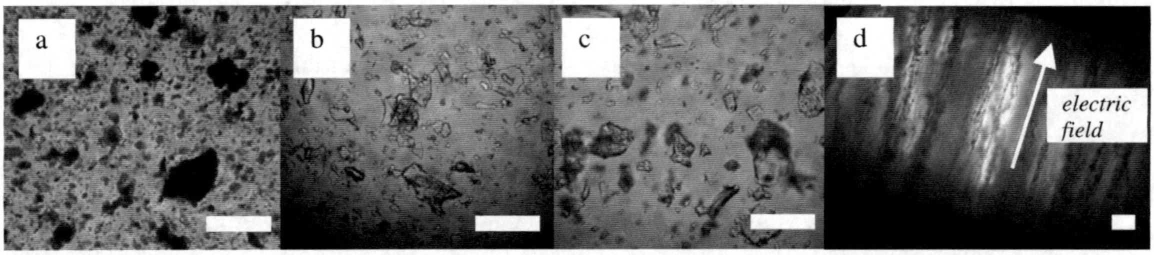

Fig.1 Optical micrographs of DNA particles: (a) as-received DNA, (b) purified DNA, (c) DNA/lipid complex, and (d) chain-like structure of purified DNA observed under 2kV/mm. Scale bar: 100μm.

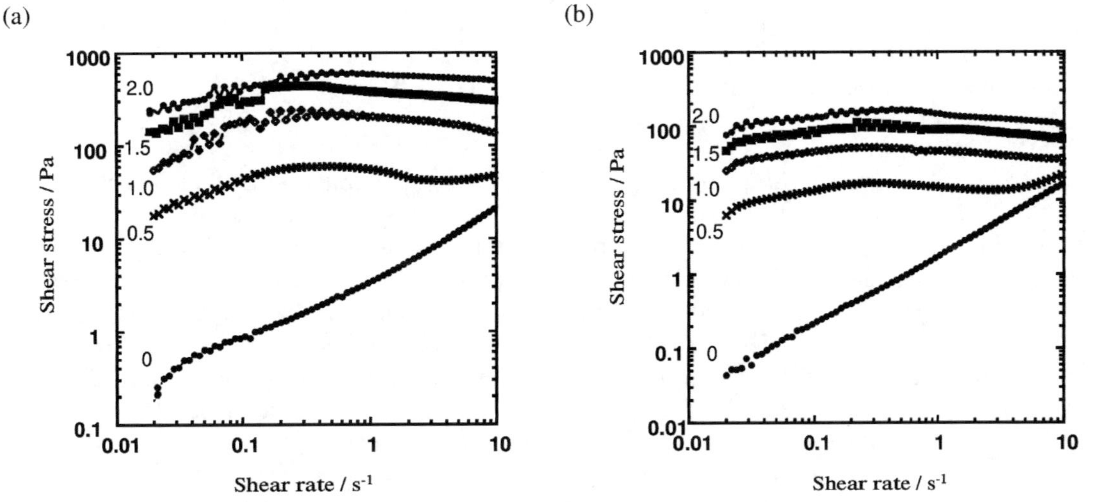

Fig. 2 Flow curves of suspensions with DNA based particles. (a) purified DNA (20wt%) and (b) DNA/lipid complex (20wt%).

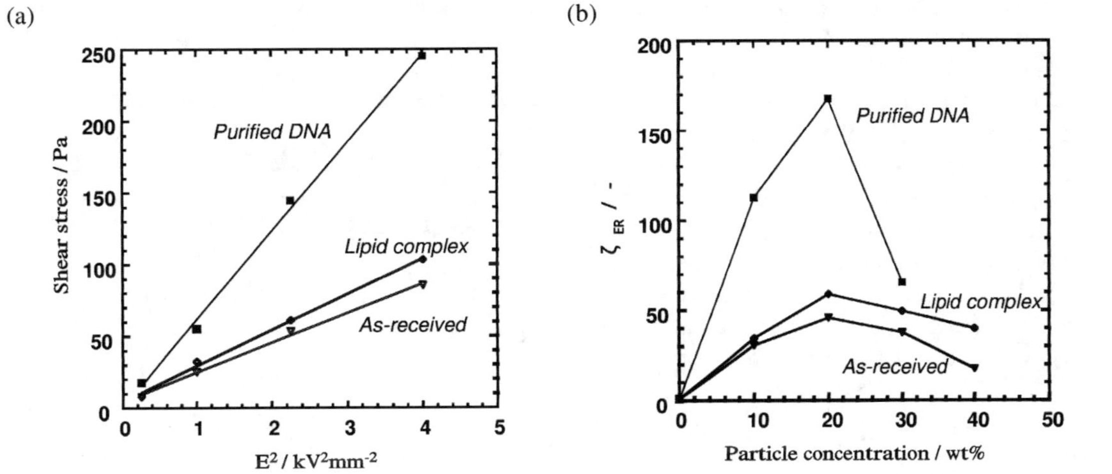

Fig. 3 Comparison of the ER effects of as-received DNA, purified DNA, and DNA/lipid complex suspensions. (a) Electric-field dependence of stress at a low shear rate (0.02 s^{-1}); (b) The ratio of stress obtained under a 2kV/mm electric field to that under no electric field measured at 1 s^{-1}.

Figure 3 shows the ER effect of as-received DNA, purified DNA, and DNA/lipid complex suspensions. In Fig. 3(a), the stress values obtained at the lowest shear rate examined were plotted against the square of the electric-field strength. These values can be used to evaluate the 'absolute value' of ER effect. The stress values are proportional to the square of field strength, and the largest stress increase was obtained with the purified DNA suspension. The large ER effect of purified DNA is due to the removal of proteins shielding the charge of DNA molecules. The lipid molecules attached to DNA have similar effect to that of proteins. It is also noticed that the relative ER effect shows a maximum at the particle concentration of 20wt%. This result is due to an increase in the shear stress in the absence of an electric field at high particle concentrations. Although the purified DNA seemed to be best from these results, the purified DNA particle was found to form sticky aggregates due to the adsorbed water from air when kept under open atmosphere. Thus the stability of this particle is not so good. It should also be noted that water affects the ER property. On the other hand, the lipid complex did not change its shape for long time, which indicates little influence of water from the air. This fact suggests that appropriate chemical modification can be used for utilization of DNA as stable particles. Further improvement is expected by using various DNA-binding substances.

3.2. LDH /DNA composite

The X-ray diffraction patterns of the particles obtained with the coprecipitation method indicated an increase in the distance between LDH layers. The increase of the layer spacing suggested that DNA molecules were intercalated into LDH layers. From the observation with optical microscopy, the average diameter of obtained LDH/DNA primary particles was about 5 μm, although the size varied with the reaction conditions. It was found that the LDH/DNA particles are in aggregated form, and the average diameter of the aggregates was typically about 25 μm. Suspensions of LDH/DNA with DMS were prepared and the dependence of ER effect on the LDH composition was examined. All samples varying Mg/Al ratio showed an ER effect: e.g., the shear stress of 20wt% suspensions under a 2kV/mm field at 1 s^{-1} for Mg/Al = 2, 3, 4 was 70, 50, and 40 Pa, respectively. The low Mg/Al ratio resulted in a larger ER effect. Although the present preliminary experiments searching for optimum conditions resulted in rather low stress enhancement under an electric field, this composite preparation method has the advantages that stable DNA-based particles can be obtained easily and that the compositions of LDH (metal species and ratio) and of composite (LDH/DNA ratio, etc.) can be widely changed. It was also found under some conditions that a 'slipping layer' of DMS by particle aggregation could be formed (Fig. 4), which decreases the apparent viscosity. The slipping occurs when the particles tend to aggregate rather than homogeneously dispersed. The shear rate and electric-field strength are also important. The control of slipping and non-slipping structures by the composition and field conditions makes it possible to vary the 'positive and negative' ER effects. From these results, LDH/DNA composites were found to be applicable to ER suspensions, and the property can be widely varied by changing the ratios of metals and DNA, and fluid composition. A variety of ER materials can be designed by use of this method.

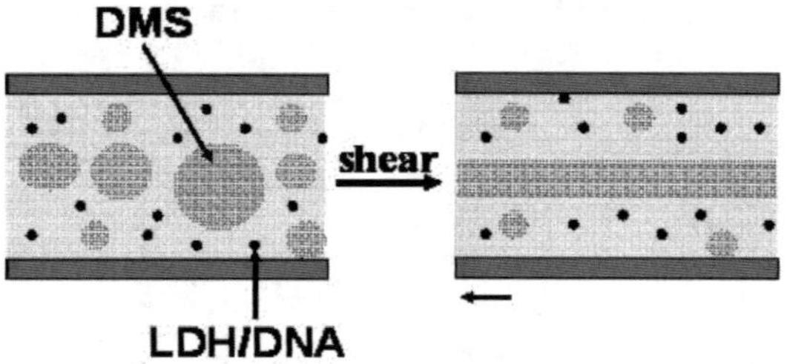

Fig. 4 Schematic illustration of slipping layer.

4. Conclusion

DNA-based particles are applicable for ER suspensions. The enhanced ER effect of purified DNA indicated the potential usefulness of DNA as a component of ER materials. Though the particle stability is a problem, chemical modification with lipid can be a solution for it. The LDH composite also indicated another possibility of the use of DNA-based ER fluids having interesting properties.

Acknowledgment

This work was partly supported by the Feasibility Study grant from JST Innovation Satellite Tokushima.

References

[1] Tanaka K and Okahata Y 1996, J. Am. Chem. Soc., 118 (44), 10679.
[2] Minagawa K and Koyama K 2005, Curr. Org. Chem., 9,1643.
[3] Minagawa K, Aoki Y, Mori, T, Tanaka M 2006, Int. J. Mod. Phys., 20 (25, 26, 27), 3987.
[4] Minagawa K, Aoki Y, Tanaka M 2007, Polym. Prepr. Jpn., 56, 2E18.
[5] Berber M R, Minagawa K, Katoh M, Mori T, Tanaka M 2008, Eur. J. Pharm. Sci. in press.

Journal of Physics: Conference Series **149** (2009) 012024 doi:10.1088/1742-6596/149/1/012024

Electrorheological flow patterns analysis

P Ponížil[1], M Stěnička[1], V Pavlínek[1], T Kitano[1] and T Dřímal[2]

[1] Tomas Bata University in Zlin, Faculty of Technology, TGM 275, 762 72 Zlin, Czech Republic
[2] Tomas Bata University in Zlin, Faculty of Applied Informatics, Nad Stranemi 4511, 760 05 Zlin, Czech Republic

E-mail: pavlinek@ft.utb.cz

Abstract. A method of image analysis of flow patterns, which are developed in electrorheological fluids, is presented. Due to the process of preparation, electrorheological samples show a radial symmetry. Numerical transformations are necessary to remove sample deformation and obtain correct radial dependence of image intensity as the function characterizing the sample image. Parameters describing the radial symmetrical image are suggested. Then the influence of electrorheological parameters on properties of final sample is studied.

1. Introduction
Electrorheological (ER) fluids are systems whose rheological properties (viscosity, yield stress, shear modulus) can be controlled by external electric field. Polarized particles are oriented in the direction of electric field and create structures (chains), which increase the rigidity of the originally liquid system.

The arrangement of the particle chains in electric field has been proved via optical microscopy. A direct optical display of ER suspensions in rotational viscometer presented in papers [1–4] indicated that particles organize into lamellar or ring structures, which are optimal for minimum energy dissipation, and this structure depends on the flow field. The aim of this paper is to describe ring ER structures, using image analysis, with respect to conditions of their development.

2. Experimental
Rotational viscometer Bohlin Gemini (Malvern Instruments, UK) in parallel plates geometry (diameter 40 mm, gap 1 mm) was used for experiments. 20 and 40 wt. % suspensions of glass beads (40 μm) in liquid silicone rubber were sheared at various shear rates in the presence of electric field (1 kV mm^{-1}) and generated circular flow patterns were fixed via curing of silicone rubber matrix.

Later, samples were scanned on commercial scanner and greyscale images with typical size 1000×1000 pixels were obtained.

3. Methods of image analysis
As can be seen in figures 1, 3 and 4, circular structures developed during the flow of ER fluids upon application of electric field significantly depend on a value of shear rate used. At low shear rates higher number of thin rings is formed. With increasing shear rate, number of rings decreases while their width rises.

The goal of presented mathematical procedure is to describe ER ring patterns by some simple numbers defined by radial dependence of darkness of the image and to correlate them with conditions of their preparation.

3.1. Image preprocessing

Suppose that $p(x,y)$ is brightness of pixel in x-column and y-row of the image. There are several problems to be solved prior analysis. Samples are relatively soft and easily deformable therefore images are slightly prolonged in some directions because of manipulation.

In the first step the centre of sample symmetry was detected as a point (x_c,y_c) minimizing the function $S(x_c,y_c)$ (equation 1):

$$S(x_c,y_c) = \sum_{x,y} \frac{(p(x_c+x,y_c+y)-p(x_c-x,y_c-y))^2}{\sqrt{x^2+y^2}} \tag{1}$$

where $p(x,y)$ is brightness of (x,y) pixel. Position of pixels $p(x_c+x,\ y_c+y)$ and $p(x_c-x,\ y_c-y)$ is symmetrical with respect to centre of symmetry (x_c,y_c). The numerator is the square of pixels in symmetrical positions. The expression in denominator is pixels distance r from centre (x_c,y_c). The number of pixels in interval $(r;\ r+dr)$ is proportional to radius r. The denominator ensures the same weight of pixels in any distance from centre. Minimizing the function $S(x_c,y_c)$ we obtain centre position for which the difference between points symmetrical with respect to it, are minimal.

If the centre is found, image is divided into 8 radial sectors and mean brightness as a function $f_i\,(r)$ (index i denotes the sector) of the real centre distance r is computed for each sector. The functions $f_i\,(r)$ and $f_j\,(r)$ can be a little shifted (due inaccuracy of centre detection) and elongated due the sample deformation.

The generalized cross-correlation function (equation 2) [5]:

$$(f_i * f_j)(h,x) = \int f_i(r) f_j(hr+x)dr \tag{2}$$

is a tool for finding the best shift x, and elongation h of function f_j, maximizing the function $(f_i{*}f_j)$ and obtain the maximal similarity of f_i and f_j. Then all functions for all sectors are averaged to obtain final $f\,(r)$ function representing radial dependence of image brightness.

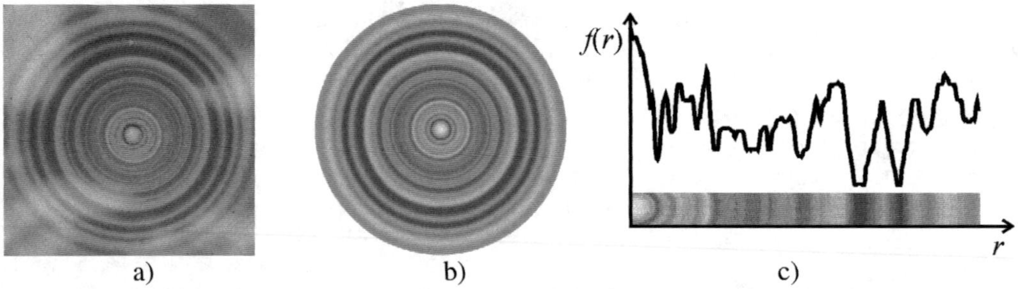

Figure 1. The original image (a), corresponding result of the image preprocessing (b) and appropriate $f(r)$ function (c).

Thanks to such image preprocessing all artifacts as noise, sample inhomogeneities, sample deformation are removed (figure 1b) and then the function $f\,(r)$ as radial intensity dependence is obtained (figure 1c).

3.2. Image quantities

After the image preprocessing some parameters for pattern characterization must be chosen. We see evident differences between samples prepared under different conditions, but we have to find some simple numbers to describe images differences. A contrast should characterize changes of intensities of image in radial direction (waviness or amplitude of $f\,(r)$ function). A period describes mean width of rings (local extremes of $f\,(r)$ function).

3.2.1. *Standard contrast* Standard contrast (amplitude of f (r) function) is inspired by standard deviation in statistics: $\sigma = \sqrt{\frac{1}{N}\sum_i^N (x_i - x)^2}$. The function $c_S = \sqrt{\frac{1}{R}\int_0^R \left(f(r) - \frac{1}{R}\int_0^R f(r)dr \right)^2 dr}$ is generalisation of this well known statististical equation. It can be seen from definition that standard contrast is sensitive to size of local minima and maxima of function, but is not sensitive to behaviour of 1st derivation of function.

3.2.2. *Linear contrast* The linear contrast is inspired by linear roughness often used in engineer mechanics. The linear contrast of function f (r) is defined as a length of function curve divided by a length of curve projection to horizontal line (equation 3):

$$c_L = \frac{1}{R}\int_0^R \sqrt{1 + f'^2(r)}\mathrm{d}r \qquad (3)$$

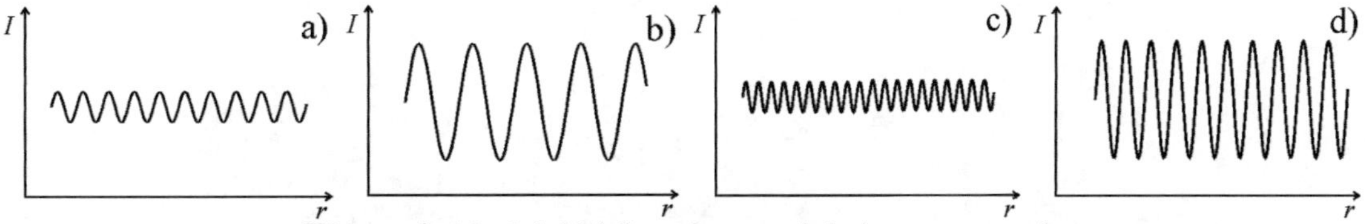

Figure 2. Model f (*r*) functions and their contrast values.

The difference between standard contrast c_S and linear contrast c_L can be explained in the figure 2. Standard contrast is low for functions 2a and 2c (low amplitude). Linear contrast is low for functions 2a and 2b (slow intensity change with growing *r*). For 2b and 2d c_S is high; c_L is high for 2c and 2d.

3.2.3. *Period* Period *P* is the mean distance between neighbouring f (r) maxima, i.e. mean gap width of neighbouring bright pattern circles. It can be replied that f (r) function is not periodic, but period is quantity describing mean density of rings.

4. Results

4.1. Reproducibility

The reproducibility of samples was tested on four samples produced at the same conditions (figure 3). It is evident that resulting images are very similar and suggested parameters have small variability.

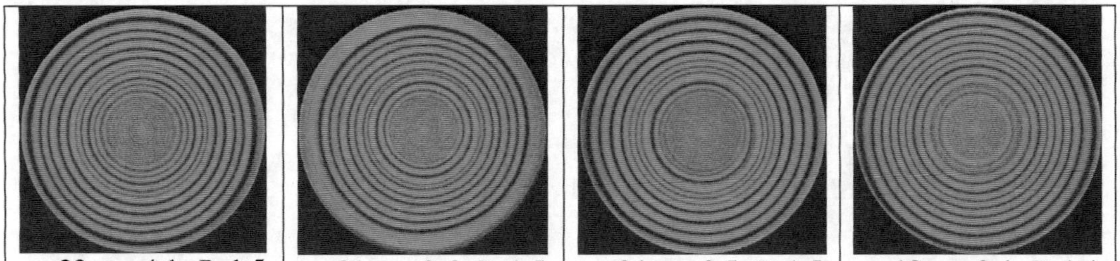

c_S=23; c_L=4.1; *P*=1.5	c_S=21; c_L=3.6; *P*=1.5	c_S=24; c_L=3.5; *P*=1.5	c_S=18; c_L=3.4; *P*=1.4

Figure 3. Four samples produced by the same conditions (concentration 40 %; shear rate γ=10 s^{-1}; time=300 s; E=1 kV mm^{-1}) and their calculated parameters (period *P* is measured in mm).

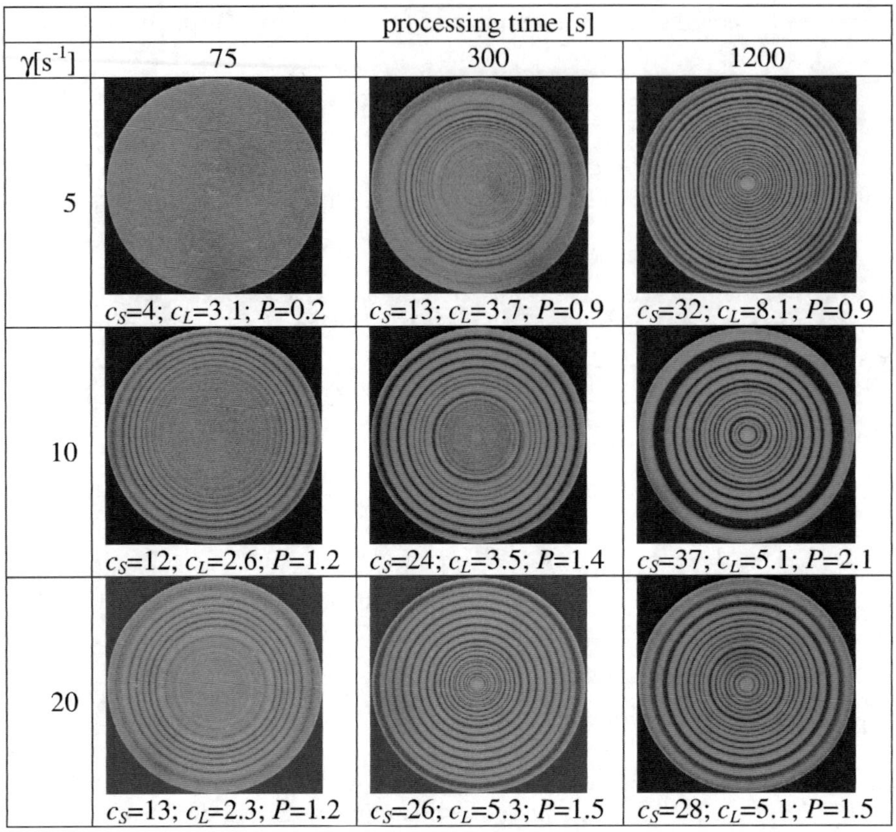

Figure 4. The influence of shear rate and processing time (concentration 40 %; E=1 kV mm^{-1}).

Figure 4 shows influence of shear rate and processing time on parameters of samples. It seems that both pattern contrasts and period increase with shear rate and processing time. The area of central unstructured part of samples is decreasing with growing time and shear rate. Probably a total plate displacement is important for structure progression. At time 1200 s and two higher shear rates a structure regression is observable (especially c_S). It seems that there is some optimal time which ensures pronounced structure progression.

5. Conclusion

The quantitative parameters for description of flow patterns, which are developed in electrorheological fluids, were suggested and verified. Growing processing time and shear rate leads to more perspicuous patterns, but the regression at higher shear rates should be properly examined.

6. Acknowledgment

The authors kindly acknowledge the support by the Ministry of Education, Youth and Sports of the Czech Republic (project MSM 7088352101) and the Grant Agency of the Czech Republic (projects 106/05/0550 and 202/06/0419).

7. References

[1] Winslow W M 1949 *J. Appl. Phys.* **20** 1137–40
[2] Hao T 2002 *Adv. Colloid Interface Sci.* **97** 1–35
[3] Henley S and Filisko F E 1999 *J. Rheol.* **43** 1323–36
[4] Filisko F E, Henley S and Quist G 1999 *J. Intell. Mater. Syst. Struct.* **10** 476–80
[5] Liu J and Iskander M 2004 *J. Comput. Civil. Eng.* **18** 46-57

Journal of Physics: Conference Series **149** (2009) 012025 doi:10.1088/1742-6596/149/1/012025

Optimal control of transient processes in an oscillating system with an electrorheological shock-absorber

G N Reizina[1], **E V Korobko**[2], **V A Bilyk**[2], **V L Efremov**[3] **and A E Binshtok**[3]

[1] National Technical University, 65 Nezavisimosti Avenue, Minsk, 220013, Belarus
[2] Luikov Heat and Mass Transfer Institute of NASB, 15 P Brovki Street, Minsk, 220072, Belarus
[3] Minsk Wheel Tractor Plant, 150 Partizanskii Avenue, Minsk, 220021, Belarus

E-mail: eva@itmo.by

Abstract. The problem of optimal control of the oscillation of a driver's seat with ER shock absorber is discussed in application to vehicles based on the principle of maximum. Coordinates of a vector of the control parameter and vertical motions of a seat are obtained. Test experimental investigations on working regimes for the driver's seat model depending on the controlling impact of the electrical signal (the intensity of the electric field) were performed.

1. Introduction

The creation of "smart" shock-absorbers and the algorithm of their control is a key issue of vibroprotective systems, especially in application to vehicles. Nowadays, an intensive development of both "smart" (electro and magnetorheological) fluids for shock-absorbers [1, 2], and methods, and devices for the control of cushioning systems of vehicles [3, 4] is carried out. For each vehicle its own algorithm of control is needed, which can provide an optimal vibroprotection. The type of electrorheological (ER) shock-absorber is determined by electrorheological properties of ER fluid at given parameters of an oscillating system.

To characterize the described single mass oscillating system a state space of vertical motions is used. The calculation of the system results in the determination of the point of extremum of functional at given physical limitations. With the aid of one of the methods of variational calculus a control parameter is determined, which minimizes the initial functional.

2. Statement of the problem

In this paper an optimal control of transition processes of driver's seat oscillations with an electrorheological shock-absorber in a vehicle is investigated. The controllable vibroprotection is shown in figure 1. The oscillation damping of the driver's seat is controlled by a control unit of viscoelastic characteristics of a shock-absorber at applying the electric impact, depended on vibro acceleration, determined by sensors. A control unit (6) creates the control signal for providing a necessary force on a shock-absorber piston, which is connected with a sprung mass. Kinematical actuation of the shock-absorber forms its vertical displacement. Oscillations of the driver's seat are described by the equation of motion:

$$m\ddot{z} + cz = S_p \Delta P(E), \tag{1}$$

and a balance of forces [5] on shock-absorber's piston is described from the equation:

$$m\ddot{z} + F(z, \dot{z}) = S_p \Delta P(E), \tag{2}$$

where the pressure drop [6] is determined from the equation of the fluid flow:

$$(V_0 / 2G)\dot{\Delta P} + k_1 \Delta P + S_p \dot{z} = k_2 E_c \tag{3}$$

with initial conditions $\dot{z}(t) = \dot{z}_0$; $\Delta P(t_0) = \Delta P_0$ for $z(t_0) = z_0$, with m – mass of the driver's seat together with the weight of a driver; $F(z, \dot{z}) = az + b\dot{z}$ – resistance force, conditional upon ER shock–absorber and consisting of elastic az and viscous $b\dot{z}$ components [7]; c – rigidity of the system without ER shock-absorber; S_p – working area of ER shock-absorber piston; V_0 – fluid volume of a working chamber of ER shock-absorber; G – complex modulus of ER fluid; k_1, k_2 – coefficients, dependent on geometrical parameters of ER shock-absorber; $\Delta P, \dot{\Delta P}$ – a drop and change of a drop in pressure over time; E_c – the intensity of electric field; $z, \dot{z}, \ddot{z}$ – vibro motion, vibro velocity, vibro acceleration of the system.

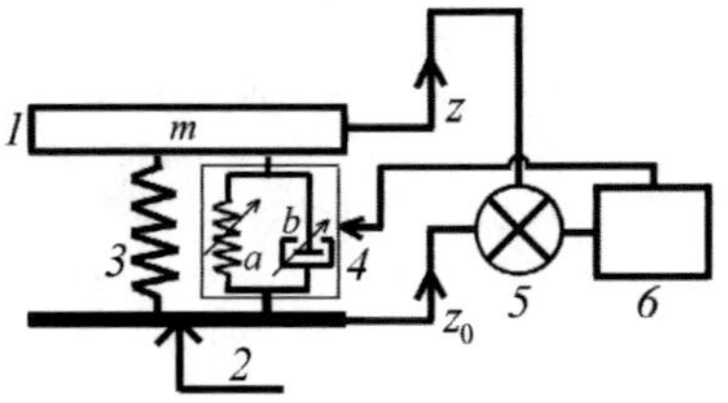

Figure 1. A scheme of a controllable vibroprotection: 1 – a sprung mass; 2 – an external impact; 3 – elastic element; 4 – ER shock-absorber; 5 – a comparator unit; 6 – a control unit.

Substituting (3) into differentiated equations (1) and (2), we get a generalized equation

$$\dddot{z} + a_1 \ddot{z} + a_2 \dot{z} + a_3 z = a_4 E_c(t), \tag{4}$$

where

$$a_1 = (bV_0 + 2Gk_1 m)/mV_0, \; a_2 = 2G(S_p^2 + bk_1)/mV_0, \; a_3 = 2Gk_1 a/mV_0, \; a_4 = 2Gk_1 S_p/mV_0.$$

Let $x_1 = z$; $x_2 = \dot{z}$, $x_3 = \ddot{z}$, $u_c = a_4 E_c(t)$, where x_1, x_2, x_3, u_c – time functions. It is needed to determine the controllable parameter for the system with the initial condition $\mathbf{X}(t_0) = [0\ 0\ 1]$ and a limitation $|u_c| \le u_{\max}$, where $u_{\max}$ – the maximum of the value of the controllable parameter u_c, which according to [8] satisfies the condition:

$$\int_0^T \ddot{z} dt \to \min, \; t \in [0, T]. \tag{5}$$

Equation (4) is transformed to the system:

$$\begin{cases} \dot{x}_1 = x_2; \\ \dot{x}_2 = x_3; \\ \dot{x}_3 = -a_1 x_3 - a_2 x_2 - a_3 x_1 + u_c. \end{cases} \tag{6}$$

System (6) in the matrix form takes the form:

$$\dot{\mathbf{X}} = \mathbf{AX} + \mathbf{B}u_c\,, \tag{7}$$

where

$$\mathbf{X} = [x_1 \quad x_2 \quad x_3]^T,\ \mathbf{A} = \begin{bmatrix} 0 & 1 & 0 \\ 0 & 0 & 1 \\ -a_3 & -a_2 & -a_1 \end{bmatrix},\ \mathbf{B} = [0 \quad 0 \quad 1]^T.$$

Hamilton function [8] to determine a controllable parameter u_c from (6) is:

$$H = \varphi_1 x_2 + \varphi_2 x_3 + \varphi_3 (-a_1 x_3 - a_2 x_2 - a_3 x_1 + u_c), \tag{8}$$

where φ_1, φ_2, φ_3 – auxiliary functions, dependent on time, according to [8] are determined from the system:

$$\begin{cases} \dot{\varphi}_1 = -\partial H/\partial x_1 = \varphi_3 a_3; \\ \dot{\varphi}_2 = -\partial H/\partial x_2 = -\varphi_1 + a_2 \varphi_3; \\ \dot{\varphi}_3 = -\partial H/\partial x_3 = -\varphi_2 + a_1 \varphi_3. \end{cases} \tag{9}$$

Necessary condition (5) is the finding of such operating parameter u_c, at which functional H accepts the maximal values during each moment of time t. A sufficient condition of minimization (5) is the correct choice of entry conditions for system (9).

To determine the parameter u_c equations (5), (8), (9) are used, providing that function H reaches extremum at each moment of time. The controllable parameter u_c is:

$$u_c = u_{\max} \operatorname{sgn}[-\varphi_3(t, z(t))], \tag{10}$$

where $u_{\max} = 9.81$ m/s^2.

Therefore, the obtained expression (10) for the controllable parameter u_c allows us to determine in the oscillating system the acceleration of a sprung mass $x_3 = \ddot{z}$ from (6), which satisfies the condition (5), i.e. creates an optimal regime of damping of oscillations of the driver's seat.

3. Results and discussion

To calculate the value of the controllable parameter u_c the following values of parameters are used: $m = 80$ kg, $b = 400$ N·s/m, $c = 2\cdot10^3$ N/m, $S_p = 5.2\cdot10^{-4}$ m^2, $V_0 = 50\cdot10^{-6}$ m^3, $G = 6\cdot10^7$ N/m^2. As a result of calculations a transient process, described by the dependence of vibro motions on time (figure 2) has been obtained for vertical oscillations $x_1(t)$ for $u_c = 0$ ($E_c = 0$ kV/mm) and $u_c \neq 0$ ($E_c = 2.5$ kV/mm). Vector of optimal coefficients of the equation (6) has the following parameters: $\mathbf{B} = (0.56; 0.24; 0.15)$.

To perform the calculation (1) we experimentally investigated a model of the driver's seat with ER shock-absorber at the condition of stimulated harmonic oscillations with the amplitude of 5 mm and the frequency of 1.6 Hz (figure 3).

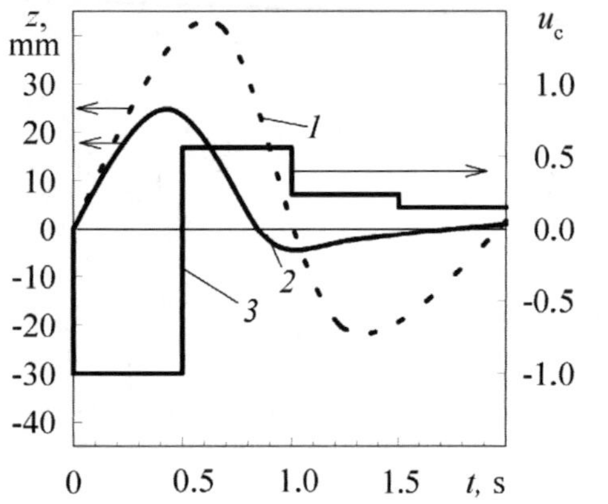

Figure 2. Dependence of vibro motion of the driver's seat on time at $E_c = 0$ kV/mm (1), $E_c = 2.5$ kV/mm (2) and the controllable parameter u_c (3) on time.

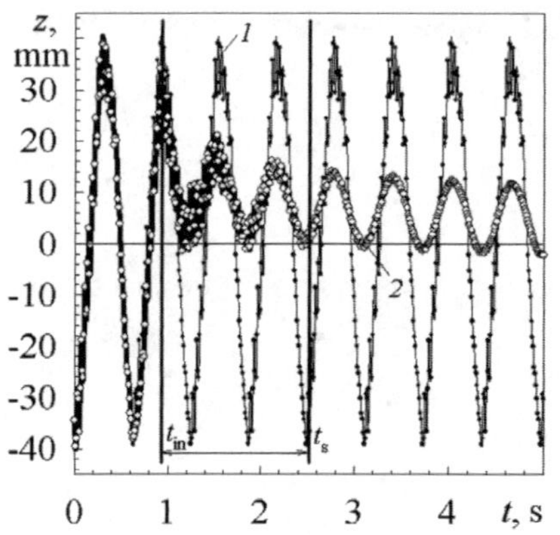

Figure 3. Dependence of vibro motion of the model of the driver's seat on time at stimulated harmonic frequencies with the amplitude of 5 mm and frequency 1.6 Hz at $E_c = 0$ kV/mm (1) and $E_c = 2.5$ kV/mm (2).

From figure 3 it is obvious that time of establishment of the constant amplitude of oscillations of the sprung mass *m* at impact of the controllable parameter u_c ($E_c = 2.5$ kV/mm) is 2-3 times smaller than that at $E_c = 0$ kV/mm which equals to 4.5 s. Calculated time of establishment of the constant amplitude of oscillations at $E_c = 2.5$ kV/mm is 1.76 s. Time of establishment t_s of the constant amplitude of oscillations at the controllable parameter u_c from the moment of switching on of the signal t_{on} from the control unit is 1.5 s, which coincides with the calculated result for 85%.

Conclusion

Suggested calculation allows us to evaluate the efficiency of damping over time of establishment of oscillation amplitudes and to determine the optimal values of the intensity of the electric field $E_c(t)$ for a maximum damping of oscillations. Considered above scheme of the oscillating system of the driver's seat is a part of a general scheme of controlling oscillations of vehicles.

References

[1] Tian Yu, Meng Y and Wen S 2004 Dynamic Responses of Zeolite-based ER fluid Sheared Between Two Concentric Cylinders *J. of Intelligent Materials Systems and Structure* **15** pp 621-626.

[2] Di K, Zhu Y, Yang X and Li C 2006 Electrorheological behavior of copper phthalocyanine-doped mesoporous TiO_2 suspensions *J. of Colloid and Interface Science* **294** pp 499–503.

[3] Barta D J, Raynauld O and Valee O 2002 Method and system for control of a variable force damper *Patent 6,438,473 B1 20/08/2002.*

[4] Choi S B, Choi J H, Lee Y S and Han M S 2003 Vibration Control of an ER Seat Suspension for a Commercial Vehicle *J. of Dynamic Systems, Measurenment and Control* **125** pp 60–68.

[5] Grigoriev P V, Zakharov U E, Ryagnov U G 1969 *Automatic control of diesel locomotive's hydrotransmission* (Moscow: Mashinostroenie).

[6] Baranov B N, Zakharov U E 1977 *Electro–hydraulic and hydraulic vibration mechanisms* (Moscow: Mashinostroenie).

[7] Shulman Z P, Husid B M, Hijinski B P, Mokeev V V, Korobko E V 1986 Electrorheological vibroisolation of oscillation systems *Preprint* (Minsk: ITMO).

[8] Pontrjagin L S, Boltianski V G, Gamkrelidze R V and Mishenko E F 1983 *Mathematical theory of optimal processes* (Moscow: Nauka).

Journal of Physics: Conference Series **149** (2009) 012026 doi:10.1088/1742-6596/149/1/012026

Dynamic Column Formation in Na-FLHC Clay Particles: Wide Angle X-ray Scattering and Rheological Studies

Z Rozynek, B Wang, M Zhou, and J O Fossum

Department of Physics, NTNU, Høgskoleringen 5, NO-7491, Trondheim, Norway

E-mail: zbigniew.rozynek@ntnu.no

Abstract. The dynamic chain and column formation of 5wt.% clay particles suspended in silicone oil has been studied using synchrotron Wide Angle X-Ray Scattering (WAXS) technique and rheometry. The anisotropic arrangement of particles described by the global order parameter S has been investigated. The WAXS data have also allowed distinguishing between the chain and column formation processes by comparison of the change of WAXS angular plots maxima with the current density growth in function of time. The saturation time t_s (after which there was no change in the system observed) was estimated. In addition, the rheological properties of the ERF have been measured including the static yield stress.

1. Introduction

Electrorheological fluids (ERFs) are liquids that solidify, or become very viscous, under an electric field. The transition from liquid into a solid-like state indicates that there is an inner ordering of the ER-constituents, which leads to a change in the rheological properties. Application of an electric field induces polarization of the suspended dielectric particles and a chain-like structure can be formed along the electric field direction [1-3]. Recently, the Giant Electrorheological (GER) fluid was discovered and it differs from the conventional ER fluids by exceeding the theoretical upper bound [4,5] of the yield stress, reaching 250 kPa at 5 kV/mm [6,7]. That discovery opens new horizons in the field of ER systems and inspires rheologists for working on further developments of such systems. In order to predict the response and behavior of ERFs under an applied electric field, it is important to gain an understandable knowledge of the physical mechanism of the chain and column formation and its dynamics. In the present study, synthetic clay particles Na-fluorohectorite (Na-FLHC) suspended in insulating, non-polar silicone oil form with time chain-like structures when subjected to an external electric field. The time needed to create columns of aggregated particles varies as the electric field (0.35, 0.5 and 0.75 kV/mm) increases. Wide Angle X-Ray Scattering (WAXS) diffraction patterns reveal changes of the direction of the dipolar moment induced in clay particles when the electric field is applied. The anisotropic arrangement of particles forming the chain can be described by the global order parameter S. The detail description of the system geometry, fitting procedure and calculations of S parameter can be found in the report of Meheust Y, et al. [8]. In-depth analysis of the system measured can be obtained using rheometry. The shear stress τ as a function of a shear rate $\dot{\gamma}$ can reveal the type of ER fluid and in case of samples of this study the shear stress is represented by the Bingham plastic rheological model [9,10]

2. Sample preparation and characterization methods

The synthetic Na-FLHC clay was purchased from Corning Inc. (New York) in a form of powder. According to supplier, its chemical formula is given as $Na_{0.6}(Mg_{2.4}Li_{0.6})Si_4O_{10}F_2$ per unit cell, where Na is an interlayer exchangeable cation. A silicone oil Dow Corning 200/100 Fluid (dielectric constant of 2.5, viscosity of 100 mPa·s and specific density of 0.973 g/cm^3 at 25°C) was used as a suspending liquid as a relatively non-polar and non-conductive medium, with a DC conductivity of the order of magnitude of 10^{-12} S/m. The preparation of the ER fluid was undertaken by the following procedure. Clay powder was crushed with a pestle and mortar, and then dried in vacuum oven for 12h at 90°C. At the same time the silicone oil was heated at 110°C for 12h. After that, the clay powder and silicone oil were mixed in glass tubes and vigorously hand-shaken for ~2min, then placed in an ultrasonic bath for 1h and again vigorously shaken for 1h in an orbital shaker. The clay concentration was assumed to be ~5 wt.% and it sedimented rather slow, however for rheological data collection the sample needed to be pre-shaken each time before measurement.

The wide angle x-ray scattering experiment (WAXS) was carried out at the European Synchrotron Radiation Facility (ESRF) in Grenoble, France. The X-ray beam with wavelength of 0.72Å and 0.3x0.3mm^2 beam size at sample was used. On the beamline BM01A a two-dimensional MAR 345 image plate detector with diameter of 345mm and a read-out time of around 40s was installed. The WAXS experiment was performed at RT for sample at rest (no liquid flow through a custom-made scattering cell). Electric fields from 0.35 to 0.75kV/mm have been chosen as a result of the following observations: for the electric filed above 0.75kV/mm the chain and column formations are too fast to be distinguished; electric fields below 0.35kV/mm however are too week for particles to form any chains.

The rheological properties of the clay suspensions were measured at RT, under DC electric fields using a Physica MCR300 rheometer.

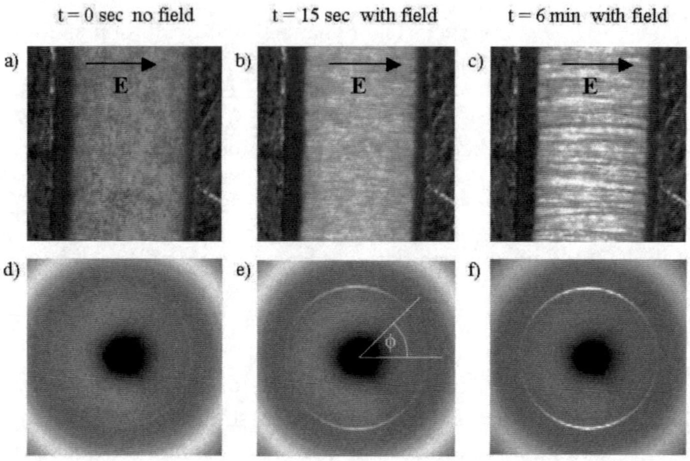

Figure 1. Optical microscope (a,b,c) and WAXS data (d,e,f)

3. Results and discussion

Normally, the Na-FLHC particles are randomly dispersed into the silicone oil (fig.1a). Microscopy images of the sample experiencing an electric field for different time are shown in fig.1b, 1c. The formation of column-like structures aligning parallel to the field is clearly observed. Many of thin chains are formed first (~1-20μm) and they attract each other resulting in creation of thicker (~50-200μm) columns. After the time t_s (which will be called here the saturation time) no major changes in the system are noticeable. Figure 1d is obtained prior to the application of an electric field and isotropic pattern can be observed, since particles are distributed randomly with not specified orientation. The two-dimensional WAXS patterns in figures 1e and 1f are becoming gradually more intensive in time, and the dependence of intensity for azimuthal angle ϕ is pronounced. This change of the 1st Bragg peak intensity in time is plotted as function of the azimuthal angle ϕ and presented in figures 2a and 2b for different electric (DC) fields.

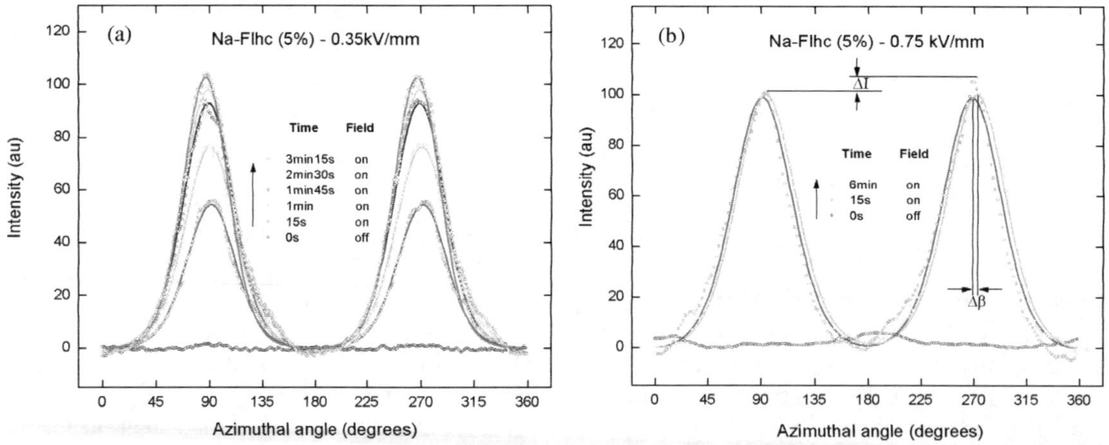

Figure 2. Angular plots of the 1st Bragg peak changing with time under an electric field

There are three possible events which can influence these WAXS pattern changes while time passes by: either clay platelets are being more orientated along the electric field or/and clay particles are forming chains or/and chains are coarsening into column-like structures in a range of the X-ray beam. These three situations can occur in different time range or they may overlap in time. By looking at figures 2 and also taking into account the calculations of S parameter (tab.1) one can conclude that the intensity does not changes due to particle orientation and it is rather dependent on the dynamic chain or/and column formation. The mean orientation distribution hardly changes and S oscillates around 0.63, 0.65 and 0.60 under applied fields of 0.35 kV/mm, 0.50 kV/mm and 0.75 kV/mm, respectively.

	Time	15s	1min	1min45s	2min30s	3min15s	6min	**Avr.**
0.35 kV/mm	**S**	0.63	0.61	0.62	0.62	0.67	0.65	0.63
0.50 kV/mm	**S**	0.62	0.62	0.64	0.64	0.68	0.67	0.65
0.75 kV/mm	**S**	0.53	0.64	0.65	0.58	0.60	0.59	0.60

Table 1. Calculated order parameters for Na-FLHC clay suspension under an electric fields of 0.35, 0.50 and 0.75 kV/mm.

The current density increases when chains are being formed. However, it does not change while chains coarsen into the column. Therefore by comparison between the current density growth and the change of WAXS angular plots maxima versus time (fig.3), both chain and column formation processes can be distinguished. This information might be of importance when one needs a self-assembly from particles to lead into the single chain formation rather than aggregated column creation. For 0.75 kV/mm and 0.50 kV/mm electric fields the current density curves overlap with WAXS data indicating the chain formation is followed by rapid column aggregation. However, for the lowest electric field current density increases faster than the WAXS records for the same field pointing out that the column formation occurs slower comparing to higher electric fields. That follows the thermal theory of coarsening [11,12], which describe the collision time of two chains as follows: $t_c \sim \rho^{3/2} \cdot k_B T^{-1/2} \cdot E^{-1}$, where ρ is a distance between two chains. Thus, when the electric field E increases (more rapid chain formation entails both the decrease of distances between chains and the increase of current resulting in the temperature T rise) the coarsening occurs more rapidly. For the different field strength, the intensity for azimuthal angle and integrated q values becomes saturated at the different time t_s which was estimated as: ~195s, ~150s and ~60s for electric fields of 0.35kV, 0.50kV and 0.75kV, respectively. One can see that the data and its fitted curves in fig. 2b are left- or right-shifted. This is due to the movement of a columnar structure that can be tilted in respect to the reference orientation while being formed.

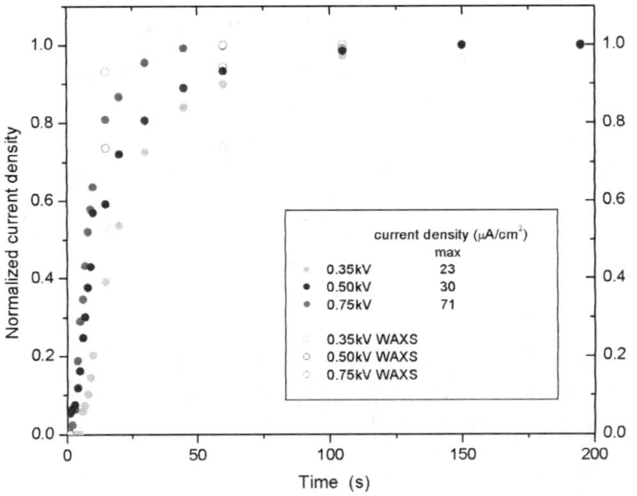

Figure 3. Leaking current density and maxima of angular plots versus time.

Fig.4 shows the yield stress τ_y of different clay particle concentration Θ as a function of electric field E. The power law with exponent α of 1.57, 1.77, 1.67 and 1.64 was used to fit the results for the samples with 5, 10, 20 and 40 wt.% of clay particles, respectively. The exponent α is commonly found in the range between 1 and 2 [13]. The yield stress for the sample with 5wt.% of clay particle was estimated as 16 Pa at 1kV/mm. whereas the yield stress of 102 Pa was measured for the sample with 40wt.% of clay particles. As one can see the yield stress τ_y and what follows the apparent viscosity η of the ER suspension is largely dependent on the particle concentration Θ and the relationship is nearly linear. Similar results have been found by many researchers [9] and the fibrillation model was used to derive that dependence. However simple and very limited, the fibrillation model can be also suggested here as the mechanism of the ER effect. In this model particles can be polarized and aligned as a dipole along the direction of the electric field. The interaction between the polarized particulates would be increased, resulting in the obvious ER effect. The particle could bear some net charge, arising from the non-uniform polarization or ionic adsorption, thus electrophoresis could contribute to the particle motion. This particle migration was observed using optical microscope and also confirmed by the experiment described below. Steady shear stress for a DC electric field was compared with a low-frequency (50Hz) periodic AC field of the same RMS voltage values. The results are shown in the inset of figure 4. The shear stress is higher when the AC power supply is used. This is due to the fact that electrically induced particle migration is minimized.

4. Conclusion

The dynamic chain and column formation of 5wt.% clay ER suspension has been studied by means of WAXS technique and rheometry. It was shown that the mean orientation distribution hardly changes in time (15sec – 6min) and the order parameter oscillates around 0.63, 0.65 and 0.60 under applied fields of 0.35 kV/mm, 0.50 kV/mm and 0.75 kV/mm, respectively. The WAXS experiments have also allowed distinguishing between the chain and column formation processes by comparison of the change of WAXS angular plots maxima with the current density growth versus time. For the lowest electric field (0.35 kV/mm) there was clear evidence that coarsening of chains into the thicker columns occurs much slower comparing to the samples at higher fields, where column formation takes place immediately after chains are being formed. The saturation time t_s (after which there was no change in the system observed) was estimated as: ~195s, ~150s and ~60s for electric fields of 0.35kV, 0.50kV and 0.75kV, respectively. The formation of chains and columns is more rapid when an applied voltage is higher. Steady shear stress for a DC electric field was compared with a low-frequency periodic field and a slight increase in the shear stress was observed due to the minimization of particle migration. In addition, the most important electrorheological properties have been measured which included both the electric field and also particle concentration dependent yield stress tests.

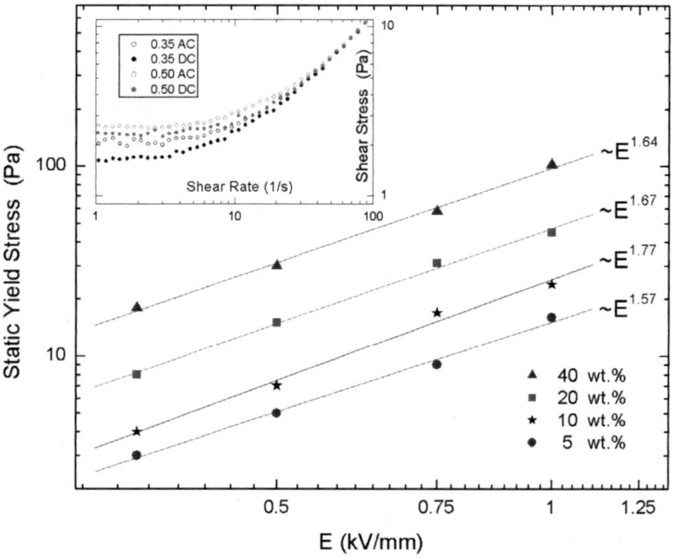

Figure 4. Log-log plot of the static stress for different clay concentrations as a function of E. Inset shows that the shear stress is higher when AC field is used due to minimizing of particle migration.

Acknowledgments

The authors acknowledge assistance from K.D.Knudsen, O.T.Buset and D.Chernyshov while performing experiments at the Swiss-Norwegian Beam Lines at ESRF. Special thanks to Y.Méheus for his interesting comments. This work was supported by the Research Council of Norway through the FRINAT Program: NFR project number 171300.

References

[1] Tao R 1991 *Physical Review Letters* **67**(3): 398-401.
[2] Espin M, et al. 2005 *Langmuir* **21**(11): 4896-4903.
[3] Fossum J, et al. 2006 *Europhysics Letters* **74**: 438-444.
[4] Ma H, et al. 1996 *Physical Review Letters* **77**: 2499-2502.
[5] Ma H, et al. 2003 *Advances in Physics* **52**: 343-383.
[6] Wen W, et al. 2003 *Nature Materials* **2**: 727-730.
[7] Huang X, et al. 2006 *Solid State Communications* **139**: 581-588.
[8] Meheust Y, et al. 2006 *Journal of applied crystallography* **39**(5): 661-670.
[9] Klingenberg D, et al. 1990 *Langmuir* **6**: 15-24.
[10] Gast A, et al. 1989 *Advances in Colloid and Interface Science* **30**: 153-202.
[11] Halsey T 1992 *Science* **258**: 761-766.
[12] Martin J, et al. 1998 *Phys. Rev. E* **57**(1): 756-775.
[13] Parmar K, et al. 2008 *Langmuir* **24**(5): 1814-1822.

The role of particle conductivity in electrorheology of suspensions of variously protonated polyaniline

M Stěnička[1], V Pavlínek[1,3], P Sáha[1], N V Blinova[2], J Stejskal[2] and O Quadrat[2]

[1] Polymer Centre, Faculty of Technology, Tomas Bata University in Zlin, TGM 275, 762 72 Zlin, Czech Republic
[2] Institute of Macromolecular Chemistry, Academy of Sciences of the Czech Republic, Heyrovsky Sq. 2, 162 06 Prague 6, Czech Republic

E-mail: pavlinek@ft.utb.cz

Abstract. Electrorheological behaviour of silicone oil suspensions of particles of polyaniline protonated to various doping level with orthophosphoric and tetrafluoroboric acids has been studied. The dynamic yield stress obtained by extrapolation of shear stress to zero shear rate using Herschel-Bulkley equation was used as a criterion of the ER efficiency. At a same molar concentration of doping acids, various protonation effects appeared and the dependences of the yield stress on the acid concentration differed. The comparison of the yield stresses with dielectric characteristics calculated from the Havriliak-Negami equation revealed that the particle conductivity unlike particle permittivity especially at higher protonation degree is dominant in polarization process. Consequently, particle conductivity or dielectric relaxation time proved to be the parameters providing the common dependences of the yield stress regardless of the way of polarization.

1. Introduction

Intrinsically conducting polymers [1–3] rank among particulate materials in electrorheology (ER) due to relatively low density, good thermal stability, and controllable electric and dielectric properties. Among them polyaniline (PANI) has been the object of many ER studies [4–7]. Its conductivity can be advantageously easily arranged by protonation to various doping levels to obtain particle suspension with a high ER performance. In present study the comparison of the ER efficiency of PANI particles protonated by two doping agents, orthophosphoric (PA) and tetrafluoroboric (TA) acids, in silicone oil suspensions has been performed.

2. Experimental

2.1. Controlled protonation of PANI with PA and TA

PANI powder was prepared by the oxidation of 0.2 M aniline hydrochloride with 0.25 M ammonium peroxydisulfate in water starting at 20 °C. The produced PANI salt was deprotonated into the PANI base by two-day immersion in an excess of 1 M ammonium hydroxide. Portions of the PANI base (0.5 g) were immersed in 100 mL of aqueous solutions of either PA or TA of various molar concentrations (C_A) for 24 h, separated, and dried. The conductivity of the samples pressed into pellets rises with the increasing C_A of the acid used (table 1). Both acids show the significant differences in achieved values of conductivity at the same C_A, obviously due to different pH of equally concentrated solutions.

Table 1. The conductivity, σ, of PANI base protonated in the aqueous solution of PA and TA of various molar concentrations, C_A.

Sample	C_A (mol L^{-1})	σ (PA) (S cm^{-1})	σ (TA) (S cm^{-1})
B	0	3.6×10^{-9}	3.6×10^{-9}
PA1/TA1	0.0001	4.7×10^{-9}	4.9×10^{-9}
PA2/TA2	0.0005	6.8×10^{-9}	4.9×10^{-9}
PA3/TA3	0.001	2.7×10^{-8}	9.0×10^{-9}
PA4/TA4	0.005	5.8×10^{-7}	1.6×10^{-8}
PA5/TA5	0.01	4.7×10^{-5}	1.2×10^{-5}
PA6/TA6	0.05	3.1×10^{-3}	2.5×10^{-4}

2.2. Suspension preparation

Suspensions (10 wt%) were prepared by mixing PANI powders with corresponding amounts of silicone oil (Lukosiol M200, Chemical Works Kolín, Czech Republic; $\eta_c = 200$ mPa s, $\sigma_c \approx 10^{-11}$ S cm^{-1}, $\varepsilon'_c = 2.6$). At first, the samples were stirred mechanically and then in an ultrasonic bath for 30 s before each measurement.

2.3. Electrorheological and dielectric measurements

Shear stress in the range of shear rates 1–100 s^{-1} and at the DC electric field strengths 0–3 kV mm^{-1} using a rotational rheometer (Bohlin Gemini, Malvern Instruments, UK) with parallel plates 40 mm in diameter and a gap of 0.5 mm was carried out at the temperature of 25 °C.

The frequency dependences of dielectric properties (ε', ε'') of 10 wt% suspensions were measured with a Hioki (3522 RCL HiTester, Japan) in the range 10^1–10^5 Hz.

3. Results and discussion

3.1. Electrorheological characterization

The typical ER behaviour of PANI suspensions at various electric field strengths is shown in figure 1. The field-off shear stress depends linearly on the shear rate and indicates a Newtonian flow. After electric field application, especially at low shear rates, the values of shear stress significantly rise by several orders of magnitude with increasing intensity of the electric field. The dynamic yield stresses (τ_0) indicate formation of the reinforced chain-like structures of organized polarized particles. A plateau region of the shear stresses appears at low shear rates due to the balance between electric and shear forces. At higher shear rates, when the hydrodynamic forces begin to dominate over the electrostatic ones, these organized particle structures are gradually destroyed and the suspension tends to return to field-off Newtonian behaviour.

3.2. Effect of PANI particle protonation

The findings suggest that the doping level of the same C_A of various acids on particle polarization may be different. Thus, at constant electric field strength τ_0 significantly depends not only on the protonation degree of PANI particles but also on the character of the acid used. At low protonation, τ_0 values achieve approximately similar constant values. At a higher protonation τ_0 of both suspensions of doped particles grew and became different. In case of PA at $C_A = 0.01$ mol L^{-1} at the highest electric field strength used, τ_0 nearly doubled in value. For TA acid, τ_0 was a little lower (figure 2). In samples doped to higher protonation at $E > 1.5$ kV mm^{-1} the current passing through the suspensions was too high, short-circuit between electrodes sets in and the measurement failed.

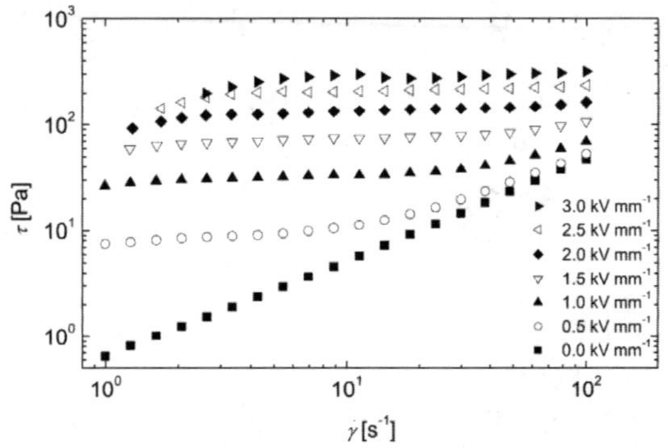

Figure 1. Double-logarithmic plot of the shear stress τ vs. shear rate γ for the suspension of the sample PA5 at various electric field strengths E.

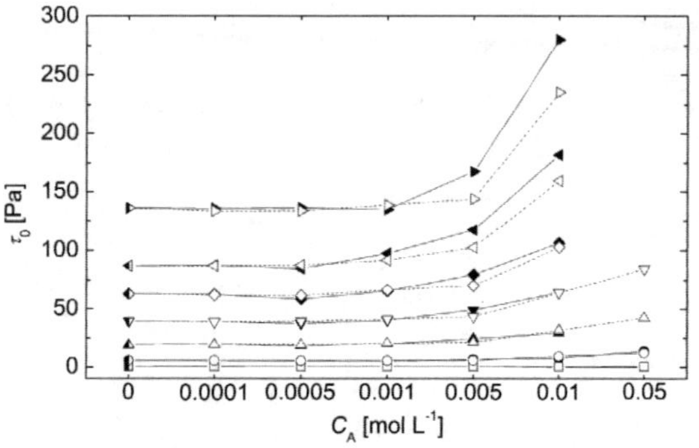

Figure 2. The dependence of τ_0 on C_A of used acids at various electric field strengths. PA (solid), TA (open), original PANI base (semi solid symbols). Electric-field strength E (kV mm^{-1}): (■□) 0, (●○) 0.5, (▲△) 1.0, (▼▽) 1.5, (◆◇) 2.0, (◀◁) 2.5, (▶▷) 3.0.

3.3. Particle polarizability and relaxation time

Dielectric characteristics of particle suspensions under this study (table 2) were calculated from the Havriliak-Negami empirical equation (1) [8]:

$$\varepsilon^* = \varepsilon'_\infty + \frac{(\varepsilon'_0 - \varepsilon'_\infty)}{(1+(i\,2\pi\,f\,t_{rel})^a)^b} \tag{1}$$

Here ε'_0 and ε'_∞ are the limit values of the relative permittivity at the frequencies below and above the relaxation frequencies, t_{rel} is a relaxation time, a is the scattering degree of t_{rel} and b is related to the asymmetry of the t_{rel} spectrum. Large values of a mean a great scattering of t_{rel}. When a differs much from zero and b significantly from unity, the t_{rel} spectrum becomes more asymmetrical. The dielectric spectra demonstrated an increase in limit permittivity ε'_0 and a steep decrease in t_{rel} with particle protonation.

Table 2. Parameters of Havriliak-Negami equation (1) for PA and TA doped PANI particles.

Sample	$\varepsilon'_{0,s}$ (PA / TA)	$\varepsilon'_{\infty,s}$ (PA / TA)	t_{rel} (s) (PA / TA)	a (PA / TA)	b (PA / TA)
B	4.62	2.66	1.48×10^{-2}	0.80	0.69
PA1/TA1	4.62 / 4.64	2.66 / 2.62	1.16×10^{-2} / 1.46×10^{-2}	0.86 / 0.84	0.60 / 0.65
PA2/TA2	4.82 / 4.58	2.67 / 2.64	9.88×10^{-3} / 1.27×10^{-2}	0.76 / 0.81	0.70 / 0.67
PA3/TA3	5.03 / 4.81	2.70 / 2.69	5.23×10^{-3} / 7.39×10^{-3}	0.70 / 0.73	0.70 / 0.74
PA4/TA4	5.10 / 4.80	2.65 / 2.61	1.27×10^{-3} / 4.47×10^{-3}	0.61 / 0.63	0.69 / 0.90
PA5/TA5	5.19 / 5.13	2.60 / 2.66	8.27×10^{-6} / 1.86×10^{-4}	0.37 / 0.54	0.97 / 0.78
PA6/TA6	5.47 / 5.32	2.64 / 2.72	3.33×10^{-7} / 3.53×10^{-6}	0.59 / 0.50	0.73 / 1.00

Because of the high conductivity of particles this quantity appeared to be more important than permittivity in the process of particle polarization and the ER efficiency could be rather controlled by the mismatch of conductivity of particles in the continuum [9]. The common dependence of τ_0 on the relative particle conductivity Γ as a ratio of conductivity of particles σ_p and that of continuum σ_c for both acids used confirmed this assumption (figure 3).

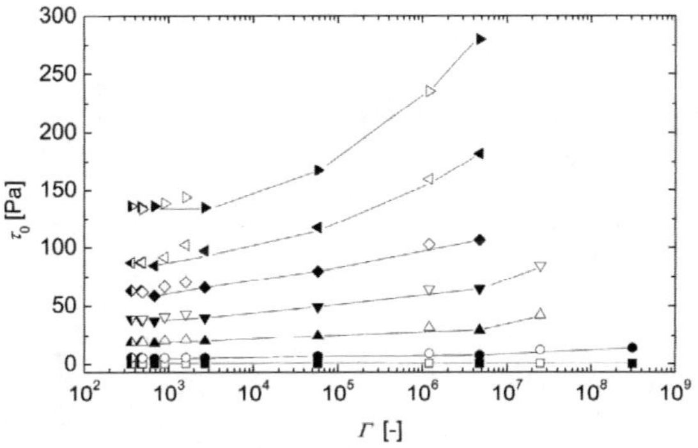

Figure 3. The dependence of τ_0 on $\Gamma (=\sigma_p / \sigma_c)$ at various electric field strengths. Points denoted as in figure 2.

For both studied suspensions, t_{rel} steeply decreased with the protonation degree of particles which indicates increasing particle mobility during restoration of the organized particle structure in the electric field. The common course of the correlations of t_{rel} with Γ regardless of the way of protonation indicated a close connection between conductivity and particle mobility in the electric field (tables 1 and 2).

Conclusions

The results of the study revealed that conductivity of PANI particles is a significant factor controlling their polarizability and thus the formation of ER structures in the electric field. The yield stress increases with conductivity of particles up to some critical value ($\approx 4.7 \times 10^{-5}$) where short-cut of the circuit prevents measurement of ER effect. The type of the acid used for PANI protonation seems to play minor role.

Acknowledgements

The acknowledgement to the Ministry of Education, Youth and Sports of the Czech Republic (MSM – 7088352101) and the Grant Agency of the Czech Republic (202/06/0419).

References

[1] Chandrasekhar P 1999 *Conducting Polymers, Fundaments and Applications* (Boston: *Kluwer Academic Publishers*)
[2] MacDiarmid A G 2002 *Synth. Met.* **125** 11–22
[3] Park S M and Lee H J 2005 *Bull. Korean Chem. Soc.* **26** 697–706
[4] Gow C J and Zukoski C F 1990 *J. Colloid Interface Sci.* **136** 175–188
[5] Hao T 2001 *Adv. Mater.* **13** 1847–56
[6] Quadrat O and Stejskal J 2006 *J. Ind. Eng. Chem.* **12** 352–61
[7] Hiamtup P, Sirivat A and Jamieson A M 2006 *J. Colloid Interface Sci.* **295** 270–8
[8] Havriliak S Jr and Havriliak SJ 1997 *Dielectric and mechanical relaxation in materials* (Munich: Hanser)
[9] Davis LC 1992 *J. Appl. Phys.* **72** 1334–40

Electrorheological fluids containing phosphorylated polystyrene-*co*-divinylbenzene

M Sukiennik, A Krztoń-Maziopa and J Płocharski

Faculty of Chemistry, Warsaw University of Technology
ul. Noakowskiego 3, 00-664 Warszawa, Poland

E-mail: janusz@ch.pw.edu.pl

Abstract. Electrorheological fluids based on ion-exchange resins were thoroughly described in literature. Most of the papers reported on application of sulphonated polystyrene-*co*-divinylbenzene as a dispersed solid phase. The aim of the presented work was to synthesise phosphorylated polystyrene-*co*-divinylbenzene (PPS/DVB) and compare properties of the obtained suspensions with a sulphonated polystyrene-*co*-divinylbenzene based fluid. PCl_3 was used as the phosphorylating agent. The prepared PPS/DVB resin powders were dispersed in a silicone oil and flow curves were recorded under DC electric field up to 3 kV/mm. It was found that yield stress values depended on the cation type in PPS/DVB: Na^+ containing resins exhibited the highest ER effect, and on the water content with the optimum concentration depending on the cation type.

1. Introduction

Electrorheological (ER) fluids comprising micron sized polarisable particles dispersed in non-polar liquids of low viscosity have been intensively investigated for a few decades. Among them a particular attention was paid to polyelectrolyte polymers, better known as ion-exchange resins [1]. Most of the performed work was, however, dedicated to electrorheological effect in suspensions of only one kind of these resins – sulphonated polystyrene-*co*-divinylbenzene (SPS/DVB). The ER effect observed in these fluids was related to humidity [2], solid phase content [3], grain size [4] as well as electric field strength and frequency [5]. One of obvious objectives of designing electrorheological fluids for practical applications is to maximize the magnitude of polarization forces within dispersed solid particles under external electric field what can be possibly achieved by modification of a kind and concentration of polar groups attached to a polymer backbone. The aim of the present work was to check whether replacing the sulphonic groups in a SPS/DVB resin by their phosphorus analogues would result in any improvement of ER properties of the related fluids.

2. Experimental

The polystyrene-*co*-divinylbenzene was synthesized using a suspension polymerisation method [6] and then polymer was phosphorylated with PCl_3 at the presence of $AlCl_3$ as a Lewis acid-type catalyst [7]. PCl_3 (188 ml) and PPS/DVB (30 g) were introduced into a three-necked flask equipped with a reflux condenser, a high speed homogeniser and a thermometer. The content was stirred at ambient temperature for 1h and $AlCl_3$ (15 g) was subsequently added. The mixture was intensively stirred at 75°C for 3 hours. Then the suspension was cooled to ambient temperature and stirred for further 24 hours. The obtained precipitate was filtered out and first washed with dioxane and then hydrolysed with water at 4°C. Acid protons in phosphinic groups were replaced by potassium, sodium or lithium cations using 10% aqueous solutions of appropriate hydroxides. The cation substituted PPS/DVB powders were filtered out and washed with distilled water to pH = 7. Then they were dried in an air drier at 80°C and/or in vacuum of approximately 10^{-5} mbar for 8 hours. For comparative reasons a sulphonated copolymer was also prepared according to well known method described in literature [8].

A few methods were applied to characterise the obtained resins. FT-IR spectroscopy was used to identify functional groups in PPS/DVB. SEM/EDS (Scanning Electron Microscopy/Energy Dispersive Spectroscopy) was used to reveal the morphology of the prepared powders and to determine concentrations of phosphorus, sulphur, carbon and oxygen in the samples. We have also examined the ionic capacity of the obtained resins with a classic titration method. Water content was determined by the Karl Fisher method directly before each ER measurement.

Aliquots of powders were dispersed in a dry silicone oil (polydimethylsiloxane) of 0.1 Pa·s viscosity. ER fluids containing 30 and 40 wt.% of solid phase were prepared. Flow curves of ER fluids were recorded using a rotational viscometer Bohlin Visco 88 equipped with a concentric cylinder geometry of 1 mm gap. The measurements were recorded under DC electric field up to 3 kV/mm. Currents drawn by the ER fluids were measured simultaneously. All measurements were taken at ambient temperature of about 23°C.

3. Results and Discussion

3.1. Characterisation of materials

Figure 1 shows the FT-IR spectrum of the raw PS/DVB and the copolymer phosphorylated with PCl_3. The band at 2326-2275 cm^{-1} is due to –P–H vibration. The absorption bands at 1402 and 1344 cm^{-1} are typical of –P–C and –P=O vibrations, respectively. The ester bands are visible at 1150-970 cm^{-1}. The wide band at 3680-3300 cm^{-1} was assigned to –OH groups. These data show that a considerable amount of phosphinic groups were attached to the polystyrene backbone.

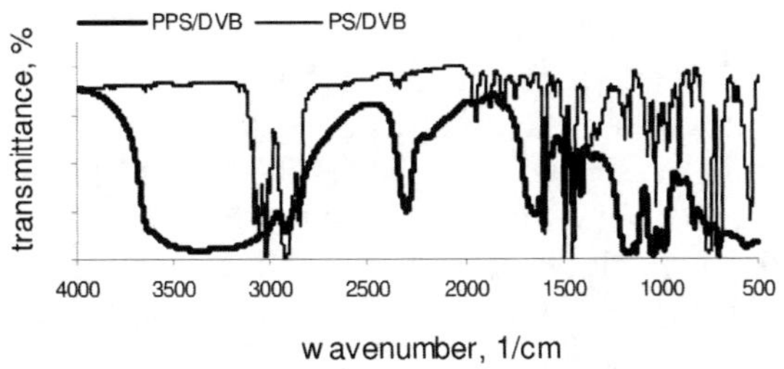

Figure 1. FT-IR spectra of raw polystyrene-*co*-divinylbenzene and the copolymer after phosphorylation.

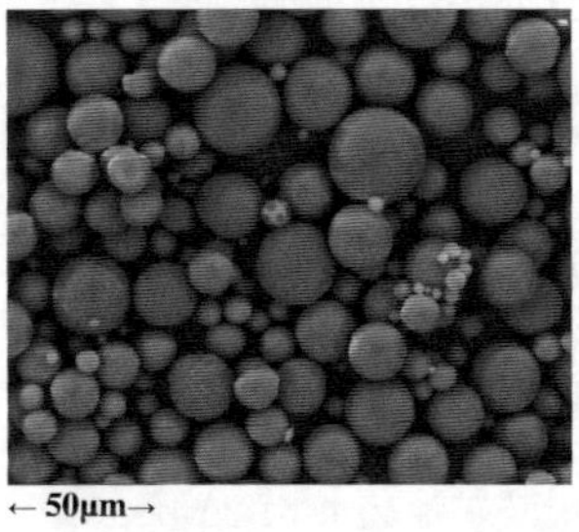

Figure 2. SEM image of PPS/DVB-Na particles.

We attempted also to determine the concentration of the polar groups in the ion-exchange resin. The EDS analysis of phosphorus showed that approximately 0.4 phosphinic group is attached to each aromatic ring. A different result was obtained from a classic titration giving only 0.19 phosphinic per one ring. This discrepancy can be easily understood after assumption that the phosphorylation process of the polymer took place mostly in a surface layer of each grain and not within a whole volume.

The EDS analysis of oxygen showed that the O/P ratio in surface layers was 2.44 suggesting that not only phosphinic but also phosphonic groups were present in the studied material. Since the phosphorylation reaction was performed in a presence of air it was possible that a simultaneous oxidation reaction took place. It seems that the concentration of the two phosphorus containing functional groups is almost equal.

A morphology of PPS/DVB-Na particles was investigated by SEM. Figure 2 shows that the obtained grains were regular and spherical in shape. Diameters of most of the particles were between 10 and 20 μm. Some small spheres – significantly below 10 μm – could be also found.

3.2. Electrorheological measurements
A set of flow curves of 40 wt.% suspensions of PPS/DVB substituted with Na^+ is presented in figure 3. The observed ER effect is strong since the yield stress amounts to about 2.6 kPa at 3 kV/mm with a reasonably low current not exceeding 6 μA/cm².

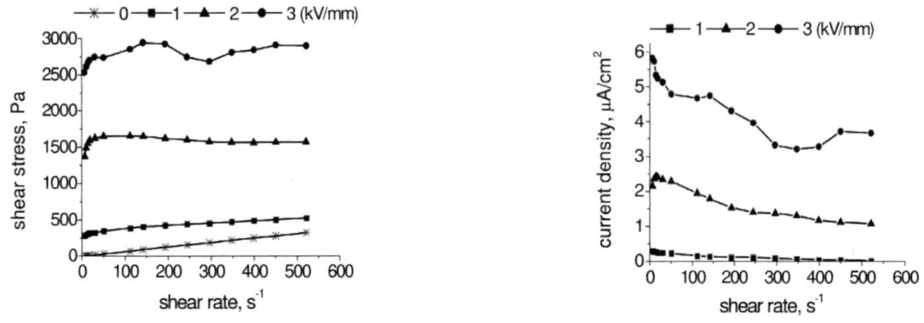

Figure 3. Flow curves and current densities of 40 wt.% PPS/DVB-Na. Water content is 1.6%.

In the case of suspensions containing sulphonated polystyrene we observed a dependence of ER properties on the cation type. We performed similar experiments for fluids based on PPS/DVB. The fluids taken for comparison had the same content of water (1.6%). High yield stresses were recorded for potassium containing fluids (2070 Pa at 3 kV/mm for 30 wt.%), but in the case of 40 wt.% suspension accompanying currents were unacceptably high and exceeding in some cases the capacity of the high voltage supply. The yield stresses of sodium containing fluids were only slightly lower (2600 Pa for 40 wt.% at 3 kV/mm) with a very acceptable level of current consumption (6 μA/cm²). In the case of lithium based fluids both yield stresses and the currents were much lower (600 Pa and 5 μA/cm² at 3 kV/mm). We believe that the sodium based fluids show optimum properties.

One of typical features of SPS/DVB based ER fluids is that their properties depend on water content [9]. Apparently this was also the case of the studied PPS/DVB based fluids (figure 4). With the exception of PPS/DVB-Li the yield stress increased with water concentration, reached the maximum value and then decreased. Possibly the PPS/DVB-Li fluids behave also in a similar way, but the eventual maximum appeared at higher concentrations of water which were not studied. The optimal values of humidity were different for K^+ and Na^+ and amounted to 1.75%, 1.6% respectively.

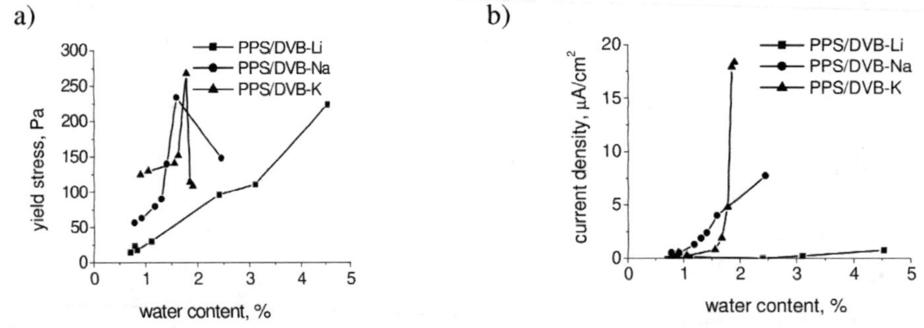

Figure 4. Dependence of yield stresses a) and current densities b) of 30 wt.% PPS/DVB suspensions on water content at 1 kV/mm. Data given for K^+, Na^+ and Li^+ substituted resins.

The results obtained for the PPS/DVB based fluids were compared with ER activity of their sulphonated analogues. The results depicted in figure 5 clearly show that the yield stress of the phosphorus fluids was almost two times higher with a significantly lower current consumption proving its higher ER activity. It is worth to note that the number of sulphur atoms per one aromatic ring (according to EDS) in the SPS/DVB resin was about 0.81 while the number of phosphorus atoms per one aromatic ring in the PPS/DVB resin was only 0.4. It means that phosphinic/phosphonic groups are considerably more effective in evoking ER activity than the sulphonic ones used to date. If we manage to obtain PPS/DVB with increased number of functional groups it will be probably possible to get ER suspensions of even higher activity.

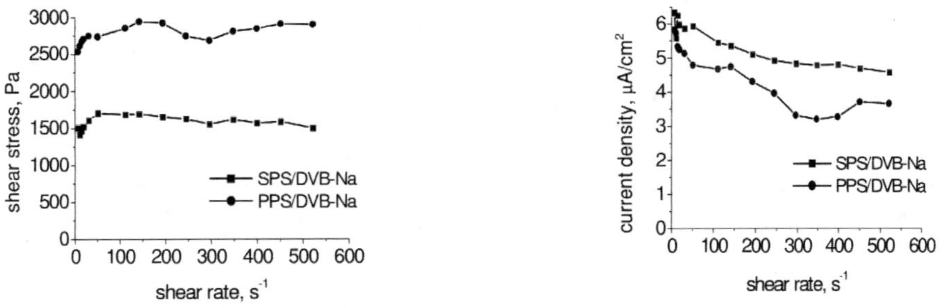

Figure 5. Flow curves and current densities of 40 wt.% PPS/DVB-Na and SPS/DVB-Na suspensions at 3 kV/mm. Water content 1.6%.

4. Conclusions

The presented findings suggest that PCl_3 is a suitable phosphorylating agent. The presence of polar groups in the product obtained by phosphorylation with phosphorus chloride(III) was confirmed by EDS and FTIR methods. The phosphorylation process takes place mostly at the surface of grains because of the high cross-linking level of the copolymer and limited diffusion of the phosphorylating agent into a bulk of a grain.

It seems most likely that suspensions of PPS/DVB can exhibit much higher ER effect provided the phosphorylation reaction efficiency is improved. We had also showed that suspension of PPS/DVB needs lower current densities in comparison to the sulphur analogue.

Summing up – suspensions of PPS/DVB exemplify new, promising materials characterized by high electrorheological activity. Prepared materials are described by following parameters: the yield stress about 2.6 kPa at 3 kV/mm, viscosity below 0.3 Pa·s (at 23°C and 0 kV/mm) and the current consumption below 10 µA/cm^2 (at 3 kV/mm). The investigated fluid can be used as a new material in ER technology.

Acknowledgments

This work was financed by Faculty of Chemistry of Warsaw University of Technology.

References

[1] Kawakami T, Aizawa R, Konishi M, Asako Y 1999 *Int. J. Mod. Phys* B **13** 1721
[2] Aizawa R, Vieira S, Nakano M, Asako Y 2001 *Int. J. Mod. Phys.* B **15** 1070
[3] Klingenberg D J, Parthasarathy M 1996 *Mat. Sci. Eng.* R **17** 57
[4] See H, Kawai A, Ikazaki F 2002 *Rheol. Acta* **41** 55
[5] See H, Kawai A, Ikazaki F 2002 *Colloid Polym. Sci.* **280** 24
[6] Florjańczyk Z 2002 *Chemia polimerów. T. 2* Oficyna Wydawnicza PW, Warszawa
[7] Trochimczuk A W 2001 *Reactive & Functional Polymers* **48** 141
[8] Aizawa R, Vieira S, Nakano M, Asako Y 2001 *Int. J. Mod. Phys.* B **15** 1070
[9] Ikazaki F, Kawai A, Kawakami T, Asako Y 1999 *Int. J. Mod. Phys* B **13** 1845

Journal of Physics: Conference Series **149** (2009) 012029 doi:10.1088/1742-6596/149/1/012029

A Study on Fixing Force Generation Mechanism of ER Gel

H Tanaka[1], Y Kakinuma[1], T Aoyama[1] and H Anzai[2]

[1]School of Integrated Design Engineering, Keio University,
3-14-1 Hiyoshi, Kohoku-ku, Yokohama 223-8522, Japan
[2]Fujikura kasei Co., Ltd., 2-6-15 Shibakouen, Minato-ku, Tokyo, Japan

h-tanaka@ina.sd.keio.ac.jp

Abstract. Electro-rheological Gel (ERG) is a new functional elastomer which changes its surface frictional and adhesive property according to the intensity of applied electrical field. This unique property is called ERG effect. The upper sliding electrode placed on the surface of ERG is fixed by the adhesive effect of ERG under electrical field. Variable fixing forces due to adhesion are generated by this effect. However, relationship between physical factors and generated fixing force has not yet been clarified. In this study, physical mechanism of fixing phenomenon is elucidated experimentally from the view point of frictional force and adhesive force. From the results, empirical equation of generated fixing force is originally derived to establish the theory of ERG effect.

1. Introduction

Electro-rheological Fluids (ERFs) are functional fluids which change their own viscoelastic properties swiftly and reversibly according to the applied electrical field intensity [1]. This property is called as the ER effect. Various devices and equipments have been developed as dampers, clutches using ERFs in a mechanical system [2]. The dispersion type ERFs are composed of ER particles and silicon oil. However, there are two defects of particle type ERFs. Firstly, separation and sedimentation of ER particles over long time reduces ER effect and stability in mechanical system using ERFs. Secondly, necessity of a mechanical seal to prevent leaking ERFs makes system complicated.

The Electro-rheological Gel (ERG) has been developed to conquer these defects of ERFs. ERG is a functional material, which changes its surface frictional and adhesive property according to the intensity of applied electrical field [3]. In this study, this unique property is named as the ERG effect to distinguish from ER effect. The estimated mechanism of ERG effect is shown in figure 1 [4]. ERG is composed of ER particles and silicone gel. ER particles polarize when electrical field is applied. The polarized ER particles attract each other and ER particles existing at the surface of ERG sink into the gel. Therefore, upper sliding electrode placed on the ERG sheet, which is pasted on the lower fixed electrode as shown in figure 2, stick to the gel under electrical field. By the adhesion between surfaces of ERG sheet and electrode, fixing force is generated when the upper electrode is forced to slide. Figure 3 shows the appearance of adhesion between the upper glass electrode and the surface of ERG sheet under various electrical fields. From this phenomenon, it is almost elucidated that the fixing force due to ERG effect is caused by adhesion of gel. However, the physical factors which are contributed to generation of fixing force have not been analyzed enough. In addition, fixing force in normal direction has not yet been evaluated.

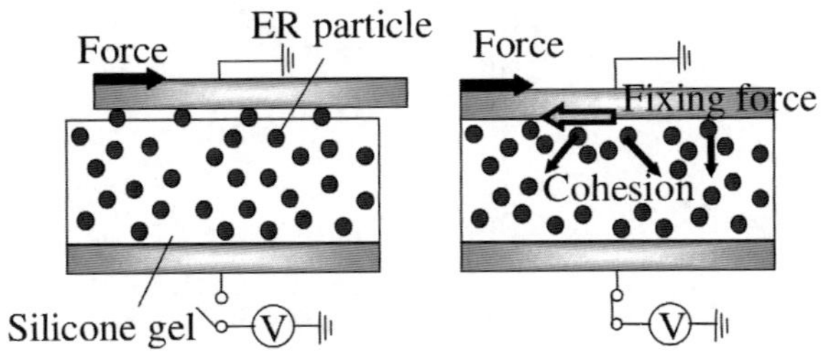

(a) Without electrical field (b) With electrical field

Figure 1. Mechanism of ERG effect

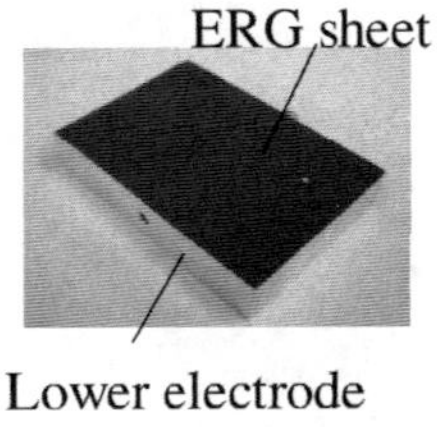

Figure 2. ERG sheet

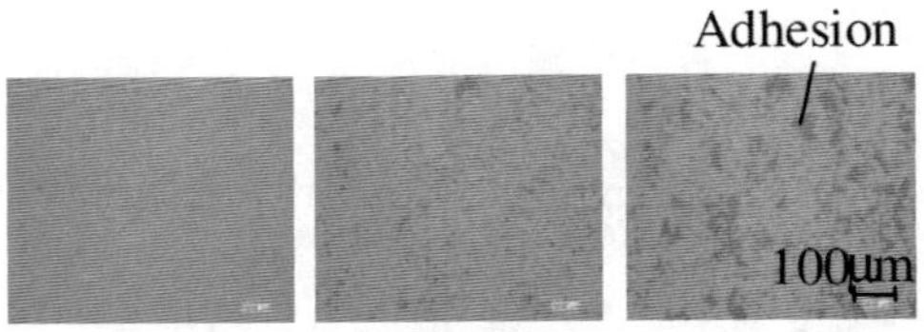

(a) E= 0.0 kV/mm (b) E= 1.0 kV/mm (c) E= 2.0 kV/mm

Figure 3. Adhesion of ERG

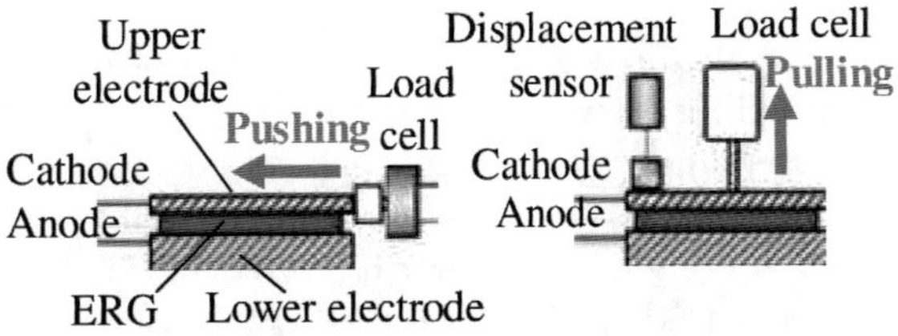

(a) Sliding direction (b) Normal direction

Figure 4. Experimental apparatus

In this study, the physical factors of sliding fixing force are experimentally analyzed from the view point of frictional and adhesive property. Based on the results of the analysis, empirical equation of generated fixing force is derived.

2. Experimental apparatus

The influence of the contact pressure and contact area between ERG sheet and upper electrode is investigated to clarify the effect of frictional force and adhesive force respectively.

The schematics of the experimental setup are shown in figure 4. ERG sheet L70×W50×T0.5mm (figure 2) is prepared. ERG sheet is sandwiched between lower fixed electrode and sliding upper electrode. The lower and upper electrode is served as the cathode and anode respectively. Figure 4 (a) shows, an upper electrode is slide aside in sliding direction by a micro screw with motor-drive equipment under various electrical fields. In contrast, a thin metal plate is lifted up in normal direction by z-axis stage via load cell as shown in figure 4 (b). Generated fixing force in both sliding and normal direction is measured by a strain gauge load cell attached to the apparatus.

3. Results

3.1. Fixing force in sliding direction

To investigate how the contact pressure influences on the ERG effect, sliding fixing forces are measured applying the various contact pressures (100, 400, 700Pa) to the center of a sliding upper electrode. The relation between the contact pressure and sliding fixing force is shown in figure 5. Sliding fixing force increases in proportion to the applied electrical field. The initial sliding fixing force increases due to increase of the applied contact pressure. However, the incremental ratio of sliding fixing force is almost constant regardless of contact pressure. Therefore, the generated sliding fixing force owing to ERG effect hardly depends on the contact pressure.

The influence of contact area between ERG and an upper electrode on the sliding fixing force under constant contact pressure is also investigated. The three samples of ERG, which have different surface area of 875mm^2 (L35×W25mm), 1750mm^2 (L70×W25mm) and 3500mm^2 (L70×W50mm), are prepared. As shown in figure 6, generated sliding fixing force increases according to increment of not

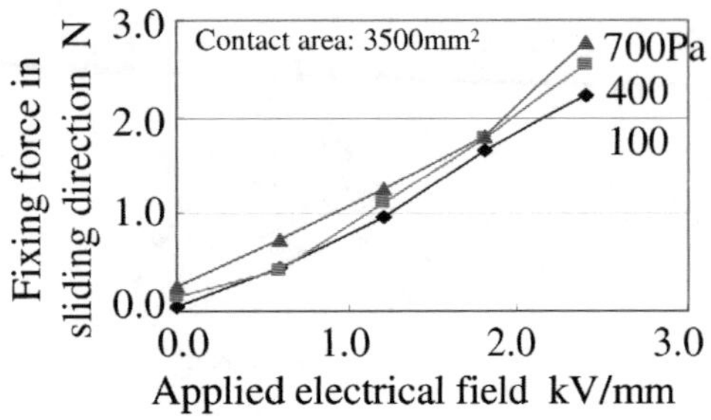

Figure 5. Influence of contact pressure on sliding fixing force

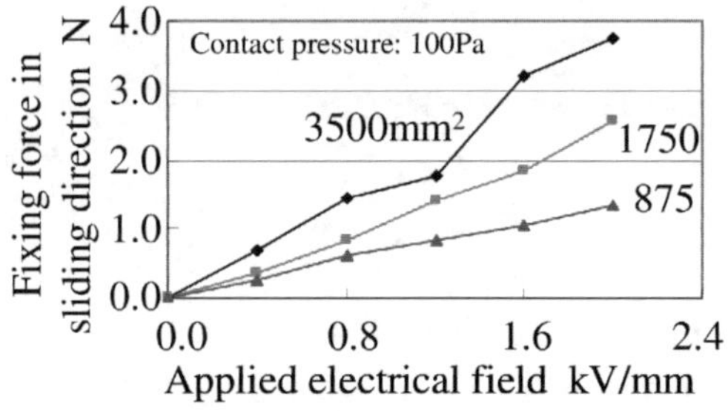

Figure 6. Influence of contact area on sliding fixing force

only the applied electrical field, but also contact area. Fixing forces per unit area in every contact area show almost steady value under constant electrical field intensity. These results indicated the sliding fixing force due to ERG effect is generated by not frictional but the adhesive effect because adhesion depends on the contact area.

3.2. Fixing force in normal direction
It is expected that the ERG generates various fixing force in normal direction since the ERG effect depends on adhesive effect between the surface of ERG and a sliding upper electrode. When the upper electrode is lifted up, the normal fixing force of ERG is measured under various electrical fields. Behaviors of generated normal fixing force are shown in figure 7. Higher fixing force was obtained under higher voltage. Up to the maximum point, the normal fixing force changed linearly with displacement since the upper electrode, which stuck to the ERG through adhesion, is subjected to an elastic force induced by the deformation of ERG. Beyond the maximum point, normal fixing force decreases gradually. This phenomenon is caused by removing on of the upper electrode from the ERG sheet. Fixing force still exists after the upper electrode separates fully from the ERG sheet. This result indicates the existence of the electrostatic force between the upper electrode and the ERG surface. To investigate the distribution of electrical field between the upper electrode and ERG surface, the numerical analysis of electrical field is carried out using the software of 'ANSYS 11.0'. Figure 8 represents the distribution of electrical field between electrodes with tiny air gap. Electrical field intensity of air gap is achieved to 15.0kV/mm theoretically while electrical field intensity of ERG is 2.0kV/mm. This analysis is conducted under the condition that dielectric breakdown does not occur. From this result, tiny air gap makes the electrical field of the air gap increase remarkably when the upper electrode apart from ERG surface completely. Consequently, this soaring electrical field generates electrostatic force.

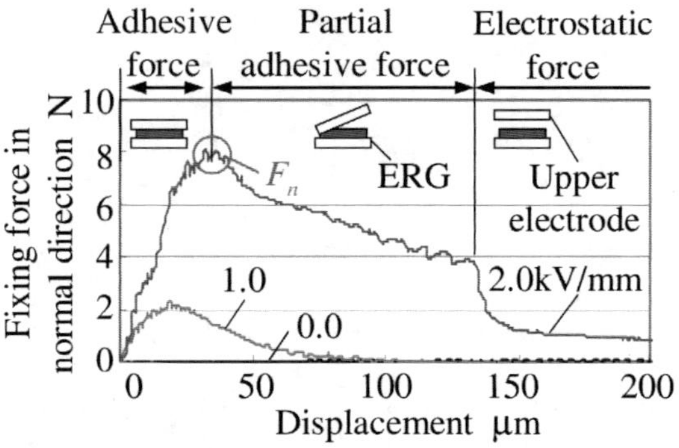

Figure 7. Behaviour of normal fixing force

3.3. Influence of surface roughness of the upper sliding electrode
The ERG effect is attributed strongly to the change of adhesive property of ERG surface. Therefore, it is possible that surface roughness of sliding electrode became potentially influential factor on ERG effect since adhesive strength depends on the contact condition between the ERG sheet and a sliding upper electrode. The correlation between surface roughness and fixing force is investigated. For this experiment, the three samples of sliding upper electrode, which have different surface roughness Ra of 200nm, 130nm, and 50nm, called as *Standard*, *Fine* and *Very Fine* respectively in this paper, are prepared. The relation between the surface roughness and fixing force in sliding direction is shown in figure 9. Generated fixing force increases with improvement of the surface roughness of sliding electrode in

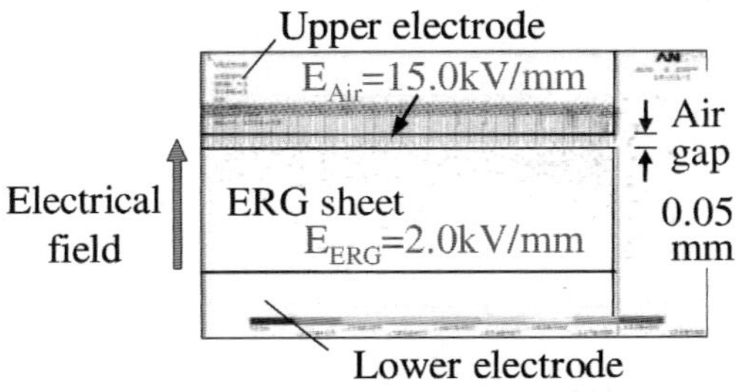

Figure 8. Results of analysis

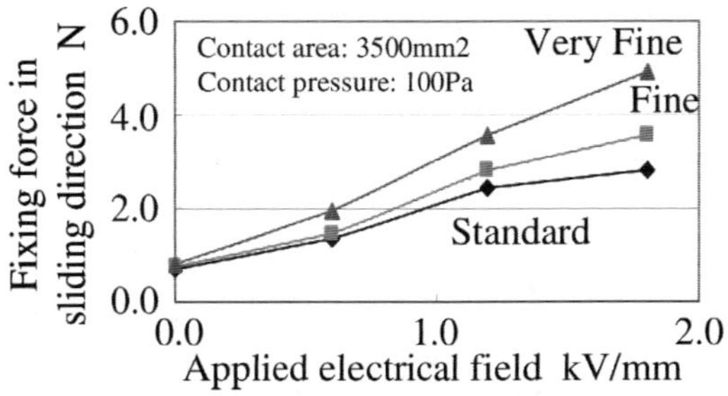

Figure 9. Influence of surface roughness of upper electrode on sliding fixing force

both directions. This result indicated that higher adhesive force due to ERG effect is obtained by making the surface of sliding electrode finer.

3.4. Theoretical model derived from the experiment analysis

In the past study, it is confirmed that gel contact area is proportional to electrical field intensity [4]. From this relation and the experimental result in this study, the empirical equation of maximum sliding fixing force F_s [N] and maximum normal fixing force F_n [N] just before sliding and detaching between ERG and upper electrode is derived as follow.

$$F_s = \mu_0 N + C\alpha SE \tag{1}$$

$$F_n = \frac{1}{2}\varepsilon SE^2 + C\alpha SE \tag{2}$$

where μ_0 is the frictional coefficient of ERG, N [N] is the normal force for applying to ERG surface, C [N/mm^2] is the adhesive force of gel per unit area, α is the increasing rate of gel contact area per unit electrical field, S [mm^2] is the contact area between the ERG surface and upper electrode, E [kV/mm] is the applied electrical field of ERG, ε [F/m] is the permittivity of ERG. α also depends on surface roughness of the upper electrode, whose magnitude is in order of 10^{-3}~10^{-4}. Maximum sliding fixing force F_s is represented as resultant force of initial friction force $\mu_0 N$ and adhesion generated by ERG effect. On the contrary, maximum normal fixing force F_n is as resultant force of electro static force $\varepsilon SE^2/2$ and generated adhesion.

4. Conclusion

In this study, relation between generated fixing force according to the ERG effect and physical factors is experimentally analyzed from the view point of frictional force and adhesive force. It clears that generated fixing force depends greatly on adhesive force due to ERG effect, although frictional force and electrostatic force make little impact on it. Based on the experimental results, the empirical equation of generated fixing force is obtained in both sliding and normal direction. In addition, impact on fixing force according to the surface roughness of a sliding upper electrode is also investigated. Finer surface roughness of a sliding upper electrode gives well fixing force.

References

[1] Winslow W M 1949 Induced Vibration of Suspension *Journal of Applied Physics* **20** 1137-1140
[2] Weinberg B, et al 2007 Design, Control and Human Testing of an Active Knee Rehabilitation Orthotic Device *Proc IEEE Int Conf Rob Autom* 2007 **9** 4126-4133
[3] Kakinuma Y, et al 2005 Basic Properties of Gel-structured Electro-rhological Fluids *International Journal of Modern Physics B* **19** 1339-1345
[4] Kakinuma Y et al 2006 Application of ER gel with Variable Friction Surface to the Clamp System of Aerostatic Slider *Precision Engineering* **30** 280-287

Electrorheology Improves Engine Efficiency

R Tao, K Huang, H Tang and D Bell

Department of Physics, Temple University, Philadelphia, PA 19122, USA

Email: rtao@temple.edu

Abstract. Improving engine efficiency and reducing pollutant emissions are extremely important. Here, we present our fuel injection technology based on physics principle that application of electrorheology can reduce the viscosity of petroleum fuels. A small device is introduced just before the fuel injection, producing a strong electric field to reduce the fuel viscosity, resulting in much smaller fuel droplets in atomization. As combustion starts at the droplet surface, smaller fuel droplets lead to cleaner and more efficient combustion. Both lab and road tests confirm that this technology improves fuel mileage significantly.

1. Introduction

Fuel atomization plays an important role in combustion efficiency and pollutant reduction. Since combustion starts at the interface between fuel and air and most harmful emissions are coming from incomplete burning, reducing the size of fuel droplets would increase the surface area to start burning, leading to a cleaner and more efficient engine.

This concept has been widely accepted as the discussions about future engine for efficient and clean combustion are focused on ultra-dilute mixtures at extremely high pressure to produce much finer mist of fuel for combustion [1-3].

Here we present our technology for efficient combustion, based on new physics principle that application of electrorheology can reduce the viscosity of petroleum fuels. A small device is thus introduced, producing a strong electric field to reduce the viscosity of petroleum fuels just before the fuel atomization. This viscosity reduction leads to cleaner and more efficient combustion. Both lab tests and road tests confirm our theory and indicate that such a device improves fuel mileage significantly. The technology is expected to have broad applications.

2. Viscosity Reduction, Atomization, and Test Results.

The principle of our device is sketched in Fig.1a. The fuel flows through two metallic meshes before it reaches the fuel injector. A voltage is applied on the two meshes to produce an electric field of around 1.0 kV/mm between the two meshes. The device consumes very low electric power, lower than 0.1 watt. In our setup, the field direction is opposite to the flow direction, which may help to provide negative charges to the fuel droplets. However, the main function of our device is to reduce the viscosity of the fuel as it passes the electric field [4,5].

Reducing the fuel viscosity improves the fuel injection. As shown in Fig.1b, the injected fuel has a pressure higher than that in the combustion chamber. The droplets are thus split, becoming smaller and smaller after they are emitted from the nozzle. If the droplets are allowed to reach the equilibrium, their radius is given by $a = 2\gamma / \Delta P$, where γ is the fuel's surface tension and $\Delta P = P_i - P_o$ is the pressure difference between the fuel's inside pressure P_i and the pressure outside the fuel P_o. However, in reality, the fuel droplets can never reach equilibrium because the viscosity acts against any deformation of the droplets. Therefore, reducing viscosity of the fuel greatly improves the fuel atomization.[6]

Proper application of eletrorheology can reduce the viscosity of liquid suspensions. According to the Krieger-Dougherty formula,[6,7] the effective viscosity of a liquid suspension η is related to the viscosity of base liquid η_0 by $\eta = \eta_0(1-\phi/\phi_m)^{-[\eta]\phi_m}$, where ϕ is the volume fraction of suspended particles, ϕ_m is the maximum volume fraction to pack particles randomly, and $[\eta]$ is the intrinsic viscosity, related to the particle shape. For example, $[\eta]=2.5$ for spherical particles.

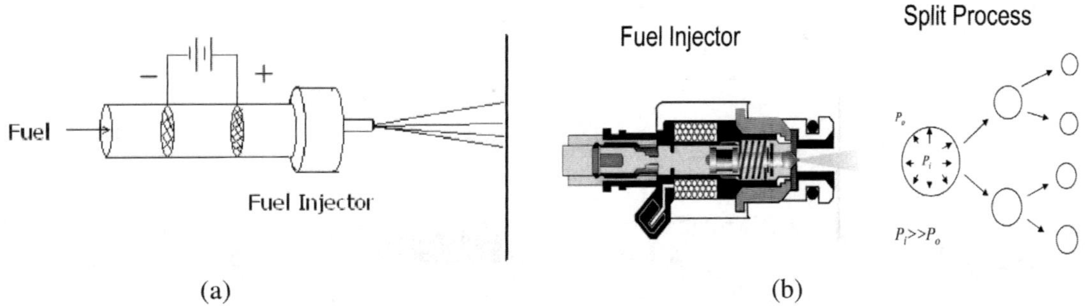

Figure 1. (a) In the device, the fuel flows through two metallic meshes before it reaches the fuel injector. A voltage is applied to produce an electric field of about 1kV/mm between the two meshes. (b) The emitted droplets split to become smaller and smaller.

Utilizing the mismatch in dielectric constant or magnetic permeability between the suspended particles and the base liquid, we can apply an electric field to aggregate the small particles into large ones. Normally, we aggregate nanoscale or sub-micrometer particles into micrometer particles. While this change in rheology does not alter ϕ, it makes ϕ_m increased as a result of increase of polydispersity and average particle size.[4,5] Hence the effective viscosity η is reduced.

Here we extend the above physics principle to refinery fuels. In fact, refinery fuels, such as diesel fuel and gasoline, are made of many different molecules. They can be regarded as liquid suspensions if we take the large molecules as suspended particles and the base liquid is made of small molecules. Under a strong electric field, the induced dipolar interaction makes the large molecules aggregate into small clusters. Similarly, this change reduces the effective viscosity of refinery fuels. The above theory was verified by our experiment. As shown in Fig. 2, after application of an electric field of 1kV/mm for about 2 seconds, the diesel oil's viscosity is reduced by about 9%. While this reduction is not permanent, it provides the opportunity to improve fuel atomization.

In our spray experiment, we used an Accel high impedance fuel injector to simulate fuel injection at engine chambers. When the device was on, the fuel took about 5 seconds to pass the electric field. The spray lasted for 4 milliseconds. The droplets were collected and analyzed for both cases, with the field and without the field. The statistical results for diesel fuel are in Fig.3. All of them are averaged over 50 tests. The repeatability was quite good with an error less than 5%. In both experiments with diesel fuel and gasoline, the current was less than 10μA, i.e. the electric power consumption is below 0.1W. For diesel fuel, the fuel pressure was 13.79 bar (200 lb/in^2) and the electric field was about 1.0kV/mm in the experiment. The electric field increased the number of droplets with diameter less than 40 μm dramatically. The number of droplets of diameter below 5 μm was increased from 5.3% to 15.3% when the device was on. The effect on diesel fuel is very significant. In the experiment with gasoline (with 20% ethanol), the fuel pressure was 7.59 bar (100 lb/in^2) and the electric field was 1.2kV/mm. The effect on gasoline is also significant. The number of droplets with diameter around 10 μm was increased from 17.6% to 20.7%, an increase of 18% when the device was on.

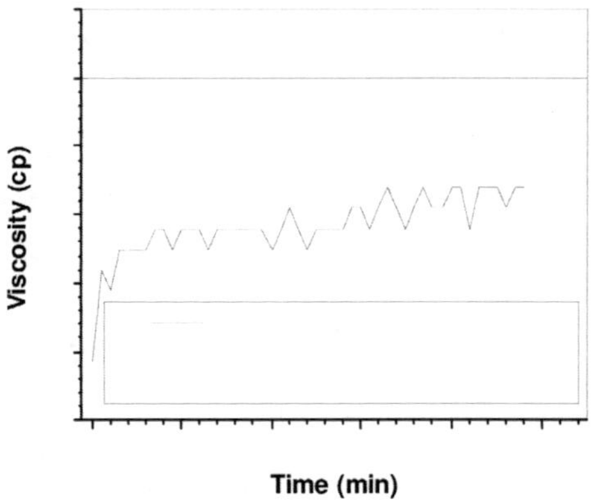

Figure 2. The diesel viscosity is reduced by 9% after application of an electric field of 1kV/mm for 2 seconds: from 4.6cp down to 4.18cp. Afterwards, the viscosity is rising to return to the original value.

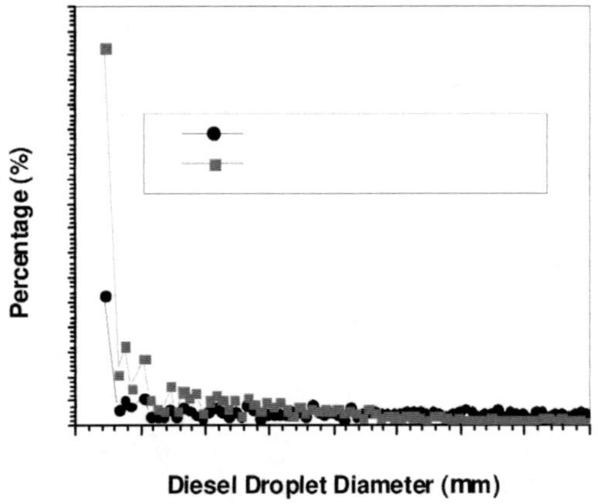

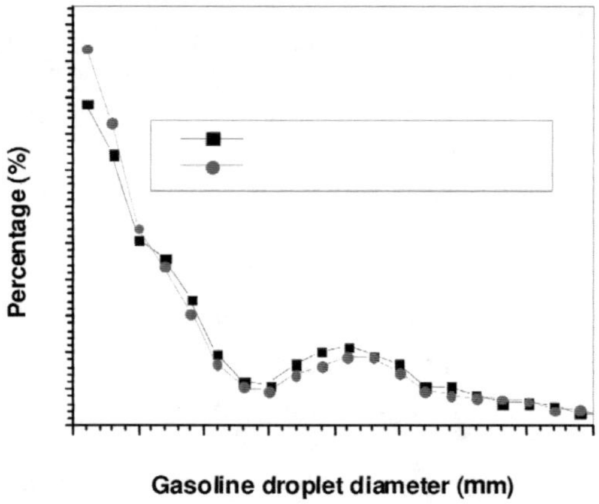

Figure 3. The size distribution of petreluem fuel in atomization with or without an applied electric field: (a) for diesel fuel and (b) for gasoline (with 20% ethanol).

Because the spray experiment suggests that diesel engines would significantly benefit from our device, we conducted extensive tests with our device on diesel engines. Our recent tests have been done on a 1993 Mercedes-Benz 300D, diesel sedan. The device on the vehicle has two mesh electrodes separated by 1cm for the diesel fuel to pass in 5 seconds. The field is about 1 kV/mm.

A continuous road tests of the Mecedes-Benz 300D for six months showed that our device increased the fuel mileage significantly. On the highway, the device increased the fuel mileage from 32 miles per gallon (mpg) to 38 mpg. In city driving, the improvement of fuel mileage was not as good as that on the highway, but was averaged at 12-15%.

Since our road tests reported significant fuel mileage improvement, we brought the vehicle to a Dynamometer for lab tests, which confirmed the road tests. The typical result is shown in Fig. 4. At a fixed fuel consumption rate close to 500g/h, the Dynamometer measured the engine output. When the device was off, the average power output was 0.3677hp. It increased to 0.4428hp after the device was turned on. This indicates that the power output was improved by about 20.4% at the same fuel consumption rate. The test was repeated for three hours and had an error within 5%.

We also did tests at high fuel consumption (power output about 40-50 hp) and at acceleration, where the improvement was even better than that at low fuel consumption. In our opinion, the low fuel consumption is more representative since the fuel mixes with air better and the different results should be the representative indication between combustion with large fuel droplets and the combustion with small droplets.

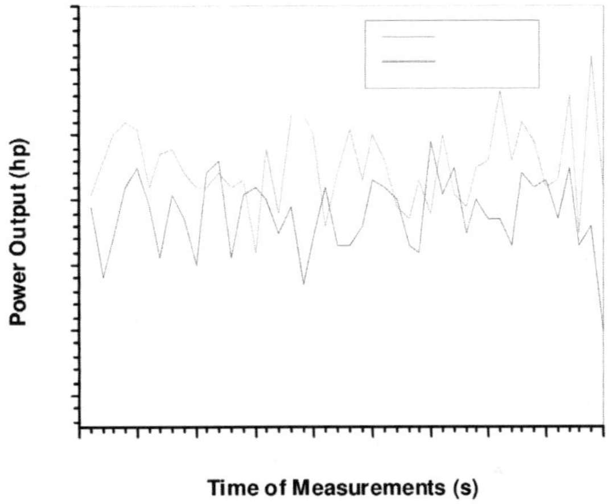

Figure 4. The lab test of Mercede-Benz 300D with a dynamometer. The average power output was originally about 0.368hp and increased to 0.443hp after the device was turned on.

3. Discussions

Our tests with the Mercedes-Benz indicate that there are optimal values for two parameters, the applied electric field strength and the time duration for the diesel fuel to pass through the electric field. The device with the Mercedes-Benz has 1kV/mm field strength and 5 seconds time duration. It is possible to have some improvement by adjusting these two parameters.

There are a couple of other alternatives to increase fuel efficiency, such as adding additives to the fuel to reduce the viscosity. Unfortunately these fuel additives are quite expensive. The extremely high fuel pressure method, as mentioned before, is still under development and not applicable to current engines. Since our technology, developed on new physics principle, consumes very small power and improves fuel efficiency significantly, we expect it will have wide applications on all types of internal combustion engines, present ones and future ones. By adjusting the values for the electric field and time duration, we could make this technology work effectively for other fuels, such as bio-diesel, kerosene, and gasoline.

Acknowledgements: This work was supported in part by RAND and STWA.

References

[1] Y. Aoyagi et al 2006, "Advanced diesel combustion using of wide range, high boosted and cooled EGR system by single cylinder engine," SAE Technical Paper, NO 2006-01-0077.
[2] M. Hiroshi et al 2004, "The potential of lean boost combustion," 2004 FISTA World Automotive Congress, May 23-27, Barcelona, Spain.
[3] J. A. Pirault & S. J. Klippenstein, *J. Phys. Chem. A*, 110, 10528-10544 (2006).
[4] R. Tao & X. Xu, *Energy & Fuels,* **20,** 2046-2051 (2006).
[5] R. Tao, Intern. J. of Modern Physics B, V. 21, N28&29, pp4767-4773 (2007).
[6] A. H. Lefebvre, *Atomization and Sprays*, (Taylor & Francis, 1989) 27-73.
[7] W. B. Russel, D. A. Saville, and W. R. Schowalter, *Colloidal Dispersion,* (Cambridge Univ. Press, Cambridge, 1991), 456-503.
[8] I. M. Krieger & T. J. Dougherty, *Tans. Soc. Rheol.* **3**, 137-152 (1959).

Measuring the Response Time and Static Rheological Properties of Electrorheological Fluids with Regard to the Design of Valves and their Controllers

S Ulrich[1], G Böhme[2] and R Bruns[3]

Faculty of Mechanical Engineering,
[1, 2] Institute of Fluid Mechanics, [3]Institute of Machine Elements and Technical Logistic, University of the Federal Armed Forces Hamburg

E-mail: stephan.ulrich@hsu-hh.de

Abstract. The rheological properties of the commercially used electrorheological fluid Rheoil 3.0 is measured using flow channels, that provide similar flow conditions to many electrorheological applications. The static dependence of the apparent viscosity on the field strength, the shear rate and the temperature as well as the response time of a change in the viscosity upon a changed electric field are measured and approximated by a mathematical model that can be used to design electrorheological valves and to estimate their dynamic performance.

1. Introduction

When designing electrorheological applications like valves, actuators or dampers the knowledge about the rheological behaviour in load conditions is essential. In particular the static and dynamic effects of the electric field strength, the shear rate and the temperature to the apparent viscosity need to be known. However, classical rheological measuring methods like rotational or cone-plate rheometry are not suitable for electrorheological fluids (ERFs) because of different flow conditions. Compared to the load conditions in technical applications these methods suffer from low shear rates and different requirements for the electric field.

In this work the static and dynamic rheological properties of the electrorheological fluid Rheoil 3.0 containing polyurethane particles in silicone oil [1], which is now available for commercial applications are measured using flow channels permitting technical relevant conditions. The test setup for the response time measurements allows measuring the response of the pressure drop of the valve to a changed electric field with a minimized influence by the hydraulic system or the inertia of the fluid. This yields a static and dynamic model of the ERF that is used to design electrorheological applications and their controllers as well as to estimate the dynamic performance [2].

2. Static Measurements

To determine the static rheological behaviour the ERF is forced to flow through a rectangular channel by a servo hydraulic test bench. This setup permits a constant flow rate without pulsations over a wide range. The gap of the flow channel is bounded by two electrodes that can be loaded with high voltage thus the flow channel acts like an electrorheological valve. By measuring the pressure drop the apparent viscosity can be calculated subjected to the electric field strength and the apparent shear rate (Fig.1). The fluid temperature was varied between 25 and 60 °C and it was found, that the ER-effect of this fluid is larger for higher temperatures (Fig. 2 right). This ERF shows some typical properties. At a zero field it shows a Newtonian behaviour with a linear relation between the shear stress and the shear rate. With rising field strength the pressure drop increases and the behaviour is more Bingham like.

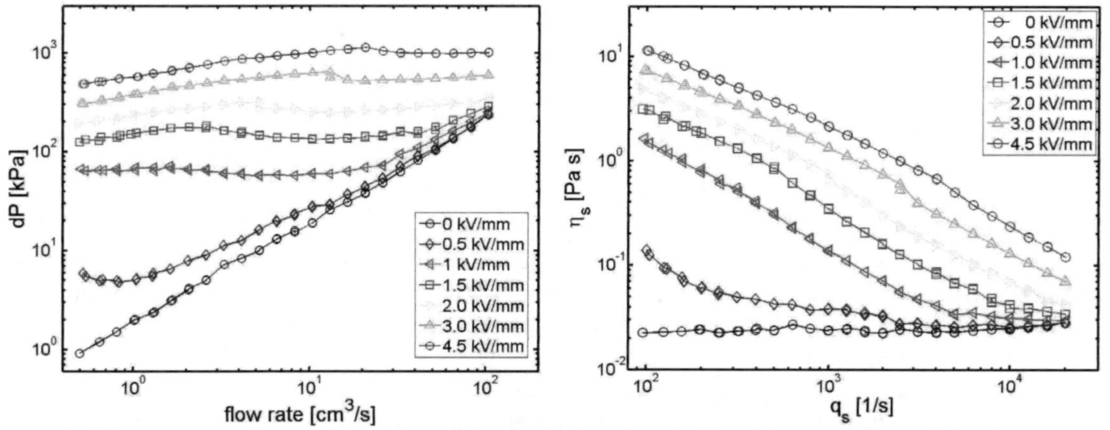

Figure 1. The pressure drop of the electrorheological valve (left) with a gap size of 1x30 mm and a length of 100 mm in dependence of the flow rate and the field strength at a temperature of 60 °C. In conjunction with the geometric dimensions the apparent viscosity and the apparent shear rate can be calculated (right).

Note that in some regions the pressure-flow rate-characteristics have a negative gradient, thus there can be no pure fluid friction with wall adhesion. Rather there must be slip with dry friction. Thus, with flow channel experiments, it is not possible to calculate the shear rate at the wall with the equation of Weissenberg-Rabinowitsch [4] in order to find a constitutive equation in the form of a shear rate-shear stress-relation. Further experiments with different geometries should be made to evaluate wall slip (Mooney relation [4]).

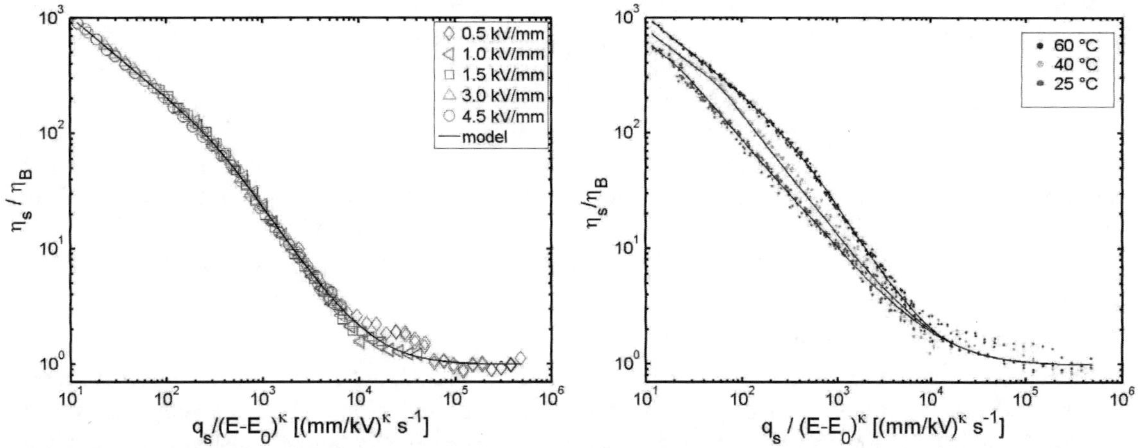

Figure 2. The viscosity ratio in dependence on the apparent shear rate scaled with the excess field strength: E_0 = 0.38 kV/mm and κ = 1.5 at a temperature of 60 °C (left). At lower temperatures the viscosity ratio decreases for equal apparent shear rate and field strength (right).

Nevertheless these results can be used to formulate a functional relation between the apparent shear rate, the electric field strength and the apparent viscosity that can be easily used to calculate the pressure drop of electrorheological valves with different geometries. Remarkably all measured values can be reduced to one single curve by scaling the apparent shear rate with some power of the excess field strength: $(E-E_0)^{\kappa}$ and the apparent viscosity with the viscosity at zero electric field (Fig.2). This curve can be approximated by a modified Carreau-Yasuda equation that is widely used for shear thinning fluids [4]:

$$\frac{\eta_s}{\eta_B} = 1 + \frac{\left(K\chi^m - \eta_B\right)}{\eta_B}\left[1 + (\lambda\chi)^{\alpha}\right]^{\frac{n-1}{\alpha}}. \qquad (1)$$

The parameters for the temperature of 60 °C are: η_B=27 mPas, m=-0.7, n=0.45, $\chi := q_s(E-E_0)^{-\kappa}$, K=140 $Pas^{m+1}kV^{\kappa m}mm^{-\kappa m}$, α=2.2, $\lambda=2.8\cdot10^{-3}\, skV^{\kappa}mm^{-\kappa}$.

3. Dynamic Measurements

Although the short response time of ERFs is one of the most propagated advantages, there are only a few publications that try to quantify the dynamic response. Additionally, these measurements are often not applicable to electrorheological valves, because of different flow conditions. Here, the dynamic performance of the ERF, i.e. the response time of the change in viscosity as a reaction of the fluid to a changed electric field, is measured with a hydraulic test set-up where two identical flow channels are switched in-line that are alternately stressed by an electric field (Fig. 3).

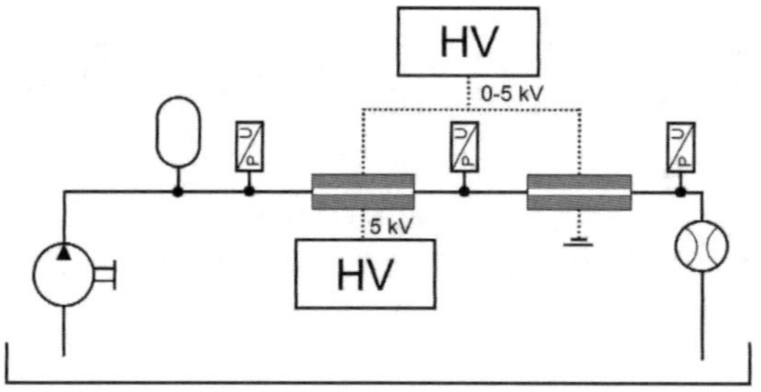

Figure 3. Schematic illustration of the test setup for response time measurements.

This facilitates measuring the pressure steps caused by the altered electric field only, without influencing the system pressure or having interferences due to the hydraulic capacity of the system. This ensures a constant flow rate to avoid inertia effects of the fluid to the dynamics. The response time of the fluid is calculated from step responses of the pressure upon steps in the electric field strength (Fig. 4). The values were received with a sampling of 20 μs and a maximum phase delay of the pressure transducers and the amplifiers of 15 μs. The step response of the pressure drop is measured in dependence on the temperature, the shear rate and the electric field strength. Although the dynamic behaviour of the ERF is non-linear, it can closely be approximated by a set of two linear first order differential equations, that can easily be used to estimate the dynamic performance of electrorheological valves:

$$\tau_1\dot{\bar{y}} + \bar{y} = a_1 u, \quad \tau_2\dot{\tilde{y}} + \tilde{y} = a_2 u \ \text{ with } \ a_1 + a_2 = 1, \quad \bar{y} + \tilde{y} = y. \qquad (2)$$

Accordingly the dynamic behaviour can be characterized by three parameters: τ_1 is the short response time that affects the fast change in the pressure to the level a_1 and τ_2 is the slow response time that affects the approach to the static value. In order to estimate the dynamic constraints of electro-rheological valves or actuators, τ_1 is the leading parameter which interestingly is neither a function of the shear rate nor of the electric field strength, but it is strongly influenced by the temperature. Thus the response time τ_1 increases from 0.9 ms at 60 °C to 12 ms at 25 °C. These results are in accordance to measurements at a rotational viscometer with a concentric cylinder geometry. Here, at a constant shear rate, the frequency of the alternating electric field is varied (Fig. 4). The frequency range where the viscosity ratio decreases seems to point out the dynamic performance of the fluid. Thus the 3 dB cut-off frequency of a simple first order dynamic model with a response time of 0.9 ms is 177 Hz which is in accordance with the frequency range where the viscosity ratio decreases for a temperature of 60 °C.

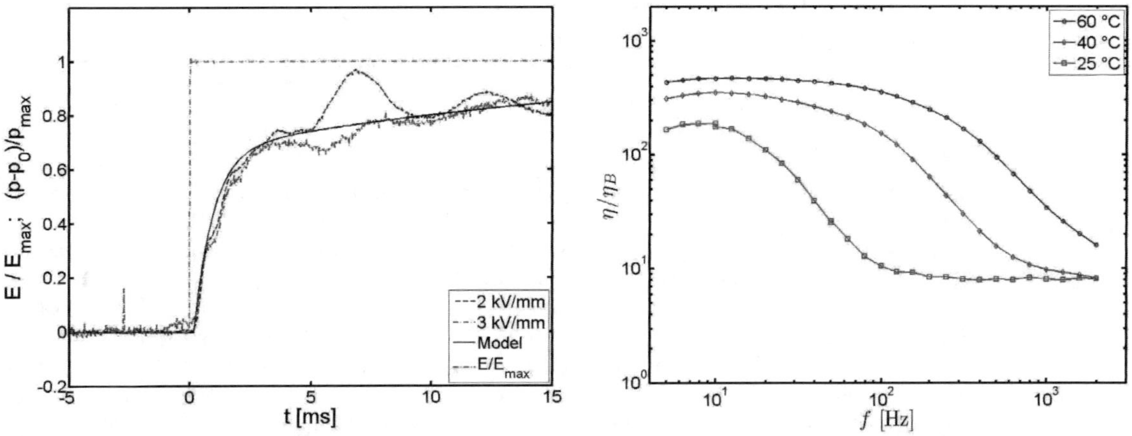

Figure 4. Normalized pressure step response at a temperature of 60 °C and an apparent shear rate of 1000 1/s (left). The parameters of the model are: τ_1 = 0.9 ms, τ_2 = 19.4 ms and a_2 = 0.67. The dependence of the viscosity ratio on the frequency of an alternating electric field with the effective field strength of 2 kV/mm for different temperatures (right).

4. Conclusions

Static and dynamic measurements were carried out for the commercially available ERF Rheoil 3.0 leading to a model of the fluid that can be used for designing electrorheological applications and their controllers as well as for estimating the dynamic performance. In particular the evidence of response time less than one millisecond shows the realizable dynamics of electrorheological valves that is much higher than the dynamics of conventional hydraulic systems. The performance of this ERF, in particular the dynamic response and its viscosity ratio is strongly related to the temperature, thus an electrorheological system should operate at about 60 °C. These results were used to design an electrorheological actuator for active vibration isolation of heavy duty applications. Further work will show if it is possible to compensate forces of a vibrating machine in a band width of 10 to 2000 Hz.

5. Acknowledgment

This research is supported by the Wehrwissenschaftliches Institut für Werk-, Explosiv- und Betriebsstoffe (WIWEB), Erding, Germany.

References

[1] N.N., product specification: RheOil 3.0, Fludicon GmbH, Darmstadt
[2] Ulrich S., Böhme G., Bruns R., The Capability of Electrorheological Actuators for Vibration Decoupling, *Actuator 08*, (H. Borgmann, ed), Bremen (2008), pp.842-845
[3] Zaun M., Elektrorheologische Ventile als Stellelemente in Zylinderantrieben, *PhD. Thesis*, RWTH Aachen, (2007)
[4] Bird R. B., Armstrong R.C., Hassager O., Dynamics of Polymeric Liquids, John Wiley & Sons, New York, (1987)

Journal of Physics: Conference Series **149** (2009) 012032 doi:10.1088/1742-6596/149/1/012032

Wide Angle Scattering Study of nanolayered Clay/gelatin Electrorheological Elastomer

B Wang, Z Rozynek, M Zhou and J O Fossum

Department of Physics, Norwegian University of Science and Technology, Høgskoleringen 5, NO-7491, Trondheim, Norway

E-mail: baoxiang.wang@ntnu.no and jon.fossum@ntnu.no

Abstract: In the general context of self-assembly of nanolayered clay, we have studied both kaolinite and montmorillonite guided assembly into chain-like structures in gelatin hydrogel. The electrorheological (ER) elastomers, containing clay particles which dispersed in gelatin/water matrix, were prepared with or without the applied DC electric field and cross-linked polymerized with the help of formaldehyde. The experimental techniques include synchrotron X-ray scattering, atomic force microscopy, optical microscopy. The aim is to produce a water-based, low-cost and environmentally friendly ER hydrogel. The wide-angle x-ray scattering (WAXS) patterns observed from clay/gelatin ER elastomers curing in the DC field are highly anisotropic and show differences clearly compared to that without curing in the field. Both clay nanolayers have preferential orientation in gelatine hydrogel along the direction of electric field.

1. Introduction

The physical properties of stimuli-responsive soft materials dramatically change corresponding to the external stimuli such as temperature, pH, electric and magnetic fields [1-3]. These materials are called as intelligent materials and have been harnessed in novel systems, for example, sensors, actuators, and biomedical applications [4, 5]. Electrorheological (ER) effects refer to the change of rheological properties of a material upon application of an electrical field [6-11]. When an ER fluid, a suspension of polarizable particles in an insulating medium (such as mineral oil or silicone oil), is subjected to an electric field, the particles polarize, which results in an anisotropic structure of aligned chains of particles along the direction of the uniaxial electrical field. ER fluids exhibit drastic changes in their rheological properties, including a large enhancement in apparent viscosity and yield stress under an applied electric field. An obvious problem with ER fluids is sedimentation of the particles [12-13]. Because of the fluidity, ER fluids must be sealed to prevent leakage, which can complicate the design. Another design problem with fluids is the electrodes, which must be fully supported by the container for the fluid, yet electrically insulated to resist high voltage. Recently, ER elastomers, cross-linked polymer gels with dispersed polarizable particles–have attracted attention. The dispersed phases filled with particles responsive to an applied electric or magnetic field include intercalated or exfoliated platelets obtained from clays, mica, or graphite, nanocages, nanotubes, dual fillers, porous particles, and nanocatalysts etc. Comparing with ER fluid, the liquid medium of ER elastomer is cross linked polymerized after particle alignment, so the anisotropic arrangement of polarisable particles is "frozen". In the gel network, as matrix's solid-like nature, particles are restricted and the congregation of particles is eliminated. Consequently, the instability of the ER effect, which is caused by the particles' congregation in ER fluid, is improved. At the same time, the interaction among the polarized particles under the applied electric field induces the elasticity change of the ER elastomer, and then the elasticity change induces a notable ER effect. The advantages of ER elastomers over ER fluids include no leakage, no sedimentation of particles and possibility to produce custom-made ER objects of exactly the right shape and size for the application (Because the shape and size of ER elastomer can be designed for the practical requirement) [14,15]. In present work, montmorillonite (MMT) and kaolinite guided assembly into chain-like structures in gelatin hydrogel under the external electric field are studied using synchrotron X-ray scattering, AFM and optical microscopy. There are several green chemistry concepts in our present strategy: the choice of friendly solvent, the selection of friendly benign and non-toxic matrix, the one-step reaction. Gelatin is a water-soluble natural polymer with a long and soft polypeptide backbone and its individual monomer contains an amino acids group. It is a commercially available and inexpensive material. In our approach, firstly water is used as a based component. The second concept is the choice of the low-cost and benign clay and gelatin. The third one is the simple and reproducible syntheses. The aim is to produce a water-based, low-cost and environmentally friendly ER hydrogel.

2. Materials and characterization methods

Gelatin and formaldehyde solution (A.R. 37%, diluted to 5% when used) were purchased from Merck Schuchardt OHG, Germany. Kaolinite is a 1:1 dioctahedral clay mineral with the ideal composition of $Al_2Si_2O_5(OH)_4$. It was used without further purification. Montmorillonite is a typical 2:1 smectites and was purchased from Southern Clay Products, Inc. as a fine gray powder. In this study, firstly 20% gelatin of aqueous solution was prepared at 70°C for 1h. Then appropriate clay (kaolinite or montmorillonite) particles (such as 1wt%, 5wt%, 10wt% etc.) were added into the gelatin/water matrix to further stir for 6h at 70°C. The clay/gelatin/water mixture will be transferred into a custom-made casting cell (Fig.1) and cured under E=1 kV/mm DC electric field for 1 h. At the same time, the temperature will be retained at 70°C during the curing process. After that the system naturally cooled down to 25°C, the clay/gelatin physical hydrogel is obtained. Then this physical hydrogel was dipped into the chemical crosslinking agent (5wt% formaldehyde solution) over a night. The chemical crosslinking clay/gelatin ER hydrogel can be obtained. For comparison, samples with the same chemical composition but in the absence of an electric field were prepared.

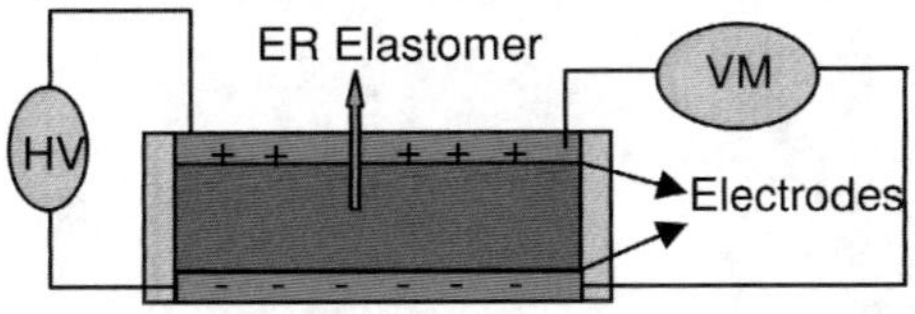

Figure 1. Schematic equipment for curing an elastomer. HV: DC power source, VM: volt meter.

The reaction of the gel chemical crosslinking is described as follows:

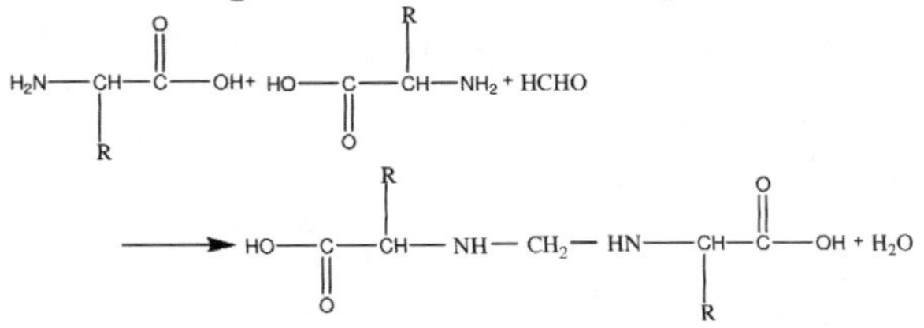

R is amino acid polypeptide macromolecule.

Relative orientations of the clay particles inside the electrorheological elastomers were determined using synchrotron X-ray scattering experiments. These experiments were performed at the Swiss-Norwegian Beamlines (SNBL) at ESRF (Grenoble, France), using the WAXS setup with a 2D mar345 detector at beamline BM01A. The X-ray beam with a wavelength of 0.72 Å and 0.3×0.3 mm^2 beam size were used. The AFM images were taken using a Multi Mode^{TM} Atomic Force Microscope (nanoscope IV) from Veeco Instruments with a contact model.

3. Results and discussion

3.1. 1:1 type clay (kaolinite) assembly in gelatin

The morphology of pure kaolinite is illustrated in Figure 2. It can be seen that kaolinite is composed of stacked small platelets (1~10μm). Furthermore, kaolinite particles are sharing

Journal of Physics: Conference Series **149** (2009) 012032 doi:10.1088/1742-6596/149/1/012032

morphology with well defined hexagonal edges and corner angles (120°) shown in large resolution image. Figure 3 shows the optical image of kaolinite/gelatin hydrogel without curing in the field. Solid-like or gel-like behaviour is observed. Kaolinite have hydroxyl group, which is beneficial for homogeneously dispersing in the gel network.

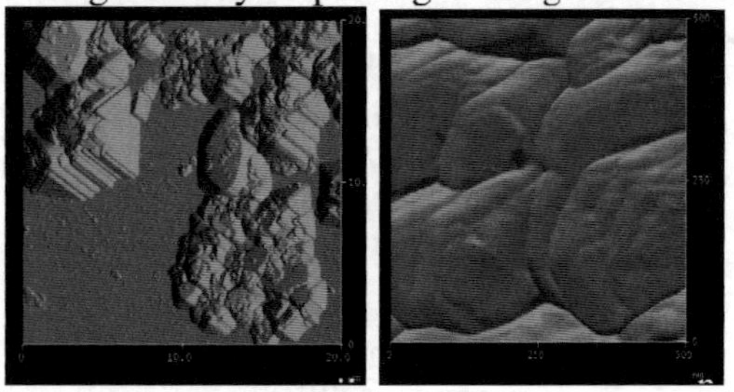

Figure 2. AFM images of pure kaolinite particles, kaolinite/gelatin the scale bar 20×20μm (left) and 500×500nm (right).

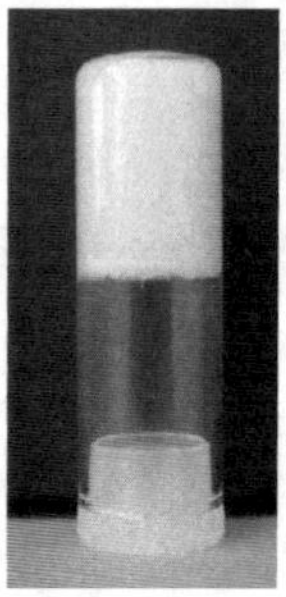

Figure 3. Optical images of hydrogel without curing in the field.

Solid kaolinite/gelatin ER elastomer is prepared by arranging suspended kaolinite particles in gelatin with an electric field and crosslinked the obtained structure with the help of formaldehyde to "freeze" the chain. Figure 4(a) shows the micrograph of kaolinite particles in the elastomer cured without the externally applied electric field, and Figure4(b) shows the elastomer cured under the field with E= 1 kV/mm. It is observed that particles are randomly dispersed and aligned dispersed in Figure4 (a) and (b), respectively. By comparing and analyzing the two elastomers, it is obvious that the particles have been assembled by the applied DC field during the curing process.

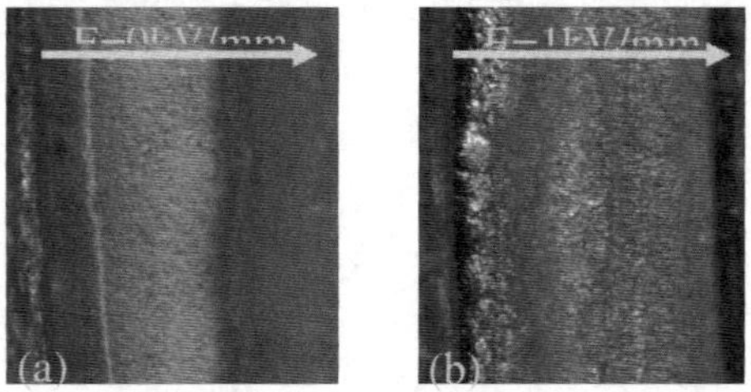

Figure 4. Optical micrographs of the kaolinite/gelatin elastomers cured in the absence/presence of the applied electric field (E=1kV/mm).

Relative orientations of the kaolinite particles inside the electrorheological elastomer were determined using synchrotron X-ray scattering experiments: chain and column formations were observed. The three-dimensional WAXS patterns obtained from kaolinite/gelatin ER elastomer are shown in figure 5.

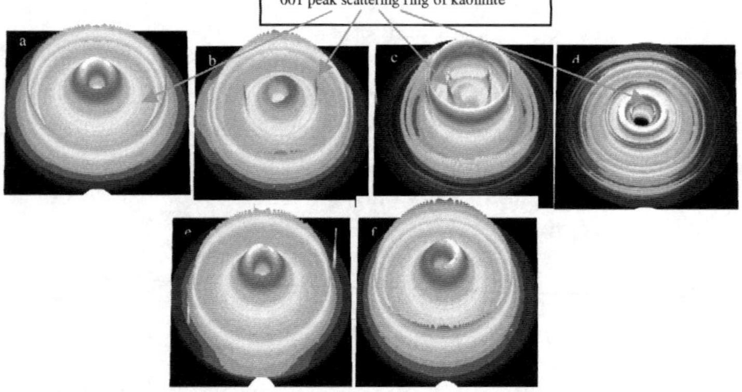

Figure 5. The three Dimensional WAXS patterns obtained from 10wt% kaolinite/gelatin hydrogel curing in a DC external electric field (a) E=0kV/mm, isotropic pattern; (b)(c)(d) E=1kV/mm, anisotropic pattern; the distance between the detector and the sample is different as follow (b) 350mm (c) 200mm (d)110mm, which more scattering rings of kaolinite/gelatin ER hydrogel can be observed; e) pure gelatin hydrogel without curing in a field; f) pure gelatin hydrogel curing in a field E=1kV/mm

The 001 scattering ring of kaolinite is indicated by arrows respectively in figure 5. Clay minerals are sheet silicates or layered silicates, whose primary particles are platelet shape. Each clay platelet is ~1 nm thick (c-direction). Kaolinite is a 1:1 dioctahedral clay mineral with the ideal composition of Al2Si2O5(OH)4. Its structure is composed of AlO2(OH)4 octahedral sheets and SiO4 tetrahedral sheets. Normally, the 001 peak of clay indicates the interlayer spacing between the two basic sheets. For the first strong peak of kaolinite in WAXS patterns represents the diffraction from the (001) crystal surface of the kaolinite, and the associated interlayer d-spacing of this peak is 0.715 nm (In figure 4) [16]. For kaolinite/gelatin hydrogel without curing in the electric field, the kaolinite particles are randomly dispersed into the gelatin matrix, so its WAXS pattern is isotropic shape due to the non-orientation of kaolinite particles (shown in figure 5a). However after curing in the electric field, kaolinite partilces will polarized and assemble or orient to respond with field. Kaolinite particle have preferential orientation in gelatine hydrogel along the direction of electric field and show an anisotropic pattern after the curing of DC electric field compared with the isotropic pattern of kaolinite/gelatin hydrogel without curing in the field (shown in figure 5b, c and d). These two patterns show clearly that the kaolinite particles assemble or orient in the direction of electric field. Likewise, the electric field should be high enough such that the electric force can overcome the effect of the gravitational force acting on the dense kaolinite particles. However□applying a DC field over E=2 kV/mm or much higher can cause the water to hydrolyze and air bubble is observed obviously. The sample will be shrunken during the curing process. The colour of sample will be changed from light gray to green for kaolinite/gelatin hydrogel. Furthermore, too weak electric field strength (below 300V/mm) is difficult to ensure particle polarized for orientation or forming chain-like structure. So we chose a medium electric field strength (E=1 kV/mm) to make sure clay assembling into gelatin matrix and in this case the hydrolyzation is not serious. The multipolar interactions from the highly polarizable particles undoubtedly contribute significantly to the ER response. The result indicates that the orientation of particle chain relative to the field is an important structure factor in the designing of the ER elastomers. Spatial arrangement influences the interparticle forces, especially those within the chains. Furthermore, pure gelatin samples with or without curing in an electric field is prepared and their three-Dimensional WAXS patterns are shown in figure 5 e) and f), respectively. Comparing with kaolinite/gelatin ER elastomer (figure5b), no kaolinite scattering ring, only gelatin scattering ring is shown in the pattern. Furthermore the anisotropic pattern (figure 5f) also indicates that pure gelatin hydrogel can respond with the electric field and oriented along the direction of electric field.

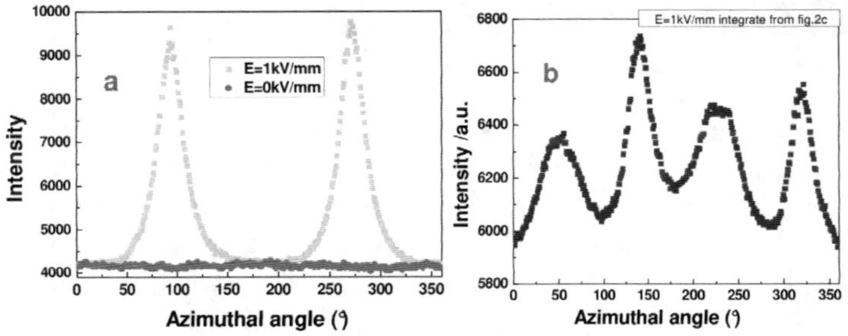

Figure 6. a) Dependence of the intensity of circular scattering rings (only integrating kaolinite 001 peak in figure 4 a and b) on the azimuthal angle for 10wt% kaolinite/gelatin composite at the different electric field; b) Dependence of the intensity of circular scattering rings (kaolinite 001 peak plus the first peak of gelatin in figure 4 c) on the azimuthal angle.

Figure 6 shows how the intensity of circular scattering rings such as those presented in figure4 a) and b) evolve as a function of the azimuthal angle Φ, between 0 and 360°. For the 10wt% kaolinite/gelatin composite, we firstly have considered the first-order Bragg peak of kaolinite (001). The scattered intensity at a given azimuthal angle is proportional to the number of particles that meet the Bragg condition for that angle, so the shapes of the scattered intensities in figure 6a provide the orientation distributions of clay particle orientations inside the chains and columnar structures. For E

= 0 kV/mm, the intensities are independent of Φ and the two-dimensional scattering pattern is isotropic because the kaolinite particles is randomly dispersed in gelatin hydrogel. For E =1kV/mm, the azimuthal positions of the maxima along the plots is anisotropic and this demonstrate that the preferred orientation of the kaolinite particles is with the lamellar stacking plane parallel to the direction of the electric field. Secondly, we have considered both the 001Bragg peak of kaolinite and the first peak of gelatin evolve as a function of the azimuthal angle Φ, between 0 and 360°. From figure 6 b), we can see that the azimuthal positions of the maxima along the plots is anisotropic and have 4 maxima. So that means both the visible diffraction rings correspond to the diffraction of kaolinite and gelatin are anisotropic. The reason is probably because the macromolecules of the matrix are polymerized anisotropically under an applied dc electric field for curing. Since the clay particles are platelets with a thickness much smaller than their lateral dimension, this suggests that they are lying in the ER bundles with one of their lateral dimensions parallel to the field; i.e., on average, the clay particles have their directors aligned perpendicular to the direction of applied field [17]. Furthermore, the gelatin matrix polarizes along their macromolecules chain parallel to the direction of the electric field, and hence in figure 5b, c and d we can see that two perpendicular directions for anisotropic rings are observed for kaolinite and gelatin separately.

3.2. 2:1 type clay (MMT) assembly in gelatin

The wide angle X-ray diffraction (WAXS) patterns of a series of MMT/gelatin hydrogel are also studied. The two-dimensional WAXS patterns obtained from MMT chains of particles dispersed in gelatin hydrogel, in the presence (a) 10wt%MMT with field(E=1kV/mm); (b) 5wt% MMT with field (E=1kV/mm) or (c) absence of a DC external electric field are shown in figure 7. The 001 scattering ring of MMT is indicated by arrows respectively in figure 7. The basic structural unit of so-called 2:1 clays (such as MMT) is a 1 nm-thick platelet consisting of two tetrahedral silica sheets sandwiching one octahedral silica sheet. The interlayer d-spacing for the peak of free MMT (001 crystal surface) is near 1nm.

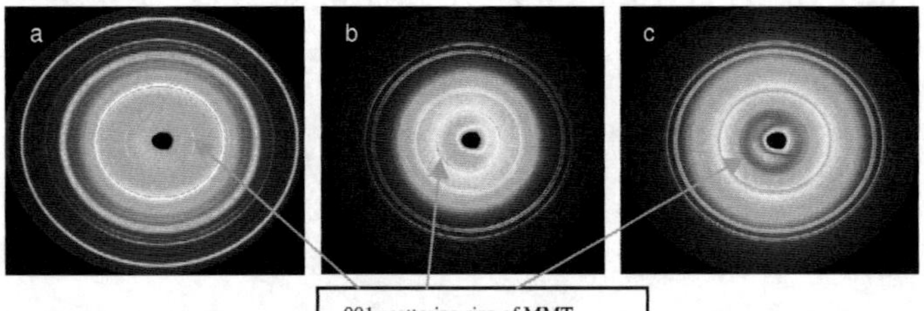

Figure 7. The 2-dimensional WAXS patterns obtained from the MMT/gelatin hydrogel, in the presence (a) 10wt%MMT with field (E=1kV/mm); (b) 5wt% MMT with field (E=1kV/mm); (c) 5wt% MMT without field (E=0kV/mm).

Similarly, the isotropic pattern of 5wt% MMT absence of a DC external electric field in figure 7(c) means that MMT particles are randomly dispersed into gelatin hydrogel and no preferential orientations for the clay aggregates. The two-dimensional WAXS patterns obtained from this MMT/gelatin hyrogel elastomer (figure7 b and c) are clearly anisotropic, reflecting the preferential orientation of the particles in the electric field. Taking advantage of the nano-layered nature of the clay crystallites, we were able to infer MMT particle orientations inside the particle bundles from the anisotropy of two-dimensional WAXS images. This study suggested that the MMT particles polarize along their silica sheets, and hence, that their directors were, on average, perpendicular to the direction of the electric field[18]. Dependence of the intensity of scattering vector for MMT/gelatin hydrogel with and without electric field for 5wt% and 10wt% MMT content are shown in figure 8, respectively. A sharp and strong peak attributable to a low degree of ordering is observed for MMT/gelatin hydrogel elastomer with electric field comparing that of MMT/gelatin hydrogel without electric field (d_{001} =10 Å, $d = 2\pi/q$). This is due to dipolar interaction between polarized MMT particles, which orientate and aggregate to forms chains and then column structures along the direction of applied electric field which also lead to the enhancement of MMT concentration. The intensity of 001 diffraction peak was increased, indicating an increase of MMT particles assembling along the direction of electric field. A chain-like structure of ER composites enhances their electrostriction response to normal and shear deformations. This makes feasible application of ER composites in stand-alone sensing systems as well as implementation of electrostriction to achieve self-sensing capabilities in ER devices.

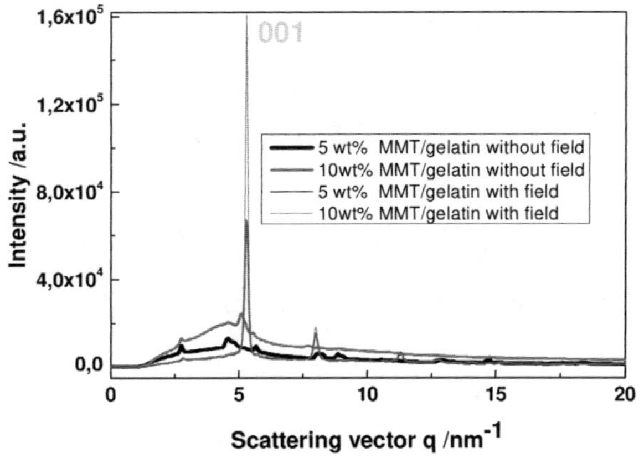

Figure 8. Dependence of the intensity of scattering vector q for 5wt% or 10wt% MMT/gelatin hydrogel with and without curing in an electric field.

4. Conclusions

A serial of clay/gelatin ER elastomers have been synthesized and characterized by wide angle -ray scattering. The resulting materials show different structures compared to uncuring clay/gelatin hydrogel. Two and three-dimensional WAXS images from bundles of clay inside the gelatin matrix exhibits a clear anisotropy, reflecting the preferential orientation of the clay particles in the field. The anisotropy of 2-dimesional diffractogram also provides the angular distribution of clay particles inside the clay assemble into gelatin hydrogel.

Acknowledgements: This work was supported by the Research Council of Norway (RCN) through the NANOMAT Program and the FRINAT Program.

References

[1] Mark J E 2003 *Prog. Polym. Sci.* **28** 1205
[2] Schmidt D F, Clement F and Giannelis E P 2006 *Adv. Funct. Mater.* **16** 417
[3] Giannelis E P 1996 *Adv. Mater.* **8** 29
[4] Gamota D and Filisko F E 1991 *J. Rheol.* **35** 399
[5] Hao T 2001 *Adv. Mater.* 13 1847
[6] Block H, Kelly J P, Qin A and Watson T 1990 *Langmuir* **6** 6
[7] Whittle M and Blullough W A 1992 *Nature* **358** 373
[8] Wen W J, Huang X X, Yang S H, Lu K Q and Sheng P 2003 *Nature Mater.* **2** 727
[9] To K and Choi H J 1998 *Phys. Rev. Lett.* **80** 536
[10] Tetsu M, Kousuke S and Kiyohito K 2004 *Polymer* **45** 3811
[11] Wang B X and Zhao X P 2005 *Adv. Funct. Mater.* **15** 1815
[12] Liu B and Montgomery T S 2001 *J. Rheol.* **45(3)** 641
[13] Martin J E and Anderson R A 1999 *J. Chem. Phys.* **111(9)** 4273
[14] Gao L X and Zhao X P 2007 *J. Appl. Polym. Sci.* **104(3)** 1738
[15] Gao L X and Zhao X P 2004 *J. Appl. Polym. Sci.* **94(6)** 2517
[16] Wang B X and Zhao X P 2002 *J. Mater. Chem.* **12** 1865
[17] Meheust Y, Knudsen K D and Fossum J O 2006 *J.Appl.Cryst.* **39** 661
[18] Fossum J O et al 2006 *Europhys. Lett.* **74** 438

11th Conference on Electrorheological Fluids and Magnetorheological Suspensions IOP Publishing
Journal of Physics: Conference Series **149** (2009) 012033 doi:10.1088/1742-6596/149/1/012033

MR-Tech: a portable smart wrist rehabilitation device

M Avraam, M Horodinca and A Preumont

Active Structures Laboratory, Université Libre de Bruxelles, Belgium

E-mail: mavraam@ulb.ac.be

Abstract. The purpose of this paper is to present a novel patent-pending, computer controlled, MR-brake actuated, muscular rehabilitation and evaluation device that can be carried to and used at patient's home (without the immediate assistance of the physiotherapist), offering similar functionalities as existing stationary devices that can be found in hospitals (i.e. isometric, isotonic and isokinetic exercise modes). The first part of the paper describes in more details these various exercise modes. The second part is devoted to the design of the machine and the third part to its control. The last part presents preliminary results obtained for the wrist pronation/supination motion.

1. Introduction

The handling of patients with joint muscular dysfunctions requires the intervention of a practitioner who currently mainly exercises his skills with the assistance of large, heavy, expensive and stationary rehabilitation devices (e.g; BIODEX, HUMAC NORM, CONTREX...) that force the patient to perform exercises in clinics or medical private practices. The objective of the device presented in this paper is precisely to make this technology available for home exercises of the wrist joint.

Muscular rehabilitation and evaluation is usually carried out by performing three main types of exercises namely *isometric*, *isotonic* and *isokinetic* [1]. The *isometric exercise* is performed statically (without any rotation of the joint) at various angular positions. The external resistance applied to the joint is always equal to the force applied by the patient. The *isotonic exercise* is performed dynamically over a predefined Range Of Motion (ROM). The resistance applied to the joint is either constant or follows a predefined pattern as a function of joint angular position (e.g. parabolic...). The *isokinetic exercise* is also performed dynamically but in that case, the resistance is applied to the joint only if a predefined angular speed is reached by the joint in order to avoid that the joint exceeds this speed value. This particular exercise mode is the only one that enables dynamic training at the maximum muscle force over the entire ROM (figure 1) since the isotonic exercise (even with a parabolic profile) does not adapt its force profile to the exact muscular capacities of the user.

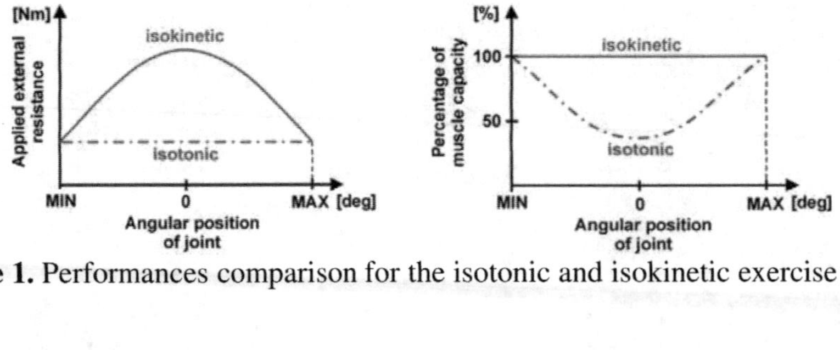

Figure 1. Performances comparison for the isotonic and isokinetic exercise modes

Journal of Physics: Conference Series **149** (2009) 012033 doi:10.1088/1742-6596/149/1/012033

2. Description of the device

2.1. System components

As can be seen on figure 2, the rehabilitation device is composed of three major subsystems: mechanical hardware, embedded electronics and user interfaces.

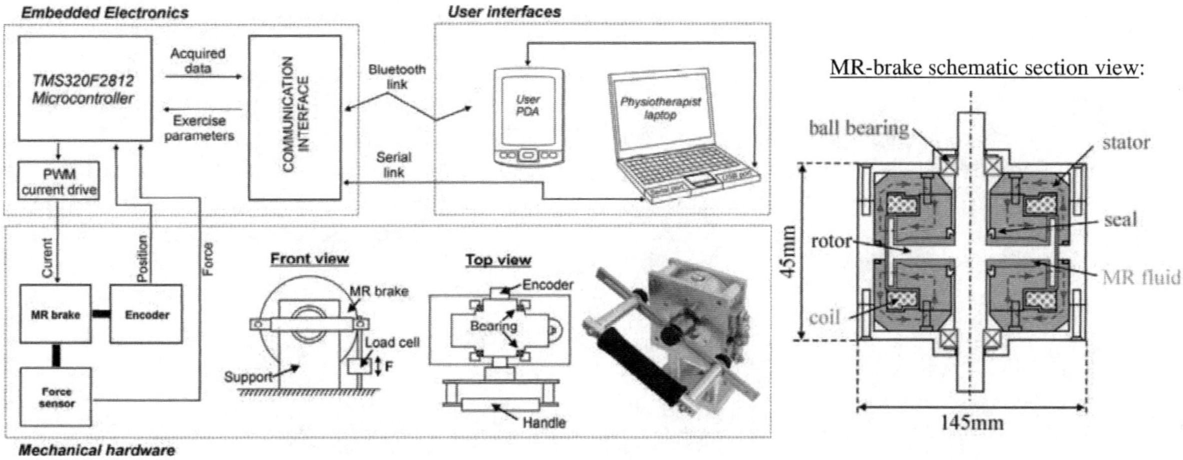

Figure 2. Schematic view of the system components

The mechanical hardware is mainly composed of a MR-brake (described in details in section 2.2.) connected to its support through two bearings. A rotational MR-brake actuation was preferred over other technologies (commercial devices use DC-motors and other authors have used ER-brakes [2][3] or linear MR damper [4]) mainly due to its high torque/volume ratio and safety of operation (more details can be found in [5]). The brake rotation relatively to its support is blocked by a load cell providing a measurement of the reaction force applied to the brake stator when a torque is applied on the rotor of the brake. An encoder is connected to the rotor of the brake providing position and speed measurements.

Embedded electronics are composed of a TI TMS320F2812 microcontroller used for control and data acquisition purposes. It is linked to an electronic board that drives the MR-fluid brake (through a a PWM current drive) and implements the interface with the sensors and the user interfaces (through Bluetooth or wired connections).

Two user interfaces are provided with the device. The first one is a laptop used by the physiotherapist to set exercise parameters and to analyse data collected during the exercise. The second one is a PDA that provides a simplified visual feedback to the patient when he is using the device and that stores exercise results for further analysis by the physiotherapist (following a data transfer between the PDA and the physiotherapist laptop).

2.2. Details on the MR-brake design

A previous study has shown the advantages of T-shaped rotor brakes over conventional drum brakes in terms of compactness [6]. Indeed, such a configuration enables to concentrate the action of the fluid as far as possible from the output axis, leading to a high output torque while maintaining a reasonably compact design (figure 2). The magnetic circuit of the brake as been optimized (using electromagnetic FE analysis) in order to achieve a maximum output torque of 28Nm while minimizing its volume and mass with the following constraints: maximum external diameter: 120mm, thickness of the MR-fluid gap: 0.4mm, thickness of the rotor T-shaped extremity: 2mm and maximum current in each coil: 0.5A.

Experiments conducted on the prototype that was built based on the previously mentioned specifications have shown a maximum output torque of 22.5Nm combined with an off-state torque of

0.4Nm. The difference between predicted and measured torque could be explained by a non-optimal filling of the fluid gap. This hypothesis seems to be confirmed by the fact that measurements of the magnetic induction field in the gap of the stator (without fluid) exhibit good correspondence with FE simulation results. The achieved torque still being compatible with physiological requirements, the brake has been kept in that state due to difficulties to clean and refill it.

3. Control strategy

A control strategy, inspired by [4], composed of three overlapping control loops, is used to control the device. The first control loop is a current loop used to compensate for the electrical time delay related to the use of coils to generate the magnetic field inside the brake. The second control loop is a torque loop used to compensate for the imperfections of the relation between the current applied to the brake and its output torque (non-linearity, hysteresis, remanence). The third loop is a motion loop used to control the motion of the shaft of the device according to the selected exercise mode (isometric, isotonic or isokinetic)(figure 3).

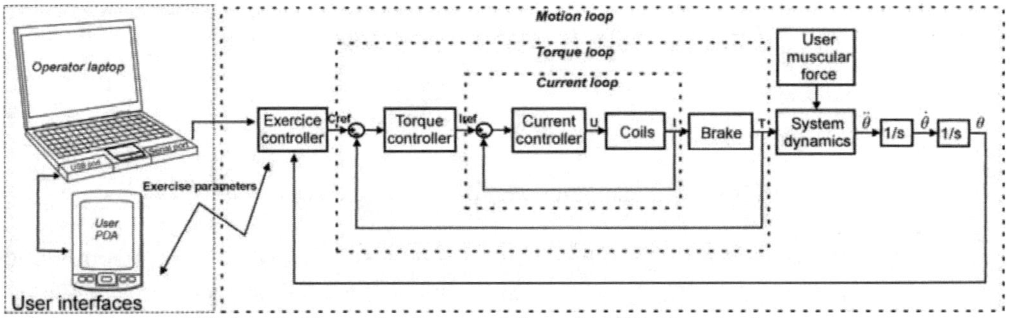

Figure 3. Control scheme of the rehabilitation device

For the isometric mode, the control is based on a position controller that implements a hard stop if the measurement position is reached and if the patient applies a force in the right direction. For the isotonic mode, when the patient moves within the measurement ROM, the reference torque, used as input of the torque loop, is simply set to a constant value or to a value varying with the angular position (following the pattern selected by the physiotherapist). For the isokinetic mode, the control is based on a speed controller that provides a reference torque to the torque loop. It should be noted that, as the speed of motion is quite slow (<180°/s), the quality of the speed signal derived from the encoder measurements is quite poor. In order to alleviate this problem, control is not based on a speed measurement but on a position measurement that is compared with a reference signal consisting of a ramp signal (corresponding to a constant speed). The difference between these two values is used as input of a PID controller. The reference position signal starts to increase/decrease from the current position as soon as the motion is initiated. The slope of the ramp signal is adjusted according to the speed to be achieved. For both isotonic and isokinetic modes, virtual end-of-motion stops have been implemented in order to limit the motion of the user to the specified ROM.

4. Experimental results

Dedicated electronics and user interfaces are still under development. Preliminary tests have however already been conducted by linking the mechanical hardware to a dSPACE PC-based acquisition board used in combination with the Control-Desk software package emulating the future user interface. Control has been implemented using MATLAB Simulink.

All the results presented here are for the wrist pronation/supination motion but similar results have been obtained for the two other wrist motions. Results obtained for the isometric exercise show that the user is perfectly blocked at the various measurement positions (0°, 30°, 60°, 90°) and that he is able to leave these positions with a very limited sticking force (figure 4). Performances of the isotonic mode have been evaluated by applying a parabolic and a constant resistive torque profile (with various maximum values: 2, 4, 6Nm) against the pronation and supination motion respectively (figure 4). A perfect behaviour can be observed. The torque peaks at the end of the ROM are due to the fact that the user motion is suddenly limited by the virtual end-of-motion stops.

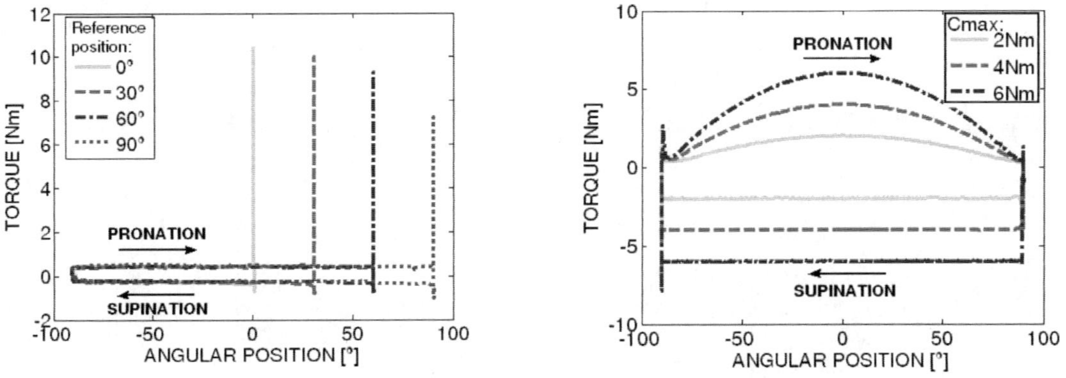

Figure 4. Torque vs position for the isometric and isotonic exercise modes

Results obtained for the isokinetic mode show that the angular speed of the pronation/supination motion can be kept almost constant whatever the torque applied by the user (figure 5), for various values of the reference speed (50°/s, 100°/s, 150°/s). Future work will be mainly focused on the statistical analysis of the reproducibility of the measurements made with the machine and of the agreement with measures obtained with existing commercial devices.

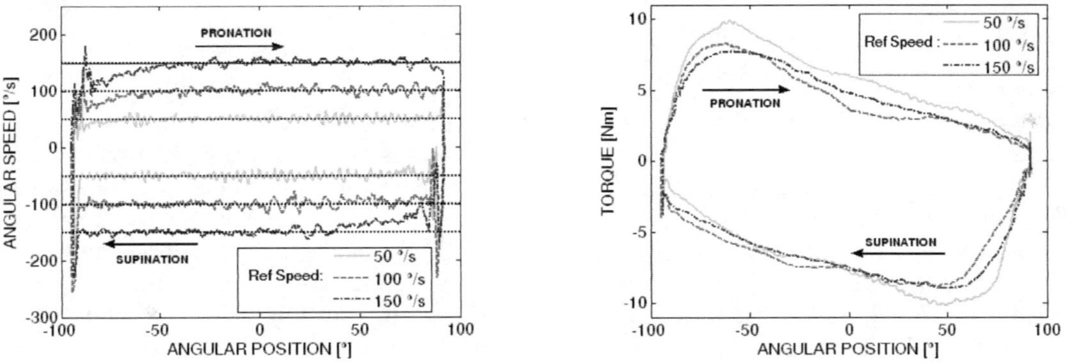

Figure 5. Speed & torque vs position for the isokinetic exercise mode

Acknowledgments

This work was supported by the Walloon Region under the FIRST Spin-off program. The MR-fluid used in the device is the AL-458, provided by the Fraunhofer-ISC, Würzburg.

References

[1] Perrin D 1993 *Isokinetic exercise and assessment* (Champaign, IL: Human Kinetics Publishers)
[2] Nikitczuk J, Weinberg B and Mavroidis C 2007 *Smart Materials and Structures* **16** 418-28
[3] Kikuchi T, Furusho J and Oda K 2003 *Proc. Int. Conf. on Robotics and Automation* (Taipei-Taiwan) 214-19
[4] Dong S, Lu K-Q, Sun J and Rudolph K 2006 *J. Intelligent Materials and Strucutres* **17** 555-61
[5] Avraam M, Horodinca M, Letier P and Preumont A 2008 *Proc. Int. Conf. on Intelligent Robots and Systems* (Nice-France) (to appear)
[6] Avraam M, Horodinca M, Letier P, Romanescu I and Preumont A 2006 *Proc. Actuator 2006* (Bremen-Germany) 836-839

Journal of Physics: Conference Series **149** (2009) 012034 doi:10.1088/1742-6596/149/1/012034

MR haptic joystick in control of virtual servo drive

P Bachman[1] **and A Milecki**[2]

[1] University of Zielona Gora, 65-246 Zielona Gora, ul. Podgórna 50, Poland
[2] Poznan University of Technology, 60-965 Poznan, ul. Piotrowo 3, Poland

E-mail: andrzej.milecki@put.poznan.pl

Abstract. In the paper the one-axis haptic joystick is proposed and described. This joystick utilizes a magnethoreological (MR) brake in order to obtain operator movement opposite force which creates feedback force. The MR brake investigations results are shown. The designed and built joystick was then connected to the computer, and used to control of prepared in Matlab/Simulink software, virtual model of electrohydraulic drive. The main equations describing this drive were shortly described. Finally the investigations stand and some results are presented.

1. Introduction

The development of computer systems enabled the creation of so called Virtual Reality (VR) environment. In this modern technology new devices used for human-computer interaction have to be applied. Applications of such technology encompass computer games, training and rehabilitation devices, minimally invasive surgery devices, teleoperation, steer by wire devices and many others. In most of these applications visual and sound signals are used by the operator as a feedback. However, it would be advantageous, to have additional feedback like feeling of the force, produced by the controlled device [1, 2, 4]. Thanks to it, the operator will have the illusion of being in direct contact with this device. The answer is a haptic device which can act as a coupling between the human sense and the controlled object. One of the most commonly used haptic devices, are joysticks. They are present on the market since more then 10 years [2, 4]. Such joysticks can be successfully applied also in communication with virtual environment giving human operator filling of the force generated by virtual drives.

In the paper the joystick with force feedback used to control of virtual electro-hydraulic servo drive is proposed. In this joystick, a brake able to produce electrically controllable movement opposite torque is applied. Unique properties of magnetorheological (MR) fluids enable to design such brakes. The paper presents the construction of joystick in which such a brake is applied. The virtual model of the electro-hydraulic drive and its connection to the joystick is described. Some aspects of control of electrohydraulic servo drive with joystick which uses MR fluid, are also shortly considered.

2. Joysticks with MR brakes

In the investigations we designed and build a joystick with MR brake (figure 1). This brake utilizes the shear mode. Its housing is movable and the rotor has to be fixed. The coil is placed in the middle of the rotor and its connection wires are placed in the middle of the rotor shaft. The torque M_o generated by an operator is in balance to the opposite torque produced by the joystick, exactly by the MR brake. The Laplace'a transfer of its output torque can be described as follows:

Journal of Physics: Conference Series **149** (2009) 012034 doi:10.1088/1742-6596/149/1/012034

$$M_d(s) = c_{d0} \cdot [s \cdot \Phi(s)] + \frac{c_{d1}}{R_d(T_d s + 1)} U_d(s) \quad (1)$$

where: M_d – joystick brake torque, Φ – angular rotor position, c_{d0} – viscosity damping coefficient (by absence of magnetic field) [N·s], c_{d1} – gain coefficient of the MR effect damping moment [N·m/A], U_d – supply voltage, R_d – brake coil resistance, T_d – total brake time constant.

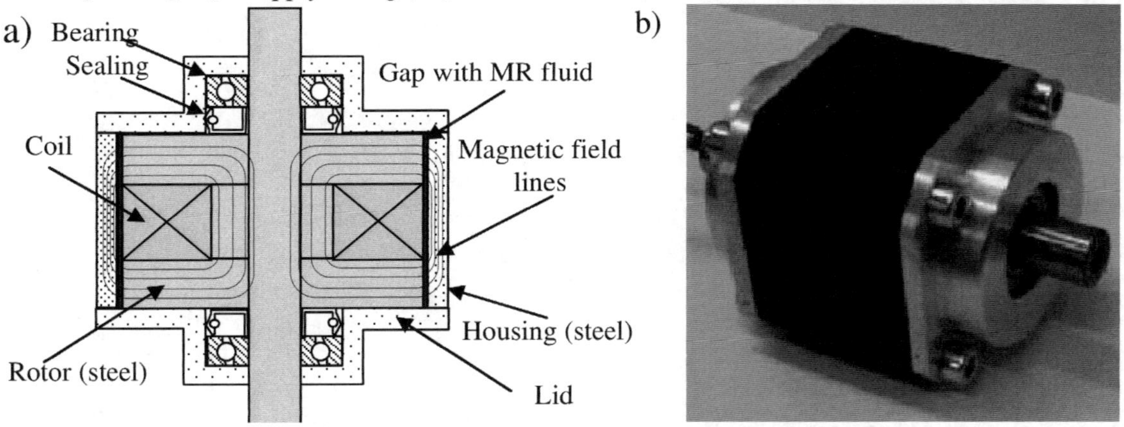

Figure 1. Design of MR brake and its photo.

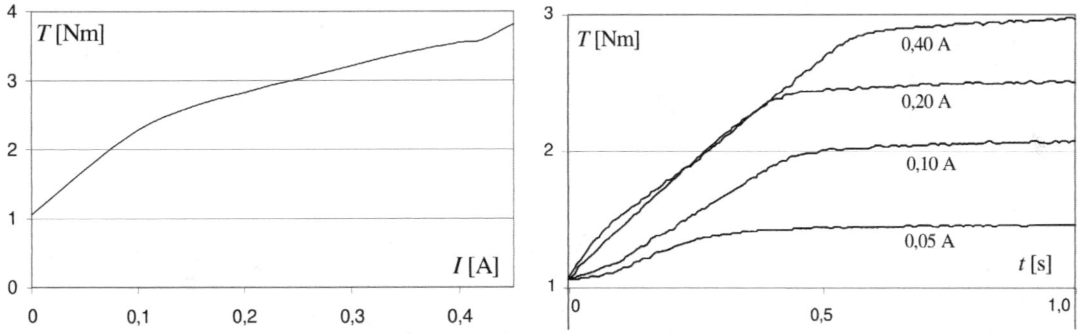

Figure 2. MR brake static and dynamic investigation results.

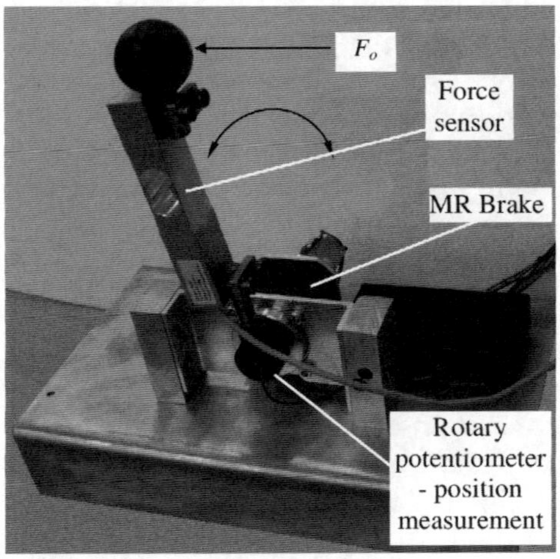

Figure 3. Joystick with MR brake.

In this equation the first component describes the non-magnetic behavior of the brake. Torque produced in this mode is proportional to the angular velocity. The second component is controlled by external signal (voltage U_d) and characterizes the brake output torque as a result of MR fluid properties changes in magnetic field.

In figure 2 investigation results of built by us MR brake are shown. The static torque reaches about 3.5 Nm when the coil is supplied by current equal to 0.4 A. However, the obtained by switching on a supply voltage, brake time constant is rather long and can be estimated at about 0.5 s.

The one axis joystick prepared for investigations is shown in figure 3. The position of its rotary arm is measured by rotary potentiometer. The joystick arm is created by tensometer transducer what enable the measurement of the force produced by an operator.

3. Virtual model of the hydraulic drive and its connection to the joystick

Electrohydraulic servo drives are widely used in many technological machines and devices [3]. Electrohydraulic drive consists of servo or proportional valve, hydraulic cylinder, position measure unit and a controller. The valve is usually regarded as a second order system. We assumed that in the drive proportional valve is used, which is described by following transfer function [3]

$$G(s) = \frac{x(s)}{U(s)} = \frac{200}{s^2 + 230s + 10000}, \tag{2}$$

where valve damping coefficient ζ_z = 1.15, natural frequency ω_z = 100 Hz and gain coefficient k_z = 0.02 (x – valve spool displacement, U – valve coil supply voltage) were specified basing on data taken from hydraulic proportional valve producer catalogue. The circuit: hydraulic amplifier-hydraulic cylinder was described as third order system (oscillatory element placed serial to an integrator). This system was described by transfer function taken as follows [3]

$$G(s) = \frac{y(s)}{x(s)} = \frac{15936437}{s^3 + 80s^2 + 159822s}, \tag{3}$$

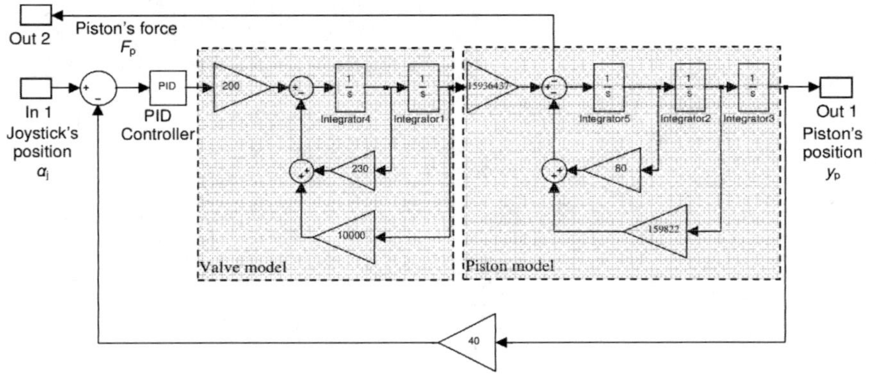

Figure 4. Hydraulic servo drive simulation model.

where y is the piston rod displacement. The parameters in this equation were calculated for following data: cylinder piston area 0.01 m^2, amplifier flow coefficient 1.0 m^2/s, dynamic friction coefficient 40 000 Ns/m, movable mass 500 kg and oil bulk modulus 10^9 Pa.

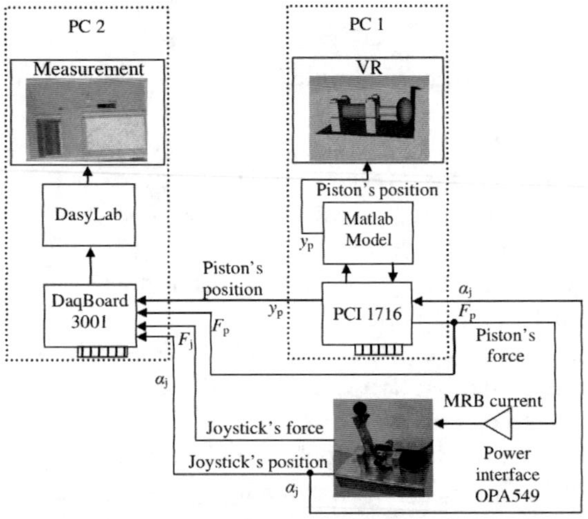

Figure 5. Measurement and investigation system's block diagram.

Basing on equations (2) and (3) the model was prepared in Matlab/Simulink software (figure 4). The drive model can communicate with the real word (here: joystick) by input and output: In 1 and Out 1 and 2, which are assigned to input/output card ports.

We used Virtual Reality Toolbox (VRLT) in Matlab software to present the movement of the cylinder piston rod and additionally to connect the joystick position measure element (In 1) and MR brake coil amplifier to the computer and to the virtual model (Out 2). In system described in this article we used as an output a servo-cylinder piston displacement signal. Available in VRLT VR Sink includes 3D library of different objects and graphic interface of the user combining simulation with virtual reality. However there are no hydraulic elements in a library and authors had to draw the hydraulic cylinder by themselves. The electrohydraulic servo drive was controlled by PID type regulator.

To the presented in figure 4 simulation model an artificial load (spring) was added. In our investigations we used two functions to describe this spring: linear $F_l = k_l \cdot y$ and non-linear $F_l = k_l \cdot y^2$. We also implemented a simulation of contact to a stiff wall, meeting and then shifting of a mass.

In figure 5 the block scheme diagram of investigation stand is shown. During the investigations we used two computers. The first of them (PC1) was used as VR system on which basing on Matlab/Simulink and VRLT, virtual reality environment is implemented. In this computer input/output card type PCI-1716 was installed and used to connect the VR model to haptic joystick output (potentiometer) and to its input (MR brake coil amplifier). The second computer (PC2) was used to collect data. In this computer the input/output card type DaqBoard 3001 was installed which was used to connect the signals taken from the joystick and the computer.

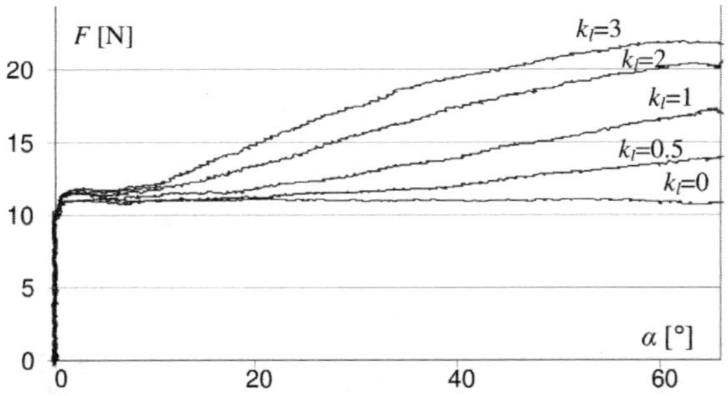

Figure 6. Results of simulation: joystick force in function of joystick angle for linear spring load.

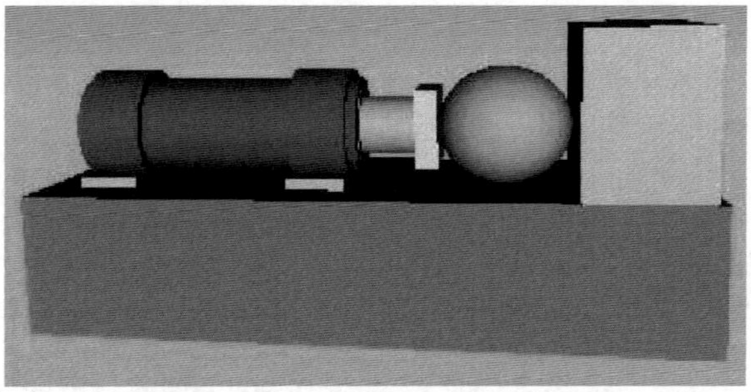

Figure 7. View of the VR screen.

In figure 6 the recorded, during virtual sphere squeezing, joystick force signal is shown. Figure 7 presents the screen of virtual reality system. The operator can watch the movement of the piston and the deformation of the virtual sphere. The force signal from the model was used to control the brake coil, thus the operator was able to feel this force on a joystick arm. As one can see, the force increases in function of joystick arm angle, what confirms the force feedback in a joystick.

4. Conclusion

The paper proposed one-axis joystick with MR brake which enable force feedback to communicate and interact with a virtual model of servo drive. The joystick was connected to the prepared virtual model of a drive. During the movement, the piston squeezed an artificial sphere. By the usage of MR brake in joystick, the impression of being in physical contact to the virtual drive and its environment was given to the human operator. Thanks to this, the operator-model interaction is more realistic.

Acknowledgments

This study was supported by Polish Ministry of Science and Education in the years 2006 to 2008 as a grant no 4 T07B 002 29.

References

[1] Adams R J and Hannaford B 1999 Stable haptic interaction with virtual environments *IEEE Transactions on Robotics and Automation*, vol. **15** 465.

[2] Harward V, 2006 Haptic synthesis *Proc. 8th Int. IFAC Symp. on Robot Control SYROCO*.

[3] Milecki A 2003 Liniowe serwonapędy elektrohydrauliczne. Modelowanie i sterowanie Poznan *Poznan University of Technology Publisher*.

[4] Nahvi A, Nelson D D, Hollerbach J M and Johnson D E 1998 Haptic manipulation of virtual mechanisms from mechanical CAD design *Proc. IEEE Intl. Conf. Robotics and Automation Leuven*.

Journal of Physics: Conference Series **149** (2009) 012035 doi:10.1088/1742-6596/149/1/012035

Performance Characteristics of Prototype MR Engine Mounts Containing LORD Glycol MR Fluids

Daniel E Barber and J David Carlson

LORD Corporation, 406 Gregson Drive, Cary NC 27511

E-mail: daniel_barber@lord.com

Abstract. LORD Corporation has recently developed glycol-based MR fluids for use in applications such as engine mounts and bushing, in which the MR fluid will contact rubber and other oil-sensitive elastomers. To demonstrate the performance characteristics of these fluids, prototype MR engine mounts were designed and their dynamic stiffness and damping were tested. In one configuration, the MR mount contained a simple MR valve and was filled with a glycol fluid containing 22% iron by volume. This mount had low dynamic stiffness and frequency-dependent damping in the off-state and higher dynamic stiffness with little damping in the on-state. The low-frequency stiffness and damping could be varied by adjusting the applied magnetic field. In a second mount configuration, the mount contained both an MR valve and an inertia track. Two such mounts were evaluated with MR fluids containing 22% and 36% iron, respectively. In the off-state, both mounts displayed low stiffness and damping as fluid flowed through the MR valve. In the on-state, the MR valve was closed and the inertia track was activated, giving characteristic frequency-dependent stiffness and damping that could be tuned by varying the strength of the magnetic field. These behaviors were observed at both low and high displacements of the mount.

1. Introduction

Over the past decade, commercial magnetorheological (MR) fluids and devices containing them have proven their value in providing dramatically improved ride comfort and performance in automotive applications [1]. There are presently more than 500,000 MR devices in use worldwide, and this number is expected to rise into the millions as more automotive platforms adopt MR fluid shock absorbers and other vibration-control technologies. Current commercial applications generally use oil-based MR fluids.

There is a growing commercial interest in applying MR technology to devices in which natural rubber or similar elastomers are a key structural component, particularly engine mounts and drive train or suspension bushings. Several US patents from LORD Corporation [2] and Delphi Technologies, Inc. [3] describe designs for MR fluid hydraulic mounts, and Choi et al. have described MR mounts for controlling vibration in passenger cars and other structures [4]. Paulstra Corporation has described the design and use of an MR suspension bushing in a twist-beam axle to give performance potentially comparable to a multi-link suspension [5]. MR fluids with either silicone oil [6] or glycol [7] as the carrier fluid have been described for use in natural rubber devices, since oil-based fluids would be incompatible with the elastomers in these devices.

Recently, LORD Corporation introduced glycol-based MR fluids for use in devices containing natural rubber or other elastomers, such as semi-active engine mounts, bushings, or similar vibration-control devices [8]. To demonstrate the potential effectiveness of such devices, prototype MR engine mounts have been developed and their performance characteristics tested in both off-state and on-state conditions. This paper describes the dynamic stiffness and damping characteristics of the MR mounts.

2. Experimental Details

LORD MR fluids with predominantly glycol carrier liquid were prepared using proprietary formulations and processes. Iron content of the various fluids ranged from 22% to 36% iron by volume.

A standard LORD hydraulic mount with inertia track and membrane decoupler (part number FL-1002-13, outer diameter 135 mm) was used as both the performance control and the prototype from which the MR mounts were designed. One MR mount was built in which the decoupler and inertia track assembly was replaced with a simple MR valve, and a second MR mount was built which had both an MR valve and an inertia track (Figure 1). The magnetic valves in both mounts had an average diameter of 46 mm, a magnetic gap of 2 mm, and an active length of 8 mm. The inertia track volume of the second mount was approximately 58 cm^3, and each mount contained approximately 500 cm^3 of the desired fluid. The elastomeric components were of the same type in all mounts.

Both types of MR mounts were filled with LORD MR fluid containing 22% iron using a vacuum-fill process. One mount with MR valve and inertia track was also filled with a LORD MR fluid containing 36% iron to assess the effect of higher fluid density, viscosity and magnetic strength. The dynamic stiffness and phase angle were measured at constant-amplitude sinusoidal displacements of either 0.1 mm or 1.0 mm and frequencies from 1-25 Hz with various applied magnetic field strengths.

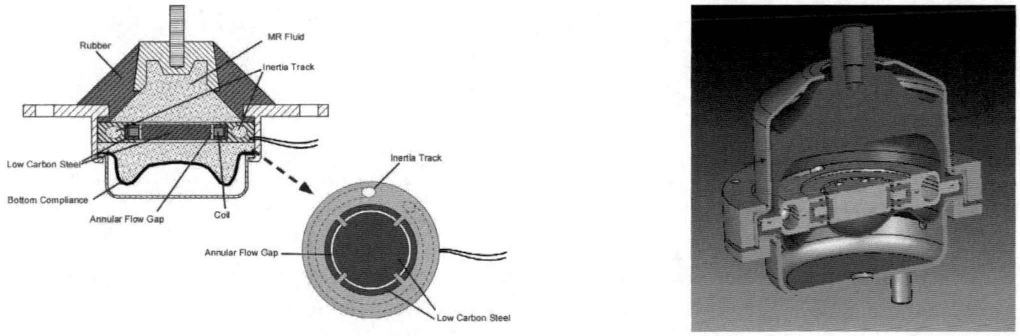

Figure 1: Schematic diagram and CAD drawing of LORD mount with MR valve and inertia track

3. Results and Discussion

3.1. MR Mount Performance Characteristics

Engine mounts are typically characterized by measuring their dynamic stiffness, K_{dyn}, and the loss angle θ as a function of vibration frequency. These parameters are related by equations (1) and (2),

$$K_{dyn} = \sqrt{(K')^2 + (K'')^2} \qquad (1)$$

$$\theta = \tan^{-1}\left(\frac{K''}{K'}\right) \qquad (2)$$

where K' is the elastic modulus and K'' is the out-of-phase modulus associated with viscous damping [9]. High values of θ correspond to high damping. Both mount configurations (MR valve only and MR valve plus inertia track) were tested at low (0.1 mm) and high (1.0 mm) displacements and various magnetic fields in the frequency range from 1 to 25 Hz.

3.1.1. *MR Mount with MR Valve Only*. Figure 2 shows the dynamic stiffness and phase angle at high (± 1.0 mm) displacement for the mount with an MR valve containing LORD MR fluid with 22% iron. Similar behavior was observed at low (± 0.1 mm) displacement. In the off-state, the mount had a lower-stiffness condition at low frequencies up to about 15 Hz and a higher-stiffness condition above this frequency, with the maximum damping occurring at the transition from low to high stiffness. This behavior is similar to that of a standard hydraulic mount (non-MR) with a simple orifice [9], so in the off-state the open MR valve acts as a fluid orifice with the shear stiffness of the rubber dominating the low-frequency behavior.

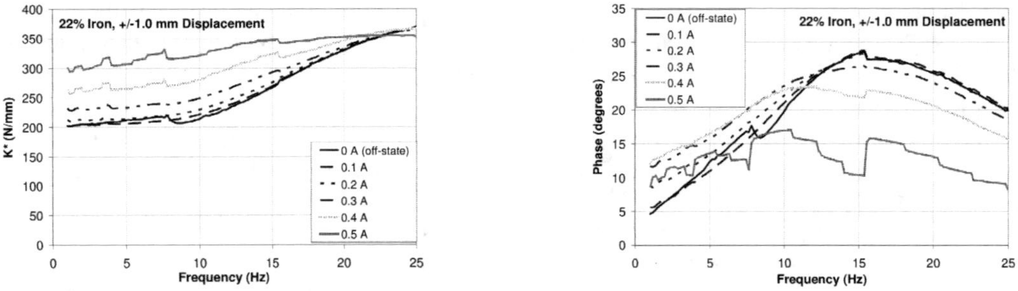

Figure 2: Dynamic stiffness (left) and phase angle (right) at ±1.0 mm displacement for mount with MR valve only and containing LORD MR fluid with 22% iron.

In the full on-state (0.5 A current), the MR valve is closed and no fluid flow occurs. The stiffness of the mount is now determined by the higher bulge stiffness of the rubber, and the mount behaves as a simple elastomeric mount with frequency-independent stiffness and low damping [9]. Variation in the low-frequency stiffness and damping was observed with varying magnetic field due to incomplete blocking of the MR valve. This behavior represents an opportunity for continuously tuning the low-frequency stiffness of such mounts. Similar behavior was observed for the multi-stage MR suspension bushing described by Piquet et al. [5].

3.1.2. *MR Mount with MR Valve and Inertia Track***.** Two mounts with this configuration were tested, each containing LORD glycol MR fluid with either 22% or 36% iron by volume. Figure 3 shows the dynamic stiffness and phase angle results for the mount containing 22% iron fluid at high (±1.0 mm) displacement. Similar results were observed at low and high displacements for a given iron content.

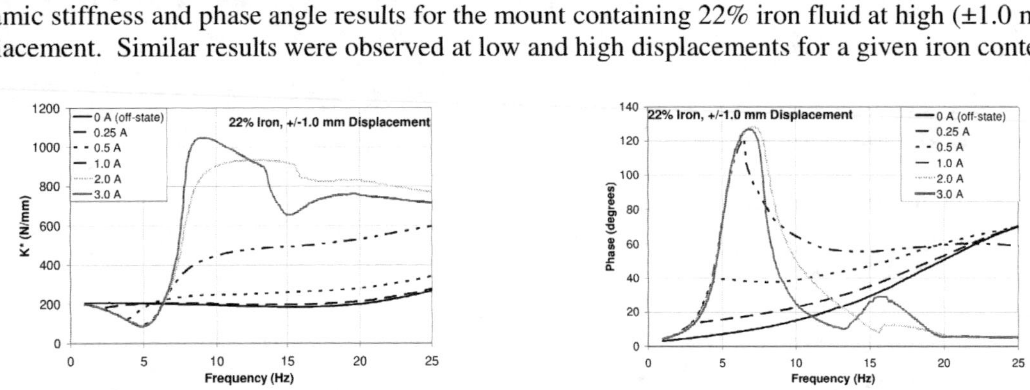

Figure 3: Dynamic stiffness (left) and phase angle (right) at high displacement (± 1.0 mm) for mount with inertia track and MR valve containing LORD glycol MR fluid with 22% iron.

For 22% iron (Figure 3), the off-state stiffness was low throughout the tested frequency range and the phase angle increased gradually from 10° or less to about 60°. For 36% iron, similar changes occurred but at somewhat lower frequencies. The curves show some similarity to those for the MR mount with valve only, although the features were somewhat broadened, possibly as a result of the additional orifice represented by the inertia track.

In the full on-state, the MR valve was essentially closed and classic inertia track behavior [9] was displayed by both mounts. The initial dynamic stiffness decreased to a minimum at a relatively low frequency (the low-stiffness "notch") and the damping reached a maximum due to resonant oscillation of the fluid mass within the inertia track. At higher frequencies the mount had high stiffness and low damping much like a simple elastomeric mount due to blocking of the inertia track by the fluid at high frequencies. As compared to the non-MR mount (Figure 4), the low-stiffness notch was much softer and the high-stiffness condition was much stiffer for the MR mounts due to the high density and high solids loading of the MR fluid as compared to glycol. The mount with 22% iron fluid had a deeper notch than the one with 36% fluid, probably due to the lower viscosity of the 22% fluid, but the high-stiffness values at higher frequencies were nearly the same. Curiously, the mount with 22% fluid at the highest current showed a second decrease in stiffness and increase in damping at about 15 Hz.

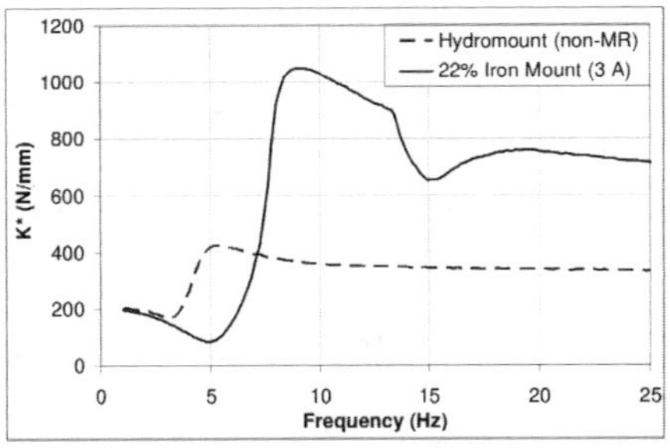

Figure 4: Comparison of dynamic stiffness at high displacement (± 1.0 mm) for standard LORD hydromount with inertia track (glycol-filled, non-MR) and MR mount with inertia track and MR valve in the on-state. MR mount was filled with 22% iron fluid.

At intermediate currents, more complex behaviors were observed due to incomplete blocking of the MR valve (Figure 3). This behavior allows some degree of tuning of the MR mount properties. For example, at currents above about 0.5 A, the low-stiffness notch was established and the higher-frequency stiffness varied over a relatively wide range.

4. Conclusion

Prototype MR engine mounts utilizing recently introduced glycol-based LORD MR fluids have been shown to give a highly tunable response. Combined with the known rapid response of the MR effect, such mounts have the potential for a high degree of vibrational motion control.

References

[1] J. D. Carlson, *Proceedings of the 10th International Conference on Electrorheological Fluids and Magnetorheological Suspensions*, World Scientific Publishing: Hackensack, NJ (2007), 389-395.
[2] J. D. Carlson, M. J. Chrzan, F. O. James., US Patent 5,398,917, Mar. 21, 1995.
[3] T. A. Baudendistel, S. G. Tewani, J. M. Shores, M. W. Long, R. E. Longhouse, C. S. Namuduri, A. A. Alexandridis, US 6,622,996 B2, September 23, 2003.
P. N. Hopkins, M. W. Long, M. O. Bodie, US 7,063,191 B2, June 20, 2006.
[4] S.-B. Choi, H.-H. Lee, H.-J. Song, J.-S. Park, *Proceedings of SPIE* 2002, *4701*, 1-8.
S.-R. Hong, S.-B. Choi, *J. Intelligent Material Systems and Structures* 2005, 16, 931-936.
[5] B. Piquet, C. A. Maas, F. Capou, SAE 2007 World Congress, Detroit, MI, April 16-19, 2007, SAE Technical Paper 2007-01-0850.
[6] V. R. Iyengar, P. N. Hopkins, M. E. Myers, US 7,070708 B2, July 4, 2006.
[7] V. R. Iyengar, T. J. Kacsandy, US 6,824,700 B2, November 30, 2004.
[8] News release, http://lord.com/Default.aspx?tabid=3879&Page=1&DocID=1134, June 9, 2008.
[9] W. C. Flower, *SAI Surface Vehicle Noise and Vibration Conference,* Traverse City, MI, May 15-17, 1985, SAE Paper 850975.

Magneto-rheological fluids redispersibility - a factorial design study of phosphate shell on carbonyl iron powder with dispersing additives

Antonio J F Bombard[1], Laís S Antunes[2], Pedro P Balestrassi[2] and Anderson P Paiva[2]

[1]Instituto de Ciências Exatas, [2]Instituto de Engenharia da Produção e Gestão
Universidade Federal de Itajubá – UNIFEI
Av BPS, 1303, Itajubá – MG, 37.500-903, BRAZIL

E-mail: bombard@unifei.edu.br

Abstract. We showed, in a previous paper, that Magneto-Rheological Fluids (MRFs) have different rheology when prepared with Carbonyl Iron Powders (CIP) phosphate (coated or uncoated). This was especially so when done without a magnetic field. This paper employs factorial design to examine the redispersibility and rheology of some MRF formulations; we use the same CIPs but with different dispersing additives. The factors are: CIP A (uncoated) or B (phosphate shell); additives with carboxylic acid or primary amine as the polar group; and n-octyl (C_8H_{17}) or n-dodecyl ($C_{12}H_{25}$) as the alkyl hydrocarbon chain (R-). CIP B was much more redispersible than CIP A, especially with amine additives; typical work values were < 5mJ @ 20 mm depth. In terms of viscosity, CIP A generated lower values, at shear rates above 100 s^{-1}. It also realized higher yield stress values (H_o = 300 kA/m) than CIP B (50% and beyond).

1. Introduction

As shown in a previous paper [1], Magneto-Rheological Fluids (MRFs) prepared with phosphate coated or uncoated carbonyl iron powders (CIP) have different rheology. This happens because Hydrophilic-fumed silica interacts with the phosphate (via hydrogen bonding) and it is especially true when a magnetic field is absent. The challenge in MRF formulation [2, 3] is redispersibility. This quality was measured, according to the test described by *Kieburg et al* [4]. In this test the work is measured by how far a steel blade, penetrates vertically downward into a test tube, at a constant speed. Each test tube contained 10 mL of an MRF formulation that was centrifuged for 15 minutes at 2000 'g'. This paper makes use of factorial design of experiments [5, 6] to study the redispersibility of some MRF formulations; the formulations include two carbonyl iron powders with different dispersing additives. The work (mJ) comprises the response variable for the factorial design; it was measured with a normal force cell built-in rheometer (Anton Paar – Physica MCR-301).

2. Experimental

The MRF formulations were made with a food grade, poly(alpha-olefin) oil. In addition to this oil (balance) the formulations consisted of the dispersing additive (0.8% w/w) and the CIP (80% w/w). A modified montmorillonite clay (0.3% w/w) was used as thixotropic agent. In each MRF formulation a high shear homogenizer (Ika – Turrax T-18) dispersed the CIP. The MRF samples were prepared in

duplicate and measured twice. More details about the CIP can be found in reference [1]. All the rheological measurements were obtained with a stress controlled rheometer Physica MCR-301.
Thus, the factorial design is 2^3 (2 levels, 3 factors). Table 1 displays the full cube of possible combinations, with duplicate samples, and randomized order.

3. Results and Discussion

Table 1. MRF 2^3 factorial design and yield stress values

Std Order	CIP	Polar Group	Alkyl	Yield 1	Yield 2	Yield 3	Yield 4	Yield mean
9	A	NH_2	Octyl	17.0	15.9	15.9	-	16.3
16	B	COOH	Dodecyl	7.1	7.6	7.1	9.1	7.7
11	A	NH_2	Dodecyl	16.8	18.0	17.4	-	17.4
4	B	NH_2	Dodecyl	10.5	10.0	10.9	11.2	10.7
2	B	NH_2	Octyl	11.2	12.2	-	-	11.7
15	A	COOH	Dodecyl	17.6	18.1	17.1	-	17.6
7	A	COOH	Dodecyl	18.1	18.5	18.1	-	18.2
1	A	NH_2	Octyl	18.5	16.0	17.5	18.4	17.6
14	B	COOH	Octyl	11.6	10.5	9.4	11.2	10.7
5	A	COOH	Octyl	15.4	13.6	14.0	-	14.3
12	B	NH_2	Dodecyl	10.3	9.8	9.7	9.0	9.7
3	A	NH_2	Dodecyl	18.5	18.2	17.1	-	17.9
6	B	COOH	Octyl	13.5	10.2	11.3	11.6	11.7
10	B	NH_2	Octyl	9.8	10.2	9.7	-	9.9
13	A	COOH	Octyl	14.0	14.2	14.0	-	14.1
8	B	COOH	Dodecyl	7.8	7.7	8.1	-	7.9

Figures 1 and 2 show the redispersibility as curves of force versus depth, for carbonyl iron powders A (without phosphate) and B (with phosphate shell). One sees first that Powder B, with its smooth force curves and absence of peaks, generally did better than A. With the exception of the octanoic acid, Powder B was nearly additive-independent. The second thing one sees is that with Powder B amines outperformed acids while with Powder A acids outperformed amines.

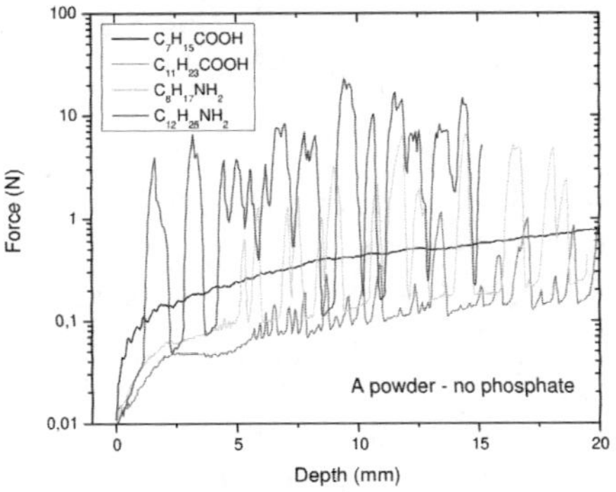

Figure 1. Force as function of blade depth, for MRF prepared with CIP A.

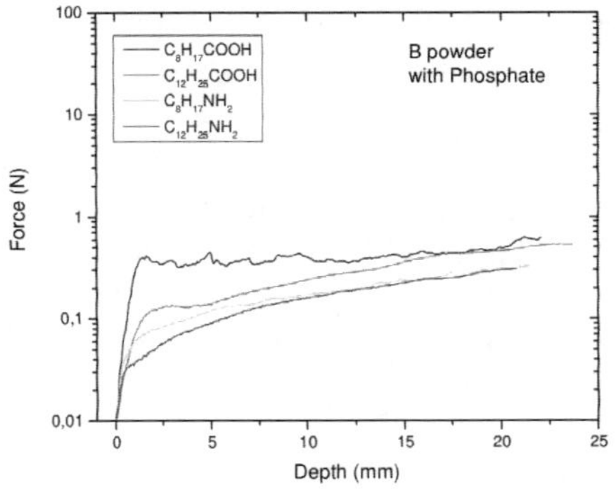

Figure 2. Force as function of blade depth, for MRF prepared with CIP B.

(Dispersing additives: 1% w/w of Octanoic Acid - Black; Dodecanoic Acid – Red; n-Octylamine - Green; or n-Dodecylamine – Blue.)

Figure 3 shows the viscosity curves (no field) for the 8 MRFs. Powder B generated MRFs with higher viscosities, above 100 s^{-1}. Figure 4 shows the shear stress versus shear rate for the 8 MRFs, under applied field H_o = 300 kA/m. It is evident that Powder A was better than B since the yield stress values of MRFs with Powder A were higher than with Powder B. Curiously, the additives seem to have an effect, though secondary, on the yield value.

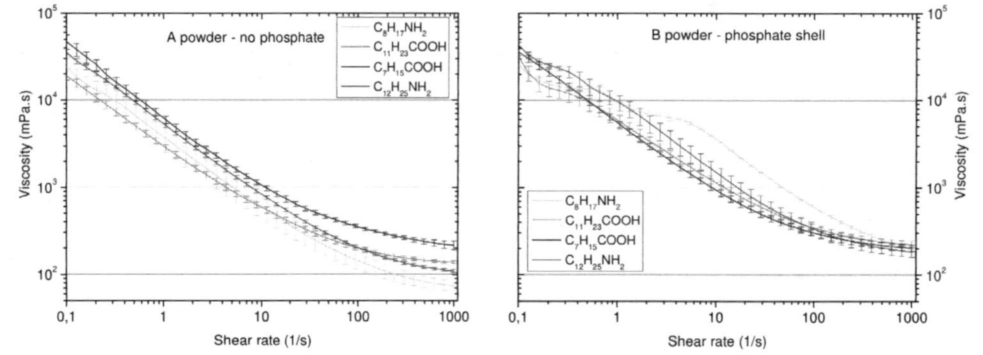

Figure 3. Viscosity curves for the MRF with CIP A (left) and B (right) without magnetic field (T = 25°C. Each point is the mean of the sample duplicate and two measurements with an SD bar. Same color legend as Figures 1 & 2.)

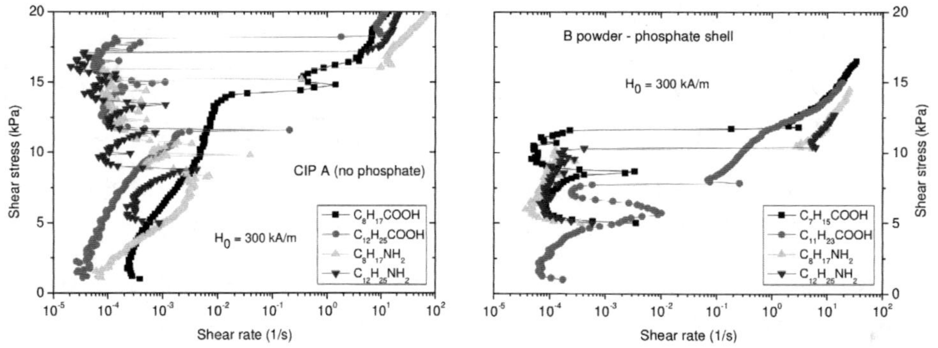

Figure 4. Flow curves for MRF prepared with CIP A (left) and B (right), measuring yield stress. (Applied field strength H_o = 300 kA/m, measured as an empty gap. Colors as before.)

Table 2 resumes the terms, coefficients, standard error, T and P-values of the factorial analysis. Boldface emphasizes relevant factors and interactions. Figure 5 shows the main effects, interactions, cube plots, and the Pareto chart for the yield stress response. Powder type, the evidence reveals, produces the main effect; phosphate coating reduces the yield stress. Polar group reveals an important effect and the HC chain length's effect is irrelevant.

Table 2. Estimated effects and coefficients for yield stress mean (coded units)

Term	Effect	Coef	SE Coef	T	P
Constant		13.331	0.1728	77.13	**0.000**
Powder Type	- 6.696	- 3.348	0.1728	- 19.37	**0.000**
HC Chain Length	0.115	0.057	0.1728	0.33	0.749
Polar Group	1.125	0.563	0.1728	3.25	**0.012**
Powder Type*HC Chain	- 2.110	- 1.055	0.1728	- 6.11	**0.000**
Powder Type*Polar Group	- 0.117	- 0.058	0.1728	- 0.34	0.744
HC Chain*Polar Group	- 0.060	- 0.030	0.1728	- 0.17	0.866
Powder Type*HC Chain*Polar Group	1.431	0.716	0.1728	4.14	**0.003**

S = 0.69 R^2 = 98.22% R^2 (adj) = 96.66%

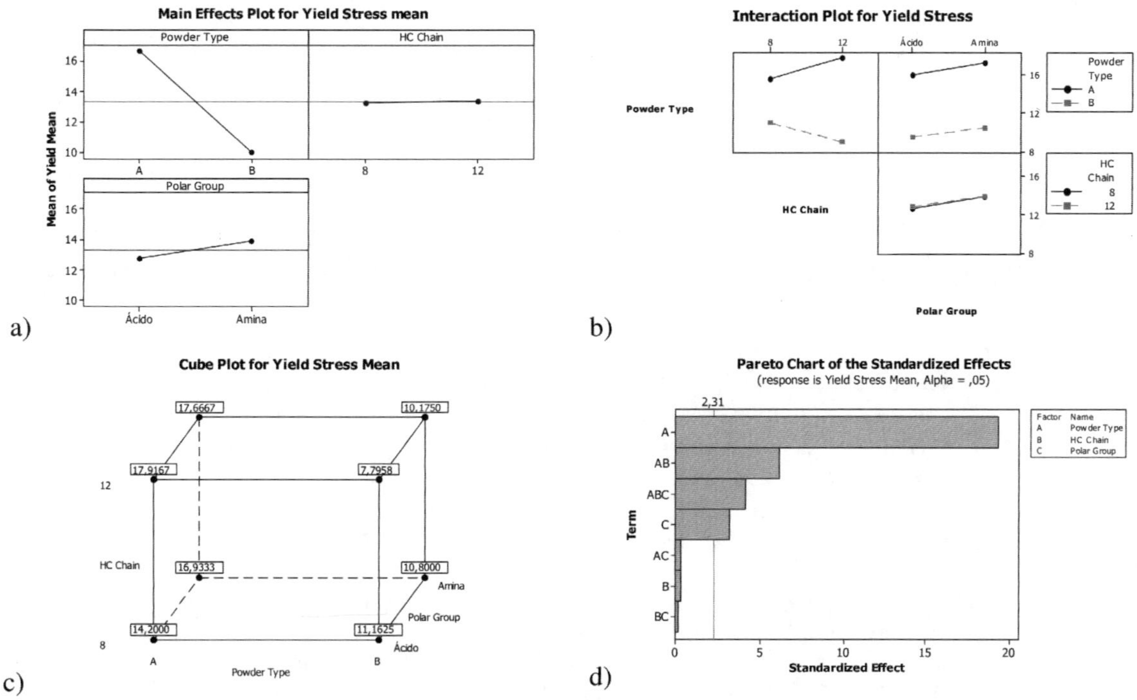

Figure 5. Factorial plots for yield stress mean analysis

a) Main effects plot; b) Interaction Plot; c) Cube Plot; d) Pareto Chart.

4. Conclusions

Carbonyl Iron Powder B (with phosphate shell) was much more redispersible than Powder A (without phosphate); this was especially true with amine additives. Powder A brought about lower viscosities, at shear rates above 100 s^{-1}. Powder A also generated higher yield stress values (H_o = 300 kA/m), at a rate of 50% and higher. It seems that additives also affected MRF behavior in magnetic fields. In the future, this aspect will be investigated using oscillatory tests.

References

[1] Bombard A J F, Alcantara M R, Knobel M and Volpe P L O 2007; "Phosphate coating on the surface of carbonyl iron powder and its effect in magnetorheological suspensions"; *Int. J. Mod. Phys. B* **21** 4858-4867

[2] Phule P P, Mihalcin M P and Genç, S 1999 *J. Materials Research* **14** 3037-3041

[3] Lopez-Lopez M T, Zugaldia A, Gonzalez-Caballero F and Duran J D G 2006 *J. Rheology* **50** 543-560

[4] Kieburg C *et al* 2007 *Proc. 10th Int. Conf. on ER Fluids and MR Suspensions -Lake Tahoe, USA* (Singapore: World Sci. Pub) pp. 101-107.

[5] Böse H 1999; "Investigations on zeolite-based ER fluids supported by experimental design"; *Int. J. Mod. Phys. B* **13** 1878-1885

[6] Box G E P, Hunter J S and Hunter W G 2005 *Statistics for Experimenters: Design, Innovation, and Discover - 2nd Edition*, (Hoboken, NJ: Wiley-Interscience)

Acknowledgments

Bombard wish to thank FAPEMIG for the Grant # CEX APQ-2676-5.02/07 as well as the financial support to attend the conference (Grant CEX-213/08). CNPq (Brazilian Council for Sci. & Tech. Development) post-doc fellowship is also acknowledged. BASF SE help was also gratefully appreciated.

Journal of Physics: Conference Series **149** (2009) 012037 doi:10.1088/1742-6596/149/1/012037

Off-state viscosity and yield stress optimization of magneto-rheological fluids: a mixture design of experiments approach

Rafael F Ierardi[1] and Antonio J F Bombard[2]
[1] Instituto de Engenharia Mecânica, [2] Instituto de Ciências Exatas
Universidade Federal de Itajubá – UNIFEI
Av BPS, 1303, Itajubá – MG, 37.500-903, BRAZIL

E-mail: bombard@unifei.edu.br

Abstract. The purpose of this paper was to optimize the blend ratio of ternary mixtures of carbonyl iron powders with different modes to get the minimum "off-state" viscosity as well as to maximize the yield stress under field in MRF. The Mixture Design of Experiments methodology was here employed considering the powders 'A' (coarse); 'B' (medium); and 'C' (fine) as mixture components and the viscosity and yield stress as mixture responses. The minimum viscosity was found over the A-C triangle edge of a response surface diagram. On the other hand, the maximum yield stress was found on the C corner (no blend). The results show the power and robustness of DOE methodology to optimize and to tailor MRF performance, according to the desired application.

1. Introduction

Magneto-Rheological Fluids (MRFs) are composite or multiphase materials. They are commonly prepared with some ferromagnetic phase dispersed in a chemically inert, non-volatile liquid. Typical materials include iron powders and mineral, silicone or synthetic oils. MRFs have the ability to change their rheological properties. When a magnetic field is applied, MRFs can change from a fluid to a near solid in a reversible way – with a relatively quick time of response (~10 milliseconds). The rheological properties are controlled by the magnetic field strength, which makes MRFs very promising for several technological devices, especially in mechatronics.

The magneto-rheological effect increases with the volume fraction of iron powder that is the active phase. However, one of the challenges in formulating good MRFs is to keep the so-called *off state* viscosity – the viscosity without magnetic field – as low as possible. Since the earlier works of *Mooney* [1], *Krieger-Dougherty* [2], *Thomas* [3], and *Frankel and Acrivos* [4] it is well established that the relative viscosity of any suspension increases exponentially with the volume fraction.

It is also well-known that mix powders with different particle sizes and/or size distribution can be advantageous. Chemical and food engineering processes, powder metallurgy and ceramics industries, etc. have used this approach with success [5]. *Farris* reported that there is substantial reduction on the relative viscosity in multimodal suspensions – when compared to unimodal suspensions – if the total volumetric fraction Φ is above 55% v/v and depending both on the size ratio and volume fraction of large particles (Φ_L/Φ) [6].

The purpose of this paper was to optimize the blend ratio of ternary mixtures of spherical iron powders with different modes to get the minimum "off-state" viscosity as well as to maximize the yield stress under field in Magnetorheological Fluids. The Mixture Design of Experiments methodology [7] was here utilized considering the powders 'A' (coarse), 'B' (medium), and 'C' (fine) as mixture components and the viscosity and yield stress as mixture responses.

2. Experimental

In order to evaluate the effect of mixing carbonyl iron powders with different particle sizes, we choose three carbonyl iron powders manufactured by BASF SE, Ludwigshafen, Germany. Table 1 summarizes some properties of these powders. The size distribution and chemical analysis, according to the manufacturer. The magnetic properties of dry, pure powders were measured with a SQUID magnetometer, Quantum Design MPMS-7. The saturation was measured at 50.000 Oe. The differential susceptibility was calculated at H = 3400 Oe, because the yield stress values of the MRF were measured at this field strength in the rheometer.

Table 1. Some properties of the BASF CIP used in the mixtures.

	Size (μm)			Magnetic properties		Analysis
CIP	D_{90}	D_{50}	D_{10}	M_s (emu/g)	dM/dH @ 3400 Oe (emu/g.Oe)	Fe (%)
A (coarse)	25	9	5	248	0.018	> 99.5
B (medium)	18	6	3.5	227	0.014	> 99.5
C (fine)	3.4	1.7	0.7	237	0.024	> 99.5

All the mixtures were prepared with a synthetic hydrocarbon oil (a polyalphaolefine). Dispersing additive as well as a thixotropic additive were used in all formulations, at the same concentrations. Each MRF was homogenized using a mechanical disperser Ika Turrax T-18, at 10,000 rpm, for 10 minutes. The total volume fraction of iron powder in all formulations is ~ 50% v/v.

The flow curve of each MRF sample was measured in a controlled stress rheometer (Anton Paar – Physica MCR 301) equipped with a magnetorheological device (Physica MRD-180) and using plate – plate geometry (20 mm diameter, 1 mm gap). The temperature was kept constant at 25°C. The profile to measure the viscosity was a shear rate ramp, increasing linearly from 1 to 1000 s^{-1}. The region between 200 – 1000 s^{-1}, visually considered linear was used to calculate the apparent viscosity, considered as the slope of the curve shear stress versus shear rate. Each experimental run was repeated three times.

The yield stress was measured through shear stress linear ramp, after applying the maximum field strength of MRD-180 (H_o ~ 3400 Oe) to the MRF sample. The shear stress value associated with a sharp drop in viscosity was taken as the yield value.

3. Results and Discussion

To obtain the data to feed the statistical analysis, apparent viscosity or yield stress values, each MRF blend was measured at least three times. Figures 1 and 2 are examples of flow curves used to measure off state viscosity and yield stress, respectively.

Table 2 resumes the proportions of each MRF blend; and the values for "off" plastic viscosity and/or yield stress values (mean and standard deviation, triple repetition, at least). Figure 3 and 4 show the mixture contour plots for apparent viscosity and yield stress, respectively.

To increase the maximum packing volume fraction and to reduce the viscosity for powders of uniform spheres with different size ratios, optimum blend ratio of binary, ternary, quaternary or quinary mixtures was well established by *Lee*: it should be the blend 73.5:26.5 of large/small powder, below 16:1 of size ratio. [8]. Besides, *Dinger and Funk* [9] – based on the seminal work of *Andreasen and Andersen* [10] – have shown that it is possible to reduce the viscosity of powder dispersions with mixtures. This, however, requires mixing at least two decades of particle size. Three or more decades are often used. The ideally infinite number of decades is obviously impossible. In such way the gross particles are the main components, and the amount of others decreases with size reduction.

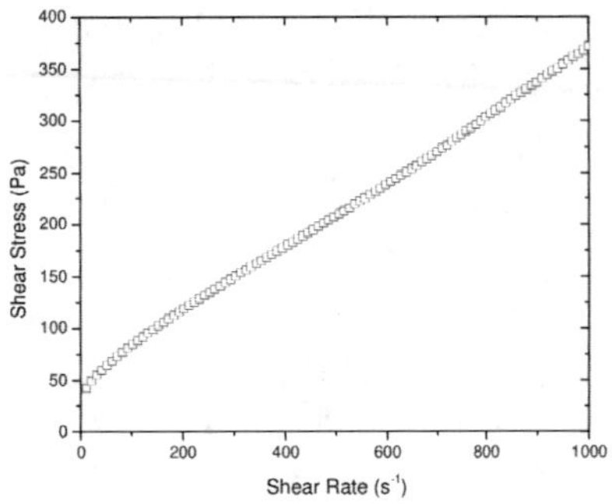

Figure 1. Flow curve of a MRF blend. Apparent viscosity, as the slope above 200 s^{-1}. No field.

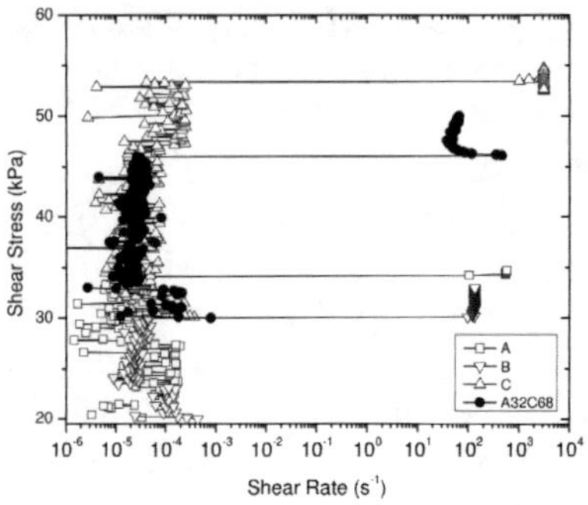

Figure 2. Flow curves under applied field (H_0 = 3400 Oe), used to measure the yield stress value.

Table 2. MRF blend proportions, apparent viscosity (no field) and yield stress @ H_o = 3400 Oe (mean values and standard deviation).

Std Order	A	B	C	Viscosity mean (mPa.s)	Viscosity sd	Yield mean (kPa)	Yield sd
1	1	0	0	-	-	33.7	0.497
2	0	1	0	-	-	30.1	0.424
3	**0**	**0**	**1**	479	47.23	**55.0**	4.92
4	0.5	0.5	0	1757	28.36	26.9	1.21
5	0.5	0	0.5	410	29.51	40.7	1.59
6	0	0.5	0.5	476	22.28	49.2	0.805
7	0.333	0.333	0.333	454	78.21	38.0	1.62
8	0.666	0.167	0.167	955	115.25	31.9	1.80
9	0.167	0.666	0.167	1026	42.12	34.6	1.14
10	0.167	0.167	0.666	346	2.89	44.7	0.702
11	0.90	0.05	0.05	926	97.69	-	-
12	0.05	0.90	0.05	1697	174.32	-	-
13	0.25	0.25	0.50	363	7.00	41.7	1.57
14	0.20	0.20	0.60	362	13.12	-	-
15	0	0.25	0.75	408	11.27	48.5	1.63
16	0.25	0	0.75	379	12.22	43.7	1.28
17	0.05	0.15	0.80	402	5.03	-	-
18	0.10	0.10	0.80	340	12.66	48.0	0.785
19	0.30	0.10	0.60	325	21.45	42.5	1.89
20	0.10	0.30	0.60	386	9.02	44.3	0.797
21	0.05	0.05	0.90	369	18.25	-	-
22	0	0.30	0.70	399	15.52	-	-
23	0.207	0.160	0.633	315	5.00	-	-
24	**0.323**	**0**	**0.677**	**269**	8.33	45.3	0.666
25	0.25	0.75	0	-	-	26.0	0.911
26	0.75	0.25	0	-	-	25.0	2.05
27	0.75	0	0.25	-	-	32.2	0.616
28	0	0.75	0.25	-	-	38.9	1.448

For powders with log-normal size distribution and mixing less than one size decade, the approach previously mentioned does not work. More recently, *He and Ekere*, through computer simulation, have suggested that for concentrated suspensions of non-colloidal particles with log-normal distribution, the viscosity decreases as the standard deviation of the particle diameters increases [11].

Journal of Physics: Conference Series **149** (2009) 012037 doi:10.1088/1742-6596/149/1/012037

Two US patents report that to maximize "on-state" force of MRF, bimodal particle distribution is better than unimodal [12,13]. We confirmed this in a previous paper [14], but this work shows that question remains open.

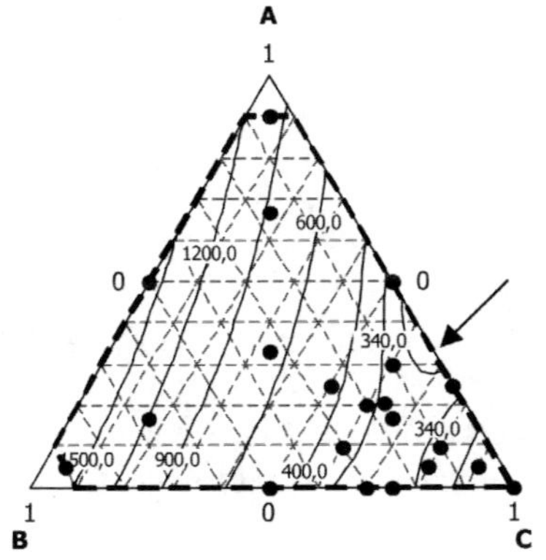

Figure 3. Mixture contour plot of mean viscosity. No field, 25°C. Arrow indicates a region where viscosity is below 300 mPa.s

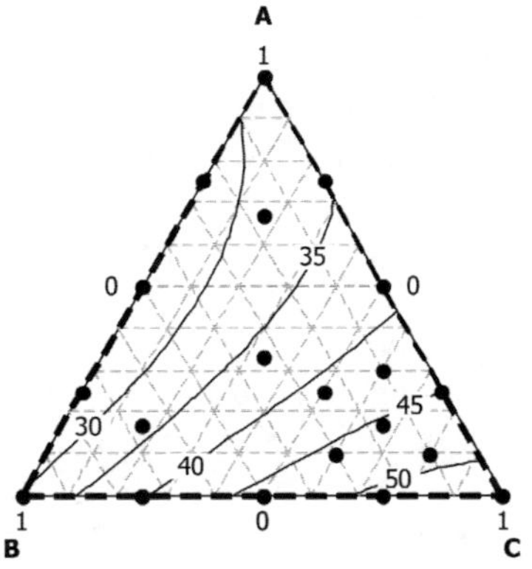

Figure 4. Mixture contour plot of mean yield stress, under applied field strength H_0 = 3400 Oe. Contour lines in kPa.

As reported by Kordonski and co-workers [15], the magnetic properties, as saturation magnetization or magnetic susceptibility could be the reason for the higher yield stress of MRF with powder C. Thus, the figure 5 shows the magnetization curves for the three powders, until the range used to measure yield values in the rheometer. One can see, from figure 5, that powder C has the lower magnetization at the field strength used to measure yield stress. But susceptibility also has a role in the yield stress. Table 2 also shows susceptibility values. Perhaps, the susceptibility of powder C, even though slightly higher than powders A and B, could be the answer for this behaviour. However, the magnetic induction B was measured through a Hall probe in the magnetorheological cell, for the MRF with powders A, B or C (blends 1, 2 or 3 of table 2), and its value was the same: B = 433 ± 2 mT. Thus, we think that the stronger structure of MRF with only powder C, under magnetic field, is due to particle size and or size distribution. Figure 6 shows the cumulative undersize curves for the powders A, B and C, measured by Low Angle Laser Light Scattering (LALLS), with a Malvern Microsizer Master.

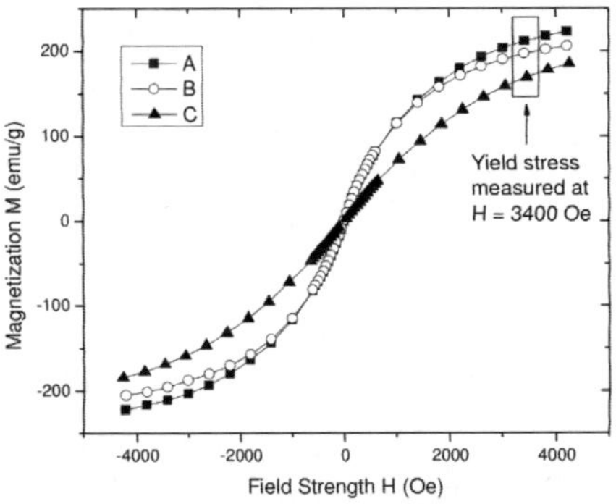

Figure 5. Magnetization curves for the pure powders used in this study. The arrow indicates the maximum field strength in the rheometer.

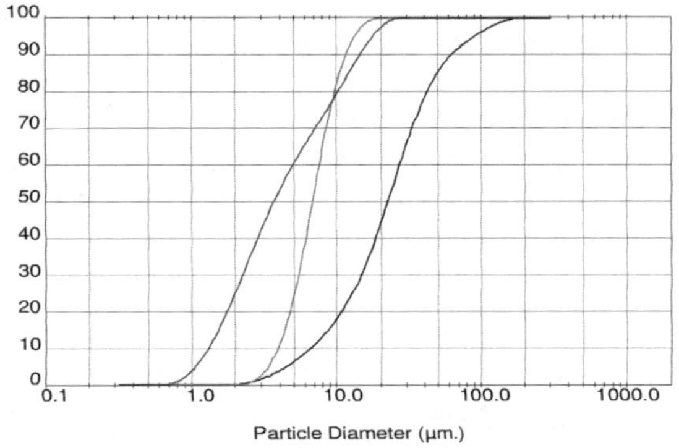

Figure 6. Cumulative undersize curves measured by LALLS, for powders A, B and C (right to left). Volume based size distribution.

According to the *Lee* [8], for a ternary blend to show reduced viscosity with relation to the pure components, the lowest size ratio between them should be 25:5:1. As the size ratio of A-B-C, taking in account the D_{50} values measured by LALLS is approximately 21:7:3, we did not find any ternary blend which minimized the viscosity.

4. Conclusions

The optimal solution for minimum off viscosity was obtained with binary blends, being the mixture of powders A-C 32:68 (MRF Blend 24th) correctly predicted as that of lower apparent viscosity. When it was prepared and measured – not by the chance, trial and error or serendipity – the apparent viscosity was ~ 270 mPa.s. Therefore, Mixture D.O.E. (Design of Experiments) was able to guide for viscosity optimization.

On the other hand, looking for maximum yield stress, the highest value (σ_y ~ 55 kPa @ 3400 Oe), was obtained with blend #3, using only 'C' powder, the finer CIP.

We found a blend which minimize the off viscosity, but not a unique MRF with the two ideal responses: minimum viscosity AND maximum yield stress. *A priori*, it seems that there is no advantage of ternary blends over binary ones.

References

[1] Mooney M 1951 *J. Colloid Sci.* **6** 162–170
[2] Krieger I M and Dougherty T J 1959 *Trans. Soc. Rheology* **3** 137-152
[3] Thomas D G 1965 *J. Colloid Sci.* **20** 267
[4] Frankel N A and Acrivos A 1967 *Chem. Eng. Sci.* **22** 847
[5] German R M 1989 *Particle Packing Characteristics* (Princeton: Metal Powder Industries Federation)
[6] Farris R J 1968 *Trans. Soc. Rheol.* **12** 281 – 301
[7] Cornell J 2002; *Experiments with Mixtures: Designs, Models, and the Analysis of Mixture Data* – 3rd edition (New York, NY: Wiley)
[8] Lee D I 1970 *J. Paint Technology* **42** 579 – 587
[9] Funk J E and Dinger D R 1993 *Predictive Process Control of Crowded Particulate Suspensions: Applied to Ceramic Manufacturing* (Norwel, USA: Kluwer Academic Pub.)
[10] Andreasen A H M and Andersen J 1930 *Koll. Zeitschr.* **50** 217 – 228
[11] He D and Ekere N N 2001 *AIChE Journal* **47** 53 – 59
[12] Foister R T 1997, "Magnetorheological fluids", *US Patent 5,667,715*
[13] Weiss K D, Carlson J D and Nixon D A 2000, *US Patent 6,027,664*
[14] Bombard A J F *et al.* 2005, *Int. J. Modern Phys. B* **19** 1332-1338
[15] Kordonski W, Gorodkin S and Zhuravski N 2001, *Int. J. Modern Phys. B* **15** 1078-1084

Acknowledgments
Bombard wishes to thank FAPEMIG for the Grant # CEX APQ-2676-5.02/07 as well as Grant CEX-213/08 to attend ERMR'08. Mr. Ierardi acknowledges CNPq (Brazilian Research Agency - Scientific Initiation Program). BASF SE help is gratefully appreciated.

Journal of Physics: Conference Series **149** (2009) 012038 doi:10.1088/1742-6596/149/1/012038

Redispersibility in magnetorheological fluids: surface interactions between iron powder and wetting additives

Antonio J F Bombard[1], Laís S Antunes[2] and Douglas Gouvêa[3]

[1,2]Instituto de Ciências Exatas - Universidade Federal de Itajubá – UNIFEI
Av. BPS, 1303, Itajubá/MG, 37.500-903, BRAZIL
[3]Dept. of Metallurgical & Materials Eng. - Escola Politécnica – USP
Universidade de São Paulo, Av. Prof. Mello Moraes 2463, São Paulo/SP, 05508-900, BRAZIL

E-mail: bombard@unifei.edu.br

Abstract. Our aim in this work was to investigate the interactions between 3 carbonyl iron powders (CIP) and different wetting additives, looking for to understand how these interactions affects the rheology and redispersibility of magnetorheological fluids (MRF). The powders were named: 'A' (uncoated), 'B' (with silica coating), and 'C' (with iron III oxide coating). The additives studied were alcohols, amines and carboxylic acids or polymeric dispersants, mainly with n-octyl or n-dodecyl hydrocarbon chains. The effect of additives concentration was also studied. We conclude that the redispersibility of MRF is strongly dependent on both carbonyl iron powder surface properties and choice and concentration of dispersing additives. The type of iron powder modification also has a significant role on the viscosity of MRF formulations additionally to the known particle size effects.

1. Introduction

Several classes of dispersing/wetting additives are discussed in literature for magnetorheological fluids (MRF) formulation [1,2,3]. However, at best of our knowledge, few papers focus on the interaction of powder surface and dispersing agents [4,5].

In this work, the effects of three carbonyl iron powders (BASF SE) with different surfaces (A – uncoated, B – silica coated and C – iron III oxide coated), and its interactions with three additives named 1 (poly-12-hydroxy-stearic acid - PHSA); 2 (1-dodecylamine) and 3 (1-dodecanol/1-tridecanol blend) on the redispersibility and rheology of 28% v/v magnetorheological fluids (MRF) was investigated. Redispersibility is a challenge in MRF formulation [6,7,8]. When the iron powder in a MRF settles and creates hard sediment, difficult to redisperse, the MRF no more works. The redispersibility can be evaluated by calculating the work done by a steel blade, which penetrates at constant speed (1 mm/s), the sediment inside a test tube. The work was calculated by integration of the area under curves of force as function of displacement. The force was measured by a load cell, attached to the blade. The sediment was formed after centrifuging 10 mL of each MRF formulation at 2000 'g' for 15 minutes, according to the test described by *Kieburg et al* [8].

2. Experimental

The MRF formulations were made with a poly(alpha-olefine) oil, food grade. Besides the oil (balance), the dispersing additives (0.8% w/w) and the CIP (80 or 85% w/w), a modified hectorite clay

(0.3% w/w) was used as thixotropic agent. A high shear homogenizer (Ika – Turrax T-18) was used to disperse the CIP in each MRF formulation. The MRF samples were prepared in duplicate, and the measurements were twice repeated.

Sedimentation stability was evaluated allowing MRF formulations to settle in test tubes during 2 weeks, at rest. Oscillatory measurements were done, with other additives, to evaluate the concentration effect.

3. Results and Discussion

Figure 1 shows the redispersibility results: the work as a function of the depth for the 9 MRF combinations. One can see that Dodecylamine was the best additive, minimizing the work, almost independent of the powder. On the other hand, PHSA was the worst. Besides, the iron powder surface also has an important role. Table 1 resumes the work values, at ~ 17 mm depth.

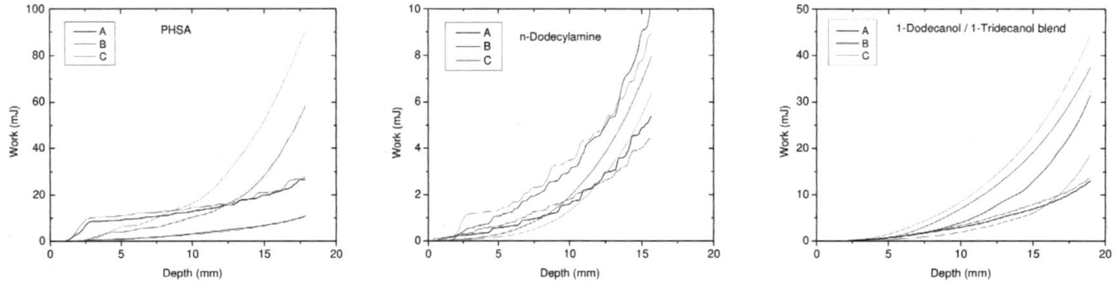

a) b) c)

Figure 1. Redispersibility test for MRF prepared with 80% w/w CIP A (black), B (blue) or C (red), and 1% w/w of different dispersing agents: a) PHSA, b) n-dodecylamine and c) dodecanol/tridecanol blend.

Table 1. Work (mJ) of the 9 MRF formulations

CIP	PHSA (17 mm)	Amine (16 mm)	Alcohol (19 mm)
A	27 ± 0.5	5 ± 0.5	13 ± 1
B	10.6 ± 0.2	9.7 ± 0.7	25 ± 6
C	74 ± 16	7 ± 1	41 ± 3

Figure 2 shows the viscosity curves without magnetic field, at 25°C. The MRF's based on powder 'C', were the more viscous, in all the cases, independent of the additive. Between powders 'A' and 'B', MRF prepared with 'B' gave lower viscosities in all shear rate range evaluated with the exception of PHSA.

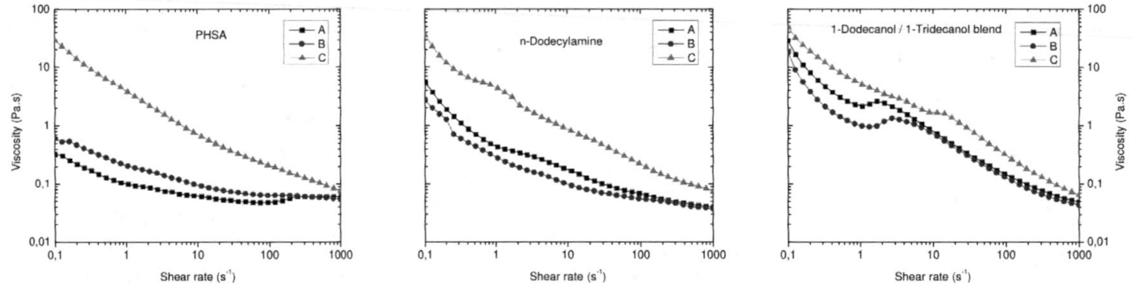

a) b) c)

Figure 2. Viscosity curves for MRF with 80% w/w CIP and 1% w/w of wetting additives: a) PHSA, b) n-dodecylamine and c) dodecanol/tridecanol blend. Without magnetic field. T = 25°C.

Journal of Physics: Conference Series **149** (2009) 012038 doi:10.1088/1742-6596/149/1/012038

Figure 3 shows the settling curves for MRF with 80% w/w iron powder 'B' and 1% w/w of PHSA, C12 alcohol, n-dodecylamine or Lauric Acid. The relevance of the different chemical groups is evident on the sedimentation stability.

Figure 4 shows the magnetic sweep at constant shear rate = 1 s^{-1}, for the MRF with powder 'B' and the additives PHSA, n-dodecylamine and dodecanol/tridecanol blend. Field strength measured in empty gap, not real field.

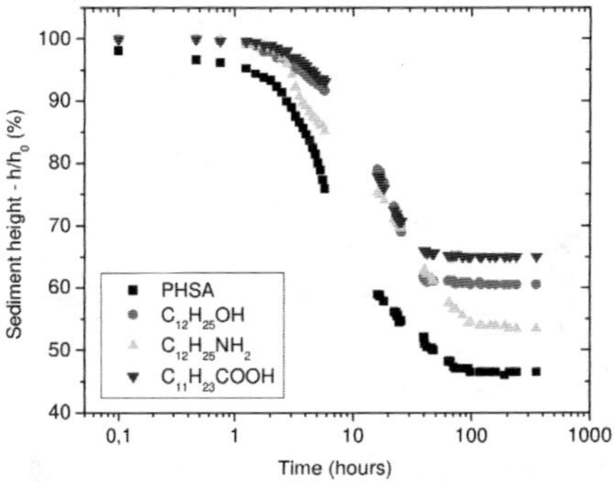

Figure 3. Settling curves for some MRF with powder 'B' at rest. Black: PHSA; Red: dodecanol; Green: dodecylamine and Blue = lauric acid.

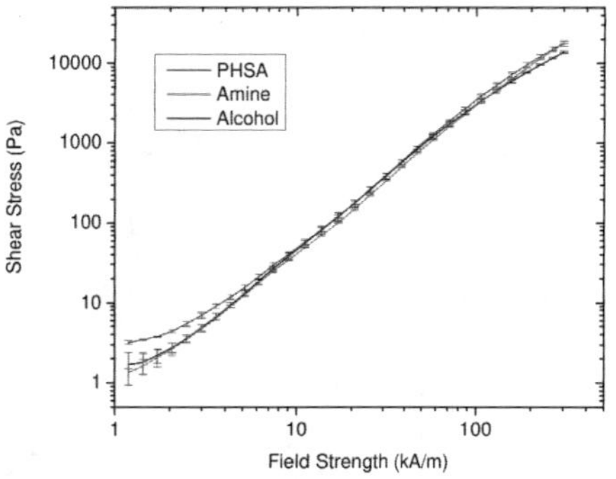

Figure 4. Magnetosweep at shear rate = 1 s^{-1}, for MRF with 3 different additives. All formulations with 80% w/w iron powder B and 1% each additive. Field strength relative to empty gap, not real field.

From figure 4, one can see that the effect of the additives on the rheological properties is weak if the field is applied. Significant changes appear only at low field region, below ~ 6000 A-m.

Then, we decided to focus only in powder 'B', and to evaluate the concentration effect of other additives, changing to n-octyl chain with carboxylic acid, besides alcohol and amine groups. Figure 5 shows the redispersibility curves for MRF with additives ranging 0.25 – 1.25% w/w.

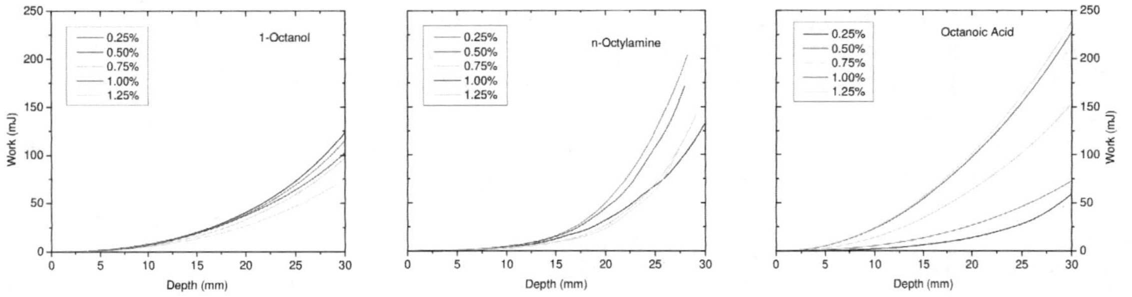

a) b) c)

Figure 5. Work versus Depth for MRF's all prepared with 85% w/w iron powder 'B' and additives with concentrations ranging from 0.25 to 1.25% w/w. a) 1-octanol; b) 1-octylamine and c) octanoic acid. Samples centrifuged @ 2000 'g' by 15 min.

One can see in figure 5a, that the work to redisperse decrease slightly with octanol concentration rises, but ranging from 75 to 125 mJ, at 30 mm depth. For the octylamine, figure 5b, the dependence is complex, changing with the concentration in a strange way, in the range from 125 to 200 mJ. For the octanoic acid, the work increases with the acid concentration until 1.00%: ~ 60 mJ with 0.25%; 75 mJ with 0.50%; 150 mJ with 0.75% and 225 mJ with 1.00 or 1.25% w/w.

Figures 6 and 7 shows respectively, the amplitude and frequency sweeps for MRF with powder 'B' and 0.5 or 1.25% w/w of the n-octyl additives: acid, alcohol or amine. It is remarkable that with alcohol, the gel-like structure is stronger than with amines, and almost independent of the concentration. On the other hand, with octanoic acid, the storage modulus G' as well as the yield value increases with the concentration.

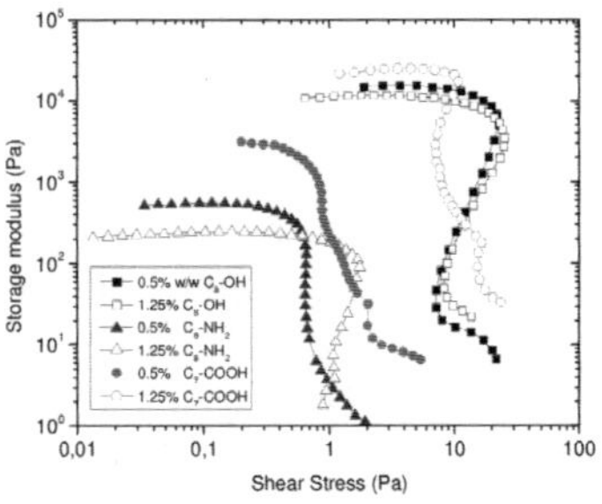

Figure 6. Amplitude sweep for MRF with 85% powder B with 0.5% (full points) or 1.25% w/w (open points) additives.

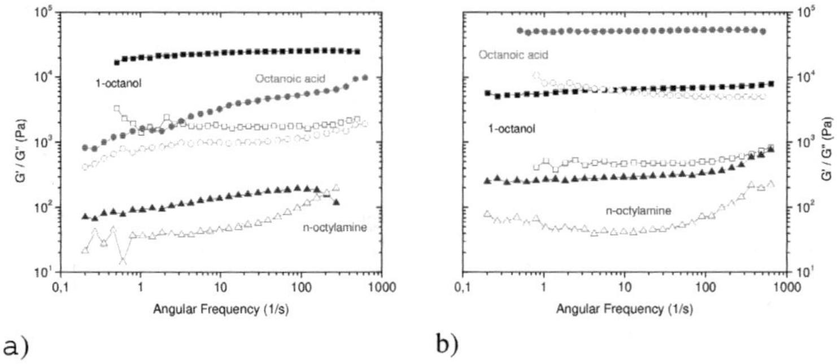

Figure 7. Frequency sweep for MRF (85% w/w CIP) prepared with a) 0.5% w/w or b) 1.25% w/w of 3 different additives. Storage and loss moduli, as full and open points, respectively. Black: 1-octanol; Red: Octanoic acid; Blue: 1-octylamine.

The paramount relevance of the additives for the redispersibility and off viscosity, as pointed by others [6,7,8] it is clear by itself. The only basic difference between the iron powders were their surfaces and between the additives, their chemical functional group. Therefore, we think it is reasonable to attribute the differences on rheology and redispersibility, to surface effects, indicating that the anchor groups in additives have specific interactions with the CIP coating, as suspected. However, more studies, including other analysis techniques are necessary to clarify it.

4. Conclusions

The redispersibility of MRF is strongly dependent on both carbonyl iron powder surface properties as well as choice and concentration of dispersing additives.

The type of iron powder modification also has a significant role on the viscosity of MRF formulations additionally to the known particle size effects.

MRF of iron powder with silica coating and 1.25% w/w 1-octanol or 0.50% w/w octanoic acid gave both the same redispersibility: work ~ 75 mJ at 30 mm depth.

References

[1] Phule P P 1999 *US Patent 5,985,168*
[2] Foister R T 1997 *US Patent 5,667,715*
[3] Muñoz B C, Adams G W, Ngo V T and Kitchin J R 2001 *US Patent 6,203,717*
[4] Johansson E and Nyborg L 2000 *Surf. & Interf. Analysis* **30**: 333-336
[5] Belyavskii S G *et al* 2006 *Protection of Metals* **42**: 244-252
[6] Phule P P, Mihalcin M P and Genç, S 1999 *J. Materials Research* **14** 3037-3041
[7] Lopez-Lopez M T, Zugaldia A, Gonzalez-Caballero F and Duran J D G 2006 *J. Rheology* **50** 543-560
[8] Kieburg C *et al* 2007 *Proc. 10th Int. Conf. on ER Fluids and MR Suspensions* -Lake Tahoe, USA (Singapore: World Sci. Pub) pp. 101-107.

Acknowledgments
Bombard wish to thanks: a) FAPEMIG for the Grant # CEX APQ-2676-5.02/07 as well as the financial support to attend this conference, Grant CEX-213/08; b) CNPq (Brazilian Gov. Research Agency) for the post-doctorate fellowship. BASF SE help is gratefully appreciated.

Journal of Physics: Conference Series **149** (2009) 012039 doi:10.1088/1742-6596/149/1/012039

Performance of Magnetorheological Fluids in a Novel Damper With Excellent Fail-Safe Behavior

H. Böse, J. Ehrlich and A.-M. Trendler

Fraunhofer-Institut für Silicatforschung ISC, Neunerplatz 2, D-97082 Würzburg, Germany

E-mail: boese@isc.fraunhofer.de

Abstract. A novel magnetorheological damper, whose magnetic circuit contains beside an electromagnet also two permanent magnets, is described and tested. Without any consumption of electric energy, a medium damping force is achieved, which ensures an excellent fail-safe behaviour in case of an electric power loss. This medium damping force can be increased or decreased, depending on the polarity of the current in the coil. Furthermore, the performance of the damper with different MR fluids was investigated. By this way, the influence of the composition of the MR fluid in terms of the concentration of iron particles between 20 and 50 vol.% on the damping force could be evaluated. Finally, the damping force after various storage times of the damper was studied and indicated a good redispersibility of the MR fluid.

1. Introduction

In a magnetorheological (MR) damper the damping force is continuously influenced by the strength of a magnetic field which controls the rheological characteristics, i.e. the flow resistance of an MR fluid. The increase or decrease of the damping force can be achieved within some milliseconds, due to the fast response of the MR fluid. A large number of different MR damper designs have been proposed so far. Such adaptable MR dampers have already been commercialized as automotive shock absorbers [1] or semi-active seat dampers in trucks and busses [2].

The magnetic field in an MR damper is usually generated with a coil in a magnetic circuit [3-5]. The circuit guides the magnetic flux through the active part of the fluid valve where the MR fluid is stiffened, causing a resistance force against the motion of the piston of the damper. This principle implies that the lowest damping force is generated when no current is generated in the coil. When a failure of the electric power supply occurs, the MR damper automatically switches to its softest damping state. This fact restricts the selection of the damping characteristics because the powerless state must fulfill the requirements of an appropriate fail-safe behavior.

An objective of this work is to propose a novel MR damper with a magnetic circuit containing an electromagnet and additionally permanent magnets in order to achieve an improved fail-safe behavior. Another objective is the evaluation of the performance of the novel MR damper filled with MR fluids having different compositions and corresponding properties.

2. Design of the MR damper

The novel MR damper has a piston which contains the magnetic circuit with a coil and two permanent magnets. A scheme of the piston and the magnetic flux in the circuit in different operation modes are shown in Figure 1. The permanent magnets generate a magnetic base field which remarkably solidifies the MR fluid in the MR valve without any electric power supply. In addition, the coil generates a magnetic field which strengthens or weakens the magnetic field of the permanent magnets in the active MR gap as requested, depending on the polarity of the current in the coil. The magnetic flux density was simulated with a Finite Elements Method (FEM). A particular advantage of the magnetic circuit is its capability to reduce the magnetic flux density in the MR valve to nearly zero. This allows to drastically reduce the damping force, caused by the permanent magnets, and to achieve a large range of damping forces. More details of the magnetic circuit were described in [6]. A demonstrator of the novel MR damper was manufactured and is shown in Figure 2. It contains a gas accumulator for the compensation of the volume change of the piston rod in the damper due to the motion of the piston.

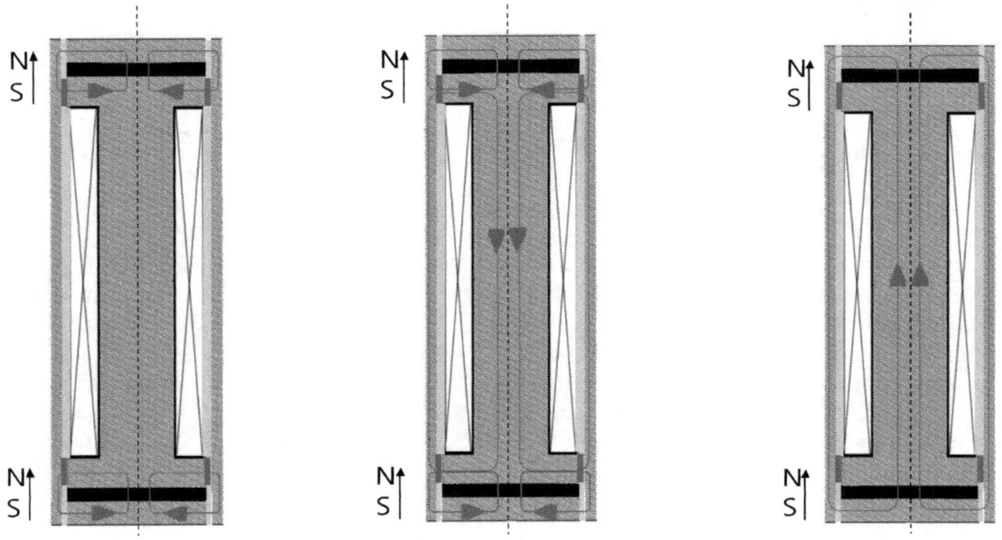

Figure 1. Scheme of the piston of the novel MR damper with the magnetic circuit in three different operation modes: without coil current (left), with positive coil current (middle) and with negative coil current (right). The annular MR fluid gap is green and the MR valves and magnetic flux lines are red.

Figure 2. Photographs of the novel MR damper: piston with electric connection (left) and damper with pneumatic connection to accumulator (right)

3. MR fluids used in the damper

The characteristics of the MR damper were evaluated by using various MR fluids with different profiles of properties. This procedure offers the possibility to compare the performance of the MR fluids and to study the influence of their composition on the damping characteristics. The MR fluids mainly differed in their concentrations of the magnetizable iron particles, which determine the MR effect, i. e. the enhancement of the shear stress in the magnetic field. The performances of three MR fluids with iron concentrations of 20, 36 and 50 vol.%, respectively, were compared. Figure 3 shows the flow curves without magnetic field and the dependence of the shear stress on the magnetic flux density for the three MR fluids. The data was measured in shear mode with a rheometer MCR 300 from Anton Paar, equipped with a magnetorheological cell.

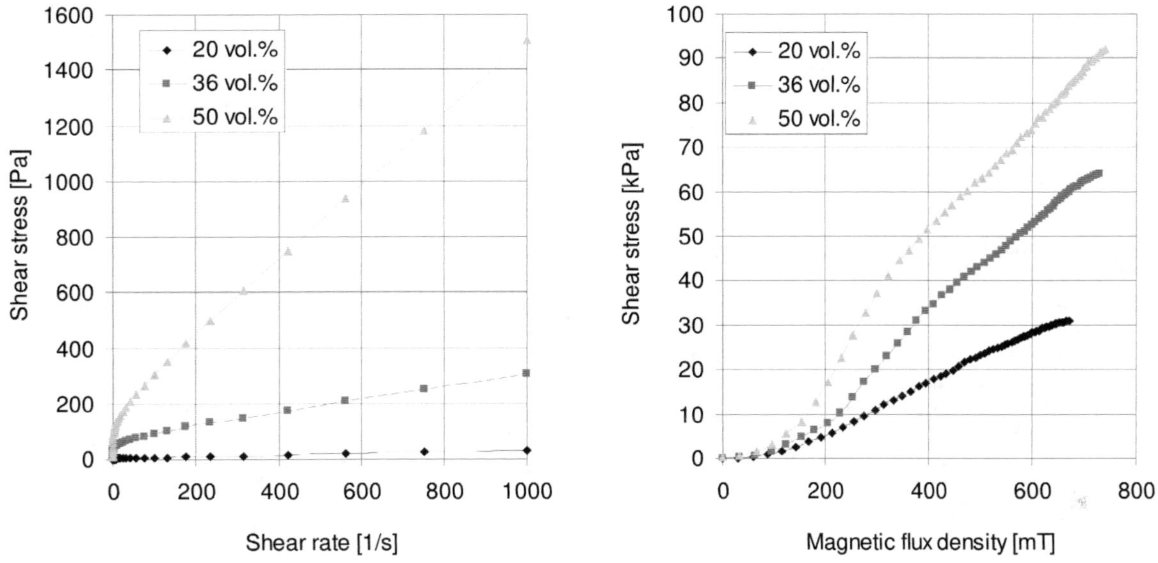

Figure 3. Properties of MR fluids with different particle concentrations (in vol.%): flow curves without magnetic field (left) and shear stress vs. magnetic flux density at 25 °C and shear rate 100 s^{-1}

4. Mechanical evaluation of the MR damper

The damping characteristics were measured on a dynamic testing machine with a sinusoidal load with an amplitude of 8 mm and a frequency of 1 Hz. The damping force was recorded for the use of the three MR fluids at various coil currents.

More detailed investigations were performed with the MR fluid containing 36 vol.% iron particles. With this MR fluid, the influence of the storage time in the damper on the damping performance at the first strokes after defined periods of storage and after defined cycles of work was studied. These experiments yield information on the storage stability and the redispersibility of the used MR fluid under near-application conditions.

5. Results and discussion

The results of the force measurements on the MR damper containing the MR fluid with 36 vol.% iron particles are shown in Figure 4. Without any current in the coil, indicating the power-less damping state, a damping force of about 1400 N was measured. The lowest damping force was determined to be about 400 N for a negative current of -3 A and the highest force 4000 N for a positive current of +3 A. These results demonstrate the excellent fail-safe behavior of the MR damper in case of a power-failure.

Journal of Physics: Conference Series **149** (2009) 012039 doi:10.1088/1742-6596/149/1/012039

Corresponding measurements were also performed with the damper containing the other MR fluids with 20 and 50 vol.% of iron particles, respectively. For the MR fluid with 50 vol.% iron particles in the damper, the maximum applicable current was restricted to 0 A, because at higher currents the damping force could destroy the accumulator in the damper. Figure 4 shows a summary of the measured damping forces, using the three MR fluids, in dependence on the coil current.

As expected, the investigations with the three MR fluids containing 20, 36 and 50 vol.% iron particles showed damping forces which, on one hand, increase with the rising iron particle content. On the other hand, the switching factor, i. e. the ratio of the maximum damping force with the strongest magnetic field to the minimum damping force with the weakest field, increased with a decreasing concentration of iron particles in the MR fluid. It was concluded that a medium iron particle content could be a good compromise for this MR damper.

Moreover, the measurements on the damper with the MR fluid containing 36 vol.% iron particles showed a good performance in terms of the stability of the damping force even after storage times of up to 36 days (see Figure 5). The initial increase in the damping force after storage of the damper, observed directly after the start of the measurement, disappeared after 1 minute of operation of the damper. This behavior emphasizes the good redispersibility of the MR fluid used in the damper.

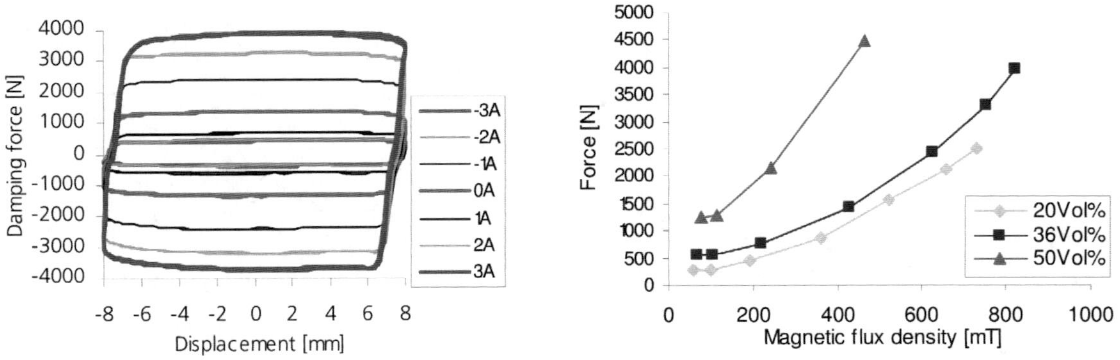

Figure 4. Damping force of the novel MR damper with the 36 vol.% MR fluid measured with a mechanical testing machine at different coil currents (left) and dependence of the damping force on the magnetic flux density for the three MR fluids with different iron contents (right)

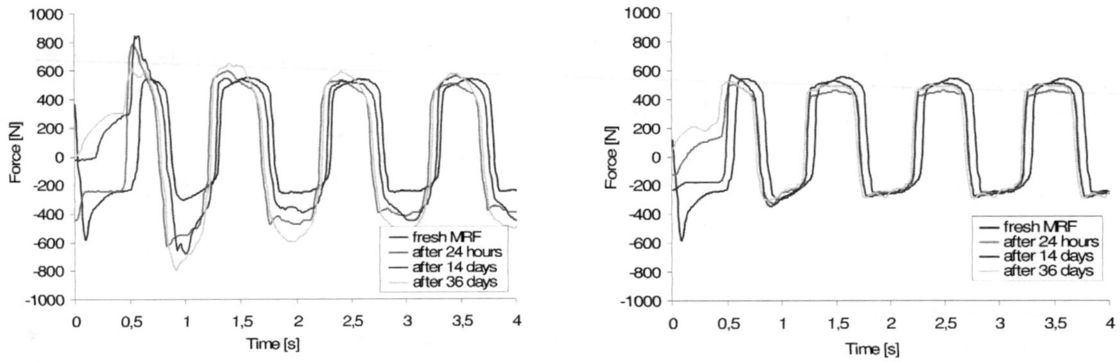

Figure 5. Damping force of the novel MR damper with the 36 vol.% MR fluid after various storage times: immediately after insertion into the testing machine (left) and after 1 min of operation (right)

6. Conclusions

A magnetorheological damper with excellent fail-safe behavior was developed and tested. Due to the integration of an electromagnet and permanent magnets, the damper generates a medium damping force without electric energy consumption, which can be increased or decreased by the electromagnet. This operation principle allows a broad variability of the damping force, as shown by this study.

The novel damper design with its special magnetic circuit offers a huge potential for a large number of products in automotive, industrial and other applications, like adaptive shock absorbers and semi-active vibration dampers. A prerequisite is the availability of an MR fluid whose properties are adapted to the damper design and working mode. The performance of a special MR fluid with a medium content of iron particles showed also very promising results in terms of storage stability and redispersibility.

Acknowledgments

Financial support for this work from the European Community is gratefully acknowledged.

References

[1] http://www.sae.org/automag/globalview_05-00/03.htm
[2] http://www.lord.com
[3] S.J. Dyke, B.F. Spencer, Jr., M.K. Sain and J.D. Carlson: An Experimental Study of MR Dampers for Seismic Protection, Smart Materials and Structures 7 (1998) 693-703
[4] U. Lange, S. Vassileva, L. Zipser: Controllable magnetorheological dampers for shock and vibration. Proceedings of Actuator 2002 – 8th International Conference on New Actuators (2002) 339-342
[5] M. Mao, W. Hu, Y.-T. Choi, N. M. Wereley: A magnetorheological damper with bifold valves for shock and vibration mitigation, J. Intelligent Mater. Syst. Struct. 18 (2007) 1227-1232
[6] J. Ehrlich, H. Böse: Novel magnetorheological damper with outstanding fail-safe characteristics. Proceedings of Actuator 2008 – 11th International Conference on New Actuators (2008) 495-498

Experimental analysis of clustering structures in magnetic and MR fluids using ultrasound

M A Bramantya[1], M Motozawa[2], H Takuma[1], M Faiz[1] and T Sawada[1]

1 Department of Mechanical Engineering, Keio University, 3-14-1 Hiyoshi, Kohoku-ku, Yokohama 223-8522, Japan

2 Department of Mechanical Engineering, Tokyo University of Science, 2641 Yamazaki, Noda, Chiba 278-8510, Japan

E-mail: sawada@mech.keio.ac.jp

Abstract. The formation of clustering structures in magnetic and MR fluids has an influence on ultrasonic propagation. We propose a qualitative analysis of these structures by measuring properties of ultrasonic propagation. Since magnetic and MR fluids are opaque, the non-contact inspection using this ultrasonic technique can be very useful for analyzing the inner structures of magnetic and MR fluids. We measured ultrasonic propagation velocity in a hydrocarbon-based magnetic fluid and MR fluid precisely. Based on these results, the clustering structures of these fluids were analyzed experimentally in terms of elapsed time dependence, effect of external magnetic field strength and angle, and hysteresis phenomena. A comparison of ultrasonic velocity propagation between magnetic and MR fluid was discussed.

1. Introduction

Magnetic fluid and magneto-rheological fluid are typical magnetic functional fluids that change their rheological properties when an external magnetic field is applied. The main difference between these two fluids is the inner particle size. The particles in a magnetic fluid are about 10 nm in diameter, while those in an MR fluid are in a range of 100 nm to 10 μm. When an external magnetic field is applied to magnetic or MR fluids, some of the inner particles coagulate and form a clustering structure. Because these clusters have a strong influence on the properties of these two fluids, it is important to analyze these clustering structures with regard to applications of magnetic functional fluids. However, because both fluids are opaque, it is difficult to analyze them by normal optical methods.

Ultrasonic propagation velocity and attenuation in magnetic and MR fluids change with the application of an external magnetic field [1]. It seems that the formation of clustering structures influences the ultrasonic propagation. Therefore, we propose a qualitative analysis of these structures by measuring properties of ultrasonic propagation. Because this ultrasonic technique can be applied to an opaque fluid, it can be useful for analyzing the inner structures of magnetic and MR fluids.

Several theoretical [2] and experimental [3] studies have been conducted to investigate the properties of ultrasonic propagation in relation to the inner structures in magnetic [4] and MR fluids. However, the inner structures are complicated and are still not clear. In this study, we measured ultrasonic propagation velocity in each of a hydrocarbon-based magnetic fluid and MR fluid precisely.

Based on these results, the clustering structures of these fluids are analyzed and the application of non-contact inspection in these fluids by ultrasonic techniques is discussed.

2. Experimental apparatus

Figure 1 shows a block diagram of the experimental apparatus. The ultrasonic measurement scheme is based on the pulse method. Ultrasonic propagation velocity in the test fluid can be calculated using this apparatus.

Figure 2 shows the details of the test cell. Two 5 mm diameter ceramic oscillators function as ultrasonic transmitter and receiver respectively. The ultrasonic frequency is 2 MHz. The test cell is made of transparent acrylic. The temperature of the test fluid in the square test cell (32 × 32 mm) is controlled and kept constant by a temperature control unit. The temperature control unit uses water as a liquid temperature control. The magnetic field is applied by an electromagnet that can be varied from 0 to 550 mT. A DC power supply is used for the electromagnet. The angle ϕ between the direction of the magnetic field and the direction of ultrasonic wave propagation can be adjusted freely between 0° and 180°. A digital oscilloscope is used to stimulate and measure the receiving ultrasonic wave. The properties of the magnetic and MR fluids used are shown in Table 1. These properties were measured at 40°C.

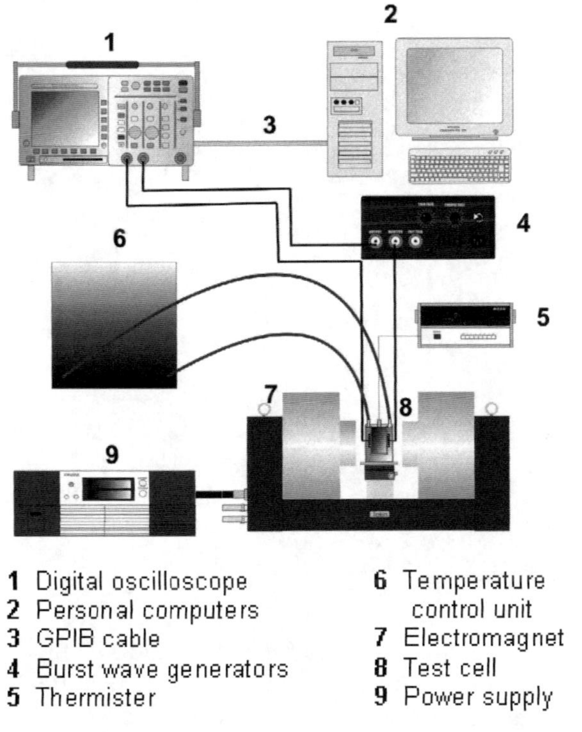

Figure 1. Experimental apparatus

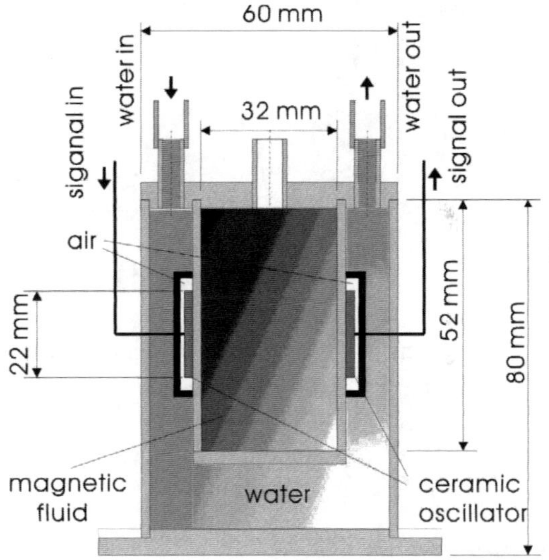

Figure 2. Detail of the test cell

Table 1. Properties of magnetic and MR fluids at 40°C.

	Magnetic Fluid	MR Fluid
Manufacturer	Ferrotec	LORD Co.
Serial name	EXP04019	MRF132-DG
Particle material	Ferrite	Ferrite
Mean particle size	10 nm	10 μm
Volume fraction	0.19	0.32
Carrier liquid	Hydrocarbon oil	Hydrocarbon oil
Viscosity	2.18±0.2 mPa.s	92±15 mPa.s
Density	1.049×10^3 kg/m^3	2.98~3.18×10^3 kg/m^3

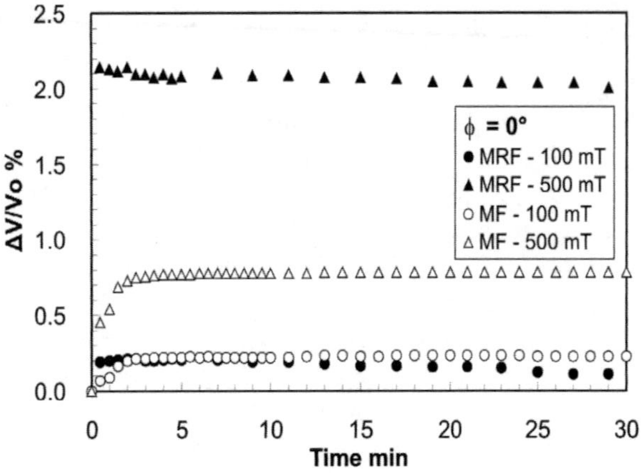

Figure 3. Elapsed time dependence of ultrasonic propagation velocity in magnetic and MR fluids

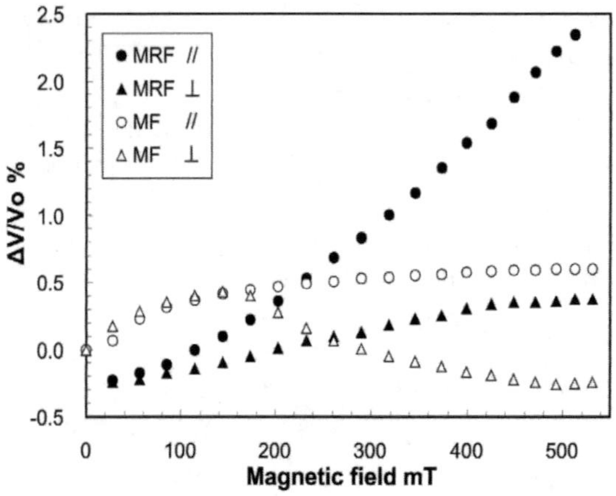

Figure 4. Change of ultrasonic propagation velocity under an external magnetic filed in magnetic and MR fluids

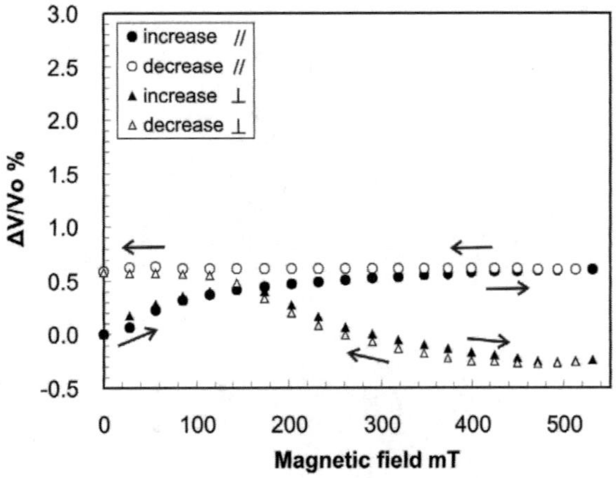

Figure 5. Hysteresis of ultrasonic propagation velocity on magnetic fluid

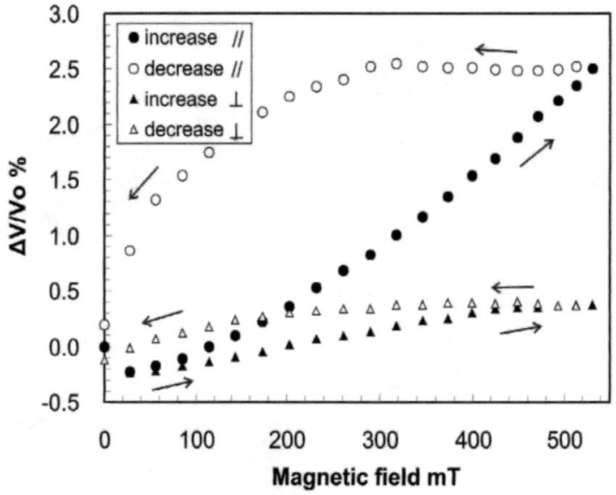

Fig. 6 Hysteresis of ultrasonic propagation velocity on MR fluid

3. Result and discussion

Figure 3 shows the change in ultrasonic propagation velocity in the magnetic fluid (MF) and MR fluid (MRF) versus the elapsed time of applying the magnetic field. In this experiment, the magnetic field intensity is kept at 100 mT and 500 mT, the angle ϕ is 0° and the temperature of the test fluid is kept at 25°C. Measurements are performed over 30 minutes. The change in ultrasonic propagation velocity is expressed by $\Delta V/Vo$. Here, ΔV is defined by $\Delta V = V - Vo$ where V and Vo are ultrasonic propagation velocities with and without an external magnetic field, respectively. Average value of Vo was founded 1092.34 m/s for MF and 803.79 m/s for MR fluid in this experiment.

The ultrasonic propagation velocity in the magnetic fluid increases with time while the magnetic field is applied. The change becomes almost constant after about 5 minutes. On the contrary, the ultrasonic propagation velocity in the MR fluid instantly increases. These changes seem to be caused by differences in cluster formation and indicate growth of clustering structures over the time the magnetic field is applied. The clusters seem to continue to grow until the ultrasonic propagation velocity becomes constant. On the other hand, robust clusters seem to form almost instantly in the MR fluid and we cannot detect the growing process of the clusters based on these measurements. It seems that the clusters grow very fast, in advance of the first measurement at 30 seconds and become stable after 30 seconds.

Figure 4 shows that ultrasonic propagation velocity in both magnetic and MR fluids changes with the application of an external magnetic field for $\phi = 0°$ (//) and $\phi = 90°$ ($\perp$). When the magnetic field intensity increases, the ultrasonic propagation velocity also increases. It seems that cluster formation is thicker when a higher magnetic field is applied. For the MR fluid, the positive gradient of the 0° curve is higher than that of the 90° curve. This clarifies that the direction of cluster formation is parallel with the direction of the magnetic field. An ultrasonic wave has more difficulty propagating across the 90° cluster formation than the 0° case.

On the other hand, the magnetic fluid has unique curves when the external magnetic field is applied at 0° and 90°. Both curves are nearly the same for values below 200 mT but very different for values above 200 mT. It seems that cluster formation has not yet fully developed when under the smaller magnetic field and only primary clusters are formed. Then secondary clusters appear under the higher magnetic fields. Again, ultrasonic propagation velocity at 90° is lower than at 0°.

An interesting hysteresis phenomenon appears when the magnetic field is increased to and decreased from a certain value, 530 mT in this experiment. The magnetic field is applied using the following process: the magnetic field intensity is increased by 12 mT every minute until 530 mT, and there after, the magnetic field intensity is decreased by 12 mT every minute.

In the magnetic fluid, during the increasing process, the change rate $\Delta V/Vo$ also increases as shown in figure 5. This phenomenon appears to be caused by the formation of chain-like clusters in proportion to the magnetic field intensity. However, $\Delta V/Vo$ remains at nearly same value during the decreasing process. It seems that the chain-like clusters do not break instantly, similar to the situation in our previous reports on water-based magnetic fluids [5]. The cluster formation remains in the magnetic fluid even though the external magnetic field becomes zero.

Hysteresis for the magnetic fluid also changes when the angle ϕ is changed. For the angle $\phi = 90°$, hysteresis only appears at small magnetic filed intensity, below 150 mT. For the angle $\phi = 0°$, hysteresis appears aver almost the whole range of magnetic field intensity. This indicates that the hysteresis phenomena can be better studied at the angle $\phi = 0°$. Hysteresis for the MR fluid is shown in Figure 6. The hysteresis curves are quite different in comparison with the magnetic fluid. Even though there is hysteresis with the MR fluid, the curve returns to the initial value. It seems that cluster formation in the MR fluid breaks more quickly and eventually becomes the same as the initial conditions after the external magnetic field intensity is zero. The effect of angle ϕ is similar to that for the magnetic fluid.

4. Concluding remarks

A precise experimental investigation of the ultrasonic velocity propagation in magnetic and MR fluids under a uniform external magnetic field was conducted. The ultrasonic propagation velocities in magnetic and MR fluids change according to the magnetic field intensity, interval time and angle of magnetic field ϕ. These results seem to be related to Brownian motion and clustering of the particles in the magnetic and MR fluids under an external magnetic field. In order to understand these clustering phenomena better, it is necessary to investigate the Brownian motion of magnetic particles, the time and the size of cluster formation. It is also necessary to consider the influence of sedimentation of the magnetic particles.

References

[1] Motozawa M, Sawada T and Matsumoto Y 2007 *Int. J. Modern Physics B* **21** 4914-21
[2] Taketomi S 1986 *J. Phy. Soc. Japan* **55** 838-844
[3] Skumiel A, Hornowski T and Jozefczak A 2000 *Ultrasonics* **38** 864-867
[4] Jeyadevan B and Nakatani I 1999 *J. Magn. Magn. Mater.* **201** 62-65
[5] Sawada T, Nishiyama H and Tabata T 2002 *J. Magn. Magn. Mater.* **252** 186-188

Journal of Physics: Conference Series **149** (2009) 012041 doi:10.1088/1742-6596/149/1/012041

The response of excited magneto-rheological fluid along field direction

Y T Chan[1], K P Liu[1], P L Wong[1] and W A Bullough[2]

[1] City University of Hong Kong, Department of Manufacturing Engineering and Engineering Management, Hong Kong, China

[2] Sheffield University, Mechanical Engineering Department, UK

E-mail: meplwong@cityu.edu.hk

Abstract. This paper describes an experimental study on the characteristics of an excited magneto-rheological fluid in the field direction using a self-developed two-plate rheometer. The effects of B-fields and shearing are investigated. The normal force can be built up by B-fields and further increased by imposing an additional shear action. The rate at which the normal force changes in response to an increasing B-field and shearing is quite rapid. In addition, the shear-enhanced normal force can be sustained even when the shear is stopped. An explanation based on chain rotation and slippage effects is proposed.

1. Introduction

Major studies on the characterization of magneto-rheological fluids (MRFs) have focused mainly on field-dependent shear responses in the direction perpendicular to that of the magnetic field. Engineering applications that utilize field-dependent rheological properties have evolved in the areas of active control damping [1], polishing/grinding processes [2, 3], flexible fixtures [4] and medical devices [5]. The response of MRF along the field direction (i.e., the normal direction) has recently attracted attention. De Vicente et al. [6] carried out studies on the normal force generated by an excited MR fluid with a commercial controlled-stress parallel-plate rheometer. The MRF tested had 50% volume fraction of carbonyl iron particles in silicone oil. The relative susceptibility and saturation magnetization of the particles were 131 and 1990 kA/m, respectively. They claimed that the normal force can only be generated where two criteria are satisfied: a) the B-field must exceed a critical value; and b) the MRF must be under shear strain. The normal force will increase with the shear strain up to a maximum value, then decrease with any further increase in the shear strain. Liu et al. [7] observed that the excited MRF did not rise until the B-field applied had been increased to around 13 mT. See and Tanner [8] reported that the normal force of MRF can be generated by a B-field even where there is no shear action. The home-made specimen MRF was a suspension of 80% by weight 4 to 6 μm carbonyl iron powder in a paraffin oil. Tests with gaps of 0.8, 0.9 and 1.0 mm were performed with a commercial controlled-strain rheometer. The effect of different gap sizes was reported not significant. The normal force increases with a magnetic field as $B^{2.6}$ and if a shear is imposed, the normal force will decrease. The normal force will continue to fall with an increase in the shear strain until it eventually settles at a steady value. The value at which it settles depends on the magnitude of the shear rate and is lower at higher shear rates. These phenomena, whereby the normal force is increased by a B-field and reduced by a shear, are respectively attributed to stronger elongated aggregates and the breaking of MR particle chains under shear strain. Laun et al. [9] recently proposed that the relationship between the normal force and the B-field can be expressed as $F_N \alpha B^{2.4}$, and found that the normal force can be increased with shear. Their tests were performed with constant shear rates and a fixed gap of 0.8 mm width using a commercial controlled-torque rheometer. The specimen MRF was a hydrocarbon-based suspension of 50% micron-sized magnetisable carbonyl iron particles.

This paper presents the results of a recent study on the normal forces measured in excited MR fluids using a self-constructed rheometer. The first set of experiments was mainly aimed at analyzing the B-field effect; no shear was imposed. The other set of experiments was conducted under steady shearing conditions.

2. Experiments

2.1. Materials

The MR fluid tested, MRF-140CG from Lord Corp., mainly consisted of 3 components [10] that included micron-sized magnetizable particles, hydrocarbon-based oil and other additives, such as suspending agent, thixotropes, anti-wear and anti-corrosion additives, and friction modifiers. Its viscosity was 0.280±0.07 Pa-s and particle concentration was 85.44% by weight. The particle size was in the approximate range of 0.88 to 4.03 μm, as measured by SEM.

2.2. Test rig

A parallel-plate rheometer was constructed to measure the characteristics of the MR fluids, as shown in Fig. 1. Fig. 2 is a schematic diagram of the rheometer.

Figure 1 Photograph of the self-constructed rheometer.

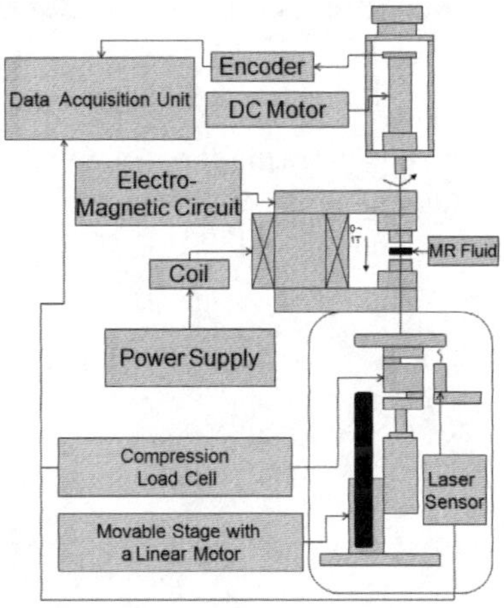

Figure 2 Schematic diagram of the self-constructed rheometer.

The setup included five major parts: a fixture for the MRF, a mechanical input, an electromagnetic circuit, a compression load cell and a data acquisition system. The fixture for the MRF consisted of two parallel plates. The shear rates specified in this paper were calculated on the basis of the radius of the plate, 20.5 mm. The lower plate was attached to a linear motor that adjusted the size of the gap between the two plates. There was also a laser sensor used to monitor the movement of the lower plate and ensure a gap of a constant width during tests. Shear actions were exerted by a D.C. servo motor with a 7700 gear head attached to the shaft of the upper plate. An optical encoder was used to measure the speed of the motor. An electromagnetic circuit with 3,700 turns of coil providing about 4 mT to 1 T was designed using ANSYS and a magnetization curve of MRF-140CG. The radius of the magnetic

poles was 20 mm. This was smaller than the plates for 1 mm in diameter used to facilitate an evenly distributed field inside the MR fluid. A gaussmeter was used to measure the B-field. Because the gaussmeter probe inserted directly into the MR fluid could not be used to measure the real B-field inside the MR fluid [9], it was inserted into the gap between the upper pole and the upper plate.
A calibration was carried out before the experiments were conducted. The values of B-fields at a position measured by the gaussmeter and calculated using ANSYS, respectively, were quite close. The ANSYS results showed that the B-field inside the MR fluid was about 35% higher than the value measured at the aforementioned position. Hence, the B-fields specified in this paper are the adjusted values from the measurements. Fig. 3 shows a typical FEA result, illustrating the difference between the B-fields in the air gap and the MR fluid. The spatial homogeneity of the magnetic field across the MR fluid is also shown in this figure. A compression load cell was fixed underneath the shaft of the lower plate to measure the normal force of the excited MRF. The major measurements, i.e., the normal force, were calibrated with standard weights before the experiments were carried out. The calibrated results are shown in Fig. 4. Signals from both the load cell and the incremental encoder of the motor were captured by two Agilent 34970A data acquisition units and recorded in a computer for subsequent analyses.

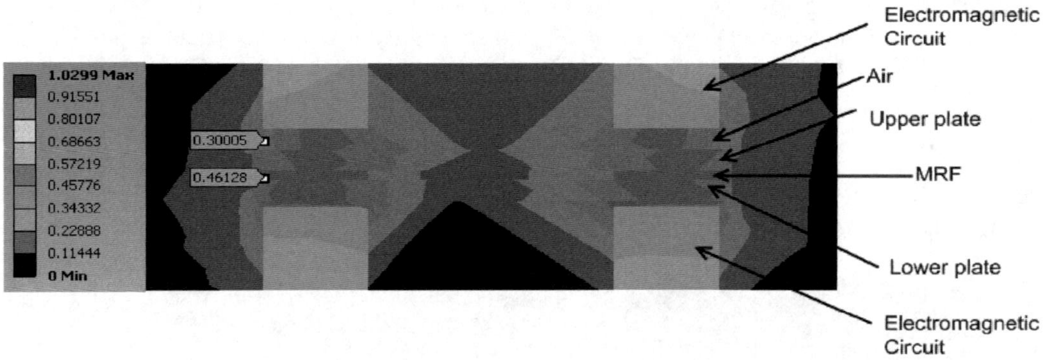

Figure 3 FEA results showing the difference between the B-fields in the air gap and the MR fluid, and the spatial homogeneity of the field across the MRF.

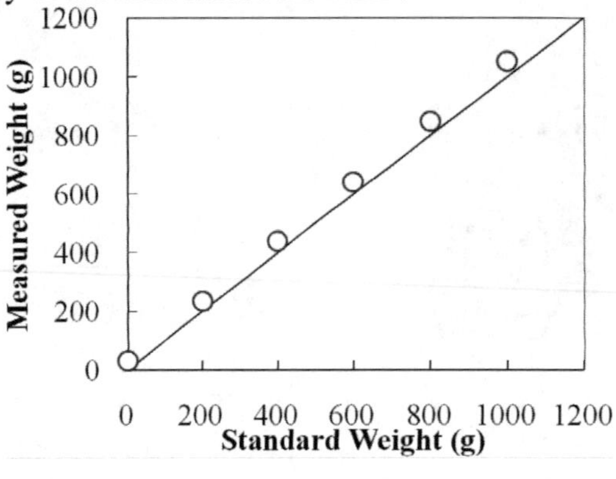

Figure 4 Calibration of normal force measurements using standard weights.

2.3. Measuring procedures
Two sets of experiments were performed to study the responses of the MRF along the field direction. One set of experiments was carried out with a steadily rotating upper plate to impose shear on the specimen fluid and the other set of experiments was conducted without shear. The specimen MR fluid contained in its original bottle was shaken vigorously and de-gassed in a vacuum before use. A controlled volume of MR fluid was injected onto the lower plate using a clean syringe. The lower plate was then elevated to form a gap of 1 mm between the plates, which was monitored by the laser sensor.

Pre-shearing produced by rotating the upper plate with 2.2 s^{-1} was applied to the MR fluid for 10 minutes to ensure good dispersion. To start with, the readings for the load cell under zero field and zero shear strain conditions were recorded for the first 3 mins. The desired field was then applied and the normal force data were captured from the load cell with a sampling rate of 10 Hz for 27 mins. The captured data were then averaged and compared with the B-fields. The experiment was carried on without turning off the B-fields for the shear test. The normal force was first recorded for 3 mins. A shear rate of 0.1 s^{-1} was subsequently imposed for 22 mins. The motor was then turned off, i.e., there was no shear, and the normal force was again recorded for another 5 mins. After the experiments, the MR fluid was demagnetized by applying an inverse impulse B-field. The experiment was repeated for other B-fields of 270, 405, 540, 675 and 810 mT.

3. Results and Discussion

3.1. Normal force of excited MR fluid

Figure 5 depicts the changes in the normal force as the B-field was increased. The trend is similar to the results of See et al. [8]. The particles aggregated due to the field effect and formed a columnar structure, pushing the two plates apart. The laser sensor showed that there was an increase of about 20 μm in the size of the gap under 600 mT. The normal force attained in low magnetic fields, i.e., < 200 mT, was not quite stable and repeatable. However, under relatively high fields, i.e., > 300mT, the normal force results were quite stable. This indicates that the internal structure under low B-fields is non-deterministic. Normal force measurement data can be fitted by the power law $\sigma_{normal} \propto B^{2.245}$. See et al. [8] and Laun et al. [9] also modeled their results using power-law curves with respective exponents of 2.6 and 2.4. The theoretical model developed by Shkel et al. [11] has an exponent of 2. This significant increase in the normal force of excited MRF without shearing is genuine. However, de Vicente et al. [6] reported that the normal force of excited MRF without shearing is too small to be measured. The different conclusions reached in these two studies may be due to the differing field strengths and MR fluids used.

3.2. Response of the normal force with steady shear

The normal responses of MRF under different B-fields and a constant shear rate of 0.1 s^{-1} are shown in Fig. 6. For the initial 180 s., the specimen MRF was under a constant B-field only, with no shear. Once shear was applied at the 180th second, the normal force increased rapidly by about 100%. This phenomenon is similar to that reported by de Vicente et al. [6]. Nevertheless, it does not accord with the results reported by See and Tanner [8], in which the normal force declined only under shearing action. Figure 6 shows that the normal force was roughly constant throughout the shearing period. While the shear was stopped at the 1,500th second, the normal force remained constant. Laun et al. [9] have recently reported similar results in which although there was a significant increase in the normal force under shear strain, the normal force dropped once the shear was stopped. De Vicente [6] used an analogue of a theory developed for ER fluids by Martin and Anderson [12] to explain the increase in the normal force of excited MRF due to shear. When a particle chain is tilted and forms an angle, θ, with the field direction, a resisting torque is generated. The torque is a function of $\sin(2\theta)$. When the aggregates/chains of the excited MRF are tilted, magnetic torques are generated to resist the further increase in the shear strain. As a result, this leads to an upward force. For large shear strain values, i.e., large tilted angles ($2\theta > \pi/2$), the increased normal force will drop. However, as shown in Fig. 6, the increased normal force remained almost constant in our experiment. This can be explained by the occurrence of slippage at the interface between the plate and the MR fluid. Hence, when the shear was applied solely by rotating the top plate, the MR particle chains became slightly deformed and experienced a resisting torque that led to an increase in the normal force. As the shear action continued, the tilted angles of the chains increased with the shear strain. Once the slippage occurred, the tilted angles became constant and there was no further increase in the normal force. The normal force remained unchanged even with a further increase in the shear strain. All the tests were under a constant shear rate of 0.1 s^{-1}. Based on the typical values of the relevant parameters including the magnetic flux density (1T), the magnetic field inside the MRF (H_o=1 A/m), the viscosity (0.28 Pa-s), μ_r ($4\pi\times10^{-7}$ Vs/Am), μ_{oil} (1), the Mason number [13] was calculated to be much less than unity which thus shows that the structure of the MRF remained intact. During the last 5 mins. of the experiment, the shear was stopped. The normal force still remained unchanged because the MRF structure had not changed. The normal force vanished only when the B-field was turned off, i.e., the MRF structure relaxed.

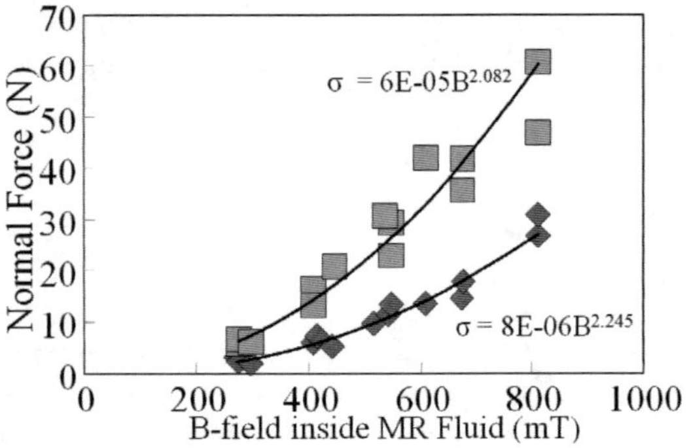

Figure 5 Normal force of the excited MRF fitted with power law curves.
■: 0.1 s^{-1}, ♦: 0 s^{-1}

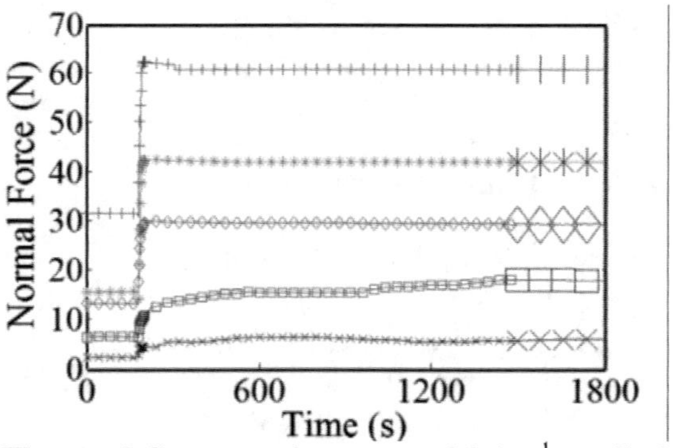

Figure 6 Constant shear rate of 0.1 s^{-1} applied at 180 s. Shear was stopped for the last 5 mins. (large signs)

4. Conclusion

The phenomenon whereby the normal force of an MRF is increased using B-fields is confirmed with a power law having an exponent of about 2. The normal force can be increased further by imposing shear actions on the excited MRF. The amount of the increase due to shear is quite significant. In the experiments carried out, a 100% increase was recorded under a shear rate of 0.1 s^{-1}. The shear-enhanced normal force can be sustained even when the shear is stopped. Hitherto, no firm conclusions can be drawn to explain the different results produced in similar studies. More comprehensive studies should be carried out to elucidate the mechanism.

Acknowledgments

The authors would like to thank Dr. Wangwen of Shanghai University for his valuable advice. The work described in this paper was fully supported by a grant from the Research Grants Council of the Hong Kong Special Administrative Region, China [Project No. CityU 121606].

References

[1] Carlson J D, Cantanzarite D M, St Clair K A 1996 *Pro. 5th Int. Con. on ERF and MRS* 20-28
[2] Kordonski W I and Golini D 1999 *Proc. 7th Int. Con. on ERF and MRS* 682-692
[3] Jha S and Jain V K 2004 *Int. J. Machine Tools & Manufacture* **44** 1019-1029
[4] Rong Y, Tao R and Tang X 2000 *Int. J. Advanced Manufacturing Technology* **16** 822-829.
[5] Flores G, Sheng R and Liu J 1999 *Proc. 7th Int. Con. on ERF and MRS* 682-692.
[6] De Vicente J, González-Cabellero F, Bossis G and Volkova O 2002 *J. Rheol.* **46**(5) 1295-1303.
[7] Liu X H, Wong P L, Wang W and Liu K P 2008 *Proc. ERMR08, Dresden.*
[8] See H and Tanner R 2003 *Rheol.Acta* **42** 166-170
[9] Laun H M, Gabriel C and Schmidt G 2008 *J. Non-Newtonian Mech.* **148** 47-56
[10] Lord Corp. 2008 Datasheet of MRF-140CG http://www.lordfulfillment.com/upload/DS7012.pdf
[11] Shkel Y M and Klingenberg D J 2001 *Int. J. Modern Phys. B* **15** (6 & 7) 795-802
[12] Martin J E and Anderson R A 1996 *J. Chem. Phys.* **104** (12) 4814-4827
[13] Martin J E 2000 *Phys. Review E.* **63** 011406

Applications of magnetorheological brakes in manual control of lifting devices and manipulators

M Chciuk[1], A Milecki[2] and A Myszkowski[2]

[1] University of Zielona Gora, 65-246 Zielona Gora, ul. Podgórna 50, Poland

[2] Poznan University of Technology, 60-965 Poznan, ul. Piotrowo 3, Poland

E-mail: andrzej.milecki@put.poznan.pl

Abstract. The article is aimed to design and testing of joystick with force feedback used in direct, human control of lifting device. The paper starts with the basic description of designed and tested by us MR rotary brake. Some initial laboratory investigations results of such brakes are presented. The usage of MR brakes in 2 axis joystick is proposed. Such, built by as joystick, is described. It was used as Human-Machine Interface in active control of lifting device. The designed and built 2 axis manipulator with electrohydraulic drive is described. In the paper, the based on PC with input/output card, control system of mentioned above joystick with MR brake and manipulator is described. Finally the control algorithm is proposed.

1. Introduction

Nowadays most devices like working machines, cranes, lifts and manipulators are controlled by human operators which utilize different joysticks and pedals. The natural feedback between the device and the operator is realized by visual observation. Because by this way not all features of the process can be noted, it can happen that the controlled device can meet unforeseen obstacles what can lead to damages or other problems. The illusion of presence can be enhanced by haptic cues which can significantly improve the flow of information from the slave site (device), to the remote operator for many tasks requiring dexterity. The aim of haptic interfaces is to render the mechanical features of controlled objects on the user, thus allowing a feeling of contact with real objects [1, 2]. In this paper we propose the usage of MR brakes in haptic joysticks in order to obtain controlled force (torque) feedback from the device. In such joysticks the produced by this brake movement opposite force can be proportional to the device movement resistance. This force can be controlled electronically by the controller according to the device opposite force [3, 4].

The article presents two-axis jostick with MR brakes used for manual control of two-axis technological lifting device. In the paper the joystick and manipulator construction, their connection to the controller and investigation results are presented. A new design of two-axis hydraulic lifting manipulator is also described. This manipulator is able to handle of different elements.

2. The construction of two-axis joystick with MR brakes and a manipulator

In figure 1, the construction of designed by us, two-axis joystick with MR rotary brakes is presented. The joystick consists of two arms (4, 6) interconnected by a joint (5). The stators of two MR rotary brakes (1, 7) are assembled to the aluminum body (2), which is placed on a heavy foundation. The left

Journal of Physics: Conference Series **149** (2009) 012042 doi:10.1088/1742-6596/149/1/012042

MR brake rotor (12) is connected directly to the bottom arm (6). The second rotor (9) is connected to the cogwheel (10) which is coupled with the upper arm through the tooth belt (3) and the second cogwheel. In the both rotors electromagnetic coils (14) are placed. Gaps between body and rotors are filling with MR fluid. When the coils are supplied, they generate magnetic field which lines closes by working gaps and housing. As a result MR fluid changes its viscosity and as a final effect, a braking torque. The rotors rotation angles (movement of arms) are measured by potentiometers (8, 13).

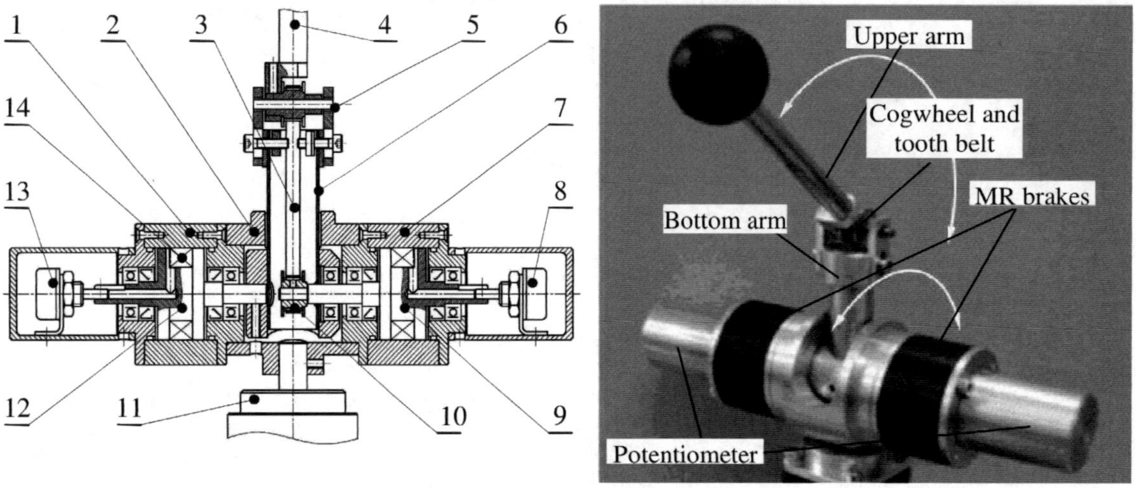

Figure 1. Two-axis joystick with MR brakes and its photo.

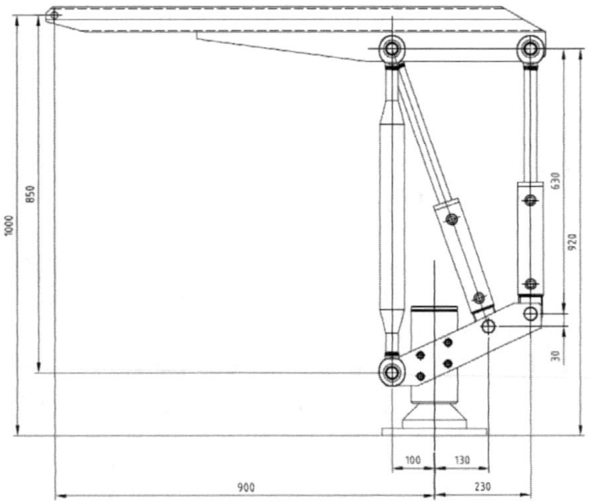

Figure 2. Two-axis manipulator.

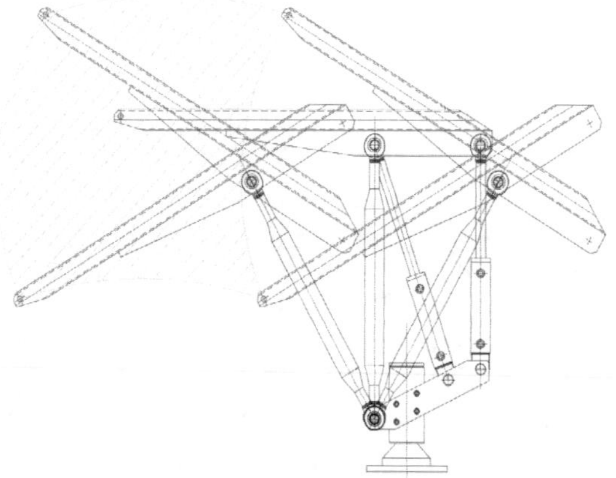

Figure 3. Manipulator working area.

In figure 2 the designed and built two-axis hydraulic manipulator is shown. It is based on not commonly known kinematics, which consists of to two arms connected to the body beam and moved by two linear hydraulic cylinders. We used cylinders which had following parameters: stroke 195 and 300 mm, piston diameters 40 mm, supply pressure 25 MPa. At the end of the upper arm (main arm) a gripper can be installed. All mentioned above elements are assembled together by usage of ball-and-socket joints. The length of hydraulic cylinders can be changed according to control signals. The actual position of a manipulator is measured by two rotary encoders placed in the joints which connect both arms and the second arm to the body. In figure 3 the manipulator working area is shown. It is 900 mm width and 800 mm high. The manipulator was able to produce forces till 10 kN at its end point.

3. The structure of control system

In figure 4 the block scheme diagram of the whole system is presented. It consists of a two-axis joystick with MR brakes and amplifiers, control computer with input/output card and hydraulic lifting device. The hydraulic cylinders in this device are controlled by proportional valves, which are connected to its electronic control cards. This cards control the proportional electromagnets according to signals send from the control computer. The angle positions of joysticks arms are measured by potentiometers. Their output signals are given to A/D converter in input/output card, transformed to digital form and finally send to the computer. The angle positions of lifting device arms are measured by incremental encoders. Their outputs are connected to special up/down counters in the input/output card. The state of these counters represents the measured manipulator joints angle position. It can be easy read by the computer. Additionally the force on the manipulator arm is measured; directly using force transducer (weight measure unit) or indirectly by measurement of pressure differences in cylinder chambers. Pressure sensors are mounted in cylinders chambers and their output signals can be also send by input/output card to the computer.

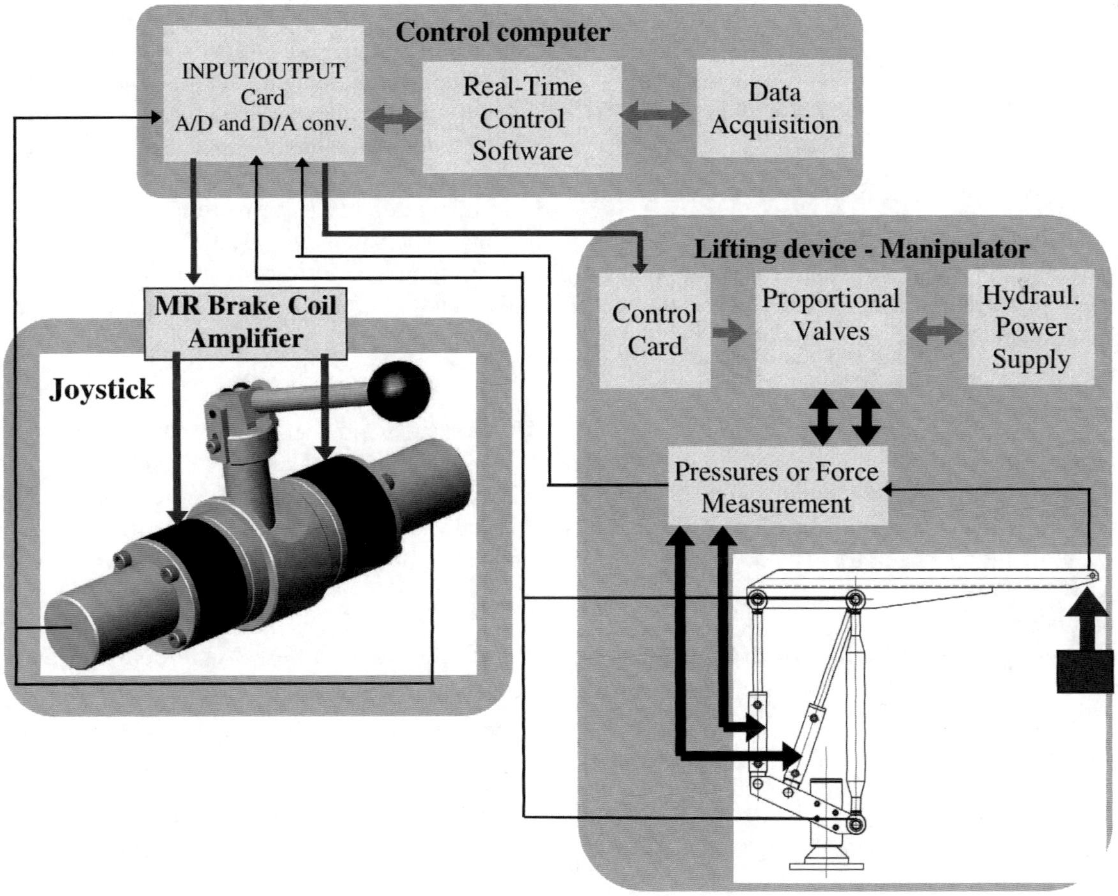

Figure 3. Scheme block diagram of the manipulator controlled by haptic joystick.

4. The control algorithm

In the computer the control card type RT-DAC was used which enables the connection of a joystick, sensors and drives to the computer which controls the whole system. The block structure of the control system is shown in figure 4. To the card analog inputs joystick arm angle positions α_1 and α_2, measured by potentiometers, which produces voltage signals (*PotX* and *PotY*), are connected. To the same type of input load force F_l signal is connected (*SenS*). These signals are transformed in AD converter. The measured by encoders, manipulator arms angle positions Θ_1 and Θ_2 (*EncX* and *EncY*), are connected to the computer.

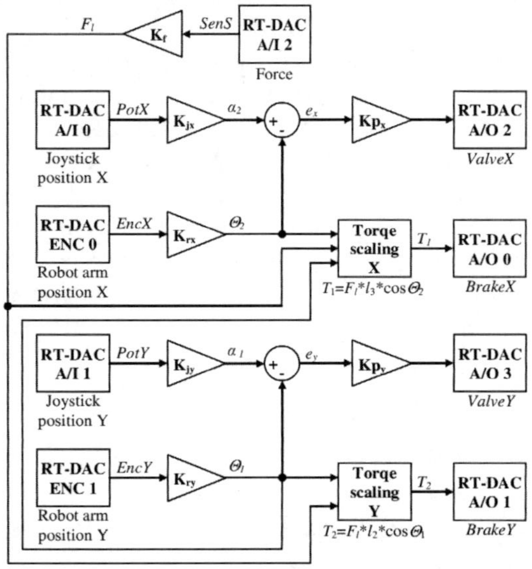

Figure 4. Structure of connections and control algorithm.

The information send to the computer enables the control of hydraulic cylinders velocity by sending voltage signal to proportional valve control cards (*ValveX* and *ValveY*). These signals are calculated basing on the difference between the joystick and manipulator arm positions. The signals sent to the joystick MR brakes coils are calculated according to the load force, arm angle and its angle position. The computer sends to the MR brake coils control voltages (*BrakeX* and *BrakeY*). These signals are sent trough D/A converter. By this way, the impression of being in physical contact to the lifting device and its environment is transformed to the human operator. Because both, the kinematics of the manipulator and a joystick and the applied measurement elements (rotary potentiometers in joystick and rotary encoders in manipulator) are different, and moreover there is no straight relation between the joystick arms positions and hydraulic cylinders displacements the control program is not ease. In our investigations we used Matlab/Simulink software, which give us the flexibility and ability to rapid change of control algorithm and its structure.

5. Conclusion

The paper proposed two-axis joystick with MR brakes which enable force feedback to communicate and interact with a lifting device. This device and a joystick were controlled by usage of a computer with input/output card and by special, worked out by us, algorithm. By the usage of MR brakes in joysticks, the impression of being in physical contact to the device and its environment was given to the human operator. Thanks to this, the manipulator can be controlled more accurately.

Acknowledgments

This study was supported by Polish Ministry of Science and Education in the years 2006 to 2008 as a grant no 4 T07B 002 29.

References

[1] Harward V, 2006 Haptic synthesis *Proc. 8th Int. IFAC Symp. on Robot Control SYROCO.*

[2] Nahvi A, Nelson D D, Hollerbach J M and Johnson D E 1998 Haptic manipulation of virtual mechanisms from mechanical CAD design *Proc. IEEE Intl. Conf. Robotics and Automation Leuven.*

[3] Scilingo P, Bicchi A, De Rossi D and Scotto A 2002 A magnetorheological fluid as a haptic display to replicate perceived compliance of biological tissues. *IEEE-EMBS Conference on Microtechnologies in Medicine & Biology Lyon*

[4] Zhuang Y and Canny J 2000 Haptic interaction with global deformations, *Proc. IEEE Robotics and Automation Conference*, IEEE, vol. 3.

Effect of storage and ball milling on the sedimentation and rheology of a novel magnetorheological fluid

H B Cheng[1], P Hou[1], Q J Zhang[1] and N M Wereley[2]

[1]State Key Laboratory of Advanced Technology for Materials Synthesis and Processing, Wuhan University of Technology, Luoshi Road 122, Wuhan, 430070, P.R.China

[2]Department of Aerospace Engineering, University of Maryland, 3180 Martin Hall, College Park, MD 20742, USA

E-mail: zhangqj@whut.edu.cn

Abstract. In this study, the sedimentation stability and rheology of a new magnetorheological fluid formulation were experimentally investigated before and after being stored for 24 months, as well as after being milled. The results reveal that, after undergoing 24 months of steady storage, the off-state viscosity and field-induced yield stress of the MR fluid remained constant, and that its flowability and redispersibility were maintained. After being subjected to ball milling for 2 hours at 200 rpm and 300 rpm, the field-induced yield stress was also maintained.

1. Introduction

Magnetorheological (MR) fluids are suspensions composed of magnetizable particles, nonmagnetic liquid and additives, and show a unique ability to undergo rapid, nearly completely reversible, significant changes in their apparent viscosity upon application of an external magnetic field [1]. Typically, this change is manifested by the development of a yield stress that monotonically increases with applied magnetic field. This yield stress controllability has stimulated considerable research in developing controllable shock absorbers, clutches, engine mounts, alternators, power steering pumps, control valves, and artificial joints. However, MR fluids are thermodynamically unstable systems which tend to settle and may even form a hard "cake" at the bottom of its container in time, which limits their application in systems where the MR fluid remains quiescent for long intervals. Therefore, investigating the long-term stability of MR fluids is very important for civil engineering [2] and single use systems like energy absorbers for crashworthiness systems such as helicopter seats. In this work, we focus on how off-state viscosity and field-induced yield stresses of a novel MR fluid are affected by long-term storage and by ball milling (simulating in use shear thickening), to evaluate the stability and durability of the MR fluid.

2. Experimental

The samples used in this study were (1) freshly mixed MR fluid (MRF-140WUT, Wuhan University of Technology, China)[3], which is composed of carbonyl iron 82 wt%, silicone oil 15 wt% and additives 3 wt%, (2) MR fluid after being stored for 24 months, and (3) MR fluids after ball milling.

Ball milling was used to simulate the durability, or in-use shear thickening, of the MR fluid. The process was carried out under vacuum conditions, using a stainless steel pot (height 7 cm, ID 7.5 cm) and stainless steel balls (diameter 6 mm and 10 mm), the weight ratio of balls to MR fluid was 5; The rotation rate of the pot was 200 and 300 rpm., respectively. Magnetorheological characterization was carried out using a parallel-plate advanced rheometric expansion system (ARES, TA Co.) equipped with a magnetic field supplier at room temperature.

3. Results and discussion

3.1. Dispersion Stability

The samples of MR fluid were transferred to 10 ml graduated test tubes and placed in still storage. The volume of the supernatant liquid can be obtained by observing the phase boundary between the supernatant liquid and the concentrated suspension, until reaching an asymptotic value. The sedimentation ratio is defined as:

$$\text{Sedimentation Ratio}\,(\%) = \frac{\text{volume of the supernatant liquid}}{\text{volume of the entire suspension}} \times 100\% \qquad (1)$$

Based on experimental observations, sedimentation ratios were calculated according to Eq. (1), and plotted as a function of time in figure 1. Sedimentation ratio is inversely proportional to the dispersion stability of the MR fluids. Figure 1a shows the dispersion stabilities of the fresh MR fluid (□ #0) and the redispersed MR fluid following 24 months of storage (■ #1). Figure 1b shows the dispersion stability of the stored MR fluid after ball milling for 2 hours at 200 rpm (□ #2) and 300 rpm (• #3) , respectively. In figure 1, sample#1, #2 and #3 were made according the procedures as follows: at first the stored MR fluid was stirred to create a homogeneous suspension as sample #1, and then portions of sample #1 were milled with a ball mill for 2 hours at the rotation rate of 200 and 300 rpm, respectively, thus obtaining sample #2 and #3. From figure 1a, the sedimentation ratio of the novel MR fluid was not affected by storage, and both the fresh and stored MR fluids exhibited nearly identical sedimentation behavior. The segregation phenomenon occurred slowly, and reached equilibrium after about 105 days. The sedimentation ratio increased at the beginning correspondingly fast, and then slowly, finally kept constant, the maximum is about 11%. Moreover, the novel MR fluid can be easily redispersed after 24 months steady storage through a weak stirring force (300 rpm×10 minutes). Comparing the sedimentation ratios of the MR fluid before and after ball milling (samples #1, #2 and #3 in figure 1b.), it was found that the sedimentation ratio decreased by about 77% after milling, and extent of the reduction increased with the rotation rate of the pot. These effects are the result of non-spherical particles and superfines produced after milling (see fig. 3 below), and are analogous to methods of improving MR fluid stability of by introducing non-spherical particles [4] and/or nanoparticles [5].

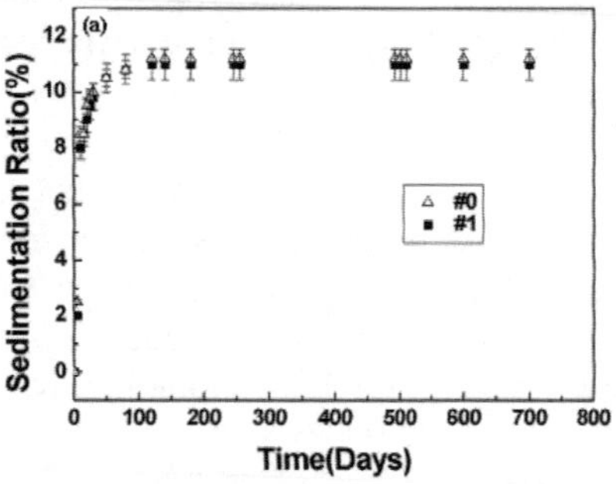

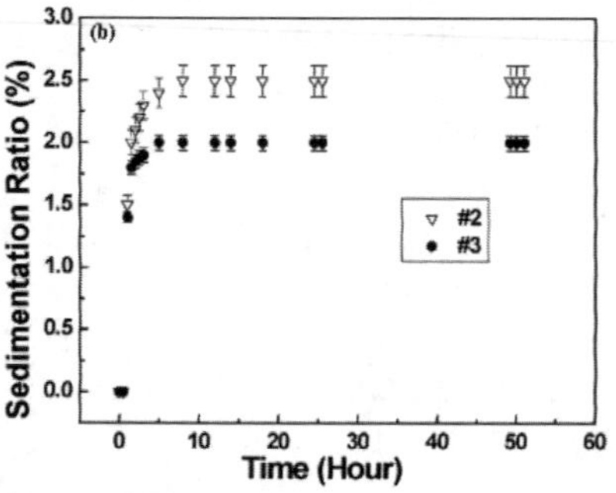

Figure 1. Sedimentation ratio versus time. (a): □ #0 fresh MR fluid; ■ #1 redispersed MR fluid after 24 months of storage; (b): □ **#2** milled 2 hours at 200 rpm; • **#3** milled 2 hours at 300 rpm

3.2. Rheological properties analysis

3.2.1. *Off-state viscosity.* Figure 2a shows the off-state viscosity of fresh MR fluid (□ #0) and after 24 months of storage (■ #1). Both MR fluid samples exhibited typical shear thinning behavior, decreasing monotonically with shear rate. The difference between the two viscosity curves is small. Even though the MR fluid was placed in static storage for 24 months, its off-state viscosity remained constant after being redispersed. This result indicates that this MR fluid is stable for long-term storage and no irreversible sedimentation and agglomeration takes place during the storage process. Figure 2b shows the off-state viscosities of MR fluids (#2 and #3) after milling at different rotation rates as a function of shear rate. Compared with that of the original sample free of milled (#1), the viscosities averagely increase by 230%, whether the rotation rate is 200rpm or 300rpm. That is to say, ball milling will increase the viscosity of MR fluids, which is in agreement with the results of the above stability test and results of Ulicny [6]; it is also similar to the so-called in-use-shear-thickening [8]. Increased viscosity may be due to the increase in aspect ratio of the particles [8], because some magnetic particles were volumetrically reduced after milling, and also partly due to the increase of

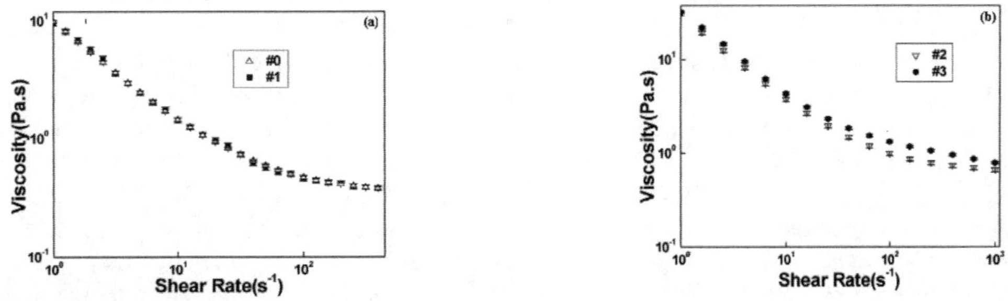

Figure 2. Off-state viscosity versus shear rate. (a): □ #0 fresh MR fluid; ■ #1 redispersed MR fluid after 24 months of storage; (b): □ **#2** milled 2 hours at 200 rpm; • **#3** milled 2 hours at 300 rpm

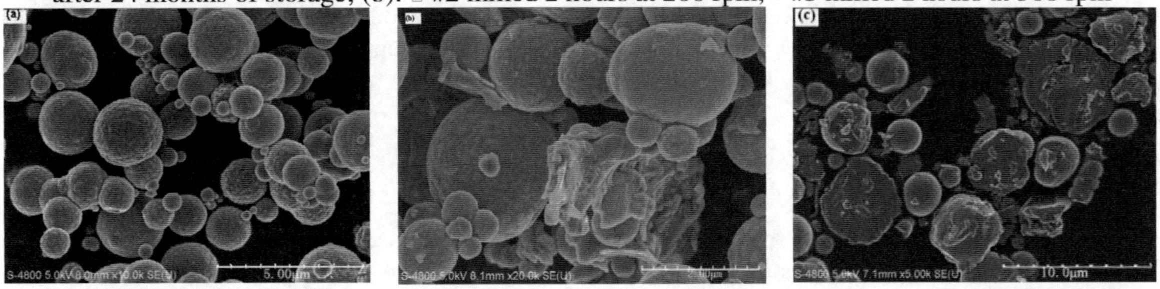

Figure3. SEM for different ball milling rates. (a): unmilled; (b): milled 2 hours at 200 rpm; (c): milled 2 hours at 300 rpm. Scale bars indicate 5μm (a), and 10 μm (b and c), respectively.

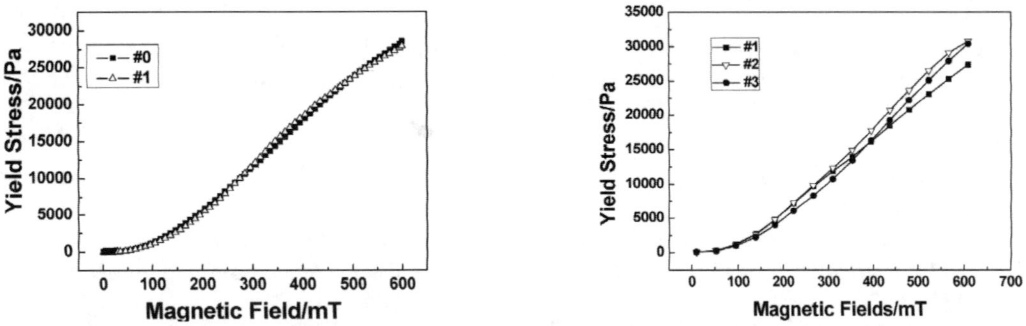

Figure 4. Yield stress versus field strength. (a): □ #0 fresh MR fluid; ■ #1 redispersed MR fluid after 24 months of storage; (b): □ **#2** milled 2 hours at 200 rpm; • **#3** milled 2 hours at 300 rpm

superfine particulates spalled from the surface of the carbonyl iron particles after milling. Since even a small quantity of carbonyl iron particles are replaced with superfines, the off-state viscosity would be increase [7]. To validate this hypothesis, photomicrographs of the particles in MR fluid before and after miling were observed in an SEM□Hitachi S-4800 FESEM□and shown in figure 3 (a-c). It was found that the larger particles were at first worn down and some non-spherical abrasion particles and superfine particles appeared after milling (seeing figure 3 b and c). Ulicny [6] obtained similar results in a durability test of an MR fan clutch. Interestingly, this MR fluid did not become too dense to flow, but presented better flowability even after being milled 2 hours at 300 rpm. That is to say, the MR fluid did not fail after being subjected so severe wear; the yield stress measurements below support this conclusion.

3.2.2. *Field-induced yield stress.*

The field-induced yield stress of the MR fluid samples were tested with magnetic flux density ranging from 0 to 0.6T, at the shear rate of 1 s^{-1} and shown in figure 4. Figure 4a shows the relationship of magnetic flux density and the field-induced yield stresses of the MR fluid before and after 24 months of storage. Interestingly, the yield stress of the MR fluid is nearly identical for the fresh and stored MR fluids. Together with the above demonstration of the stability of MR fluid viscosity before and after 24 months of storage, we conclude that the MR fluid has very stable properties even after being stored for 24 months and redispersed. Figure 4b shows the field-induced yield stresses of MR fluid samples before and after milled increase. Under weak magnetic field (lower than 0.45T); the yield stresses of the three samples have exhibit similar trend of increasing yield stress with magnetic field. The results may imply that the field-induced shear stress of the MR fluids is not very sensitive to the shape of particles under weak field, or the amount of broken particles is not enough to induce large change of particles-chain in MR fluids under weak field. For magnetic flux density greater than 0.5T, yield stress of the two milled samples #2 and #3 exceeded that of the unmilled sample #1. This behavior is analogous to the results observed when adding nanometer lithium magnesium silicate into MR fluids [9]; it further validates that one cause of increasing viscosity is the superfine particles produced in the milling process.

4. Conclusions

The novel MR fluid formulated here was proven empirically to have a high level of stability. There was measurable change in neither the viscosity nor yield stress between freshly mixed MR fluid and fluids subjected to 24 months of quiescent storage. In addition, there was no measurable worsening of the field-induced yield stress after ball milling at 300 rpm for 2 hours, although viscosity increased because of the increase of weight fraction of superfines and non-spherical as the result of milling.

References

[1] Ginder J M, 1998, Behavior of magnetorheological fluids. *MRS Bull.* **23** 26
[2] Dyke S J, Spencer B F, Jr, Sain M K and Carlson J D. 1998, *Smart Materials and Structures*, **7** 693-703.
[3] Cheng H B et al. 2006 *China Pat.* ZL200610124728.0
[4] Pu H T, Jiang F J, Yang Z L, Yan B and Liao X, 2006, *Journal of Appllied Polymer Science*, **102**, 1653-1657.
[5] Ngatu G T and Wereley N. 2007 *2007IEEE Transactions on Magnetics*, **43(6)** 2474-2476
[6] Ulicny J C, Balogh M P, Potter N M and Waldo R A. 2007, *Materials Science and Engineering A*. **443** 16–24
[7] Carlson, J.D. 2002. *Journal of Intelligent Material Systems and Structures*, **13(7–8)** 431–435.
[8] Kwon T M, Jhon M S and Choi H J. 1998, *Journal of Molecular Liquids.* **75** 115-126.
[9] Cheng H B, Zhang J, Feng J,Guan J G, Zhang Q J and Qu W L. 2006, *Journal of functional material* **7(37)** 1166-68

Journal of Physics: Conference Series **149** (2009) 012044 doi:10.1088/1742-6596/149/1/012044

Preparation and Characterization of Composite Magnetic Particles and Aqueous Magnetorheological Fluids

H B Cheng[1], J M Wang[1], Q J Zhang[1] and N M Wereley[2]

[1]State Key Laboratory of Advanced Technology for Materials Synthesis and Processing, Wuhan University of Technology, 430070, Wuhan, P.R.China
[2]Department of Aerospace Engineering, University of Maryland, 3180 Martin Hall, College Park, MD 20742, USA

E-mail: wereley@umd.edu

Abstract. Composite magnetic particles (CMPs) with carbonyl iron (CI) core and N-glucose ethylenediamine triacetic acid (GED3A) shell were prepared by an in-situ chelating reaction, where the CI surface was coated with GED3A. And novel kind of aqueous magnetorheological (MR) fluids were prepared with thus obtained CMPs. The properties of CMPs, including morphology, structure, and magnetic behaviours were characterized, and the magnetorheological properties of the MR fluids were analyzed via a strain-controlled rheometer. It was found that the MR fluids exhibited high shear stress and quick response rate to magnetic field. Furthermore, the dispersion stability and the anti-oxidation property were substantially improved as revealed by stability test.

1. Introduction

Aqueous magnetorheological (MR) fluids are composed of water as base liquid and magnetic particles as dispersed phase [1-3], and have many potential applications [4, 5]. Especially used in magnetorheological finishing, they are favorable to the dispersion of fine abrasive in MR fluids and have the function of cooling and accelerating the hydrolysis of the polishing material surface to enhance the polishing efficiency [6]. However, there are still problems which have so far not been resolved. Most aqueous MR fluids are easily oxidized, and are prone to settle and agglomerate in a relatively short period of time [6, 7]; since they are commonly made with reduced iron powder or carbonyl iron particles (CIPs) [1-3], the density of CIPs is about 7.86g/cm^3 much larger than that of water. These disadvantages limit their practical application. In this work, we developed a preparation of composite magnetic particles (CMPs) and an aqueous MR fluid to minimize oxidization and agglomeration, and to increase sedimentation stability. The morphology, chemical surface composition, magnetic properties and anti-oxidization ability of the obtained CMPs were investigated. The magnetorheological behavior of the aqueous MR fluid based on the CMPs was also experimentally studied using a commercial parallel-plate type rheometer with a magnetic induction iron yoke.

2. Experimental Methods

The CIPs were added to N-glucose ethylenediamine triacetic acid (GED3A) solution in a flask, stirring at 40□ for 24 hours, and then the product was washed with water, dried at 60□ in a vacuum oven for 8 hours to yield CMPs.

The aqueous MR fluids were prepared according to the patent CN 200610124727.6 [3]. At first, the previously obtained CMPs were dispersed in pure water with the composition in volume of 30% CMPs, 0.5% additives (including anti-volatile agent and nanometer lithium magnesium silicate as a thixotropic agent), and 69.5% water. Then the mixture was blended with a high-speed emulsifier at the rate of 10,000 rpm for 15 minutes. The obtained aqueous MR fluid was labelled as sample #a. With the same method but with the CMPs substituted by the same volume of CIPs, the CIPs based MR fluids were also prepared for comparison, which are referred to as contrast aqueous MR fluids and labelled as sample #b.

The morphology, magnetic properties and anti-oxidization ability of the CMPs were investigated by SEM, VSM and acid erosion test, respectively. The MR properties were measured using a parallel-plate advanced rheometric expansion system (ARES, TA Co.) equipped with a magnetic field supplier at room temperature.

3. Results and Discussion

3.1 Morphology and Structure of CMPs

The characteristic morphologies of CMPs (a) and raw CIPs (b) observed using the scanning electron micrograph (SEM) are shown in Figure 1. The CMPs and CIPs were typically spherical in shape and of varying diameter. This property could enhance the development of columnar structure in MR fluids under an applied magnetic field and shear flow [4]. Compared with the smooth surface morphologies of the raw CIPs, the CMPs have thick and rough surface. This type of morphology is in good agreement with that of PEG/CI core-shell structured magnetic particle [5], and reveals that the CMPs are composed of a GED3A coating a CIP core.

3.2. Magnetic properties of the CMPs

Because the coating layer of the CMPs consists of nonmagnetic organic molecules GED3A, the magnetic properties of CMPs should change as the thickness of the GED3A layer increases. Thus, the difference in magnetic properties between CMPs and CIPs can be used to judge further whether GED3A is coated on the surface of CIPs and estimate the thickness of the coating layer. The magnetization curves of CMPs and CIPs were measured by VSM at room temperature and incorporated together in Figure 2 (a) and (b). The magnetization loop of the CMPs is very similar to that of CIPs, and the hysteresis of CMPs is nearly the same as that of CIPs. The coercivities for CMPs and CIPs are all very low, 4.17 Oe and 4.07 Oe, respectively. The saturation magnetization Ms of CMPs is as high as 182.2emu/g, and a little lower than but close to that of CIPs (186.7emu/g). According to the reduced Ms value, it was estimated that the composition of GED3A in the outer layer of CMPs is about 2.4 wt% according to the mixing law [8]. These results imply that the obtained CMPs have favorable soft magnetic properties and are very suitable to be used as the dispersed phase in MR fluids. This composition of CMPs, when used in preparing MR fluids, can not only supply a quick response to magnetic field and a robust control system, but also induce high sheer stress at the external magnetic field, since the yield stress increases with an Ms^2 dependence [9].

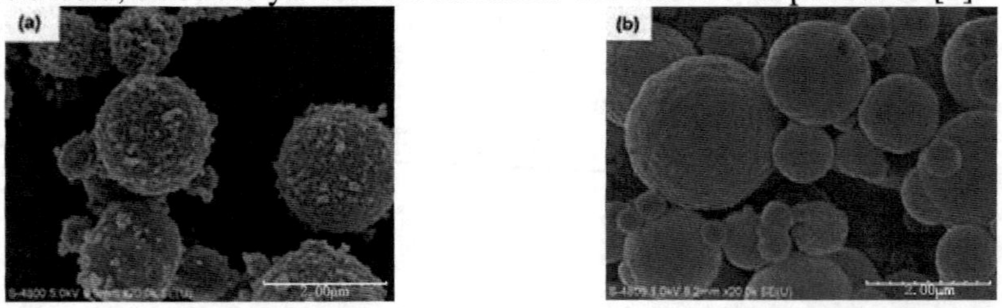

Figure 1. SEM images of CMPs (a) and CIPs (b) (×20k). Scale bars indicate 2μm

3.3. Dispersion Stability

Figure 3 shows the sedimentation ratio as a function of time for the MR fluids (sample #a and sample #b), measured via recording the amount of supernatant liquid after placed in still storage for given time. The sedimentation ratio is defined as:

$$\text{Sedimentation Ratio}\,(\%) = \frac{\text{volume of the supernatant liquid}}{\text{volume of the entire suspension}} \times 100\% \tag{1}$$

From Figure 3, it can be seen that the settling rate of the novel aqueous MR fluids is much smaller than that of the contrast MR fluids, and the sedimentation of the former started much later than the latter. Moreover, the novel aqueous MR fluids can be easily redispersed after a long period of time (3 months), using only weak stirring (300rpm×5minutes). However, the contrast MR fluids had difficulty homogenizing even after strong stirring (500rpm×10minutes). The results revealed that the dispersion stability of the novel aqueous MR fluids was much better than that of CIPs based aqueous MR fluids, and at the same time implied that the hydrophilic organic coatings of the CMPs may play an important role in stabilizing the MR fluids.

3.4. Magnetorheological properties

Figure 4 shows the changes of shear stress of the MR fluids as a function of shear rate under five different magnetic field strengths: 0 KA/m, 79.6 KA/m, 159.2 KA/m, 238.8 KA/m, and 318.4 KA/m, respectively. The shear stress of sample #a is smaller than that of sample #b at lower magnetic field strength such as 0KA/m and 79.6KA/m, then the difference becomes less quickly as further increasing magnetic field strength such as at 318.4 KA/m. It means that sample #a can provide larger tuning capacity of damper force when used in MR damper. For sample #a the shear stress abruptly increases from 10 to 2000 Pa, while the applied magnetic field changed from 0 to 79.6KA/m, and reaches approximately 20 kPa at 318.4 KA/m. Also, the shear stress under 79.6KA/m to 318.4 KA/m fluctuates at limited amplitude with increasing shear rate, which is probably due to remnant magnetization of the magnetic particles and high-volume fraction of magnetic particle in water [10]. These results show that the MR fluids exhibit high shear stress and quick responsive rate to magnetic field, and are insensitive to shear rate.

3.5. Oxidation Resistance

An acid erosion test is an effective and direct method to evaluate the oxidation resistance of MR fluids. Figure 5 shows the effect of HCl on the shear stresses of MR fluids. The measurements were performed under 318.4 KA/m and $10s^{-1}$ shear rate 3 hours after HCl was added. In Figure 5, observe that when the HCl concentration is 0.002mol/L and 0.1mol/L, the shear stress of the contrast MR fluid (sample #b) decreased by 11.5% and 48.6%, compared with that of the original sample free of HCl, but the shear stress of the MR fluid (sample #a) decreased only by 0.5% and 3.6%, respectively, compared with the original sample #a free of HCl.. That is to say, although both of the shear stresses of sample #a and #b decreased with the increase of the HCl concentration, but the reduction in shear stress value of sample #b was greater than that of sample #a for the same HCl concentration. This result demonstrates that sample #a has better anti-acid corrosion performance. This demonstrates that the organic coating surrounding CMPs prevents contact of the inner CI core with hydrogen ion in water in order to mitigate corrosion.

Conclusions

CMPs were prepared with a CIP core and a GED3A coating. These CMPs, which remain higher saturation magnetization and lower coercivity compared to CIPs, are suitable for preparing aqueous MR fluids. A novel aqueous MR fluid based on the obtained CMPs was produced and exhibited high shear stress and quick responsive rate to magnetic field, especially better dispersion stability and the anti-oxidation property.

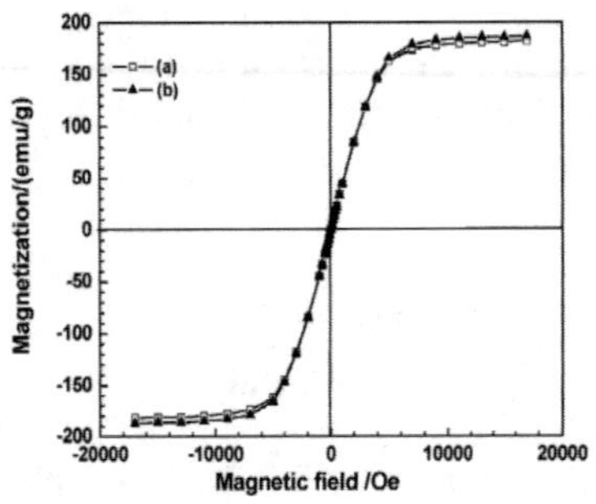

Figure 2. Magnetization curves of CMPs (a) and CIPs (b)

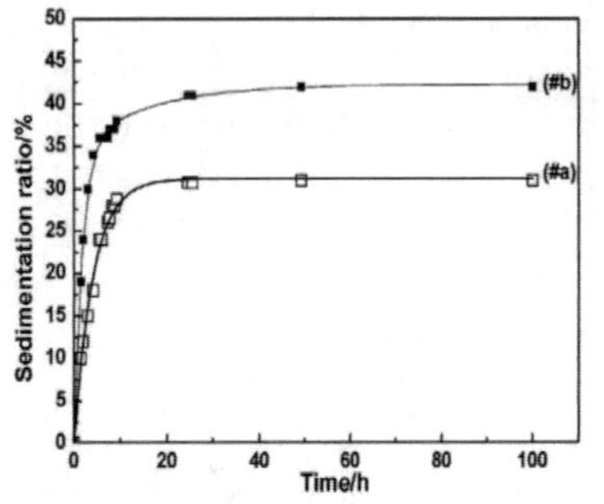

Figure 3. Sedimentation ratio versus time for the MR fluid (#a) and the contrast MR fluid (#b)

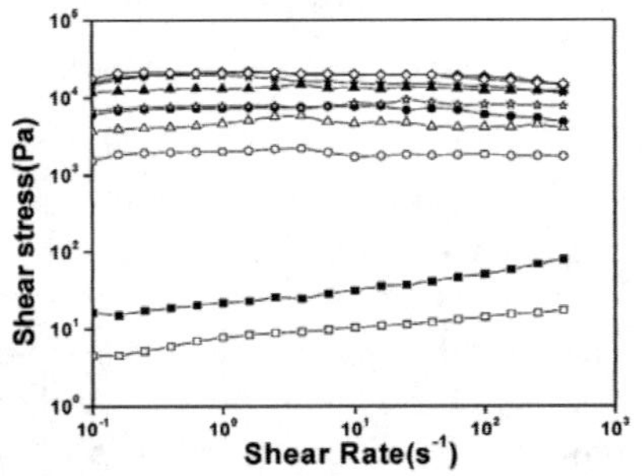

Figure 4. Shear stress versus shear rate for sample under different magnetic field strength. □ ■0KA/m, ○ ●79.6KA/m, □ ▲159.2 KA/m, □ □ 238.8 KA/m, □ □ 318.4 KA/m; hollow points for sample #a and black points for sample #b.

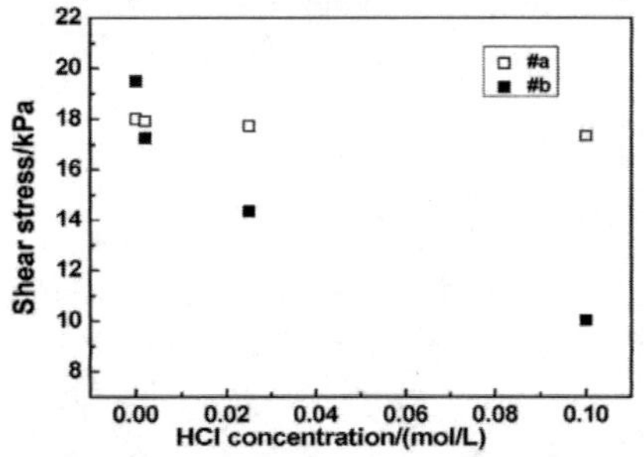

Figure 5. Shear stress versus the concentrations of HCl for sample #a and #b, Testing conditions: 318.4 KA/m, shear rate: 10s^{-1} and temperature: 25□

References

[1] Carlson J D, 2000 Aqueous Magnetorheological Material. *U.S. Pat.* 613263
[2] Cheng Haibin, Zhang Qingjie, Gao Wenxin 2006 A Stable Aqueous Magnetorheological Fluid and a process for preparation thereof. *China Pat.* ZL200610124727.6
[3] Tang L, Liu Q, Zhang P. 2006 Preparation and Properties of Water-based Magnetorheological Fluid. *Journal of Functional Materials.* **37(4)** 543-545
[4] Flores G A, Liu J. 2002 Embolization of Blood Vessels as a Cancer Therapy Using Magnetorheological Fluids. *Journal of Intelligent Material Systems and Structures.* **13(10)** 641-646
[5] Kordonski W I, Jacobs S D. 1996 Magnetorheological Finishing. *International Journal of Modern Physics B.* **10** 2837~2848
[6] Jacobs S D, Arrasmith S R. 1999 Development of New Magnetorheological Fluids for Polishing CaF2 and KDP. *LLE Review.* **80** 213-219
[7] Li Lichun, Guan Jianguo, Cheng Haibin and Tao Jianqing. 2005 Preparation of Poly (ethylene glycol)-coated Carbonyl Iron Prticles for Aqueous Magnetorheological Suspension. *Acta Phys.-Chim. Sin.* **21(7)** 817-821.
[8] Rosensweig R E. 1985 *Ferrohydrodynamics.* (Cambridge: Cambridge University Press) p344
[9] Ginder J M, Davis LC, and Elle L D. 1999 Rheology of Magnetorheological Fluids: Models and Measurements. *Int. J. Mod. Phys. B* **10** 3293-3303
[10] Choi H J, Jang I B, Lee J Y, Pich A, Bhattacharya S and Adler H J. 2005 Magnetorheology of Synthesized Core-shell Structured Nanoparticle. *IEEE Trans. Magn.* **41(10)** 3448-3450

Journal of Physics: Conference Series **149** (2009) 012045 doi:10.1088/1742-6596/149/1/012045

Magnetorheological fluids based on amorphous magnetic microparticles

H Chiriac[1], G Stoian[1] and M Lostun[1]

[1]National Institute of Research and Development for Technical Physics, 47 Mangeron Boulevard, Iaşi, Romania

Email: gstoian@phys-iasi.ro

Abstract. Magnetorheological fluids were prepared using FeSiB particles with an irregular flattened shape and also spherical Fe particles commercially available. In this paper we point out that particles oxidation and particles size distribution can influence the MR fluid performance and measurements.

1. Introduction

Magnetorheological fluids are suspensions of magnetically polarizable particles in a magnetically neutral fluid, usually mineral or silicone oil. The main characteristic of magnetorheological fluids is their ability to reversibly change from a linear viscous fluid with free flow to a semi-solid material with controllable yield stress when placed in magnetic field. When magnetic field is applied, the dispersed particles in the fluid form oriented chains parallel to magnetic field and the result is an increase of the shear stress perpendicular to the magnetic field direction. Due to their ability to modify their properties (viscosity, shear stress) in an applied magnetic field, magnetorheological fluids are widely used in semi-active controllable fluid dampers for vehicles, machineries, for buildings protection against vibrations. [1]

Yield stress is very important in applications and depends on magnetic field applied, volume fraction of solid magnetic material and is limited by magnetic saturation of the magnetic material. Usually, spherical magnetic particles are used as they are easier to be obtained in this form. The preferred magnetic material is Fe due to high saturation magnetization [2]. We presumed that the magnetic particles blocking in magnetic field may be improved using non-spherical magnetic particles.[3] Our final goal is to study contribution of the particle size and shape to the performance of the MR fluids, and the FeSiB flattened particles are made for this purpose [5], [6], [7]. In this first stage of the study we only point out that particles oxidation and particles size distribution can influence the MR fluid performance and measurements.

2. Experimental

2.1. Particle preparation and characterisation

We prepared magnetorheological fluids with two types of magnetic particles: Fe and FeSiB particles.

FeSiB particles were obtained from amorphous microwires as they have convenient shapes. We used magnetic amorphous magnetic microwires prepared by an in-rotating-water quenching technique [4]. FeSiB wires were cut to pieces of millimeter sizes so that they could be milled in a Retsch planetary ball mill. The purpose was to obtain particles of different sizes and shapes corresponding to different milling times. The particles were imaged by mean of SEM that revealed the particles dimensions and shapes. The wires were milled and then sieved by means of a sieve shaker from Retsch. The dimensional ranges of the particles used were < 20, 20-40, 40-60, 60-80 micrometers. SEM images show that they have irregular flattened shape as seen in figure 1.

a)

b)

Figure 1. SEM images of the FeSiB particles a) 20-40 microns b) 40-60 microns

Fe particles used were commercially available spherical Fe particles (mesh 325) from Sigma-Aldrich. The powder was sieved and we obtained the following dimensional ranges: < 20, 20-40 and 40-60 microns. VSM measurements were performed for all the particles.

2.2. Magnetorheological fluid preparation and characterisation
All the magnetorheological fluids were prepared in volume concentrations of 10%, by mixing the mineral oil with the stabilizer (Al stearate) first and then with the magnetic particles using a grinding mortar. The fluid was homogenized afterwards by mean of a stirrer. Flow (viscosity) curves were measured in and out of magnetic field using a rheometer Physica MCR101 fitted with a magnetorheological module which can apply a magnetic field up to 1T corresponding to a 5A current. The measurements were performed using a parallel-plate system with a diameter of 20 mm at a gap of 1 mm and at 20^0C for all tests. We applied different magnetic fields and flow curves were plotted for each magnetic field. The yield stress was calculated by the rheometer software for each flow curve according to Bingham model.

3. Results and discussions

Magnetorheological fluids were prepared in volume concentration of 10% using FeSiB particles with different dimensional range. Variation of the yield stress with applied field in the MR cell was plotted for each MR fluid based on the measured flow curves. The values of the yield stress for each magnetic field were calculated by the rheometer software according to Bingham model. As seen in figure 3, a direct dependence between the particles size and the maximum yield stress was found. Further VSM magnetic measurements shown in figure 2 revealed that the saturation magnetization of the particles was greater as the particles were bigger. This could mean that the particles were affected by oxidation. The degree of oxidation was larger as the particles were smaller due to their larger surface in contact with air. Smaller particles were more affected by oxidation leading to lower saturation magnetization and thus to lower yield stress. Our EDS analysis reveals that the particles were indeed oxidized. Thus it cannot be clearly separated the influence of the oxidation from the influence of the dimension of the particles. Further investigations have to be made taking into consideration the particles oxidation.

Saturation magnetization for each particle dimensional range is given in Table 1. It can be seen also in Figure 3 that although the dimensional ranges of 20-40µ and 40-60µ have notable different saturation magnetizations they present almost the same yield stresses. Starting with this observation and other results with the same type of MR fluids we assumed that there could be another factor that alters the resulting yield stress and that would be the particle size distributions.

Journal of Physics: Conference Series **149** (2009) 012045 doi:10.1088/1742-6596/149/1/012045

Table 1. Saturation magnetization for each FeSiB particle dimensional range

Particle size range	Ms(emu/g)
<20 μ	149.69
20-40 μ	161.32
40-60 μ	172.20
60-80 μ	174.17

For a better understanding of the particle size distributions we made MR fluids using Fe particles. The Fe particles were Mesh325 commercially available from Sigma-Aldrich. The particles were sieved identically as in the case of the FeSiB particles. Thus we obtained Fe particles with size ranges of < 20, 20-40 and 40-60 microns. Over 60 microns there was a negligible quantity after sieving.

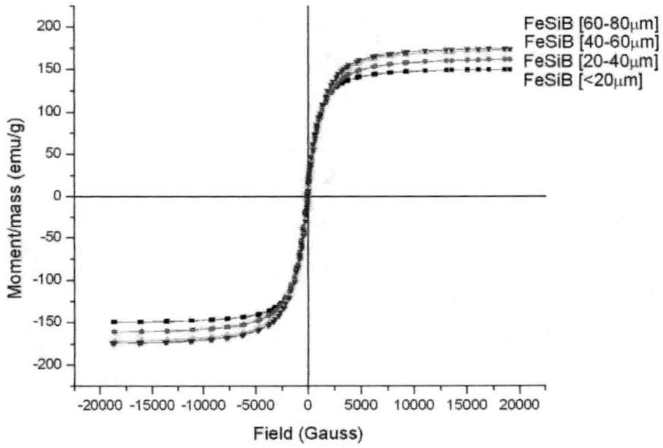

Figure 2. VSM magnetic measurements for the FeSiB particles with different size ranges

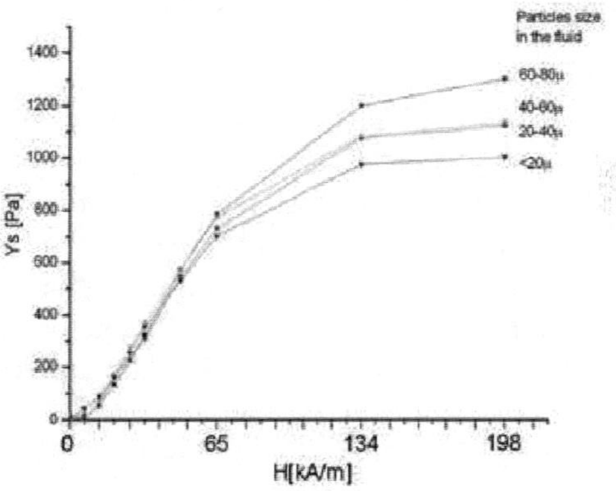

Figure 3. Yield stress vs. applied field in the MR cell for the MR fluids with FeSiB particles with different sizes

MR fluids were prepared using the same procedure as in the case of FeSiB particles. The magnetization curves are shown in figure 4. Yield stress dependence with applied field has been plotted after flow curves measurements. The result is shown in figure 5. We note that the MR fluid with particles in range of 20-40μ presents maximum yield stress higher in the case of ranges <20μ and 40-60μ.

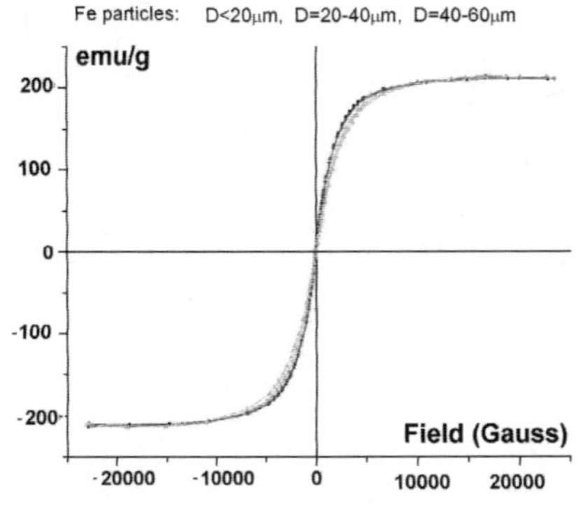

Figure 4. VSM magnetic measurements for the Fe particles with different size ranges

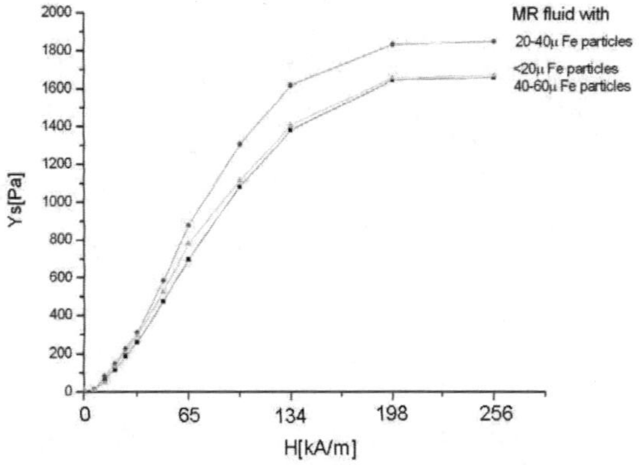

Figure 5. Yield stress vs. applied field in the MR cell for the MR fluids with Fe particles

The size distribution has been measured with a Microtrac S3500 particle size analyzer. A maximum in the range 20-40 μ, and an ascending respectively descending slope distribution in the ranges <20μ and 40-60μ were observed. We presume that there is a link between the particle size distribution and yield stress curves shown in figure 5. This could mean that a strictly ascending or descending slope distribution has a negative role for the performance of the MR fluid. Also, it is probably that a more uniform particle size distribution leads to more homogenous and strong structure in the MR fluid when magnetic field is applied. Further investigations have to be done.

4. Conclusions

Magnetorheological fluids were prepared using FeSiB particles with an irregular flattened shape and also spherical Fe particles commercially available. FeSiB particles in four dimensional ranges were obtained by milling magnetic amorphous microwires and sieving the resulting powder. Rheological measurements were performed for each MR fluid: flow curves having the magnetic field as parameter and yield stress vs. applied magnetic field curves were calculated using Bingham model. In order to highlight the influence of the particle dimension and shape on the performance of the MR fluid it is necessary to eliminate or keep constant the other influencing factors. We found that two factors were influencing our measurements: particles oxidation and particles distribution. The degree of oxidation was higher as the particles were smaller leading to lower magnetic saturation and hence to lower yield stresses in magnetic field. The influence of the particle distribution was better evidenced using commercially available Fe particles with a known particle distribution and sieved in the same dimensional range as the FeSiB particles. Our hypothesis is that a strictly ascending or descending slope-like distribution has a negative effect on the performance of the MR fluid.

Further experiments have to be done regarding the role of the size distribution to the MR fluid performances. In order to study the particles dimension and shape contribution to the properties of a MR fluid we must take into account the above mentioned influences.

References

[1] Deshmukh S S and McKinley G H 2007 *Smart Mater. Struct.* **16** 106-113

[2] Genc S and Phule P P 2002 *Smart Mater. Struct.* **11** 140-146

[3] Bell R C, Karli J O, Vavreck A N, Zimmerman D T, Ngatu G T and Wereley N M 2008 *Smart Mater. Struct.* **17** 015028

[4] Chiriac & Lupu; in: *"Properties and Applications of Nanocrystalline Alloys from Amorphous Precursors"* Edited by Bogdan Idzikowski, Peter Svec, Marcel Miglierini, North Atlantic Treaty Organization Scientific Affairs Division; Published by Springer ISBN 1402029632, 9781402029639

[5] Bell R C, Miller E D, Karli J O, Vavreck A N and Zimmerman D T 2007 "Influence of particle shape on the properties of magnetorheological fluids" *International Journal of Modern Phisics* B, **21** 5018-25

[6] de Gans B J, Duin N J, van den Ende D and Mellema J 2000 "The influence of particle size on the magnetorheological properties of an inverse ferrofluid" *J. Chem. Phys.* **113** 2032-42

[7] Lopez-Lopez MT, Vertelov G, Bossis G, *et al.* 2007 "New magnetorheological fluids based on magnetic fibers" *J. Mater Chem.* **17** 3839-44

Theoretical analysis and evaluation of an optimally controlled full-car vehicle model with a variable-damping semi-active vehicle suspension forced by measured road inputs

Y Ercan

TOBB University of Economics and Technology, Sogutozu, 06560 Ankara, Turkey

E-mail: yercan@etu.edu.tr

Abstract. This study aims to obtain the optimal control algorithm for a full-car model with a variable-damping semi-active suspension, such as a magnetorheological damper, by solving the linear quadratic regulator problem, and then to evaluate the system performance if the control inputs are constrained and delayed, and the vehicle is subjected to measured road inputs. A seven-degree of freedom full-car vehicle model was considered, and the state equations of the system were obtained in bilinear form. An integral performance index involving a weighted combination of the mean squares of average sprung mass acceleration and suspension deflections was defined. Trade-off curves were obtained between the sprung mass acceleration and suspension deflections of the optimally controlled system which is subjected to a measured road profile input. Performance of the optimally controlled system was compared to the performance of the corresponding optimum passive suspension system. For the vehicle parameters and the road input profile considered in this study, a reduction of 6.4% in the average vertical acceleration and 2.8% in the average suspension deflection was achieved by the semi-active suspensions. The response of the system to an initial condition has shown that its transient oscillations are damped out effectively by the semi-active suspension.

1. Introduction

A set of suitable parameters, normally the coefficients of the dampers between the sprung mass and the unsprung masses, are changed in semi-active suspensions to optimize the system performance. Dampers with variable coefficients, such as magnetorheological dampers, are now commercially available, and the use of semi-active suspensions in vehicles has become economically feasible.

Semi-active suspensions were first proposed more than 30 years ago by Karnopp and Crosby [1, 2]. Many studies have been carried out since then, including those employing linear quadratic Gaussian (LQG) control of ¼- and ½-vehicle models, and other control schemes with and without preview [3-15]. Some of the previous studies have been specifically on suspensions with magnetorheological dampers [16]. A review of the major studies may be found in Hrovat [17].

The linear quadratic state regulator optimization yields the optimal linear quadratic Gaussian control if the disturbance input to the system is white noise and the control inputs are not constrained. However, the road profile inputs to a vehicle are not white noise, and the control inputs to a vehicle with a semi-active suspension with variable damping are constrained. Moreover, the control inputs can be applied only with a certain time delay caused by processing of data and the response time of controllers. As a result, performance of a vehicle suspension system using an algorithm based on the optimal control solution of the LQG problem will be affected by these deviations from the underlying assumptions. The purpose of this study is to obtain the optimal control algorithm for a full-car model with a variable damping semi-active suspension by solving the linear quadratic state regulator problem, that is the LQG problem, and evaluate its performance when the vehicle is subjected to measured road inputs, and the control inputs are constrained.

2. System model and the method of obtaining the optimal control algorithm

A full-car vehicle model with 7 degrees of freedom as shown in figure 1 is considered in this study. Here u_1, u_2, u_2, u_4 are coefficients of the variable dampers, which may be adjusted to control the system behaviour. z_{fl1}, z_{fr1}, z_{rl1}, and z_{rr1} represent the disturbance inputs from the road. The road inputs from the left and right tracks are different from each other. On the other hand, the inputs from the same track at the front and at the back of the vehicle are the same in shape, but there is a time difference between them that is determined by the length L and speed V of the vehicle.

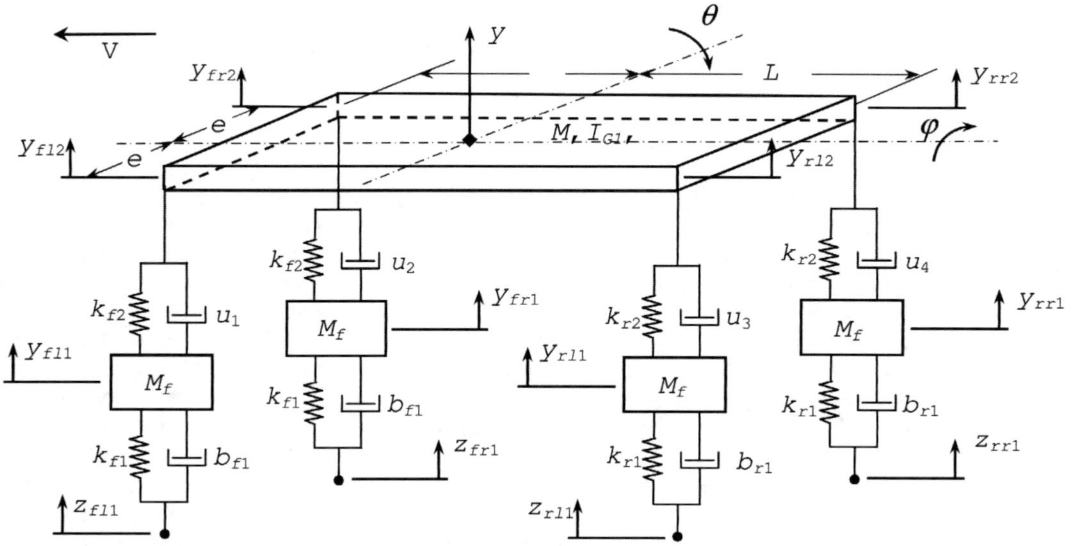

Figure 1. System model.

The state equations of the system, which are not given here to save space, were obtained. The state vector **x** has 14 elements. The dynamic suspension clearances $x_1 = y_{fl2} - y_{fl1}$, $x_5 = y_{rl2} - y_{rl1}$, $x_9 = y_{fr2} - y_{fr1}$, $x_{13} = y_{rr2} - y_{rr1}$, tire deflections $x_3 = y_{fl1}$, $x_7 = y_{rl1}$, $x_{11} = y_{fr1}$, and their rates of change are defined as the state variables. The state equations involve terms in which state and control variables multiply each other. Such systems are called bilinear. It can be shown that the solution of the state regulator problem for bilinear systems may be obtained from the Riccati equation. In the present study, the state regulator problem is solved first. Then, performance of the resulting optimally controlled system is evaluated for constrained control inputs, and in the presence of disturbance inputs which are the measured road elevations.

A quadratic performance index was assumed in this study as follows, involving the average squared heave accelerations of the points on the sprung mass, squares of suspension deflections and tire deflections.

$$P = \int_0^{t_f} [(\ddot{y}^2)_{av} + \rho_1 (x_1^2 + x_5^2 + x_9^2 + x_{13}^2) + \rho_2 (x_3^2 + x_7^2 + x_{11}^2 + y_{rr1}^2)]dt \qquad (1)$$

It was shown that the four variable damping coefficients u_1, u_2, u_3 and u_4 are not all needed to minimize the selected performance index. Therefore, u_4 was set to a constant value u_{4s}, and the remaining control variables u_{1o}, u_{2o}, and u_{3o} were solved in terms of the state variables by assuming the

Journal of Physics: Conference Series **149** (2009) 012046 doi:10.1088/1742-6596/149/1/012046

final value in the performance $t_f = \infty$, and making use of the corresponding Riccati equation. However, rather than solving the algebraic Riccati equation, differential Riccati equation was integrated in the negative time direction starting from zero final condition until its steady state solution is reached.

3. Performance of a vehicle with a semi-active suspension

Performance of a 7-ton vehicle with a semi-active suspension traveling at a speed of 90 km/h was investigated. However, coefficients of the variable dampers were constrained in applying the optimal control algorithm such that 2000 N·s/m < u_i < 40000 N·s/m (u_i = 1 to 3), assuming that magnetorheological dampers are used in the system. A randomly selected road profile from The University of Michigan Transportation Research Institute (UMTRI) World Wide Web site was used to disturb the system. Performance of the optimal passive system for the same vehicle was used as a benchmark to evaluate the performance of the semi-active system.

Simulations showed that tire deflections are hardly effected by the semi-active control, and they are within acceptable ranges. Therefore, the weighting coefficient ρ_2 was taken zero in the simulations, and the optimal trade-off curve between the average sprung mass vertical acceleration and the average suspension deflections, shown in figure 2, was obtained by changing the values of ρ_1.

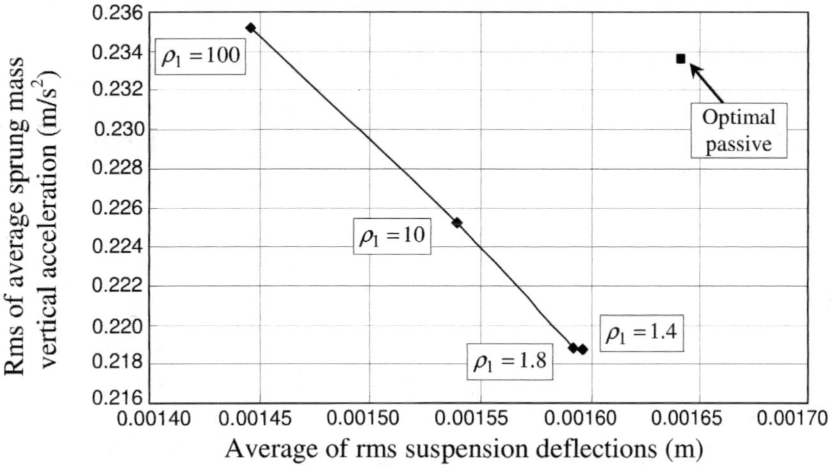

Figure 2. Sprung mass acceleration versus suspension deflection trade-off curve.

Minimum sprung mass acceleration was obtained for $\rho_1 = 1.4$, which is 0.2187 m/s^2 as compared to 0.2336 m/s^2 for the optimal passive suspension. The corresponding average rms suspension deflections are 1.596×10^{-3} m and 1.642×10^{-3} m respectively. Hence, a reduction of 6.4% in the rms average vertical acceleration and 2.8% in the average rms suspension deflection is achieved by the semi-active suspensions. The simulations also have shown that either the maximum or the minimum value of the variable damping coefficient is used most of the time; hence, the control is almost like a bang-bang control.

Figure 3 shows the responses of the semi-active system for $\rho_1 = 1.4$, and of the passive system to an initial condition. It is noted that both the average rms vertical acceleration and center of gravity deflection of the sprung mass are reduced effectively by the semi-active suspensions as compared to the passive suspension.

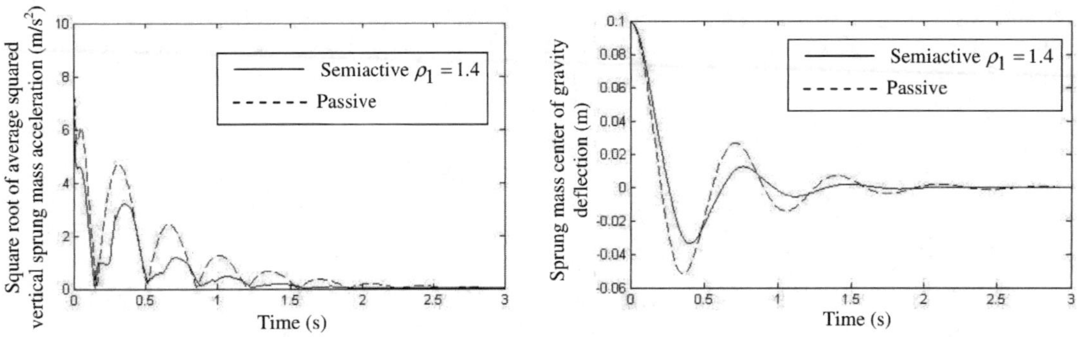

Figure 3. Responses of semi-active and passive systems to an initial condition ($\rho_2 = 0$).

4. Conclusions

A full-car model of a vehicle equipped with variable-damping type semi-active suspensions was developed in this study and bilinear form of state equations was obtained. An analytical study on optimal control of the system has shown that only three variable dampers are sufficient to minimize a performance index consisting of average sprung mass acceleration and suspension deflections. The optimal control algorithm obtained was applied to a vehicle with magnetorheological dampers and moving on a road with a specified profile, and the performance the system was evaluated for constrained control inputs. The results show that rms average sprung mass vertical acceleration is reduced by 6.4% and the average rms suspension deflections by 2.8% by semi-active suspensions as compared to the optimal passive system. The transient oscillations of the system were shown to be effectively damped out by the semi-active suspensions.

5. References

[1] Karnopp D C, Crosby M J and Harwood R A 1974 Vibration control using semi-active force generators *ASME J. of Engineering for Industry* **96** 619

[2] Crosby M J and Karnopp D C 1973 The active damper – A new concept for shock and vibration control *Shock and Vibration Bull.* **H**

[6] Hac A and Youn I 1993 Optimal design of active and semi-active suspensions including time delays and preview *ASME J. of Vibration and Acoustics* **115** 498

[8] ElMadany M M and Abduljabbar Z 1991 Alternative control laws for active and semi-active automotive suspensions - A comparative study *Computers and Structures* **39** 623-29

[9] Hwang Y R and Wang S C 2003 Computer control algorithm for semi-active suspensions based on a two-wheel vehicle model *International Journal of Computer Applications in Technology* **17** 37-46

[10] Tseng H E and Hedrick J K 1994 Semi-active control laws – Optimal and sub-optimal *Vehicle System Dynamics* **23** 545-69

[12] Elbeheiry E M 1998 A method for preview vibration control of systems having forcing inputs and rapidly switched dampers *J. of Sound and Vibration* **214** 269-83

[13] Elbeheiry E M 2001 Suboptimal bilinear control methods applied to suppressing car vibrations *J. of Vibration and Control* **7** 279-306

[14] Ercan Y 2006 *Proc. The 10th Int. Conf. on Electrorheological Fluids and Magnetorheological Suspensions ERMR 2006 (Lake Tahoe)* (Hackensack: World Scientific) p 431

[16] Gordaninejad F, Evrensel C A, Yeo M S, Liu Y and Karakaş ES 2003 Study of a quarter model HMMWV suspension system using a magnetoheological fluid damper *Proc. of SPIE - The Int. Society for Optical Engineering*, **5056** 506-13

[17] Hrovat D 1997 Survey of advanced suspension developments and related optimal control applications *Automatica* **33** 1781-817

Journal of Physics: Conference Series **149** (2009) 012047 doi:10.1088/1742-6596/149/1/012047

Magnetorheological characteristics of carbon nanotube wrapped carbonyl iron particles

Fei Fei Fang, Byung Il Choi and Hyoung Jin Choi

Department of Polymer Science and Engineering, Inha University, Incheon 402-751, Korea

E-mail: hjchoi@inha.ac.kr

Abstract. Carbonyl iron (CI) based magnetorheological (MR) fluid exhibits serious dispersion defect in general due to large density mismatch between CI particles and continuous medium, which restricts further MR application. Thus, various strategies were explored either to reduce the density or to prevent CI particle aggregation. Among them, polymer coating technology becomes more prevalent due to favorable morphology obtained and effective decrease in density by introducing polymeric shell; nevertheless, coating polymer on the surface of CI particles is always influenced by the selected grafting agent, mole ratio of reactant or the temperature of reaction. In this work, considering self-assembling trend of carbon nanotube (CNT) which exhibits similar density with polymer but better magnetic property due to the iron catalyst, we constructed a dense nest composed of CNT on the surface of CI particles by using 4-aminobenzoic acid (PABA) as a grafting agent under sonication. Thickness and morphology of the CNT nest were found to be related with sonication duration via SEM/TEM images. MR performances (yield stress behavior, shear viscosity) of the CI/CNT particles based MR fluid were investigated via controlled shear rate and controlled shear stress methods. Finally, sedimentation observation was checked to be improved.

1. Introduction

Magnetorheological (MR) fluids consisting of magnetic particles dispersed in nonmagnetic fluids are being regarded as smart/intelligent engineering materials, since they are free-flowing liquids under normal condition with out magnetic field but immediately become solid-like when exposed to an external magnetic field [1-5]. During this process, the rheological properties (yield stress, apparent viscosity, and storage modulus et. al) are altered. Therefore, MR fluids, along with their electrically analogous electrorheological (ER) fluids, attract considerable attentions in widespread engineering applications of designing damper, torque transducer or polishing devices [6-10]. Exploring superior MR materials which can promote their further engineering application is in progress, and various magnetic species or alloys have been investigated [11-16]. So far, much attention has been focused on carbonyl iron (CI) particles (CD grade, BASF Germany, average particle size: 4.25 μm; density: 7.91 g/cm^3) which are considered as an excellent candidate for MR fluid due to the superior magnetic properties as well as their appropriate size [17-20]. Nevertheless, CI based MR fluid has not been widely applied in designing MR devices as expected because serious sedimentation problem caused by large density mismatch between CI particles and medium oil affects the operation of MR test as well as the re-dispersion. Thus, plentiful efforts (introducing sub-micro additives or polymer coating technology) have been paid on preventing CI particles' contact or decreasing density of CI particles to improve the sedimentation rate indirectly. A mass of sub-micro fillers (carbon nanotube, graphite nanotube, fumed silica et al) have been introduced into CI based MR suspension, in which the interspaces of CI particles are occupied, by which way restrains the direct contact of CI particles, consequently sustains the stability of MR fluid [21-24]. Additionally, when exposed to a magnetic field, the produced chain-like structure undergoes a chain rupture and reformation process, adding fillers can donate to repairing the deformed chain structure under an external shear field. However, we have to regard the influence of size dimension, morphology of additives as well as the affinity with CI particles on improving this problem. Therefore, modifying CI particles to reduce the density mismatch with continuous oil comes to front. Coating polymer on the surface of CI particles has become popular due to the produced favorable core-shell structure along with the apparently decreased density for CI/polymer composite particles [17, 25, 26]. However, coating polymers on the surface of CI particles is not a facile process in spite of using versatile functional grafting agents. Failing to select an optimal grafting agent may lead to a very thin coating layer even a failure in coating. In addition, the mole ratio among monomer, initiator or stabilizer as well as the reaction temperature also plays a critical role in affecting the final shell product. Taking into account of the above mentioned complicated procedures, we have to seek a new strategy based on the coating technology. Concurrently, carbon nanotubes (CNT) have been extensively studied as filler materials for fabricating polymer/CNT nanocomposites which exhibit synergistic mechanical, electrical, and thermal properties. Chemical functionalization of CNT with carboxyl/amino group (COOH/NH_2-CNT) or physical adsorption of dispersants onto the wall of CNT has been developed to enhance the dispersibility of CNT in solvent [27]. In addition, CNT exhibits unique self-assembling trend [28], therefore, in this work we tried to construct a dense nest of CNT on the surface of CI particles with the aid of using 4-aminobenzoic acid (PABA) as a grafting agent. Anther two principal reason of choosing CNT as the shell material for coating CI particles may lie in two main aspects. On the hand, the density of CNT is similar with that of polymer; on the other hand, the existence of magnetic iron catalyst within CNT wall may contribute to better magnetic properties, which does not occur in polymer coating systems. The thickness and morphology of CNT nest were investigated by alternating sonication duration. MR performances (yield stress behaviour, viscosity) of CI/CNT particles based MR fluid were studied via controlled shear rate and controlled shear stress methods. Finally, sedimentation of the produced particles was checked.

2. Experimental

2.1. Preparation of CI/CNT composite particles

Initially, the grafting agent, 4-aminobenzoic acid (PABA) was dispersed in distilled-water at 60°C for 2h. Raw CNT was chemically pre-treated to obtain COOH-CNT [29]. Then, a certain amount of CI particles (CD grade, BASF, Germany) were dispersed in this solution and underwent sonication for 15 minutes to modify the surface. After removing the excess PABA by washing with di-water, the PABA modified CI particles were added to reactor, in which a dispersion of COOH-CNT (0.1 wt% of CI particles) in di-water was predispersed. The reaction was kept with sonication at room temperature under vigorous stirring in order to avoid sedimentation due to the large density of CI particles. At first, the color of reaction dispersion is absolute black due to the dispersed CNT in di-water, as time goes on, the liquid turns to be light brown, at last becomes transparently brown. Meanwhile, CI particles change its color from initial silver grey to complete black which suggests the successful wrapping of COOH-CNT. Therefore, the majority COOH-CNT was adsorbed to the surface of CI particles, leaving minority COOH-CNT with very short chain dispersed in water which may cause the transparently brown liquid because sonication process will cut COOH-CNT long chains into very small parts. This assumption was proved by holing the transparent brown liquid in vial still for about 1 month, after which the liquid became completely transparent but left a very thin black layer at bottom which was considered to be short COOH-CNT. The fabricated CI/CNT particles were separated from the residual COOH-CNT dispersion using a magnet and washed with di-water and methanol separately.

Finally, the sample was dried in vacuum oven at 60°C for 24 hours. Here, in order to obtain densely coated CNT nests and perfect spherical profile along the surface of CI particles, different sonication duration (12, 24 hours) was performed with preserved other reaction condition.

2.2. Characterization of CI/CNT composite particles

Both surface morphology and cross section view of fabricated CI/CNT particles were detected by Scanning Electron Microscope (SEM, S-4300, Hitachi Japan) and Transmission Electron Microscope (TEM, Philips CM200), respectively. The sampling for TEM view was done by an initial moulding of particles in epoxy bath, a following nano-scaled cutting by using an ultramicrotome (UMT) as well as a final dropping on a copper grid. Density of the synthesized CI/CNT particles was examined by using pyconometer. In order to prepare MR fluid, the fabricated CI/CNT particles were dispersed in lubricant oil (Yubase 8, SK Corp. Korea) with a particle concentration of 20vol%. Magnetorheological measurements (controlled shear rate/CSR mode and controlled shear stress/CSS mode) were performed using a rotational rheometer (MCR 300, Physica, Stuttgart, Germany) with the magneto-rheological device (MRD 180, Physica, Stuttgart, Germany). A parallel-plate measuring system with a diameter of 20 mm and a gap of 1 mm was made of non-magnetic metal to prevent the occurrence of radial magnetic force components on the shaft of the measuring system. The magnetic field direction was set to be perpendicular to the flow direction. To control temperature of the device and the sample, a water circulator was adopted to the MRD 180, and the temperature was set at 25 °C for all tests.

3. Results and discussion

3.1. Morphology of CI/CNT particles and influence of sonication duration on morphology

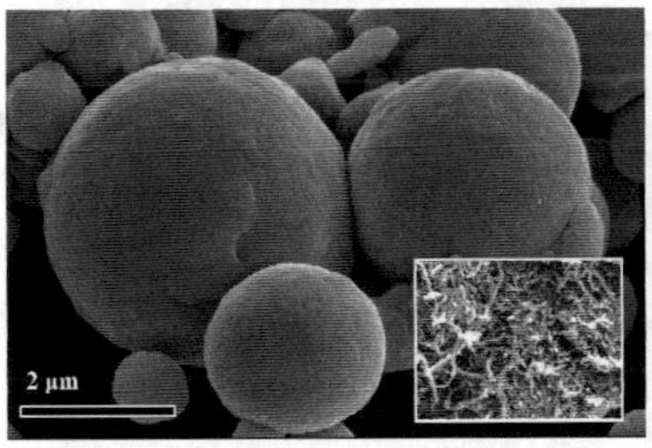

Figure 1. SEM images of pure CI particles and COOH-CNT (the inset picture).

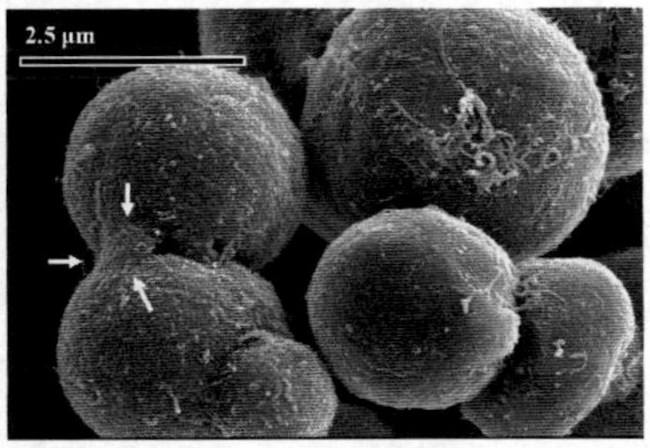

Figure 2. SEM image of fabricated CI/CNT particles with mild sonication for 12 hr.

Figure 1 indicates the morphology of pristine CI particles and COOH-CNT (the inset picture) used in this experiment. It is clear that most CI particles are spherical and polydisperse with considerably smooth surface. Chemically treated COOH-CNT with an average diameter of 15 nm gives serious aggregation. After immersing PABA modified CI particles in COOH-CNT dispersion with sonication, obvious change in surface appearance is visible as indicated in Fig. 2. A dense layer composed of huge CNT is found to wrap CI particles leading to a considerably rough surface compared with that of the pure CI particles. As pointed with white arrow head, there seems to be a nest-like shell which connects two adjacent CI particles. Innumerable CNT piled together thus formed a firm network overspreading the whole surface of CI particles. Besides this, compared with the compactly aggregated CNT nest, there is some single CNT standing at the surface towards to the outside. It is worthy to notice that the length of CNT here seems to be shorter than that of pure COOH-CNT, which may be attributed to the sonication. In order to track the role of PABA as grafting agent, an additional experiment without employing PABA was made. Unfortunately, there is no any apparent CNT layer around the CI particle, consequently confirming that PABA take crucial function in joining hydrophilic CI particles and CNT networks. On knowing that PABA possesses a carboxyl group (COOH) and an amino group (NH_2), we assume that an amide functional group may form between PABA modified CI and COOH-CNT, besides this, the non-covalent bond appeared between adjacent –COOH functionalized CNT may also help producing the network of CNT [28, 30]. Anyway, a further study in either FT-IR or XPS spectra is necessary to confirm this prediction.

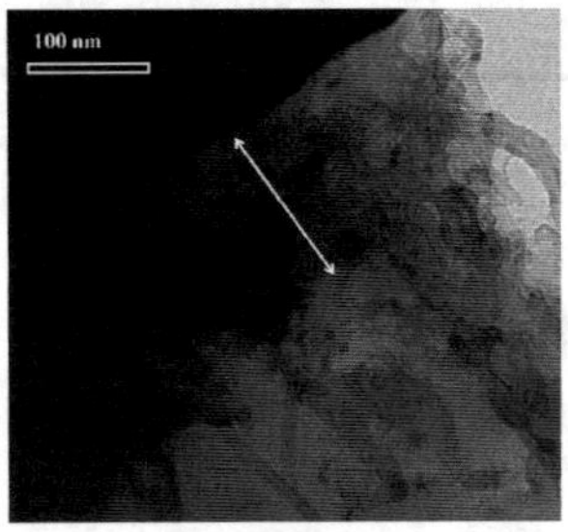

Figure 3. TEM image of fabricated CI/CNT particles with mild sonication for 12 hr.

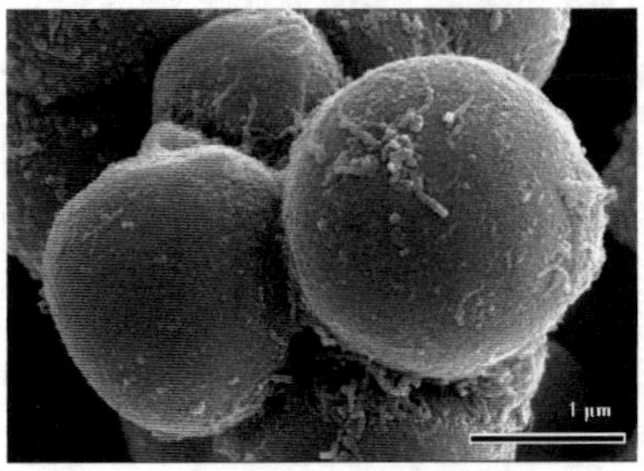

Figure 4. SEM image of fabricated CI/CNT particles with mild sonication for 24 hr (other reaction condition is same).

In order to observe the internal structure within CNT nest as well as the coating thickness, we examined the cross section view for an individual CI/CNT particle via TEM as indicated in Fig. 3. The black block represents CI core particle, while the rest fibril part is CNT. There is no apparent leakage between CI particles and CNT shell. The CNT close to the surface of CI particles seems to aggregate probably due to the overlap of multi-layer of CNT shell. Therefore, in this region, dense CNT nest was constructed, consequently revealing that the coating thickness of CI particles was about 100 nm as pointed in the picture. The CNT extending out over the surface exhibits relative straight shape with an average size of about 11 nm. This result is well coincided with that of Fig. 1. In order to coat CNT nest as dense as possible on the surface of CI particles, the effect of sonication duration was examined. When the COOH-CNT is dispersed in di-water without using any other dispersant, sonication will contribute to break CNT agglomerates leading to homogeneous CNT dispersion as well as a chemical banding between carboxyl and amino group [31, 32]. Thus, we set sonication duration for 12 hr and 24 hr to observe the difference. Figure 4 is a SEM image taken for the CI/CNT particles sonicated for 24 hours. Unfortunate, compared with the apparent nest structure found by sonicating for 12 hr, the expected more dense CNT nest did not appear. We find, although the surface becomes rougher than pure CI particle which is possibly caused by the initial coating of CNT, plenty of short CNT segments rather than long ones were observed. This may be attributed to the vigorous long term-sonication which may cut the long CNT into short ones, by which way the formed CNT nest gets broken up. The interspace of adjacent CI particles was filler of short CNT segments. Thus, the CNT within the interspaces may suffer less influence from sonication than that of loaded on the surface of CI particles. In this experiment, the chemical interaction produced between PABA modified CI and COOH-CNT as well as the noncovalent bond appeared within –COOH functional groups of adjacent CNTs was assumed to be the driving force to construct CNT network [21]. Although we can not decide the optimal sonication duration, excess sonication does have a negative effect on coating.

3.2. The MR characterization of CI/CNT particles based MR fluid

MR characterization was performed at magnetic field strengths ranging from 0 to 343kA/m. Yield stress was investigated via two modes: controlled shear rate (CSR) and controlled shear stress (CSS). The CSR mode is the most common way to obtain a yield stress by shearing the sample over a range of shear rate, then plotting the shear stress as a function of shear rate as shown in Fig. 5 [33, 34]. The intersection on the stress axis is then taken as dynamic yield stress demonstrating that any stress below this is insufficient to cause the sample to flow [35]. In this system, the dynamic yield stress is extrapolated to be about 0.89 kPa, 3.5 kPa and 11.8 kPa, corresponding to the strength of magnetic field of 86, 171 and 343kA/m. In addition, when the magnetic field is present, all of shear stress curves shift upwards and represent a wide plateau range over the whole region of applied shear rate which may be caused by the robustly formed columns via strong dipole-dipole interaction among the adjacent magnetic particles. Compared with the above dynamic yield stress, static yield stress was also investigated via CSS approach which starts testing in rest state and increases the stress to a critical value at which the sample starts to flow [36]. Figure 6 describes the change of viscosity as a function of shear stress, in which viscosity initially holds its value at low shear rate region then falls abruptly and finally approaches to a constant value at high shear stress. The static yield stress develops below which there is no real macroscopic flow. From the point marked with arrow head, all curves indicate about 4 decade drop in viscosity. Thus, the value of this point can be considered as static yield stress which is about 1.02 kPa, 5.31 kPa and 13.9 kPa separately. Actually, the sample is undergoing creep behavior below this stress; however we just assume that it is static.

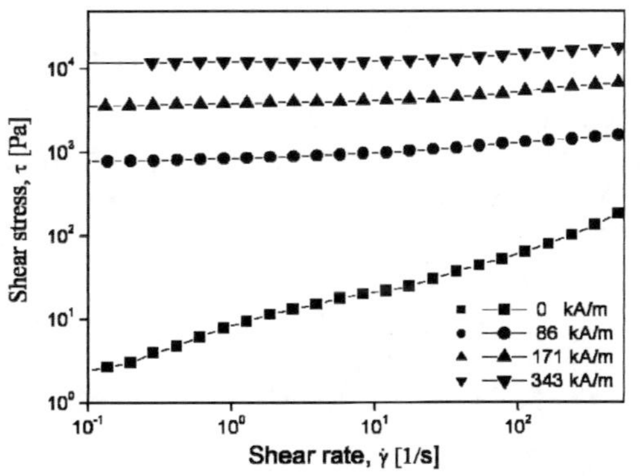

Figure 5. Shear stress curves (CSR mode) for CI/CNT particles based MR fluid under various magnetic field strength.

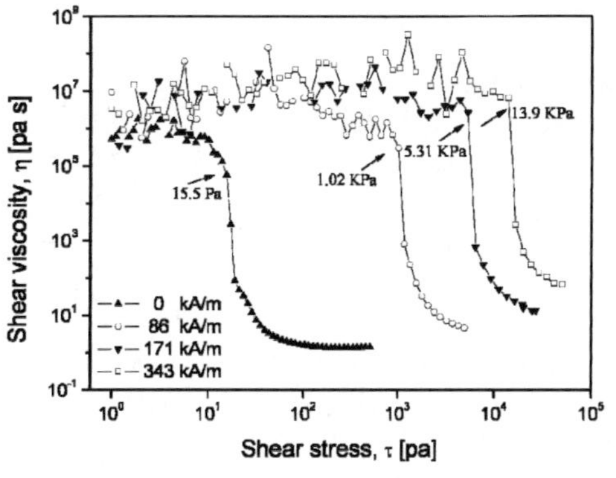

Figure 6. Shear stress curves (CSS mode) for CI/CNT particles based MR fluid under various magnetic field strength.

3.3. Investigation on sedimentation problem

Finally, sedimenation problem was examined. The density for synthesized CI/CNT particles is 6.42 g/cm^3 which is much smaller that pure CI particles (7.91 g/cm^3). Obviously, the density mismatch between dispersed particles and medium oil (0.85 g/cm^3) is reduced, thus an improved sedimentation rate should be expected [17, 26, 37]. Figure 7 indicates the sedimentation ratio for the CI/CNT suspension and pure CI suspension as a function of time. In this method, the settling of macroscopic phase boundary between concentrated suspension and supernatant liquid was observed. We notice that initially, CNT wrapped CI particles settle down more slowly than that of pure CI particles at the same time duration. As time goes by, after most particles settle completely, compared with the nearly transparent supernatant of pure CI suspension, CNT wrapped CI suspension exhibits obscure phase, in which many CI/CNT particles still suspended in the medium oil. Actually, when we prepared the MR fluid and operated MR test, the CI/CNT particles dispersed more easily in medium and homogeneously re-dispersed MR fluid can also be achieved by momently mild shaking. The improved sedimentation and easy redispersion can be attributed to the reduced density along with the rough surface. Compared with the smooth surface of pure CI particles, the rough surface of CI/CNT particles is available to produce much friction force with medium oil, evidently, this may help to postpone the settling of particles, thus reduce the sedimentation rate.

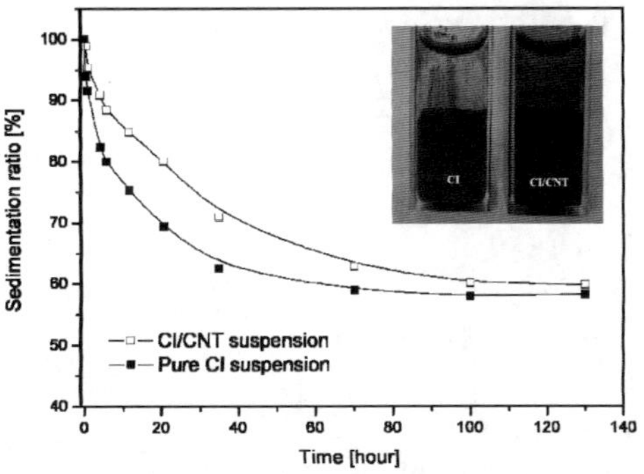

Figure 7. Sedimentation ratio as a function of time for CI suspension (closed symbol) and CI/CNT suspension (open symbol)

4. Conclusions

In this work, we focused on fabricating unique spherical CI/CNT particles with densely coated CNT network by using PABA as grafting agent basing on the CNT's self-assembling principle. The perfect core-shell morphology as well as the coating thickness was confirmed via SEM/TEM view. Long-term sonication was not favored to construct much denser CNT nest on the surface of CI particles. Although the possible mechanism of forming CNT nest was assumed, further physical characterization should be continued. In the case of MR characterization, both dynamic yield stress and static yield stress were investigated. Finally, sedimentation was observed to be improved by the reduced density along with easy redispersity. Working on coating more densely CNT shell on the surface of CI particles which leads to improved sedimentation is in progress.

Acknowledgments
The authors make a grateful acknowledgment to Mr. Hwang Sun Jae for the valuable assistance with the transmission electron microscopy (TEM) investigations. This study was supported by a grant from the Fundamental R&D Program for Core Technology of Materials funded by the Ministry of Commerce, Industry and Energy, Republic of Korea (2007).

References
[1] Brand H R and Pleiner H 2000 *Phys. Rev. Lett.* **86** 1385
[2] Lita M, Popa C, Velescu C and Vekas L N 2004 *IEEE Trans. Magn.* **40** 469
[3] Bica I and Choi H J 2008 *Int. J. Mod. Phys. B* **22** 5041
[4] Ulicny J C and Golden M A 2007 *Int. J. Mod. Phys. B* **21** 4898
[5] See H, Joung C and Ekwebelam C 2007 *Int. J. Mod. Phys. B* **21** 4945
[6] Phule P P and Ginder J M 1999 *Int. J. Mod. Phys.* B **13** 2019
[7] Bica I 2006 *J. Ind. Eng. Chem.* **12** 501
[8] Bossis G, Khuzir P, Lacis S and Volkova O 2003 *J. Magn. Magn. Mater.* **258-259** 456
[9] Goncalves F D and Carlson J D 2007 *Int. J. Mod. Phys. B* **21** 4832
[10] Bose H 2007 *Int. J. Mod. Phys. B* **21** 4790
[11] Samouhos S and McKinley G 2007 *J. Fluid Eng.-T ASME* **129** 429
[12] Kanno H, Shimada K, Ogawa J and Inoue N 2007 *Int. J. Appl. Electrom.* **25** 109
[13] Margida A J, Weiss K D and Carlson J D 1996 *Int. J. Mod. Phys.* B **10** 3335
[14] Yang Y, Li L, Chen G and Liu E 2008 *J. Magn. Magn. Mater.* **320** 2030
[15] Bell R C et al 2007 *Int. J. Mod. Phys. B* **21** 5018
[16] Fang F F, Park B J, Choi H J and Jhon M S 2007 *Int. J. Mod. Phys. B* **21** 4981
[17] Jang I B, Kim H B, Lee J Y, You J L, Choi H J and Jhon M S 2005 *J. Appl. Phys.* **97** 10Q912
[18] de Vicente J et al 2004 *Rheol. Acta* **44** 94
[19] Tang X, Zhang X, Tao R and Rong Y 2000 *J. Appl. Phys.* **87** 2634
[20] de Vicente J, López-López M T, González-Caballero F and Durán J D G 2003 *J. Rheol.* **47** 1093
[21] Lim S T, Cho M S, Jang I B and Choi H J 2004 *J. Magn. Magn. Mater.* **282** 170
[22] Lim S T, Choi H J and Jhon M S 2005 *IEEE Trans. Magn.* **41** 3745
[23] Fang F F, Jang I B and Choi H J 2007 *Diamond Relat. Mater.* **16** 1167
[24] Wu W P, Zhao B Y, Wu Q, Chen L S and Hu K A 2006 *Smart Mater. Struct.* **15** N94
[25] Cho M S, Lim S T, Jang I B, Choi H J and Jhon M S 2004 *IEEE Trans. Magn.* **40** 3036
[26] Fang F F and Choi H J 2007 *Phys. Stat. Sol. (a)* **204** 4190
[27] Moniruzzaman M and Winey K I 2006 Macromolecules **39** 5194
[28] Sano M, Kamino A, Okamura J and Shinkai S 2002 *Nano Lett.* **2** 531
[29] Sun Y P, Fu K, Lin Y and Huang W 2002 *Acc. Chem. Res.* **35** 1096
[30] Yu F, Yao K, Shi L, Wan W, Zhong Q, Fu Y and You X 2007 *Chem. Mater.* **19** 3412
[31] Liu J et al 1998 *Science* **280** 1253.
[32] Jeong S H, Lee O J and Lee K H 2002 *Chem. Mater.* **14** 1859
[33] Wereley N M et al 2006 *J. Intel. Mat. Syst. Str.* **17** 393
[34] Rankin P J, Horvath A T and Klingenberg D J 1999 *Rheol. Acta* **38** 471
[35] Wollny K, Läuger J and Huck S 2002 *Appl. Rheol.* **12** 25
[36] Uhlherr P H T, Guo J, Fang T N and Tiu C 2002 *Korea-Aust Rheol. J.* **14** 17
[37] Lee M A, Fang F F and Choi H J 2007 *Phys. Stat. Sol. (a)* **204** 4186

Journal of Physics: Conference Series **149** (2009) 012048 doi:10.1088/1742-6596/149/1/012048

Non-dimensional Modeling and Experimental Evaluation of a MR Squeeze Mode Rheometer

Alireza Farjoud[1], Ryan Cavey[1], Mehdi Ahmadian[1] and Chandra Namuduri[2]

[1] Center for Vehicle Systems and Safety (CVeSS), College of Engineering, MC-0901, Virginia Tech, Blacksburg, VA 24060, USA

[2] GM R&D Center, Mail Code 480-106-390, 30500 Mound Road, Warren MI 48090, USA

Email: ahmadian@vt.edu

Abstract. Squeeze mode flow in MR fluids has recently become an interest to many researchers. In squeeze mode, MR fluids can generate a wide range of force associated with a small displacement. The generated force is controllable by means of a variable magnetic field. This is promising for applications that require variable forces along short strokes. There are not many published data on the behavior of MR fluids in squeeze mode. Many of the basic properties of MR fluids in squeeze mode are still unknown. The presented research here focuses on modeling and testing MR fluids in squeeze mode. A novel squeeze mode rheometer was designed and built. The results show that MR fluid can deliver a force that is comparable in magnitude to the force in shear mode. A non-dimensional mathematical model was developed and validated experimentally.

1. Introduction

Shear and valve mode flow of MR fluids are the most extensively explored modes in MR research and many commercial products are available based on shear or valve mode. The basic properties of MR fluids in shear and valve mode are completely understood and there are numerous literature available on modeling and simulation of MR fluids in these modes. On the other hand, there is very little literature concerning the squeeze mode flow of MR fluids and their basic properties in compression [1]. However, the number of publications on squeeze mode flow is increased recently [2, 3]. In this paper, the authors present basic research on MR fluids in squeeze mode. MR fluid is squeezed and its behavior and performance are evaluated. Having the knowledge of the basic properties of MR fluids in squeeze mode is essential for the design of devices based on this mode.

In squeeze mode, the MR fluid is placed between two plates (commonly referred to as poles). In the presence of a magnetic field, the poles move toward each other and squeeze the MR fluid as shown in Figure 1. As the magnetic field is turned on, iron particles line up in the direction of the magnetic field and form chains. These chains increase the resistance of MR fluid to flow.

2. Non-dimensional mathematical model

The pressure in the MR fluid when squeezed is known to be due to three factors [4]:

- Pressure due to the viscosity of the MR fluid.
- The pressure due to formation of chains and their resistance to flow (MR effect).

- The pressure due to the inertia of the MR fluid. If the amount of MR fluid is small and the relative velocity of the poles is low, the inertial effect can be neglected.

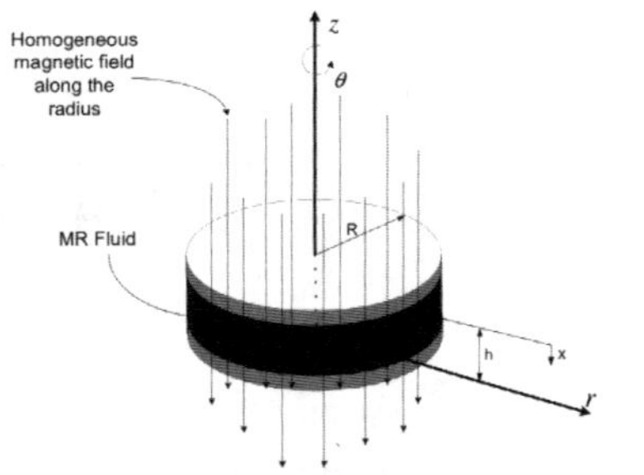

Figure 1 MR fluid in squeeze mode

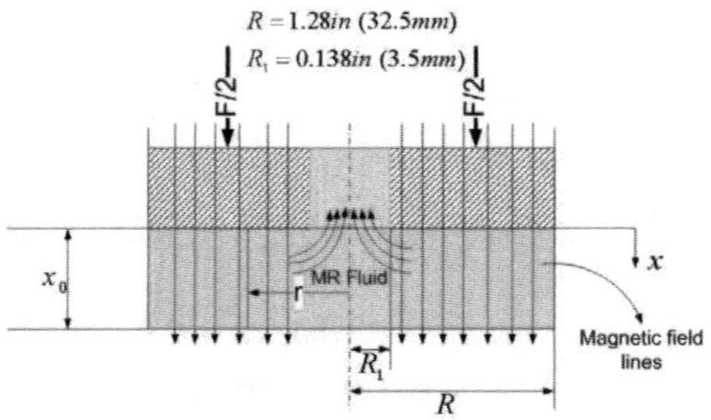

Figure 2 Schematics of the squeeze mode rheometer configuration

The rheometer designed and built at CVeSS has the schematics shown in Figure 2 to squeeze the MR fluid. A piston with a hole in the center is pushed down in a cylinder filled with MR fluid, squeezing the MR fluid. The excess of the MR fluid escapes from the center hole to compensate the volume decrease caused by movement of the piston.

Neglecting the inertial effects, the total force, F, can be calculated as [5]:

$$F = \frac{3\pi\eta\dot{x}}{2(x_0 - x)^3}\left[R^4 - R_1^4 - 4R_1^2R^2\ln\left(\frac{R}{R_1}\right)\right] + \frac{2\pi\tau_y}{3(x_0 - x)}\left(R_1^3 + 2R^3 - 3R^2R_1\right) \tag{1}$$

Where, τ_y is the yield stress of MR fluid, η is the viscosity of the MR fluid, R is the radius of the piston, R_1 is the radius of the center hole, x_0 is the initial gap size, and $\dot{x}$ is the relative velocity of the piston.

Equation (1) can be used to obtain the squeeze force as a function of the gap size ($h = x_0 - x$). The first term on the right hand side is the force generated by the viscosity of the MR fluid and the second term is the force caused by the magnetic effect of the MR fluid.

Transforming equation (1) to the non-dimensional form gives the ability to better understand the behavior of the MR fluid and make some means to compare its performance in squeeze mode with other flow modes. Rearranging equation (1) into its non-dimensional form results:

$$\psi = \frac{A}{Bn\zeta^3} + \frac{B}{\zeta} \tag{2}$$

where, $\zeta = \frac{h}{R}$ is the non-dimensional gap size. The Bingham number, $Bn = \frac{R\tau_y}{\eta\dot{x}}$, can be considered as the ratio of the MR effect to the viscous effect. The non-dimensional force, $\psi = \frac{F}{\pi R^2\tau_y}$, is the ratio of the squeeze force to the yield stress multiplied by the area of the pole. It can be used as a measure for comparison between the squeeze mode force and direct shear mode. The denominator of ψ can be interpreted as the force acting on the same area and the same MR fluid to yield the MR fluid in direct shear mode. A and B are geometric constants which depend on the geometry of the rheometer.

3. Test results

A large number of tests have been done on MRF-120RD from LORD, with the squeeze mode rheometer. Tests have been carried out for different initial gap sizes and different magnetic field densities. The velocity of the piston has been chosen to be very small assuring a quasi-static condition. The results show that MR fluid is capable of delivering a wide range of force in squeeze mode. Figure 3 shows the force versus displacement plot. The initial gap is 0.1in (2.5mm) and the MR fluid is squeezed to a gap size of 0.05in (1.3mm). The coil current is increased from 1Amp to 3 Amps (corresponding to 0.3T to 0.7 Tesla in the MR fluid gap). As can be seen, MR fluid could deliver a maximum force of approximately 700lb (3100N) at 3Amp. A further increase in current did not considerably increase the maximum force. The yield stress of the MR fluid does not increase significantly as the magnetic field density is increased beyond 0.7 Tesla in the gap.

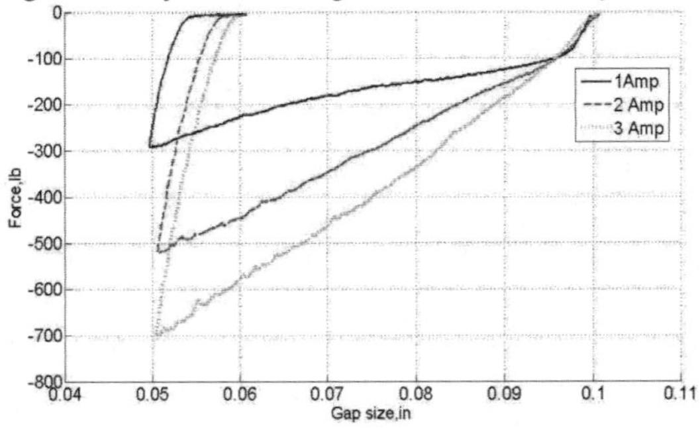

Figure 3 Force vs. displacement plots for MRF-120RD

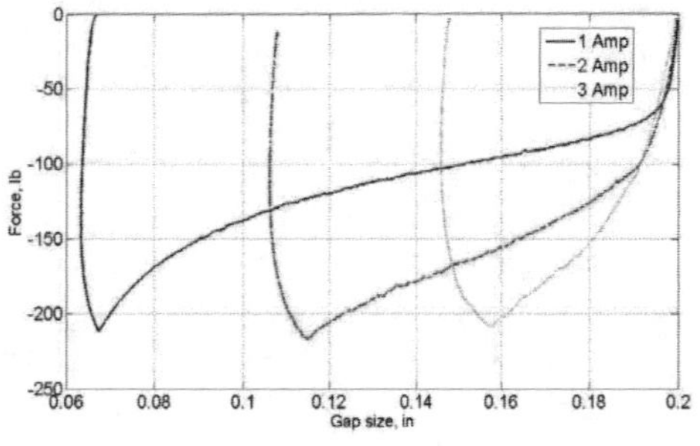

Figure 4 Force vs. gap size plots for MRF-120RD in force controlled mode

One of the behaviors seen was the clumping effect. At high currents the iron particles are trapped in the magnetic field forming chains, but the carrier fluid is free to flow. Therefore, as the piston moves, the iron particles stay in the gap and the carrier fluid leaves the gap leaving iron particles behind. After a few repetitions of tests, the amount of iron particles is considerably high in the gap and a large percentage of the carrier fluid has left the gap. This makes the volumetric percentage of the MR fluid increase and consequently, increases the force. Similar behavior is also seen by other researchers [2].

Another important property of MR fluids in squeeze mode, shown in Figure 3, is the inability to withstand tensile forces. After the minimum gap is reached, the piston contracts at the same rate of approach but the force rapidly decreases from its maximum value to zero.

Test results shown in Figure 3 are based on displacement control tests of the rheometer. If the force is used as the input to the rheometer (force controlled mode), the MR fluid will flow based on the force that is exerted. With the force controlled mode the yielding behavior of MR fluid is more obvious in the beginning of the tests. As the force is increased, the MR fluid is resisting the movement of the piston without yield. This can be identified as a high slope region in the force-displacement curves. Once the MR fluid is yielded, the slope is decreased and the MR fluid flows. shows the results of the forced controlled tests on MRF-120RD. For the force controlled tests, the initial gap was 0.2in (5mm) and the final gap depends on the current and the amount of force exerted. The force was increased from 0lb to approximately 200lb (890N) with a constant rate. At higher currents the MR fluid is more resistive and the piston moves a shorter distance.

The mathematical model introduced in the previous section agrees with the test results, especially the force controlled tests because they better illustrate the pre-yield phase of the MR fluid in squeeze. At 1amp the average magnetic field density is 0.3Tesla. At this magnetic field density, the properties of MR fluid can be approximated as: $\eta = 6.09\times10^{-6}$ Reyn (0.042 Pa.s), $\tau_y = 1.9$ Psi (13 kPa). The average velocity of the piston is 0.0135 in/sec (0.3mm/sec). Using these parameters along with the non-dimensional model, a comparison between the test data and the mathematical model is obtained as shown in Figure 5.

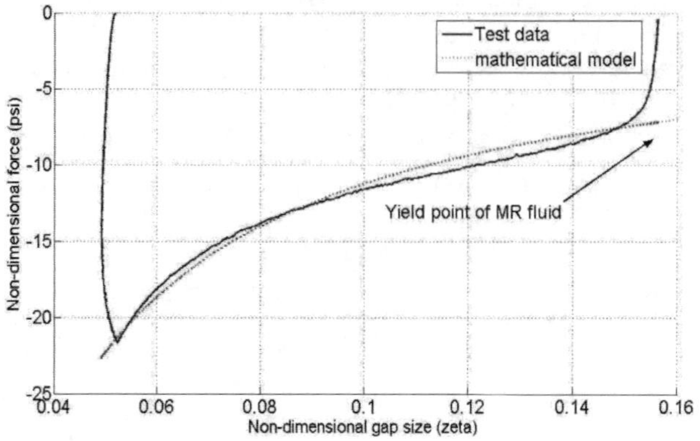

Figure 5 Comparison between test data and mathematical model for MRF-120RD at 1Amp

As shown in Figure 5, the magnitude of the non-dimensional force at the yield point of MR fluid in squeeze mode is approximately 7. This shows that the amount of force required to yield the fluid in squeeze mode is almost 7 times larger than the force needed to yield the same MR fluid in direct shear mode when $\zeta = 0.16$. As the non-dimensional gap size decreases, the magnitude of the non-dimensional force increases.

4. Conclusion

It is shown that the tested MR fluid is capable of carrying a wide range of force in squeeze mode. This force range is achievable along a short stroke. The amount of the force achieved depends on the magnetic field density and gap size. The clumping effect was observed in squeeze mode testing. The design of MR devices which work on squeeze mode requires a solution to reduce the clumping effect. Clumping causes the carrying liquid to separate from iron particles reducing the repeatability. The clumping behavior is currently being studied at CVeSS.

The mathematical model presented is able to illustrate the basic behavior of MR fluids in squeeze mode and can be used for design purposes. The mathematical model is in agreement with the test results. However, the force controlled test results have a better match with the mathematical model. At force controlled tests, MR fluid can resist movement of the piston before yielding. While in displacement controlled tests, the piston starts to move right after the test is started and MR fluid has to flow.

References

1. Carlson, J.D. and M.R. Jolly, *MR fluid, foam and elastomer devices.* Mechatronics, 2000. **10**(4-5): p. 555-69.
2. Mazlan, S.A., N.B. Ekreem, and A.G. Olabi, *Apparent stress-strain relationships in experimental equipment where magnetorheological fluids operate under compression mode.* Journal of Physics D: Applied Physics, 2008. **41**(9): p. 095002 (6 pp.).
3. Vieira, S.L., et al., *Behaviour of MR fluids in squeeze mode.* International Journal of Vehicle Design, 2003. **33**(1-3): p. 36-49.
4. Jolly, M.R. and J.D. Carlson. *Controllable squeeze film damping using magnetorheological fluid.* 1996. Bremen, Germany: AXON Technologie Consult GmbH.
5. Phillips, R.W., *Engineering applications of fluids with a vairable yield stress.* 1969, University of California, Berkeley: California.

Journal of Physics: Conference Series **149** (2009) 012049 doi:10.1088/1742-6596/149/1/012049

Novel Active Vibration Absorber with Magnetorheological Fluid

T. Gerlach, J. Ehrlich, H. Böse

Fraunhofer-Institut für Silicatforschung ISC, Neunerplatz 2, D-97082 Würzburg, Germany

E-mail: thomas.gerlach@isc.fraunhofer.de

Abstract. Disturbing vibrations diminish the performance of technical high precision devices significantly. In search of a suitable solution for reducing these vibrations, a novel concept of active vibration reduction was developed which exploits the special properties of magnetorheological fluids. In order to evaluate the concept of such an active vibration absorber (AVA) a demonstrator was designed and manufactured. This demonstrator generates a force which counteracts the motion of the vibrating body. Since the counterforce is generated by a centrifugal exciter, the AVA provides the capability to compensate vibrations even in two dimensions. To control the strength of the force transmitted to the vibrating body, the exciter is based on a tunable MR coupling. The AVA was integrated in an appropriate testing device to investigate its performance. The recorded results show a significant reduction of the vibration amplitudes by an order of magnitude.

1. Introduction

A multitude of technical high precision devices generate unintended vibrations while they are operating. By reducing these disturbing vibrations, the performance of the devices can be significantly enhanced. An already known approach to solve this problem is the semi-active damping of vibrations with magnetorheological (MR) fluids [1]. In MR dampers, the influence of a magnetic field on the rheological properties of the MR fluid is exploited to control the damping force [2-4].

Another general technical solution to minimize disturbing vibrations is based on an active vibration absorber (AVA). In such a device, an active force is generated by an appropriate actuator which acts against the motion of the vibrating body. If the strength of the counter-force equals that of the force which causes the disturbing vibration, the forces compensate one another and the vibration is canceled.

The key question is how to generate a force of correct strength and phase to counteract the disturbing vibration. Piezoelectric actuators [5] have received much attention in this respect, but are expensive and show only small displacements. The use of a centrifugal exciter is a well known solution for generating a periodic force, because it is a light, small and cost-effective device which can produce very high forces at low power consumption. However, a problem of this approach is that the strength of the generated force increases quadratically with the rotation frequency of the exciter. Thus, the generated force at each tuned rotation frequency is fixed and cannot be changed independently.

A possible technical solution for an adjustably controllable centrifugal exciter is described in a US patent [6]. Furthermore, various real-time controlled techniques for reducing disturbing vibrations were developed for a multitude of technical applications, i.e. in the helicopter industry [7, 8].

Journal of Physics: Conference Series **149** (2009) 012049 doi:10.1088/1742-6596/149/1/012049

This paper addresses a novel concept of controllable vibration reduction, in which a centrifugal exciter is connected to a force-modifying transmitter based on magnetorheological fluids. This combination leads to an active vibration absorber, where the MR fluid modifies the force of the centrifugal exciter to the required strength.

2. Concept and realization of the active vibration absorber

In order to prove the feasibility of the novel concept, a demonstrator of the active vibration absorber with a centrifugal exciter and an MR fluid for the reduction of vibrations was designed and manufactured. A scheme of the demonstrator is presented in Figure 1.

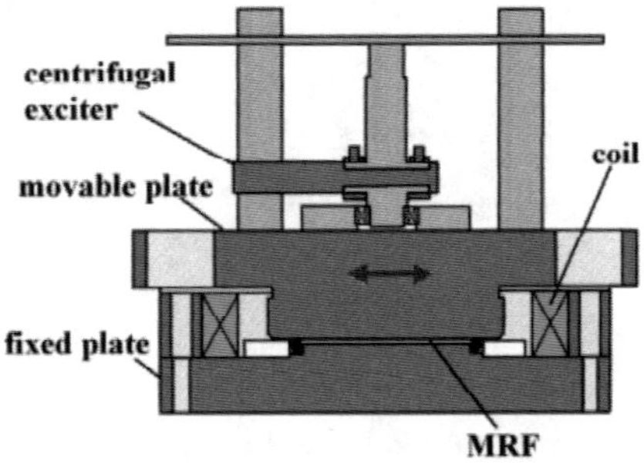

Figure 1. Scheme of the active vibration absorber

Figure 2. Picture of the active vibration absorber

The assembly can be attached to a vibrating body in order to minimize the amplitude of its vibrations. The demonstrator consists of two main components. The first component is a centrifugal exciter which generates a rotating force. The second component is a tunable MR coupling device which enables a controlled transmission of the force generated by the centrifugal exciter to the vibrating body.

The tunable coupling device exploits the MR shear mode. The force is transmitted between two parallel plates which are moveable with respect to each other, parallel to their surfaces. The first plate is fixed to the vibrating body and the second plate is connected to the centrifugal exciter. The gap between the two plates contains the MR fluid. The magnetic circuit with the coil for generating the requested magnetic field is integrated in the tunable MR coupling device (see Figure 1). The centrifugal exciter generates a high rotating force whose amount quadratically increases with the rotation frequency.

The magnetic field strength and the phase of the exciter must be tuned in such way, that the force which is transmitted to the vibrating body compensates the vibration force. Due to the broad variability of the shear stress of MR fluids, the counterforce transmitted by the MR coupling device can be adapted to meet a large range of requirements in terms of force and frequency.

Since the counterforce is generated by a centrifugal exciter, this vibration absorber provides the opportunity to compensate vibrations even in two dimensions. This is advantageous for a multitude of applications where rotating machines cause vibrations in the plane of the rotation. Figure 2 shows a photograph of the demonstrator of the novel AVA.

3. Experimental evaluation

3.1. Design of the testing device

In order to study the performance of the active vibration absorber, the manufactured demonstrator was integrated in a testing device, which is able to provide vibrations within the range between 15 Hz and 30 Hz. Figure 3 shows a scheme of this testing device.

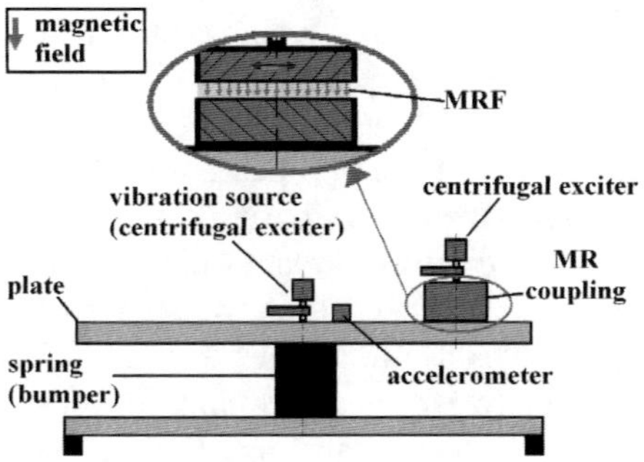

Figure 3. Scheme of the testing device for the active vibration absorber

Figure 4. Picture of the testing device for the active vibration absorber

The disturbing vibration force for the testing is generated by a second centrifugal exciter fixed on the same plate to which the active vibration absorber is attached. While the AVA is turned off, the disturbing force will cause an oscillation of the plate. By turning on the centrifugal exciter of the AVA, the second centrifugal force is generated. If the force transmitted to the plate, connecting the two exciters by the tunable MRF coupling, has the same value but the opposite orientation, the two generated forces compensate each other and the disturbing vibration of the system is drastically reduced.

In order to investigate, whether the tunable MRF coupling will be able to transmit only a diminished amount of energy, the exciter of the disturbing vibration of the test rig has to generate a lower centrifugal force than the one in the active vibration absorber. By modulating the magnetic field strength of the tunable MRF coupling, only that portion of the force has to be transmitted which is necessary to achieve a maximum vibration reduction. The vibration amplitudes are recorded by an accelerometer which is attached to the plate. Figure 4 reveals the experimental set-up of the testing device.

3.2. Investigations and results

A self-manufactured MR fluid with an iron particle concentration of 50 vol.% in silicone oil as carrier liquid was used for the investigations with the AVA. This selection was motivated by the intention to transmit high forces by the MR fluid.

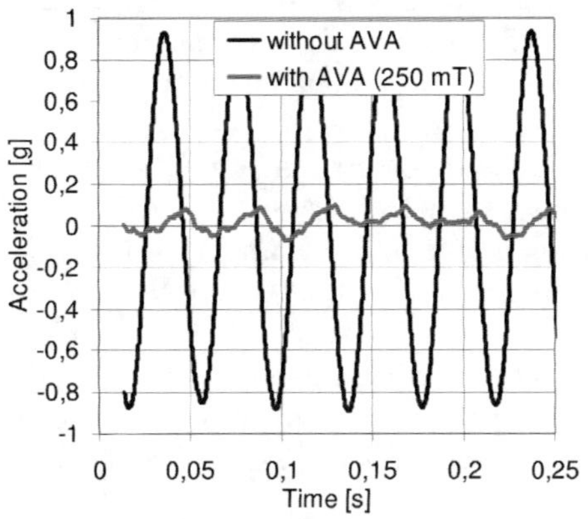

Figure 5. Reduction of the vibration amplitude by the AVA at 25 Hz

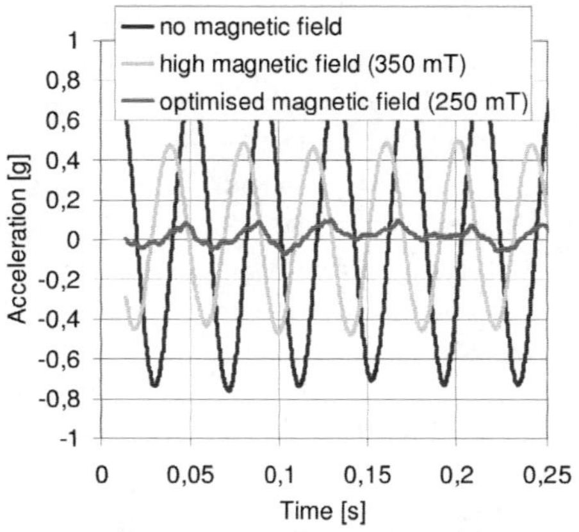

Figure 6. Influence of the magnetic field strength on the vibration amplitude at 25 Hz

For the sake of simplicity, the following discussion is limited to vibrations in only one dimension. The aim of a first test series was to study the performance of the demonstrator at vibration frequencies in the range of 19 to 28 Hz. Within the tested frequency range, a high reduction of the vibration amplitudes was achieved. Figure 5 presents the recorded amplitudes at a frequency of 25 Hz by way of example.

The results demonstrate that a significant reduction of the vibration amplitudes can be achieved with the active vibration absorber. Figure 5 depicts that a significant decrease of the vibration amplitude by an order of magnitude is possible without an appreciable effort of electronic control.

Furthermore, it is has been demonstrated that it is possible to modify the force transmitted by the MR coupling by varying the magnetic field strength inside the MR fluid. If the magnetic field is optimized, the disturbing force and the counter-force nearly equalize each other and the vibration amplitude is reduced significantly. If the magnetic field is too weak, an insufficient force is transmitted by the MR coupling and the amplitude of the vibration increases. If the strength of the magnetic field is too large, an excessive force is transmitted and the vibration amplitude also increases. A comparison of these different conditions in terms of magnetic field strength is given in Figure 6.

4. Conclusions

A novel concept of an active vibration absorber (AVA) was developed, which combines the efficiency of force generation by a centrifugal exciter with the controllability of force transmission by MR fluids. With an experimental set-up, it could be demonstrated that the amplitude of a vibration can be decreased by an order of magnitude. This principle can be adjusted to a large range of different working conditions, while the power consumption is very low.

Moreover, the concept of an active vibration absorber with an MR fluid provides a powerful tool for the effective reduction of disturbing vibrations, even in two dimensions. By reducing these unintended vibrations, the performance of large number of technical high precision devices can be significantly enhanced. This approach has a large potential for many applications in transportation and industry, where disturbing vibrations are caused by rotating engines.

5. Acknowledgement

The authors of this paper would like to thank Massimo Ippolito from SEQUOIA Automation and Angelo Merlo from Ce.S.I. for the stimulating discussions.

References

[1] M. Mao, W. Hu, Y.-T. Choi, N. M. Wereley: A magnetorheological damper with bifold valves for shock and vibration mitigation, J. Intelligent Mater. Syst. Struct. 18 (2007) 1227-1232

[2] J. Ehrlich, H. Böse: Novel magnetorheological damper with outstanding fail-safe characteristics. Proceedings of Actuator 2008 – 11th International Conference on New Actuators (2008) 495-498

[3] S. J. Dyke, B.F. Spencer, Jr., M.K. Sain and J.D. Carlson: An Experimental Study of MR Dampers for Seismic Protection, Smart Materials and Structures 7 (1998) 693-703

[4] U. Lange, S. Vassileva, L. Zipser: Controllable magnetorheological dampers for shock and vibration. Proceedings of Actuator 2002 – 8th International Conference on New Actuators (2002) 339-342

[5] A. Illgen, V. Wittstock: Piezo based active vibration absorber for gear box noise reduction in wind turbines, Fraunhofer-Institut für Keramische Technologien und Systeme-IKTS, Dresden: Piezocomposite Applications. CD-ROM: September 27-29, 2007, Dresden, ISPA2007

[6] G. J. Orzal, Gregory: Adjustably controllable centrifugal vibratory exciter, US Patent 4568218

[7] M. Jolly, S. Hildebrand, R. Altieri, M.Ferguson, D. Ivers: Helicopter vibration control system and rotary force generator for canceling vibrations, US Patent 7448854

[8] I. Reed, H. Wilmer: System for controlling higher harmonic vibrations in helicopter rotor blades, US Patent 5314308

An alternate operation mode for MR fluids—Magnetic Gradient Pinch

F D Goncalves and J D Carlson

LORD Corporation, 406 Gregson Drive, Cary, NC USA

Email: fernando_goncalves@lord.com

Abstract. MR fluids have been employed in a wide variety of applications, providing viable solutions to many engineering challenges. Controllability in MR fluid devices is achieved through one of three commonly recognized operating modes—valve mode, direct-shear mode, and squeeze mode. This paper presents evidence of an alternative valve configuration for controlling MR fluids. This valve design represents a considerable departure from the commonly recognized modes. Specifically, this novel valve design replaces the conventional narrow gap with a relatively large circular orifice, and it changes the way in which the magnetic field interacts with the fluid. With these changes, we present evidence that this novel valve design may offer functionality not currently available in conventional MR fluid valves.

1. Introduction

Today's MR fluid devices achieve controllability through proper valving of the fluid using one of the three commonly recognized operating modes—valve mode, direct-shear mode, and squeeze mode. Regardless of the operating mode, all current MR fluid devices rely on creating a more or less uniform magnetic field across the thin layer of fluid in the channel and perpendicular to the flow direction. This has profound practical implications on the geometry and size of the flow channel. Specifically, the narrow dimension of the flow channel or gap must be kept small to minimize the power required for generating the magnetic field and to allow a large on-state force. Yet the gap must not be too small so as to make the off-state viscous force too large. Moreover, the size of the particles in the MR fluid must be carefully chosen to avoid jamming in the narrow gap.

This study presents evidence of an alternative valve configuration for controlling MR fluids, namely, Magnetic Gradient Pinch or MGP [1]. Initial motivation for the MGP valve was to find a way to enable the use of low-cost iron powder in MR fluid devices. The carbonyl iron powder that is used today is quite expensive and the goal was to design a valve that could allow the use of larger, coarser particles that are significantly less expensive than carbonyl iron.

The idea is to radically change the way in which one attempts to valve MR fluid such that the normally accepted MR fluid design paradigms and constraints are significantly altered to allow the use of coarser, low-cost particles. Rather than the very narrow rectangular flow channel of a normal MR fluid valve, the MGP valve is a relatively large circular aperture through which large particles can easily pass. The MGP valve also changes the way in which the magnetic field interacts with the fluid.

2. Magnetic Gradient Pinch valve

Conventional MR fluid valves rely on the ability to generate a field-dependent yield stress throughout the entire active fluid volume. The fluid essentially becomes a solid and no flow occurs until the pressure becomes sufficiently large to yield the fluid near the walls at which point the entire body of fluid moves as a plug. The greater the magnetic field, the larger the yield strength and the more pressure required to move the plug, as shown in figure 1.

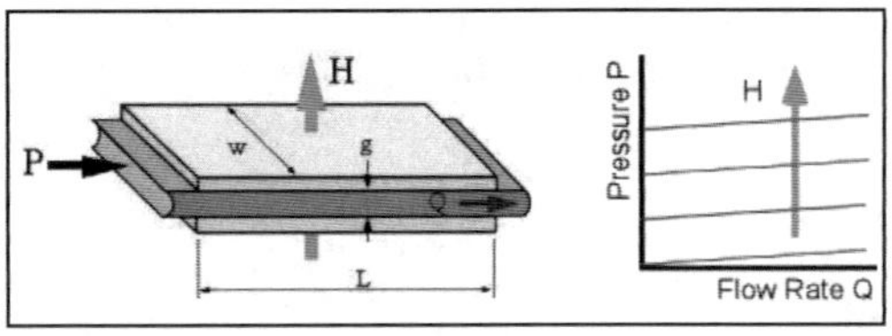

Figure 1. MR fluid valve showing key geometric features and resulting pressure.

The fundamental difference between a conventional MR fluid valve and the Magnetic Gradient Pinch (MGP) valve is in the design of the magnetic circuit. The MGP valve replaces the rectangular flow channel by a simple orifice and arranges the poles axially, as shown in figure 2. Arranging the poles axially along the flow path and separating the poles by a non-magnetic spacer generates a highly non-uniform magnetic field.

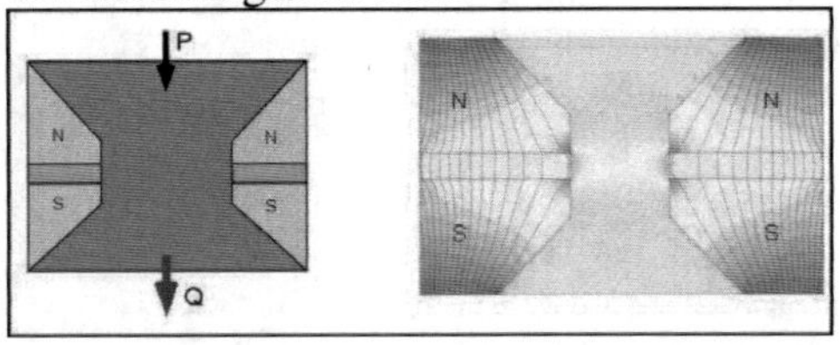

Figure 2. The Magnetic Gradient Pinch valve has poles arranged axially along the flow channel (left) which generates a highly non-uniform magnetic field (right).

The basic premise for the MGP valve is that the flow of MR fluid through a circular orifice can be controlled with a strong magnetic field gradient. Rather than solidifying the fluid throughout the valve, the non-uniform field only solidifies MR fluid near the wall. The overall magnetic field strength will control the inward distance that such solidification occurs, thus effectively controlling the orifice diameter. This type of control has been identified as the MGP pinch mode.

A second, more extreme, mode of control was discovered while exploiting another benefit of the MGP valve. The MGP valve replaces the rather narrow rectangular flow channel by a relatively large (2 mm) orifice. Such a large orifice enables the use of much coarser iron in the fluid formulation. Without the fear of spontaneous jamming, a fluid with iron particles on the order of 100 μm can be used in the MGP valve. While testing with these coarse-iron fluids, it was observed that a strong enough magnetic field could initiate controllable, and reversible, jamming in the valve. This type of controlled jamming was identified as the MGP jamming mode.

Preliminary studies with an MGP valve identified two modes of control—the pinch mode, and the reversible jamming mode. The details of each mode will be highlighted below, along with supporting experimental data.

3. Experimental investigation

An MGP valve was built to evaluate the controllability of this novel valve design. A piston driven rheometer was used to drive fluid through the valve. Several fluids of varying concentrations and particle size distributions were tested. Experimental testing identified two modes of control for the MGP valve—pinch mode, and jamming mode.

3.1. Pinch Mode

The MGP pinch mode is a continuously variable control mode in which the application of a magnetic field changes the effective diameter of the MGP orifice. As indicated in figure 3, a magnetic field mediated change in effective orifice diameter will change the slope of the pressure/velocity relationship. This relationship can be approximated by the Wuest equation for flow through a sharp

Journal of Physics: Conference Series **149** (2009) 012050 doi:10.1088/1742-6596/149/1/012050

edged orifice [2]. This type of control is unique to MGP valves and is very different from the normal MR fluid case where the slope remains constant.

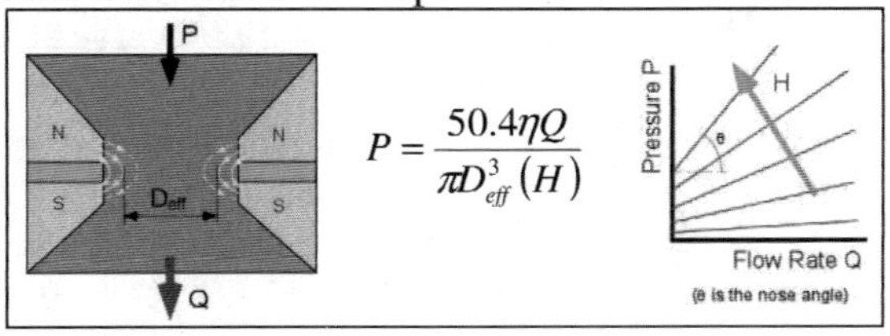

Figure 3. Magnetic Gradient Pinch valve illustrating the MGP pinch mode (left) and the resulting pressure profile (right). P is pressure, η is viscosity, Q is flow rate, D is diameter, H is field strength.

Preliminary experiments conducted on an MGP valve and have validated the MGP pinch effect. As shown in figure 4, the slope of the pressure versus flow rate curve increases as applied field increases, indicative of a change in the effective diameter of the flow orifice. To quantify the pinch effect, the data is curve fit and the slope is extracted from each curve. From the slope of each curve, and the nominal valve geometry, an effective diameter can be found. As shown in figure 4 (right), the increase in slope corresponds to a reduction in effective diameter of greater than 50% from the off-state to the 3000 mA test.

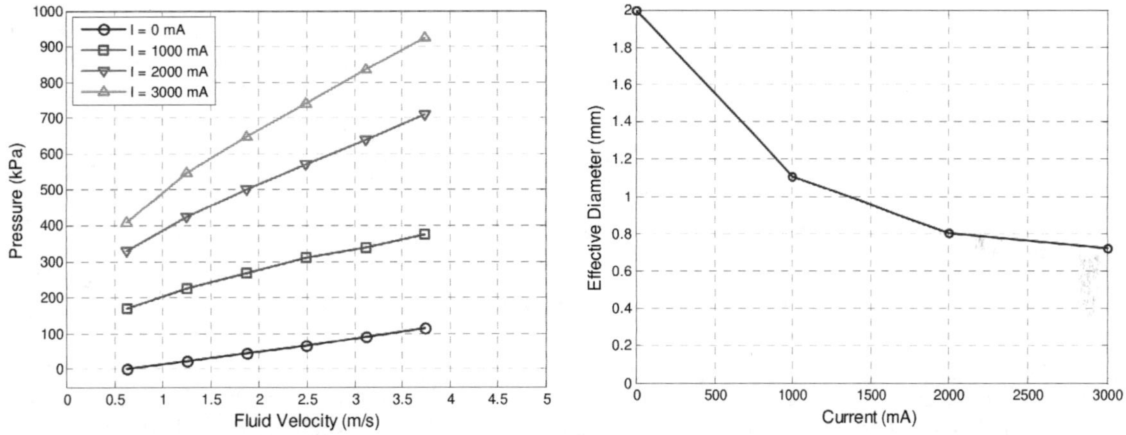

Figure 4. MGP pinch effect showing slope increase (left) and reduction in effective diameter (right).

3.2. Jamming Mode

Another MGP operating mode was discovered while testing with a coarse-particle MR fluid. It appeared that a strong enough magnetic field could initiate jamming in the valve. At a critical field strength, the granular nature of the coarse-particle fluid would set up a jam in the valve causing the pressure to climb quite suddenly as illustrated in figure 5. We normally make every effort to avoid jamming in conventional MR fluid devices because it is normally irreversible. An exciting feature of the MGP valve is that jamming is indeed reversible. Once the magnetic field is removed, fluid flow resumes and the pressure drops. This reversible jamming concept appears to be a completely new operating regime for MR fluids and devices.

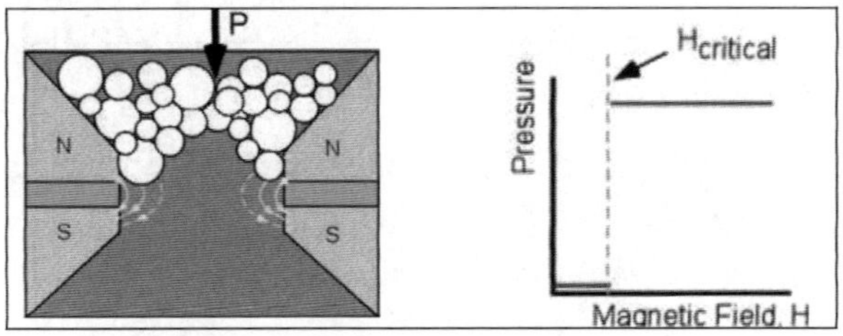

Figure 5. MGP reversible jamming mode.

Journal of Physics: Conference Series **149** (2009) 012050 doi:10.1088/1742-6596/149/1/012050

The MGP reversible jamming mode has been studied with fluids which contain relatively large iron particles. Unlike the very narrow rectangular flow channel of a normal MR fluid valve, the MGP valve is a relatively large circular aperture through which large particles can easily pass. Particle sizes ranged from 45–150 microns. Fluid was passed through the valve at various operating currents while the pressure was monitored. At some critical field strength, the measured pressure climbed suddenly indicating that the valve had jammed. Figure 6 shows the peak pressure measured during the experiment versus the applied current. For low currents (i.e. less than 700 mA) the pressure remains relatively low; once the current reaches 700 mA, the pressure peaks to 20 MPa (nearly 3000 psi). The pressures observed in the MGP valve operating in the jamming mode are an order of magnitude higher than pressures normally observed in ordinary MR valves.

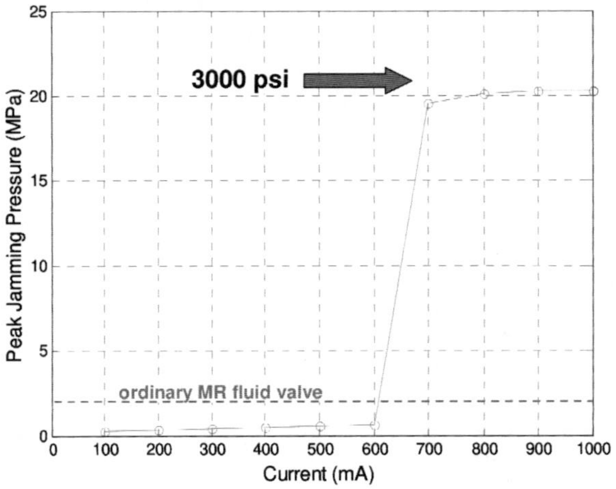

Figure 6. Sudden onset of jamming at critical field strength.

4. Discussion

The Magnetic Gradient Pinch valve is a novel MR fluid valve that may expand the solution space for MR technology. This study has presented evidence of new functionality through the two identified modes—pinch mode and reversible jamming mode. A particularly advantageous feature of the circular MGP valve compared to a normal MR fluid valve is the potential to have a much lower viscous off-state. The viscous flow loss of a single 2 mm MGP orifice is considerably smaller than the flow loss experienced with the narrow gap (~0.5 mm) found in many MR fluid valves. This may be paradigm shifting with regard to desirable or necessary fluid properties.

The MGP jamming mode represents a significant leap in functionality for MR technology. Jamming has historically been considered a significant problem in ordinary MR valves because jamming in such valves is often irreversible and leads to poor device performance. However, with the MGP valve, jamming reversibility has been validated and this mode of control may provide solutions to challenges unmet by conventional MR technology. Many applications may benefit from the truly enormous pressures achievable with the MGP jamming mode. Such applications may include small, high-force devices or controllable locking devices. Another key benefit of the MGP jamming mode is that it may enable the use of extremely low cost iron powder in MR fluids. Coarse, water-atomized iron particles have been shown to perform well in the MGP jamming mode.

References

[1] Carlson J D, Goncalves F D, Catanzarite D M and Dobbs D R, 2008 Controllable magnetorheological fluid valve, devices, and methods WIPO 2008/024957A1

[2] Merritt H E 1967 *Hydraulic Control Systems* (New York: John Wiley & Sons, Inc.) p 44

Journal of Physics: Conference Series **149** (2009) 012051 doi:10.1088/1742-6596/149/1/012051

Magnetic Properties of Carbonyl Iron Particles in Magnetorheological Fluids

S R Gorodkin, R O James and W I Kordonski

QED Technologies International, 1040 University Avenue, Rochester, NY 14607 USA

E-mail: gorodkin@qedmrf.com

Abstract. Knowledge of the magnetic properties of dispersed magnetic particles is a prerequisite to the design an MR fluid with desired performance. A term *specific susceptibility* is introduced for characterization of particle susceptibility. The study was performed with the Bartington MS2B magnetic susceptibility system on small samples volume. Specific magnetic susceptibility of iron particles was found to be a linear function of median particle size. Structural change in the fluid, including particle organization, led to susceptibility drift and may affect fluid performance. It was shown that susceptibility data can be used for evaluation of the concentration of carbonyl iron particles in MR fluids.

1. Introduction

Dispersed magnetic particles are the main components of MR fluid and knowledge of their magnetic properties is required to design MR fluids with the desired performance [1 - 3]. Here we report initial susceptibility data for dispersed particles of commercial powders. In MR fluid applications, it is desirable to trace the drift or aging of particle properties (e.g., due to particle oxidation/abrasion/agglomeration while the fluid is in use) that might be responsible for MR fluid performance change. We show that this can be detected by simple magnetic susceptibility measurements.

Reported methodology is based on a small scale (sample volume) and low cost technical means. Two basic cases are considered: first, dilute fluid samples, where individual particles mainly contribute to the magnetic properties of the system, and second, neat concentrated fluids, for which particle interactions and ordering may introduce complex effects.

2. Problem configuration

The Maxwell-Garnett equation for magnetic susceptibility of a suspension of non-interacting spherical particles has the following form [4]: $\chi = 3\varphi\chi_p/[3+\chi_p(1-\varphi)]$, where χ_p is the susceptibility of the particles and φ is the volume fraction of the particles. For dilute suspensions, when φ is a low number, the equation can be reduced to $\chi = \varphi[3\chi_p/(3+\chi_p)]$. This equation can be further transformed to the linear form

$$\chi = \chi_S\varphi \quad \text{or} \quad \chi_S = \chi/\varphi \tag{1}$$

where the slope $\chi_S = 3\chi_p/(3+\chi_p)$ is named as specific *magnetic susceptibility* for convenience.

Hence, one can evaluate the specific susceptibility χ_s from the measured sample susceptibility χ and the particle volume fraction φ of the fluid samples and then use this value of χ_s as a measure for the particle susceptibility χ_p.

3. Experimental: magnetic materials and techniques

Carbonyl iron powders (CIP) of different grades and median particle size, D_{50}, were received from BASF (Germany) and ISP (USA). The powders are polydisperse and particles are spherical in shape as observed with optical microscope and scanning electron microscopy images. According to manufacturer's specification CIP chemical composition includes > 97% w/w of Fe and the density is 7500 kg/m^3. In one case, the BASF EW iron powder of D_{50} ~ 3.5 μm was fractionated to produce samples of large particle sizes with D_{50} up to ~ 4.5 μm.

The measurement of the initial magnetic susceptibility was performed with a Bartington MS2B system [5]. Particle sizing for slurries of CIP, involved in the study, was carried out using an acoustic absorption spectrum method (Colloidal Dynamics, Acoustosizer ™). Otherwise, CIP manufacturers' particle size data were used. The concentration of CIP in samples was determined by weight using mass balance, or moisture measurements for water based fluid samples.

4. Results and discussion

4.1. Experiments with diluted samples

The standard 10 ml cell was used. The discussed below results for specific susceptibility were obtained using diluted (< 2 % v/v of CIP in glycerol) fluid samples. The inert and viscous glycerol carrier retarded the sedimentation of dense CIP enabling consistent, multiple measurements. The average values were calculated from the 10 measurements on each sample.

Typical experimental characteristics for three different types of CIP are shown in the figure 1 where magnetic susceptibility values are plotted as a function of concentration for each particle type. The solid line is calculated using equation (1) at $\chi_p = 4$ and well coincides with experimental data for BASF HQ CIP. At the same time, each type of CIP is characterized by a different slope χ_s, which suggests that particles have different susceptibility.

All results of testing are listed in table 1. Even though various commercial CI powders may differ due to some difference in component compositions and internal structure (mechanically hard and soft materials), a significant trend in the specific susceptibility - a dependence on particle size is immediately apparent. This trend is shown in the figure 2 and data for all samples are fitted by the following empirical linear equation: $\chi_s = 0.668D_{50} + 0.961$. Thus, knowing the median CIP size (reported in μm) one can estimate the specific susceptibility of the particles.

4.2. Experiments with concentrated MR fluid

The small 0.27 ml special cell was used in this set of experiments to overcome the problem with the limit of instrument output signal. A large volume (~ 1 liter) of fresh MR fluid with CIP concentration of more than 35% v/v was sheared continuously, by pumping through a closed loop, to simulate real application conditions while small portions were taken periodically, and susceptibility measurements were performed.

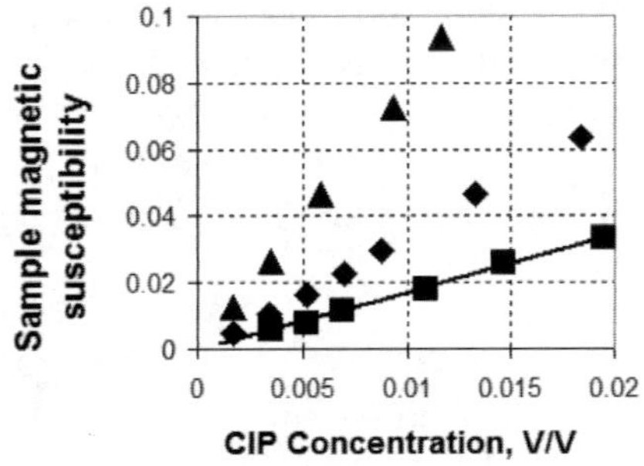

Figure 1. Magnetic susceptibility vs. concentration: (▲) ISP S-1000; (◆) BASF EW; (■) BASF HQ; (line) by Maxwell-Garnett equation (1) at $\chi_p = 4$.

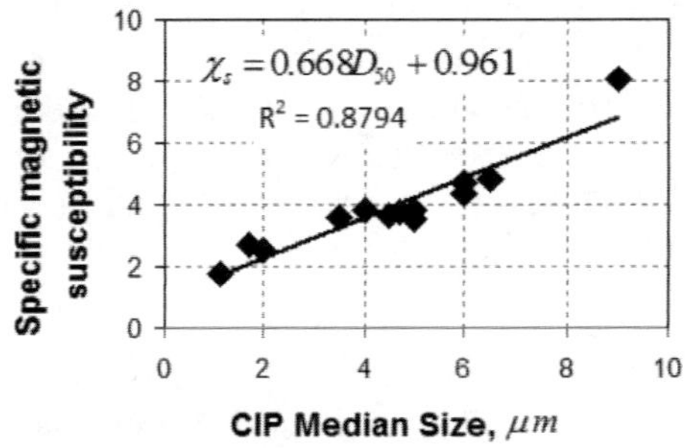

Figure 2. Specific magnetic susceptibility of different grades CI vs. particle size with linear fit (line).

Table 1. Specific magnetic susceptibility χ_S of various types of CIP and their particle size

CI powder ID	Size (D50, μm)	χ_S (SI unit)	CI powder ID	Size (D50, μm)	χ_S (SI unit)
BASF HQ ^{h}	1.1	1.77	BASF CC ^{s}	5	3.83
BASF HF ^{h}	1.7	2.68	ISP S-1281 ^{h}	5	3.55
BASF HS ^{h}	2	2.53	ISP R-1521 ^{s}	6	4.38
BASF EW ^{h}	3.5	3.58	BASF OR ^{h}	6	4.68
BASF EW $*^h$	4.5	3.63	BASF CL ^{s}	6.5	4.8
ISP S-1701 ^{h}	4.7	3.77	ISP S-1000 ^{h}	9	8.03
BASF OS ^{h}	4	3.82			

^{h} Mechanically hard; ^{s} mechanically soft; * fractionated.

It was noticed that the susceptibility increased by 4 to 5 % from the initial values within first 30-50 hrs and leveled afterwards. This drift seemed to be related to particle and fluid "structural changes" that occurred in the fluid during use.

The ability of agglomerated/structured CI particles to affect the susceptibility was confirmed in separate tests. Small samples were briefly placed in the field of a permanent magnet. In order to assess the effect of particle organization, two types of samples based on BASF EW CIP were tested: (i) using viscous liquid carrier, glycerol, and (ii) using a rigid composite carrier, polyester resin (Castin' Craft Casting Resin with Catalyst [6]).

As it shown in the figure 3, particles mobility and structuring results in increased susceptibility (by ~12%). The effect was persistent and was not completely diminished by vibratory agitation out of field (left). At the same time the magnetic field did not affect the susceptibility for particles locked in the rigid composite sample (right). A similar effect in permeability measurements has been reported elsewhere [1].

An attempt was made to use equation (1) for analyzing the concentration of neat (concentrated) MR fluid with CIP of defined specific susceptibility by measuring fluid sample susceptibility. This approach could be considered as a simple method for probing fluid concentration for various purposes. Four fluid samples were prepared using CIP with χ_s of 3.9 and "*true*" concentration of 0.38, 0.42, 0.44 and 0.46 v/v. Samples were subjected to shear to account for the initial drift in the susceptibility (see previous section). The sample susceptibility χ was measured for each composition (figure 4), and CIP concentration was computed using equation (1) and susceptibility data. Finally, the error Δ = *(true concentration – computed concentration) / true concentration* was calculated. Both computed concentration and error Δ are plotted in the figure 4. Up to CIP concentration of 0.44 v/v the error Δ doesn't exceed 1%. Thus, the simple and quick susceptibility measurements can be used to determine the volume concentration of MR fluid samples, at least for a practical range of volume concentration of $0.38 < \varphi < 0.44$ v/v.

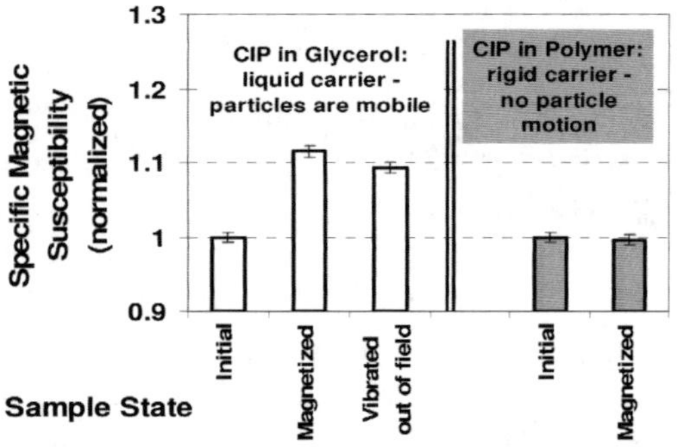

Figure 3. Normalized by initial specific magnetic susceptibility for MR fluid samples of different states based on liquid carrier (left) and rigid carrier (right).

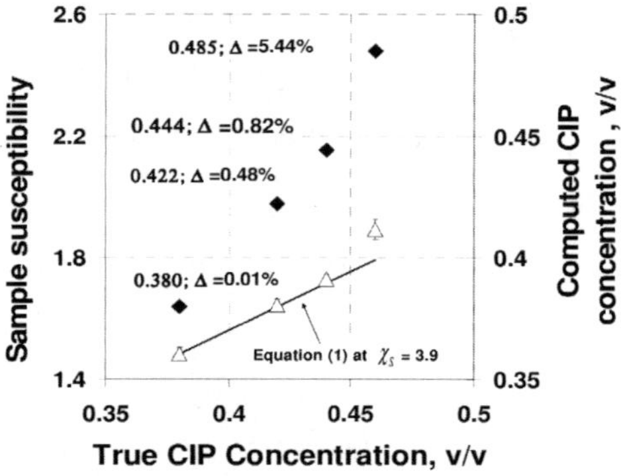

Figure 4. Sample magnetic susceptibility (open triangles) and computed CIP concentration (solid diamonds) with the error Δ as a function of the true CIP volume concentration.

5. Summary

- Bartington MS2B magnetic susceptibility system is a convenient tool to evaluate initial susceptibility of carbonyl iron; it requires only a small sample volumes providing quick reproducible measurements;
- Specific magnetic susceptibility, χ_s, of iron particles is found to be a linear function of median particle size. The following empirical equation can be used to define specific magnetic susceptibility if the particle diameter, D_{50}[μm] is known: $\chi_s = 0.668D_{50} + 0.961$;
- Structural change, including particle organization such as translation, aggregation and chaining, occurring in the MR fluid during use led to susceptibility drift and may affect fluid performance;
- The linear relationship between sample susceptibility and specific magnetic susceptibility can be used for screening of the particles concentration for a fluid with well defined specific susceptibility.

Acknowledgements
We thank Prof. S. Jacobs at University of Rochester for scanning electron microscopy and for access the Colloidal Dynamics Acoustosizer™ equipment for measurement of particle size.

References
[1] De Vicente J, Bossis G, Lacis S and Guyot M 2002 *JMMM* **251** 100
[2] Bombard A, Knobel M, Alcantara M and Joekes I 2002 *JIMSS* **13** 471
[3] Heider F, Zitzelberg A and Fabian K 1996 *Physics of the Earth and Planetary Interiors* **93** 239
[4] Bregar V and Pavlin M 2004 *Journal of Applied Physics* **95** 6289
[5] http://www.bartington.com
[6] http://www.eti-usa.com

Journal of Physics: Conference Series **149** (2009) 012052 doi:10.1088/1742-6596/149/1/012052

Temperature effect on the magneto-rheological behavior of magnetite particles dispersed in an ionic liquid

C Guerrero-Sanchez[1,2,3], A Ortiz-Alvarado[1] and U S Schubert[1,2,4]

[1] Laboratory of Macromolecular Chemistry and Nanoscience, Eindhoven University of Technology, P. O. Box 513, 5600 MB Eindhoven, The Netherlands
[2] Dutch Polymer Institute, P. O. Box 902, 5600 AX Eindhoven, The Netherlands
[3] Ioniqa Technologies, Eindhoven, The Netherlands
[4] Laboratory of Organic and Macromolecular Chemistry, Friedrich-Schiller-University Jena, Humboldtstr. 10, D-07743 Jena, Germany

E-mail: u.s.schubert@tue.nl, c.guerrero@ioniqa.com

Abstract. The magneto-rheological properties of iron(II, III) oxide particles dispersed in the ionic liquid 1-butyl-3-methylimidazolium hexafluorophosphate were investigated in the temperature range from 25 °C to 76 °C. The experimental results have revealed that the apparent viscosity of the dispersion slightly changes with the temperature when a constant magnetic field is applied and its value mainly depends on the shear rate and the strength of the magnetic field. The viscosity of the dispersion remains practically unmodified with both the temperature and the magnetic field intensity as the magnetic saturation of the material is reached; in this regime the viscosity will only depend on the applied shear rate. In contrast, the yield stress values of the dispersion as well as the corresponding shear stress *vs.* shear rate curves have shown and inversely proportional behavior with temperature for a constant magnetic field. Moreover, a power law model was utilized to predict the temperature dependence of the yield stress under the presence of a constant magnetic field.

1. Introduction

The use of ionic liquids (ILs) has been recently proposed for the preparation of magneto-rheological fluids (MRFs) with outstanding properties [1-6]. This is mainly due to the interesting and intriguing properties shown by ILs (*e.g.,* negligible vapor pressure and flammability, outstanding chemical and physical stabilities in a broad range of temperature, possibility of tuning their properties, etc.) [7]. Thus, it is thought that ILs will play an important role in the preparation of advanced MRFs in the coming years, which might find novel applications (and/or improve the existing ones) in several fields of science and engineering. To our best knowledge, there are only a few reports in the open literature addressing the effect of the temperature (T) on the magneto-rheological properties of MRFs and magnetic fluids [8-10]. Indeed, the lack of investigations in this particular aspect may be related to the fact that, so far, MRFs have been prepared in very specific liquid carriers (e.g. hydrocarbon, silicon and mineral oils) [4]. Due to their remarkable thermal stability and negligible vapor pressure, ILs might be considered as suitable liquid carriers to perform detailed studies of MRFs in a broad range of temperature. Based on these facts, this contribution addresses the magneto-rheological properties of iron(II, III) oxide (magnetite) particles dispersed in the hydrophobic

IL 1-butyl-3-methylimidazolium hexafluorophosphate (BMI-PF_6) in the temperature range from 25 °C to 76 °C.

2. Experimental part

The MRF investigated in this work was prepared and characterized following similar procedures as reported elsewhere [4]. Iron(II, III) oxide (magnetite) powder (Aldrich, <5 μm, 98%, density 4.8-5.1 g cm^{-3} (25 °C)) was dispersed in BMI-PF_6 (obtained from Solvent-Innovation GmbH as a kind gift). BMI-PF_6 is an hydrophobic IL with a density and viscosity values of 1.37 g cm^{-3} (20 °C) and 211 mPa s (25 °C), respectively. MFR used in this study had a content of 12.5% wt of magnetite particles dispersed in BMI-PF_6. The dispersion process was performed in a cylindrical polyethylene container using polyethylene stirring paddles. The mixing process was achieved by mechanical stirring at a rate of 2400 rpm for 15 min at 20 °C. Magneto-rheological measurements of the prepared MRF were carried out at the desired temperatures under steady shear and at several shear rates using a Physica MCR 301 rheometer (Anton Paar) coupled with a commercial magneto-rheological device (MRD180/1T magneto-rheological cell). The homogeneous magnetic field was oriented perpendicular to the shear flow direction. A circular parallel plate measuring system (PP20/MR made of nonmagnetic metal to prevent the occurrence of radial component of magnetic forces on the shaft of the measuring system) with a diameter of 20 mm and a gap of 1 mm between the plates was used. Before performing the magneto-rheological measurements, the prepared MRF was additionally and vigorously shaken in order to ensure the homogeneity of the dispersion (*e.g.* no supernatant clear layer formation (sedimentation of the dispersed particles) was observed for at least two days).

3. Results and discussion

The first investigations in this study correspond to the rheological measurements of the liquid carrier (BMI-PF_6) of the prepared MRF. Figure 1 reveals that the BMI-PF_6 IL has a Newtonian behavior in the investigated range of temperature. Furthermore, Figure 1 also shows that the viscosity (η) of BMI-PF_6 obeys an Arrhenius model at the investigated experimental conditions.

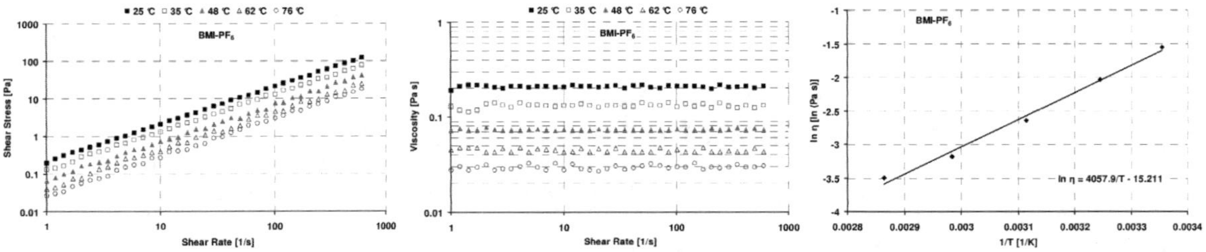

Figure 1. Temperature effect on the rheological characteristics of the hydrophobic ionic liquid 1-butyl-3-methylimidazolium hexafluorophosphate (BMI-PF_6).

The flow curves of the investigated MRF at several temperatures and under the influence of a magnetic field of different intensities (H) are displayed in Figure 2. Note that the shape of the curves resembles more to a pseudo-plastic system with a yield stress than to the Bingham model [11] conventionally used to describe MRFs under the presence of a magnetic field. This is also demonstrated in Figure 2 (bottom right) for a selected case of the obtained experimental data. In this specific case, two pseudo-plastic models (*e.g.* Ostwald (power law) and Carreau-Yasuda) [11], and the Bingham model are fitted and compared to experimental data. From this latter plot, it is clear that both pseudo-plastic models are more suitable to predict the experimental data than the Bingham model; similar findings were obtained for the rest of the obtained experimental data. Yield stresses of the MRF at the investigated conditions (different temperatures and magnetic field intensities) were estimated by using a yield stress analysis method incorporated in the software package of the utilized commercial rheometer. This analysis method estimates the yield stress by calculating the flexion point in a stress-strain curve in a logarithmic plot. The evaluation does a regression on the input data and

Journal of Physics: Conference Series **149** (2009) 012052 doi:10.1088/1742-6596/149/1/012052

checks for the point with the largest distance to the regression curve; this point is taken as the yield point. The yield stresses of the MRF obtained by this latter method are displayed in Figure 3.

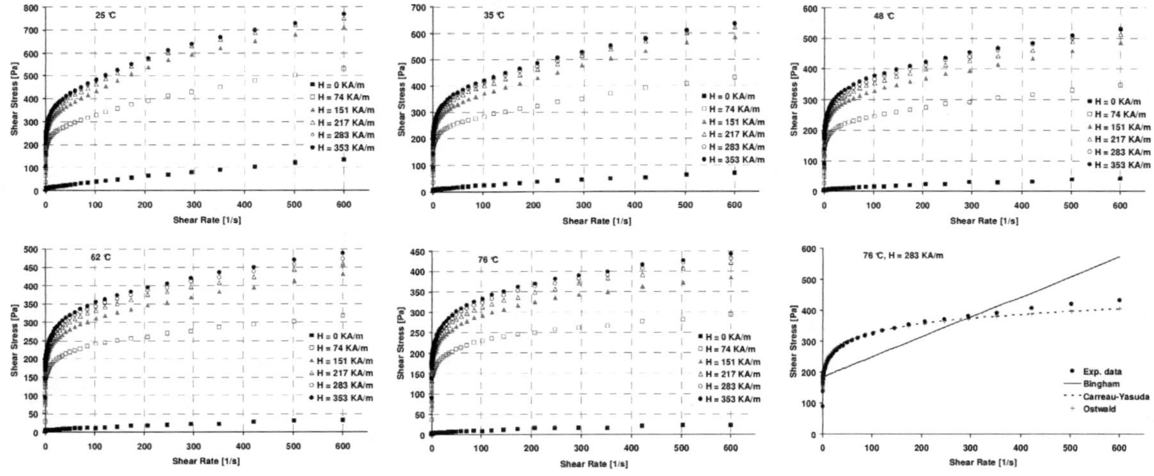

Figure 2. Flow curves of the investigated magneto-rheological fluid (MRF) at different temperatures and intensities of a magnetic field (H). Bottom right plot compares the predictions of two pseudo-plastic models (Ostwald and Carreau-Yasuda), and the Bingham model with the experimental data obtained for the investigated MRF at 78 °C and H = 283 KA m^{-1}.

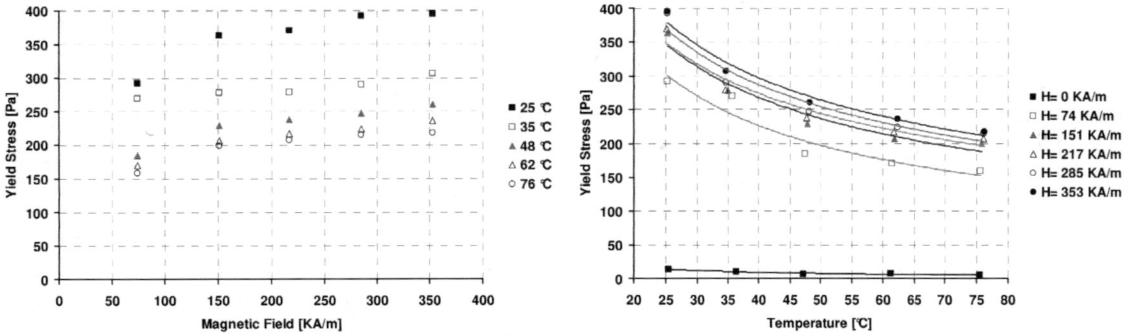

Figure 3. Yield stresses of the investigated magneto-rheological fluid (MRF) as a function of the temperature and of the magnetic field intensity (H).

As observed in the plots of Figures 2 and 3, the shear stress (τ) vs. shear rate ($\dot{\gamma}$) curves as well as the estimated yield stress (τ_o) values of the investigated MRF reveal an inversely proportional behavior with temperature for a fixed magnetic field intensity. The curves shown as solid lines in the plot on the right hand side of Figure 3 correspond to fittings of the obtained data with a power law model as represented by equation (1)

$$\tau_o = kT^n \tag{1}$$

where τ_o, k, and T are the yield stress, a fitting constant, and the temperature, respectively. The results in the plot on the right hand side of Figure 3 reveals that the temperature dependence of the yield stress in a MRF under the influence of a constant magnetic field can be qualitatively approached by using the power law model of equation (1).

In addition, the experimental results of this study have revealed that the apparent viscosity of the investigated MRF slightly changes with the temperature when a constant magnetic field is applied, and its value mainly depends on the shear rate and the strength of the magnetic field. Furthermore, the viscosity of the MRF remains practically unmodified with both the temperature and the magnetic field intensity as the magnetic saturation of the material is reached; in this regime the viscosity will only depend on the applied shear rate. These effects can be observed in the plots of Figure 4.

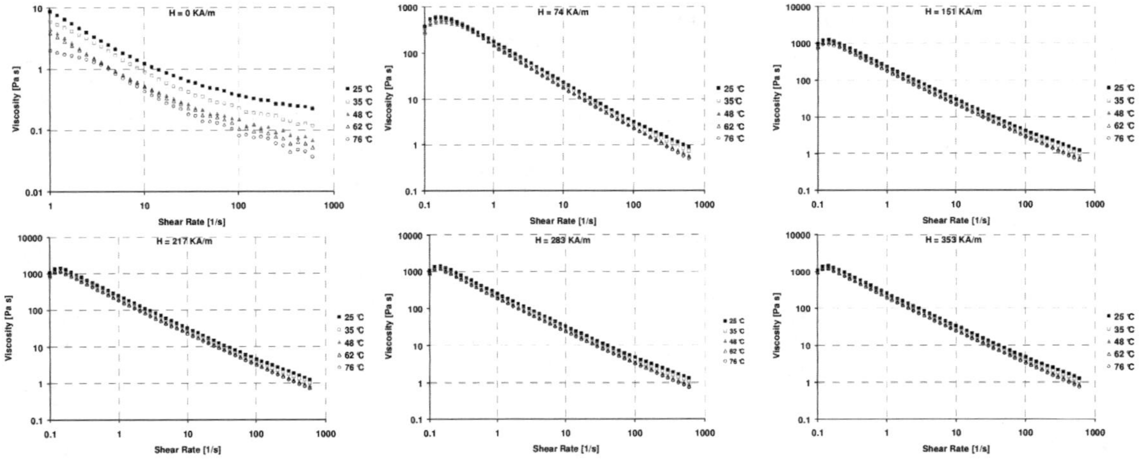

Figure 4. Viscosity curves of the investigated magneto-rheological fluid (MRF) as a function of the shear rate at different temperatures and intensities of a magnetic field (H).

4. Conclusions

In this contribution, the magneto-rheological properties of magnetite particles dispersed in the hydrophobic IL BMI-PF_6 were investigated in the temperature range from 25 °C to 76 °C. The use of ILs for the preparation of MRFs is very convenient to perform detailed temperature dependence magneto-rheological studies in this kind of dispersions due to the fact that ILs are in liquid state in a broad range of temperature, and they have a remarkable thermal stability and negligible vapour pressure.

Acknowledgement

This research is supported by the Technology Foundation STW, applied science division of NWO and the technology programme of the Ministry of Economic Affairs. The authors also thank the Fonds der Chemischen Industrie and the Dutch Polymer Institute for financial support.

References

[1] Guerrero-Sanchez C, Rasa M and Schubert U S *WO/2008/055645*
[2] Kieburg C, Pfister J, Gabriel C, Oetter G, Laun M and Lochtman R *WO/2008/034820*
[3] Dodbiba G, Park H S, Okaya K and Fujita T 2008 *J. Magn. Magn. Mater.* **320** 1322
[4] Guerrero-Sanchez C, Lara-Ceniceros T, Jimenez-Regalado E, Rasa M and Schubert U S 2007 *Adv. Mater.* **19** 1740
[5] Guerrero-Sanchez C, Lara-Ceniceros T, Jimenez-Regalado E, Rasa M and Schubert U S 2007 Ionic liquids as carriers of magnetorheological fluids (5P7) *11th Int. Conf. on Magnetic Fluids*, July 23-27
[6] Altin E, Gradl J and Peukert W 2006 *Chem. Eng. Technol.* **29** 1347
[7] Deetlefs M and Seddon K R 2006 *Chim. Oggi* **24(2)** 16
[8] Kordonsky V I, Shulman Z P, Gorodkin S R, Demchuk S A, Prokhorov I V, Zaltsgendler E A and Khusi B M 1990 *J. Magn. Magn. Mater.* **85** 114
[9] Weiss K D and Duclos T G 1994 *Int. J. Mod. Phys. B* **8** 3015
[10] Chooi W W and Oyadiji S O 2005 *Int. J. Mod. Phys. B* **8** 1318
[11] Macosko C W 1994 *Rheology: Principles, measurements, and applications* VCH Publishers

Journal of Physics: Conference Series **149** (2009) 012053 doi:10.1088/1742-6596/149/1/012053

Optimal Design of a Magnetorheological Fluid Suspension for Tracked Vehicle

S H Ha[1], S B Choi[1], E J Rhee[2] and P S Kang[2]

[1]Smart Structures and Systems Laboratory, Department of Mechanical Engineering, Inha University, Incheon 402-751, Korea

[2]R&D Center, Hyundai Rotem Company, Gyeonggu-do, 437-718, Korea

E-mail: seungbok@inha.ac.kr

Abstract. This paper presents optimal design of controllable magnetorheological suspension system (MRSS) for a tracked vehicle. As a first step, a double-rod type MRSS is designed on the basis of the Bingham model of commercially available MR fluid, and its damping characteristics are evaluated with respect to the intensity of the magnetic field. Subsequently, the governing equation of motion of the MRSS featuring the MR valve is established Then, the optimization problem to find optimal geometric dimensions of the MRSS is formulated by considering an objective function which is related to damping torque and control energy. The first order optimization method intergrated with a commercial finite element method(FEM) software is adopted to obtain optimal solution of the system. The performance characteristics of the optimized MRSS are then evaluated and compared with initial ones.

1. Introduction

Recently, various semi-active suspension systems featuring MR fluid have been proposed and successfully applied in vehicle suspension systems [1]. However most of previous researches have focused on wheeled vehicles such as passenger vehicles [2]. In general, tracked vehicles are exposed to severe working environment such as rough road and complex ride behaviors. Therefore, high performance suspension systems are required for tracked vehicles. The primary function of the tracked vehicle's suspension system is to isolate the vehicle body and its contents from terrain induced vibrations to permit high vehicle speed and steering stability [3]. Consequently, the main contribution of this paper is to show how to design the magnetorheological suspension system (MRSS) which is incorporated with the optimized MR valve. A new double-rod type MRSS is modeled and two-coil annular MR valve is designed. The optimization is undertaken using ANSYS and solutions such as damping torque are presented.

2. Modelling of MRSS

One of representative passive type suspension systems for tracked vehicle is in-arm type hydropneumatic suspension unit (ISU). Some benefits of the ISU include weight savings, greater load levelling capability and easiness of height control. In this work, the passive double-rod-type ISU is modified in order to devise the MRSS and its schematic configuration is shown in figure 1. It is seen that an MR valve is attached to the manifold part of the passive ISU. The manifold of the MRSS consists of the MR valve and two-coil annular valve. Above a certain pressure difference between actuating and accumulating cylinders, the MR valve needs to be activated by the magnetic field to achieve desired damping torque. To facilitate calculation of damping torque of the proposed MRSS, consider the coordinate system shown in figure 2. When the MR fluid flows through the MR valve, the pressure drop can be expressed by

$$\Delta P_e = \Delta P_{MR} + \Delta P_{mv} \tag{1}$$

where ΔP_{MR} is the pressure drop due to the field dependent yield stress of the MR fluid, and ΔP_{mv} is the pressure drop due to the viscosity of the MR fluid.

Figure 3 shows a simplified structure of the two-coil annular MR valve constrained specific volume. The valve geometry is characterized by the valve height L, the valve radius R, the valve housing thickness d_h, the MR channel gap h, the iron flange thickness t and the distance of between upper coil and lower coil a. When the electric current is applied to the coil, a magnetic circuit appears as shown in the figure 3. The pressure drop of the yield stress and viscosity are calculated by [4]

$$\Delta P_{MR} = 2ct\tau_t / h + ca\tau_a / h \tag{2}$$

$$\Delta P_{mv} = 6\eta L Q_0 / \pi h^3 R_1 \tag{3}$$

where c is a coefficient which depends on the flow velocity profile, η is the base viscosity, Q_o is the flow rate through the MR valve and R_1 is the average radius of the annular duct.

The pressure drop ΔP_0 due to the orifice can be also obtained as

$$\Delta P_0 = \rho g h_1 \tag{4}$$

where ρ is the density of the MR fluid, g is the acceleration due to gravity and h_1 is the head loss including the major and the minor losses of the orifice. Thus, the damping torque at the rotational center of the wheel arm is expressed by

$$T_d = l_{ob} \sin(87 - \alpha + \beta) F_d / \cos(\beta) \tag{5}$$

where F_d is the damping force at the actuating piston given by

$$F_d = \pi r_p^2 (\Delta P_e + \Delta P_0) \tag{6}$$

3. Optimization of MRSS

In this study, ANSYS integrated with an optimization tool is used to obtain optimal geometric dimensions of the MR valve of MRSS in order to minimize an objective function. The optimal objective is to minimize the valve ratio defined by the ratio of the viscous pressure drop to the field-dependent pressure drop of the MR valve. This ratio has a large effect on the characteristics of the MR

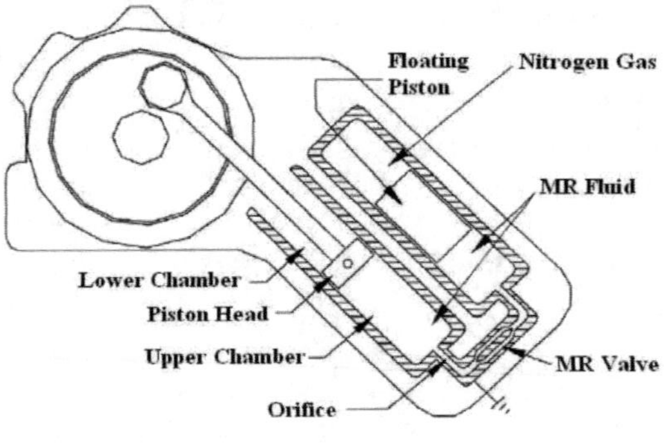

Figure 1. Schematic configuration of MRSS

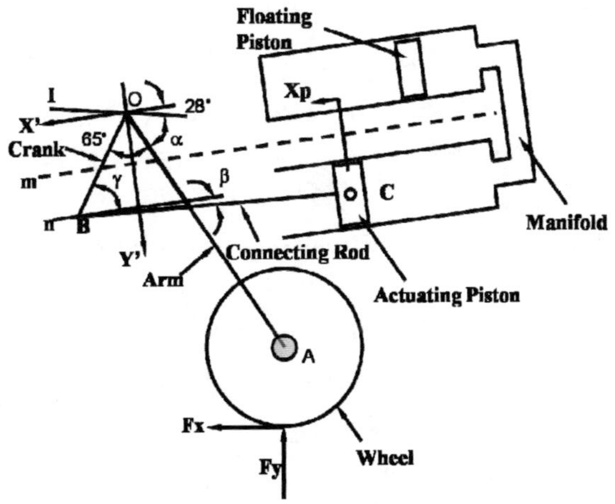

Figure 2. Coordinate system of MRSS

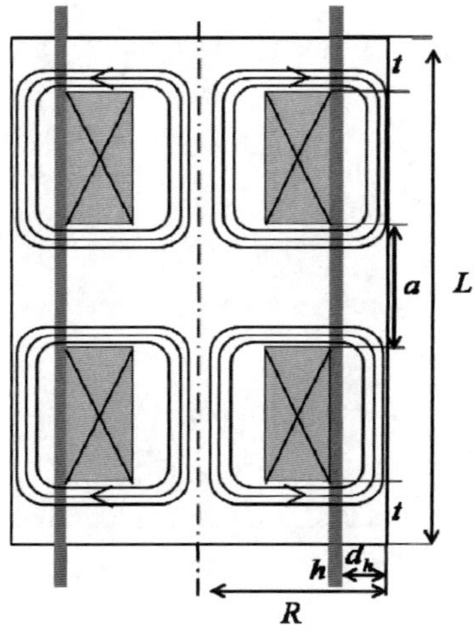

Figure 3. Schematic diagram of MR valve

valve. It is desirable that the valve ratio takes small value. From equations (2) and (3), the valve ratios of the two-coil MR valve is calculated by

$$\lambda = \Delta P_{mv} / \Delta P_{MR} = 3\eta L Q_0 / \pi h^2 R_1 c(t\tau_t + 0.5a\tau_a) \tag{7}$$

The yield stress of the MR fluid caused by the magnetic circuit is calculated from the approximate polynomial curve of the yield stress. This curve can be obtained as [5]

$$\tau(kPa) = 52.962B^4 - 176.51B^3 + 158.79B^2 + 13.708B + 0.1442 \tag{8}$$

where τ is the yield stress caused by the applied magnetic field and B is the magnetic flux density of the applied magnetic field. To achieve optimal design parameters of the MR valve, the first-order method of the ANSYS optimization tool is used. The ANSYS optimization tool transforms the constrained optimization problem into the unconstrained one via penalty functions. The dimensionless, unconstrained objective function is formulated as follows;

$$\lambda^{(*)}(y,p) = \lambda_1 / \lambda_0 + \sum_{i=1}^{n} P_y(y_i) + q\sum_{i=1}^{m} P_g(g_i) \tag{9}$$

where λ_1 is the objective function, λ_0 is the reference objective function, q is the response surface parameter, P_x is the exterior penalty function and P_g is the extended-interior penalty function.

For the initial iteration, the search direction of design variables is assumed to be the negative of the gradient of the unconstrained objective function. The direction vector is calculated by

$$d^{(0)} = -\nabla\lambda^{(*)}(y^{(0)},1) \tag{10}$$

The values of design variables in the next iteration is calculated by line search parameter s_j.

$$y^{(j+1)} = y^{(j)} + s_j d^{(j)} \tag{11}$$

In the subsequent iterations, the procedures are similar to those of the initial iteration except that the direction vectors are calculated according to the Polk-Pibiere recursion formula as follows;

$$d^{(j)} = -\nabla\lambda^{(*)}(y^{(j)},q_k) + u_{j-1}d^{(j-1)} \tag{12}$$

$$u_{j-1} = \{\nabla\lambda^{(*)}(y^{(j)},q) - \nabla\lambda^{(*)}(y^{(j-1)},q)\}^T \nabla\lambda^{(*)}(y^{(j)},q) / \left|\nabla\lambda^{(*)}(y^{(j-1)},q\right|^2 \tag{13}$$

4. Results and discussion

The valves are constrained in a cylinder of radius $r = 23mm$ and height $L = 120mm$. The base viscosity of the MR fluid is assumed to be constant, $\eta = 0.092Pas$ and the flow rate of the MR valves is $Q_0 = 3\times10^{-4} m^3 s^{-1}$. In this study, the valve gap is chosen as $1mm$. The current density applied to the coils can be approximately calculated by

$$J = I / A_w \tag{14}$$

where I is current applied to the coils and A_w is the cross section of the coil wire. To evaluate power consumption of the valve coils, N is calculated by the following equation.

$$N = I^2 R_w \tag{15}$$

where R_w is the resistance of the coil wire which can be approximately calculated as follow;

$$R_w = L_w r_w = V_c r_w / A_w \tag{16}$$

where L_w is the length of the coil wire, r_w is the resistance per unit length of the coil wires and V_c is the volume of all coils of the MR valve. The valve core radius R_c, the iron flange thickness t and the valve housing thickness d_h are calculated by the following equations.

$$R_c = (1/2)\cdot\{\sqrt{2R^2 - (W+h)^2} - (W+h)\} \tag{17}$$

$$t = R_c / 2 \tag{18}$$

$$d_h = R - (W + h + R_c) \tag{19}$$

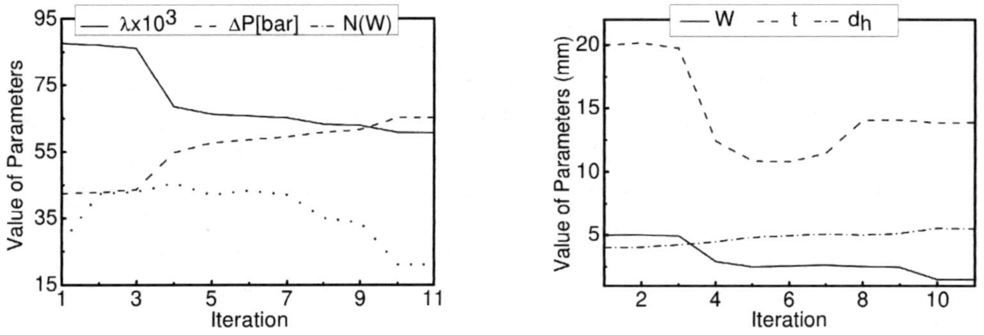

(a) valve ratio, pressure drop and power consumption (b) design variables

Figure 4. Optimization results of the MR valve

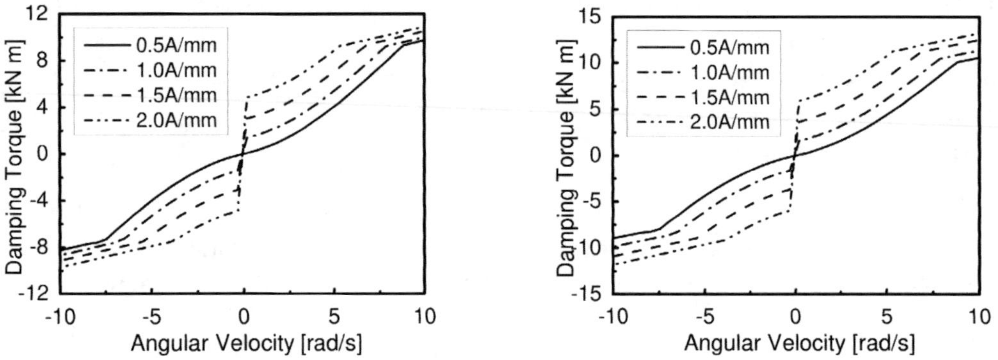

(a) initial (b) optimized

Figure 5. Damping Characteristic of the MRSS

Figure 4 shows the optimal solution of a two-coil annular MR valve which is constrained in the specific volume when a current of 2.0A is applied to the valve coils. The valve ratio, pressure drop and power consumption at these initial values are $\lambda_0 = 0.087$, $\Delta P_{\text{int}} = 42.47bar$ and $N_o = 24.51W$, respectively. Optimal design variable are determined by $t_{opt} = 13.88mm, W_{opt} = 1.5mm$ and $d_{h,opt} = 5.53mm$. At these optimal design variables the power consumption is $N_{opt} = 21.08W$ and the difference between the magnetic flux density of the outer and inner duct is 0.0398T which is small and acceptable.

As shown in figure 5, the damping torque of MRSS is compared between initial one and optimized one. It is evident that damping torque is increased as the magnetic field increased. It is also observed that the damping torque of optimized MRSS is larger than initial ones for the same control input. This means that the optimized MRSS has good performance in terms of energy efficiency.

5. Concluding remarks

In this work, a controllable MRSS has been proposed for tracked vehicles. A double-rod-type MRSS was devised and its damping characteristics were analysed. An optimization procedure based on the ANSYS has been developed in order to find the optimal geometry of MR valve employed in the MRSS. Optimal solutions have been obtained and the performance of the MRSS such as damping torque has been analysed and presented.

References

[1] Yao G Z, Yap F F, Chen G, Li W H and Yeo S H 2002 *Mechatronics* **12** 963
[2] Saxon N L and Meldrum W R 1997 *SAE technical paper* **970386**
[3] Hoogterp F B Saxon N L and Schihl P J 1993 *SAE technical paper* **930847**
[4] Nguyen Q H, Han Y M, Wereley N M and Choi S B 2007 *Smart Materials Structures* **16** 2242
[5] Lord Corporation 2003 MR fluid product bulletins online

Journal of Physics: Conference Series **149** (2009) 012054 doi:10.1088/1742-6596/149/1/012054

Adaptive Magnetorheological Seat Suspension for the Expeditionary Fighting Vehicle

G J Hiemenz[1], W Hu[2] and N M Wereley[2]

[1]Techno-Sciences, Inc., 11750 Beltsville Dr. Ste. 300, Beltsville, MD 20705 USA

[2]Aerospace Engineering, University of Maryland, College Park, MD 20712 USA

E-mail: greg@technosci.com, wereley@umd.edu

Abstract. The Expeditionary Fighting Vehicle (EFV) is an amphibious vehicle designed to operate through harsh conditions and at much higher speeds than its predecessors. These unique capabilities and broadly varying operational conditions lead to a complex design and human factors scenario for the forward seating positions that cannot be solved using conventional passive seat suspension systems. Injurious shock loads transmitted to the occupants when traversing over water in high sea states and/or at high speeds, as well as harmful shock and vibration transmitted to the occupants when the vehicle is travelling over land, pose a threat to occupant health and significantly limit mission duration. In this study, a semi-active magnetorheological (MR) seat suspension is developed which adapts to broadly varying operational conditions, as well as occupant weight, to provide optimal protection of EFV occupants. It is shown that this MR seat suspension system will reduce the shock and vibration transmitted to the occupant by up to 33% and 65%, respectively, as compared to the existing passive suspension.

1. Introduction

Figure 1. EFV in Water

The Expeditionary Fighting Vehicle (EFV), the newest United States Marine Corps (USMC) amphibious vehicle, is designed to operate over harsh off-road terrain as well as in oceans and rivers. Travelling over water (Figure 1), the EFV is capable of much higher speeds (3x) than its predecessors, which has led to high shock loads being transmitted to the occupants when operating in high sea states. These shock loads are particularly problematic in the forward seating positions for the driver and troopcrew commander. Harmful shock and vibration may also be transmitted to the occupants when the vehicle is travelling over land and traversing rough terrain. These operating conditions, combined with the requirement to handle both ballistic and UNDEX shock, make for a complex design and human factors scenario. The transmission of such high and repetitive shock loads into the seats drastically increases the risk of occupant pelvic or spinal injury. Moreover, health risks, fatigue, and discomfort are also associated with lower amplitude vibration that may be transmitted into the seat during normal operating conditions. Furthermore, the fact that these seats are occupied by USMC personnel of widely varying weights complicates the design solution. Current attempts at addressing these issues with passive seat suspensions have been unsuccessful because passive systems can only be optimally tuned for one excitation level and occupant weight. A semi-active magnetorheological seat suspension was developed in this study to adapt to these different vibration and shock spectra. The key technologies behind this system are retrofit-capable MR fluid shock absorbers and a compact control electronics module featuring a control algorithm developed specifically for this EFV seat application. MR shock absorbers have the advantage that their damping levels can be adjusted automatically in real-time with low power control signals, and accomplish this controllability with no more moving parts than conventional hydraulic dampers, so that MR dampers are highly reliable. While passive suspensions, such as those currently installed in the EFV seat, can be designed to mitigate either water-borne shock loads or lower level land-mode excitations, passive dampers cannot be optimally tuned to mitigate both simultaneously. In contrast, the MR suspension system can be adaptively tuned to simultaneously mitigate both shock and vibration. In the present study, a retrofittable semi-active MR vehicle seat (SAMVS) system for the EFV was developed and semi-active control performance was analyzed for both land and water operations.

2. System Design

2.1. Loading and Evaluation Criteria

There were 5 shock conditions for which this system was designed to mitigate. In order of increasing severity, these are: 1) underwater explosive (UNDEX) shock (52.8 *g*, 5ms half-sine pulse), 2) ballistic shock (19 *g*, 15ms half-sine pulse), 3) typical wave shock (6-8 *g*, 50ms half-sine pulse), 4) basic shock (19 *g*, 15ms terminal saw-tooth pulse), and 5) maximum wave shock (16 *g*, 50ms half-sine pulse). Since the maximum wave shock is the most severe, this excitation was used for determining the maximum MR damper force necessary to prevent end-stop impact (damper bottom-out). Shock performance was evaluated using two metrics: 1) peak lumbar loads, and 2) the ISO 2631-5 shock "dosage". Peak lumbar loads should be kept as low as possible and should at all times be kept below the injury thresholds presented in JSSG-2010-7 [1]. The ISO 2631-5 shock dosages are typically used to calculate health effects and occupant exposure durations [2]. For vibration excitation, PSD data generated using MIL-STD-810F Method 514.5, Annex A, Category 20 [3] were utilized. These PSD profiles consist of a wideband random noise superimposed with swept narrowbands. From this PSD data, eight sample vertical acceleration time histories were determined for inclusion into the design analysis. For performance evaluation for these vibration cases, ISO 2631-1 was utilized. The acceleration at the occupant interface was filtered to give a frequency-weighted acceleration time history per the ISO standard. A simple RMS of the weighted acceleration was used as a vibration performance metric [4]. Similarly to the ISO 2631-5 shock dosage, performance between design options was compared by showing percent changes in this ISO 2631-1 weighted RMS acceleration value.

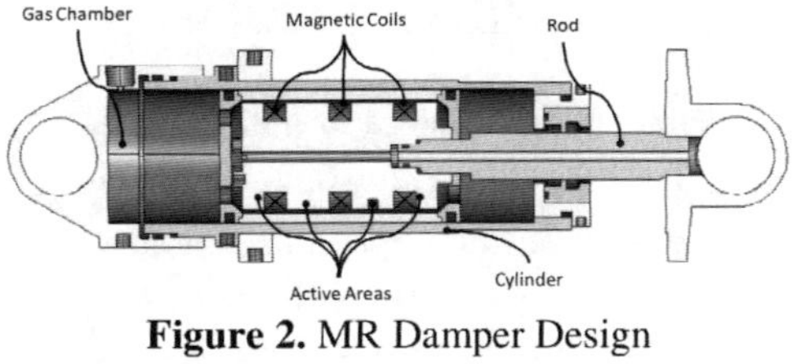

Figure 2. MR Damper Design

Figure 3. Fabricated MR Damper

2.1. MR Damper Design and System Integration

An MR damper was designed to provide an on-state force, F_{on}, of 6 kN at a piston velocity, v, of 1.7 m/s with a dynamic range (ratio of on-state to off-state force), of 3.52; thus, an off-state viscous damping, C_o, of 1,000 N-s/m. This MR damper (Figure 2) was designed to use Lord Corporation's MRF-132 fluid, have a fully extended length of 10 inches and an outer cylinder diameter of 2.8 inches such that it can be retrofitted into the existing EFV seat. This damper was then fabricated using 6061 aluminum and anodized as shown in Figure 3.

Journal of Physics: Conference Series **149** (2009) 012054 doi:10.1088/1742-6596/149/1/012054

2.2. Semi-Active Control

A multi-mode control algorithm has been developed specifically for this EFV seat application. This control algorithm has two primary modes, one which is intended for normal, low amplitude excitation and one which is intended to optimally mitigate shock loading. The objective of the first mode is to combine the benefits of a highly damped and lightly damped system to both suppress the seat suspension resonance while optimally isolating higher frequency vibrations. When a shock condition occurs, real-time sensor feedback automatically triggers the controller into the 2nd (shock mitigation) mode. In this mode, the MR dampers are optimally controlled to provide the minimum amount of load possible without causing harmful end-stop impact. In this shock mode, the recoil of the suspension is also controlled to return the occupant to the static position as quickly as possible while keeping rebound accelerations to a minimum and preventing an extensional end-stop impact. It should be noted that this controller will automatically switch between these modes based upon real-time sensor feedback. Furthermore, the control gains for all control modes automatically adapt for occupant weight (calculated by the controller), thus providing optimal protection for all occupants.

3. Results and Discussion

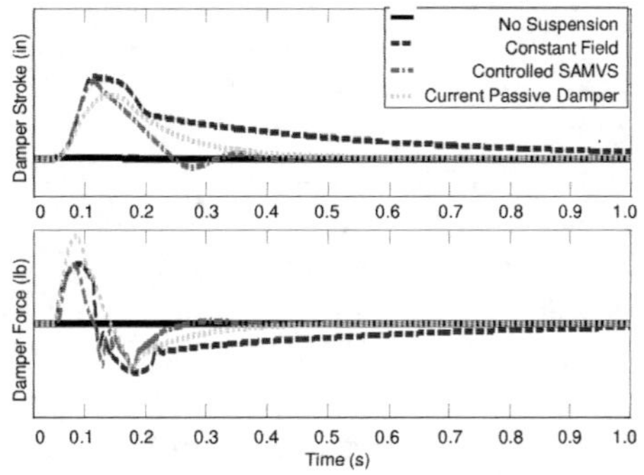

Figure 4. Response of Damper to Maximum Wave Condition

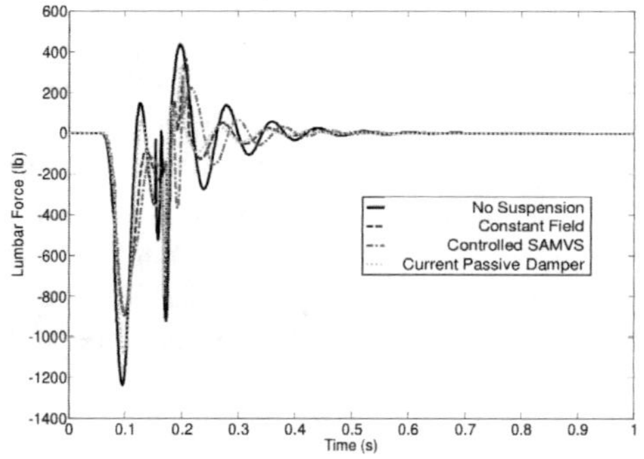

Figure 5. Occupant Lumbar Load Response to Maximum Wave Condition

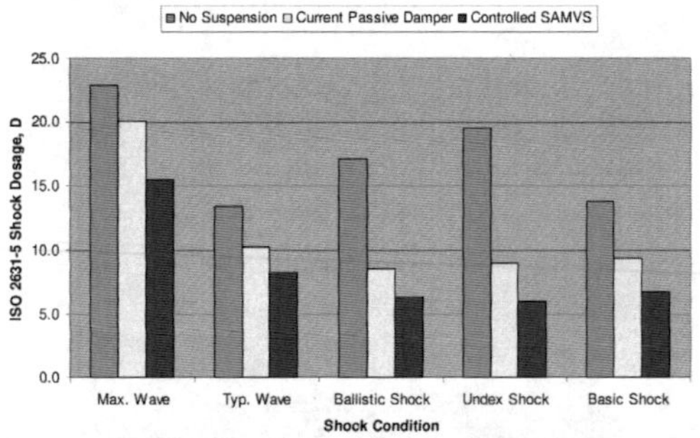

Figure 6. Shock Performance Comparisons

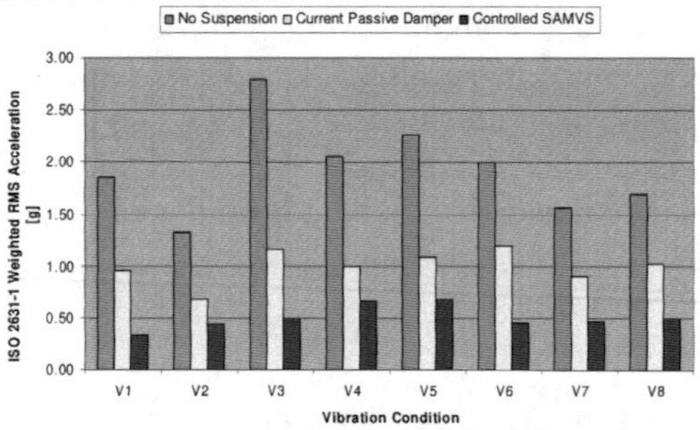

Figure 7. Vibration Performance Comparisons

For design and performance evaluations, a mathematical model of a seated human was coupled with the nonlinear model of an MR seat suspension. The biodynamic model used for this analysis is described in detail in Ref. [5] and has been updated for the inertial properties of the EFV seat.

Figure 4 shows the analytical response of the MR damper to the max wave impact condition discussed above for four different conditions. The first of these assumes that the seat has no suspension. The second condition includes an MR damper simply turned on to a constant applied magnetic field level (thus uncontrolled). The third condition utilizes an MR damper that is semi-actively controlled using the adaptive control algorithm discussed above. Finally, the fourth condition simulates the current passive suspension being used for the EFV seat. In Figure 4, the upper plot shows that the each of the conditions with suspensions utilize approximately the same amount of stroke at onset of the shock. These data also show, however, that the controlled

SAMVS system returns to its static equilibrium position (zero inches) faster than the other suspended conditions. This is because the MR damper can lower its force upon recoil (as shown in the bottom plot) and allow the occupant to return to the normal seated position sooner, which is important for line-of-sight concerns. Figure 5 presents the analytical occupant lumbar load response to the maximum wave impact condition for these same four conditions. Figure 5 also shows that the controlled SAMVS system provides significantly lower lumbar loads than the existing passive suspension, both in compression (negative) at the onset of the shock, and upon recoil. The controller is optimal in the sense that it fully utilizes available suspension stroke for all conditions, while also providing an optimal recoil.. The ISO 2631-5 shock dosages were calculated for the five design shock cases discussed above. Figure 6 shows that the theoretical controlled SAMVS system provides significantly reduced shock dosages over both a seat with no suspension and the existing seat with a passive suspension which is tuned for shock. The controlled SAMVS system reduces the ISO 2631-5 shock dosage by up to 33% over the passive suspension because the controller optimally adjusts the damper to mitigate the harsh compression and recoil. Similarly, the ISO 2631-1 weighted RMS accelerations were calculated for the eight design vibration load cases discussed above. Figure 7 shows that the theoretical controlled SAMVS system significantly reduces vibration transmitted to the occupant, and can reduce land-mode vibration transmitted to the occupant by up to 65% as compared to the existing passive suspension, because the MR damper is turned off during low amplitude excitations to provide very low transmissibility. The MR damper is only turned on to mitigate the suspension resonance and to ensure that end-stop impacts are avoided when larger amplitude bumps occur. The passive suspension is fixed at a high damping level for shock so that it provides poor vibration isolation.

4. Summary

This study has demonstrated the feasibility and benefits of the SAMVS system. The key technologies behind the SAMVS system are retrofit-capable magnetorheological (MR) fluid dampers and a control algorithm developed specifically for the EFV seat application. While passive suspensions, such as those currently installed in the EFV seat, can be designed to mitigate either the water-borne shock loads or the lower level land-mode excitations (but not both), the SAMVS system enables a unique single solution for both shock and vibration environments. This study has shown that, with a real-time controller developed specifically for this application, the SAMVS system can reduce the shock dosages applied to the occupant by up to 33% in water mode and reduce the vibration transmitted to the occupant by up to 65% as compared to the current passive suspension system.

5. Acknowledgements

This research was supported by a MARCOR Phase I SBIR (Topic No. N06-103) to Techno-Sciences, Inc. The research team acknowledges Lord Corporation for their MR fluid support.

References

[1] JSSG-2010-7, "Crew Systems – Crash Protection Handbook", U.S. Department of Defense Joint Services Specification Guide, 1998.

[2] ISO-2631-5, "Mechanical vibration and shock - Evaluation of human exposure to whole-body vibration, Part 5: Method for evaluation of vibration containing multiple shocks," 1st Edition, 2004.

[3] MIL-STD-810F, Method 514.5, Vibration, 2000.

[4] ISO-2631-1, "Mechanical vibration and shock - Evaluation of human exposure to whole-body vibration, Part 1: General Requirements," 2nd Edition, 1997.

[5] Hiemenz G J, Choi Y T and Wereley N M, *Journal of Aircraft*, AIAA, Vol. 44, No. 3, 2007, 1031

Journal of Physics: Conference Series **149** (2009) 012055 doi:10.1088/1742-6596/149/1/012055

Preparation and Characterization of MR fluid consisting of Magnetite Particle coated with PMMA

M K Hong, B J Park and H J Choi

Department of Polymer Science and Engineering, Inha University, Incheon 402-751, Korea

E-mail: hjchoi@inha.ac.kr

Abstract. To improve physical characteristics of MR fluid, we synthesize magnetite nanoparticles (MNP) in shape of magnetic core- poly(methyl methacrylate) (PMMA) polymeric shell by atomic transfer radical polymerization. Prepared MNPs-PMMA which possesses a density lower than that of pure MNPs were characterized by FT-IR, XPS and TEM. Synthesized MNP-PMMA based MR fluid which was dispersed in non-magnetic medium lubricant oil (Yu-base 8) with 10 vol% was prepared. Rheological properties such as shear viscosity, shear stress and storage modulus were measured by both static and dynamic tests using a rotational rheometer with various external magnetic field strength applied.

1. Introduction

Magnetorheological (MR) fluids, which consist of either ferro- or ferri-magnetic particles with a size on the order of sub to several microns dispersed in a nonmagnetic fluid medium, are considered as smart particle suspensions because they have readily controllable characteristics by an external magnetic field. When exposed to an external magnetic field, the fluid transforms from a freely-flowing liquid-like state to a solid-like state as quickly as a flash, because the dispersed magnetic particles attract each other to form fibril structures aligning along the applied magnetic field direction, consequently inhibiting the fluid flow. That makes a change of rheological properties of the MR fluids such as an improvement of yield stress, shear viscosity and storage modulus with external magnetic field strength [1-3]. In addition, soft magnetic materials such as magnetite (Fe_3O_4) are more reversible to control rheological properties of the MR fluids than hard magnetic materials due to be relatively easy to be magnetized and demagnetized. These characteristics of the MR fluids allow their engineering applications such as active controllable dampers, position controllers, clutches and shock absorbers. However, their some problems which are sedimentation and abrasion by high density and rigid surface of magnetic particles need to be improved for practical engineering applications [4-6]. Generally, dispersants have been added in MR fluids to improve dispersion stability. In addition, either magnetic particles were coated with polymers or monodispersed polymer cores were coated with magnetic particle to lower the density of magnetic particles [7-9]. In this study, to improve sedimentation and abrasion problems caused by high density and rigid surface of magnetic particles, we synthesized PMMA/magnetite nanoparticle composites by atomic transfer radical polymerization (ATRP) method [10-14]. Synthesized PMMA/magnetite composites (MNP-PMMA) in shape of magnetite core - PMMA shell were observed to possess lower density and be softer than pure magnetite particles. Some drawbacks of the MR fluids such as sedimentation and abrasion of magnetic particles are improved by using PMMA/magnetite composites.

2. Experimental

2.1. Materials

Ferric chloride hexahydrate ($FeCl_3·6H_2O$, 98%), ferrous sulfate heptahydrate ($FeSO_4·H_2O$, 99%), oleic acid (OA), methanol, 3-chloropropionic acid (CPA, 98%), 2,2-dipyridyl (bpy, 99%) and copper(I)

bromide (CuBr, 98%) were purchased from Aldrich, USA, while methy mathacrylate (MMA), n-hexane and ammonium hydroxide (NH_4OH) were from DC Chemical Co,. LTD, Korea.

2.2. Synthesis of magnetic Fe_3O_4 nanoparticles

The magnetite nanoparticles (MNP) were produced by co-precipitation of an aqueous Fe^{3+}/Fe^{2+} solution with a base [15]. Ferric chloride hexahydrate ($FeCl_3{\cdot}6H_2O$, 98%) and ferrous sulfate heptahydrate ($FeSO_4{\cdot}H_2O$, 99%) were dissolved in distilled water and heated to 60 ºC. Then ammonium hydroxide and oleic acid (OA) (Aldrich, USA) were added rapidly under stirring for 30min. Temperature was then decreased to room temperature (25 ºC). After 6h, the precipitates were washed several times with distilled water and methanol. After being dried in the vacuum oven at 80 ºC, the dark brown powder was obtained.

2.3. Immobilized initiator on MNPs surface

Prepared MNPs which were stabilized by OA were mixed with hexane. The mixture was sonicated for 1 hour in order that the MNPs were well dispersed in hexane. Initiator CPA was added in the solution with stirring and sonication. After 1 day, the precipitate was separated from mixture using a centrifuge (at 5000rpm, for 15min). They were washed with hexane several times to remove the excess initiator (CPA). After being dryied in the vacuum oven at 80 ºC, the dark brown MNPs-CPA in which the initiator CPA was immobilized on the surface of MNPs were obtained.

2.4. Surface initiated atoimic transfer radical polymerization of MMA

Prepared MNPs-CPA were dispersed in toluene and then were sonicated for 1 hour. MMA monomer, CuBr and bpy were added in the mixture and then the mixture was sealed in glass reactor under vacuum. The final mixture was purged with nitrogen for 1 hour and was agitated with a mechanical stirrer at 300 rpm speed. Surface initiated ATRP was carried out at 80 ºC with stirring for 1day. After reaction, the precipitate was washed with toluene using a centrifuge and a magnet. After washing sevral times, the precipitate were dried in vacuum oven at 80 ºC. The red brown powder, PMMA coated MNPs with core (MNPs)–shell (PMMA) structure, was obtained.

2.5. Characterizaiton

Fourier transform infrared (FT-IR) spectra were obtained by using VERTEX 80v vacuum FT-IR spectrometers (Bruker Optics, USA). The spectrum was taken between $400cm^{-1}$ and $4000cm^{-1}$ and 16scan at resolution of $8cm^{-1}$. X-ray photoelectron spectroscopy (XPS) was performed with Axios (Phillips, USA). Transmission electron microscopy (TEM) was performed with Philips CM200 (Phillips, USA) at 120kV. Rheological properties of the MR fluid were characterized by using MCR300 rheometer (Physica, Germany) with the MR device (MRD 180). The steady shear rotation test was perform with the shear rate range from 10^{-2} s^{-1} to 10^{3} s^{-1} and the dynamic test was perform with the frequency range from 10^{-1} Hz to 10^{1} Hz.

3. Results and discussion

3.1. Preparation MNPs-PMMA particle

Figure1 shows the FT-IR spectra of (a) MNPs-OA, (b) MNPs-CPA, (c) MNPs-PMMA and (d) pure PMMA. In curves (a), the strong absorption band between $590cm^{-1}$ and $630cm^{-1}$ was corresponded to the Fe-O bond of Fe_3O_4 and the strong band around $3400cm^{-1}$ was attributed to O-H bond. It was conformed that the MNPs were synthesized by co-precipitation of an aqueous Fe^{3+}/Fe^{2+} solution. Two peaks around $2930cm^{-1}$ and $2850cm^{-1}$ were attributed to the asymmetric CH_2 stretch and the symmetric CH_2 stretch, respectively. There are the picks around $1660cm^{-1}$ which were characteristic of C=O double bond stretching. These results indicate that the oleic acid has been coated on the surface of MNPs. In curve (c), the strong absorption band around $1090cm^{-1}$ was corresponded to the C-O-C ether stretch. It was conformed that PMMA has been coated on the surface of MNPs.

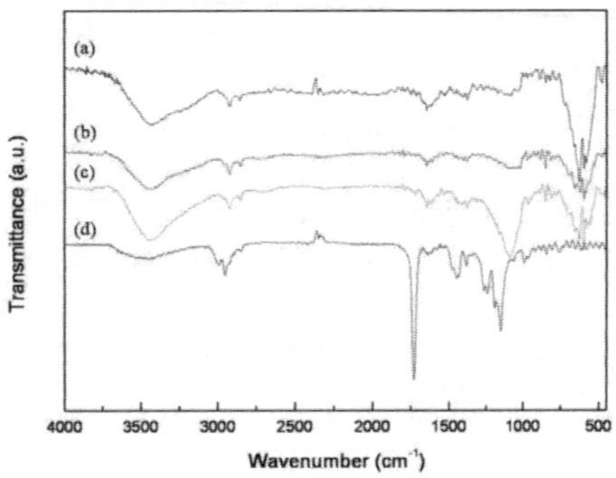

Figure 1. FT-IR spectra of (a) MNPs-OA, (b) MNPs-CPA, (c) MNPs-PMMA and (d) pure PMMA

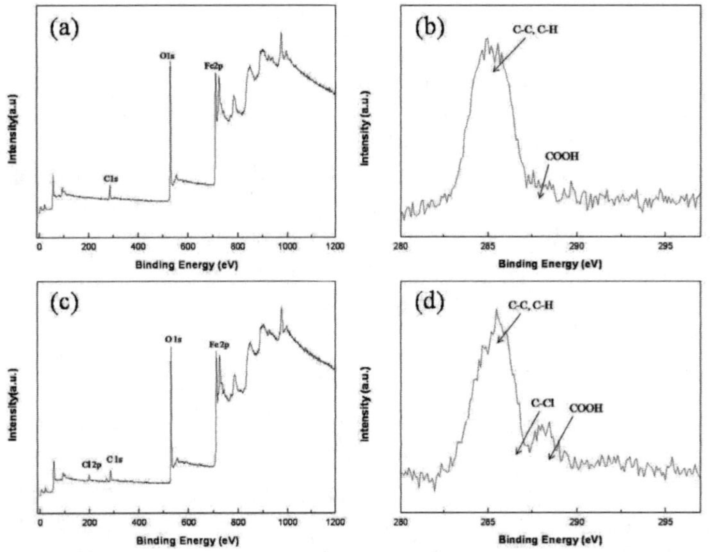

Figure 2. XPS wide scan of (a) MNPs-OA and (c) MNPs-CPA and XPS spectra of C 1s core level of (b) MNPs-OA and (d) MNPs-CPA.

As shown in Figure 2 (a), there are some picks at 284eV, 530eV and 710eV which are attributed C 1s, O 1s and Fe 2p respectively. In figure (b) which is the C 1s core level spectrum of MNPs-OA, there are two picks at 284.9eV and 287.8eV that are attributed to C-C/C-H and COOH. It means that the oleic acid is coated onto MNPs. New pick was observed at 200eV in figure (c) that is attributed to Cl 2p. As shown figure (d) which is the C 1s core level spectrum of MNPs-OA, three pick are observed at 285.4eV, 286.8eV and 288eV which are attributed to C-C/C-H, C-Cl and COOH. It was conformed that initiators are immobilized on the surface of MNPs by observing the C-Cl.

Figure 3 is TEM images of MNPs (a), (d) MNPs-OA, (b), (e) MNPs-CPA and (c), (f) MNPs-PMMA. As shown in figure 3 (a) ~ (c), MNPs-OA, MNPs-CPA and MNPs-PMMA have practically spherical shape and uniform particle size of 8~15nm. It is difficult that the PMMA coating layers of MNPs-PMMA were found, because of only small contrast difference between PMMA and background and particles are severely aggregated. In addition, as shown in Figure 3 (d) ~ (f), we could find more clearly that MNPs-PMMA aggregates are smaller than others. The MNPs which have very high surface energy are aggregated to reduce their surface energy. Since the MNPs were coated with PMMA, surface energy of the MNPs was reduced. It was also confirmed that the MNPs-PMMA was less aggregated than MNPs-OA and MNPs-CPA as given in Fig. 3.

3.2. Rheological properties of MR fluids

MR fluids (10 vol%) was prepared by dispersing synthesized MNP-PMMA in non-magnetic medium oil of Yu-base 8, and then their rheological properties were measured by both static and dynamic tests using a rotational rheometer. Figure 4 shows their relation which was obtained by the rotation test, between shear rate and shear stress for different external magnetic field strength. MR fluids act as like general suspensions without a magnetic field of which the shear stress increases with a shear rate. However, once the external magnetic field is applied to MR fluids, MR fluids behave like a Bingham fluid with yield stress.

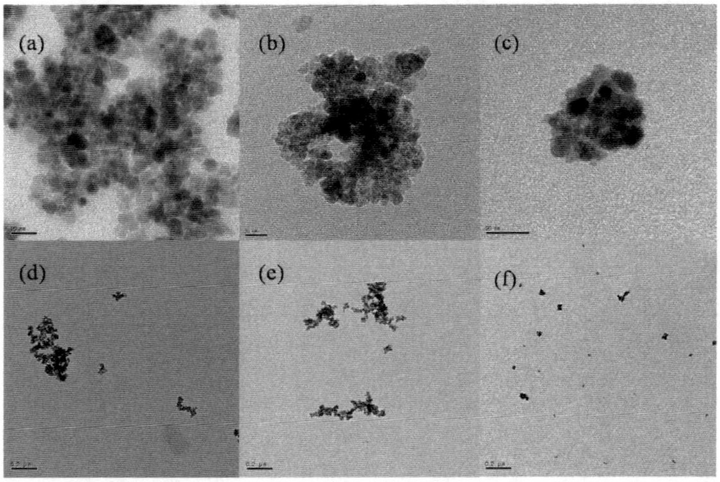

Figure 3. TEM images of (a),(d) MNPs-OA, (b),(e) MNPs-CPA and (c),(f) MNPs-PMMA.

This is the reason why the magnetic particles moving freely in the medium make particle chains or particle cluster by polarization force between particles, which is induced by an applied external magnetic field. As shown in Fig. 4, the shear stress at low shear rate increased with augmenting external magnetic field between 100mA and 1999mA. The polarization force between particles increases with increasing external magnetic field strength. Increased polarization force between particles makes strong particle chains or particle clusters in the MR fluid. Yield stress at low magnetic field drastically increases with increased magnetic field strength. On the other hand, at high magnetic field the yield stress rarely increases with magnetic field strength. At low applied fields, the shear stress of the MR fluids increases with increased external magnetic field. In intermediate fields, the contact or polar region of each particle saturates, thus reducing the rate of increase of the shear stress with increasing field. At high fields, the shear stress reaches a limiting value due to completely saturated particles [16]. In addition, the shear stress is independent of shear rate until critical shear rate, because the particle chains which are formed by an external magnetic field are maintained by repeated creation and destruction of the particle chain until the critical shear rate reaches. On the other hand, shear stress is dependent of shear rate above the critical shear rate because the destruction of particle chains is superior to their creation above critical shear rate region.

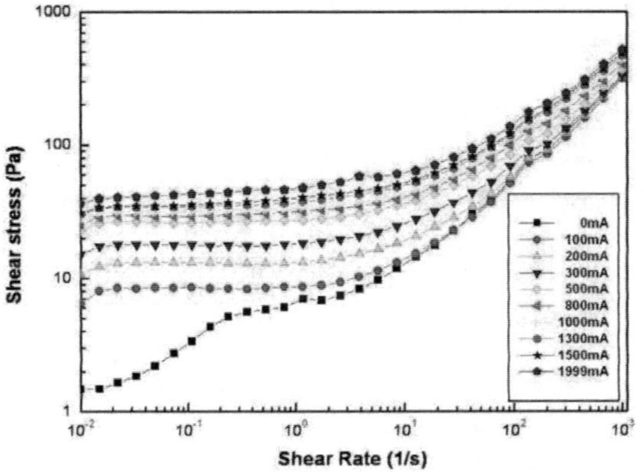

Figure 4. Shear stress of MR fluid with various external magnetic field strength.

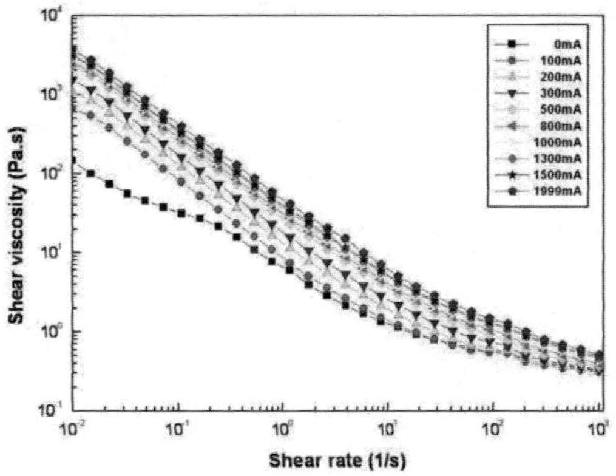

Figure 5. Shear viscosity of MR fluid with various external magnetic field strength.

Figure 5 shows the relation between shear viscosity and shear rate. It is confirmed that shear viscosity at low shear rate increases as the external magnetic field strengths increase. In addition, shear thinning behaviour of which shear viscosity decreases with increased shear rate is shown. Figure 6 represents the storage modulus with frequency. It was carried out with increasing frequency at linear viscoelasic region.

Journal of Physics: Conference Series **149** (2009) 012055 doi:10.1088/1742-6596/149/1/012055

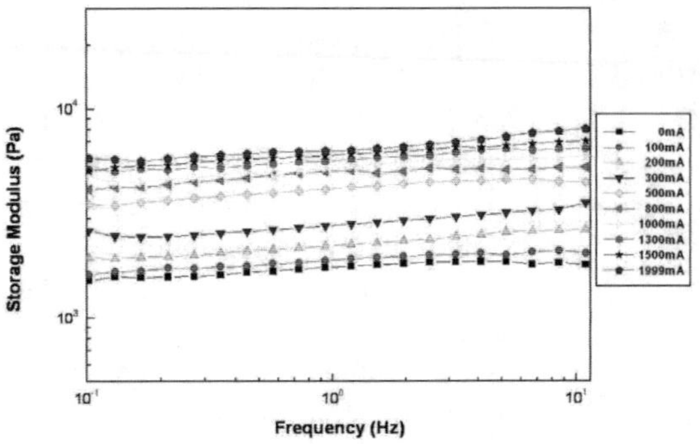

Figure 6. Storage modulus of MR Fluid with various external magnetic field strength.

Before the test, we found the linear viscoelastic region of the MR fluid through amplitude sweep test in which the storage modulus is measured with changing the strain at a fixed frequency (10Hz). The applied external magnetic field induces polarization force between particles, and induced polarization force between particles becomes strong with external magnetic field strength, so that storage modulus increases with magnetic field strength. As shown figure 6, storage modulus of the MR fluid as similar as shear viscosity and shear stress were increased with external magnetic field strength but the increase of storage modulus was insignificant in high magnetic field strength. Storage modulus are observed to be independent of frequency, indicating that the MR fluid changes from liquid like to solid like mainly by the applied external magnetic field.

4. Conclusion

The MNPs-PMMA composites, which are magnetite core-PMMA shell type, were synthesized by the ATRP. They were observed to possess quite uniform particle size and less aggregation compared with MNPs-OA and MNPs-CPA. Rheological properties of the MR fluids prepared with synthesized MNP-PMMA and non-magnetic medium Yu-base 8 with 10 volume% were measured by both static and dynamic tests using a rotational rheometer with various external magnetic field strength. Shear stress, shear viscosity and storage modulus of the MR fluids increased with applied external magnetic field strength.

References

[1] Ulicny J C and Golden M A 2007 *Int. J. Mod. Phys. B* **21** 4898
[2] See H, Joung C and Ekwebelam C 2007 *Int. J. Mod. Phys. B* **21** 4945
[3] Bica I and Choi H J 2008 *Int. J. Mod. Phys. B* **22** 5041
[4] Lim S T, Cho M S, Jang I B, and Choi H J 2004 *J. Magn. Magn. Mater.* **282** 170
[5] Jang I B, Kim H B, Lee J Y, You J L, and Choi H J 2005 *J. Appl. Phys.* **97** 10Q912
[6] Genc S and Phule P P 2002 *Smart Mater. Struct.* **11** 140
[7] Bell R C, Miller E D, Karli J O, Vavreck A N and Zimmerman D T 2007 *Int. J. Mod. Phys. B* **21** 5018
[8] Tang X, Zhang X, Tao R and Rong Y 2000 *J. Appl. Phys* **87** 2634
[9] Goncalves F D and Carlson J D 2007 *Int. J. Mod. Phys. B* **21** 4832
[10] Lim S T, Choi H J, and Jhon M S 2005 *IEEE Trans. Magn.* **41** 3745
[11] Cho M S, Lim S T, Jang I B, Choi H J and Jhon M S 2004 *IEEE Trans. Magn.* **40** 3036
[12] Choi H J, Jang I B, Lee J Y, Pich A, Bhattacharya S and Adler H J 2005 *IEEE Trans. Magn.* **41** 3448
[13] Fang F F, Jang I B and Choi H J 2007 *Diam. Relat. Mat.* **16** 1167
[14] Fan Q L, Neoh K G, Kang E T, Shuter B and Wang S C 2007 *Biomaterials* **28** 5426
[15] Marutani E, Yamamoto S, Ninjbadgar T, Tsujii Y, Fukuda T and Takano M, 2004 *Polymer* **45**, 2231
[16] Hu B, Fuchs A, Huseyin S, Gordaninejad F and Evernsel C 2006 *Polymer* **47** 7653
[17] Banert T and Peuker U A 2006 *J Mater Sci* **41** 3051
[18] Chen Y H, Liu Y Y and Lin R H 2008 *J Appl Polymer Sci* **108** 583
[19] Massart R 1981 *IEEE Trans. Magn.* **17** 1247
[20] Ginder J M, Davis L C and Elie L D 1996 *Int J Mod Phys B* **10** 3293

Journal of Physics: Conference Series **149** (2009) 012056 doi:10.1088/1742-6596/149/1/012056

A Design Strategy for Magnetorheological Dampers Using Porous Valves

W Hu, R Robinson and N M Wereley

Smart Structures Laboratory, Alfred Gessow Rotorcraft Center, Department of Aerospace Engineering, University of Maryland, College Park, MD 20742, USA

Email: wereley@umd.edu

Abstract. To design a porous-valve-based magnetorheological (MR) damper, essential design parameters are presented. The key elements affecting the damper performance are identified using flow analysis in porous media and an empirical magnetic field distribution in the porous valve. Based on a known MR fluid, the relationship between the controllable force of the damper and the porous valve characteristics, i.e. porosity and tortuosity, is developed. The effect of the porosity and tortuosity on the field-off damping force is exploited by using semi-empirical flow analysis. The critical flow rate for the onset of nonlinear viscous damping force is determined. Using the above design elements, an MR damper using by-pass porous valve is designed and tested. The experimental damper force and equivalent damping are compared with the predicted results to validate this design strategy.

1. Introduction

In conventional MR devices, flow channels are straight and electromagnetic coils are configured so that the flow streamlines are normal to magnetic field. This requires that the magnetic coil be positioned in a bobbin that is concentric to a tubular flux return. The magnetic coil is usually immersed in the MR fluid, which can reduce the life of the coil and lead to difficulties in extracting heat generated by the coil and damping action. This magnetic coil configuration limits the active length of the flow channel to be far less than the valve length. Configuration of the flow channel and flux return is also constrained by magnetic field saturation and precise MR valve flow gaps. In comparison, an MR porous valve configuration, in which MR fluids are forced to flow through voids within a packed bed of particles, has the advantage of a larger active volume. In this case, the active length can be as long as the MR valve. Shulman [1] proposed spiral channels or packed beds of particles placed inside a solenoid. In [2], a hydraulic ram was used to quasi-statically investigate the resisting force as MR fluids flowed through porous media in which the magnetic field was parallel to the average flow. Cook et al. [3] experimentally evaluated the dynamic performance of an MR damper exploiting a bypass porous valve. However, a reliable design methodology for porous-valve-based MR damper has not yet been proposed since the flow mechanisms of the MR fluid in the porous valve are not well understood.

A design strategy for developing a porous-valve-based damper is described in this paper. Firstly, based on the Darcy's law of the flow in a porous media [4], the effect of the porosity and tortuosity of the valve on the flow resistance is described, in which the condition for the onset of a nonlinear viscous force term is included. The effect of the nonlinear viscous force on the performance of the adaptive damper is evaluated. By using an empirical magnetic field analysis for the MR fluid flowing through the porous valve, the effect of the porosity and permeability of the filler material on the field

Journal of Physics: Conference Series **149** (2009) 012056 doi:10.1088/1742-6596/149/1/012056

distributions is presented. These analytical tools can be explored for damper performance evaluation. Using experimental data [3], the analytical damper behavior with different porous media structures is validated. Finally, based on the known material properties of an MR fluid and determined design parameters, an MR damper using a bypass porous valve is developed and the predicted damper performance is validated by the experimental results of the damper.

2. Damper configuration and testing setup

To verify the design strategy for porous valves, an MR damper with a bypass porous valve is developed, and the damper performance, i.e. damping capacity and controllability, is experimentally evaluated. As shown in Figure 1, the damper mainly consists of a hydraulic cylinder connected to a bypass porous MR valve. The bypass valve was fabricated around a stainless steel tube with an inner diameter of 9.4 mm and a length of 100 mm. This tube can be packed with filler materials, such as steel or plastic spheres and rods, and a magnetic coil with 50 ohms of resistance is wrapped around the outside of the bypass valve to provide magnetic filed in the valve. The piston diameter in the hydraulic cylinder is 38 mm (1.5 in) and the double-ended rod diameter is 25.4 mm (1 in). Sinusoidal displacement excitations (with amplitude at 12.7 mm and frequency up to 7 Hz) were applied to the damper on an MTS dynamic testing machine, and the applied current ranged from 0 to 1 A. The measured force and displacement were used to reconstruct hysteresis behavior, compute equivalent damping and other characteristics of the damper, i.e. yield force and viscous damping.

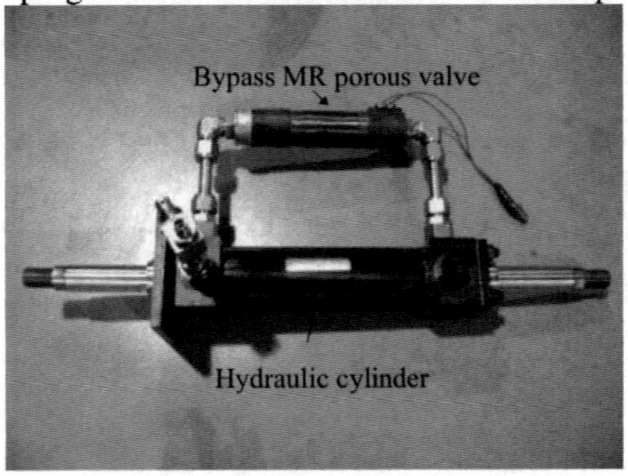

Figure 1. MR damper with porous valve

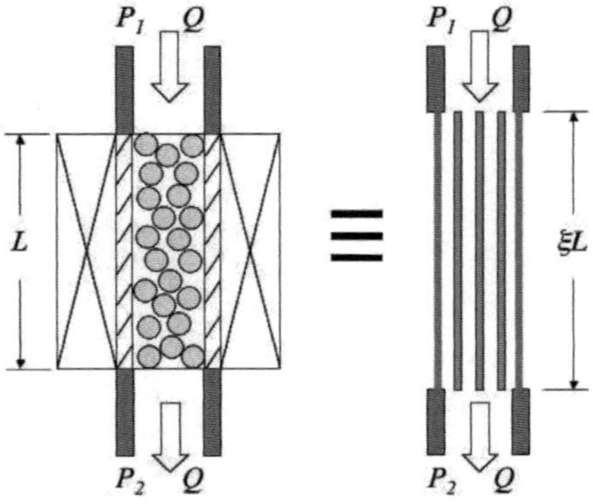

Figure 2. Equivalent flow analysis model

3. Design parameters for porous MR valve

Since the bypass valve is randomly packed with fillers, the flow of the MR fluid through the valve containing a porous media is evaluated using a semi-empirical methods. Three major parameters related to the porous media are used for fluid behavior characterization in this design strategy, that is, porosity, tortuosity and Reynolds number. The porosity, ε, is defined as the ratio of volume of the void space to the volume of the whole space in the valve. The tortuosity, ξ, is defined as the ratio of the length of an imaginary flow path in the porous media to the whole length of the valve. The Reynolds number, Re, is introduced to evaluate the onset condition of a nonlinear viscous force. To derive the flow motion, the flow path in the porous media is considered as a multiple-pipe system, in which a set of identical cylindrical tortuous channels is grouped in parallel as shown in Figure 2. For each channel, the ratio of the internal fluid volume to the internal wetted surface area is defined as a hydraulic radius, R_h, which can be related to the volume and surface area of the filler and the porosity as:

$$R_h = \frac{V_p}{S_p}\frac{\varepsilon}{1-\varepsilon}, \quad V_p : \text{volume of filler}, \quad S_p : \text{surface area of filler} \tag{1}$$

The diameter and length of each channel are then given as $4R_h$ and ξL, respectively. The behavior of the MR fluid flowing through a cylindrical channel is well described by a chain model [2]. The behavior of the MR fluid in a uniform magnetic field (of intensity *H*) can be represented by the Bingham model [5]. By combining the nonlinear Darcy's law, the pressure difference between the inlet and outlet of the porous valve as a function of the flow rate, *Q*, as shown in Figure 2 can be empirically obtained by:

$$\Delta P = P_1 - P_2 = \frac{8\xi L}{3\pi R_h}\tau_y \,\mathrm{sgn}(Q) + \frac{2\eta\xi^2 L}{R_h^2\varepsilon}\frac{Q}{A}\left[1+\left(\frac{1-\varepsilon}{\varepsilon}+\frac{\varepsilon-\varepsilon_0}{\beta}\right)\frac{\mathrm{Re}}{\xi\,\mathrm{Re}_{\mathrm{cr}}}\right] \tag{2}$$

where τ_y, η and ρ are the yield stress, viscosity and density of the MR fluid, respectively, and *A* is the cross-sectional area of the porous valve. In equation (2), the first term is a controllable resistance determined by the yield stress of the MR fluid, and the second term includes a linear viscous resistance and a nonlinear viscous resistance. The Reynolds number is defined as $\mathrm{Re} = (\rho 4 R_h Q)/(\varepsilon A \eta)$ such that the nonlinear viscous term in equation (2) is similar to a nonlinear force derived from inertial terms in the Navier-Stokes equation [4]. The constant critical Reynolds number, $\mathrm{Re}_{\mathrm{cr}}$, and nonlinear shape constants ε_0 and β are empirically determined, and $\mathrm{Re}_{\mathrm{cr}}$ represents the onset of the nonlinear viscous force.

In addition, based on an empirical equation [6], the yield stress of the MR fluid is determined to be:

$$\tau_y = 271700\psi^{1.5239}\tanh\left(6.33\times10^{-6}H_f\right) \tag{3}$$

where, Ψ is the volume fraction of the iron particles in the MR fluid. The magnetic field transverse to the MR fluid, H_f, is known to be [2]:

$$H_f = \kappa(\xi)\frac{\mu_s-\mu}{\varepsilon(\mu_s-\mu_f)}H, \quad \text{where } \mu = \left[\varepsilon\mu_f^{1/3} + (1-\varepsilon)\mu_s^{1/3}\right]^3 \tag{4}$$

In equation (4), μ_s and μ_f are the magnetic permeability of the fillers and MR fluid, respectively, and μ is an average magnetic permeability. External magnetic field, H, is known to be a function of the coil number and valve length of the porous valve, and the corrector κ is used to represent the effect of the tortuous channel on the orientation of the magnetic filed applied to the MR fluid, and then is a function of the tortuosity. Notably, the magnetic field transverse to the MR fluid is affected by the permeability of the filler materials and the porosity and tortuosity of the porous valve.

To design an MR damper with a porous valve, maximum damper force at a maximum piston velocity and controllable range (the ratio of the total force to the viscous force) are usually given as design objectives, and the MR fluid is chosen in advance. In the current design strategy, optimized ε and ξ are determined to maximize an achievable yield stress by using equations (3) and (4), and then R_h can be known by using (1). To reduce the effect of the nonlinear viscous force on the controllable range, the valve area, *A*, will be chosen by maintaining a lower Re than $\mathrm{Re}_{\mathrm{cr}}$. Thus, the valve length, *L*, and the piston area, A_p, can be determined from the required force and controllable range by using equation (2).

4. Parameter identification and validation

To determine ε_0, β and $\mathrm{Re}_{\mathrm{cr}}$, measurements of the MR damper with 3.5 mm magnetic steel beads in the porous valve [3] were used, such that the analytical force obtained by using equation (2) matched the experimental force with the same displacement excitation. Figure 3 shows an example of the force-velocity hysteresis comparison between analytical and experimental results. To validate the identified parameters, the experimental data of the MR damper with 2.0 mm and 5.5 mm magnetic steel beads in the porous valve [3] were compared with the analytical results. Equivalent damping, which was obtained by equating the dissipated energy over a cycle to the energy dissipated by an equivalent viscous damper, was used to characterize the damping performance and to evaluate the analytical and measured results. As shown in Figure 4, the equivalent damping as a function of the peak velocity for three different beads are grouped together, and the experimental results match the analytical results quite well.

Based on the design strategy and the determined parameters, an MR damper with a porous valve was developed as shown in Figure 1. To validate the design strategy, the predicted damper performance is compared with the experimental results as shown in Figures 5 and 6, where the porous media in the valve is 3.5 mm and 5.5 mm magnetic steel beads, respectively.

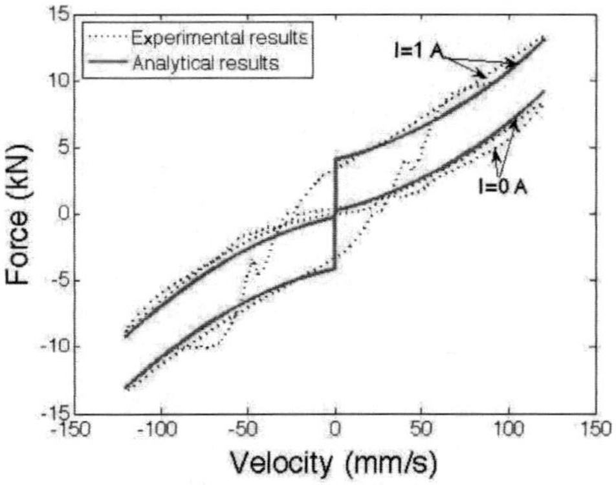

Figure 3 Force-velocity hysteresis evaluation
Bead diameter: 3.5 mm; amplitude:12.7 mm; 1.5 Hz

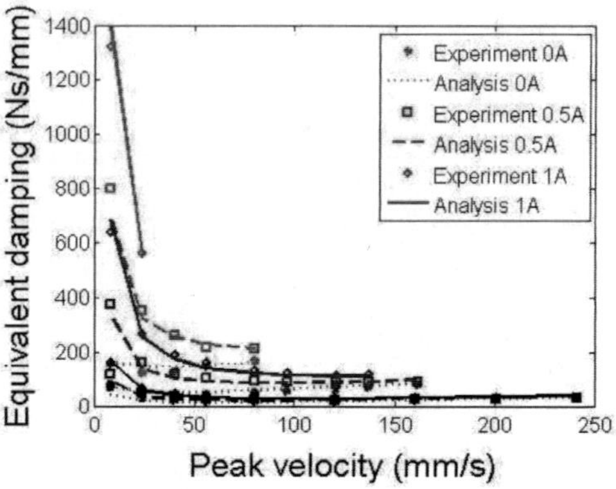

Figure 4 Damping evaluation using design strategy
Bead diameter: 2.0 mm (red), 3.5 mm (blue), 5.5 mm (black)

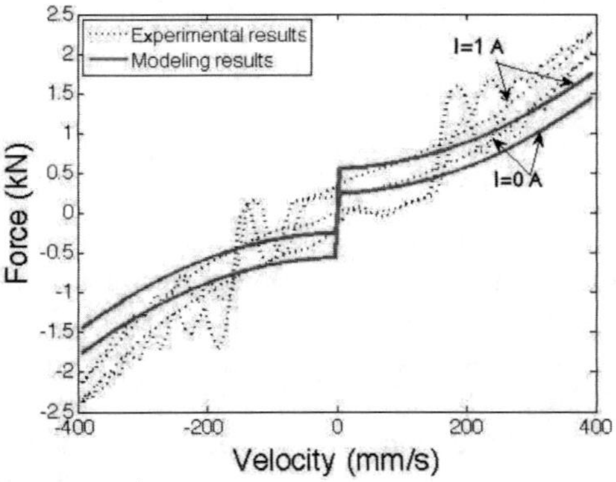

Figure 5 Force-velocity hysteresis prediction
Bead diameter: 5.5 mm; amplitude 12.7 mm; 5 Hz

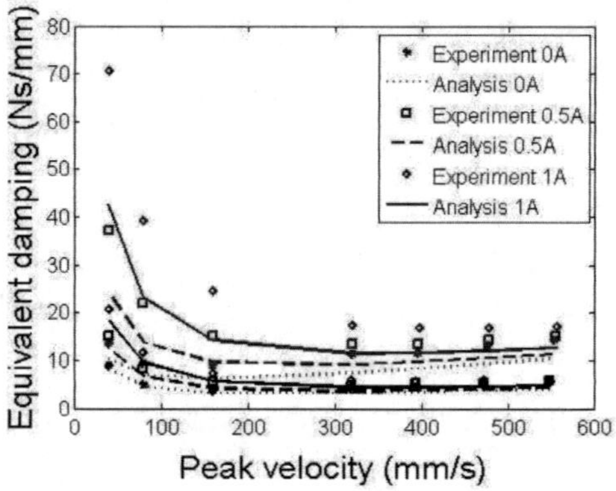

Figure 6 Damping prediction using design strategy
Bead diameter: 3.5 mm (blue), 5.5 mm (black)

5. Conclusions

A design strategy for the MR porous valve was developed. The effect of the nonlinear viscous force on the damper performance was evaluated. The parameters of the analytical model were empirically determined, and the design strategy was validated using experimental data.

References

[1] Shulman Z 1996 *Magnetic Fluids and Applications Handbook* (New York: Begell House) pp 188-229.
[2] Kuzhir P, Bossis G, Bashtovoi V and Volkova O 2003 *Euro. J. Mech. B/Fluids* **22** 331-43
[3] Cook E, Hu W and Wereley N M 2007 *J. Intel. Mater. Systems and Struct.* **18** 1197-203
[4] Bear J 1988 *Dynamics of Fluids in Porous Media*, (New York: Dover Publications)
[5] Wereley N M and Pang L 1998 *Smart Mater. Struct.* **7** 732-43
[6] Carlson J D 2004 Proc. 9th Int. Conf. on Electrorheological Fluids and Magnetorheological Suspensions (Singapore: World Scientific) pp 531-38

Journal of Physics: Conference Series **149** (2009) 012057 doi:10.1088/1742-6596/149/1/012057

Numerical Analysis of Microstructure Formation of Magnetic Particles and Nonmagnetic Particles in MR Fluids

Y Ido[1], T Inagaki[2] and T Yamaguchi[1]

[1] Department of Engineering Physics, Electronics and Mechanics, Nagoya Institute of Technology, Gokiso-cho, Showa-ku, Nagoya 466-8555, Japan
[2] Toyota Motor Corporation, 1, Toyota, Toyota 471-8571, Japan

E-mail: ido.yasushi@nitech.ac.jp

Abstract. Microstructure formation of magnetic particles and nonmagnetic particles in MR fluids is investigated using the particle method simulation. Nonmagnetic sphere particles are rearranged in the field direction due to the chain-like cluster formaion of magnetic particles. In the contrast, the nonmagnetic spherocylinder particles are not sufficiently rearranged in the field direction by using the cluster formation of sphere magnetic particles.

1. Introduction

Magentic particles of micronsize form clusters in MR uids in the presence of magnetic eld [1]. Microstructure formation of suspended particles in MR uids is very interesting problem in physics, chemistry, and engineering. Nonmagnetic particles in MR uids are also rearranged by applying magnetic eld due to the cluster formation of magnetic particles. Magnetic intelligent compound (MAGIC) is one of applications using this phenomenon [2]. In the previous papers [3,4], we have reported that the e ects of the volume fraction of particles on the distribution of suspended particles in MR uids containing nonmagnetic particles. In this paper, we report the microstructure formation of both magnetic particles and nonmagnetic particles in MR uids. We investigate the in uence of both size and shape of particles (both of magnetic particles and of nonmagnetic particles) in the presence of uniform steady magnetic eld.

2. Numerical method

A three-dimensional system to be simulated is based on a magnetic pole model for magnetic particles and the simple Stokesian model. We use a hard particle picuture without full hydrodynamic e ects through the medium such as many-body hydrodynamic interactions, lubrication force and buoyancy. In our simulations, particles are sphere or spherocylinder particles. Figure 1 shows the analytical model of our simulations. When the magnetic particles are sphere, we use the dipole model, while the poles model is used in case of the spherocylinder magnetic particles. All particles are arranged rondomly in the inital state. Uniform steady magnetic eld is applied in the z-direction and periodic boundary condition is imposed in all directions.

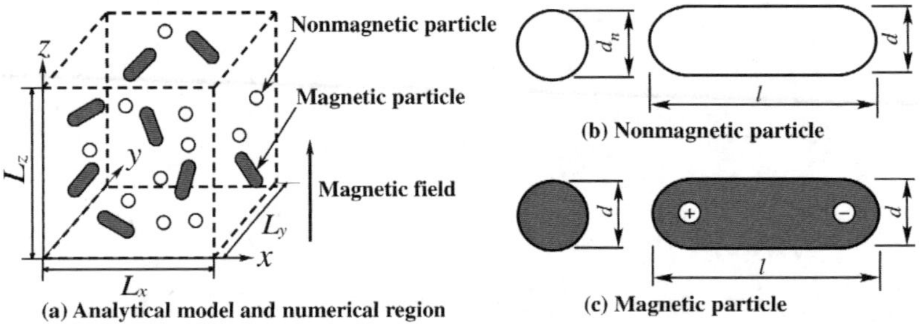

Figure 1. Analytical model of suspended particles in MR uids

The motion of the ith particle having mass m_i at time t and position $\boldsymbol{r}_i$ is described by using the following equations.

$$m_i \frac{d^2 \boldsymbol{r}_i}{dt^2} = \boldsymbol{F}_i - \frac{k_B T}{D_i^t} \frac{d\boldsymbol{r}_i}{dt} + \boldsymbol{R}_i, \tag{1}$$

$$I_i \frac{d\boldsymbol{\Omega}_i}{dt} = \boldsymbol{T}_i - \frac{k_B T}{D_i^r} \boldsymbol{\Omega}_i + \boldsymbol{N}_i, \tag{2}$$

where k_B is the Boltzmann constant, T is the absolute temperature, D_i^t and D_i^r are the di usion coe cients for translational motion and rotational motion [5], I_i is the moment of ineratia, $\boldsymbol{R}_i$ and $\boldsymbol{N}_i$ are Brownian force and torque which are independently calculated by using Box-Müller method, and $\boldsymbol{\Omega}_i$ is the angular velocity. The total force $\boldsymbol{F}_i$ and the total torque $\boldsymbol{T}_i$ for the magnetic particles are given by the following equations:

$$\boldsymbol{F}_i = \sum_{j(\neq i)} \boldsymbol{F}_{ij}^M + \sum_k \boldsymbol{F}_{ik}^{rep}, \qquad \boldsymbol{T}_i = \sum_{j(\neq j)} \boldsymbol{T}_{ij}^M + \boldsymbol{T}_i^{field}, \tag{3}$$

while the total force and the total torque for the nonmagnetic particles are given by the following equations:

$$\boldsymbol{F}_k = \sum_i \boldsymbol{F}_{ik}^{rep} + \sum_l \boldsymbol{F}_{kl}^{rep}, \qquad \boldsymbol{T}_i = \boldsymbol{0} \tag{4}$$

where subscripts i and j are the numbers of magnetic particles, subscripts k and l are the numbers of nonmagnetic particles. The magnetic force $\boldsymbol{F}_{ij}^M$ is the dipole-dipole interaction in case of spherical magnetic particles, while $\boldsymbol{F}_{ij}^M$ is the magnetic pole interaction force in case of the spherocylinder magnetic particles. The short-range repulsive force $\boldsymbol{F}_{ij}^{rep}$ is based on the DLVO theory. The torque $\boldsymbol{T}_{ij}^M$ is the magnetic torque due to the dipole interaction and $\boldsymbol{T}_i^{field}$ is the magnetic torque due to the interaction between the magnetic moment of the particle and applied magnetic eld.

The simple Euler method with a time step 10 $^{8}t_0$ is used to integrate the above equations (1) and (2), where $t_0 = m/3\ \eta d$, d is the diameter of magnetic particles, η is the viscosity coe cient.

We simulate microstructure formation of interacting magnetic particles and nonmagnetic particles. Simulations are conducted on three di erent aspect ratios of the spherocylinder particles: l/d =1(spherical particle), 2 and 3, and six di erent ratios of diameter of spherical nonmagnetic particles to that of spherical magnetic particles: 0.5, 0.8, 1.0, 1.2, 1.5 and 2.0. The volume fraction of magnetic particles is 0.16 and that of nonmagnetic particles is 0.16. We assume that the MR uid is composed of manganese zinc ferrite particles (diameter: 3 m, density: 2.5×10^3kg/m^3, saturated magnetization: 0.223T) and myristic myristate (denisty: 8.3×

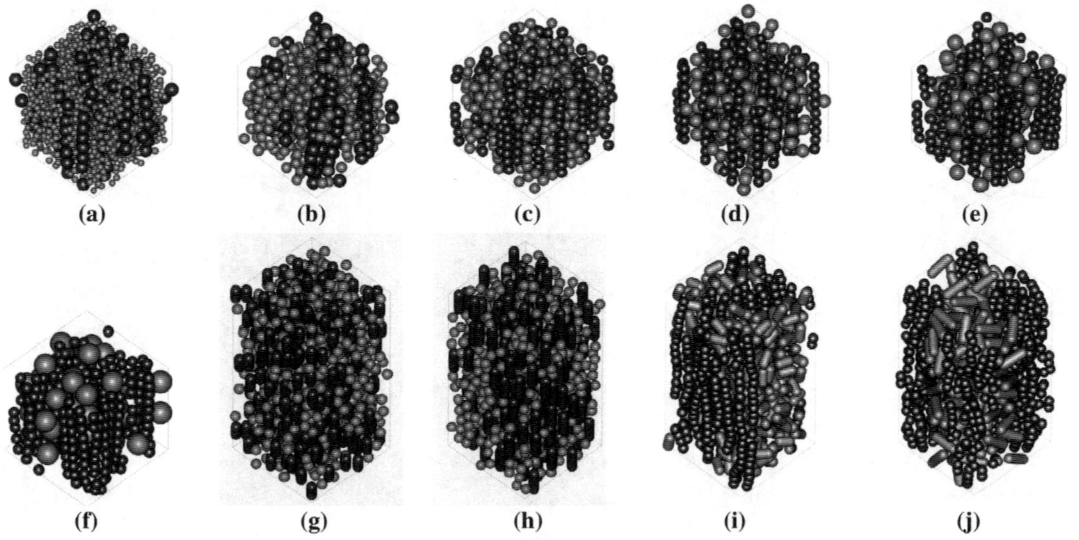

Figure 2. Snapshots of bird's eye views of distribution of suspended particles in the presence of magnetic eld. The dark particles are magnetic particles and the light particles are nonmagnetic particles. The ratio of the diameter of nonmagnetic particles to that of magnetic particles is (a) 0.5, (b) 0.8, (c) 1.0, (d) 1.2, (e) 1.5 and (f) 2.0. The aspect ratio of the spherosylinder particles is (g,i) 2.0 and (h,j) 3.0, respectively.

10^2 kg/m^3, viscosity: $1.5\times$ $10^{\ 4}$ Pa s, melting point: 40°C) and the nonmagnetic particles are made of silicon carbide (density: $3.2\ \times 10^3$kg/m^3). The applied magnetic eld and the temperature are assumed to be constant, 70mT and 70°C, respectively. In our simulations, the dependence of temperature on the distribution of particles is neglected.

3. Results and discussion

Figure 2 shows the snapshots of distribution of suspended particles in MR uids after 10^5 time steps. As we can see in gure 2, both spherical magnetic particles and spherical nonmagnetic particles form chain-like clusters in the presence of magnetic eld, while any organized structures of nonmagnetic particles cannot be seen in cases of the spherocylinder particles.

In order to analyze the distribution of particles statistically, we de ne the modiied contact coe cient C_v [1,4]. The contact coe cient is the ratio of the number of contacts between the target particle and the other particles in the eld direction against the number of contacts when all particles form chain-like clusters (e.g. if all magnetic or nonmagnetic particles form chain-like clusters, i.e., all magnetic or nonmagnetic particles are members of chain clusters, thus $C_v = 1$). The number of contacts is counted when the surface distance between the target particle and the neighbour particle is within the cut-o distance r_c. In case of the spherocylinder particles, the distance between the magnetic poles is less than $d + r_c$ and the axis of the particle is inclined within 30° to the eld direction, we count as contact in the eld dirtection. In each cases shown in gure 2, C_v becomes almost stable after 10^5 steps.

Figure 3 illustrates the contact coe cients of spherical magnetic particles ($r_c = 0.1d$) and spherical nonmagnetic particles ($r_c = 0.5d$), respectively. Figure 4 demonstrates the contact coe cients of spherocylinder particles and spherical particles. From gures 3(1) and 4(1), we have found that magnetic particles are rearranged in the eld direction and form chain-like clusters in almost all cases, because the di erences between the initial C_v and the nal C_v are

Journal of Physics: Conference Series **149** (2009) 012057 doi:10.1088/1742-6596/149/1/012057

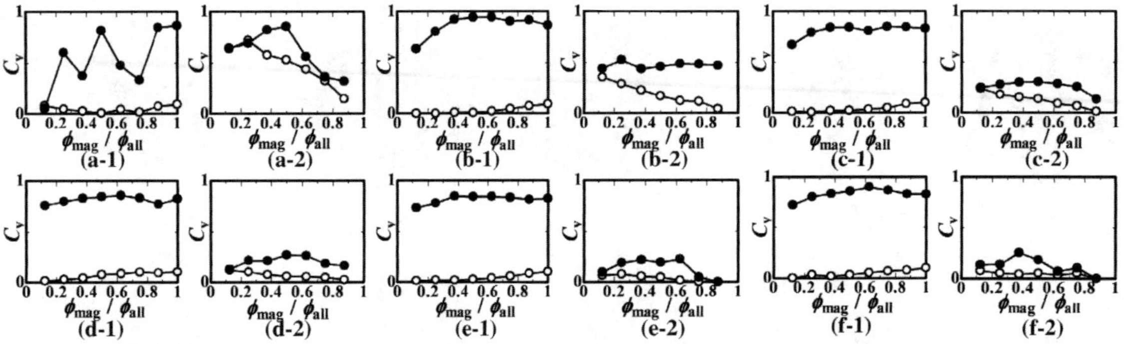

Figure 3. The modi ed contact coe cients. The ratio of diameter of nonmagnetic particles to that of magnetic particles is (a) 0.5, (b) 0.8, (c) 1.0, (d) 1.2, (e) 1.5 and (f) 2.0. The particles are (1) magnetic particles and (2) nonmagnetic particles. The open circles: initial state, and the closed circles: nal state, respectively.

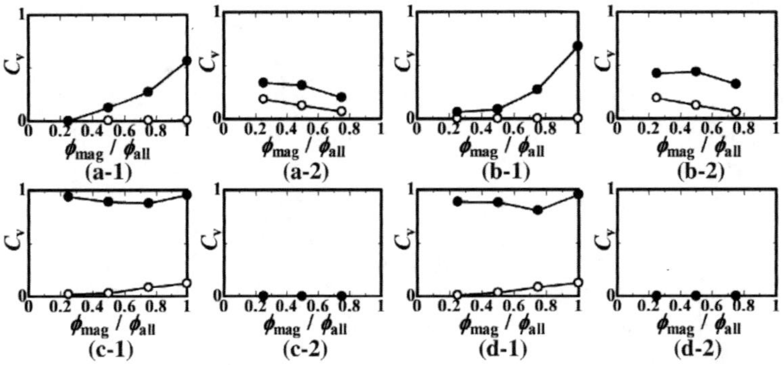

Figure 4. The modi ed contact coe cients. The aspect ratio of the spherocylinder particles is (a,c) 2.0 and (b,d) 3.0, respectively. The spherocylinder particles are (a,b) magnetic particles and (c,d) nonmagnetic particles. The open circles: initial state, and the closed circles: nal state.

large. When all particles are sphere, from gure 3(2), nonmagnetic particles are also rearranged in the eld direction. The increase of C_v is the largest when the ratio of diameter is 0.8. Thus, slightly small nonmagnetic particles are suitable in order to form chain-like clusters of nonmagnetic particles in MR uids. From gure 4(2), using spherocylinder magnetic particles is e ective for rearranging spherical nonmagnetic particles in the eld direction, while cluster formation of spherical magnetic particles cannot rearrange spherocylinder nonmagnetic particles in the eld direction.

4. Concluding remarks

Performing the simple particle method analysis, we have simulated various ordering processes and microstructure formation of both magnetic and nonmagnetic particles in MR uids. When spherical nonmagnetic particles are slightly small comparing with spherical nonmagnetic particles, more nonmagnetic particles are rearranged in the eld direction. Using spherocylinder magnetic particles is e ective to arrange spherical nonmagnetic particles in the eld direction. In the contrast, cluster formation of spherical magnetic particles is not useful to rearrange spherocyliner nonmagnetic particles in the eld direction.

Acknowledgments
This research was partially supported by Japan Society for the Promotion of Science, Grant-in-Aid for Scienti c Research (C) 18560163.

References
[1] Mohebi M, Jamasbi N and Liu J 1996 *Phys. Rev. Lett.* **54** 5407
[2] Umehara N 2003 *J. Magn. Magn. Mat.* **252** 341
[3] Zhu Y, Umehara N, Ido Y and Sato A 2006 *J. Magn. Magn. Mat.* **302** 96
[4] Ido Y, Inagaki T and Umehara N 2008 *Magnetohydrodynamics* **44** 83
[5] Torre J G and Bloomfield V 1981 *Quart. Rev. Biophys.* **14** 81

Experimental Verification of Sensing Capability of an Electromagnetic Induction System for an MR Fluid Damper-based Control System

H J Jung[1], D D Jang[1], S W Cho[2] and J H Koo[3]

[1]Dept. of Civil & Environmental Engineering, KAIST, 305-701, Guseong-dong, Yuseong-gu, Daejeon 305-701, Korea

[2]Samsung SDS Co., Ltd., Yeoksam-dong, Gangnam-gu, Seoul 135-918, Korea

[3]Dept. of Mechanical & Manufacturing Engineering, Miami Univ., Oxford, Ohio 45056, USA

E-mail: hjung@kaist.ac.kr

Abstract. This paper investigates the sensing capability of an Electromagnetic Induction (EMI) system that is incorporated in a vibration control system based on MR fluid dampers. The EMI system, consisting of permanent magnets and coils, converts reciprocal motions (kinetic energy) of MR damper into electrical energy (electromotive force or emf). According to the Faraday's law of electromagnetic induction, the emf signal, produced from the EMI, is proportional to the velocity of the motion. Thus, the induced voltage (emf) signal is able to provide the necessary measurement information (i.e., relative velocity across the damper). In other words, the EMI can act as a sensor in the MR damper system. In order to evaluate the proposed concept of the EMI sensor, an EMI system was constructed and integrated into an MR damper system. The emf signal is experimentally compared with the velocity signal by conducting a series of shaking table tests. The results show that the induced emf voltage signal well agreed with the relative velocity.

1. Introduction

The magnetorheological (MR) fluid dampers have received considerable attention for vibration mitigation of structures in the field of civil engineering because of their attractive features such as mechanical simplicity, high dynamic force range, and low operating power [1]. The semi-active control system based on MR fluid dampers should include a feedback control system including an external power supply, a controller, and sensors, which may cause maintenance and cost problems, particularly for large-scale civil engineering structures, such as high-rise buildings and cable-stayed bridges. In an effort to solve the abovementioned problems, the smart passive system, which consists of an MR fluid damper and an electro-magnetic induction (EMI) system, has been recently developed by the authors [2]. The EMI system consisting of permanent magnets and a coil is capable of generating power by converting mechanical energy (i.e., reciprocal motions of an MR fluid damper) into electrical energy (i.e., electromotive force or induced voltage) based on the Faraday's law of induction [3]. As such, the EMI was studied as an alternative power source for MR dampers to eliminate an external power supply. In a recent study, the authors demonstrated the powering ability of the EMI system by operating an MR damper device using the harvested energy [4].

In addition to the energy harvesting capability, the EMI system can provide a sensing capability. This is because the induced voltage (emf) from the EMI system is linearly proportional to the relative velocity across the MR fluid damper according to the Faraday's law. This principle, in fact, has been used in commercial, inductive velocity transducers. The EMI's sensing capability can be useful in the MR fluid damper-based control systems because some control algorithms require the measurement information on the response related to the relative velocity of the damper. As such, this study examines the sensing capability of the EMI system experimentally using a prototype EMI system.

2. Electromagnetic Induction System

Figure 1 shows the schematic of the EMI system augmented in an MR damper and the prototype of the large-scale EMI system installed in a shaking table.

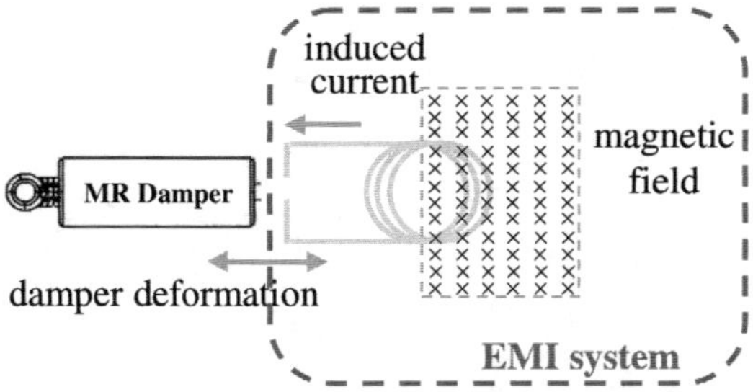

(a) Schematic of EMI system

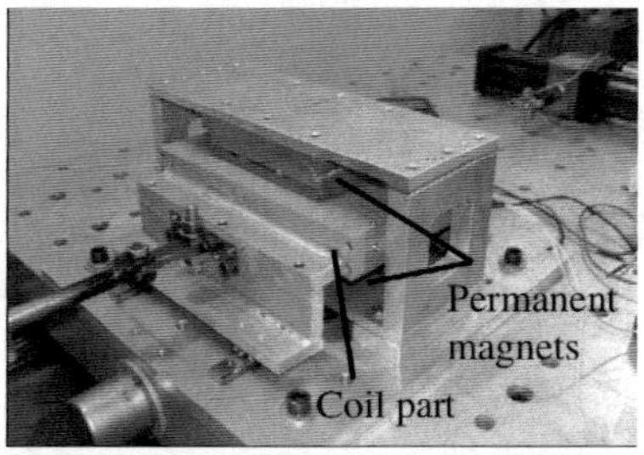

(b) Prototype of large-scale EMI system

Figure 1. The Electromagnetic Induction System

The electromotive force (or the induced voltage) produced by the EMI system can be expressed as

$$\varepsilon = -N\frac{d\Phi}{dt} = -N\frac{d(BA)}{dt} = -NB\frac{dA}{dt} = -NBw\frac{dx}{dt} \tag{1}$$

where ε is the induced electromotive force (emf) in volts; N is the number of turns of coil; Φ is the magnetic flux in webers; B is the magnetic field; A is the area of cross section; w is the coil width; and x is the relative displacement of the permanent magnet. Negative sign in equation (1) is the direction of the induced current.

As shown in equation (1), the induced emf (ε) in a closed loop is equal to the negative value of the time rate of change of magnetic flux through the loop, and it is also proportional to the relative velocity across the MR fluid damper. According to the Faraday's law, therefore, the EMI system can be considered as a velocity sensor for the MR fluid damper. It might be used in the MR fluid damper-based semi-active control systems that require the relative velocity information to implement them. Figure 2 shows the block diagram of the semi-active control system using the EMI device as a velocity sensor as well as a power supply.

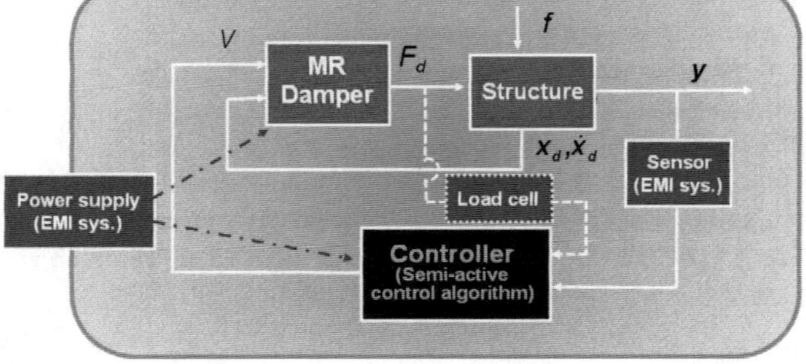

Figure 2. Block diagram of MR fluid damper-based semi-active system using EMI system

3. Experimental Validation

In order to validate the EMI system's sensing capability, a series of shaking table test are carried out under various loading conditions. Figure 3 compares the relative velocity of the MR damper along with the induced voltage signal generated from the EMI under several harmonic input excitations. In this experiment, the relative velocity of the damper is calculated by differentiating the measured displacement from the shaking table. As seen from Figure 3, the time history responses of the relative velocity of the MR fluid damper match well with those of the induced voltage from the EMI system.

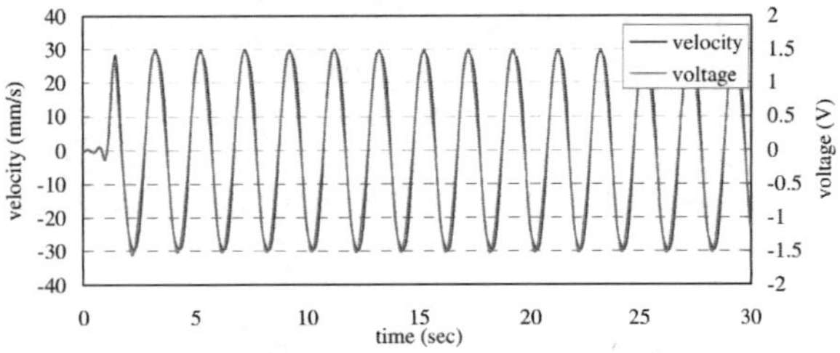

(a) 0.5Hz of exciting frequency

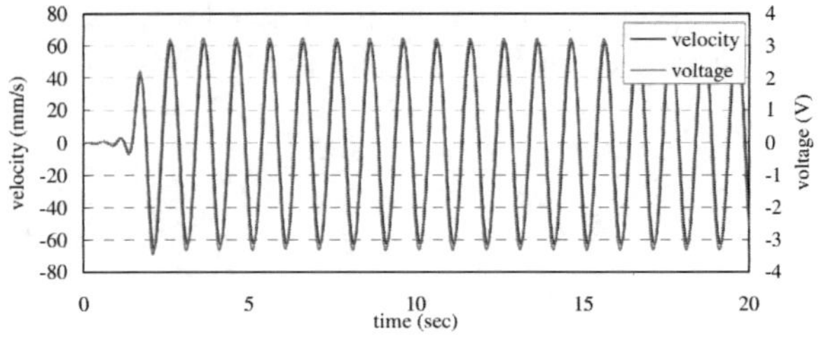

(b) 1.0Hz of exciting frequency

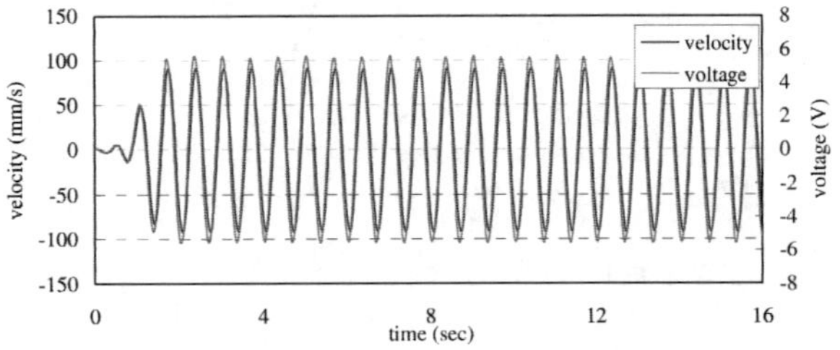

(c) 1.5Hz of exciting frequency

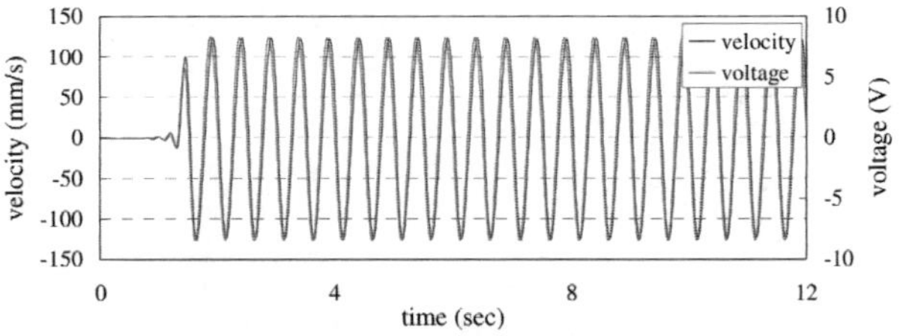

(d) 2.0Hz of exciting frequency

Figure 3. Time history responses of the induced voltage from the EMI system and the relative velocity of the MR fluid damper with various harmonic input frequencies

To further characterize the relationship between the relative velocity across the MR fluid damper and the induced voltage from the EMI system, their peak amplitude values are compared. As shown in Figure 4, the trend line of the data points is linear, indicating that the induced voltage signal is linearly proportional to the velocity signal. This result suggests that the EMI system may be used as a velocity sensor.

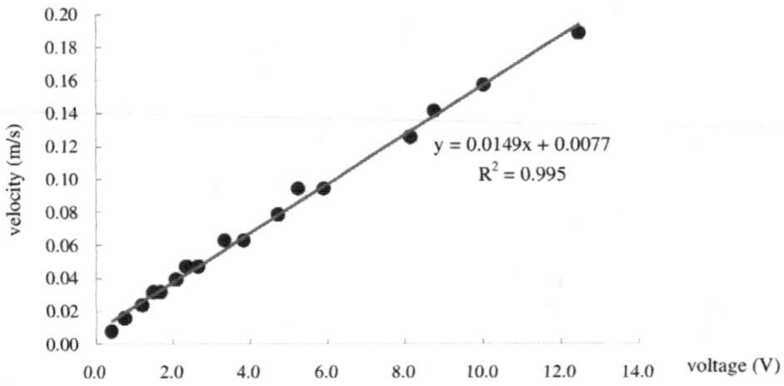

Figure 4. Relationship between the peak relative velocity and the peak induced voltage

In addition to the harmonic inputs, scaled historical earthquake excitations (El Centro and Kobe earthquakes) are used to further validate the EMI's sensing capability. Before presenting the results, it is noteworthy that some semi-active control algorithms currently being used in the MR damper-based control systems, such as the maximum energy dissipation algorithm, often require just a direction of the relative velocity across the damper (i.e., its sign change), not exact velocity amplitudes. In other words, it is sufficient to implement such control algorithms if the sign (or phase) of the induced voltage from the EMI system coincides with that of the relative velocity across the damper. Thus, the degree of agreement between signs of the induced voltage and the relative velocity, under the two scaled earthquake records, are compared in Table 1.

Table 1. Degree of agreement between signs of the voltage and the velocity (%)

Earthquake	$\lvert\dot{x}\rvert > 5\,\mathrm{mm/s}$	$\lvert\dot{x}\rvert > 10\,\mathrm{mm/s}$	$\lvert\dot{x}\rvert > 30\,\mathrm{mm/s}$
El Centro	98.55	100	100
Kobe	95.77	95.52	100

As shown in Table 1, the degree of the sign agreement is slightly velocity dependent, yet the signs of the induced voltage and the relative velocity match well, over 95 % at lower velocities. Moreover, the signs completely match when the velocity is relatively large (i.e., 30 mm/s). The results indicate that the EMI system can be used as a "relative velocity" sensor for the MR damper-based semi-active control system, employing phase based control algorithms (for example, the maximum energy dissipation algorithm).

4. Conclusions

The sensing capability of the EMI system, incorporated in the MR fluid damper-based semi-active control system, is experimentally investigated. After constructing a prototype EMI system, it is augmented to an MR damper system installed in a shaking table. A series of shaking table experiments were conducted to compare the induced emf (or voltage signal) from the EMI and the velocity signal under various harmonic and scaled historic earthquake loading conditions. The experimental results validated that the linear relationship between the induced emf and the velocity signal, as expected based on the Faraday's law of induction. Thus, the results indicate that the EMI system may be considered as a velocity sensor, albeit more thorough investigation should be required to use it as a velocity sensor. The results further show that the EMI system can be used as a "velocity-sign" sensor because the sign change of the emf signal agrees well with that of the velocity signal. This characteristic of the emf signal might be particularly useful for control algorithms that only require sign change of the velocity signals.

Acknowledgements

The work presented in this paper was supported by the Ministry of Construction and Transportation in Korea through the Construction Core Technology Research and Development Project (Grant No.: C105A1000021-06A0300-01910). This support is gratefully acknowledged.

References

[1] Dyke, S.J., Spencer, B.F., Jr., Sain, M.K. & Carlson, J.D. 1996. Modeling and control of magnetorheological dampers for seismic response reduction. *Smart Materials and Structures.* 5: 565~575.

[2] Cho, S.W., Jung, H.J. & Lee, I.W. 2005. Smart passive system based on magnetorheological damper. *Smart Materials and Structures.* 14: 707-714.

[3] Jung, H.J., Choi, K.M., Spencer, B.F., Jr. & Lee, I.W. 2006. Application of some semiactive control algorithms to a smart base isolated building employing MR dampers. *Structural Control and Health Monitoring.* 13(2-3): 693-704.

[4] Choi, K.M., Jung, H.J., Lee, H.J. & Cho, S.W. 2007. Feasibility study of an MR damper-based smart passive control system employing an electromagnetic induction device. *Smart Materials and Structures.* 16: 2323-2329.

Journal of Physics: Conference Series **149** (2009) 012059 doi:10.1088/1742-6596/149/1/012059

Compact MR fluid clutch device for human-friendly actuator

T Kikuchi, K Ikeda, K Otsuki, T Kakehashi and J Furusho

Osaka University, 2-1 Yamadaoka, Suita, Osaka 565-0871, Japan

E-mail: kikuchi@mech.eng.osaka-u.ac.jp

Abstract. In this paper, we describe a design method and experimental results of a newly developed MR Fluid clutch which has a multi-layered disks and micro-size (50 micro meters) gaps of MR Fluid. The micro-size gap works for the reduction of magnetic resistance, amount of power supply and size of the total system. Static torques of the device was predictable with conventional magnetostatic analyses. Additionally, dynamic test shows that its response time is about 20 milliseconds.

1. Introduction

Due to the increasing number of the aged population, there are strong demands for life-support robot systems, for example power-assist systems, to aid ADL (Activities of Daily Living) of the aged people. In order to develop safe and feasible robot systems that will contact human directory, we need to develop human-friendly actuators to keep firm safety at hardware level. Pneumatic power has been used in many conventional human-friendly actuators. However, pneumatic actuators have low elasticity and in many case it causes difficulties for control. Additionally, pneumatic actuators basically need compressors that cause to enlarge total systems.

In our previous researches [1], we have proposed MR Fluid Clutch-type Actuators (MRFCA). An MRFCA has three components; actuation part, reduction part and torque transmission part [1]. The MRFCA can perform high speed torque response depending on the rapid response of MR Fluids [2], and high safety thanks to the interposition of the MR Fluid Clutches (MRFC) [3] ~ [5]. From the view point of safety for human, the MRFCA is suitable for human-machine-coexistent systems like the life-support robots.

For further advancements of the MRFC, we need to improve torque / mass ratio and reduce amount of the power supply. Kavlicoglu B., et al. [4] have applied the multi-plate (multi-layered disk) structure for a high-torque MRFC. However, conventional MRFCs have millimetres-sized gap (0.5~1.0mm) of the MRF layers and it causes a high magnetic resistance. High magnetic resistance, at the same time, need high power supply for sufficient generation of the magnetic field to excite effective viscosity change of MRFs.

In this study, we describe a design method and experimental results of a newly developed MRFC which has a multi-layered disks and micro-size (50 micro meters) gaps of MRF layers. Application of the micro-size gaps works for the reduction of the size and the level of power-supply.

2. Basic Structure of Compact MRF clutch with Multi-Layered Disks and micro-size-Gaps

A conceptual drawing of a compact MRF clutch (CMRFC) is shown in Fig.1. An electric coil is rolled round an output shaft and it generates magnetic fluxes shown by dashed lines in the drawing. Multi-layered disks are fixed on the input / output shaft. The MRF is filled between these disks.

MRFs have basically low relative magnetic permeability (3.5~10) [2], and that is significantly lower than ferromagnetic materials that will be used for a magnetic circuit. For example, the relative magnetic permeability of iron is more than 1000 at least. Therefore, MRF layers are most influential sources of the magnetic resistance that will cause large energy-loss.

In many previous reports [3] ~ [5], millimeter-size (0.5~1.0mm) gaps of MRF layers have been applied. On the other hand, in this study, we utilize micro-size-gaps (50μm) to reduce the magnetic resistance of the MRF layers.

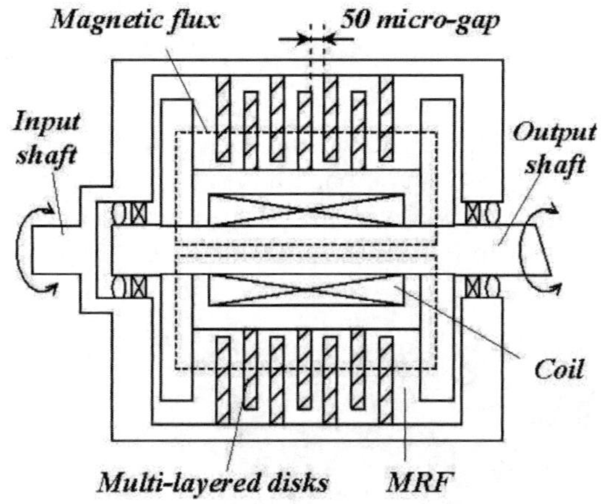

Figure 1. Basic structure of a compact MRF clutch (CMRFC) device with multi-layered disks

Figure 2. CMRFC (upper: front view, lower: rear view)

3. Development of A Compact MRF Clutch (CMRFC)

We formulated the way to estimate output torques of the CMRFC. The analysis flow for the estimation is shown as follows;

(1) Geometric design of a CMRFC with 3-D CAD software,
(2) Magnetostatic analysis for the estimation of the magnetic flux density with CAE (FEM) software,
(3) Decision of a yield stress of the MRF depending on the results of the process (2),
(4) Calculation of a maximum torque depending on the results of the process (3) and other size-parameters.

In the process (2), we need to input nonlinearity data (B-H curve) of ferromagnetic materials (silicon steal etc.) and the MRF. Commercially produced MRF (140CG, Lord Corp.) was used as an MRF in this study.

In the process (3), we referred to the characteristics data between the yield stress and the magnetic flux density of 140CG, which is presented by Lord Corp. [6].

Figures 2 show a developed compact CMRFC. Specification data of this clutch is shown in table 1. Multi-layered disks are fixed with a gap of 50μm accurately. The MRF is filled between these gaps completely. Figure 3 shows a cross-sectional view of the CMRFC. In this figure, black broken line means a loop of the magnetic flux.

Table 1. Specifications of CMRFC

Total thickness [mm]	32
Outer diameter [mm]	52
Number of disks	9
Number of MR fluid layer	18
Turning number of coil	191
Idling torque [Nm]	0.15
Maximum torque [Nm]	6.0
Weight [g]	237

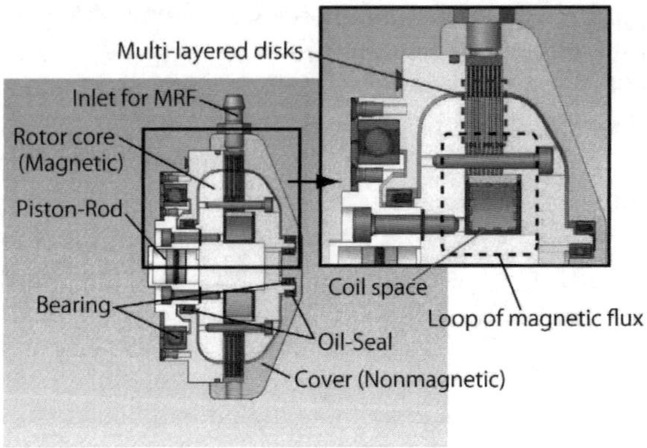

Figure 3. Sectional view of CMRFC

4. Experimental Results

In this session, we evaluate torque characteristics of the CMRFC comparing with the estimation mentioned above. An experimental setup is shown in Fig.4. Input part (body) of the clutch was fixed on a turning table and rotated by constant speed (1rad/s). A lever-arm of 250 mm was fixed on the output shaft of the clutch. We measured transmission torque by a force sensor contacting this arm with application of electric current to the coil built in the clutch.

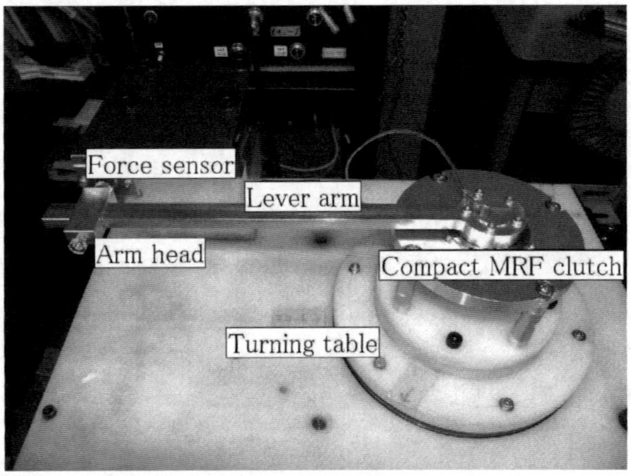

Figure 4. Experimental setup

At first, static torque tests were conducted. The experimental results are shown in Fig.5. White circles mean transmission torque under the static condition (peak torque). Black squares mean the results of the analysis mentioned in the previous session. As shown in this figure, the experimental result under the static condition indicates good similarity to the analytic result. Therefore, we can predict the maximum torque of the CMRFC with the suggested method.

Secondly, we also conducted dynamic torque tests (step-response). Time constant of step-response was about 20 milliseconds as shown in Fig.6. This results shows that the developed CMRFC has a better response time than that of conventional powder clutches. But it is not sufficient for servo-control. We have to improve dynamic characteristics with dynamic analyses in the next stage.

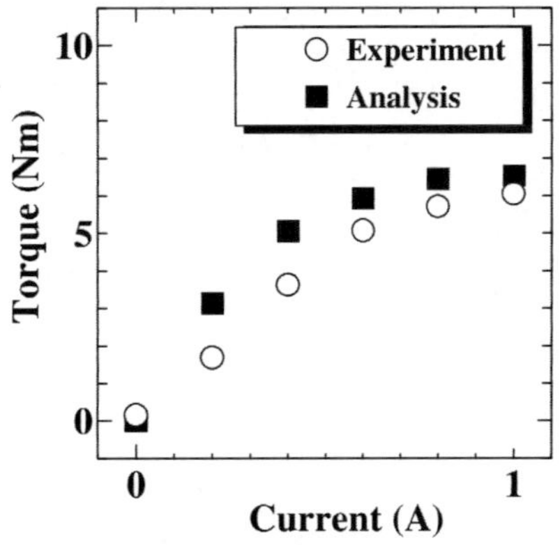

Figure 5. Static torque of CMRFC

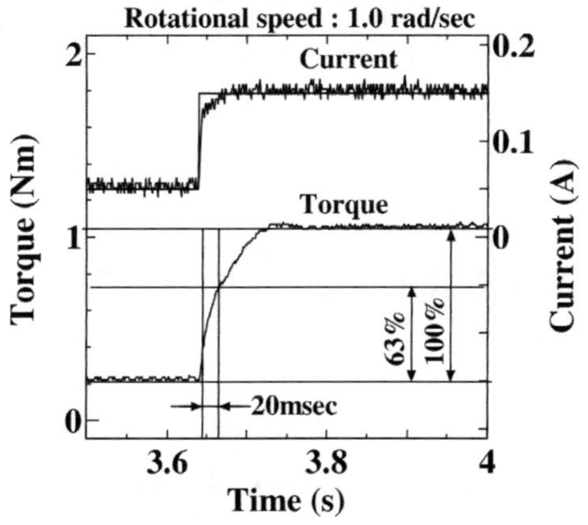

Figure 6. Step response of CMRFC

5. Conclusion

In this paper, we suggested a Compact MR Fluid Clutch (CMRFC) which has a multi-layered disks and micro-gaps (50μm) structure. Design method and experimental results were described. With conventional magnetostatic analyses and geometrical calculation methods, we could estimate static torques of the CMRFCs. As experimental results, we developed a CMRFC with compact body (237g), high torque (6Nm) and rapid response (20ms).

Acknowledgments

A part of this work was financially supported by a JAPAN Grant-in-Aid for Scientific Research on Priority Areas, No.438 "Next-Generation Actuator Leading Breakthroughs".

References

[1] Furusho J., Kikuchi T., 2007, A 3-D Rehabilitation System for Upper Limbs "EMUL", and a 6-DOF Rehabilitation System "Robotherapist", and Other Rehabilitation System with High Safety, Rehabilitation Robotics (Edited by Sashi S Kommu), Chapter 8, I-Tech Education and Publishing, 115-136

[2] Carlson J. D., 2004, MR Fluid and Devices in The Real World, Proceedings of the ninth international Conference on Electrorheological Fluids and Magnetorheological Suspensions, 531-538

[3] Lee U., Kim D., Hur N. and Jeon D., 1999, Design Analysis and Experimental Evaluation of an MR Fluid Clutch, Journal of Intelligent Material Systems and Structures, 10, 701-707

[4] Kavlicoglu B., Gordaniejad F., Ecrensel C. A., Liu Y., Kavlicoglu N. and Fuchs F., Heating of a High-torque Magnetorheological Fluid Limited Slip Differential Clutch, 2008, Journal of Intelligent Material Systems and Structures, 19, 235-241

[5] Lin J., Asaoka H., Sakaguchi M., Zhang G. and Furusho J., 2000, Applying Study Connected with Development of New Actuator Using MR Fluid, Proceedings of the 2000 Japan-USA Flexible Automation Conference, 413-416

[6] Lord Corp.: http://www.lord.com/

Leg-robot with MR Clutch to realize virtual spastic movements

T Kikuchi, K Oda, S Yamaguchi and J Furusho

Osaka University, Yamadaoka, Suita, Osaka 565-0871, Japan

E-mail: kikuchi@mech.eng.osaka-u.ac.jp

Abstract. In this study, we propose a leg-robot with an MR clutch to realize virtual haptic control for spastic movements of brain-injured patients. This system can be used in the practical training for trainees of physical therapy. Additionally, we will study to figure out the physiological mechanism of spastic movements of human with the process to simulate patient-like spastic motion by this robot. In this paper, basic structure and mechanism of the leg-robot with the MR clutch are explained. Finally, experimental results of some kinds of haptic control for spastic movements are described.

1. Introduction

"Spasticity" is a well know syndrome, most commonly arising after stroke [1]. Spasticity is defined as a velocity-dependent increase in resistance to muscle stretch. In severe case, spasticity can cause reduced flexibility, posture, and functional mobility, as well as joint pain, contracture, and difficulty with positioning for comfort and hygiene. In stroke, an increase occurs in both tonic and phasic reflexes. Loss of upper motor neuron control causes disinhibited alpha and gamma motor neuron activity and heightened sensitivity to class 1a and 2 muscle spindle afferents [2]. Consequently, monosynaptic and multi-synaptic spinal reflexes become hyperactive. In many cases, "clasp-knife phenomenon" and "clonus" appear with spasticity. However, theoretical models for such phenomenon are not unveiled.

Clinical diagnosis methods for spasticity depend on skills of physical therapists. Our purpose of this study is to develop a haptic controllable leg-robot for education of physical therapists. Our targets of this study are spastic movements of an ankle joint. Therefore, this leg-robot was developed to realize virtual haptic force and movement of its ankle joints. At same time, this robot must be developed to have same level of the mass (inertia) characteristics of human. In conventional educational setting of physical therapists, students can not have sufficient experiences to diagnose real patients with spasticity. However, by using our leg-robot, they can practice their techniques as much as they want to do. It will cause upskilling for therapists.

In this study, an MR fluid clutch works as a safety mechanism and a haptic generator of the system. The MR fluid [3] is one of functional fluids which can change its viscosity with application of a magnetic field. And they respond with a good repeatability and good rapidity. In many researches, such functional fluid devices are utilized for therapy / rehabilitation application [4], [5]. However, almost all application has targeted trainings or assists for patients. The system that enables trainings of therapeutic treatments for therapist is novel application of the functional fluids device. In this paper, basic structure and control method of the leg-robot with MR clutch are described. Additionally, two types of spastic movements are conducted with this system and suggested control methods.

2. Structure of leg-robot with MR clutch

Figure 1 shows the basic structure of our leg-robot to realize ankle-spastic movement. This system has an MR fluid (MRF) clutch between a user (e.g. therapist) and an actuator. The system with clutch-driving mechanism has several merits by comparing with direct-driving mechanism as follows;

1) The MRF clutch works as a torque limiter with a maximum transmittable torque of the clutch. Additionally, we can disconnect transmitting torque instantly when the system become uncontrollable state for the motor.

2) When the system performs a spastic movement that does not require active movement, e.g. spasticity and clasp-knife phenomenon, we can use this MRF clutch as a controllable brake without actuating the motors. In this case, this system becomes passive haptic devices with high safety.

3) When the clutch is in free-state, the system does not transmit inertias and friction force of the motor side. This makes it possible to realize weak force like relaxed joint. An MR brake (RD-2085-02, LORD Corp.) was used as a clutch by rotating its body in this prototype.

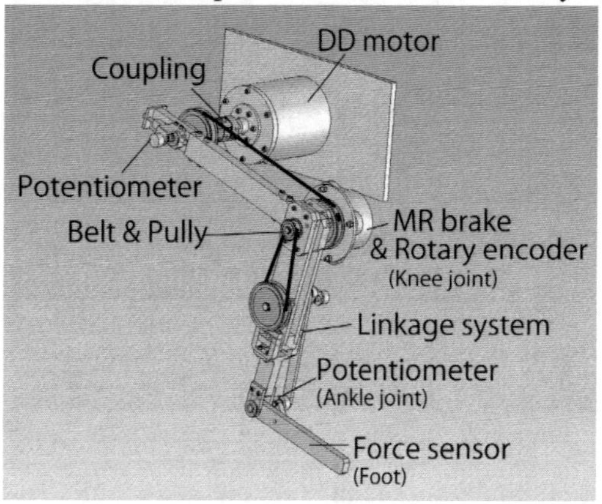

Figure 1. Basic structure of leg-robot

3. Control method

3.1. Clasp-knife phenomenon

A flow chart for “clasp-knife phenomenon” is shown in figure 2. In this process, the DD motor is servo-locked at initial position. If an operator pushes up the foot with the angular velocity of ω toward dorsiflexion, the robot outputs resistant torque depending on the level of ω. If ω is larger than ω_{clasp} [rad/s], the output torque is controlled with “clasp-knife mode” framed by broken lines of bottom side of figure 2. In clasp-knife mode, the robot fixed its ankle angle with maximum braking torque of T_{max} [Nm] (12Nm) for a time limit of t_{clasp} [s]. If the time passed over t_{clasp}, the braking torque becomes zero instantly. This algorithm is proposed by a physical therapist to express a patient-like behavior. If ω is not larger than ω_{clasp}, the output torque is controlled with “spasticity mode”, in which the resistant torque is proportional to the rotational peed of ω.

3.2. Ankle clonus

A flow chart for “ankle clonus” is shown in figure 3. In this process, the motor is servo-locked initially. However, in “ankle clonus mode”, the DD motor is rotated with constant velocity. If an operator pushes up the foot with the angular velocity of ω [rad/s] toward dorsiflexion, the robot outputs resistant torque depending on the level of ω. If ω is larger than ω_{clonus}, the output torque is controlled with “clonus mode” framed by broken lines of bottom side of figure 3. In clonus mode, the robot outputs oscillatory torque with on-off control between +3deg and -3deg with T_{max} [Nm] and 0. The operator must apply force toward dorsiflexion in order to continue the ankle clonus mode. If applied force is not sufficient to keep that, control mode becomes “spasticity mode”. In the spasticity mode, the robot outputs resistant torque depending on ankle's rotational velocity with a proportional coefficient α [Nms].

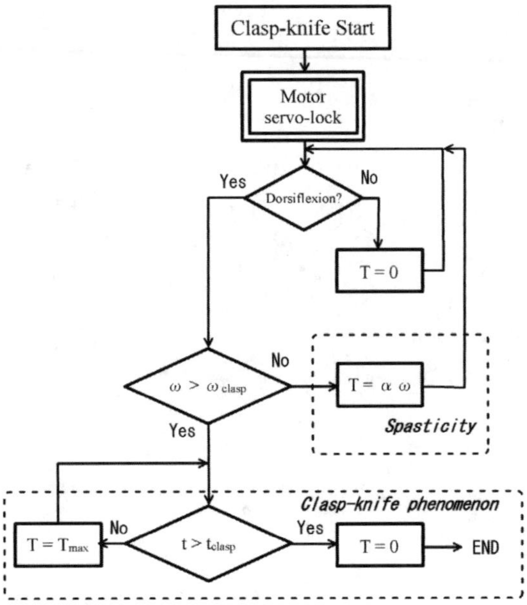

Figure 2. Control chart for clasp-knife phenomenon

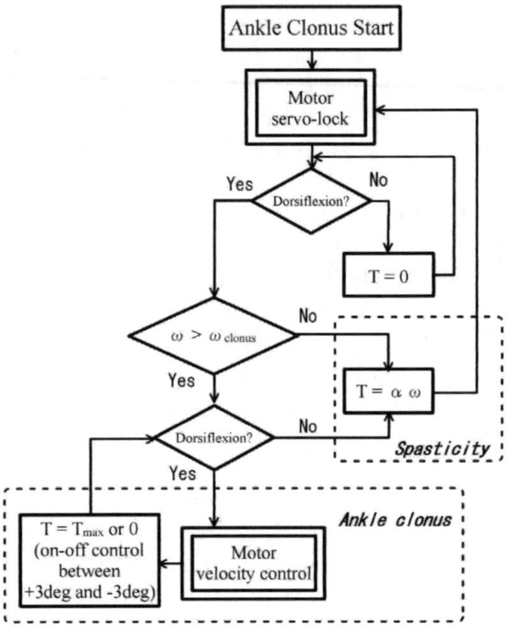

Figure 3. Control chart for ankle clonus

4. Experimental Result

The operator is a well experienced physical therapist. A scene of experiment is shown in figure 4. All variables were decided based on his subjective evaluations.

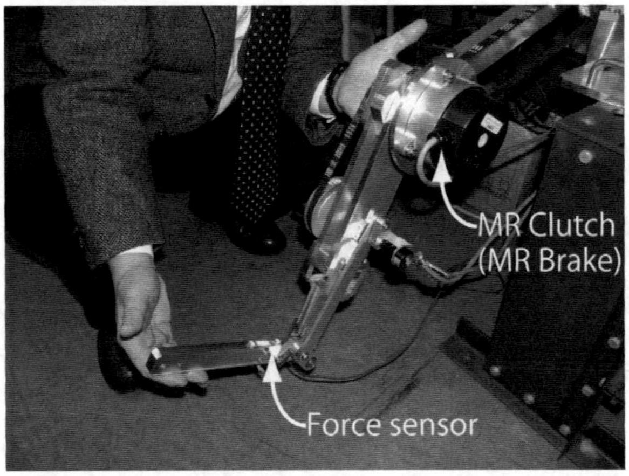

Figure 4. Scene of experiment

4.1. Clasp-knife phenomenon

Figure 5 shows an experimental result of "clasp-knife phenomenon" control. α, ω_{clasp} and t_{clasp} are set to 2.35Nms, 0.64rad/s and 2.3s, respectively. The ankle angle was stopped during the "clasp-knife mode" and after 2.3 seconds, it was released instantly. A rapid peak that appears at the end of the period shows that the foot hit a stopper.

4.2. Ankle clonus

Figure 6 shows an experimental result of "ankle clonus" control. α and ω_{clonus} are set to 2.35Nms and 0.21rad/s, respectively. The angle moves in the oscillatory manner in the "clonus mode". However, the oscillation does not continue if operators do not keep lightly pushing the foot.

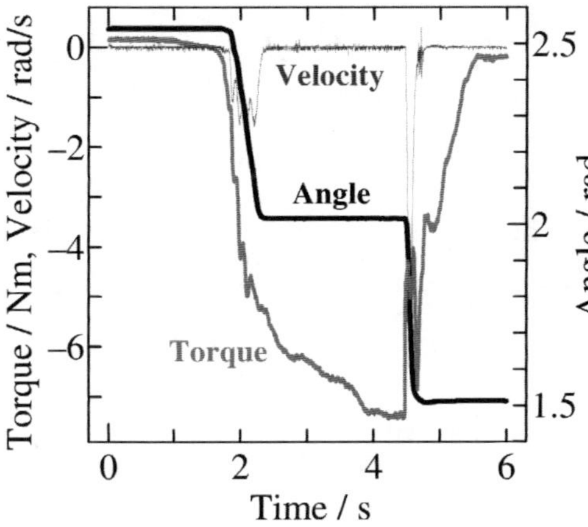

Figure 5. Experimental result of clasp-knife phenomenon

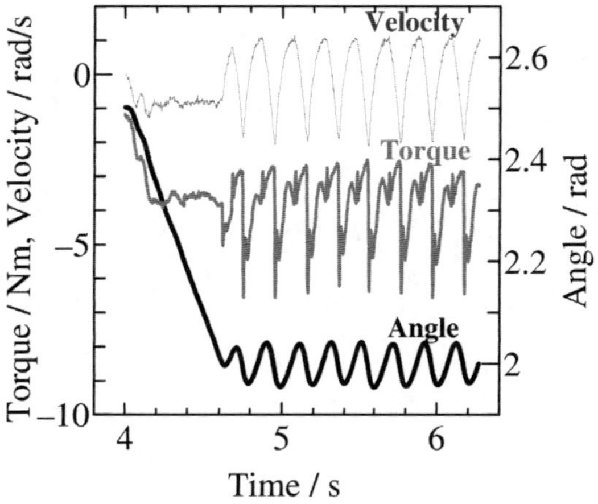

Figure 6. Experimental result of ankle clonus

5. Conclusion

In this study, we proposed a leg-robot with an MR clutch to realize several kinds of haptic characteristics of spastic movements. This robot system has high safety depending on the clutch-type actuation mechanism. Two types of the control method for spastic motions (clasp-knife phenomenon and ankle clonus) were proposed and conducted experimentally. Experimental results show good reality for these spastic motions.

Acknowledgments

This work was financially supported by a JAPAN Grant-in-Aid for Scientific Research, No.19760174.

References

[1] Volker D and Thomas S 2007 Spastic movement disorder: impaired reflex function and altered muscle mechanics, Lancet Neurology, **6**, 725-7333

[2] Lance JW 1981 Disordered muscle tone and movements, Clinical and Experimental Neurology, **18**, 27-35

[3] Carlson J D 2004 MR Fluid and Devices in The Real World, Proceedings of the ninth international Conference on Electrorheological Fluids and Magnetorheological Suspensions, 531-538

[4] Weinberg B, Nikitchzuk J et al. 2007 Design, Control and Human Testing of an Active Knee Rehabilitation Orthotic Device, Proceedings of the 2007 IEEE International Conference on Robotics and Automation, 4126-4133

[5] Dong S, Lu K et al. 2005 Rehabilitation Device with Variable Resistance and Intelligent Control, Medical Engineering & Physics, **27**(3), 249-255

Journal of Physics: Conference Series **149** (2009) 012061 doi:10.1088/1742-6596/149/1/012061

An Experimental Study on the Ultra-precision Polishing of Quartz Crystal Using MR Fluids and Micro Abrasives

D W Kim[1,2], J W Lee[1], M W Cho[1] and S B Choi[1]

[1] Micro Manufacturing Systems Laboratory, Department of Mechanical Engineering, Inha University, Incheon 420-751, KOREA

[2] Smart Structures and Systems Laboratory, Department of Mechanical Engineering, Inha University, Incheon 420-751, KOREA

E-mail: kdw525@inha.ac.kr

Abstract. This paper presents ultra-precision MR polishing results of quartz crystal, which has been widely used in many applications, such as piezo-electric transducer, surface acoustic wave (SAW) filters and SAW resonators, etc. It is known that smooth surface with sub-nanometer roughness is needed for higher-frequency application. The MR fluids, used for the polishing, consist of DI water based carbonyl iron (CI), nonmagnetic polishing micro abrasives, and required amount of stabilizers. In the process, mixed fluids were supplied into the gap between a rotating wheel (with electromagnetic field) and the workpiece. Then, the micro abrasives contained in the fluids perform material removal action from the workpiece. Such material removal mechanism in the MR polishing is considered as a process governed by the Bingham flow in the contact zone. In this study, material removal characteristics and generated surface roughness of the quartz crystal specimens using the MR polishing process were investigated through a series of experiments. The surface roughness variations of the polished specimens were investigated by changing imposed polishing conditions, such as wheel speed, magnetic field intensity. As a result, very fine surface roughness of Ra=0.770nm was obtained.

1. Introduction

In this study, variations of the machined surface characteristics of the quartz crystal glass were investigated in a polishing process using MR fluids containing micro alumina abrasives. The quartz crystal has been used in many applications such as optical devices, piezo-electric transducers, surface acoustic wave (SAW) filters and SAW resonators. It is known that ultra-precise surfaces of sub-nanometer roughness with very few surface damages are required for higher frequency applications. The MR polishing method is considered as a potentially effective solution to solve the existing problems arising in traditional polishing methods, such as pressure control, pad wear, subsurface damage, and micro crack propagation problems.

MR fluids, which are known as controllable smart materials, are suspensions of very small magnetic additives, such as carbonyl iron (CI), and nonmagnetic fluids such as mineral oils or water[1-3]. Flow properties of the fluids such as viscosity and flow rate can be easily changed using externally imposed magnetic fields[4]. Such special characteristics of the MR fluids can be applied to polishing processes for nonmagnetic parts, such as aspheric lenses, ceramics, etc. In the MR polishing process, the surface roughness and material removal rate of the workpiece are determined by the process parameters such as relative velocity, intensity of the imposed magnetic field, gap size, and used abrasives, etc. Thus, the objective of study is to investigate the effects of the process parameters on the surface roughness in the polishing of quartz crystal glass using the MR fluids and micro abrasives. A series of experiments were performed, and the machined surface qualities were measured using noncontact type measuring instrument. As a result, ultra-precise surface roughness of Ra=0.770nm was obtained using Al_2O_3 (alumina) micro abrasives.

2. Polishing principle using MR fluids

Under the imposed magnetic field, the MR fluid can compose chain-like structures via magnetization of the particles in the fluid. To use the MR fluid as a polishing media, proper abrasives should be added to the fluid: these abrasives adhere to the MR particles as shown in the Figure 1 [5]. A schematic diagram of the MR polishing mechanism is shown in the Figure 2 [6]. As shown in (a), the MR fluid is supplied to the gap between a workpiece and a moving wall (vertically rotating wheel in this study). For the polishing purposes, proper abrasive slurry is incorporated into the fluid, which is supplied to the narrow gap between the wheel and workpiece. Then, the abrasives which are adhered to the MR particles can remove required parts of the workpiece surface.

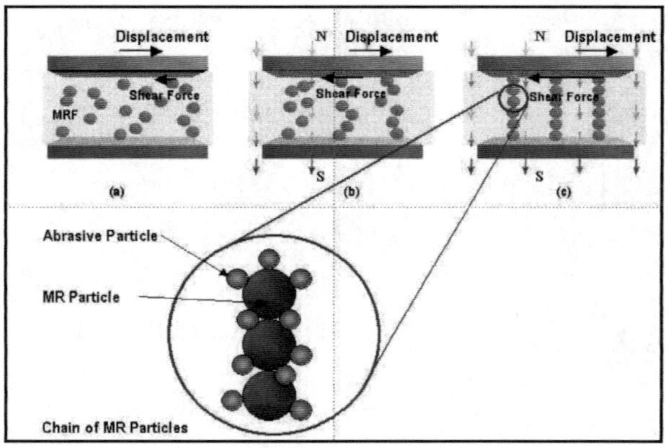

Figure 1. Chain cluster formation of MR particles and abrasives.

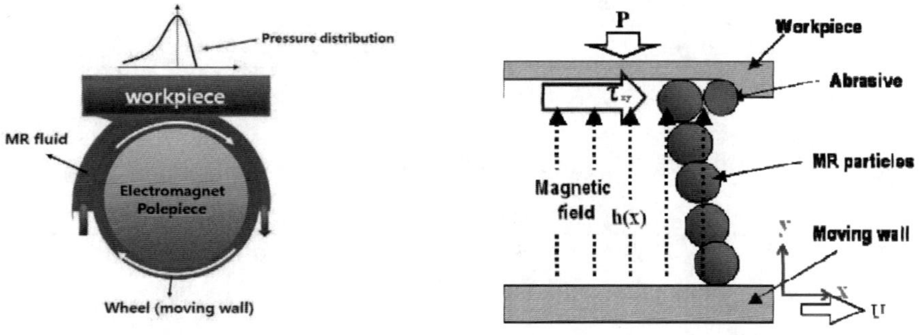

(a) MR polishing method. (b) Shear direction at polishing spots.

Figure 2. Applied MR polishing method.

3. Experimental setup

Experimental setup for the MR polishing is shown in Figure 3. The MR polishing equipment is composed of a rotating wheel with electromagnetic field and a workpiece fixture, and moving stages. A SEM image of the prepared MR fluid and abrasive slurry mixture is shown in Figure 4(a). Figure

Journal of Physics: Conference Series **149** (2009) 012061 doi:10.1088/1742-6596/149/1/012061

4(b) illustrates the particle size distribution of the used alumina powder for the polishing. In the Table 1 and 2, compositions of the used MR fluid and abrasive slurry are listed. The experiments were performed by changing the wheel rotating speeds (50~300 rpm) and magnetic field intensities to investigate the surface roughness variations.

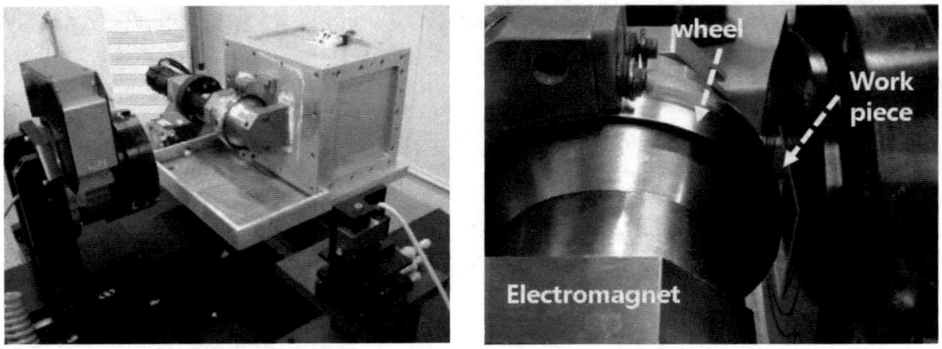

Figure 3. Experimental setup for the MR polishing.

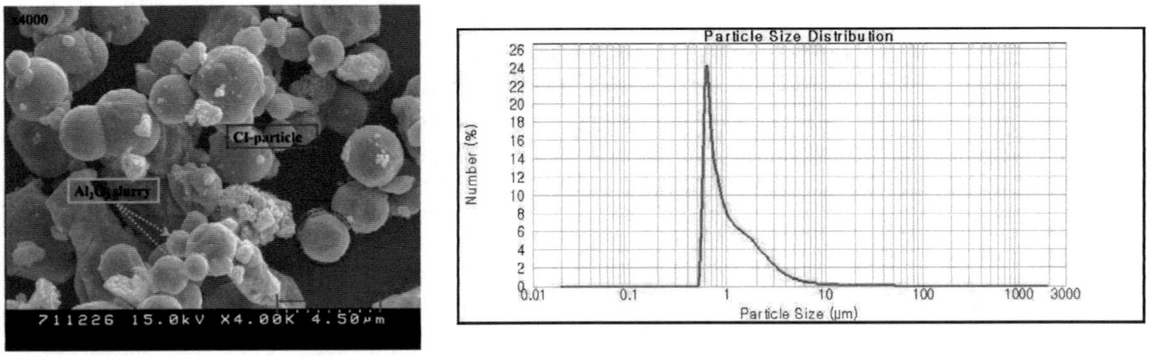

(a) SEM image. (b) Particle size distribution.

Figure 4. SEM image and particle size distribution of the alumina abrasive.

Table. 1 Composition of the used MR fluids.

CI-powder	**DI- Water**	**Na_2CO_3**	**Glycerin**
50%	48%	1%	1%

Table. 2 Composition of the used abrasive slurry.

	Al_2O_3 Abrasive	**DI- Water**
Composition	17.3%	82.7%

4. Results and discussions

Machined surfaces were measured using a noncontact type three dimensional measuring instrument (Zygo, NV-6200). Surface roughness variations according to the wheel speed and magnetic field intensity variations are shown in the Figure 5. It can be observed that the surface roughness improves with the increase of the wheel rotating speeds, and with the decrease of the imposed magnetic field intensities. As the results of the experiments, very fine surface roughness of Ra=0.770nm was obtained as shown in Figure 6. Such results insist that the MR polishing method can be an effective method for the polishing of quartz crystal glass.

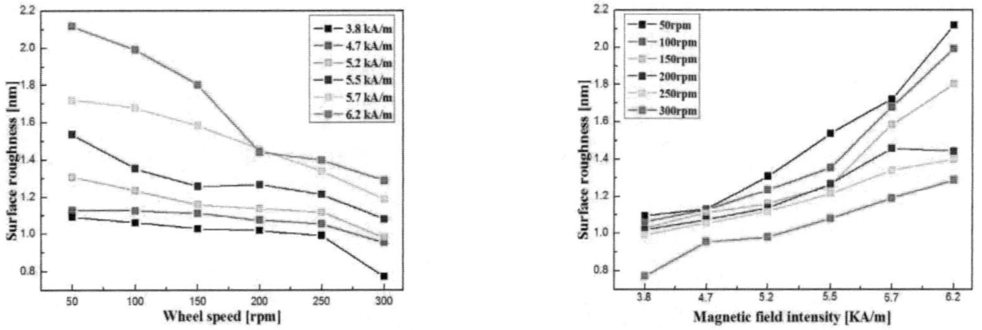

(a) According to the wheel the speed variations. (b) According to the magnetic field variations.

Figure 5. Surface roughness variations according to the applied polishing conditions.

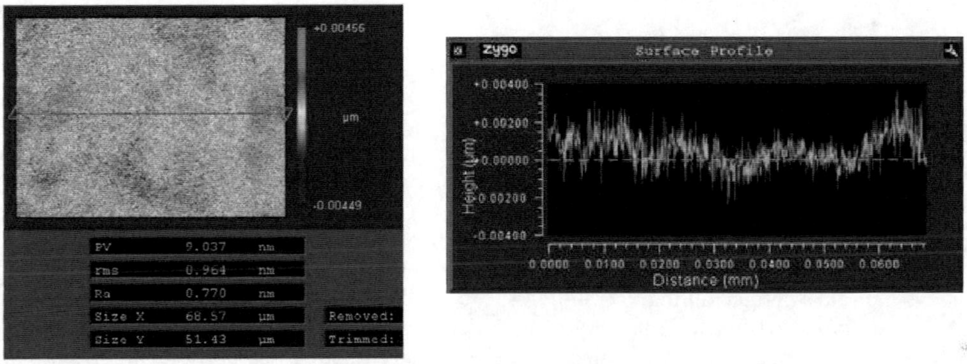

Figure 6. Measured results of the machined surfaces (Ra=0.770nm when the wheel speed=300rpm and the magnetic field intensity= 3.8KA/m).

5. Conclusions

In this study, a series of experiments were performed to investigate the ultra-precision polishing characteristics of quartz crystal glass using the MR fluid and polishing slurries with micro alumina abrasives. Surface roughness variation tendencies and polished results of the quartz crystal specimens were investigated by changing the applied process parameters such as the imposed magnetic field intensities and the rotating wheel speeds, etc. The results of this study can be summarized as follows.

- Surface roughness of the specimens improves with the increase of the rotating wheel speeds, and with the decrease of the imposed magnetic field intensities.
- Very fine surface roughness of Ra=0.770nm was obtained.
- MR polishing can be a very effective method for the polishing of quartz crystal glass.

References

[1] Kordonski W I and Jacobs S D 1996 *Int. J. Modn. Phys.* B **10** 2837-48
[2] Kordonski W I and Golini D 1999 *Int. J. Modn. Phys.* B **14, 15 & 16** 2205-12
[3] Kim D W, Cho M W, Seo T I and Shin Y J 2008 Sensors **8** 222-235
[4] Kordonski W I and Golini D 1999 *J. Int. Mate. Sys. and Strus.* **10-Sep** 683-689
[5] Shorey A B, Jacobs S D, Kordonski W I and Gans R F 2001 *Applied Optics* **40** 20-33
[6] Shorey A B, Kordonski W I, Gorodkin S R, Gorodkin S D, Jacobs S D, Gans R F, Kwong K M and Farny C H 1999 *Rev. sci. instum.* **11** 4200-06

Journal of Physics: Conference Series **149** (2009) 012062 doi:10.1088/1742-6596/149/1/012062

Smart Mouse: 5-DOF Haptic Hand Master Using Magneto-Rheological Fluid Actuators

K H Kim[1], Y J Nam[1], R Yamane[2] and M K Park[1]

[1] Graduated School of Mechanical Engineering, Pusan National University, Busan 609-735, Korea
[2] Kokushikan University, 4-28-1 Setagaya, Setagaya-ku, Tokyo 154-8515, Japan

E-mail: mkpark1@pusan.ac.kr

Abstract. This paper is concerned with a haptic hand master intended to display force feedback at the fingertip of the human user. The haptic hand master, named 'Smart Mouse' has two significant differences from previous other hand masters: actuators and structure. Five passive actuators featured with magneto-rheological fluid are used to ensure the interface safety and the control stability. In order to eliminate the unnecessary reaction force and reduce muscular fatigue during operating, the mouse-like structure is adapted. Also, two assumptions are proposed for the simplicity of its kinematics and manufacturing; one is that the fingertips lies on a plane during grasping the objects, and the other is that the equilibrium point of the fingertip reaction forces is in the object. Due to these characteristics, the human hand operating the smart mouse has the kinematic configuration similar to a 5-DOF parallel manipulator.

1. Introduction

Human hands enable dexterous manipulations such as handling tools, fabrications, and drawings. Especially, multi-fingered manipulations are highly adaptable to extreme environments including nuclear reactors, underwater or space. Therefore, haptic devices with multi-finger inputs are attracting much attention in these days. Generally, such haptic devices are named the 'haptic hand master'. For past decade, several haptic hand masters have been developed and now in practice [1-6]. The conventional haptic hand masters can be generally divided into two categories: the ground-based type and the body-based type. Ground type devices are fixed to its environment such as desk, ceiling or wall. The weight of the devices can be compensated. However, because they are mainly using a pen or a ball, it is difficult to develop devices with multi-finger inputs. On the other hand, the body-based type haptic devices have their ground on the operator's body. Since the shape of the device is very much like a glove, the operator can manipulate it intuitively and portably. However, the operator should bear the weight of device, leading to physical fatigue in long-period use. Therefore, the authors have considered that its mobility on the plane and implementability of the multi-fingered inputs make the mouse-like shape device more attractive. According to the locations of the actuators, the haptic hand masters can be again classified into two categories: the exoskeleton type and the endoskeleton type. Most of the existing exoskeleton type devices provide forces onto the fingertips by pulling them backwards, either having the actuators integrated in the exoskeleton or remotely located. On the other hand, the actuators in the endoskeleton

Journal of Physics: Conference Series **149** (2009) 012062 doi:10.1088/1742-6596/149/1/012062

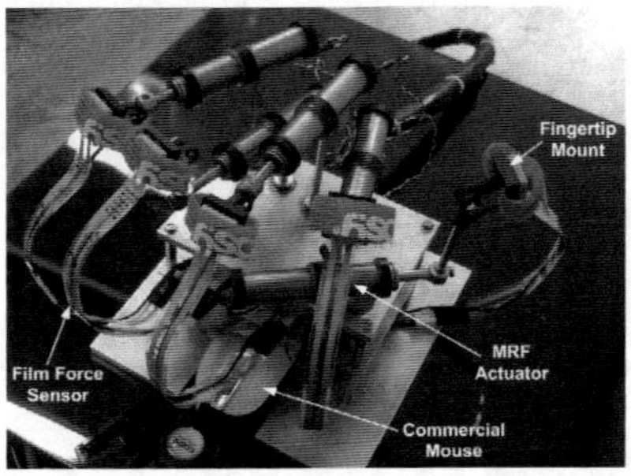

Figure 1. External appearance structure of the smart mouse

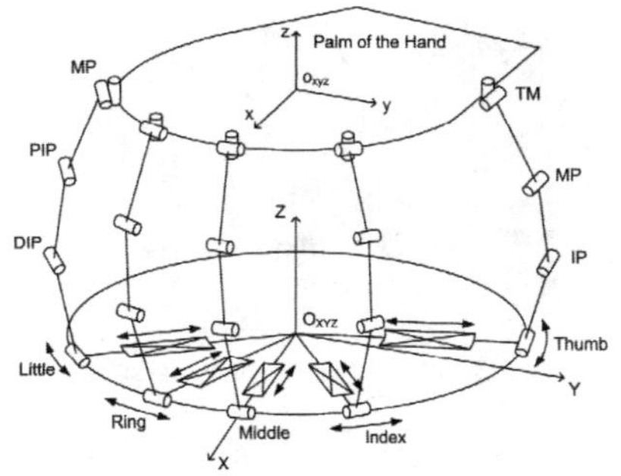

Figure 2. Schematic configuration of the human hand manipulating the smart mouse

type devices provide forces onto the fingertips by pushing or resisting against them. Unfortunately, both types do not feel free from unrealistic constraint forces that disturb the operator. For the selected concept in this paper, its actuators are not placed on the human body such as the palm of the hand or the wrist, but on the mouse-like haptic device. This consideration can make them independently attached to the corresponding fingers without any constrained forces. In the conventional haptic hand masters, force feedback mechanisms can be divided into two categories: the active type and the passive type. The active type displays force feedback using actuators such as electromagnetic motors or hydraulic/pneumatic cylinders. However, this type gives potential danger to the operator because the actuators can move independently regardless of the operator's intentions. On the other hand, brakes and crutches are mainly applied for the passive type. Actuators in this category are not activated until the operator manipulators the devices. Therefore, it is possible to display a large force feedback without danger. Obviously, force patterns that can be displayed by the passive force mechanism are limited. Nevertheless, this type seems to be more promising for the proposed haptic hand master, in the viewpoints of the interface safety and the haptic stability. Consequently, this work proposes the new type device, which is called the 'Smart Mouse' (figure 1). The proposed device has the following characteristics. (i) The mouse-like shape for mobility and portability. (ii) The endoskeleton-type attachment of the actuators for haptic fidelity. (iii) The passive force feedback for interface safety and haptic stability.

2. Mechanisms of the Smart Mouse

In order to develop the mechanism of the smart mouse, the following anatomical features of the human hand in grasping a real object are considered; (i) Because fingertips are mainly used for the dexterous tasks, only the positions and reaction forces of each fingertip are required to be measured. (ii) Obviously, humans have a certain functional position. The most appropriate position for grasping or manipulating of the human hand can be considered in the case that all of the fingertips are placed on a certain plane. (iii) The reaction forces from the object act against the finger's flexion, which is generated by the muscles and tendons in the hand and the forearm. Here, it can be sensibly assumed that these reaction forces are normal to the contact surface. (iv) Statically stable grasping can be achieved only when the equilibrium point of the reaction forces is within the object, but not on the body such as the palm of the hand or the wrist. In these anatomical viewpoints, the kinematic model of the smart mouse is proposed as shown in figure 2. Due to these characteristics, the human hand in operating the smart mouse has the kinematic configuration similar to a 5-dof spatial parallel manipulator. In order words, it can be considered that the plane on which the actuators are placed is the base platform of the manipulator, the palm of the hand is the moving platform, and each of the

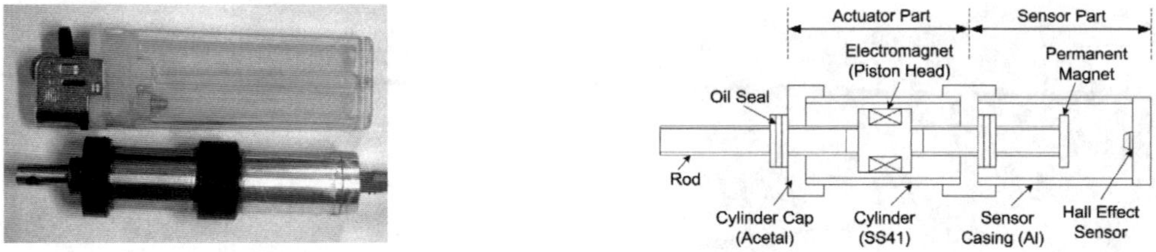

Figure 3. External appearance of the MR fluid actuator and its schematic configuration

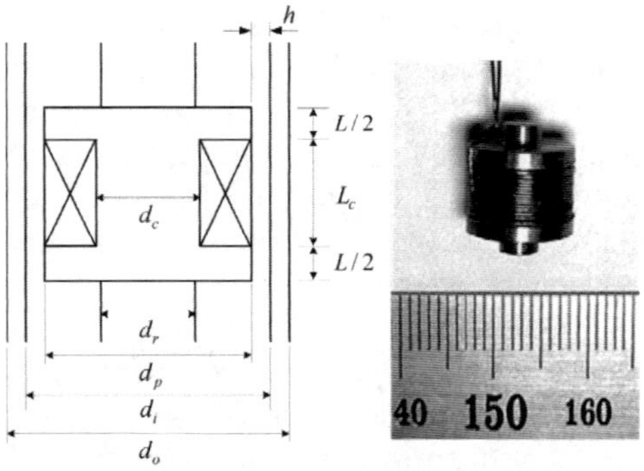

Parameters	Symbol	Value
Desired Max. Force	$F_{out,\ max}$	15N
Available Max. current	I_{max}	3A
Gap for MR fluid	h	1mm
Coil turn number	N	60
Inner diameter of cylinder	d_i	13mm
Outer diameter of cylinder	d_o	15mm
Diameter of rod	d_r	5mm
Diameter of magnetic core	d_c	7.4mm
Diameter of piston head	d_p	11mm
Length of magnetic pole	L	2.6mm
Length of magnetic core	L_c	5.4mm
Coil diameter	-	0.35mm

Figure 4. Design specification of the MR fluid actuator

fingers is the connecting limb with four passive revolute joint. Here, the actuators play a role to control the position of the 1-dof revolute joint on the plane as well as to provide the reaction force to the fingertip. For 5-dof spatial parallel manipulator with five actuators, the position/orientation of the moving platform and the positions of all the passive joints can be analytically determined from the given position of all the actuator. Based on this fact, the spatial configuration of the human hand in operating the smart mouse can be obtained only by using five actuators.

3. Structure of the Smart Mouse

3.1. Actuator: Miniature MR Fluid Actuator

For the passive force feedback of the smart mouse, the magneto-rheological (MR) fluid actuator is adopted. MR fluid is a non-colloidal solution which is composed of ferromagnetic particles with micrometers in diameter dispersed in a nonconductive carrier fluid. The MR fluid has the characteristic that their rheological properties are continuously and reversely changed within several milliseconds solely by applying or removing external magnetic fields. Due to this feature, devices using the MR fluid are simple in construction, high in power and low in inertia, compared to conventional actuation methods. Especially, MR actuators can provide much safety in equipment interacted directly with human, due to its inherent stabilizing effect as a passive device. The MR fluid actuator proposed in this work and its schematic configuration are shown in figure 3. This actuator is designed with the mechanic and electromagnetic viewpoints. This is composed with a cylinder, two piston rods and a piston head. The ferromagnetic material for the cylinder and piston head is mild steel, SS41. The piston rods are made with stainless steel which does not influenced by magnetic field. The MR fluid is MRF-132AD composed by LORD Corporation. When the piston head moves, the MR fluid is moved through the annual gap between the cylinder and the piston head. When an external current is supplied to the electromagnetic coil, the corresponding magnetic field is applied to the MR fluid in the gap. Then, the dynamic yield stress of the MR fluid is changed depending on the magnetic field intensity, and the resultant output force is activated in the opposite direction to the motion of the piston head.

Therefore, the reaction force of the MR fluid actuator can be continuously and reversely controlled by adjusting the applied coil current. The full stroke of the actuator is 20mm and its weight is about 52g. The further specifications and design parameters are shown in figure 4.

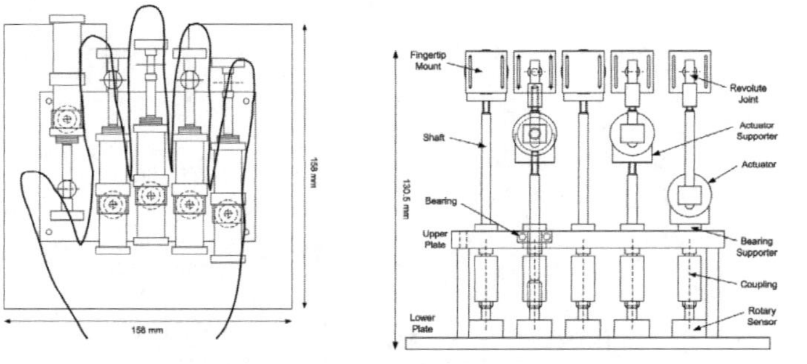

Figure 5. CAD drawings of the smart mouse (left: top view, right: front view)

3.2. Sensor

As shown in figure 3, the displacement sensor for the actuator consists of a sensor casing, a Hall-effect sensor, and a permanent magnet. When the permanent magnet attached to one end of the piston rod is close to the Hall-effect sensor, the magnetic flux density between them is increased. Therefore, the displacement of the piston head can be measured with Hall-effect sensor which outputs the voltage corresponding to the magnetic flux density. The sensor case is made of aluminum which does not make the magnetic field disturbed. In a preliminary experiment, it was observed that the output voltage of the Hall-effect sensor is nearly linear to the displacement of the piston head. The position of each fingertip in operating the smart mouse is different depending on the size and shape of the grasping object. For this consideration, the actuators are installed on the base platform with the revolute joint. In order to obtain all positions of the finger joints from the actuator displacements, the rotational positions of the actuators are required to be known. Therefore, the revolute joint for the actuator is connected with the rotary potentiometer which can output the maximum voltage of 5V with respect to the maximum rotational angle of 270 degree. For measuring the reactive force applied to the fingertip and achieving the feedback loop in the force control, the film-type force sensor (Flexiforce®) is installed on the other end of the piston rod. This sensor is characterized by its electrostatic capacity depending on the applied pressure.

3.3. Body

Figure 5 shows the CAD drawing of the proposed device. In order to avoid the mechanical interference between the actuators during rotating, adjacent actuators are installed on the different layer to each other, and therefore the device has three layers. However, the actuator on lower layers has an extension rod in such a way that all fingertips can be placed on the top layer. Considering that the direction of the reactive force for the thumb is mostly opposite or perpendicular to other four fingers during grasping, the rotational joint of the thumb actuator is positioned at the opposite side to the human hand for its functional position. Instead, the structures of the four fingers other than the thumb are almost the same as the human hand, and are distributed parallel to each other. At the lowest part, a real computer mouse is adopted to provide the ground position of the smart mouse in the virtual environment.

4. Performance Evaluations

4.1. MR Actuator

An experimental system was constructed for the performance evaluation of the MR fluid actuator. A motor-cam system operates the actuator, and its reaction force and displacement are measured with the load cell and the laser displacement sensor, respectively. For the static characteristic, the coil current at is applied a specified level and MR fluid actuator is oscillated in a sinusoidal waveform. Figure 6 shows that the reaction force of the proposed MR fluid actuator can be effectively controlled with the coil current.

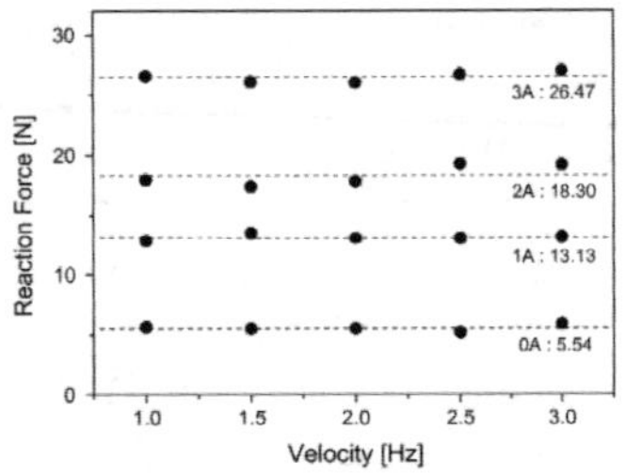

Figure 6. Static response of the MR actuator

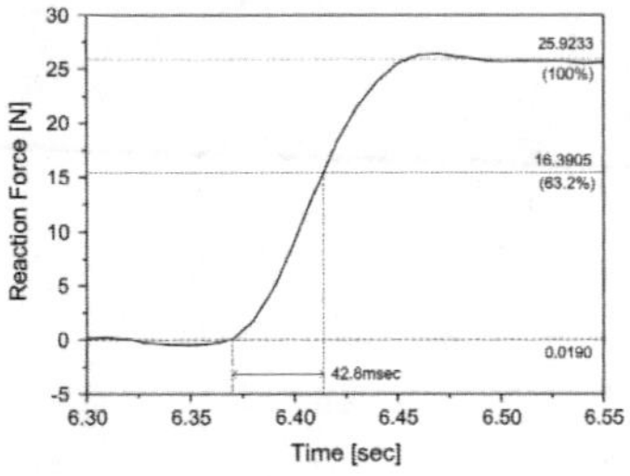

Figure 7. Step response of the MR actuator

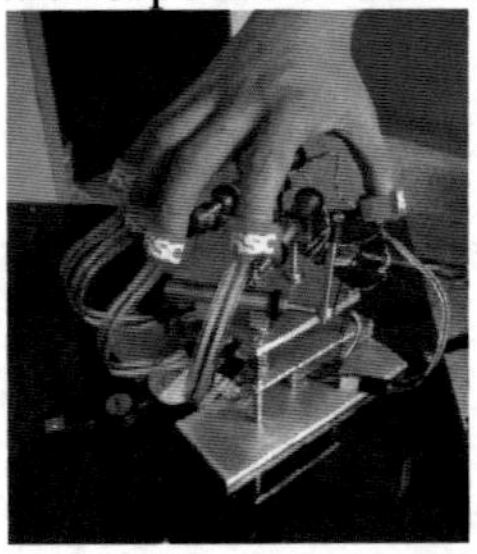

Figure 8. Manipulation of the smart glove

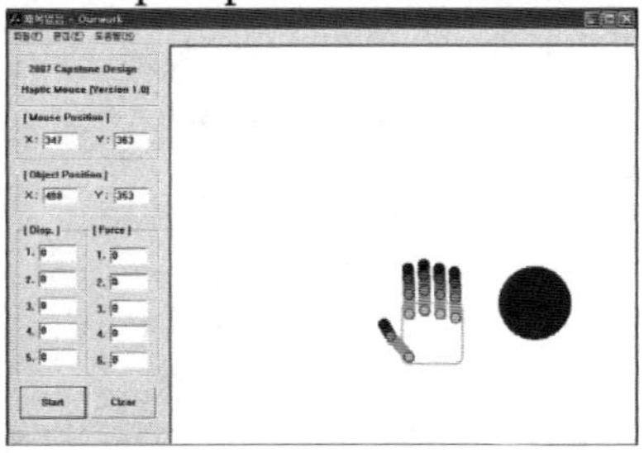

Figure 9. Virtual Environment

Considering that general human finger generates 15~20N, the MR fluid actuator (the maximum output of 26N at 3A) is applicable to the smart mouse. Figure 7 shows the output force histories of the MR fluid actuator with respect to the step input current of 3A. As the actuator is compressed by the motor-cam system (1rpm), the current is input. In general, actuators using MR fluid display a dynamic behavior similar to the first-order linear system where the time constant is defined as the time required for the output to reach 63.2% of its saturated value after applying the input. Based on this fact, the time constant of the proposed MR fluid actuator is found by about 42.8msec. Considering that most of the commercial available MR fluid actuator's time constant is about 50msec, it can be seen that the proposed MR fluid actuator has good response characteristic. [7] Therefore, the applicability of the proposed actuator to smart mouse is verified.

4.2. Interface Environment

In order to evaluate the haptic performance of the smart mouse, 2D virtual environment is constructed under Visual C++ environment. Figure 8 shows the human hand in manipulating the smart glove interacted with the virtual environment, and figure 9 shows the virtual environment consisting of the simply made virtual hand and the virtual object with three different shapes (circle, triangle or quadrangle) and sizes (10, 12 or 15 cm in enveloped diameter) which is selectable on demands. Even though the spatial kinematics of the smart mouse can be actually determined, its top-viewed configuration is reflected on the virtual hand for the convenience of the implementation. The interaction of the smart mouse with the virtual hand on computer is achieved by using the digital signal processor with five DA channels for controlling the reactive force of the actuators and fifteen AD channels for measuring the Hall-effect sensors, the rotary potentiometers and the force sensors. Therefore, the human user can move the virtual object as well as perceive the touch sensation by manipulating the smart mouse. For further interaction with the virtual environment, the human user can use the visual information displayed at the left side of the viewer, such as the position of the smart mouse, the actuator displacements and the reaction forces in real time. From several subjects volunteering for evaluating the haptic performance, it is drawn that the smart mouse can reflect the motion of the human hand and display the voluminous feeling and touch sensation quite effectively.

5. Conclusion

The haptic hand master, called 'Smart Mouse' has two significant differences from the previous haptic hand masters: actuators and structure. Five passive actuators featured with magneto-rheological fluid are used to ensure the interface safety and the control stability. In order to eliminate the unnecessary reaction force and reduce muscular fatigue while operating, the mouse-like structure is adapted. Also, two assumptions are employed for the simplicity of the kinematics and manufacturing; one is that the fingertips lies on a plane while grasping the objects and the other is that the equilibrium point of the fingertip reaction forces is in the object. Due to these characteristics, the human hand operating the smart mouse has the kinematic configuration similar to a 5-dof parallel manipulator.

6. References

[1] Ueda Y and Maeno T 2004 "Development of a Mouse-Shaped Haptic Device with Multiple Finger Inputs," *Proc. of IEEE/RSJ Int. Conf. on Intelligent Robots and Systems* Sendai Japan 2886-91

[2] Shields B L, Main J A, Peterson S W and Strauss A M 1997, "An Anthropomorphic Hand Exoskeleton to Prevent Astronaut Hand Fatigue During Extravehicular Activities," *IEEE Transactions on Systems, Man, and Cybernetics-Part A* **27** 5 668-73

[3] Koyama T, Yamano I, Takemura K and Maeno T 2002, "Multi-Fingered Exoskeleton Haptic Device Using Passive Force Feedback for Dexterous Teleoperation," *Proc. of IEEE/RSJ Int. Conf. on Intelligent Robots and Systems* Lausanne Switzerland 2905-10

[4] Nakagawara S, Kajimoto H, Kawakami N, Tachi S and Kawabuchi I 2005, "An Encounter-Type Multi-Fingered Master Hand Using Circuitous Joints," *Proc. of IEEE Int. Conf. on Robotics and Automation* Barcelona Spain 2667-72

[5] Winter S H and Bouzit M 2006, "Testing and Usability Evaluation of the MRAGES Force Feedback Glove," *Proc. of Int. Workshop on Virtual Rehabilitation* New York NY 82-7

[6] Bouzit M, Burdea G, Popescu G and Boian R 2002, "The Rutgers Master II-New Design Force-Feedback Glove," *IEEE/ASME Transactions on Mechatronics* **7** 2 256-63

[7] Y J Nam and M K Park 2007, "Performance Evaluation of Two Different Bypass-type MR Shock Dampers," *J. of Intelligent Systems and Structures* **18** 707-17

Effects of nonmagnetic interparticle forces on magnetorheological fluids

D J Klingenberg[1], **C K Olk**[2], **M A Golden**[2] **and J C Ulicny**[2]

[1] Chemical and Biological Engineering Dept. and Rheology Research Center, University of Wisconsin, Madison, WI 53706, USA

[2] General Motors R&D and Planning, Warren, MI 48090-9055, USA

E-mail: `klingen@engr.wisc.edu`

Abstract. Effects of nonmagnetic interparticle forces on the on- and off-state behavior of MR fluids are investigated experimentally and with particle-level simulations. Suspensions of iron particles in an aliphatic oil are modified by surface-active species. The modifications significantly alter the off-state properties, but have little impact on the field-induced stresses. Simulations show similar behavior. Off-state rheological properties are strongly influenced by van der Waals forces and modifications of the short-range repulsive forces. Field-induced stresses are less sensitive to the nonmagnetic forces.

1. Introduction

Much research has illustrated that the field-induced behavior of magnetorheological (MR) fluids can be explained largely in terms of magnetostatic forces and their competition with hydrodynamic forces. However, other properties, such as off-state viscosity, sedimentation, redispersability, and durability, are strongly influenced by forces other than magnetic forces. Off-state viscosity of particulate suspensions in general depends strongly on colloidal forces [1]. Sedimentation can be controlled by various additives, some of which may act by inducing colloidal interactions between particles [2, 3]. van der Waals forces, as well as remnant magnetization, can influence the stability and redispersability of suspensions [4]. Ulicny et al. [5] found that surface active additives significantly reduced the off-state drag in an MR clutch, and improved the durability of MR fluids. Surfactants can also help to disperse particles [6]. These observations are consistent with the added chemicals altering the nonmagnetic interparticle forces.

Here we investigate the effects of nonmagnetic interparticle forces on the on- and off-state rheological properties of MR suspensions. Specifically, we experimentally investigate the effects of surface-active additives on rheological properties. We also employ particle-level simulations to investigate the effects of van der Waals forces and modifications of short-range repulsive forces on rheological properties.

2. Experimental

The MR fluids were prepared using two iron powders, with average diameters of 2 and 8 μm (BASF HS and CM, respectively). The base liquid was Mobil SHF 21—a polyalphaolefin (PAO) of molecular weight of approximately 280. Three MR suspensions were prepared, each composed of 30 ± 1 vol% iron particles in PAO. One was composed of only the small diameter (HS) particles, one composed of only the large diameter (CM) particles, and one composed of a bimodal mixture of equal masses of each type of particle. These suspensions are referred to as the "nontreated" suspensions.

A second set of MR fluids were prepared by modifying aliquots of the MR fluids described above. To a portion of each of the fluids, lithium 12-hydroxy stearate (LiHS; City Chemical) and zinc-o,o-di-n-butylphosphorodithiolate (ZBPD; Flexsys) were added, in ratios of 3.47×10^{-3} g LiHS/g MR fluid and 2.78×10^{-4} g ZBPD/g MR fluid. The resulting suspensions are referred to as "treated" suspensions.

Off-state rheological properties (no applied magnetic field) were measured using a Haake RS150 rheometer with serrated parallel disks. The shear stress was measured as a function of shear rate, as the shear rate was first increased from 0 to 400 s^{-1}, and then decreased from 400 to 0 s^{-1}. Results were reproducible after the first such shear rate "loop."

Rheological properties in the presence of a magnetic field were measured using an Anton Paar MCR501 magnetic rheometer. After placing the sample between parallel disks, the shear stress was measured as a function of shear rate, with the shear rate increased in steps from 0 to 400 s^{-1}. Experiments were repeated for various magnetic field strengths. Shear stress-shear rate data were subsequently fit with a Bingham model to extract the yield stress and plastic viscosity.

3. Simulations

Particle-level simulations were used to investigate effects of various forces on MR fluid behavior. We used the point-dipole simulation method that we have employed previously [7, 8] modified by adding van der Waals forces, and altering the short-range repulsive forces to account surface modifications.

van der Waals forces are represented by the near-field form [9] with a small separation cutoff $h_{\min}$,

$$\boldsymbol{F}_{ij}^{\mathrm{vdw}} = \begin{cases} \frac{A}{24}\frac{\sigma}{h_{ij}^2}\boldsymbol{e}_r & \text{for } h_{ij} > h_{\min}, \\ \frac{A}{24}\frac{\sigma}{h_{\min}^2}\boldsymbol{e}_r & \text{for } h_{ij} \le h_{\min}, \end{cases} \tag{1}$$

where A is the Hamaker coefficient, σ is the particle diameter, and $h_{ij} = r_{ij} - \sigma$ is the separation between the surfaces of spheres i and j. Here we use a cutoff of $h_{\min}/\sigma = 10^{-3}$. Short-range repulsive forces are modeled with the short-ranged function

$$\boldsymbol{F}_{ij}^{\mathrm{rep.}} = -F_0 \exp\left[-\kappa(h_{ij} - h^*)/\sigma\right]\boldsymbol{e}_r, \tag{2}$$

where

$$F_0 = \begin{cases} \frac{3}{16}\mu_0\sigma^2\beta_M^2 H_0^2 & \text{for linear magnetization,} \\ \frac{\pi}{48}\mu_0\sigma^2 M_s^2 & \text{for saturated magnetization,} \end{cases} \tag{3}$$

is the magnitude of the magnetostatic force, $\kappa = 10^3$ for all results reported here, and h^* represents a steric barrier. We include h^* to model the effect of adsorbed or grafted species (i.e., $h^* \approx 2\ell$, where ℓ is the thickness of a densely grafted layer). Analogous expressions are employed to describe interactions between the particles and the bounding surfaces.

The ratio of hydrodynamic to field-induced forces defines a dimensionless shear rate, $\widetilde{\dot{\gamma}} = 3\pi\eta_c\dot{\gamma}/F_0$, which is set to 10^{-3} for the results presented here. The dimensionless shear rate is related to the Mason number by $\widetilde{\dot{\gamma}} = 32\mathrm{Mn}$. The ratio of van der Waals and field-induced forces is

$$\widetilde{F}_0^{\mathrm{vdw}} = \frac{A/24\sigma}{F_0} = \begin{cases} \frac{2A}{9\pi\mu_0\sigma^3\beta_M^2 H_0^2} & \text{linear magnetization,} \\ \frac{2A}{\pi\mu_0\sigma^3 M_s^2} & \text{saturated magnetization.} \end{cases} \tag{4}$$

Typical values for $\widetilde{F}_0^{\mathrm{vdw}}$ are estimated using the parameter values $A = 50 \times 10^{-20}$ J [9], $\mu_0 M_s = 2$ T, and $\beta_M = 1$. For $\sigma = 1.0$ μm, $\widetilde{F}_0^{\mathrm{vdw}}$ is roughly 1.0×10^{-7} for large magnetic field strengths (saturated magnetization) and increases with decreasing field strength; for a flux density of 0.1 T, $\widetilde{F}_0^{\mathrm{vdw}} \approx 4 \times 10^{-6}$. Although these values for $\widetilde{F}_0^{\mathrm{vdw}}$ are much less than 1, we show below that van der Waals forces can significantly impact the field-induced stresses in this range.

Simulations were performed by randomly placing 150 spheres in a periodic simulation cell of dimensions $(L_x, L_y, L_z) = (10\sigma, 5\sigma, 5\sigma)$, corresponding to a particle volume fraction of $\phi = 0.314$. Shear flow is simulated to a strain of 5. The shear stress, calculated in a conventional manner [8], becomes steady after a strain of 1. Results reported here are averaged over the strain interval $\gamma = 1$–5, and over 5 different initial configurations for each set of simulation parameters.

4. Results

The field-induced yield stress is plotted as a function of flux density in Fig. 1, for both the nontreated and treated suspensions. The surface treatments have little effect on the field-dependent yield stress. This is in contrast to results for thicker coatings where the field-induced stress is significantly decreased by the coating [10, 11, 12].

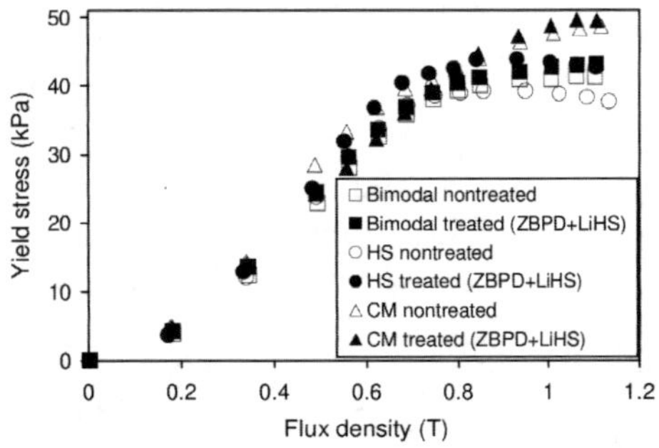

Figure 1. Yield stress vs flux density for various monodisperse and bimodal suspensions (30 vol%), with and without surface treatments (ZBPD + LiHS).

Off-state rheological properties were also measured for these systems. Shear stress-shear rate data were fit with the Bingham equation. The yield stresses and plastic viscosities obtained from these fits are presented in Figs. 2 and 3, respectively. The surface treatments significantly reduce the off-state yield stress as well as the plastic viscosity.

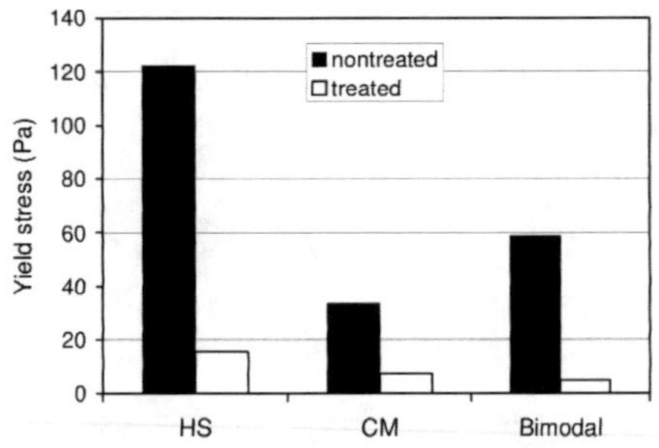

Figure 2. Off-state yield stress for various nontreated and treated suspensions.

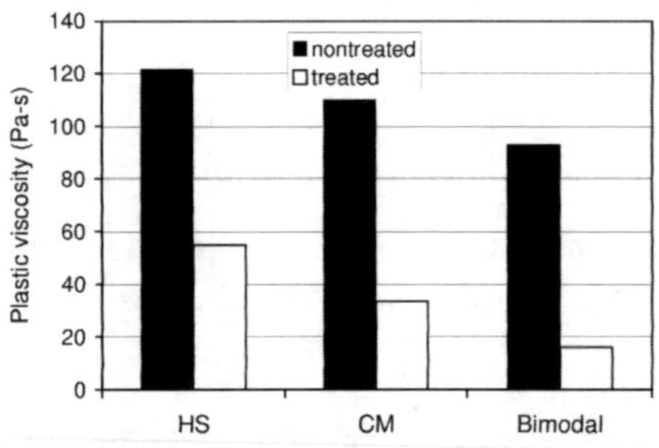

Figure 3. Off-state plastic viscosity for various nontreated and treated suspensions.

Simulation results for the on-state properties are illustrated in Fig. 4, where the average dimensionless shear stress, $\langle \tilde{\tau} \rangle = \langle \tau \sigma^2 / F_0 \rangle$, is plotted as a function of $\widetilde{F}_0^{\mathrm{vdw}}$ for several values of h^*. The results for $h^* = 0$ show that the field-induced stress is independent of $\widetilde{F}_0^{\mathrm{vdw}}$ for small $\widetilde{F}_0^{\mathrm{vdw}}$, and then increases with $\widetilde{F}_0^{\mathrm{vdw}}$ for larger values. For $\widetilde{F}_0^{\mathrm{vdw}} = 1.5 \times 10^{-6}$, the stress is 26% larger than in the absence of van der Waals forces. Thus, van der Waals forces significantly impact the rheology for small values of $\widetilde{F}_0^{\mathrm{vdw}}$; however, values of $\widetilde{F}_0^{\mathrm{vdw}} \gtrsim 10^{-6}$ are typically only achieved for relatively small magnetic field strengths or small particle diameters. The effect of steric barrier thickness h^* on the stress is also illustrated in Fig. 4 (for $\widetilde{F}_0^{\mathrm{vdw}} = 10^{-6}$). Increasing h^* causes the stress to decrease. For $h^*/\sigma = 10^{-3}$, the field-induced stress is reduced to that obtained in the absence of van der Waals forces. A decrease in stress with increasing h^* is expected, since larger interparticle separations result in weaker van der Waals forces. As mentioned above, such trends are observed experimentally, where sufficiently thick coatings can cause a decrease in the field-induced stresses [10, 11, 12].

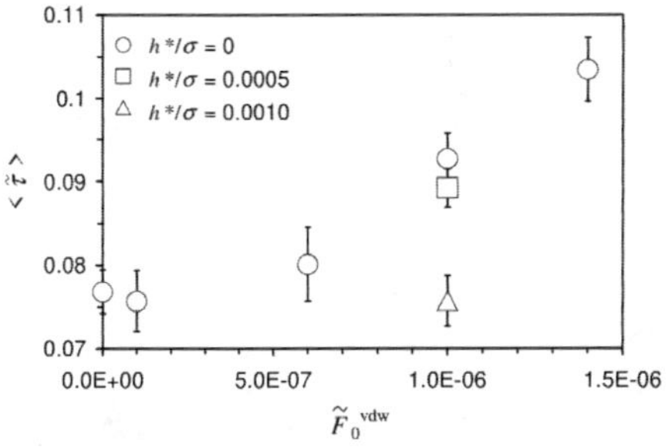

Figure 4. Field-induced shear stress vs $\widetilde{F}_0^{\mathrm{vdw}}$ for several values of h^*.

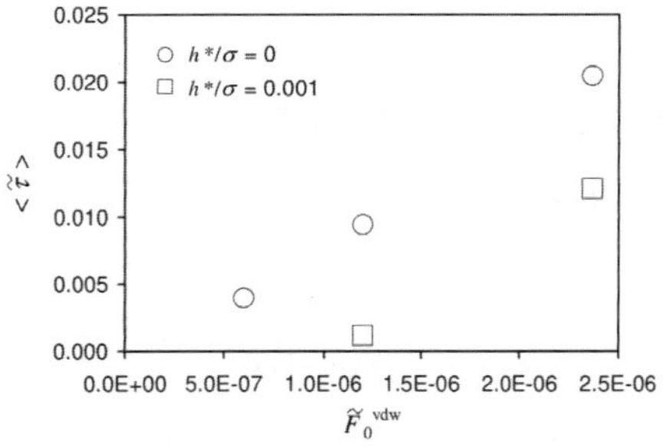

Figure 5. Off-state shear stress vs $\widetilde{F}_0^{\mathrm{vdw}}$ for several values of h^*.

The stress is much more sensitive to van der Waals forces and the steric barrier in the off state, as illustrated in Fig. 5 where the shear stress is plotted as a function of $\widetilde{F}_0^{\mathrm{vdw}}$ for two values of h^* (for the off state, $\widetilde{F}_0^{\mathrm{vdw}} \equiv A/24\sigma F_0$ is scaled by the repulsive force magnitude F_0). In contrast to the field-induced stress, the off-state stress (at low shear rates) approaches zero as $\widetilde{F}_0^{\mathrm{vdw}} \rightarrow 0$. The effect of the barrier is more pronounced as well, as the stress is reduced by 88% as h^*/σ is increased from 0 to 10^{-3} (for $\widetilde{F}_0^{\mathrm{vdw}} = 1.2 \times 10^{-6}$).

5. Conclusion

Effects of nonmagnetic interparticle forces on the on- and off-state behavior of MR fluids are investigated experimentally and with particle-level simulations. Surface modifications significantly alter the off-state properties, but have little impact on the field-induced stresses. Simulations show similar behavior. Off-state rheological properties are strongly influenced by van der Waals forces and modifications of the short-range repulsive forces. Field-induced stresses are less sensitive to the nonmagnetic forces.

References

[1] Russel W B, Saville D A and Schowalter W R 1989 *Colloidal Dispersions* (Cambridge; Cambridge University Press)
[2] Ulicny J C, Golden, M A, Namuduri C S and Klingenberg D J 2005 *J. Rheol.* **49** 87–104
[3] Goncalves F D, Koo J and Ahmadian M 2006 *Shock and Vibration Digest* **38** 203-19
[4] Phule P P, Mihalcin M P and Genc S 1999 *J. Mat. Res.* **14** 3037–41
[5] Ulicny J C, Smith A L, Golden M A and Hayden C A 2005 U.S. Patent 6,881,353 B2
[6] Weiss K D, Carlson J D and Nixon D A 2000 U.S. Patent 6,027,664
[7] Heine M C, de Vicente J and Klingenberg D J 2006 *Phys. Fluids* **18** 023301
[8] Parthasarathy M and Klingenberg D J 1995 *Rheol. Acta* **34** 430–39
[9] Israelachvili J N 1991 *Intermolecular and Surface Forces* (London; Academic Press)
[10] Cho M S, Lim S T, Jang I B, Choi H J and Jhon M S 2004 *IEEE Trans. Mag.* **40** 3036
[11] Jang I B, Kim H B, Lee J Y, You J L, Choi H J and Jhon M S 2005 *J. Appl. Phys.* **97** 10Q912
[12] Fang F F and Choi H J 2008 *J. Appl. Phys.* **103** 07A301

Magnetorheological measurements with consideration for the internal magnetic field in samples

W Kordonski and S Gorodkin

QED Technologies International, 1040 University Ave., Rochester, NY 14607 USA

E-mail: kordonski@qedmrf.com

Abstract. The magnetically induced yield stress in a sample of suspension of magnetic particles is associated with formation of a field-oriented structure, the strength of which depends on the degree of particles magnetization. This factor is largely defined by the actual magnetic field strength in the sample. At the same time it is common practice to present and analyze magnetorheological characteristics as a function of the applied magnetic field. Uncertainty of an influence function in magnetorheology hampers interpretation of data obtained with different measurement configurations. It was shown in this paper that rheological response of magnetorheological fluid to the applied magnetic field is defined by the sample's actual (internal) magnetic field intensity, which, in turn, depends on sample geometry and field orientation all other factors being equal. Utilization of the sample's actual field as an influence function in magnetorheology allows proper interpretation of data obtained with different measuring system configurations. Optimization of the actual internal field is a promising approach in designing of energy efficient magnetorheological devices.

1. Introduction

The magnetorheological effect constitutes a change in rheological properties of a suspension of magnetic particles, also known as a magnetorheological (MR) fluid, upon the application of a magnetic field and is commonly associated with the magnetically induced fluid *yield stress*. The yield stress appears due to the magnetization of particles followed by their magnetic dipole-dipole interaction and formation of a field oriented structure. In shear flow such a magnetized fluid exhibits elevated *effective viscosities* as well as *normal stresses.*

Historically, these rheological characteristics are measured, presented and analyzed as a function of applied magnetic flux density (B) or applied magnetic field strength ($H=B/\mu_0$), where μ_0 is permeability the vacuum [1,2]. At the same time, an effective magnetization of particles and therefore the magnitude of measured rheological parameters are defined by the actual magnetic field intensity in the sample. In general, the internal magnetic field intensity (H_i) in the sample is commonly presented as follows:

$$H_i = H - DM \tag{1}$$

where H is the applied magnetic field intensity, D is a demagnetization factor and M is material magnetization.

The demagnetization factor is introduced to account for the magnetic field created by the material magnetization and directed opposite to the applied field. The magnitude of the demagnetization factor depends on sample geometry (aspect ratio) as well as the field and the sample mutual orientation. For example, it may change from 0 to 1 when a high aspect ratio magnetic plate is oriented along or perpendicular to the magnetic field. Taking into account the relatively high MR fluid magnetization, the influence of the demagnetization factor on a sample's internal magnetic field intensity and consequently on rheological response of magnetorheological fluid to the applied magnetic field can be significant. As it is shown in this presentation, utilization of the internal magnetic field intensity in samples as an influence function in magnetorheology allows one to avoid misinterpretation of rheological data and provides basis for generalization of data obtained with different measurement configurations. Optimization of the actual internal field can also be a promising approach in designing of energy efficient magnetorheological devices.

2. Problem configuration

A configuration with shear flow oriented normal to a magnetic field is utilized in most MR fluid applications as well as in magnetorheological measurements. Usually, the sheared layer is thin, has relatively high aspect ratio and the field-shear orientation is realized in two ways as depicted in figure 1. The configuration B_z is characteristic to such application as magnetorheological finishing (MRF®) [3] and the configuration B_y is widely used in magnetically controlled clutches, dampeners, etc [4]. It is reasonable to expect that MR fluid magnetization as well as rheological response on applied magnetic field for configurations B_z and B_y will differ due to effect of the demagnetization factor on the actual (internal) field in the sample. A comparative study of these two cases was performed both numerically and experimentally.

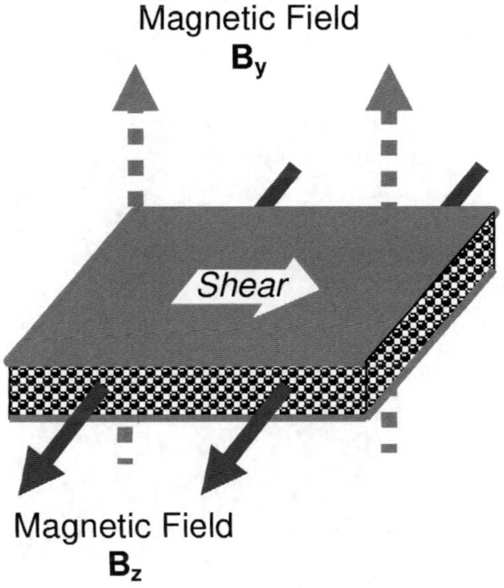

Figure 1. Shear flow and magnetic field mutual orientations.

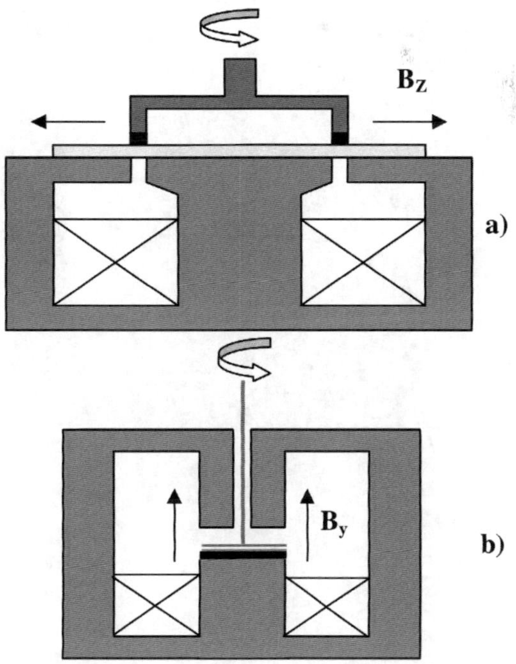

Figure 2. Schematics of configurations for magnetic modeling and rheological measurements.

By virtue of the fact that adequate instrumental measurements of the magnetic field intensity inside the very thin samples (thickness of 0.2 mm) are unrealizable, the internal field was computed using finite element analysis (FEA). As discussed below, the 3D magnetic modelling and the sample's internal field intensity calculations were performed in strict compliance with the design of experimental hardware and the sample geometry used in rheological measurements. The experimental hardware constituted two rotational rheometers schematically depicted in figure 2, where the schematic a) renders the configuration B_z and the schematic b) renders the configuration B_y as shown in figure 1. Black areas in figure 2 represent samples of MR fluid and arrows show magnetic field orientation.

MR fluid relative magnetic permeability required for simulation was derived from measurements of the initial magnetic susceptibility of the fluid. The measurements were conducted with a Bartington MS2B magnetic susceptibility system. An assumption was made that the susceptibility remains constant for the initial part of the magnetization curve that is in the range of relatively low magnetic field intensities where modeling was performed. All computations and rheological measurements were conducted for the relative magnetic permeability of 2.8, which corresponds to the silicon oil-based MR fluid with 45 vol. % concentration of ~3.5 micron in size carbonyl iron particles (BASF).

3. Magnetic field modelling

A commercial software MagNet developed by Infolytica was used for simulations of magnetic field in the magneto-rheometer magnetic gap formed by pole pieces of electromagnet. First and foremost the results of magnetic modelling were verified with measurements of magnetic flux density in the empty gap using the Hall probe. A reasonable agreement between computed and experimental data was obtained.

Figure 3a shows an exemplary working space (solid 3D model) corresponding to the configuration B_y with high demagnetization factor of the sample. In this case MR fluid sample is defined in the form of thin disk (aspect ratio of 100) placed between electromagnet poles in such a way that its plane is normal to the magnetic field. An example of a computed map of magnetic field intensity distribution in the electromagnet's gap containing a sample of MR fluid is shown in figure 3b.

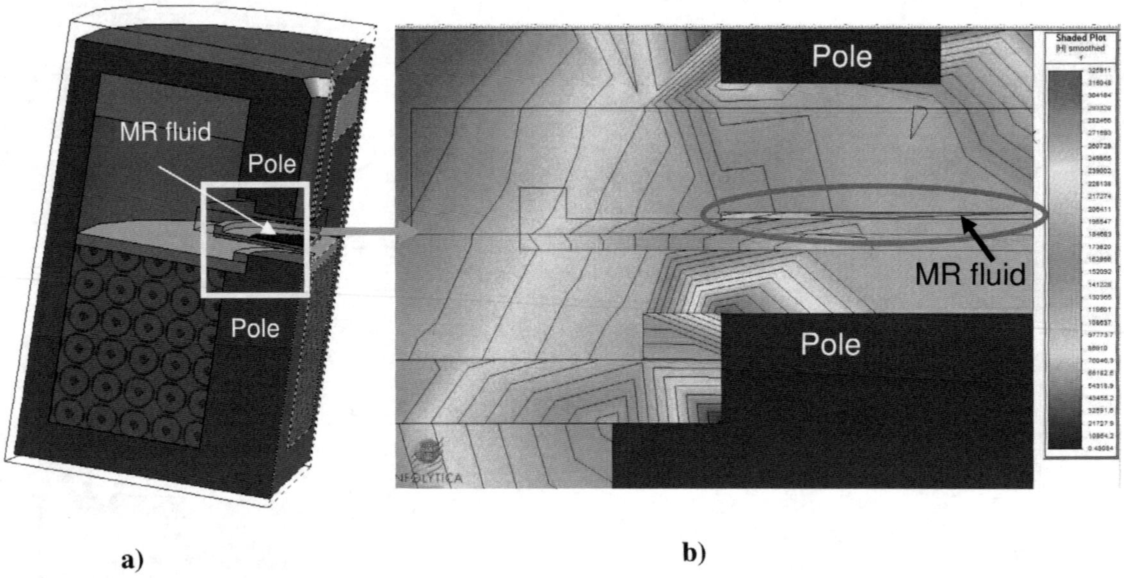

Figure 3. Working space and a map of field intensity distribution for the configuration B_y.

It can be seen that the field strength (blue color) in the sample of MR fluid is significantly lower than the strength of applied field (light green color). In this particular case the field intensity was ~ 50 kA/m and ~ 120 kA/m respectively. In the case of configuration B_z (the model is not shown here) the sample in the form of a thin ring was placed in a radial tangential fringing magnetic field (see fig. 2a).

Some numerical results of calculations of the sample's internal field intensity along with applied field intensity (in legend defined as air) for configurations B_y and B_z are presented in figure 4.

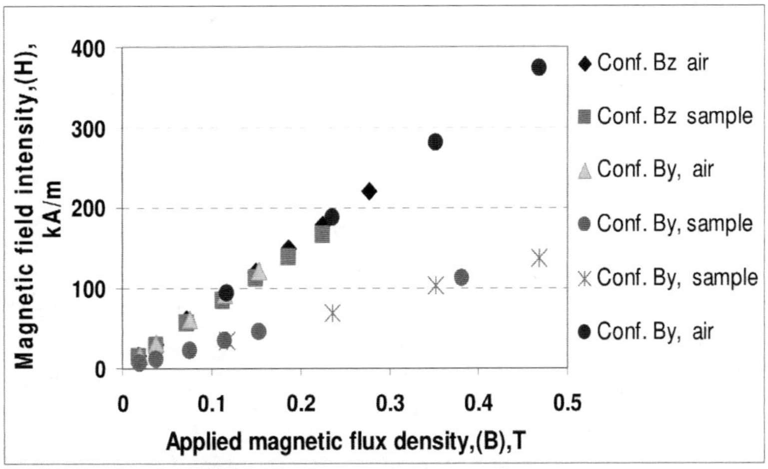

Figure 4. Results of calculations of the intensity of magnetic field in the gap for the configurations B_y and B_z.

In the case of configuration B_z, with low demagnetization factor, the field intensity inside the sample is practically equal to the intensity of surrounding (applied) field for the whole range of applied magnetic flux densities. At the same time, the intensity of the sample internal field for configuration B_y, all factors being equal, is much lower than intensity of applied field.

2. Rheological measurements

Rheological measurements were performed with a commercial Anton Paar magnetorheometer Physica MCR 301. A standard plate-plate configuration with rotor diameter of 20 mm and working gap height of 0.2 mm was used to analyze the configuration B_y as schematically shown in Fig 2b. To render case B_z, an attachment in the form of a magnetic unit with a radial fringing magnetic field and ring-plate geometry was used (see figure 2a). In both cases the Hall probe was arranged in the vicinity of the sample in order to monitor the applied magnetic flux density. A special anti-slip coating was applied to the working surfaces to provide not unreasonable and repeatable results. A proper calibration of units with standard Newtonian fluid provided unambiguous measurements. Yield stress measurements were conducted both in constant shear stress and constant shear rate regimes.

Results of the yield stress measurements shown in figure 5 demonstrate that the magnitude of yield stress in case B_z significantly exceeds the yield stress for configuration B_y at the same applied magnetic flux density. This experimental result is in a qualitative agreement with results of field modeling discussed above and supports the assumption that MR fluid yield stress is defined by the strength of the internal magnetic field in the sample. By this means, it is safe to say that MR fluid rheological response to applied magnetic field depends not only on applied field intensity and fluid magnetic properties but also on distinctive features of the actual conditions at which sample is magnetized. The use of the intensity of internal field in the sample as an influence function in magnetorheology allows obtaining unambiguous presentation of data derived from different measurement configurations. As it is shown in figure 6, all yield stress measurements for both B_y and B_z configurations are reasonably well rationalized when the internal field of the sample is used as the influence function. Alternatively, taking into account that the magnetization of a typical MR fluid saturates at the field intensity of ~ 240 kA/m [5], one would expect that optimization of the internal magnetic field in MR fluid, that is conditions at which MR fluid is magnetized in different magnetorheological devices or processes, may result in improvement of applications performance and efficiency.

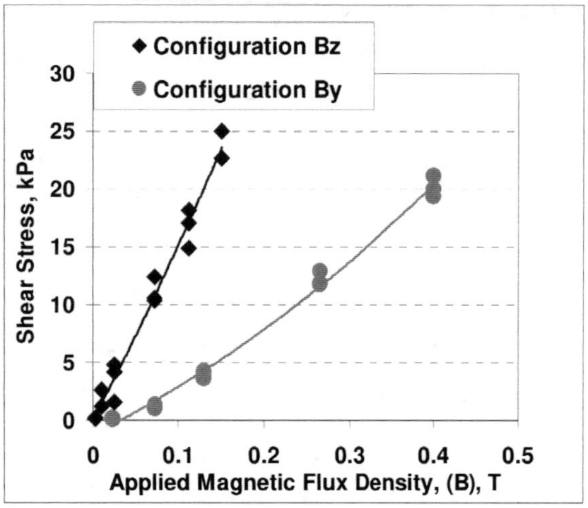

Figure 5. Effect of applied field on the yield stress for configurations B_y and B_z.

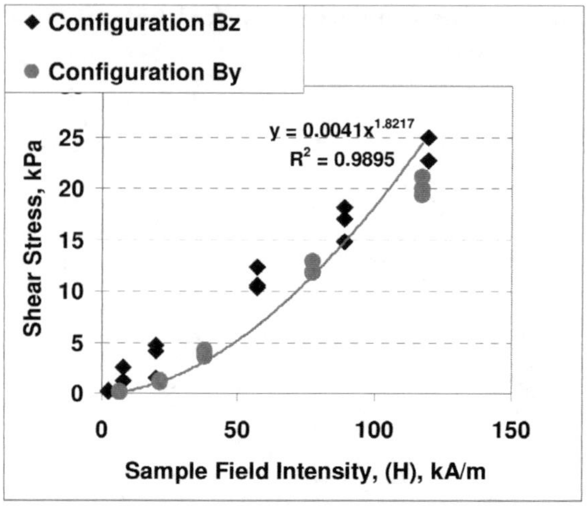

Figure 6. Generalization of rheological measurements with the intensity of magnetic field in the sample.

3. Summary

Numerical modeling of magnetic field in the air gap containing a sample of MR fluid was performed for two configurations of magneto-rheometers, which design differed markedly in the geometry of the sample and the shear-field orientation. Results of modeling show that the actual magnetic field intensity in the sample depends among other factors on sample geometry and sample orientation with respect to the magnetic field. Since the strength of field induced structure in MR fluid depends on the internal field intensity, rheological measurements demonstrate expected dependence of the yield stress on magneto-rheometer configuration at the all other conditions being equal. Utilization of the actual field in the sample as an influence function in magnetorheology allows generalization of data obtained with different measuring system configurations as well as their proper interpretation. Optimization of the internal magnetic field may result in design of energy efficient devices and processes based on the use of MR fluids.

Acknowledgement

The authors appreciate contributions from Bob James and Arpad Sekeres at QED Technologies

References

[1] Lemaire E and Bossis G 1991 Yield stress and wall effects in magnetic colloidal suspensions *J. Phys. D: Appl. Phys.* **24** 1473-77

[2] Laun H, Gabriel C and Schmidt G 2008 Primary and secondary normal stress differences of magnetorheological fluid (MRF) up to magnetic flux densities of 1T *J. Non-Newtonian Fluid Mech.* **148** 47-56

[3] Kordonski W and Golini D 2002 Multiple application of magnetorheological effect in high precission finishing *J.Intelligent Material Systems and Structures* **13** 401-4

[4] Kieburg C , Oetter G, Lochtman R, Gabriel C, Laun H and Peister J 2006 High performance magnetorheological fluids tailored for A 700 Nm automotive 4-wheel-drive clutch *Proc. 10th Int. Conf. on Electrorheological Fluids and Magnetorheological Suspensions* (Lake Tahoe, USA: World Scientific) p 101-7

[5] Ping Zhang Qi Liou and Long Tang 2004 Synthesis and properties of practical MR fluid *Proc. 9th Int. Conf. on Electrorheological Fluids and Magnetorheological Suspensions* (Beijing, China: World Scientific) 94-9

Journal of Physics: Conference Series **149** (2009) 012065 doi:10.1088/1742-6596/149/1/012065

Rheological properties of magnetoelectrorheological fluids with complex disperse phase

E V Korobko, M A Zhurauski, Z A Novikova and V A Kuzmin

A. V. Luikov Heat and Mass Transfer Institute of Academy of Sciences of Belarus, 220072, 15 P. Brovka st., Minsk, Belarus

E-mail: eva@itmo.by

Abstract. The results of experimental investigations of mechanical properties of magnetoelectrorheological fluids – adaptive media, which respond to both electric and magnetic fields – are presented. The influence of the concentration of components of complex disperse phase on the rheological sensitivity is investigated. All compositions have shown a synergetic effect.

1. Introduction

Dielectric fluid dispersed systems, having a capability to change their rheological properties under the influence of both electrical and magnetic fields have appeared for the first time in the paper of S. Gorodkin, W. Kordonsky et al, 1991, and were called magnetoeletrorheological fluids (MERFs) [1]. The formation of the structure in such dispersions, being a result of interactions of both ferromagnetic cores of particles and surface charges, gives a strengthening, synergetic effect. Theoretical discussion of this effect was reported in papers [2, 3]. E. Postrekhin and L. Zhou [2] have suggested a formula to evaluate the effect of both fields: $\tau = 0.5(\varepsilon_0 a_c E^2 + \mu_o a_\mu H^2)$, where a_c, a_μ – are the functions of the dielectric constant and magnetic permeability respectively.

Despite of perspective flexible use of two channels of controlling hydromechanical characteristics of MERF, not more than twenty publications, dedicated to this problem, have appeared for the last 15 years [2 – 12]. Apparently, it is due to the complicity of the studied phenomena both from experimental and theoretical points of view. MERF (or lately they have been also called electromagnetic rheological fluids [2, 3, 7]) – are multi component systems, having a structure dependent both on complex peculiarities of physical-chemical phenomena at phase interfaces and on peculiarities of polarization and magnetization of the filler material. Therefore, the main problems of effective MERF development are associated with initial problems of search and creation of the dispersed filler, which should be structurally sensitive to the simultaneous impact of two fields.

In the present time, the creation of commercially valuable compositions, that have a good response in each of the fields, give a total substantial increase of shear stresses in two fields, and also provide a noticeable above summary effect, has not been yet finished, and the best of the compositions still need the application of high-cost and labor consuming technologies of obtaining.

We have set a goal to develop an effective MERF on the basis of accessible materials without complex preliminary preparation stages.

2. Materials and methods

In the present paper we report on the results of the experimental investigation of rheological characteristics of three compositions of magnetoelectrorheological fluids based on the complex dispersed phase, composed of two fillers. All fluids had a strong magnetic filler (C_1) – carbonyl iron (CI). The particles of carbonyl iron are a complex formation in the form of onion like structures from elementary Fe and its compounds with carbon, nitrogen and oxygen in the quantity not more than 3 %, distributed in the form of concentric sublayers. We have used carbonyl iron of the brand BASF EW in as received condition, the particles of which had the size of 2 – 8 μm and had a double dielectric coating of SiO_2 and were characterized by electric resistance > 10^{10} Ohm·cm. The second filler (C_2) in each composition was a powder in as received condition of α-Fe_2O_3, or iron oxide γ-Fe_2O_3, known after a certain treatment as filler for MERF [1]; or SiO_2 in the form of aerosil, activated by amine, also known as active filler for electrorheological fluids. The transformer oil served as dispersion medium. By preparation of MERF, we have mechanically mixed two powders of dispersed phases, then we have added a dispersion medium. In experiments the fluids with the total volume concentration of dispersed phase 10 % were used. The compositions of MERF with the following composition of two phases were studied: C_1=0%, C_2=10%; C_1=2.5%, C_2=7.5%; C_1=5%, C_2=5%; C_1=7.5%, C_2=2.5%. To investigate the role of carbonyl iron the results were compared to the data of analogical measurements with the content of only C_2=5% in MERF.

The rheological experiments were performed on a special coaxial-cylindrical viscometric bell-type cell serving as an attachment to torque meter of the viscometer RV-12 manufactured by HAAKE. The cell is similar to that described in [8]. The force lines of the magnetic and electric fields are parallel one to another and normal to the shear.

The measurements of shear stress τ have been carried out at varying: magnetic field intensity H = (0 – 100) kA/m, electric field strength E = (0 – 1.8) kV/mm, shear rate $\dot{\gamma}$ = (2 – 575) s^{-1}.

3. Main Results

Magnetoeletrorheological fluids behave similarly in linear viscoplastic medium with a small yield stress in the absence of fields. The essential shear stress enhancement occurs under action of the electric and magnetic fields. The dependencies of shear stress on the electric field strength without and with magnetic field of maximum intensity for MERF with C_1=0 and C_2=5% for α-Fe_2O_3 (n1), SiO_2 (n2), γ-Fe_2O_3 (n3) and with C_1=5% and C_2=0 (n4) are displayed in Figure 1. It is seen that for the first and third compositions the electrorheological effect is insufficient in comparison with the bigger increase of τ in electric field for the fluid sample n2. In magnetic field, however, the sample n2 is not sensitive, while the filler γ-Fe_2O_3 (n_3) in two fields gives the noticeable increase $\Delta\tau = \tau - \tau_o$, which is due to the both magnetorheological and electrorheological responses in almost equal share. Here τ_o is the shear stress at E = 0 and H = 0. At E = 0 in magnetic field with intensity 100 kA/m, $\Delta\tau$ = 85 Pa, and at additional impact of the electric field of E = 1 kV/mm, we get $\Delta\tau$ = 150 Pa.

By the presence of carbonyl iron in the composition of MERF, the response on the magnetic field and the impact of both fields is noticeably increased. The dependencies of τ for MERF-1, containing carbonyl iron at volume concentration 5% and 5% of α-Fe_2O_3 vs. intensity of magnetic field are shown in Figure 2. It is seen, that the magnetic field influence is more essential for this MERF than that of the electric field, because of significant magnetic sensitivity of carbonyl iron. MERF-2 with the same concentration of aerosil gives larger increment of shear stress in the electric field than ones with α-Fe_2O_3, since aerosil has a higher electrorheological activity (Figure 3). The MERF-3 with γ-Fe_2O_3 of the same concentration has bigger values of shear stress in the absence of the external fields because of a needle like shape of γ-Fe_2O_3 particles (Figure 4). Under influence of the electric and magnetic fields τ for MERF-1, MERF-2 are the biggest.

The proportion between components concentrations of the disperse phase does not exert essentially on the relative shear stress increment under the electric field. The carbonyl iron concentration increase caused enlargement of the relative shear stress under the magnetic field influence and under combined

Journal of Physics: Conference Series **149** (2009) 012065 doi:10.1088/1742-6596/149/1/012065

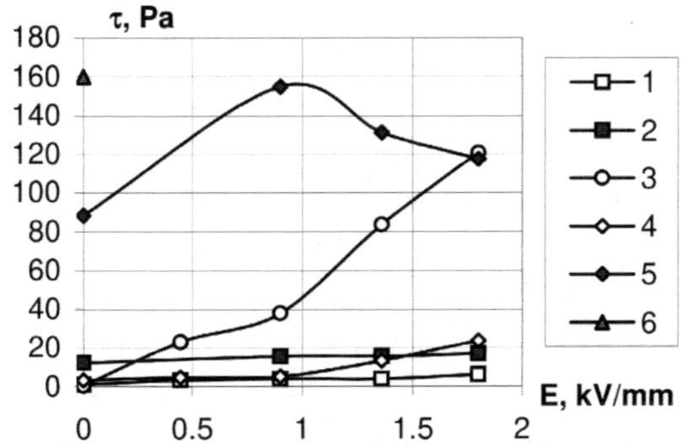

Figure 1. Shear stress of MERFs vs the electric field intensity ($\dot{\gamma}$ = 2.2 s^{-1}). 1, 2 – n1; 3 – n2; 4, 5 – n3; 6 – n4. 1, 3, 4 – H=0; 2, 5, 6 – H=100 kA/m.

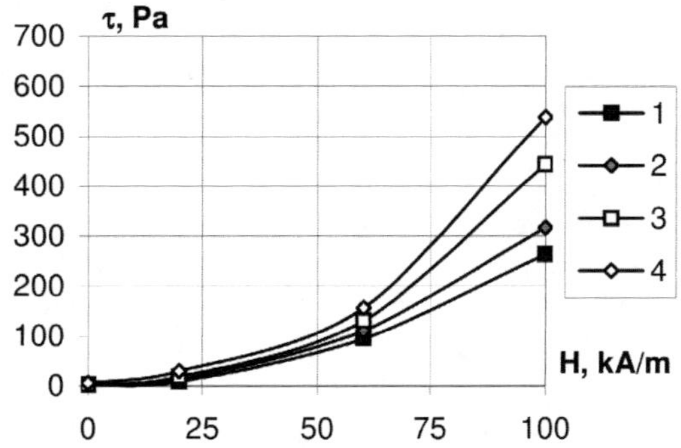

Figure 2. Shear stress of MERF-1 vs the magnetic field intensity ($\dot{\gamma}$ = 2.2 s^{-1}). 1 – E=0; 2 – E=0.9 kV/mm; 3 – E=1.36 kV/mm; 4 – E=1.8 kV/mm.

action of the both fields.

The concentration dependencies of relative shear stress increment of MERF on the basis of carbonyl iron and α-Fe_2O_3 are shown in Figure 5. Analogous dependencies occurred for MERFs on the base of γ-Fe_2O_3 and aerosil. Only for fluids on the basis of γ-Fe_2O_3 the relative shear stress increased practically linear with the growth of carbonyl iron concentration in the magnetic field and under combined action of the fields, while for fluids on the basis of α-Fe_2O_3 and aerosil this increase is more significant for the carbonyl iron concentration range up to 0.025 (Figure 5).

For every MERF with two types of disperse particles the synergistic effect is revealed in the most of shear rate range. The dependence of its index $\xi = \dfrac{\Delta\tau_{EH}}{\Delta\tau_E + \Delta\tau_H}$ on the shear rate for MERF-3 on the basis of carbonyl iron and γ-Fe_2O_3 (C_1=C_2=5 vol. %) is shown in Figure 6. As seen from the Figure, the synergistic effect is more significant for the low shear rate.

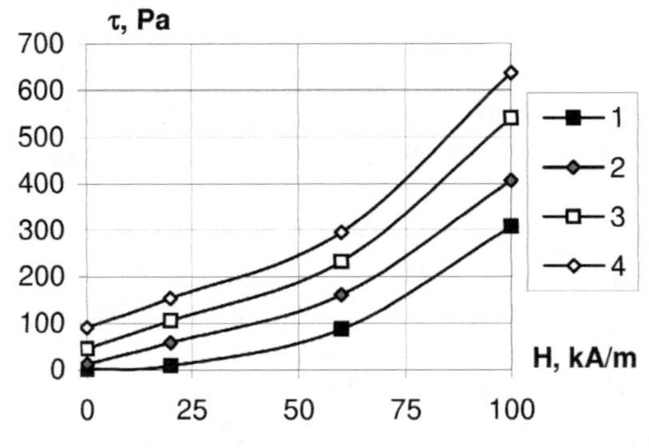

Figure 3. Shear stress of MERF-2 vs. the magnetic field intensity ($\dot{\gamma}$ = 2.2 s^{-1}). Notations are the same as in Figure 2.

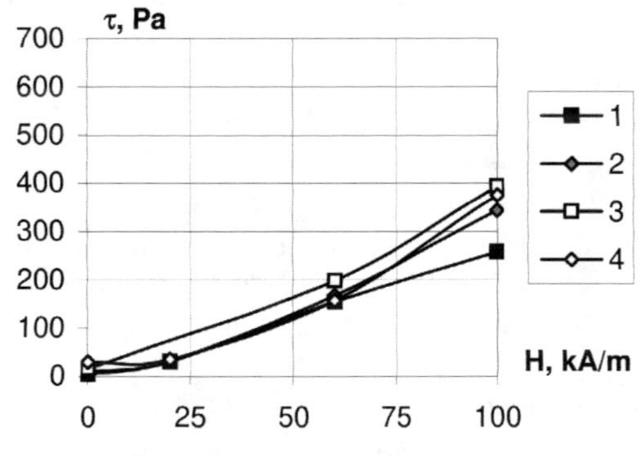

Figure 4. Shear stress of MERF-3 vs. the magnetic field intensity ($\dot{\gamma}$ = 2.2 s^{-1}). Notations are the same as in Figure 2.

Journal of Physics: Conference Series **149** (2009) 012065 doi:10.1088/1742-6596/149/1/012065

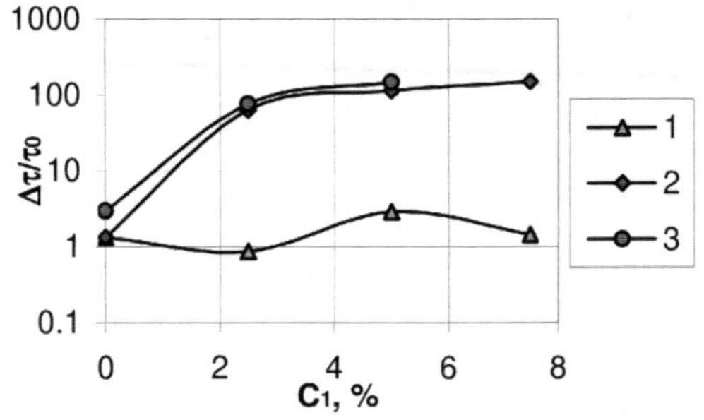

Figure 5. The dependence of relative shear stress increment of MERF with α-Fe_2O_3 and CI on the CI concentration. $\dot{\gamma}$ =36 s^{-1}. 1 – E=1.36 kV/mm; 2 – H=100 kA/m; 3 – E=1.36 kV/mm; H=100 kA/m.

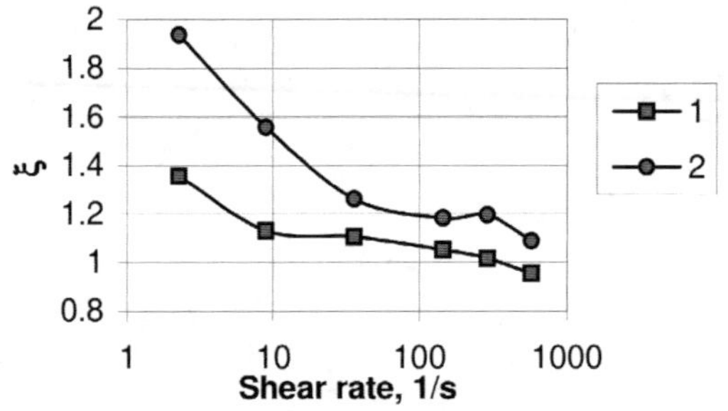

Figure 6. The dependence of the synergistic effect on the shear rate for MERF-3 under combined action of two fields. E=1.36 kV/mm, 1 – H=60 kA/m; 2 – H=100 kA/m.

4. Conclusions

The results of the experimental investigations of rheological properties of MERF on the basis of carbonyl iron, iron oxides and aerosil are presented. Optimum compositions of MERFs with a two-component solid phase are obtained, one of the components (CI) having magnetorheological and the other one – electrorheological properties. The investigated fluids show a high increase of the absolute value of the shear stress under action of the electric and magnetic fields due to the presence of carbonyl iron as filler. Minimal content of CI in MERF with the filler α-Fe_2O_3, giving the maximal effect, is 2.5 vol. %. All the compositions have shown the synergetic effect. Its value depended on the ratio of concentrations of the filler and shear regimes, and also on the material of the solid phase. The possibility of the effective control of the properties using two independent physical channels will allow us to apply these materials to many devices and technologies (hydraulic systems, heat exchangers, vibroprotector devices, etc.).

References

[1] Gorodkin S, Kordonsky W, Medvedeva E 1991 *Problems of heat and mass transfer-91* Minsk, HMTI NAS Belarus pp 48–51
[2] Postrekhin E V, Zhou L W 1998 *Proc. 6th Int. Conf. on ER Fluids (Yonezawa, Japan, July 22 – 25, 1997)* ed. by M Nakano, K Koyama pp 563–570
[3] Huang J P et al 2004 *Journal of Chemical Physics* **121** 7526
[4] Jian-Guo Guan, Jun Huang, Su-Ling Zhao, Run-Zhang Yua 2000 *Proc. 7th Int. Conf. on Electrorheological Fluids and Magnetorheological Suspensions (Honolulu, Hawaii, USA, July 19 – 23, 1999)* ed. by R Tao pp 53–63
[5] Minagawa K, Watanabe T, Koyama K, Sasaki M 1994 *Langmiur* **10** 3926
[6] Koyama K 1996 *Proc. 5th Int. Conf. on ER Fluids, Magneto-Rheological Suspensions & Associated Technology (Sheffield, UK, 10 – 14 July 1995)* ed. by W Bullough pp 245–250
[7] Takeda H, Matsushita K, Masubuchi Y, Takimoto J.-I, Koyama K A 1998 *Proc. 6th Int. Conf. on Electro-Rheological Fluids, Magneto-Rheological Suspensions and Their Applications (Yonezawa, Japan, July 22 – 25, 1997)* ed. by M Nakano, K Koyama pp 571–577
[8] Kordonsky W, Gorodkin S, Medvedeva E 1994 *Proc. 4th Int. Conf. on ER Fluids, Feldkirch, Austria, July 20 – 23, 1993,* ed. by R Tao, G D Roy pp 23–36
[9] Medvedeva E V 2005 *International Journal of Modern Physics* B **19** 1402
[10] Fujita T, Jeyadevan B, Yamaguchi K, Nishiyama H 1999 *Powder Technology* **101** 279
[11] Wen W, Sheng P 2003 *Physica B* **338** 343
[12] Wen W et al 1997 *Appl. Phys. Lett.* **71** 2529

Journal of Physics: Conference Series **149** (2009) 012066 doi:10.1088/1742-6596/149/1/012066

Direct numerical simulation of MR suspension: the role of viscous and magnetic interactions between particles

Sandris Lācis and Didzis Goško

Department of Physics, University of Latvia, Zellu street 8, Riga, LV-1002

E-mail: sandris.lacis@lu.lv

Abstract. A numerical method is developed with aim to simulate the magnetorheological (MR) suspension taking into account realistic magnetic forces. The MR suspension is described by spherical particles with nonlinear magnetic properties suspended in a shear flow. Inertia effects, Brownian motion and buoyancy forces are neglected. The hydrodynamic interaction between close particles is taken into account approximately. Results of some test simulations are presented.

1. Introduction

The behavior of magnetorheological suspensions is governed by magnetic and hydrodynamic interactions between particles. The direct experimental observation of micron size particles inside a non-transparent suspension is almost impossible. Use of numerical simulation methods seems to be an appropriate tool to solve this problem. However even for present computational resources numerical simulations are still limited to intermediate number of particles (few hundreds or thousands) and use of appropriate simplifications.

Pappas andKlingenberg [1] demonstrated that processes inside MR suspensions could be explained by means of direct numerical simulation of particle system placed in fluid flow. Reasonable simplifications were used to describe the hydrodynamic forces and the effect of the magnetic field. Formation of lamellar structures are observed at the developed flow regimes. Similar numerical simulations were carried out also in [2], [3].

Another recent paper [4] contains similar approach to direct numerical simulation based on use of the Rotne-Prager Yamakawa tensor.

More accurate simulation of a flow of rigid particle suspension [5] is limited to two dimensional case. Three dimensional case requires extremely high computational resources.

This study aims at three dimensional direct numerical simulation of the MR suspension flow under the action of magnetic field. We are following main ideas of [1], concentrating on more accurate account for the hydrodynamic interactions between neighbor particles and the more realistic model of magnetic interactions.

2. Method

Particles reside in a box shaped space enclosed by two opposed walls in xy planes according to Figure 1. The space between walls is filled with carrier fluid. One wall is still, but the other wall is moving with velocity U_0 in the direction of x-axis causing a shear flow of a carrier fluid. The walls are cosidered to be ferromagnetic, this is obtained by introducing ghost particles [1]. The "computational box" with dimensions $L_x \times L_y \times L_z$ is periodic in x- and y-directions.

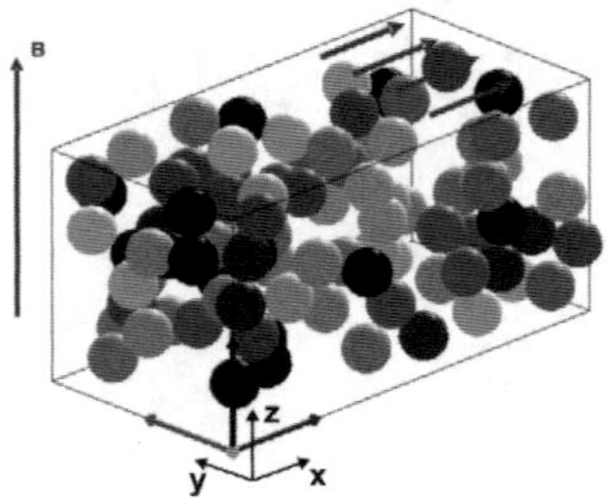

Figure 1. Periodic computational domain.

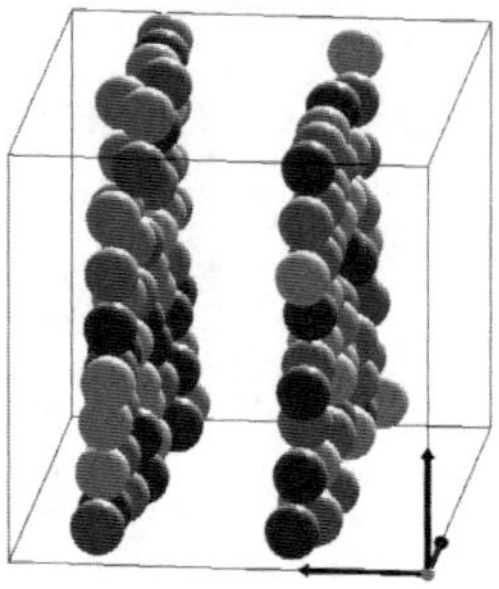

Figure 2. Lamellar structure, $\varphi = 0.1$.

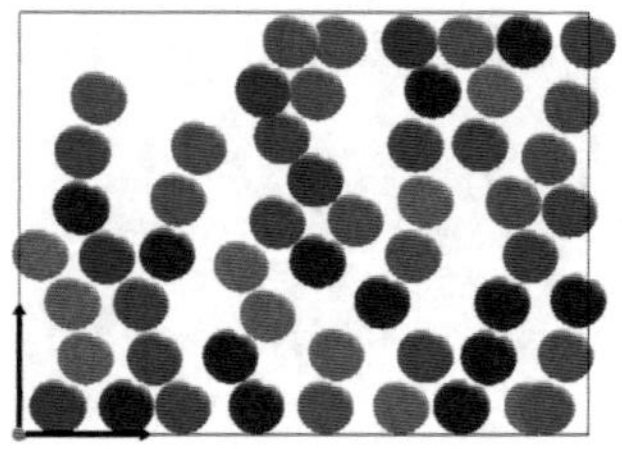

Figure 3. Lamellar structure: side view, $\varphi = 0.1$.

Nondimensional equations are introduced using the reference values. The reference time $\tau = \frac{L_z}{U_0} = \frac{1}{\dot{\gamma}}$ is introduced by the rate of shear flow $\dot{\gamma}$. Here L_z is the length of the box in the z-direction (perpendicular to shear flow) and U_0 is the velocity of the moving upper wall. So the dimensional time t is related to the nondimensional one $\hat{t}$ as $t = \hat{t}\tau$. In a similar way other nondimensional quantities are introduced. For distance measures $R = \hat{R}r_0$ and for velocity $v = \hat{v}v_0$. Here reference quantities are r_0 (the radius of a spherical particle) and the reference velocity $v_0 = \frac{r_0}{\tau} = r_0\dot{\gamma}$. Nondimensional force is derived from Stokes drag: $F_0 = 6\pi\eta r_0 v_0$ and thus the dimensional force is $F = \hat{F}F_0$ (η is the fluid viscosity).

Main magnetic reference value is the saturation magnetization M_S of magnetizable spheres (particles). Nondimensional magnetization $\hat{M} = M/M_S$, magnetic field intensity $\hat{H} = H/M_S$. The Frolich-Kennelly law [6] is used to describe nonlinear magnetization of particles, beside M_S so called initial magnetic permeability μ_{in} characterizes the magnetic material properties. Mason number $M_n = \frac{9\eta\dot{\gamma}}{2\mu_0 M^2}$ characterizes ratio of viscous and magnetic forces, here μ_0 is magnetic constant.

To calculate the velocity of the i-th sphere, we use the force balance equation for nondimensional quantities assuming that all inertial effects are negligible:

$$\sum_{j=1,\, j\neq i}^{N} \vec{F}_{m,ij} + \sum_{i=1,\, i\neq j}^{N} \vec{F}_{c,ij} + \sum_{j=1.\, j\neq i}^{N} \vec{F}_{hd,ij} + \vec{F}_{v,i} = 0. \tag{1}$$

In this equation $\vec{F}_{m,ij}$ is magnetic force between i-th and j-th particles, $\vec{F}_{c,ij}$ is collision force (repulsive force, introduced to prevent overlapping of particles [1]), $\vec{F}_{hd,ij}$ is hydrodynamic force and $\vec{F}_{v,i}$ is the Stokes drag. The Stokes drag is in form

$$\vec{F}_{v,i} = -R_i\left(\vec{v}_i - \vec{u}_i\right), \tag{2}$$

where R_i is the diameter, $\vec{v}_i$ is the velocity of i-th particle and $\vec{u}_i$ is the velocity of ambient flow at the location of i-th particle. Hydrodynamic forces between near particles (as shown in [7]) can be calculated by means of expression

$$F_{hd} = \eta (X_{11}^A - X_{12}^A) U.$$

Here U is half of the relative velocity between spheres. The expressions X_{11}^A and X_{12}^A for different motion modes can be found in [7].

Equations (1) are solved iteratively for all system of particles. Iteration scheme is build expressing velocity $\vec{v}_i$ from the equation (1). Knowing velocities $\vec{v}_i$ positions of all particles are advanced.

At dipolar approximation magnetic forces are calculated using the following expression:

$$\vec{F}_{m,ij} = \frac{1}{M_n \hat{R}^5} \left[2(\vec{m} \cdot \vec{\hat{R}})\vec{m} + (\vec{m} \cdot \vec{m})\vec{\hat{R}} - 5(\vec{m} \cdot \vec{\hat{R}})(\vec{m} \cdot \vec{\hat{R}})\vec{\hat{R}} \right]. \tag{3}$$

Here $\vec{m} = \vec{M}/M$ (unit vector) and $\vec{\hat{R}}$ is nondimensional vector between centers of both particles. Denoting angle between vectors $\vec{m}$ and $\vec{\hat{R}}$ as θ we have $\vec{\hat{R}} = \hat{R}(\cos\theta\,\vec{m} + \sin\theta\,\vec{k})$, where $\vec{k}$ is the unit vector perpendicular to $\vec{m}$. Thus we arrive at

$$\vec{F}_{m,ij} = -\frac{1}{4M_n \hat{R}^4} \left[(3A_1 \cos\theta + 5A_3 \cos 3\theta)\,\vec{m} + (B_1 \sin\theta + 5B_3 \sin 3\theta)\,\vec{k} \right], \tag{4}$$

where $A_1 = A_3 = B_1 = B_3 = 1$. If nonlinear magnetization is taken into account the expression (4) provides good approximation, but values of A_1, A_3, B_1, B_3 depend on field value and distance $d = \hat{R} - 2$ between particles, so they are calculated by finite element method, solving separate magnetic field problem for two spherical particles. Here we discuss only the case of $\mu_{in} = 100$ and $\hat{H}_0 = 0.1$, values of coefficients are given in Table 1.

In the simplest case, ambient flow velocity can be given as linear shear flow profile depending on the distance from the walls. However in more general situations the Stokes flow problem should be solved, taking into account volume force created by Stokes drag of particles. So called SIMPLE algorithm [8] was selected as fluid flow solution method.

3. Results

Table 1. Coefficients A_1, A_3, B_1, B_3 at the magnetic field value $\hat{H}_0 = 0.1$.

d	A_1	A_3	B_1	B_3
0.00	11.67	2.53	19.28	3.33
0.02	10.53	2.49	16.60	3.11
0.04	9.39	2.44	13.93	2.89
0.08	7.61	2.31	10.71	2.51
0.10	6.89	2.21	9.58	2.31
0.14	5.76	2.01	8.03	2.07
0.20	4.67	1.79	6.60	1.82
0.30	3.57	1.57	5.13	1.58
0.50	2.47	1.32	3.54	1.32
1.00	1.63	1.13	2.16	1.13

Calculations where carried out for different volume fractions φ of particles inside MR suspension at $M_n = 0.00003$. If hydrodynamic interaction is limited only to Stokes drag, formation of lamellar structures is reproduced like in [1], see Figure 2, 3.

Effect of using both improved hydrodynamic and magnetic forces is depicted on Figures 4, 5, 6. If magnetic field is perpendicular to the gap between walls, the concentration of particles near both wall is observed (Figure 4). Once some particle comes close to the wall, it is attracted by ghost particle (ferromagnetic wall) and there is no force to tear it away. Structure inside the central region of the gap is quite complex, formation of long chains of particles is not present. If the orientation of the magnetic field is parallel to the walls, particles are slightly repulsed from walls due to the dipolar nature of magnetic forces, what could be observed in Figure 5 and Figure 6. Both figures certify that also in this case there is no formation of long chains of particles in MR suspension. If ambient fluid flow calculation is included, rise of effective viscosity of MR suspension is observed. The accuracy of viscosity calculation is still not sufficiently good due to some numerical instabilities in the computational algorithm which should be fixed.

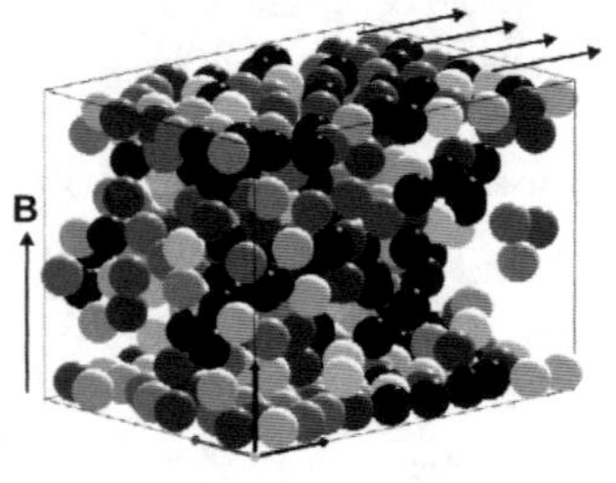

Figure 4. Magnetic field in Z-direction, $\varphi = 0.3$

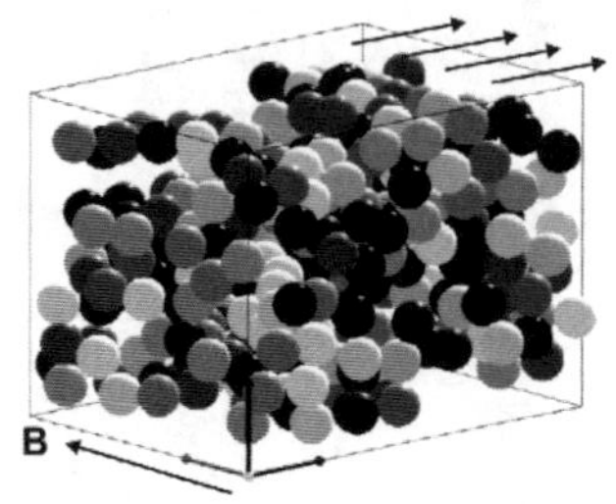

Figure 5. Magnetic field in Y-direction, $\varphi = 0.3$

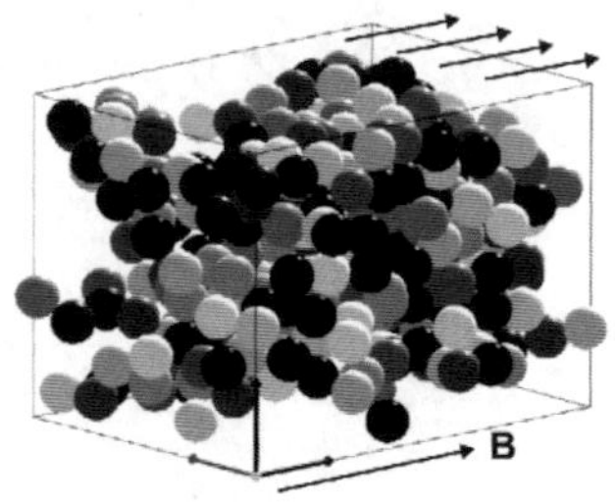

Figure 6. Magnetic field in X-direction, $\varphi = 0.3$

4. Conclusions

The results of simulation show that introduction of more realistic magnetic interaction force between particles change particle behavior in the flow. The hydrodynamic force between close particles prevents particles from fast movements caused by strong magnetic forces. In such a way the hydrodynamic interaction exerts a stabilizing influence on the numerical algorithm suppressing oscillations due to the repulsive force. Numerical method is considerably fast, after small improvements it could be used also for systems of few thousand particles.

5. Acknowledgments

The authors would like to acknowledge financial support of Latvian Science Council Project 04.1290 and ERDF project Nr. VPD1/ERAF/CFLA/05/APK/2.5.1./000038.

References

[1] Pappas Y and JKlingenberg D 2006 *Rheol.Acta* **45** 1435
[2] Klingenberg D J, van Swol F and Zukoski C F 1991 *J. Chem. Phys.* **94** 6160
[3] Tamura H and Doi M 1992 *J. Phys. Soc. Jpn.* **61** 3984
[4] Joung C and See H 2007 *J. Cent. South Univ. Technol.* **14** 271
[5] Avino G D, Maffettone P L, Hulsen M A and Peters G W M 2008 *J. Non-Newtonian Fluid Mech.* **150** 65
[6] Jiles D 1998 *Introduction to Magnetism and Magnetic Materials* (New York: Chapman & Hall)
[7] Kim S and Karilla S J 1991 *Microhydrodynamics: Principles and Selected Applications* (Stoneham: Butterworth-Heinemann)
[8] Versteeg H and Malalasekra W 2007 *An Introduction to Computational Fluid Dynamics: The Finite Volume Method* (New York: Prentice Hall)

Journal of Physics: Conference Series **149** (2009) 012067 doi:10.1088/1742-6596/149/1/012067

Magnetorheological fluid (MRF) in oscillatory shear and parameterization with regard to MR device properties

H M Laun[1], C Gabriel[2], and Ch Kieburg[2]

[1]Senior Technology Consultant, Robert-Blum-Strasse 40, D-68199 Mannheim
[2]BASF SE, Carl-Bosch-Strasse, D-67056 Ludwigshafen

E-mail: claus.gabriel@basf.com

Abstract. A mechanical analogue, consisting of a *Bingham* element (friction element and damper arranged in parallel) and elastic spring in series, is used to capture the measured amplitude dependences of the storage and loss moduli of an MRF in oscillatory shear. Model predictions for the time-dependent shear stress and resulting moduli at agitation frequency for various amplitudes are discussed. The application to plate-plate geometry, imposing a linear radial shear strain profile, is addressed.

1. Introduction

We present measurements of the storage and loss moduli of a magnetorheological fluid (MRF) versus shear amplitude in a plate-plate magnetorheometer and demonstrate the applicability of a pertinent mechanical analogue to capture the main features. The MRF is a concentrated suspension of spherical micron-sized carbonyl iron particles (CIP) in low viscosity Newtonian base oil. Submitted to a magnetic field, the ferromagnetic particles get magnetized and magnetic particle-particle interaction causes the CIP to arrange in strings parallel to the external field. This structure formation causes an increase of the MRF's yield stress τ_Y, depending on the magnetic flux density *B*. The controllability of the flow properties by an external magnetic field is used in MR devices like dashpot dampers (for shock absorbers) and clutches. Here, the performance is mainly controlled by steady shear flow properties of the MRF, approximated by a *Bingham* model,

$$\tau = \tau_Y(B) + \eta_B(B)\dot{\gamma}, \qquad (1)$$

where τ is the shear stress, $\dot{\gamma}$ the shear rate, and η_B the Bingham viscosity. The parameterization is useful to link MRF flow parameters to MR device performance. MR bushings, MR mounts as well as dashpot dampers for controllable damping of oscillatory motions, require additional information about the MRF response in oscillatory shear at various amplitudes. While measurements with imposed sinusoidal shear strain $\gamma = \hat{\gamma}\sin\omega t$ ($\hat{\gamma}$ shear strain amplitude, ω angular frequency) are easily performed by commercial magnetorheometers, a complex MRF response is observed: The resulting oscillatory shear stress strongly depends on the applied shear amplitude, deviates from a sinusoidal wave, and yields a predominantly elastic response at very small shear amplitudes.

2. Experimental

Measurements were performed on a rotational rheometer MRC501 equipped with a MRD180/1T magnetocell (*Anton Paar*) with modifications to allow online flux density measurements using a Hall probe [1]. The geometry is plate-plate with radius *R* = 10 mm and gap *h* = 0.3 mm, the flux density vector being perpendicular to the plane of shear. For sinusoidal shear, the rheometer software provides a storage and loss modulus $G'(\omega)$ and $G''(\omega)$, respectively, by decomposing the oscillatory shear stress wave into in-phase and off-phase components:

$$\tau = \hat{\gamma}\left[G'(\omega)\sin \omega t + G''(\omega)\cos \omega t\right] \tag{2}$$

While Eq. (2) gives an adequate representation of the actual shear stress in the linear viscoelastic regime at small amplitudes, the response becomes nonlinear at high amplitudes and harmonics occur:

$$\begin{aligned}\tau &= \hat{\gamma}\left[G'(\omega,\hat{\gamma})\sin \omega t + G''(\omega,\hat{\gamma})\cos \omega t\right] \\ &+ \hat{\gamma}\sum_{k=1}^{\infty}\left[a_k(\omega,\hat{\gamma})\sin[(2k+1)\omega t] + b_k(\omega,\hat{\gamma})\cos[(2k+1)\omega t]\right]\end{aligned} \tag{3}$$

The rheometer software, however, filters out the harmonics at $(2k+1)\omega$ and provides the amplitude-dependent moduli $G'(\omega,\hat{\gamma})$, $G''(\omega,\hat{\gamma})$ solely for the agitation frequency ω.

Figure 1 shows measured moduli versus shear amplitude. A linear viscoelastic behavior - both moduli being amplitude-independent - is found for amplitudes $\hat{\gamma} \le 0.0005$ (Note: the off-state storage modulus is about four powers of ten smaller). In the small amplitude regime, the MRF responds predominantly elastic, i.e. $G' >> G''$. With increasing amplitude, the loss modulus first increases while the storage modulus remains fairly constant. Finally, both moduli exhibit a strong decrease and $G'' > G'$ indicates a predominantly viscous response. If plotted versus shear stress amplitude $\hat{\tau}$ (not shown by a figure), both moduli exhibit a drastic drop at the value of the steady shear yield stress $\tau_Y(B)$. **Figure 2** shows the unfiltered shear stress signal versus time for $\hat{\gamma} = 0.01$ (shear stress wave close to sinusoidal), and for $\hat{\gamma} = 1$ (saw tooth shape indicating the generation of many harmonics).

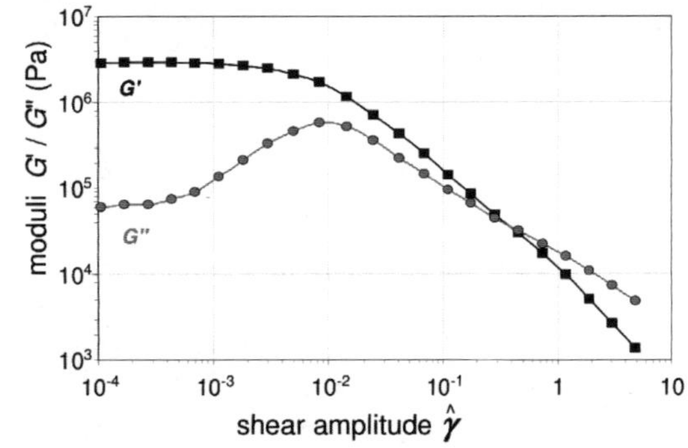

Figure 1. Storage modulus G' and loss modulus G'' versus shear amplitude output of the magneto-rheometer at $B = 0.39$ T, $\omega = 1.9$ rad/s, and 25 °C.

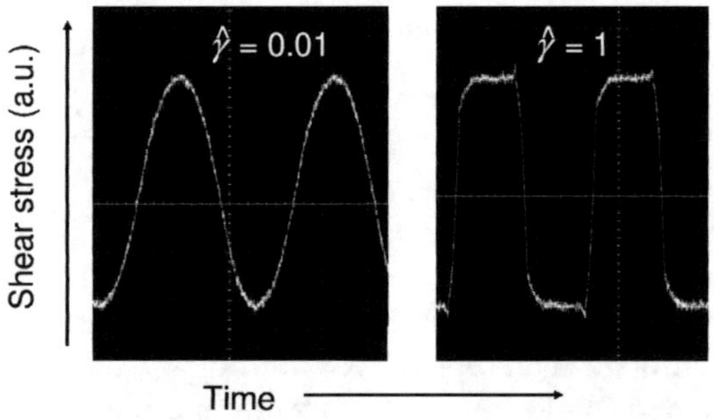

Figure 2. Unfiltered shear stress signals from the magnetorheometer for two different shear amplitude levels (parameters as in Figure 1).

3. Simple model

Figure 3 schematically depicts the structure of the MRF under the action of a magnetic field perpendicular to the plane of shear. At rest, the particles form strings and create a static normal stress [2]. Submitted to a small shear strain, the strings experience an essentially affine deformation with the macroscopic shear, i.e. a slight tilt without string rupture. The magnetic particle-particle interaction in the tilted position causes a torque which tries to reorient the strings parallel to the flux density vector, thus causing an elastic reverse force on the moving plate. This is the origin of the elastic response at very small strains. If the acting shear amplitude exceeds the yield stress, however, the strings start to rupture and achieve in part a less tilted arrangement, which means a loss of elastic memory and increased dissipation due to particle movements relative to the base oil. The MRF behavior in large amplitude oscillatory shear thus shifts from predominantly elastic to mainly viscous.

A simple mechanical analogue to capture such a behavior is the model shown in **Figure 4**. It consists of a *Bingham* element, i.e. a damper of viscosity η_B parallel to a friction element with yield stress τ_Y, arranged in series to an elastic spring of modulus G_0 [3]. The *Bingham* element only

Journal of Physics: Conference Series **149** (2009) 012067 doi:10.1088/1742-6596/149/1/012067

contributes to the deformation γ as long the shear stress τ, acting on both the spring and the *Bingham* element, exceeds the yield stress. In this case, spring and damper define a relaxation time $\lambda \equiv \eta_B / G_0$.

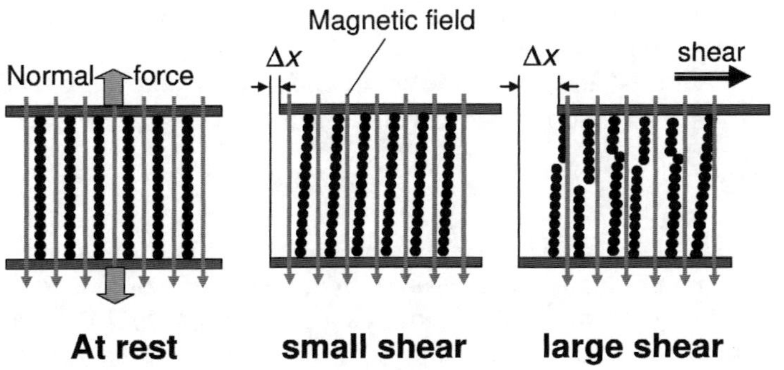

Figure 3. Structural model with CIP strings parallel to external field at rest (left), slightly tilted but intact strings at small shear (middle), and tilted but ruptured strings for large shear (right).

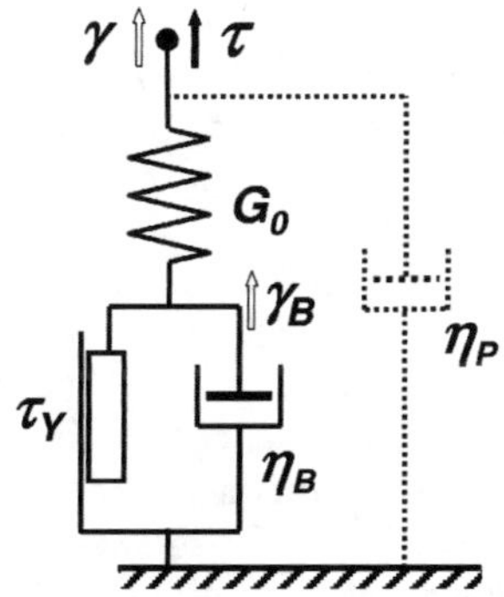

Figure 4. Mechanical analogue to represent the structural model (with optional viscosity η_P).

As soon as the absolute value of the acting stress τ becomes smaller than τ_Y, the friction element turns rigid again and the deformation of the Bingham element γ_B, reached up to that instant, remains constant. The following set of equations enables to calculate the time-dependence of the shear stress:

$$\tau = \begin{cases} G_0(\gamma - \gamma_B) & \text{for } |\tau| \le \tau_Y \\ \tau_Y sign(\gamma) + \eta_B \dot{\gamma} - \lambda \dot{\tau} & \text{for } |\tau| > \tau_Y \end{cases} \quad (4)$$

The differential equation in the second line of Eq. (4) yields for imposed sinusoidal shear

$$\tau(t) = C\exp(-t/\lambda) + \tau_Y sign(\gamma) + \hat{\gamma}[g'\sin\omega t + g''\cos\omega t], \quad (5)$$

where the integration constant C is determined by the value of the shear stress at the change of mode, while the abbreviations g'' and g' stand for $g'' = G_0\omega\lambda/(1+\omega^2\lambda^2)$ and $g' = \omega\lambda g''$, respectively.

4. Model predictions

Figure 5 shows examples for the shear stress versus time for $\omega = \pi$ rad/s, compared to a purely elastic response $\tau_e = \gamma G_0$ for $\gamma_B = 0$. In general, the solution to Eq. (4) is composed by subsequent regimes: Once the elastic shear stress exceeds τ_Y, the shear stress follows Eq. (5) until $\tau = \tau_Y$ is met again. Here, the Bingham element has reached a shear strain γ_B^0. The latter remains constant in the subsequent elastic phase given by the first line of Eq. (4). From this regime on, the shear stress is periodic in time. Due to the finite γ_B^0, however, the elastic response is phase shifted compared to τ_e,

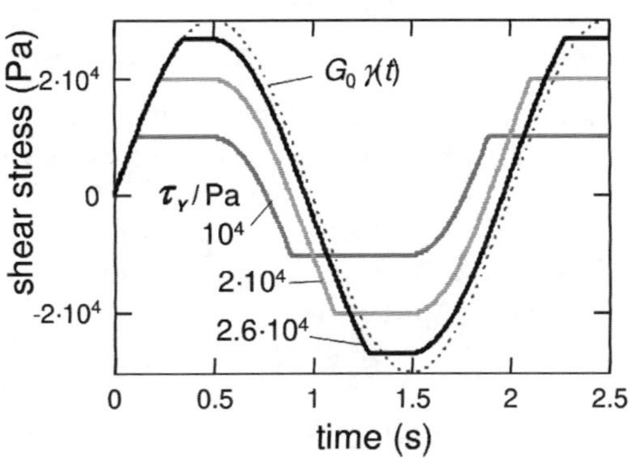

Figure 5. Shear stress versus time for various τ_Y values ($G_0 = 3\cdot10^6$ Pa, $\eta_B = 10^3$ Pas, $\eta_P = 0$).

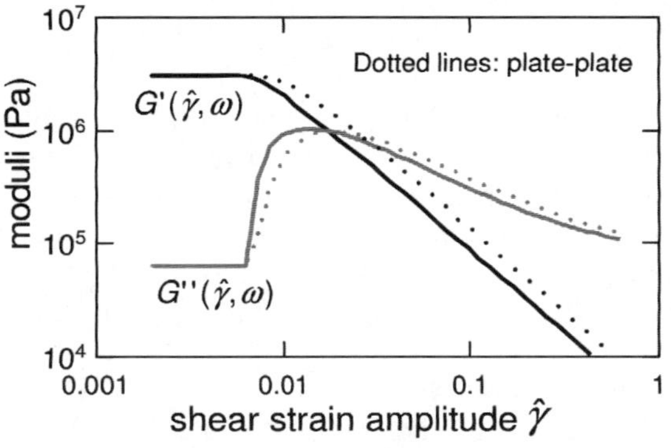

Figure 6. Storage and loss modulus versus shear amplitude ($\tau_Y = 2\cdot10^4$ Pa, $\eta_P = 2\cdot10^4$ Pas).

the phase shift increasing with the ratio $\hat{\gamma}G_0/\tau_Y$. When the shear stress becomes smaller than $-\tau_Y$, the friction element opens again and finally reaches a value $-\gamma_B^0$, which enters in the subsequent elastic response. If an optional damper of viscosity η_P (dotted lines in Figure 4) is arranged in parallel, the response of the *Bingham* element with spring remains unchanged, but a stress component proportional to the shear rate and η_P is added to the total shear stress.

From the shear stress versus time, the moduli G' and G'' for the agitation frequency ω (base mode) follow as (Note: Integration limits are chosen such that only the periodic shear stress signal is used)

$$G'(\hat{\gamma},\omega)=\frac{1}{\hat{\gamma}}\int_{t=\pi/\omega}^{3\pi/\omega}\tau(\hat{\gamma},\omega,t)\sin(\omega t)dt \text{ and } G''(\hat{\gamma},\omega)=\frac{1}{\hat{\gamma}}\int_{t=\pi/\omega}^{3\pi/\omega}\tau(\hat{\gamma},\omega,t)\cos(\omega t)dt. \quad (6)$$

An example is shown in **Figure 6** (model parameters chosen to fit the moduli from Figure 1). Once the shear amplitude exceeds the yield stress, G' monotonously decreases with increasing amplitude, due to the cut-off of the stress amplitude. Below the yield stress, the model yields G''=0 for $\eta_P = 0$. Above the yield stress, G'' shoots up, passes through a maximum and finally decreases less strongly than G' with increasing amplitude. Noteworthy, the high G'' level is mainly caused by the distinct phase shift of shear stress with regard to the imposed shear strain. Large values of the *Bingham* viscosity η_B also contribute to G'', but mainly at high shear amplitudes. The optional viscosity η_P is required to provide a G''-plateau in the limit of small amplitudes. From the experimental view, it might be possible that part of the G'' plateau at small amplitudes may be due to the limited phase angle resolution of the magnetorheometer for small G″/G′. In summary, typical features of the measured moduli are captured by the simple model, while a fit of the measured shape is not achieved. In addition, it is necessary to take into account the radial shear strain profile in the plate-plate measurement (see below).

5. Conversion from physical to engineering parameters

To predict the performance of MR devices, it is necessary to convert the physical model parameters $G_o, \tau_Y, \eta_B, \eta_P$ into the geometry-dependent engineering parameters $\tilde{G}_o, \tilde{\tau}_Y, \tilde{\eta}_B, \tilde{\eta}_P$, which e.g. allow to relate a force to a displacement. A disk clutch of radius R and gap h is chosen as example. This corresponds to the situation in the plate-plate geometry. The **Table** summarizes relations between torque M and rim shear stress τ_R as well as angular displacement φ and rim shear strain γ_R, respectively, in the linear viscoelastic regime. They yield a conversion factor f, applicable to the modulus and optional viscosity.

$M=\tau_R\frac{\pi R^3}{2}$	$\frac{M}{\varphi}=f\frac{\tau_R}{\gamma_R}$	$\tilde{G}_0=fG_0$	$\tilde{\tau}_Y\approx f\tau_Y$
$\varphi=\gamma_R\frac{h}{R}$	$f=\frac{\pi R^4}{2h}$	$\tilde{\eta}_P=f\eta_P$	$\tilde{\eta}_B\approx f\eta_B$

Table. Relations between physical and engineering model parameters (example plate-plate geometry).

Only approximate values may be given for the parameters of the Bingham element, due to the linear radial shear amplitude profile in the plate-plate gap. The resulting plate-plate moduli $G'_{PP}(\hat{\gamma}_R,\omega)$ and $G''_{PP}(\hat{\gamma}_R,\omega)$ for a given rim shear amplitude $\hat{\gamma}_R$ may be calculated as

$$G'_{PP}(\hat{\gamma}_R,\omega)=4\int_0^1 x^3G'(x\hat{\gamma}_R,\omega)dx \text{ and } G''_{PP}(\hat{\gamma}_R,\omega)=4\int_0^1 x^3G''(x\hat{\gamma}_R,\omega)dx. \quad (7)$$

The moduli from Eq. (7) are contained in Figure 6 as dotted lines. Caused by the radial amplitude profile, the moduli are smeared out and somewhat shifted to higher amplitudes. The shape is thus closer to the measurement in Figure 1, obtained by using plate-plate magnetorheometry. This is an example for the complications caused by strain or stress profiles in MR devices. Finally, it should be mentioned that the storage modulus of the MRF may contribute to a shift of the resonance frequency of an MR device. The applicability of a related model to a dashpot damper is demonstrated in [4].

References

[1] Laun, H.M., Schmidt, G., Gabriel, C., Kieburg, Ch., Rheol. Acta 47 (2008) 1049-1059
[2] Laun, H.M., Gabriel, C., Schmidt, G., J. Non-Newtonian Fluid Mech. 148 (2008) 47-56
[3] Hong, S.R., Choi, S.B., Choi, Y.T., Wereley, N.M., Int. J. of Vehicle Design 33 (2003) 17-20
[4] Hu, W., and Wereley, N.M., Journal of Guidance, Control, and Dynamics 31 (2008) 479-489

Journal of Physics: Conference Series **149** (2009) 012068 doi:10.1088/1742-6596/149/1/012068

Performance Evaluation on Vibration Control of MR Landing Gear

D Y Lee[1], Y J Nam[1], R Yamane[2] and M K Park[1]

[1]Graduate school, Pusan National University, Busan 609-735, Republic of Korea
[2]Kokushikan University, 4-28-1 Setagaya, Setagaya-ku, Tokyo 154-8515, Japan

E-mail: ldy5577@yahoo.co.kr, mkpark1@pusan.ac.kr

Abstract. This paper is concerned with the applicability of the developed MR damper to the landing gear system for the attenuating undesired shock and vibration in the landing and taxing phases. First of all, the experimental model of the MR damper is derived based on the results of performance evaluations. Next, a simplified skyhook controller, which is one of the most straightforward, but effective approaches for improving ride comport in vehicles with active suspensions, is formulated. Then, the vibration control performances of the landing gear system using the MR damper are theoretically evaluated in the landing phase of the aircraft. A series of simulation analyses show that the proposed MR damper with the skyhook controller is effective for suppressing undesired vibration of the aircraft body. Finally, the effectiveness of the simulation results are additionally verified via HILS (Hardware-in-the-loop-simulation) method.

1. Introduction

A magneto-rheological (MR) fluid is a non-colloidal solution which is composed of ferromagnetic particles with micrometers in diameter dispersed in a non-conductive carrier fluid [1]. MR fluids have the characteristic that their rheological properties can be continuously and reversely changed within several milliseconds solely by applying or removing external magnetic fields. Due to this feature, MR fluids are increasingly being considered in a large variety of devices such as shock absorbers, vibration insulators, brakes or clutches [2]. Compared to conventional actuation methods, devices using the MR fluid can be made simple in construction, high in power, and low in inertia. Therefore, the use of the MR fluid actuators is very effective for improving the function and performance of conventional control systems. Recently, the authors have proposed an electromagnetic design methodology for the magneto-rheological (MR) damper [3]. In order to improve the performance of the MR damper, the magnetic field should be effectively supplied to the MR fluid. For this purpose, two effective approaches were proposed: one is to shorten the magnetic flux path by removing the unnecessary bulk of the yoke in order to improve the static characteristic of the MR damper, and the other is to increased the magnetic reluctance of the ferromagnetic material by minimizing its cross-sectional area through which the magnetic flux passes in order to improve the dynamic and hysteretic characteristics. The superiority of the proposed design methodology to a conventional one was verified through the magnetic field analysis and a series of basic experiments. As further study, this paper is concerned with the applicability of the developed MR damper to landing gear system for attenuating undesired vibration in the landing phase.

2. MR damper

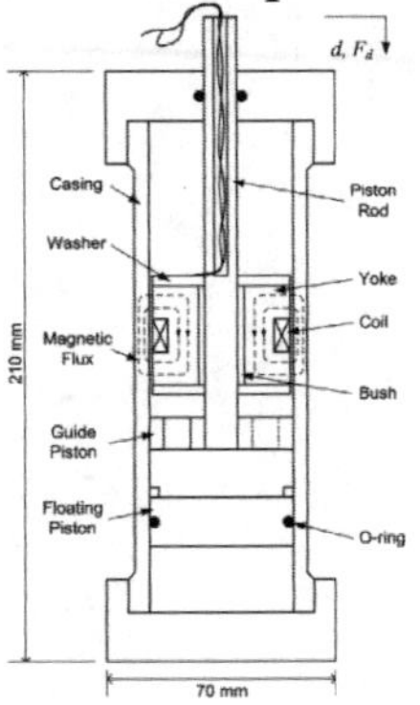

Fig. 1. Schematic configuration of the MR damper

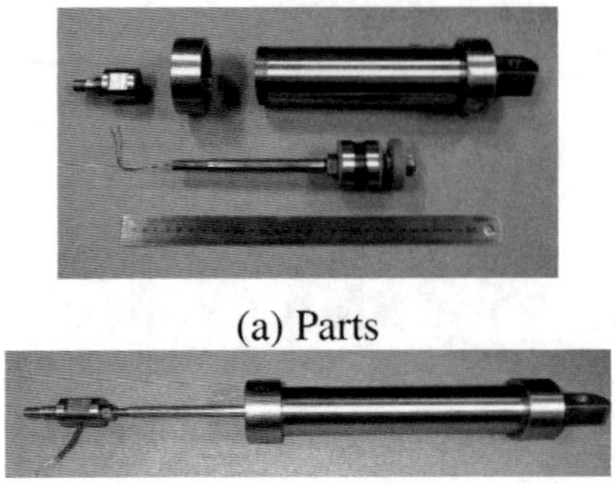

(a) Parts

(b) Assembly

Fig. 2. External appearance of the MR damper

Fig. 1 depicts the schematic configuration of an MR damper considered in this work. This MR damper consists of a MR fluid, a casing, a gas accumulator, and a piston head in which the magnetic circuit is installed. The casing (SS41) is fully filled with the MR fluid (MRF-132AD) and divided into upper and lower chambers by the piston head. When the piston head moves, the MR fluid is moved through the annual gap between the casing and the piston head from one chamber to the other, where the MR fluid is exposed to an applied magnetic field. In order to compensate for changing the fluid volume occupied by the piston rod in the upper chamber, the gas accumulator is located in the lower chamber. The bush and the washer (MC nylon) are employed for preventing leakage of magnetic field, and the guide piston with the porous structure (MC nylon) are installed to keep the gap size constant. When an external current is supplied to the electromagnet coil, a magnetomotive force is generated in the magnetic circuit and the corresponding magnetic field is applied to the MR fluid. Then, the dynamic yield stress of the MR fluid is changed depending on the magnetic field intensity, and the resultant damping force is activated in the opposite direction to the motion of the piston head. In the absence of a magnetic field, the damping force occurs only due to the viscosity of the MR fluid itself. However, in the presence of a magnetic field, an additional damping force occurs due to the MR effect of the MR fluid. Therefore, the damping force of the MR damper can be continuously and reversely controlled by adjusting the applied coil current. The external appearances of the manufactured dampers are shown in Fig. 2, and its full stroke is 80 mm (ruler scale: 30cm). Now, the experimental model of the MR damper is developed based on the results of performance evaluations obtained from the previous research. The magnetic field-dependent damping characteristic of the MR damper can be effectively described by the Bingham plastic model. Although an MR fluid behavior actually exhibits some departures from the Bingham model due to its viscoelastic property, it has been proved by numerous researchers that this simple model is very useful for the design and modeling of MR fluid-based devices [4][5]. Then, the constitutive equation of the MR damper can be expressed as

$$\tau \dot{F}_d + F_d = F_{MR} + F_\eta + F_g + F_f \qquad (1)$$

The above equation is written in SI units. The first term in right hand side is the damping force due to the magnetic field-dependent characteristic of the MR fluid (the MR effective force), and was found by F_{MR} = (426.8 I - 3.833) sgn(v) ≈ 426.8 I sgn(v), where I is the coil current and v is the excitation velocity of the piston head. Especially, it is notable that the MR effective force of the considered MR damper is nearly linear to the coil current. The second term represents the viscous damping force due to the viscosity of the MR fluid itself, and was obtained as $F_\eta = 702.2\ v$. Also, by assuming that the gas in the accumulator undergoes a polytropic process during operations, the gas spring force due to the gas pressure was obtained as $F_g = 78.54\ [11634\ /\ (11634 - 78.54\ d)]^{1.4}$, where d is the displacement of the piston head. The frictional force due to the oil seals and the guide piston was found by F_f = 63 sgn(v). Finally, the differential term in the left hand side of (1) represent the dynamic characteristic of the MR damper. In general, the damping force of the MR damper for the step input current exhibits the dynamic behavior similar to a 1st linear system. Note that the characteristic time constant τ is dependent on the excitation velocity of the MR damper as well as the field-induced condition whether is field-on or field-off; $\tau = \tau(v)$.

3. Simplified skyhook control

In the literature, several control schemes have been developed to improve ride quality of active or semi-active damper systems. One of the most straightforward, but effective approaches is the skyhook controller initially introduced by Karnopp and Cosby [6]. The main advantages of this controller are its simple implementation and easy understanding of the relationship between design and performance. Especially, this controller is applicable for both a semi-active system and an active system. These merits have made the skyhook controller adopted in a large number of real applications and used as the reference control scheme in the literature. At this control, the damping force of the actuator is given proportionally to the vertical velocity of the upper mass. However, most previous researches have found the proportional gain by the trial and error method. In this chapter, a simplified skyhook controller with its control gain determined analytically is proposed for the vibration control of the aircraft landing gear system. In general, a landing gear system featuring the controllable damper can be constructed by a simplified two rigid body model, where one rigid body is the aircraft and the other is the landing gear. Its schematic diagram is shown in the first column of Table 1, where the effect of the tire deflection is neglected. Then, the dynamic governing equation of the landing gear system can be expressed as

$$\begin{cases} m_1\ddot{x} = -k_1(x-y) - F_d + L_f - m_1 g \\ m_2\ddot{y} = -k_1(y-x) - k_2(y-z) + F_d - m_2 g \end{cases} \quad (2)$$

where m_1 is the mass of the aircraft body, m_2 is the mass of the landing gear, x is the displacement of the aircraft body, y is the displacement of the landing gear, z is the surface profile amplitude, k_1 is the linearized stiffness coefficient of the additional coil spring, k_2 is the linearized stiffness coefficient of the tire, F_d is the damping force of the active damper, and g is the gravitational acceleration. Also, the lift force can be described as $L_f = [1.2 - 0.9\tanh(3t)](m_1 + m_2)g$, where $t \geq 0$ is the time in seconds.[7] To obtain the vibration transmissibility of the upper mass, linearizing (2) and rearranging in the Laplace domain yield

$$[m_1m_2s^4 + (k_1m_1 + k_2m_2 + k_1m_2)s^2 + k_1k_2]\delta x(s) + (m_2s^2 + k_2)\delta F_d(s) = k_1k_2\delta z(s) \quad (3)$$

where s is the Laplace variable, and δ represents the variation of the corresponding variable at the static equilibrium state. Similarly, the vibration transmissibility of the 2-dof skyhook model shown in the second column of Table 1 can be expressed as

$$[m_1m_2s^4 + cm_2s^3 + (k_1m_1 + k_2m_2 + k_1m_2)s^2 + c(k_1 + k_2)s + k_1k_2]\delta x(s) = k_1k_2\delta z(s) \quad (4)$$

Then, through the comparison between (3) and (4), the desired damping force can be obtained as follows

$$F_d(s) = c[1 + H(s)]s\,x(s) \quad (5)$$

where $H(s) = k_1/(m_2s^2 + k_2)$. With the above control input, the 2-dof general model has the identical dynamic behavior to the 2-dof skyhook model. For the simple implementation of the controller (5), the following assumption is introduced; the mass of the lower body is much smaller than that of the upper body, and therefore its effect is negligible. Then, the resultant damping force, which is called the simplified skyhook controller in this work, is given by

$$F_d(s) \approx F_d(s)\big|_{m_2=0} = c[1 + H(0)]s\,x(s) \quad (6)$$

With the aid of the above assumption, the 2-dof skyhook model can be simplified to the 1-dof skyhook model with two coil spring serially connecting to each other and the ideal skyhook damper, as shown in the third column of Table 1. Then, the skyhook damping coefficient can be analytically determined as $c = 2\zeta\sqrt{m_1k}$, where $k = k_1k_2/(k_1 + k_2)$. On the other hand, the validity of the introduced assumption can be investigated in the frequency domain, as follows.

Journal of Physics: Conference Series **149** (2009) 012068 doi:10.1088/1742-6596/149/1/012068

$$H(s)/H(0) = 1/(\hat{s}^2 + 1) \tag{7}$$

where $\hat{s} = s/\sqrt{k_2/m_2}$. Fig. 3 shows the effect of the lower mass on the vibration control performance of the simplified skyhook controller. From this figure, it can be seen that the simplified skyhook controller (6) give the similar control performance to the ideal one (5) within the frequency bandwidth less than the natural frequency of the landing gear $\omega_2 = \sqrt{k_2/m_2}$. In real situation, the natural frequency of the aircraft body $\omega_1 = \sqrt{k_1/m_1}$ is much smaller than that of the landing gear. In other words, the lower mass has little effect on the performance of the skyhook controller intended to attenuate the vibration of the aircraft body. Therefore, it is concluded that the simplified skyhook controller is useful for the landing gear system, in terms of the practicality and implementability of the algorithm rather than its perfection.

Table 1. Schematic diagrams of the vibration systems

2-dof general model	2-dof skyhook model	Simplified model	1-dof skyhook model
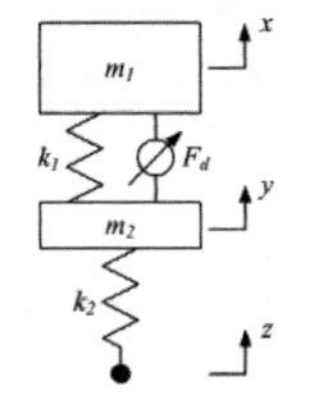	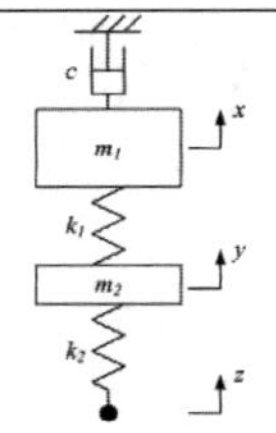	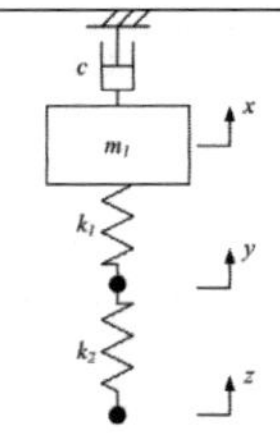	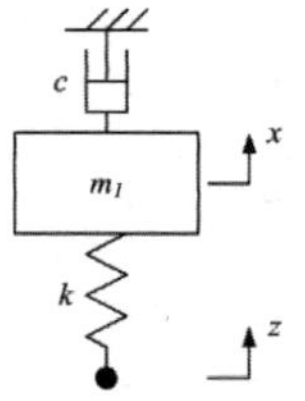

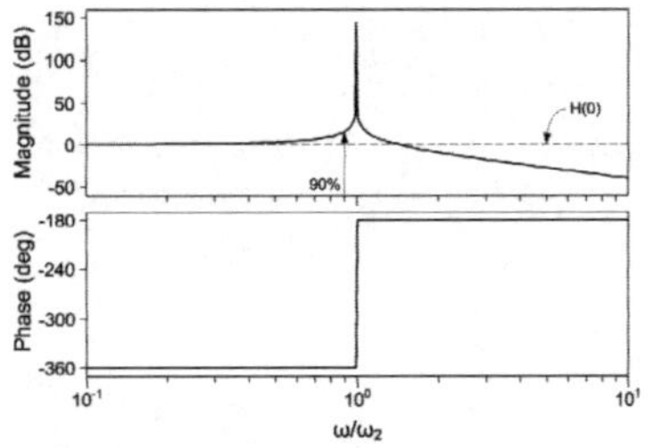

Fig. 3. Relative transmissibility

4. Simulation result

The vibration control performance of the landing gear system featuring the MR damper is investigated through a series of simulations. The specifications of the commercial aircraft 'Boeing 707' are used for simulations; m_1 = 68,210 kg, m_2 = 1,140 kg, k_1 = 11,250,000 N/m, k_2 = 8,461,000 N/m, and c = 105,400 Ns/m [7]. The sink velocity of the aircraft is set to 0.76 m/s, which can be considered as the practical condition that the aircraft is landing softly [8]. Also, the feasibility of the proposed skyhook controller is investigated through the comparison with the Min-Max controller which is a kind of bang-bang controls. Considering the damping force of the passive-type damper in the real aircraft, the capacity of the MR damper is scaled by 18.5. The input coil current is limited to 5.0 A.

On the other hand, the simplified skyhook controller (6) is designed in the active actuator manner. However, since the MR damper is a semi-active actuator, the following semi-active condition should be superposed on the input coil current;

$$I = \begin{cases} I & for \quad \dot{x}(\dot{x} - \dot{y}) \geq 0 \\ 0 & for \quad \dot{x}(\dot{x} - \dot{y}) < 0 \end{cases} \tag{8}$$

This semi-active condition ensures the increase in energy dissipation of the MR landing gear system. It is well known that semi-active dampers can attenuate vibrations as well as active dampers without additional cost and complication required in the active damper system.

Fig. 4 shows the vibration control performance of the landing gear system according to the controller. It can be seen that compared with the passive damper, the MR damper gives better performances in attenuating the displacements. Also, the accelerations of the upper mass are most significantly reduced by employing the simplified skyhook controller to the MR landing gear system. The prime purpose of adopting the skyhook controller is to improve the ride comfort of the aircraft by minimizing the acceleration of the upper mass. Since this strategy does not pay attention to the vibration of the lower mass, the maneuverability of the vehicle might be deteriorated due to excessive vibrations of the lower mass. Obviously, the ride quality and the drive stability are two conflicting criteria in the vibration control of the landing gear system. For the reason, the use of the MR damper increases the acceleration of the lower mass. Nevertheless, the skyhook controller gives better performance in attenuating the acceleration of the lower mass than the Min-Max controller. Also, it is notable that the skyhook controller requires less energy than the Min-Max controller. Therefore, it is concluded that the MR damper featuring the simplified skyhook controller is quite effective to improve the vibration control performance of the aircraft landing gear system in the landing phase.

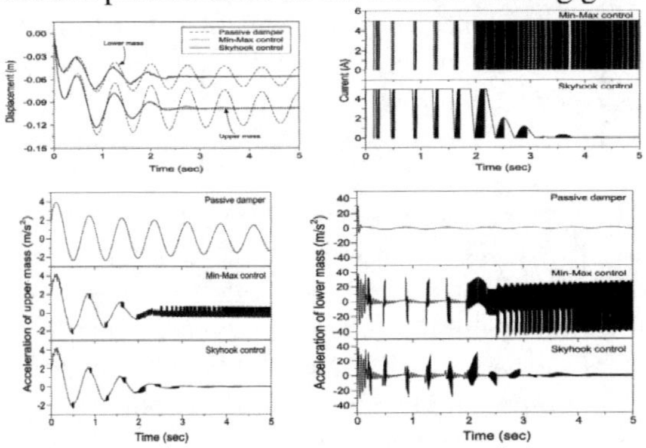

Fig. 4. Control performance of the landing gear system according to the controllers

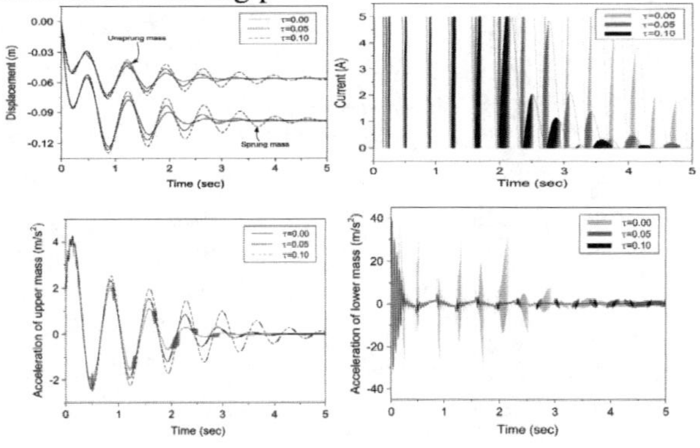

Fig. 5. Control performance of the landing gear system according to the time constants of the MR damper

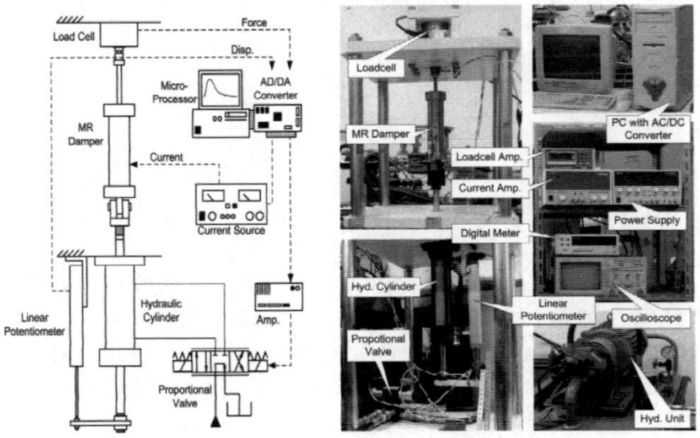

Fig. 6. Apparatus of the HILS for the performance evaluation of the landing gear system

Fig. 5 shows the vibration control performance according to the characteristic time constant of the MR damper featuring the simplified skyhook controller. When the time constant is small, the displacements of the upper and lower masses as well as the acceleration of the upper mass are reduced. However, its opposite is true in the case of the acceleration of the lower mass. In other words, fast response of the MR damper can deteriorate the vehicle stability of the aircraft. If a low-pass filter is employed to adjust the response time of the MR damper, then the relative importance of the ride comport versus the vehicle stability can be controlled considering the operating condition of the landing gear system. Among many other variations of the skyhook controller proposed to overcome the drawback of the original skyhook controller, the groundhook and hybrid controller have been paid much attention in the literature. The groundhook controller is similar to the skyhook controller, except that it is intended to improve the vehicle stability by minimizing the acceleration of the lower mass. The hybrid controller combines the concept of skyhook and groundhook controller to take advantage of the benefits of both, where a weighing factor is employed to determine the relative ratio between both controllers. Note that the time constant of the MR damper plays a similar role to this weighting factor. Therefore, the time constant of the MR damper can be considered as one control gain which is controllable with the low-pass filter.

5. Experimental results: HILS

In this chapter, HILS (Hardware-in-the-loop simulation) is used to evaluate the vibration control performance of the MR landing gear system. In general, the theoretical analysis is rather difficult to precisely predict the performance of the system, since many real situations are often neglected or approximated during its modeling and emulating. Therefore, the HILS method is frequently adopted in order to overcome this limitation. Especially, this method has major advantages such as easy modification of system parameters and wide application of operating conditions [9][10]. Fig. 6 shows the apparatus of HILS constructed in this study. The MR damper is excited by a hydraulic cylinder at the given relative displacement, which is calculated from the difference in displacement between the sprung and unsprung masses. Here, a proportional hydraulic valve is employed for the hydraulic cylinder to track its desired displacement even in the presence of the varying damping force of the MR damper. The coil current input to the MR damper can be adjusted by the current amplifier which is controlled by the personal computer with the DC converter. During operations, the damping force and displacement of the MR damper are measured by a loadcell mounted on the base frame and a linear potentiometer connected with the hydraulic cylinder, respectively. The sensor signals are transmitted to the personal computer via the AD converter with its sampling frequency of 1000 Hz. Fig. 7 shows the results of the HILS conducted under the same conditions to the simulations. It can be seen that the HILS results are well agreed with the simulation ones. However, there is little difference in the displacement. It is considered that it results from several assumptions introduced in the modeling of the MR damper. In all the simulations, the time constant of the MR damper is assumed to be constant, but it can be varied according to the excitation velocity or current-induced condition as mentioned above. Also, MR fluid behavior actually exhibits some departures from the Bingham model due to its viscoelastic and hysteresis property. Nevertheless, we can conclude that the proposed MR damper featuring the simplified skyhook controller is effective for attenuating undesired vibration of the landing gear system, with the viewpoint of the practicability and implementability rather than the perfection.

6. Conclusions

With the help of the simplified skyhook controller proposed in this work, the applicability of the MR damper constructed as Pusan National University to the aircraft landing gear system was investigated through a series of simulations. The obtained results can be summarized as follows; (i) the MR damper featuring the simplified skyhook controller is effective for attenuating undesired vibration of the landing gear system. (ii) The control gain of the skyhook controller can be analytically obtained with the assumption that the mass of the landing gear system is much smaller than the mass of the aircraft body and therefore is negligible. (iii) The time constant of the MR damper is one important factor determining the vibration control performance of the MR landing gear system, and plays a role similar to the weighting factor in the hybrid controller combining the function of the skyhook and groundhook controllers. (iv) Using the HILS method, it is verified that the proposed MR damper featuring the simplified skyhook controller is effective for attenuating undesired vibration of the landing gear system.

References

[1] P. Phule, Smart Materials Bulletin, 2, 7-10 (2001).
[2] M. R. Jolly, J. W. Bender, and J. D. Carlson, J. of Intell. Mater. Syst. Struct., 10, 5-13 (1999).
[3] Y. J. Nam and M. K. Park, J. Intell. Mater. Syst. Struct., (2007) (submitted).
[4] Y. J. Nam and M. K. Park, J. Intell. Mater. Syst. Struct., 18, 707-718 (2007).
[5] G. J. Stelzer et al J. Intell. Mater. Syst. Struct., 14, 743. (2003)
[6] D. C. Karnopp and M. J. Cosby U.S. Patent 3,807,678 (1974).
[7] C. C. Tung, J. Penzien, and R. Hornonjeff, NASA CR-119 (1964).
[8] Y. T. Choi and N. M. Wereley, Journal of Aircraft, 40(3), 432-439 (2003).
[9] Lee, H. S. and Choi, S. B., 2000, J. Intell. Mater. Syst. Struct., Vol. 11, pp. 80
[10] Hanselmann, H., 1993, SAE International Congress and Exposition, SAE paper 930207.

Ride Comfort Characteristics with Different Tire Pressure of Passenger Vehicle Featuring MR Damper

H G Lee[1], K G Sung[2] and S B Choi[3]

[1] Department of Automotive Engineering, Daeduk College, Daejeon 305-715, Korea
[2] Department of Mechanical Engineering, Yeungnam College of Science & Technology, Daegu 705-703, Korea
[3] Smart Structures and Systems Laboratory, Department of Mechanical Engineering, Inha University, Incheon 402-751, Korea

E-mail: seungbok@inha.ac.kr

Abstract. This paper presents ride comfort characteristics of a quarter-vehicle magneto-rheological (MR) suspension system with respect to different tire pressure. As a first step, controllable MR damper is designed and manufactured based on the optimized damping force levels and mechanical dimensions required for a commercial mid-sized passenger vehicle. After experimentally evaluating dynamic characteristics of the manufactured MR damper, the quarter-vehicle suspension system consisting of sprung mass, spring, tire and the MR damper is constructed in order to investigate the ride comfort. After deriving the equations of the motion for the proposed quarter-vehicle MR suspension system, vertical tire stiffness with respect to different tire pressure is experimentally identified. The skyhook controller is then implemented for the realization of quarter-vehicle MR suspension system. Ride comfort characteristics such as vertical acceleration RMS of sprung mass are evaluated under bump road condition and presented in time domain.

1. Introduction

The vehicle dynamic characteristics such as ride comfort and steering stability can be normally improved by suspension systems [1-2]. So far, passive oil damper, which provides design simplicity and cost-effectiveness, is widely employed for conventional vehicles. However, the performance limitation is inevitable due to uncontrollable damping force. Recently, active damper using a motor and electro-servo hydraulic valve is gradually used to achieve ride comfort as well as steering stability. However, design complication and high cost prevent the popularization of advanced active suspension systems. Therefore, alternative mechanisms for vehicle suspension systems have been studied to replace the passive and active dampers. For the last decade, the research work on ride comfort and steering stability of a vehicle system using semi-active suspension has been significantly increased. The semi-active suspension system using magnetorheological (MR) damper can offer a desirable performance in the active mode without requiring large power consumption and expensive hardware. MR fluid, whose rheological properties can be rapidly changed by applying a magnetic field to the fluid domain, has been used in many devices such as shock absorbers, valves, engine mounts and clutch/ brake systems. Recently, Choi et al. [3] manufactured a MR damper for a small-sized passenger vehicle and presented its control characteristics of the damping force. On the basis of this work, they extended their research and evaluated control performance of the proposed MR damper via hardware-in-the-loop simulation [4].

The main contribution of this work is to present ride comfort characteristics of a quarter-vehicle suspension system equipped with continuously controllable MR damper with respect to different tire pressure. As a first step, controllable MR damper is designed and manufactured based on the optimized damping force levels. After experimentally evaluating dynamic characteristics of the manufactured MR damper, equation of the motion for the proposed quarter-vehicle suspension system consisting of sprung mass, spring, tire and the MR damper is derived in order to investigate the ride comfort. Subsequently, vertical tire stiffness with respect to different tire pressure is experimentally identified. Ride comfort characteristics such as vertical acceleration RMS of sprung mass are evaluated under bump road condition.

2. MR Damper

The schematic configuration of the MR damper proposed in this work is shown in figure 1. The MR damper is composed of the cylinder, piston and gas chamber. The floating piston is also used in order to compensate the volume induced by the motion of the piston. The cylinder and the gas chamber in the MR damper are fully filled with the MR fluid and the nitrogen gas, respectively [5].

In the absence of the magnetic field, the MR damper produces the damping force only caused by the fluid viscous resistance. However, if a certain level of the magnetic field is supplied to the MR damper, the MR damper produces additional damping force owing to the yield stress of the MR fluid. This damping force of the MR damper can be continuously tuned by controlling the intensity of the magnetic field. Therefore, the damping force of the proposed MR damper can be written as follows:

$$F_D = k_e x_p + c_e \dot{x}_p + F_{MR} \tag{1}$$

where k_e is the effective stiffness due to the gas pressure, c_e is the effective damping due to the fluid viscosity, x_p is the excitation displacement, and F_{MR} is the field-dependent damping force. The controllable damping force F_{MR} can be expressed by

$$F_{MR} = (A_p - A_r)2\frac{cL_p}{h}\tau_y(B)\cdot \mathrm{sgn}(\dot{x}_p) \tag{2}$$

where c is a coefficient which depends on flow velocity profile, L_p is the length of the magnetic pole, h is the gap between outer and inner cylinder and $\tau_y(B)$ is the yield stress caused by the magnetic flux density. A_p and A_r represent piston and piston rod areas, respectively; and $\mathrm{sgn}(\cdot)$ is a signum function.

The size and the level of required damping force is determined so that the MR damper can be applicable to a commercial passenger vehicle. In order to obtain optimal geometric dimensions of the proposed MR damper, commercial FEM software is adopted [5]. Figure 2 presents the damping force characteristics of the optimal MR damper with respect to the piston velocity at various magnetic fields. This is obtained by calculating the maximum damping force at each velocity.

3. Quarter-Vehicle Model

In this study, a quarter-vehicle MR suspension system is established to evaluate control performances of the manufactured MR damper with optimal parameters. Figure 3 shows the quarter-vehicle model

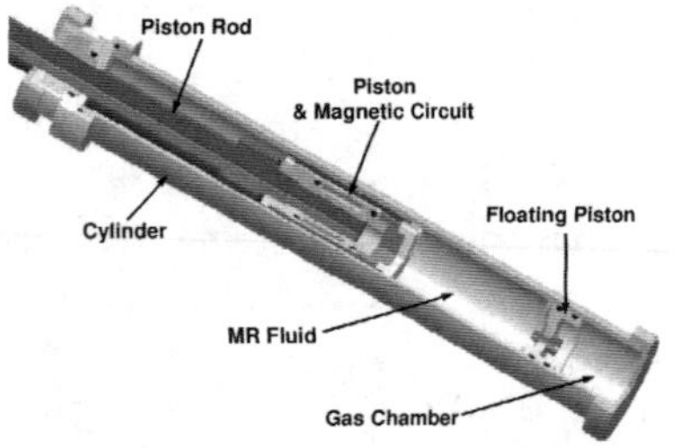

Figure 1. Schematic configuration

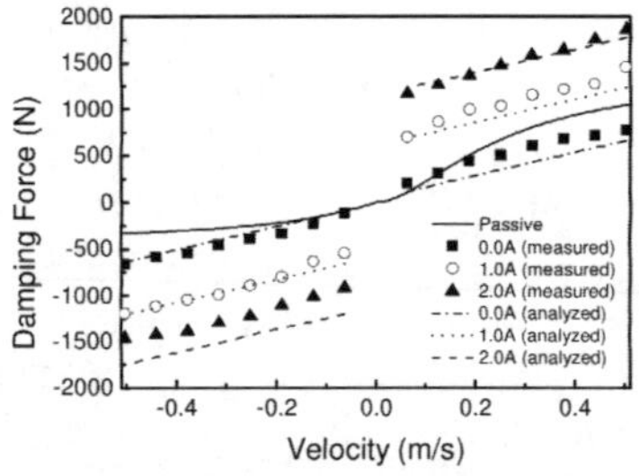

Figure 2. The field-dependent damping force

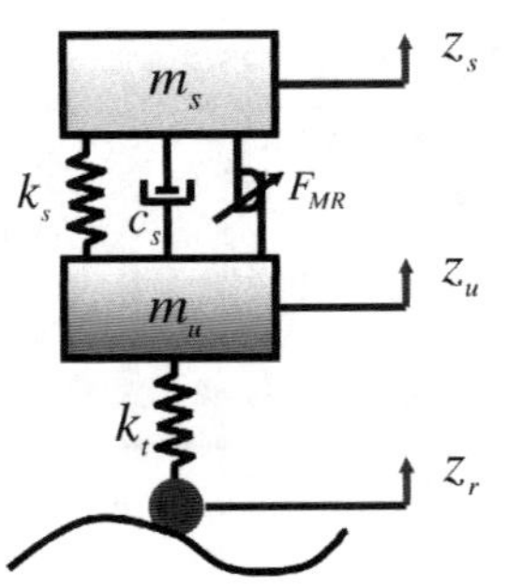

Figure 3. Quarter vehicle model of MR suspension

Figure 4. Stiffness test of the tire

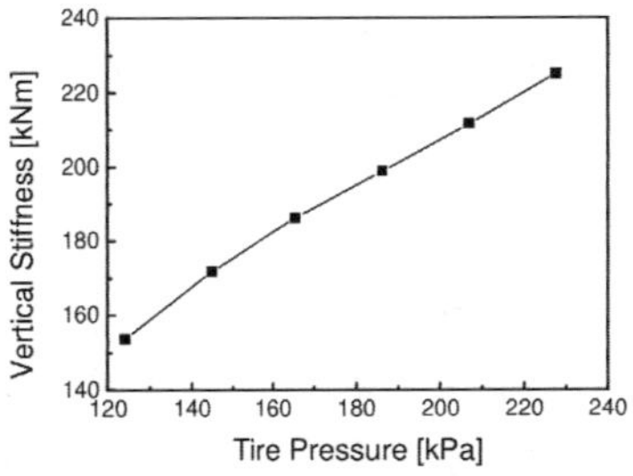

Figure 5. Vertical stiffness of the tire

of the semi-active MR suspension system which has two degree-of-freedom. m_s and m_u represent the sprung mass and unsprung mass, respectively. The spring for the suspension is assumed to be linear and the tire is also modeled as linear spring component. Now, by considering the dynamic relationship, the state space control model is expressed for the quarter-vehicle MR suspension system as follows:

$$\dot{x} = Ax + Bu + Lz_r \quad (3)$$
$$\dot{y} = Cx$$

where

$$x = [z_s \quad \dot{z}_s \quad z_u \quad \dot{z}_u \quad F_{MR}]^T$$

$$A = \begin{bmatrix} 0 & 1 & 0 & 0 & 0 \\ -\frac{k_s}{m_s} & -\frac{c_s}{m_s} & \frac{k_s}{m_s} & \frac{c_s}{m_s} & -\frac{1}{m_s} \\ 0 & 0 & 0 & 1 & 0 \\ \frac{k_s}{m_u} & \frac{c_s}{m_u} & -\frac{k_t + k_s}{m_u} & -\frac{c_s}{m_u} & \frac{1}{m_u} \\ 0 & 0 & 0 & 0 & -\frac{1}{\tau} \end{bmatrix}, \quad B = \begin{bmatrix} 0 & 0 & 0 & 0 & \frac{1}{\tau} \end{bmatrix}^T$$
$$L = \begin{bmatrix} 0 & 0 & 0 & \frac{k_t}{m_u} & 0 \end{bmatrix}^T$$
$$C = [1 \quad 0 \quad 0 \quad 0 \quad 0]$$

where k_s is the total stiffness coefficient of the suspension including the effective stiffness k_e of the MR damper in equation (1). c_s is the damping coefficient of the suspension and it is assumed to be equal to c_e. k_t is the stiffness coefficient of the tire. z_s, z_u and z_r are the vertical displacement of sprung mass, unsprung mass and road excitation, respectively. In addition, τ is the time constant of the MR damper and u is the control input.

The photograph of vertical stiffness test for the tire is shown in figure 4. Figure 5 presents the vertical stiffness characteristics of the tire with respect to different tire pressure. It is clearly observed that the vertical stiffness is increased as tire pressure increases. The vertical stiffness of 153kNm at tire pressure 124kPa is increased up to 225kNm at 227kPa.

Among many potential candidates for control algorithms, the skyhook controller is adopted in this work. It is well known that the logic of the skyhook controller is simple and easy to implement to real field [5]. The desired damping force is set by

$$u = C_{sky} \cdot \dot{z}_s \quad (4)$$

where C_{sky} is the control gain which physically indicates the damping coefficient. In above equation, control input (u) directly represents the damping force of F_{MR}.

4. Performance Evaluation

Ride comfort characteristics of the quarter-vehicle MR suspension system with respect to different tire pressure are evaluated under bump road excitations [3]. Figure 6 and 7 presents ride comfort comparison of the quarter-vehicle MR suspension system under bump road excitation with constant vehicle velocity of 3.08km/h. As shown in figure 6, it is obvious that unwanted vibrations induced from the bump excitation have been well reduced by adopting the MR damper associated with the

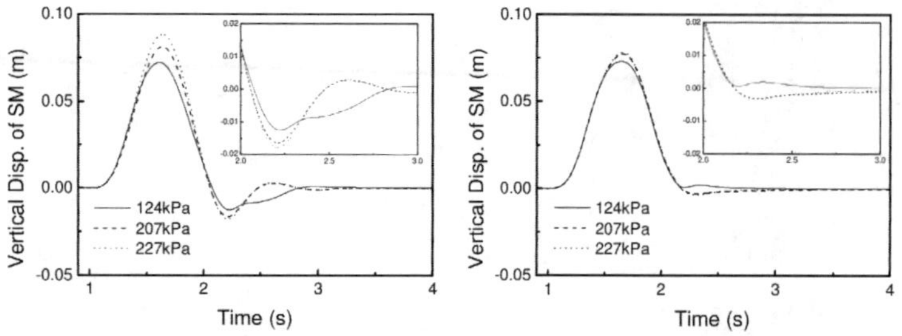

(a) uncontrolled (b) controlled

Figure 6. Bump responses of the quarter-vehicle MR suspension system

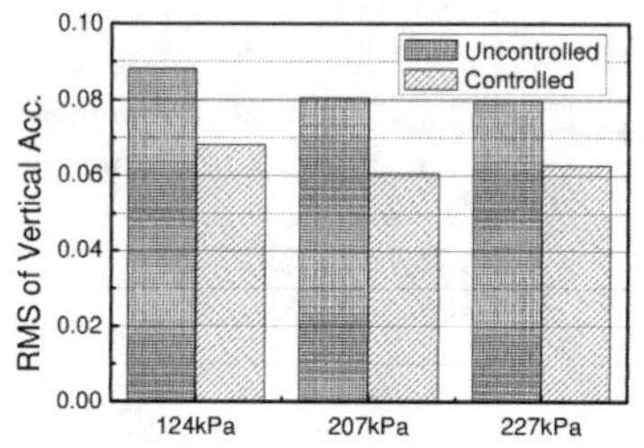

Figure 7. Ride comfort comparison of vertical acceleration RMS

skyhook controller. It is also observed from the vertical displacement of sprung mass that the control performance of the controlled case is better than that of the uncontrolled case independent of tire pressure. The RMS (root mean square) of vertical acceleration ($\ddot{z}_s$) shown in figure 7 indicate that ride comfort characteristics of a vehicle system can be substantially improved by employing the MR damper and skyhook controller even though tire pressure is changed during driving.

5. Conclusion

In this work, ride comfort characteristics of a quarter-vehicle magneto-rheological suspension system with respect to different tire pressure was undertaken. The MR damper was designed and manufactured based on the optimized damping force levels and mechanical dimensions required for a commercial mid-sized passenger vehicle. After experimentally evaluating dynamic characteristics of the manufactured MR damper, the quarter-vehicle suspension system consisting of sprung mass, spring, tire and the MR damper was constructed in order to investigate the ride comfort. After deriving the equations of the motion for the proposed quarter-vehicle MR suspension system, vertical tire stiffness with respect to different tire pressure was experimentally identified. The skyhook controller has been then implemented for the realization of quarter-vehicle MR suspension system. Ride comfort characteristics such as vertical displacement and acceleration RMS of sprung mass were evaluated under bump road condition. The results presented in this work are quite self-explanatory justfiying that the MR damper with the controller can provide imprioved ride comfort independent of tire pressure.

Acknowledgments

This work was supported by Acceleration Research (Center for ER Fluid Technology and Application) of MOST/KOSEF. This financial support is gratefully acknowledged.

References

[1] T.D. Gillespie, *Fundamentals of Vehicle Dynamics*, Society of Automotive Engineers, Warrendale, PA, 1992.

[2] E.M. Elbeheiry, D.C. Karnopp, M.E. Elaraby and A.M. Abdelraaouf, "Advanced ground vehicle suspension systems - a classified bibliography", *Vehicle System Dynamics* **24**, 231-258 (1995).

[3] H.S. Lee and S.B. Choi," Control and response characteristics of a magnetorheological fluid damper for passenger vehicles", *Journal of Intelligent Material Systems and Structures* **11**, 80-87 (2000).

[4] S.B. Choi and K.G. Sung, "Vibration control of magnetorheological damper system subjected to parameter variations", *International Journal of Vehicle Design* **46**, 94-110 (2008).

[5] K.G. Sung, Y.M. Han and S.B. Choi, "Geometric Optimization of Controllable Magnetorheological Shock Absorber for Commercial Passenger Vehicle", *Proceedings of the SPIE's 11th Annual Symposium on Smart Structures and Materials*, 9-13 March, 2008, San Diego, California.

Journal of Physics: Conference Series **149** (2009) 012070 doi:10.1088/1742-6596/149/1/012070

Theoretical and experimental analysis of MR valve

Z Leicht[1], H Urreta[1], A Sanchez[1], A Agirre[1], P Kuzhir[2] and G Magnac[3]

[1]IDEKO, S. COOP. Arriaga Kalea 2, 20870 Elgoibar, Spain
[2]LPMC, University of Nice – Sophia Antipolis, 06108 Nice Cedex2, France
[3]Cedrat Technologies, 15 Chemin de Malacher - Inovallée - 38246 Meylan, France

E-mail: zleicht@ideko.es

Abstract. The properties of magnetorheological (MR) fluid can be rapidly varied by the application of a magnetic field. This behaviour allows the designer to construct a machine that's quality can be changed in action, according to the variation of the surround and to the expectations. The commercial use of MR fluid is already not limited in dampers and breaks. Thanks to the advantageous quality - that requires low voltage - is on the increase. Using the MR fluid in a valve, the pressure drop can be adjusted by the intensity of the magnetic field, without moving parts. In this work a MR valve has been designed, that can supply a hydrostatic bearing lubricated with magnetic fluid. Its behaviour has been simulated with three models. The analytical model based on the Bingham law of the magnetic fluid flow, the Buckingham model (Bingham modified) and the dimensional model suggested by Lord Corporation, the manufacturer of used MR fluid, MRF 122 2EG. The results of the simulations are compared with the experimental data.

1. Introduction

In this work a novel MR valve has been designed and manufactured to adjust the flow rate and the pressure drop of the MR fluid feeding any magnetic device, such as a magneto-hydrostatic bearing, end application of the project. The experimental results have been compared to a model based on the Non-Newtonian viscous theory, since the MR fluid behavior differs from the Newtonian viscous equation. According to the model, the MR fluid possesses a yield stress below that the fluid behaves like a semi-solid, or even a solid. Just after the shear stress exceeds this value is allowed of flow. The MR fluid used during the experiment is a product from Lord Corporation (MRF 122 2EG).

2. Model for non-Newtonian behaviour

The MR fluids in general case behave corresponding to the Newtonian constitutive equations. Applying a magnetic field the characterization of the fluid alters into a non-Newtonian fluid. There are many different theories presented in the works of Carlson and Jolly [1,2], Wereley [3] and Goncalves [4] prescribing the material with distinct behavior from Newtonian fluids. The Bingham model ensures a base to analyze the MRF-122-2EG from Lord Corporation. This fluid possesses a yield stress τ_0. The fluid with shear stress below this value behaves like a solid. If the shear stress exceeds the yield stress the fluid starts to flow. This way, Bingham model of the shear stress acting on the MR fluid is given in the equation (1):

$$\tau = \left(\tau_0(H) + K\,\dot{\gamma}\right)\operatorname{sgn}(\dot{\gamma}). \tag{1}$$

Where H is the field intensity, K is the viscosity with no magnetic field applied and $\dot{\gamma}$ is the shear rate.

3. Magnetorheological valve

In a MR valve, the MR fluid flows through the duct between inflow core (B_1) and the outflow core (B_2) as it is depicted in figure 1.a. When the power of coil (C) is turned on and magnetic field is exerted on MR fluid, the flow through the relief valve will change its state into semi-liquid or extremely into solid. Only when the supply pressure gets high enough to offset the yield stress, the fluid can flow through the valve again. Thus, this type of relief valve can regulate the pressure of a MR fluid system under a certain flow rate.

The test bench configuration is depicted in figure 1.

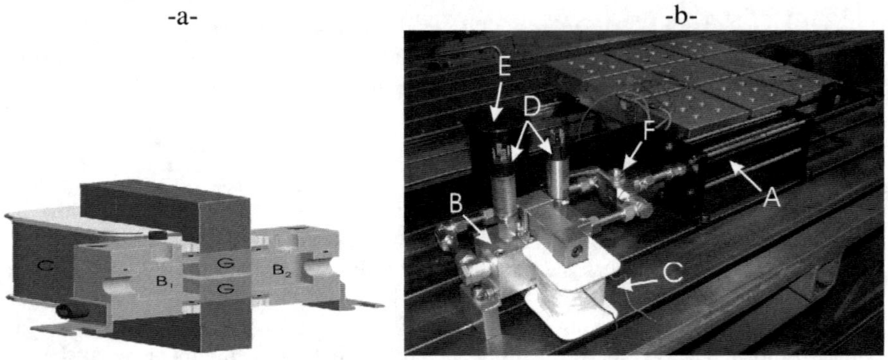

Figure 1. The profile of the MR valve (a) and construction of the test bench (b).

The cylinder (marked with A in the figure 1.b) contains the MR fluid. Its moving supports the constant flow rate to the valve (B). With the electromagnet (C) the pressure drop of the valve can be changed. The inflow and outflow pressures are measured by the sensors (D). The fluid after crossing the valve is collected in the tank (E). After the measurement switching the flap (F) the fluid returns to the cylinder, and the measurement can be repeated.

The pressure drop generated by the gap treads between the inflow core (B_1) and the outflow core (B_2). The piece G of the figure 1a determines the geometry of the gap. Different flow conditions can be examined by replacing this part, and thanks to the test bench design, it can be changed rapidly.

In the figure 2.a a typical velocity distribution of MR fluid is depicted.

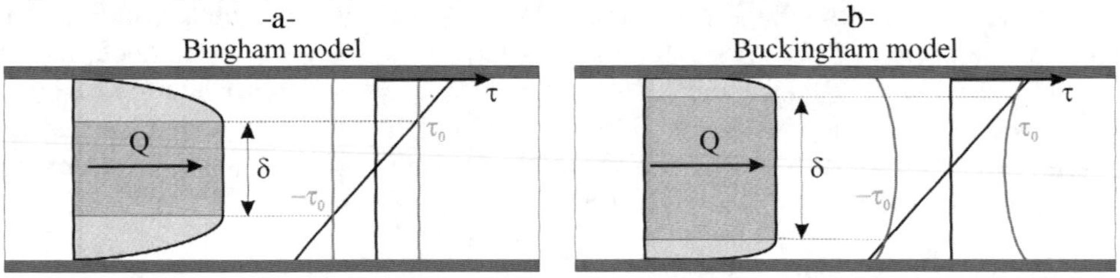

Figure 2. Velocity and shear stress distribution across the gap in the case of (a), Bingham model, and (b), Buckingham model.

The plug thickness δ according to the Bingham model equations is

$$\delta = -\frac{2\tau_0}{d\,p/d\,x}. \tag{2}$$

The pressure drop, supposing a duct length L and a linear pressure gradient:

$$\Delta p = \frac{2\tau_0}{\delta} L. \tag{3}$$

To achieve a more accurate model, the Buckingham theory is applied [5]. It supposes a non-linear magnetic flux distribution across the gap. In the case of relatively large gap, the value of the flux density increases at the walls. With this, the equation (3) is modified in the value of $\tau_0 = f(h)$, being (h) across distance in the gap. Through this larger yield stress, the plug thickness also became larger, as it is depicted in figure 2.b.

4. Comparison the results of different models

The test material MRF-122-2EG (Lord Corporation), is a hydrocarbon-based fluid with high yield stress and fast response time. The geometry of the valve is: length, L=30mm and width, W=28mm. Three different gaps has been examined, namely 1, 2 and 3 mm in the thickness. An iron core and a coil, inducing the magnetic field by the applied current intensity, have generated the magnetic field intensity into the MR fluid.

Three different models have been compared in this work: The solution advised by Lord Corporation (a dimensional solution), Bingham model and the modified Bingham model, named Buckingham model. Bingham solution doesn't fit to experimental data in any of analyzed conditions, this way, yield stress variation proposed by Buckingham model is mandatory to get right results in this kind of MR valves, see figures 3, 4 and 5.

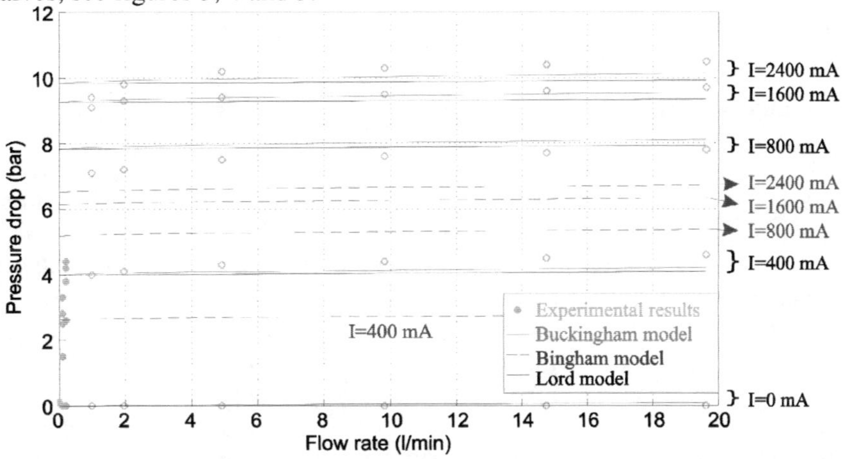

Figure 3. Pressure drop as a function of the flow rate h=3 mm

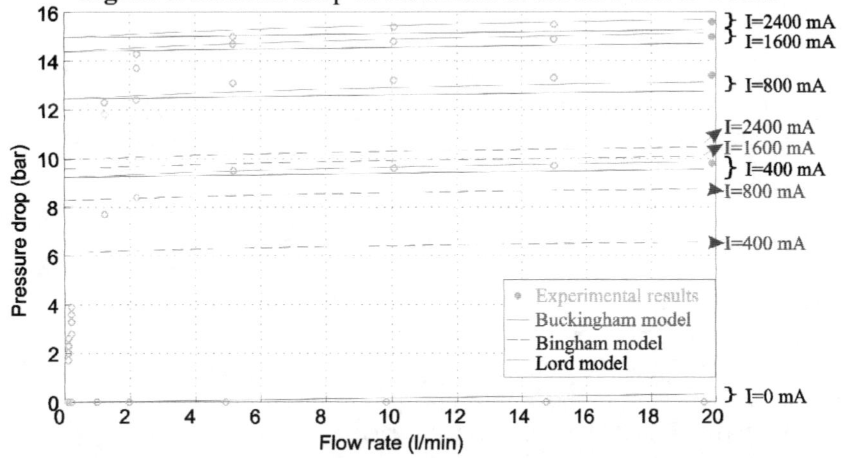

Figure 4. Pressure drop as a function of the flow rate h=2 mm

Comparing Buckingham model and dimensional one (suggested by Lord), the approximation given by fluid's manufacturer agree with experimental results at relative large gaps 2 and 3 mm (figures 3 and 4), meanwhile at small gap (1mm, figure 5) the pressure drop is oversized. That is because the Lord solution supposes a non-linear flux distribution independently to the thickness, and on the other hand, Buckingam theory considers the dependence to the gap's size in the effect of flux density. This way, when the gap thickness is relatively small (1mm in this case) the differences between both models appear, and the solution of Buckingham model agrees with experimental data due to it takes account flux inequality across the valve's gap, providing more accurate solutions.

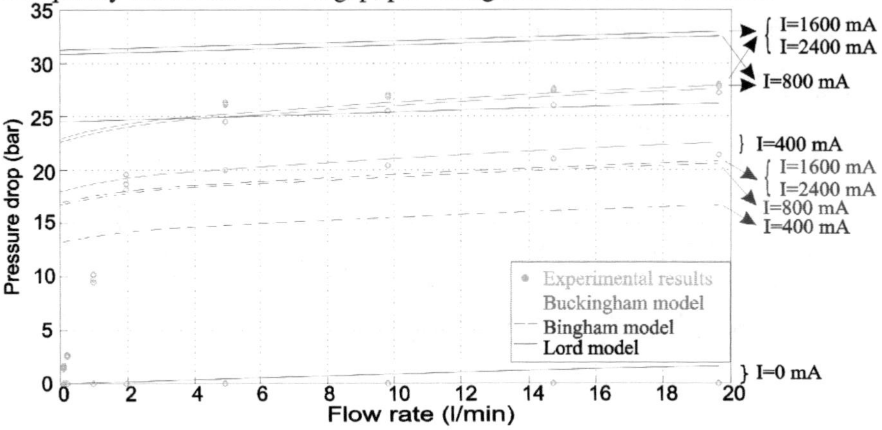

Figure 5. Pressure drop as a function of the flow rate h=1 mm

5. Conclusion

The MR valve designed and manufactured in this work has two favorable features: In one hand, the wide working range in the pressure drop and flow rate, and since the pressure drop is adjusted by the magnetic field intensity, the valve works without moving parts, thus diminishes the chance of failure. And on the other hand, the simple design of magnetic poles provides a quick and low cost geometry modification to try different design's solutions in the test bench. Modeling the valve, i.e. describing the flow-rate in function of pressure drop, three cases have been analyzed. With large gap, the simple and dimensional model proposed by Lord provides enough right results. However with small gap, the dimensional model it's not enough accurate and analytical solutions (based in Bingham theory) are required. Bingham model doesn't fit to this kind of valves due to non-linear magnetic flux distribution. This way, Buckingham model's proposed modification to Bingham theory is applied, and its results fit accurately to experimental data.

References

[1]Jolly, M. R., Bender, J. W., and Carlson, J. D., "Properties and applications of commercial magnetorheological fluids," Journal of Intelligent Material Systems and Structures, Vol. 10, No. 1, p. 5, 1999.

[2]Jolly, M. R., Carlson, J. D., Munoz, B. C., "A Model of the Behavior of Magnetorheological Materials," *Smart Materials and Structures*, Vol. 5, p. 607-614, 1996.

[3]Wereley, N. M., Pang, L., "Nondimensional Analysis of Semi-Active Electrorheological and Magnetorheological Dampers Using Approximate Parallel Plate Models," *Smart Materials and Structures*, Vol. 7, p. 732-743, 1998.

[4]Goncalves, F. D., "Caracterizing the Behavior of Magnetorheological Fluids at High Velocities and High Shear Rates", Dissertation, Blacksburg, Virginia, p.20-25, (2004)

[5]Mikulowski G., Batterby C.D., "Flux Density Variation in Magnetorheological fluid devices" (Poland, Eccomas Thematical Conference on Smart Sstructures and Materials), http://smart.ippt.gov.pl/pdf/2007_ECCOMAS-gdansk_GM-DB.pdf, 2007.

Journal of Physics: Conference Series **149** (2009) 012071 doi:10.1088/1742-6596/149/1/012071

Characterization of sedimentation and high magnetic field flow behavior of some magnetorheological fluids

Marin Lita[1], Adelina Han[1] and Daniela Susan-Resiga[1]

[1]National Center for Engineering of Systems with Complex Fluids, "Politehnica" University of Timisoara, Romania

E-mail adealina@yahoo.com

Abstract: Preparation and characterization of an improved sedimentation rate MR fluid is presented. Three different commercial additives were used for dispersion/stabilization of iron particles with sizes between 4-6 μm in synthetic oil carrier. The structural and magnetorheological properties of the sample (MRF-LM5) were compared with the same characteristics measured for a commercial MR fluid (MRF-132DG, LORD Co., USA) of similar saturation magnetization. The sedimentation rate of particles measured by X-ray transmission in MRF-LM5 is twice slower than in MRF-132DG, while the MR effect is higher for the commercial MR sample.

1. Introduction

One of the important characteristics of MR fluids is their significant gravitational sedimentation rate. Slowing down the particle settling velocity is an actual research task and involves various solutions concerning the morphology and size range of magnetic particles, as well as the use of various additives. A good compatibility of additives with the carrier and the iron particles reduces sedimentation rate and irreversible agglomeration of particles. In order to ensure long-term controllable and reproducible MR effect, the carrier liquid with additives has to keep its stability, while the suspended iron particles should be magnetically soft [1-4].

In this work a laboratory prepared MR fluid, with improved sedimentation behavior is investigated and compared with the properties of a commercial MR fluid (MRF-132DG, Lord Co., USA).

2. Experiments and discussion

2.1. MRF samples

For the preparation of MRF-LM5 samples micropowder of iron (99.5% Fe, 80g) with particle sizes between 4-6 μm [5] was used. Three different carrier liquids (20g) were experimented (Castrol oil, transformer oil and silicon oil), which are stable in the temperature interval - 40 °C, + 130°C. Also, several additives were considered to prevent sedimentation and coagulation, as well as to facilitate dispersion of particles (liquid and/or powder); finally there were selected three additives denoted Thix (1.5g), HT (1g) and Pur 8050 (3g). These additives are commercial products (Q Speciality) designed for structuring the carriers and there were selected taking into account their characteristics and the observed improvements of the sedimentation stability of MRF samples. The density of the selected MRF-LM5 sample was 2.75 g/cm^3. The MRF-132DG fluid, product of Lord Co.(USA), is a

Journal of Physics: Conference Series **149** (2009) 012071 doi:10.1088/1742-6596/149/1/012071

suspension of micron-sized (mean size 8.87 μm), magnetizable particles in a hydrocarbon carrier fluid. This sample has the following characteristics [6]: density 2.98-3.18 [g/cm^3], solid content by weight 80.98 [%].

2.2. Dimensional and morphological analysis of particles in MRF-LM5

In order to determine the shape and mean size of iron particles of the powder dispersed, several MRF-LM5 type samples were analyzed.

These were deposited on 300 mesh supports for transmission electron microscopy investigations (Tesla BS-613; U = 50 kV, I = 80mA). Investigations performed at 1000X magnification resulted in TEM images (Fig. 1) which show that the iron particles in suspension are of spherical shape and have close sizes. Data analysis referring to more than 400 particles conducted to the mean size $\bar{d} = 4.32 \mu m$ and to the histogram of relative frequency (Fig. 2).

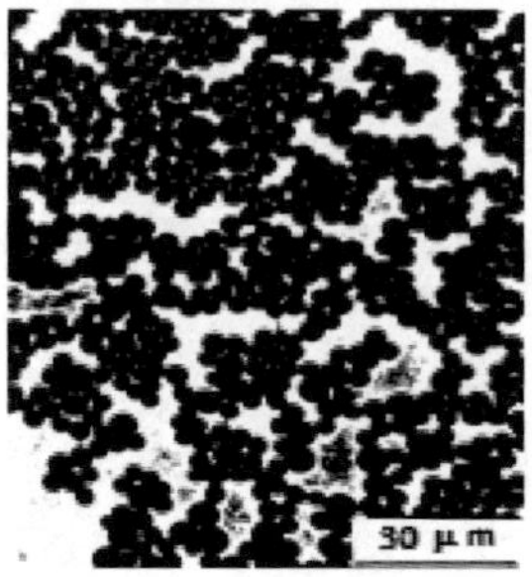

Figure 1. Shape, size and distribution of particles in a sample of MRF-LM5

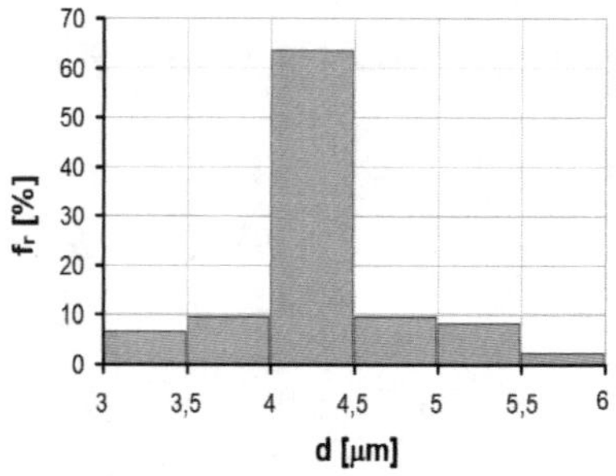

Figure 2. Relative frequency histogram of particle sizes in MRF-LM5

For the determination of the state of internal tensions the iron powder was supposed to X-ray diffraction investigations using a „Dron 3" type diffractometer with molybdenum radiation ($\lambda_{\alpha1,\alpha2}$= 0.71 Å, U = 40 kV, I = 30 mA) in the angular interval 2θ ∈ (14°, 65°), the detector having the speed of 1°/min (Fig 3).

In Fig. 3.a the spectral lines of the doublet $\lambda_{k\alpha1}$, $\lambda_{k\alpha2}$ are splitted already at small 2θ angles (2θ = 35°) and at large 2θ angles are almost completely splitted. That means that the particles of this sample do not have internal stresses and, consequently, they have the characteristics of a soft magnetic material. A similar structural behaviour is specific also to the commercial sample MRF-132DG (Fig. 3.b.)

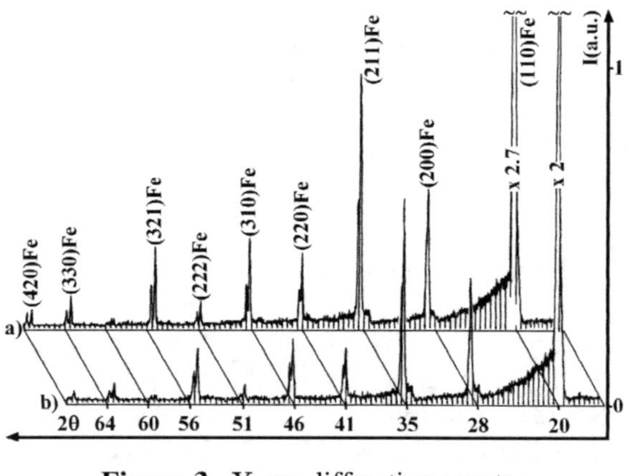

Figure 3. X-ray diffraction spectra
a) MRF-LM5; b) MRF-132DG

2.3. Analysis of gravitational settling by X-ray transmission

For the analysis of gravitational sedimentation of particles the set-up of the „Dron-3" X-ray diffractometer was used, with the X-ray radiation generated by the molybdenum tube.

To perform the sedimentation rate measurement, each sample was first homogenised and after it was introduced in a rectangular Plexiglas sample holder having the dimensions 9x4x400mm. The filled sample holder was placed in vertical position in front of the X-ray detector, the radiation passing through the 4 mm thick layer of MRF. For having the maximum radiation intensity I_t, transmitted through the MRF layer at the bottom of the sample holder, the detector was rotated in the angular interval 2θ ∈ (-0.5°, +0.5°) with the speed of (1/2)°/min.

The X-ray transmission experiment was repeated four times, at one week time intervals, keeping the same experimental conditions for each sample. The variation of particle concentration is represented in Fig. 4 by the dependence of the maximum intensity of the transmitted radiation as a function of time. Also, in Fig 4 the mass density (ρ) of the samples is represented after each time interval.

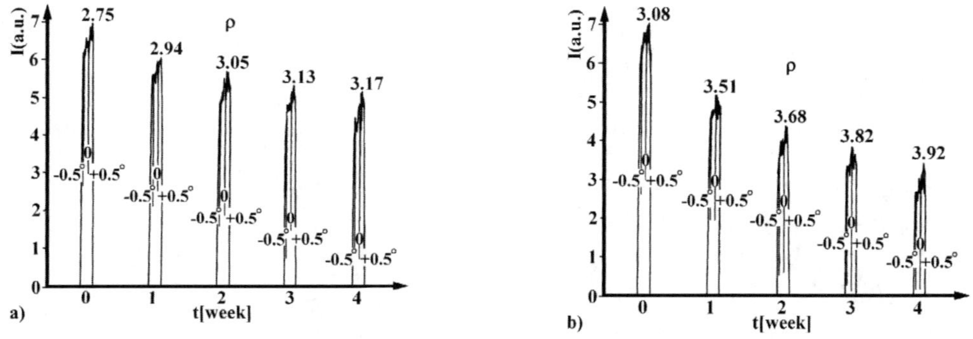

Figure 4. Gravitational sedimentation at 1, 2, 3 and 4 weeks time intervals (ρ - the corresponding density values) a) MRF-LM5; b) MRF-132DG

By comparing the transmitted X-radiation intensities and the corresponding density values for the two types of MR fluids analyzed, it follows that the laboratory prepared sample MRF-LM5 has better gravitational sedimentation behaviour than the commercial sample MRF-132DG.

2.4. Magnetic and magnetorheological characterization

The full magnetization curves of MRF samples were determined by vibrating sample magnetometry (VSM 880, DMS(USA)). The samples have close values of saturation magnetization, 620 kA/m for MRF-LM5 and 660 kA/m for MRF-132DG.

Magnetorheological measurements were performed using a PHYSICA MCR300 rheometer with plate-plate MR cell (MRD180), having 20 mm diameter and 0.5 mm gap size. Oscillatory (Amplitude Sweep) and rotational (Flow Curve) tests were performed at the temperature T = 20 °C, in the absence of the field (B=0) and in applied field (B = 0, 64.8, 125, 187, 253, 379 and 502 mT).

To describe the behavior of the two MR fluid samples, two analytical models were used (Fig 5). In the absence of the field the model of Carreau-Yasuda [7], while for the flow behaviour in magnetic field the model developed by Susan-Resiga proved to be adequate [8]:

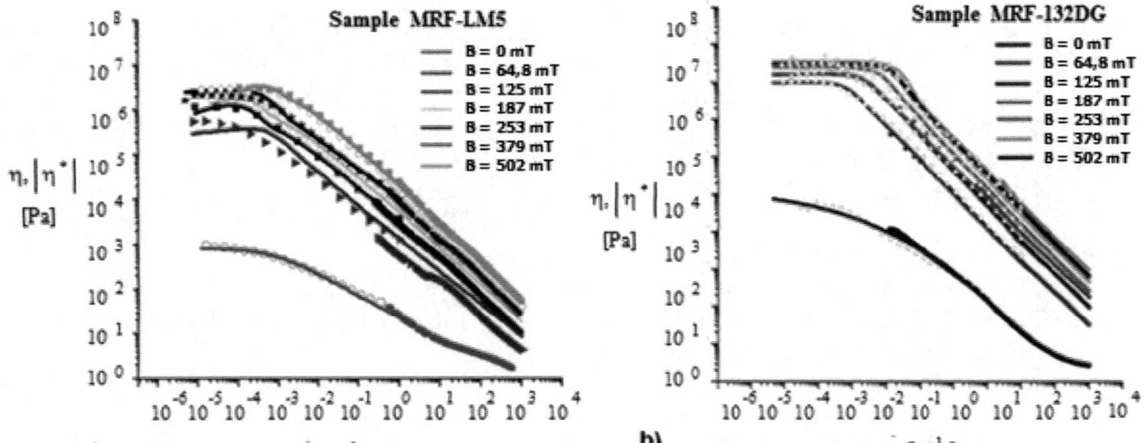

Figure 5. Viscosity curves at T=20°C: a) MRF-LM5; b) MRF-132 DG

$$\eta = \eta_0\left[1 - \tanh\left(\dot{\gamma}/\dot{\gamma}^*\right)\right] + \left[\left(\tau_0/\dot{\gamma}\right) + \left(\tau_1/\dot{\gamma}\right)\left(\dot{\gamma}/\dot{\gamma}^*\right)^{1-n}\right]\tanh\left(\dot{\gamma}/\dot{\gamma}^*\right) \quad (1)$$

where: η_0 - zero field viscosity; $\dot{\gamma}$ - shear rate; $\dot{\gamma}^*$ - transition shear rate between two different flow behaviours; τ_0 - threshold shear tension; τ_1 - fit parameter; n - pseudo-plastic exponent.

The flow diagrams (Fig.5) evidence that the two MR fluid samples have quasi-Newtonian behaviour at low shear rate; such behaviours can be observed for various systems, including nanocomposites [9]. At high shear rate, both samples have pseudo-plastic (shear thinning) properties, the change in flow behaviour being observed at $\dot{\gamma} = 10^{-3}\ s^{-1}$. Increasing the shear rate the magnetic field induced particle structures are progressively destroyed. Both samples show large MR effect under the action of applied magnetic field in the interval B = 0-502 mT. The relative increase of viscosity $\Delta\eta/\eta_0$ (Fig. 6) evidence that the observed MR effect is larger in the case of the commercial sample MRF-132DG.

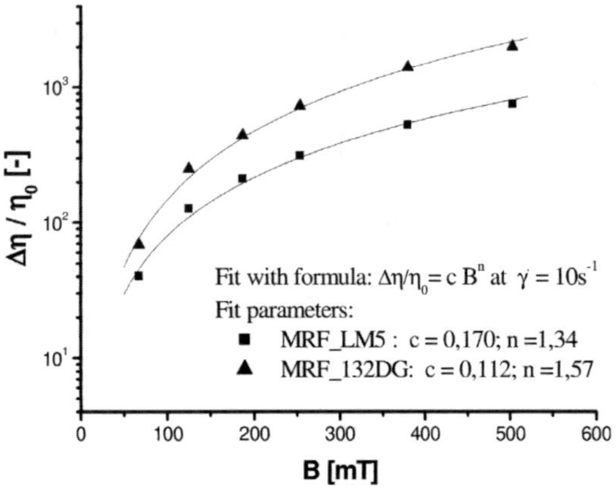

Figure 6. MR effect

3. Conclusion

Sedimentation rate of particles in the case of the sample MRF-LM5 is twice slower than for the commercial sample MRF-132DG, of similar magnetic properties. The flow properties of samples in zero magnetic field are fitted by the Carreau-Yasuda model, while their flow behaviour in magnetic field is well described by a newly developed model [8].

Acknowledgement: The present research has been supported by the Romanian National Authority for Scientific Research through the CEEX- C2-M1-1185, C64/2006-2008 „i-SMART-flow" project.

References:

[1] Bossis G., Volkova O., Lacis S., Meunier A., Magnetorheology: Fluids, Structures and Rheology, in: Odenbach S. (Ed.) - Ferrofluids. Magnetically Controllable Fluids and Their Applications, Springer-Verlag Berlin Heidelberg 2002, pp.202-232.
[2] Liţă M., Popa C.N., Velescu C., Vekas L., Investigations of a Magnetorheological Fluid Damper. IEEE Transactions on Magnetics, March 2004, Vol. 40, nr.2, p.469-472.
[3] Sung Taek Lim; Min Seong Cho; Jang, I.B.; Hyoung Jin Choi; Jhon, M.S. - Magnetorheology of carbonyl-iron suspensions with submicron-sized filler. IEEE Trans. Magn. 40, 3033 (2004).
[4] I. B. Jang, H. B. Kim, J. Y. Lee, J. L. You, and H. J. Choi - Role of organic coating on carbonyl iron suspended particles in magnetorheological fluids. J. Appl. Phys. 97, 10Q912 (2005).
[5] Carl Roth GmbH + Co: www.carl-roth.de/website/en-de/agb.jsp?lang=en-de&catId=DE
[6] Lord Cooporation: www.lord.com
[7] Yasuda, K., Armstrong, R. C. & Cohen, R. E. 1981 Shear-flow properties of concentrated linear and star branched polystyrenes. Rheologica Acta 20, 163-178.
[8] Daniela Susan-Resiga, A Rheological Model for Magneto-Rheological Fluids, Journal of Intelligent Material Systems and Structures(accepted).
[9] Yang H. Hyun, Sung T. Lim, Hyoung J. Choi and Myung S. Jhon - Rheology of Poly(ethylene oxide)/Organoclay Nanocomposites. Macromolecules 34, 8084 (2001)

Modelling of the B-field effect on the free surface of magneto-rheological fluids

X H Liu[1, 2], P L Wong[1], W Wang[2] and K P Liu[1]

[1] City University of Hong Kong, Department of Manufacturing Engineering and Engineering Management, Hong Kong, China
[2] Shanghai University, Department of Mechanical Engineering and Automation, Shanghai, China.

E-mail: meplwong@cityu.edu.hk

Abstract. The free surface of magneto-rheological fluids (MRFs) can be disturbed by an external magnetic field to form peaks along the field direction. The protruding volume is largely related to the magnitude of the applied B-field. This paper presents a theoretical analysis of the effects of this type of magnetic field on the protruding height (volume) of MRFs. A theoretical model of the protruded volume of MRFs is used according to the Euler equation and the boundary conditions for equilibrium and based on an ideal ellipsoidal shape. The surface profile due to the presence of a magnetic field is also experimentally analysed.

1. Introduction

Magneto-rheological fluids (MRFs) are characterised by fast switching between the liquid and semi-solid states upon the application of a B-field. The response of excited MRFs in the field direction has recently been studied by several research groups [1-6]. The magnetisation of magnetic fluids that have similar structures to MRFs but with lower particle concentrations and smaller particle sizes has been reported, in which the fluids formed different free surface patterns due to effects such as the Rosensweig normal field instability [7,8]. Based on the equilibrium of the surface profile in normal field instability, magnetic fluids can be pushed up to form a featured surface profile by increasing the magnetic field, which occurs once the field exceeds a critical value. This surface profile remains even when the local field is removed [9]. The equilibrium height achieved and the growth of the excited fluids due to the application of the magnetic field have been investigated both theoretically and experimentally [8]. The solutions in the theoretical studies are largely approximations achieved by assuming the profiles to have ideal ellipsoidal shapes [7, 11]. However, there is good agreement between the theoretical and experimental results in terms of the critical value of the magnetic field [10]. Theoretically, the excited free surface profile is periodic when presented as a stable state of the Rosensweig phenomenon [12, 13], and a similar approach is adopted in this study of the response of MRFs along the normal direction of an applied magnetic field. Comparison with previous studies on the normal field instability at the interface of ferrofluids [13] reveals that the main differences displayed by MRFs are the particle size and strength and also the initial susceptibility χ_0. In this study, $\chi_0 = 5.05$ of MRF is adopted, compared with $\chi_0 \approx 1$ used in the ferrofluid studies, and the aim is to determine the effects of a magnetic field on the protruding height (volume) of MRFs. The Navier-Stokes equation, the Maxwell equation and the Young-Laplace equation are solved according to the Euler equation and the boundary conditions for stable cases to obtain the size of a single MRF peak of assumed ellipsoidal shape. The surface profile due to the presence of a magnetic field is also experimentally analysed.

2. Experiment

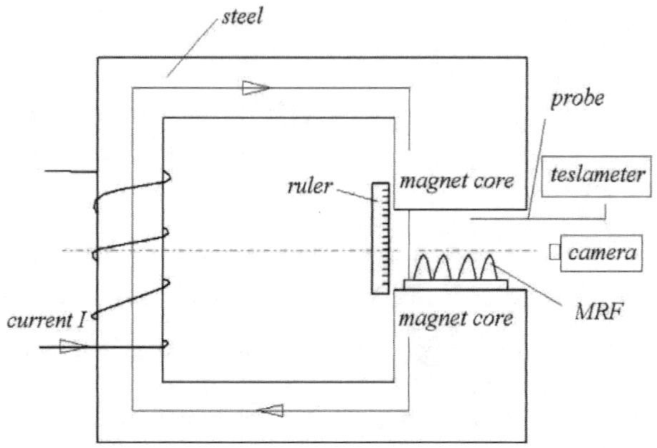

Fig. 1 Schematic diagram of the set up.

The experimental set up is shown schematically in Fig. 1. The specimen MRF was initially contained in a cylindrical plastic ring of about 6 mm in depth and 13 mm in diameter that was placed in the gap between the magnetic poles. Before switching on the B-field, the free surface of the fluid was flat. When the applied B-field was increased to a certain level, the MRF free surface became wavy and peaks formed. The height and base diameter of the protruding MRF peaks were measured using a scale and a camera focused on the vertical axis. The magnitude of the external B-field in the gap was obtained using a teslameter. The specimen MRF was MRF-132AD from Lord Corp., which is a hydrocarbon-based suspension of 30-36% micron-sized iron particles with a density of 3.06 g/cm^3 and a viscosity of 0.092 Pa·s. The surface tension of the typical hydrocarbon-based oil used was 60 mN/m. The initial susceptibility χ_0 of the MRF was 5.05, which was calculated from the relationship between the magnetic induction B and the magnetic field strength H.

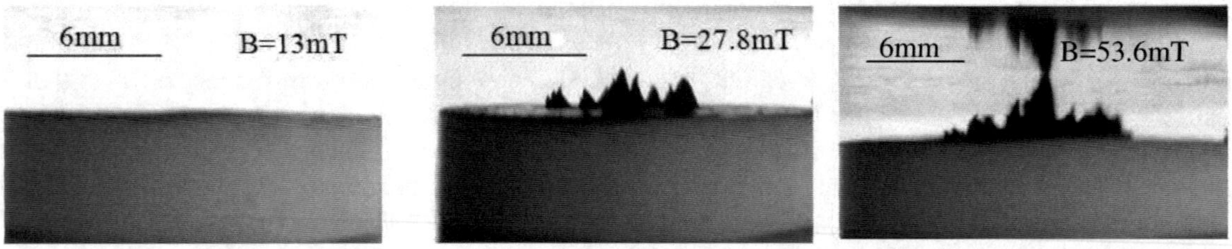

Fig. 2 Relationship between the height of the MRF peaks and the external B-field.

Fig. 2 shows the MRF protrusions under different B-fields. The height and base diameter of individual protrusions can be measured from the photos, and the protruded volume can thus be estimated by assuming its shape to be ellipsoidal. In general, the MRF starts to stretch along the magnetic field upon the application of a sufficiently large field strength, and the protrusions grow as the magnitude of the magnetic field is increased. The surface configuration of individual MRF protrusions is initially flat, but ends up being approximately ellipsoid.

3. Analysis of the MRF surface

The profile of an excited MRF is expected to be wavy with peaks of equal heights if the interfacial force, gravity and instability of the MRF under a magnetic field are all negligible. However, as shown in Fig. 2, the experimental results are different due to deviations from the ideal conditions. The final profile is therefore the result of self-adjustment by the specimen MRF to various surface instabilities under a uniform magnetic field. The complex and nonlinear interactions in MRFs lead to a peak shape with a free surface. Previous analyses of the change in the MRF peak shape consider only the interactions inside the free surface, and the effects from the surrounding fluid are minimal and usually neglected [8]. The profile of a protruding MRF peak can be modelled as an ellipsoid, as illustrated in Fig. 3, where it is assumed that 1) the protrusion is fully filled with the fluid, 2) the MRF peak is surrounded by air and 3) the magnetic field is uniform and normal to the horizontal MRF layer. This gives the embranchment $H_0 = \{0, 0, H_0\}$.

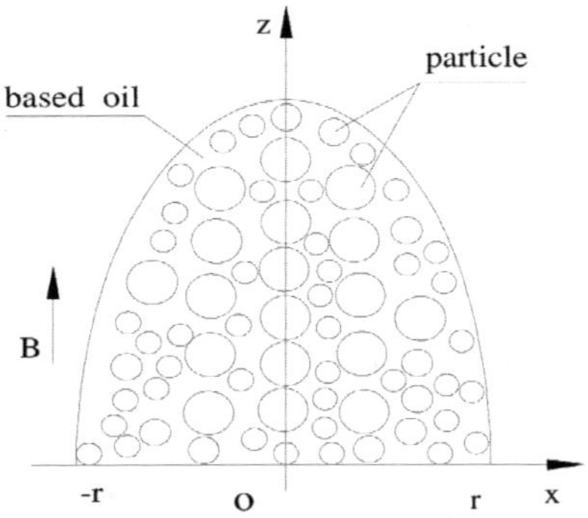

Fig. 3 Ideal profile of an MRF peak

When under a vertical uniform magnetic field, the minimum magnetisation intensity as denoted by the critical M* value can be obtained [2, 8, 9] by

$$M^* = [2\sqrt{\rho g \alpha} / \mu_0 (1 + \frac{1}{\sqrt{\mu_{abs}\mu_r}})]^{\frac{1}{2}} \tag{1}$$

The radius of its base area is

$$r^* = \pi\sqrt{\alpha} / 2\sqrt{\rho g} \tag{2}$$

The magnetisation is

$$M = \frac{\chi}{\mu_0(\chi+1)} \mathrm{B} \tag{3}$$

The system is governed by the equation of continuity and the Euler equation [10, 11]

$$\rho\left[\frac{\partial v}{\partial t} + (v \text{ grad })v\right] = -\text{grad } p + \mu_0 M \text{ grad } H - \rho g \tag{4}$$

where the magnetisation and the B-field in the magnetic force term are considered to be in the same direction. In Eqs. (1) to (4), μ_r and μ_{abs} are, respectively, the relative and absolute coefficient of magnetic permeability; μ_0 is the magnetic permeability of the free space and is equal to $4\pi \times 10^{-7}$ Vs/Am; α is the surface tension coefficient; and χ is the susceptibility of the MRF. The velocity field is denoted by v and the density of the MRF byρ. p is the pressure and g is the gravity acceleration, and M and H are the absolute values of the magnetisation and magnetic field strength in the fluid. It is assumed that in the equilibrium state the fluid and the surrounding air are at rest. In the static case, this gives

$$p = \mu_0 \int_0^H \mathrm{M d}H - \rho g z + c \tag{5}$$

where $z = \xi(x, y, t)$ denotes an unknown vertical displacement of the free surface layer and c is a constant. At $z = \xi$, if the pressure of the non-magnetic medium is set to zero, then Eq. (4) becomes

$$p = \alpha K - \frac{\mu_0}{2}\mathrm{M}^2 \tag{6}$$

where K denotes the curvature of the MRF surface. Combining Eqs. (5) and (6) and setting $\xi = h$ at $r = 0$ and $\xi = 0$ for ||r||>>0 gives

$$-\rho gh-\alpha K(h)+\frac{\chi}{2\mu_0(\chi+1)}[(B^2(h)-B_0^2)]=0 \tag{7}$$

where B_0 is the external B-field. The curvature $K(h)$ and the induction $B(h)$ at the tip of the peak can thus be estimated. The relation between the magnetic field $B(h)$ and the external magnetic field B_0 [5] is

$$\frac{B(h)}{B_0}=\frac{1+(\mu_r-1)(1-\frac{\beta}{2})}{1+(\mu_r-1)\beta(1-\frac{\beta}{2})} \tag{8}$$

where β is a geometric factor that depends on the eccentricity of the ellipsoid and is obtained from [1]. The relationship between the magnetic field $B(h)$ and the rising height of the MRF can be obtained by combining Eqs. (7) and (8).

4. Results and discussion

Fig. 4 shows that the MRF starts to increase in height when the magnitude of the B-field is raised to a certain value. The calculated critical magnetic induction value B* at the surface of the MRF is 12.62 mT and the experimental result is 13.08 mT. The radius of the peak is 2.22 mm.

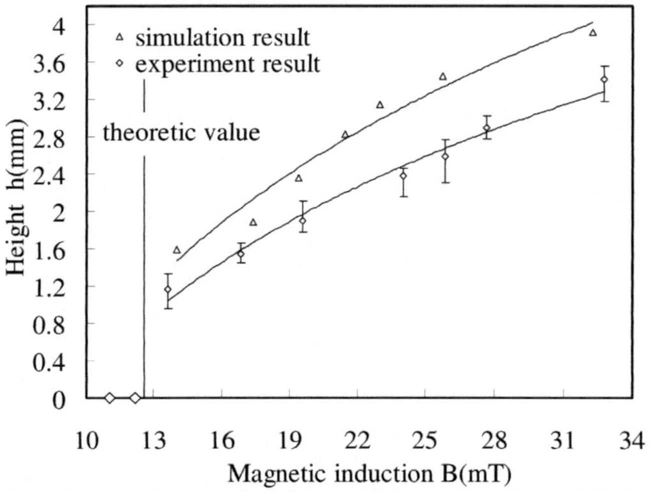

Fig. 4 Protruding height of the MRF under various B fields

The height to which the MRF protrudes is a function of the external magnetic field. The high level of correlation between the experimental and theoretical results, as shown in Fig. 4, proves the validity of the theoretical model.

According to the experimental results and the simulation, when the intensity of the external magnetic field is above the critical value B > B*, the free surface of the MRF becomes unstable and jumps abruptly to the height h > 1 mm. The height further increases at a slower rate with further increase in the external magnetic field. The volume of the MRF peaks can be approximated from the peak height by $V_{fluid} = 4N\pi hr^2/3$, and the distance to the apex can be obtained by the most instable wavelength $\lambda^* = 2\pi/k^*$ and the radius $r = \pi/2k^*$. The number of peaks N can be calculated from the wavelength. The required magnetic induction strength can thus be calculated by the model to obtain a certain protruding volume.

5. Conclusion

The surface configuration of an excited MRF is studied experimentally and theoretically, and a relationship that can be used to compute the rising height and volume of the MRF layer and the external magnetic field is developed. Although there is a quantitative difference in the experimental and theoretical results, they agree in terms of trend and the critical value of the magnetic field at which the MRF starts to increase in height. The relationship can thus be used to describe the surface configuration of MRFs under an external magnetic field. The choice of initial surface profile plays an important role in numerically solving the peak height and volume of an MRF. The selection of an ellipsoid shape is appropriate for approximating the actual configuration of an MRF.

Acknowledgments
The work described in this paper was fully supported by a grant from the Research Grants Council of the Hong Kong Special Administrative Region, China [Project No. CityU 121606].

References

[1] See H. (2003), *Rheol Acta*, 42, 86-92.
[2] Tao R. (2001), *J. of Phys.: Condensed Matter*, 13, 979-999.
[3] De Vicente J., González-Cabellero F., Bossis G., Volkova O. (2002), *J. Rheol.* 46, 5, 1295-1303.
[4] Zhang X.Z., Gong X.L. and Zhang P.Q. (2004), *J. of App. Phys.*, 96, 4, 2359-64.
[5] See H., Mackenzie S. and Chua B.T. (2006), *Korea-Australia Rheology Journal*, 18, 3, 121-126.
[6] Chan Y.T., Liu K.P., Wong P.L. and Bullough W.A. 2008 *Proc. ERMR08, Dresden.*
[7] Lange A., Langer H. and Engel A. (2000), *J. of Physica D*, 140, 294–305.
[8] Gollwitzer C. and Matthies G. (2007), *J. of Fluid Mech*, 571, 455-474.
[9] Bashtovoia V.G., Lavrovab O.A., Polevikovb V.K. and Tobiskac L. (2002), *J. of Magnetism and Magnetic Materials*, 252, 299-301.
[10] Matthiesa G. and Tobiskab L. (2005), *J. of Magnetism and Magnetic Materials*, 289, 346–349.
[11] Engel A., Lange A., Langer H., Mahr T. and Chetverikov M.V. (1999), *J. of Magnetism and Magnetic Materials*, 201, 310-31.
[12] Friedrichs R. and Engel A. (2001), *Physical Review E*, 64, 021406, 1-10.
[13] Chen C.Y. and Wu S.Y. (2007), *Physical Review E*, 75, 036310, 1-10.

Journal of Physics: Conference Series **149** (2009) 012073 doi:10.1088/1742-6596/149/1/012073

Synthesis and magnetorheology of suspensions of cobalt particles with tunable particle size

M T López-López, P Kuzhir, A Meunier and G Bossis

Laboratoire de Physique de la Matière Condensée, CNRS UMR 6622 Université de Nice, Parc Valrose, 06108 Nice Cedex 2, France

E-mail: Bossis@unice.fr

Abstract. Different samples of cobalt powder were synthesized. Particle size and shape were characterized by electron microscopy and light scattering. These measures showed that the synthesized powders consisted of monodisperse spheres with average diameters ranging between 63 and 760 nm. These powders were used for the preparation of magnetorheological (MR) fluids by dispersing them in silicone oil. It was found that particle size did not have much influence on the MR response of MR fluids for average particle diameters larger than 100 nm. On the other hand, the MR response decreased appreciably when average particle diameter was diminished below 100 nm.

1. Introduction

Suspensions of colloidal magnetic particles are complex fluids that exhibit magnetic field-dependent rheological (flow) properties. Upon magnetic field application, these particles experience attractive magnetostatic forces, which lead to the formation of particle structures aligned with the field direction. The formation of these structures, which strengthen the suspension and oppose to the flow, is the basic phenomenon underlying the magnetorheological (MR) effect [1-4]. The formation of field-induced particle aggregates, i.e. the intensity of the MR effect, will depend on the ratio of the magnetic interaction energy to the thermal energy. For two magnetic dipoles this ratio is given by [3]:

$$\lambda = \frac{\pi\mu_0\mu_f\beta^2 a^3 H_0^{\ 2}}{2kT}. \tag{1}$$

Here, a is the particle radius, μ_0 the permeability of vacuum, μ_f the relative permeability of the carrier liquid, H_0 the external magnetic field strength, k the Boltzmann constant, T the absolute temperature, and β the magnetic contrast factor. Generally, the particles have a large permeability and, therefore, $\beta\sim1$. Thus, taken $T = 300$ K and $\mu_f\sim1$, for MR fluids ($a \sim 1\ \mu$m) $\lambda = 1$ is obtained for $H =$ 0.046 kA/m, whereas for FF ($a \sim 5$ nm) the same value is obtained for $H = 368$ kA/m. Therefore, the Brownian forces normally dominate the magnetic forces in FF and, consequently, they only exhibit weak MR effect [5-6]. On the other hand, even at low magnetic field, the magnetic forces are much higher than the Brownian forces in MR fluids and, consequently, they exhibit strong MR effect [7-8].

From the experimental viewpoint, little is known about the effect of particle size on the MR properties of suspensions intermediate between FF and MR fluids. Lemaire et al. [9] measured the shear stress as a function of the shear rate for three different suspensions, made of polystyrene particles containing inclusions of magnetite, under the presence of an applied magnetic field of approximately 10 kA/m. The only difference between the suspensions was the average particle diameter: 0.5, 0.8 and 1.0 microns. They found that, for a given shear rate, the apparent shear stress slightly increased with the diameter of the particles. Rosenfeld et al. [10] investigated the MR properties of two suspensions of iron particles, the average particle size being the only difference between both of them: 30 microns and 26 nm. They found that, even at high magnetic field strength, the suspension of microparticles exhibited yield stresses up to 2 times higher than the suspension of nanoparticles. According to Equation 1, even for a particle diameter of 25 nm the parameter λ is still larger than unity for magnetic fields of interest ($\lambda \sim 10$ for $H = 100$ kA/m). Thus, from the theoretical viewpoint, the MR effect should not show any size dependence for diameters larger than 25 nm.

The difficultness of performing a rigorous experimental study about the influence of particle size on the MR properties, lies in obtaining monodisperse and well dispersed spherical particles with different average particle diameters. For instance, Kormann et al. [11] obtained a quite large yield stress (3 kPa at 0.3 Tesla) with ferrite nanoparticles of 30 nm diameter likely because of the presence of aggregates of nanoparticles. The aim of the present work is to rigorously analyze this dependence for particle diameters in the range 50 nm to 1 μm. With this objective, we synthesized different samples of cobalt powder with controlled diameter, by reduction of cobalt ions in polyols. The so-synthesized powders were used for the preparation of MR suspensions. Finally, the MR properties of these suspensions were investigated as a function of the applied magnetic field.

2. Experimental methods

Reduction of metallic ions in a liquid polyol has been extensively used for the preparation of spherical particles of magnetic and non-magnetic materials [8,12-16]. This method was used in this work to synthesize spherical cobalt particles with different average diameter. Particles size and shape were characterized by means of scanning electron microscopy (SEM) and transmission electron microscopy (TEM). In addition, the hydrodynamic diameter of the particles was obtained by light scattering measurements using a Mastersizer 2000 instrument (Malvern Instruments, UK). The magnetization curves measured on the powders with a vibrating sample magnetometer show no noticeable difference between the different sizes and we find Ms between 13900 and 14000 Gauss whatever the size. Nevertheless it is worth noting that this value is significantly lower than the one of bulk cobalt (Ms= 17900G) which means that some part of the catalyst has been included into the cobalt nanoparticles.

MR suspensions ($\phi = 0.05$) were prepared by dispersing the synthesized cobalt powders in silicone oil. Based on discussions reported in previous papers [17-18], aluminum stearate was used as dispersant. The MR properties of the suspensions were measured using a Haake RS150 controlled stress rheometer. The measuring system geometry was a plate–plate configuration of diameter 20 mm and gap 0.300 mm with a controlled temperature T=20°C. The magnetic field was applied in the vertical direction using a home-made electromagnet similar to the one sold by Anton Paar for the rheometer MCR301 , which allows reaching fields up to 580 kA/m in the measuring gap.

Table I. Average diameter (nm), measured by different techniques, of the synthesized cobalt particles.

Sample name	S1	S2	S3	S4
Electron Microscopy	760	330	83	63.6
Light Scattering	629	498	126	69

3. Results and discussion

The synthesized powders consisted of relatively monodisperse spheres of submicrometer size. Table I shows the average particle size obtained by electron microscopy and light scattering. As observed, there is a quite good agreement between the data obtained by electron microscopy and by light scattering which confirms the absence of aggregation in liquid phase.

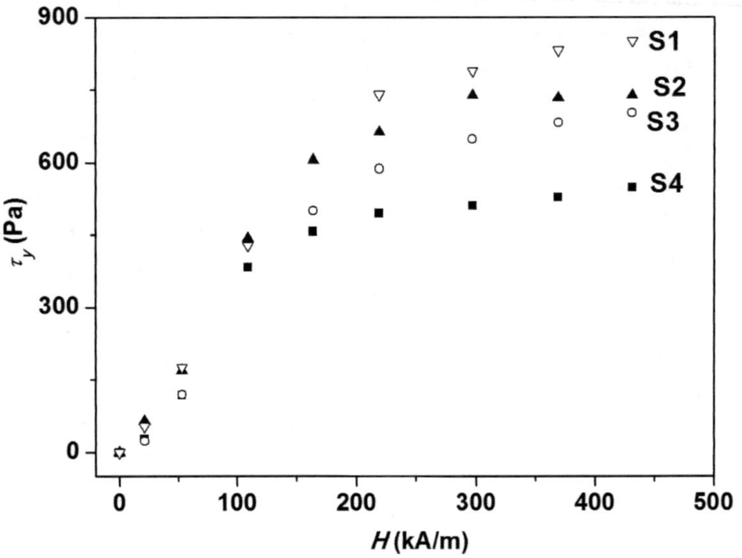

Figure 1. Yield stress, τ_y, plotted as a function of the magnetic field strength, H, for suspensions containing 5 vol.% of cobalt particles (samples S1-S4; see Table I).

The steady-shear flow of suspensions containing 5 vol.% of cobalt powders of samples S1-S4 dispersed in a silicone oil of viscosity: 0.479Pa.s and density = 0.97g/cm^3 , was investigated as a function of the applied magnetic field. The values of the shear stress were obtained as a function of the shear rate, for the different applied magnetic field strengths (up to 431 kA/m). From extrapolation at zero shear rate, of the shear stress vs. shear rate curves in the range 0.01 s^{-1} - 0.1 s^{-1} on a logarithmic scale, the values of the static yield stress were obtained. These values are plotted as a function of the magnetic field strength in Figure 1. As expected for MR fluids, the values of the yield stress shown by samples S1-S4 increase with magnetic field strength. Let us focus on the comparison of the values of the yield stress for the different samples. As observed, the yield stress decreases as the average particle diameter is diminished (see particle diameter in Table I). This decrement is negligible between S1(760nm) and S2 (330nm)except at the highest fields, but becomes quite significant for average diameter smaller than 100 nm. For example, taken as the reference, the value of sample S1 at H = 297 kA/m, the yield stress is respectively: 94 % for sample S2; 82 % for sample S3; and 65 % for sample S4. The reason for this decrement could be the Brownian motion, which is able to generate fluctuations on a large scale and promote the presence of gaps between the particles, which will decrease the strength of the columnar structures under strain and, consequently, the strength of the MR effect [9].Nevertheless, considering that, even for the smallest diameter, we have $\lambda > 1000$ for $H = 297$ kA/m it is not obvious that such a low Brownian motion can explain this size dependence; furthermore, this decrease should be less pronounced at the largest fields, which corresponds to the highest values of λ, and actually we observe the contrary. Another explanation stands on the surface energy of the aggregates, which determines their size and the condition of their rupture under strain. The energy difference produced by the breakdown of an aggregate in two smaller ones will be a decreasing function of the size of the particles as it is the case for the surface energy. The consequence is that aggregates constituted of smaller particles will break at lower strain and consequently at lower stress. This approach developed by Zubarev et al. [19] for ferrofluids could likely explain the observed behavior and will be developed in a future paper.

4. Conclusions

Different samples of spherical cobalt particles with average diameter in the range 60 nm – 800 nm, have been synthesized by reduction of ions in liquid polyols. Being dispersed in silicone oil, these particles allowed us to carry out the first systematic study on the effect of particle size on the yield stress of MR suspensions. This effect has been shown to be almost negligible for average particle diameter higher than 100 nm. For smaller diameters, a significant decrement of the MR effect is observed. Brownian motion is probably not responsible for the observed behavior but rather the change of surface energy when an aggregate is broken and form two half aggregates of the same length. Besides the academic interest of this study, it has also been evidenced that nanometric particles have large potential applications in smart MR devices, especially those having narrow flow channels, such as active hydrostatic bearings for precision machinery [20].

Acknowledgements

Eureka E! 3733 Hydrosmart project and "Conseil Régional PACA" (Biomag project) are acknowledged for their financial support. One of the authors (M.T.L.-L.) also acknowledges financial support by Secretaría de Estado de Universidades e Investigatigación (MEC, Spain) through its Postdoctoral Fellowship Program. We are also grateful to A. Zubarev for helpful discussions and V.Cagan from Versailles University for magnetization measurements.

References

[1] Odenbach, S. (ed.), "Ferrofluids," *J. Phys.-Condens. Matter* **18**, S2527-S2998 (2006).
[2] Rosensweig, R.E., "Ferrohydrodynamics," Cambridge University Press, Cambridge, 1985.
[3] Bossis, G., O. Volkova, S. Lacis and A. Meunier, in "Ferrofluids," Magnetorheology: Fluids, Structures and Rheology. S. Odenbach, ed., Springer, Berlin, 2002.
[4] Winslow, W. M., *J. Appl. Phys.* **20**, 1137-1140 (1949).
[5] Shahnazian, H. and S. Odenbach, *Int. J. Mod. Phys. B* **21**, 4806-4812 (2007).
[6] Zubarev, A.Y. and L.Y. Iskakova, *J. Phys.-Condens. Matter 18*, S2771-S2784 (2006).
[7] López-López, M.T., P. Kuzhir, S. Lacis, G. Bossis, F. González-Caballero and J.D.G. Durán, *J. Phys.-Condens. Matter* **18**, S2803-S2813 (2006).
[8] López-López, M. T., G. Vertelov, P. Kuzhir, G. Bossis and J. D. G. Duran, *J. Mater. Chem.* **17**, 3839-3844 (2007).
[9] Lemaire, E., A. Meunier, G. Bossis, J. Liu, D. Felt, P. Bashtovoi and N. Matoussevitch, *J. Rheol.* **39**, 1011-1020 (1995).
[10] Rosenfeld, N., N.M. Wereley, R. Radakrishnan and T.S. Sudarshan, *Int. J. Mod. Phys. B* **16**, 2392-2398 (2002).
[11] Kormann, C., H.M. Laun and H.J. Richter, *Int. J. Mod. Phys. B* **10**, 3167-3172 (1996).
[12] Fiévet, F., in "Fine Particles," Polyol process. T. Sugimoto, ed., Marcel Dekker, New York, 2000.
[13] Fiévet, F., J. P. Lagier and M. Figlarz, *MRS Bull.* **14**, 29-34 (1989).
[14] Sun, X. C., S. S. Kang, J. W. Harrell, D. E. Nikles, Z. R. Dai, J. Li and Z. L. Wang, *J. Appl. Phys.* **93**, 7337-7339 (2003).
[15] Sun, X. C., C. J. Thode, J. K. Mabry, J. W. Harrell, D. E. Nikles, K. Sun and L. M. Wang, *J. Appl. Phys.* **97**, 10Q901 (2005).
[16] Viau, G., F. Fiévet-Vincent and F. Fiévet, *Solid State Ionics* **84**, 259-270 (1996).
[17] López-López, M. T., A. Zugaldía, F. González-Caballero and J. D. G. Durán, *J. Rheol.* **50**, 543-560 (2006).
[18] López-López, M. T., P. Kuzhir, G. Bossis and P. Mingalyov, *Rheol. Acta* In press (2008).
[19] Zubarev, A.Y. and L.Y. Iskakova, *Physica A* **365**, 265-281 (2006).
[20] Ochonski, W.; "The attraction of ferrofluids bearings", Machine Design Nov 3, 96 (2005).

Journal of Physics: Conference Series **149** (2009) 012074 doi:10.1088/1742-6596/149/1/012074

Self-Locking MRF Latches & Dampers

G.Magnac[1], F.Claeyssen[1], K.Vial[1], R.Le Letty[1], O.Sosnicki[1], C.Benoit[1]

[1] Cedrat Technologies S.A., Meylan, France, www.cedrat.com

Email: actuator@cedrat.com

Abstract. MRF actuators are new electromechanical components using Magneto Rheological Fluids (MRF). When submitted to a high enough magnetic field, MRF switch from a liquid to a near solid body. New MRF actuators were developed in order to reach three aims: to offer a blocking force without power consumption which can be strongly reduced by applying a current, to provide an electrically-controllable resistive force over a stroke of 30 mm, to perform the control of the force in a very short time, typically in a few milliseconds. Thus, these MRF actuators can be used for two main applications: damper and latch - lock. Experiments on two versions of the actuator (a single piston rod and a feed through output axis) allow getting a blocking force around 100N, which is more than 10 times the actuator weight (its mass is 700gr). The current and electric power required to cancel the blocking force are only 1.6A and 4W. The paper further presents the design and the electromechanical properties of the Self-braking MRF Actuators for dampers & latches and new results on the control of these actuators.

1. Introduction

The MRF actuators are new electromechanical components using Magneto Rheological Fluids (MRF) (Carlson, et al, 2006). These smart fluids are characterized by their ability to change their rheological properties versus applied magnetic field. With sufficient field, they can switch from a liquid to an almost solid body. This effect is reversible. It operates in a few milliseconds. This effect can be used to create controllable damper, smart shock absorber or brakes. Most of MRF actuators (Loureiro and S.Harvin, 2006) offer a controllable braking force but do not offer a blocking force without power consumption for self-locking operations.

2. Aim

The purpose of this paper is to present the performances of the new MRF actuators which are developed to reach three aims:

- to offer a blocking force without power consumption, which can be nullified by applying a current,
- to provide an electrically- controllable resistive force over a stroke of 30 mm, with as less as possible moving parts, offering high reliability,
- to perform the control of the force in a very short time, typically in a few ms.

A fourth aim could be added for the actuators presented in this paper. They were designed, in term of blocking force and free force, to be in the range of human capability and therefore to be easily manipulated.

3. Applications

The MRF actuators have two main applications, damper and latch-lock. In some cases, the two applications can be combined to provide a smooth lock or unlock operation. An other application for some embedded instruments for example is the adaptative suspension function. The aim of such kind of suspension is to manage shocks with much better efficiency than passive hydraulic dampers. Yet another application is, combined with non controllable linear actuators as pneumatic cylinders for example, to create a fully controllable, high stroke and low cost actuators.

The self-locking feature has been obtained in the presented actuators thanks to a particular magnetic circuit, designed with FLUX software (http://www.cedrat.com/en/software-solutions/flux.html) and using knowledge acquired (Magnac, 2006) on MRF fluid magnetic and rheological properties.
The optimized damping (the first application) needs to control the energy dissipation during the piston movement. The device provides dissipated energy control in real time thanks to the current control. In a few milliseconds, the energy dissipation can be increased by 500%. A maximum damping can be maintained without power supply and so the fail safe operating can be guaranteed in several applications. Vibrations damping and shocks absorption can be optimized thanks to a closed loop. The second main application is the locking in position. The current control allows blocking the load in any positions along the stroke. The MRF actuator acts therefore as a linear electro-mechanical brake without moving parts. Whatever the piston rod position, if the motion force is lower than the maximal force damping (@0A), the control of the current allows to get the rod locked.

4. MRF actuators

Two mechanical architectures of the device are presented:

- an actuator with a single rod piston, the device allows to damp a movement versus the frame
- an actuator with a double rod piston which allows to transmit the movement through the actuator, which operates in series with the motion line.

Both devices operate in flow mode with a twin tube structure (Florian Zschunke, 2005):

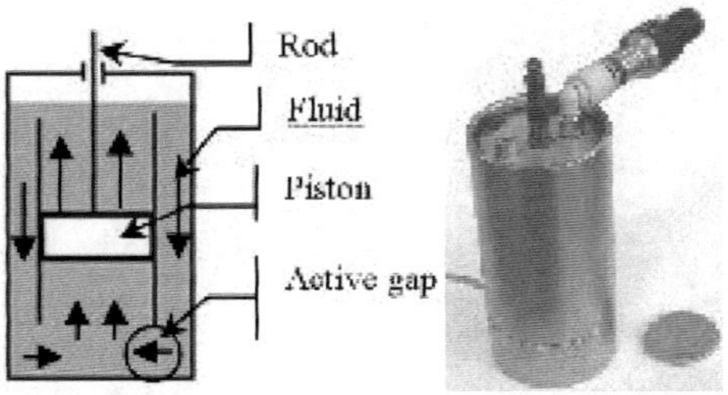

Fig. 1: Twin tube structure & Actuator

The following reasons led to the choice of the twin tube structure:

- Decoupling fluidic aspect and magnetic aspect.
- Non mobile coil, simplifying wires path and increasing reliability
- Immobile fluid gap, avoiding geometric deformations during the rod displacement
- Fast, easy and efficient replacement of the fluid gap parts, for easy customization to meet customers' requirements

The magnetic field in the fluid gap is controlled from 0 T to 0.8 T, thanks to a 0.38 cm^3 permanent magnet and a coil. The coil resistance is 1.4 Ohms with an inductance close to 36mH, this leads to an electrical time constant of 25ms. The 132DG Lord MR fluid was used.

Single rod and double rod solutions are using mostly the same structure. Thus, same behaviors between both actuators could be expected. The main difference is that the single rod actuator needs a specific component, a moss fixed in one chamber, to solve the volume variation problem. Volume variation is due to the rod movement, entering or coming out of the actuator.

5. Damper performances

A first test campaign has been performed to assess the technologies versus the theoretical expectations. First generation of prototypes reveals cavitation effects (Fig 2) during pulling movement.

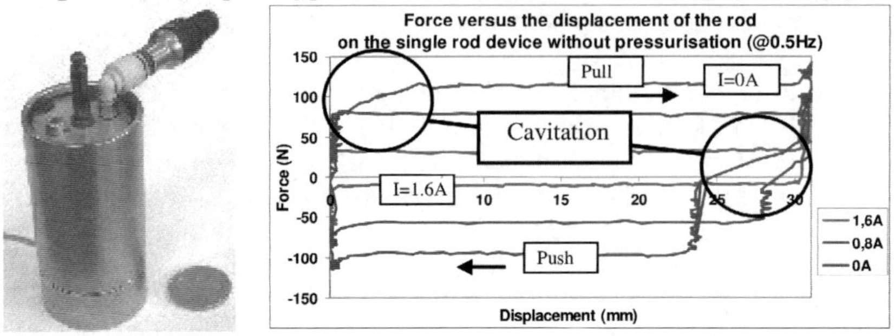

Fig. 2: Performances with cavitation

This limitation was met only with the single rod actuators. Cavitation is due to pressure drop in the MRF bellow the saturation vacuum pressure. This pressure drop appeared only in the single rod actuator because of the volume compensation.

Indeed, as in every hydraulic damper, the piston movement creates a pressure difference between its both sides. Because of the moss fixed in one chamber of the actuator, the pressure in this chamber reacts as if it was filled with a compressible fluid, whereas the other chamber is filled only with MRF, an incompressible fluid. Consequently, during pulling, there is a very low level of pressure increase in the compressible chamber, and there is an high pressure decrease in the other chamber. As soon as the pressure falls under the saturation vacuum pressure, the MR fluid evaporates and the cavitation appears.

To solve this limitation, specific solutions have been implemented and tested to be validated. To avoid cavitation issue, the single rod actuators was filled up with a pressure of 4 bars before being tested again. The results show that the cavitation is suppressed and performances become similar in either pulling or pushing operations.

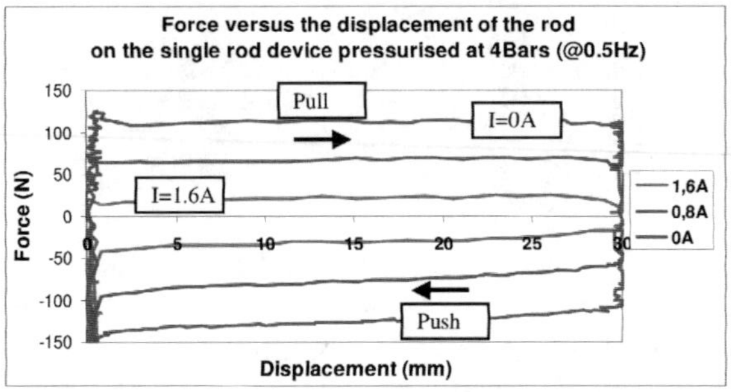

Fig. 3: Performances of the device with a single piston rod as damper without cavitation

The damper with the double rod piston (fig 4), does not need moss because it has no volume variation (excepted thermal expansion, which is negligible in this case). It presents roughly the same performances as the former prototype.

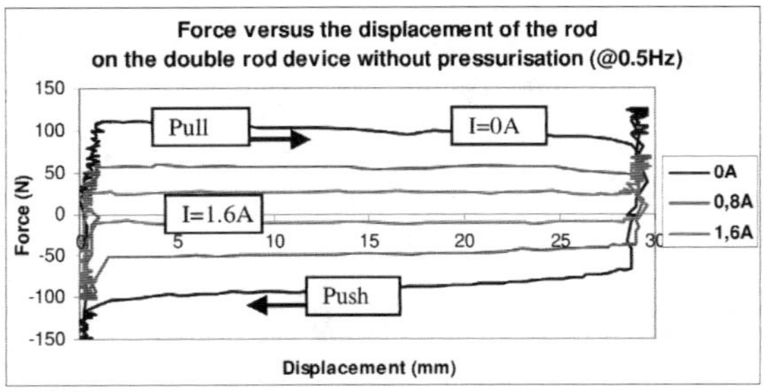

Fig.4: Damper with a double piston rod and its performances as damper

6. Latch performances

When the device is used in a latch, the rod is locked without power consumption. The moving parts are locked as long as the force applied through the load on the latch does not exceed the maximal locking force (around 100N for these prototypes). However, after experiment, latching shows limitations as slow loss of position (fig 5).

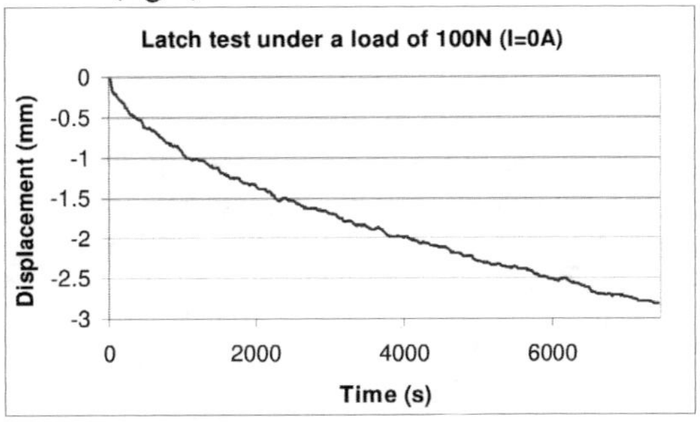

Fig. 5: Displacement versus time locking

7. Conclusion

As targeted, both developed MRF actuators offer blocking without power consumption, electrically controllable resistive force and fast control. They operate in all orientations and in both directions (pulling & pushing) with same performances. Several physical limits were solved like the cavitations, the sealing and the volume variation when the rod goes in the damper.
A test campaign has been performed to assess the technologies versus the theoretical expectations. Force versus displacement at different currents, shows the achievement: locking without power consumption, controllable forces, pulling & pushing operations with similar properties for both actuators.

8. Acknowledgements

The authors thank M.Csukai (OSEO) and M.Amiet (DGA) for supporting these works via OSEO.

9. References

Carlson, J.D. et al, 2006, "Magnetorheological and electrorheological fluid highlight", Actuator 2006 Conf Proc, pp 2 35-239
Loureiro and S.Harvin, 2006, "Design and characterisation of a low profile controllable magneto-rheological fluid based actuator for vibration mitigation", Actuator 2006
Magnac, 2006, "Characterisation of magneto-rheological fluids for actuators applications", Actuator 2006
Florian Zschunke, 2005, "Aktoren auf Basis des magnetorheologischen Effekts", Thesis from Erlangen-Nürnberg university
http://www.cedrat.com/en/software-solutions/flux.html

Journal of Physics: Conference Series **149** (2009) 012075 doi:10.1088/1742-6596/149/1/012075

Positioning magnetorheological actuator

Valery Mikhailov[1], Dmitry Borin[2], Alexey Bazinenkov[1] and Igor Akimov[1]

[1] Bauman Moscow State Technical University, 2-nd Baumanskaia st. 5, MT-11, 105005, Moscow, Russia

[2] Technische Universität Dresden, Chair of Magnetofluiddynamics, 01062, Dresden, Germany

E-mail: mikhailov@bmstu.ru

Abstract. In this work we consider a construction of a positioning magnetorheological actuator based on bellow units, as well as dynamical model, which include such elements as a magnetically hysteresis, pressure loses in hydraulic system, nonlinearity of rheological behaviour of working fluid. Two operating modes of positioning actuator are taken into account and transients are presented. Dynamical modelling shows possibility for the improvement of a real control system and ensure of submicron precision of positioning with millisecond time of response.

1. Introduction

Rapid response of magnetorheological (MR) fluids to an applied magnetic field, as well as precisely controllable yield stress, allows to use MR fluids in engineering applications such as semi-active dampers, clutches etc [1-3]. Studies on MR hydraulic power system has been performed in [4,5]. Besides, for more than ten years the use of MR effect for the control of hydraulic actuating systems in vacuum technological equipment [6, 7] and astronomical telescopes [8] is being proposed. In this work we briefly describe an actuator developed recently by us and its dynamical model which allow improvements of the positioning system's parameters.

2. Construction of the actuator

The principle idea of proposed actuator is in the use of bellow units, as a pressure chambers, and MR valves system based on Wheatstone bridge hydraulic power circuit. MR fluid is pumped from the pump to the bellow units through the input MR-valves and than pumped out through output MR valves. Objects moves relatively to the base of elastic membrane because of pressure difference in opposite hydrocylinders. Control of pressure drop is carried out due to adjustment of the magnetic field applied to MR fluid in MR valves. By means of bellow units it is possible to realize not only the linear actuating but a multicoordinate positioning system as well. The scheme of the linear actuator is shown in figure 1.

Journal of Physics: Conference Series **149** (2009) 012075 doi:10.1088/1742-6596/149/1/012075

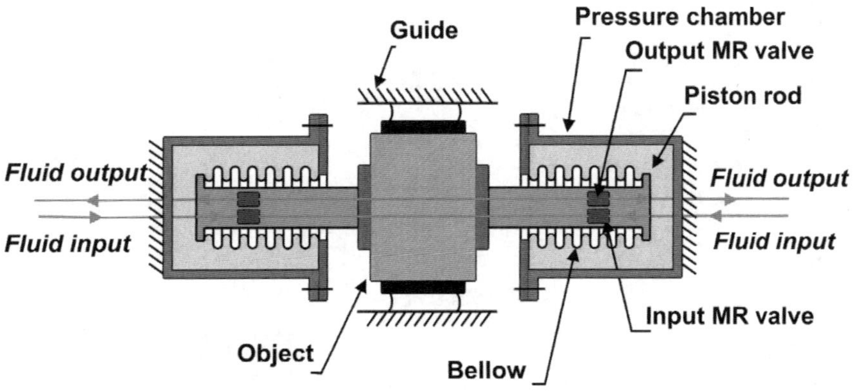

Figure 1. Scheme of the linear actuating system

3. Parameters of actuator

Experiments on actuators prototype have been performed with a control system analogous to one used in [8]. Figure 2 shows response of the actuator (red curve) on step signal (green curve). Actuator was loaded with 70 kg of external mass in direction of moving, a spherical bearing was used as a guide and measurements were made with contact inductive sensor. The set position (2 μm shift relative to initial coordinate) is achieved in 250 ms with precision of $\Delta X=0.25$ μm which is limited by sensor resolution, resistance force of guide and used control algorithms. A new control system should be developed to ensure of the actuators characteristics improvement.

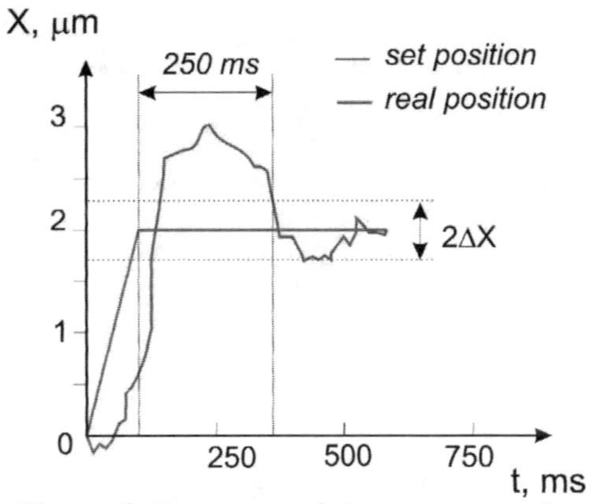

Figure 2. Response of the actuator to the set signal

4. Dynamical model of the actuator

The most important instrument for the control system analysis and synthesis is a dynamical modelling of the considered device. To define the elements of the dynamical model, one needs to find an influence of rheological and dynamical properties of MR fluid on the pressure drop on the working channel of the MR valve. Various models for some types of MR devices are already known e.g. see in [10, 11]. In [11], among others, the magnetic hysteresis effect is included in the model, as well as in a model for the valve of the MR positioning actuator [12].

Generalized functional scheme of proposed model for the case of one coordinate positioning is shown in figure 3 and consists of such elements as input and output MR valves (which are considered in model including magnetic and rheological effects), pump line, positioning object and measuring system.

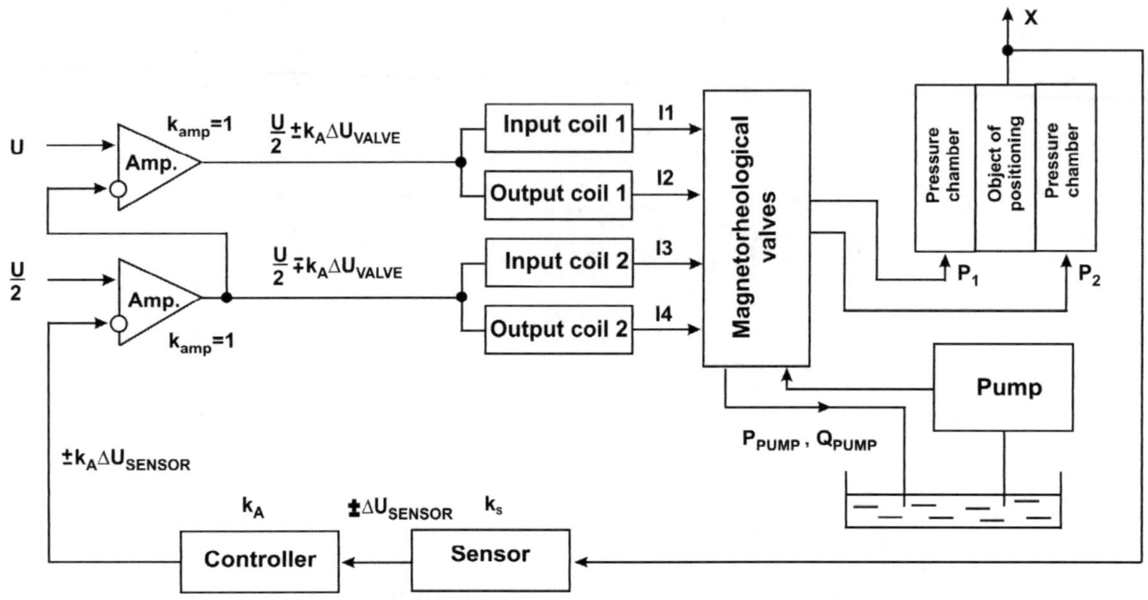

Figure 3. Functional scheme of proposed model (U – voltage, I – electrical current, P – pressure, Q – volume flow, k_S – sensors coefficient, k_A – amplifying coefficient)

To describe a rheological behavior of MR fluid in working elements of MR valves, the Bingham plastic model is usually used [13, 14]. However, in the case of positioning actuator with MR control, the different operating modes must be considered. In advance positioning mode (acceleration and rapid moving) the MR fluid is a viscous medium, because of the not high magnetic field strength and high shear rates, and the Bingham model is applicable. In adjustment positioning mode there are two regimes, namely preliminary adjustment and precise regimes. In preliminary adjustment regime, the shear rates are usually higher than 1 s^{-1} and the Bingham model is still applicable, even under action of high field strength, although the behavior of fluid is viscoplastic. In precise adjustment regime, shear rates are less than 0.1 s^{-1} and because of the strong viscoplastic behavior of MR fluid and conformably non-linearity of flow curves one must use a better model for the behavior of the MR fluids. In this case the fit for flow curves for the used by us self prepared MR fluid (20 vol. % of 5 μm particles in organic oil) can be obtained with the so-called Casson equation [15]:

$$\tau^{\frac{1}{2}} = \tau_y^{\frac{1}{2}} + (\eta \cdot \dot{\gamma})^{\frac{1}{2}}, \tag{1}$$

where τ is the shear stress, τ_y is the yield stress, η is the viscosity and $\dot{\gamma}$ is the shear rate.

In case of other MR fluids optimal fit have to be found, e.g. see [16] for the Lord Corp. fluids.

Besides, the time of aggregation of MR fluids at low shear rates must be taken into account in the model as an important element, which has a remarkable influence on dynamics of control system. Relaxation and cluster formation in MR fluids under action of magnetic field can be described with the so-called method of mechanical analogies. Namely, for the adjustment positioning mode, when fluid has viscoplastic behavior, one can consider the elements of Voigt model, where the elastic element with elasticity modulus E oscillates in viscous medium with viscosity η and the time of transient is defined as $t = \eta \cdot E^{-1}$, there both variables are magnet field and therefore control signal dependent.

5. Results of simulations

Simulation was performed by means of *Simulink* application of *MathLab* software package. We considered two operating modes separately and results of transients are presented in figures 4 and 5 for the advance positioning and adjustment positioning mode respectively. Two various regulators were chosen for each mode to ensure the stability of transients. In case of advance positioning the proportional-integral-differential and for the adjustment positioning mode the proportional-differential regulator was used. Time of response for both considered modes is 85 ms and 5 ms which are chosen for the accuracy of the positioning of 5 mm and 50 nm respectively.

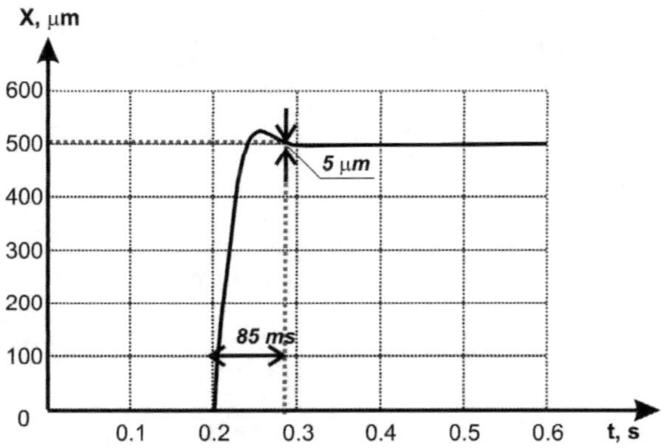

Figure 4. Transient of model for the advance positioning mode

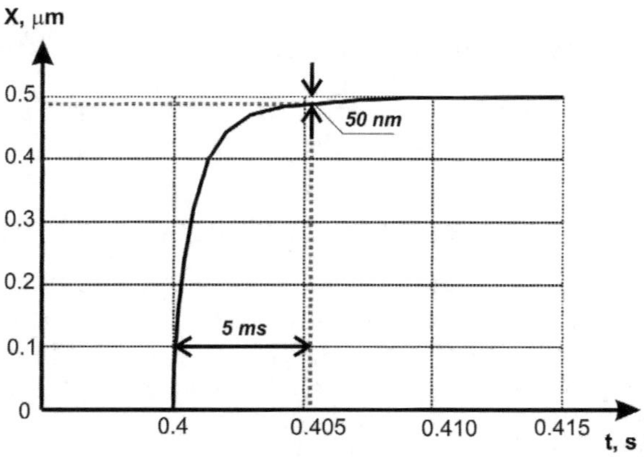

Figure 5. Transient of model for the adjustment positioning mode

6. Summary and outlooks

Construction of a positioning magnetorheological actuator based on bellow units and its dynamical model are presented. In the dynamical model of MR actuator for the analysis and synthesis of control laws two operating modes are considered. For the precise adjustment regime a non-linear model of MR behaviour must be implemented. To take into account the dynamics of structure formation in MR fluids, the Voigt mechanical model of fluid behaviour is used. Dynamical modelling shows possibility for the improvement of a real control system and ensure of submicron precision of positioning with millisecond time of response. A question of a switching between two regimes should to be at the centre of attention during the development of a program and electronic part of the control system. Results of modelling have to be experimentally verified.

References

[1] Carlson J D (2005) *Int J Mod Phys B* 19(7-9), 1463
[2] Lee U, Kim D, Hur N and Jeon D (1999) *J Intell Mater Syst Struct* 10(9), 701
[3] Gordaninejad F and Kelso S P (2000) *J Intell Mater Syst Struct* 11(5), 395
[4] Yoo J-H, Sirohi J and Wereley N M (2001) *Proc. of SPIE* 4327(22) 148
[5] Yoo J-H and Wereley N M (2004) *J Intell Mater Syst Struct* 15(11), 847
[6] Anisimov V V, Deulin E A , Khramtsov S V and Mikhailov V P (1996) *Vacuum* 47 (10), 1163
[7] Borin D, Mikhailov V, Deulin E and Zobov I (2007) *Vacuum technique and technology* 17(2),141 (in russian)
[8] Deulin E A, Mikhailov V P, Eliseev O N and Sychev V V (2000) *Proc. of SPIE*, 4003, 303
[9] Lord Corporation Engineering Note: Designing with MR Fluids (online http://www.lord.com)
[10] Jolly M R, Carlson J D and Munoz B C (1996) *J Smart Mater Struct* 5, 607
[11] Jinung An and Dong-Soo Kwon (2002) *J Intell Mater Syst Struct* 14(9), 541
[12] Borin D, Mikhailov V and Bazinenkov A (2007) *Conversation in machine buildings of Russia* 3, 37 (in russian)
[13] Wereley N (2008) J *Intell Mater Syst Struct* 19(3), 257
[14] Gordaninejad F (1999) *J Intell Mater Syst Struct* 10(8), 601
[15] Casson N (1959) in *Rheology of Disperse Systems,* ed. Mill C (London, Pergamon)
[16] Carlson J D (1997) Proc. Of Int. Conf. on ERS and MRF, 112

Geometric Optimal Design of MR Damper Considering Damping Force, Control Energy and Time Constant

Q H Nguyen[1], S B Choi[1] and K S Kim[2]

[1] Smart Structures and Systems Laboratory, Department of Mechanical Engineering, INHA University, Incheon 402-751, Korea
[2] Department of Mechanical and Automotive Engineering, Kongju National University, Chonan 330-240, Korea

E-mail: seungbok@inha.ac.kr

Abstract. This paper presents an optimal design of magnetorheological (MR) damper based on finite element analysis. The MR damper is constrained in a specific volume and the optimization problem identifies geometric dimensions of the damper that minimizes an objective function. The objective function is proposed by considering the damping force, dynamic range and the inductive time constant of the damper. After describing the configuration of the MR damper, a quasi-static modelling of the damper is performed based on Bingham model of MR fluid. The initial geometric dimensions of the damper are then determined based on the assumption of constant magnetic flux density throughout the magnetic circuit of the damper. Subsequently, the optimal design variables that minimize the objective function are determined using a golden-section algorithm and a local quadratic fitting technique via commercial finite element method parametric design language. A comparative work on damping force and time constant between the initial and optimal design is undertaken.

1. Introduction

The suppression of mechanical and structural vibration using semi-active control method has been actively worked by many researchers in last two decades. Recently, various semi-active suspension systems featuring MR fluid have been proposed and successfully applied in the real field, especially in vehicle suspension systems [1-3]. Generally, the modeling and design of MR dampers have been done based on the quasi-static assumption of the damper [4,5]. In order to accurately model MR dampers, various dynamic models considering hysteresis behavior of MR damper have been also proposed [6,7]. However, researches on optimal design in order to improve performance of the MR dampers such as damping force, dynamic range and inductive time constant are still considerably rare.

Consequently, the main contribution of this research work is to find optimal dimensions of vehicle MR dampers considering an advanced objective function which includes damping force, dynamic range and inductive time constant altogether. In this study, the cylindrical MR damper for vehicle suspension proposed by Sung [3] is considered. After describing the configuration of the MR damper, a quasi-static modelling of the damper is performed based on Bingham model of MR fluid. The initial geometric dimensions of the damper are then determined based on the assumption of constant magnetic flux density throughout the magnetic circuit of the damper [8]. Subsequently, the optimal design variables that minimize the objective function are determined using a golden-section

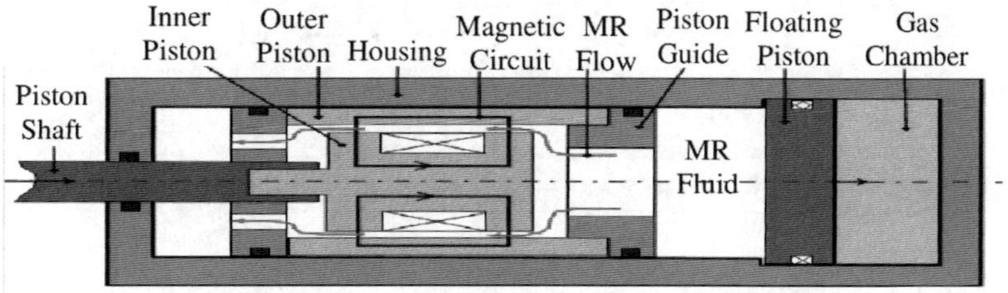

Figure 1. Configuration of the MR damper.

algorithm and a local quadratic fitting technique via ANSYS parametric design language. A comparative work on damping force and time constant between the initial and optimal design is undertaken.

2. Quasi-static modelling of the MR damper

The cylindrical MR damper for vehicle suspension proposed by Sung [3] is presented in figure 1. The outer and inner pistons are combined to form a MR valve structure which divides the MR damper into two chambers: the upper and lower chambers. These chambers are fully filled with the MR fluid. As the piston moves, the MR fluid flows from one chamber to the other through the annular duct (orifice) of the valve structure. The floating piston and gas chamber function as an accumulator to accommodate the damper rod volume as it enters and leaves the fluid chamber. By neglecting the frictional force and assuming quasi-static behavior of the damper, damping force (F_d) and dynamic range (λ_d) of the MR damper can be expressed as follows:

$$F_d = P_a A_s + c_{vis}\dot{x}_p + F_{MR}\,\mathrm{sgn}(\dot{x}_p)\,;\ \lambda_d = \frac{P_a A_s + c_{vis}\dot{x}_p + F_{MR,\max}}{P_a A_s + c_{vis}\dot{x}_p} \tag{1}$$

where gas chamber pressure P_a, the viscous coefficient c_{vis} and the yield stress force F_{MR} are given by

$$P_a = P_0\Big(\frac{V_0}{V_0 + A_s x_p}\Big)^{\gamma};\ c_{vis} = \frac{12\eta L}{\pi R_d t_d^3}(A_p - A_s)^2;\ F_{MR} = (A_p - A_s)\frac{2cL_p}{t_d}\tau_y$$

In the above, P_0 and V_0 are initial pressure and volume of the accumulator, γ is the coefficient of thermal expansion which is ranging from 1.4 to 1.7 for adiabatic expansion, x_p is the piston displacement, A_p and A_s are respectively the piston and the piston-shaft effective cross-sectional area, τ_y is the yield stress of the MR fluid induced by the applied magnetic field, η is the post yield viscosity of MR fluid, L is the length of the inner piston, L_p is the length of the magnetic pole, R_d and t_d are the average radius and gap of the annular duct, c is the coefficient which depends on flow velocity profile, $F_{MR,max}$ is the yield stress force with the maximum applied current. In this work, the commercial MR fluid (MRF132-LD) from Lord Corporation is used. The induced yield stress of the MR fluid can be expressed as a function of the applied magnetic field intensity (H_{mr}) as follows:

$$\tau_y = p(H_{mr}) = C_0 + C_1 H_{mr} + C_2 H_{mr}^2 + C_3 H_{mr}^3 \tag{2}$$

In Eq. (2), the units of τ_y and H_{mr} are *kPa* and *KA/m*, respectively. The coefficients C_0, C_1, C_2, and C_3 are respectively identified as 0.3, 0.42, -0.0012 and 1.05E-6. In order to calculate the damping force it is necessary to solve magnetic circuit of the damper. In general, for improving accuracy of the magnetic circuit solution, FEM is employed. In this study, the commercial FEM software ANSYS is used. The FE model to solve the magnetic circuit of the damper using 2D-axisymmetric coupled element (PLANE13) is shown in figure 2. It is noted that in FEM solution, the magnetic field intensity and flux density are not constant along the pole length. Therefore, the magnetic flux, the average magnetic flux density and magnetic field intensity across the MR ducts are determined as follows:

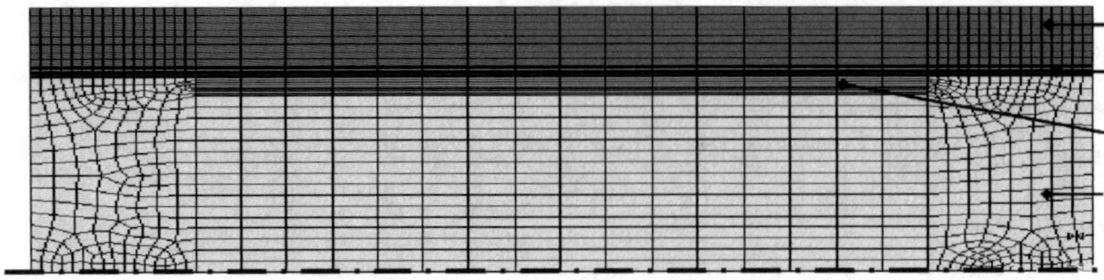

Figure 2. FE model to solve magnetic circuit of the MR damper

$$\Phi = 2\pi R_d \int_0^{L_p} B_{mr}(s)ds; \quad B_{mr} = \frac{1}{L_p}\int_0^{L_p} B_{mr}(s)ds; \quad H_{mr} = \frac{1}{L_p}\int_0^{L_p} H_{mr}(s)ds \tag{3}$$

where $B_{mr}(s)$ and $H_{mr}(s)$ are the magnetic flux density and magnetic field intensity at each nodal point on the defined path along the MR active volume where magnetic flux passes. The inductive time constant and control energy of the MR damper can be calculated as follows:

$$T = L_{in}/R_w \ ; \quad N = I^2 R_w \tag{4}$$

where L_{in} is the inductance of the valve coil given by $L_{in} = N_c\Phi/I$, R_w is resistance of the coil wire and N_c is the number of coil turns.

3. Optimization of the MR damper

In this study, FEM integrated with an optimization tool is used to obtain optimal values of design variables (DV) such as the coil width w_c, the pole length L_p, the MR orifice gap t_d and the outer piston thickness t_{op} which minimize an objective function. For vehicle suspension design, ride comfort and suspension travel are two conflicting performance indexes to be considered. In order to reduce the suspension travel, high damping force is required. On the other hand, for improving ride comfort, low damping force is expected. Thus a large dynamic range is required. Furthermore, a fast time response MR damper is also required to improve controllability of the suspension system. Taking the above requirements into account, the following objective function is proposed:

$$OBJ = \frac{F_{MR,r}}{F_{MR}} + \frac{\lambda_{d,r}}{\lambda_d} + \frac{T}{T_r} \tag{5}$$

where $F_{MR,,r}$, $\lambda_{d,r}$ and T_r are respectively the reference damping force, dynamic range and inductive time constant of the damper which determined from the solution of MR damper at initial values of design parameters (DV). Based on widely used MR dampers, the initial value of t_d is chosen by 1*mm*. The values of other DV are determined based on the assumption of constant magnetic flux density throughout the magnetic circuit of the damper as follows [8]:

$$w_c = \sqrt{R^2 - 4L_p^2} - 2L_p; \quad t_{op} = R - w_c - 2L_p \tag{6}$$

In the above, the initial value of L_p is determined such that the magnetic flux density does not exceed the saturation values of the valve core material (silicon steel) which is 1.5*Tesla* in this study. For the valve structure constrained in a cylinder of radius R=23*mm* and length L=50*mm*, the initial value DV can be determined as follows: L_p=7.75*mm*, w_c=1.5*mm*, t_{op}=5*mm*. In this study, the first order method implemented in the ANSYS optimization tool is used to find the optimal values of the design variables that minimize the proposed objective function.

Figure 3 shows the optimal solution of the MR damper based on FEM. In the optimization problem, the post-yield viscosity of the MR fluid is η=0.092*Pas*, the piston velocity is 0.4*m/s*, the coil wires are sized as 24-gauge and the maximum applied current is I=2*A*. The initial values of the

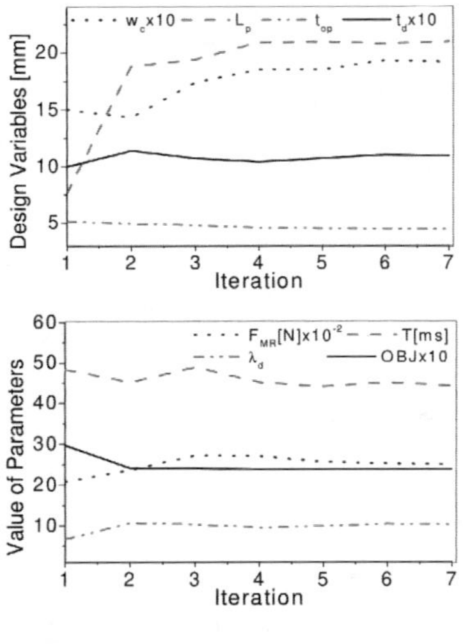

Figure 3. Optimal solution of the MR damper

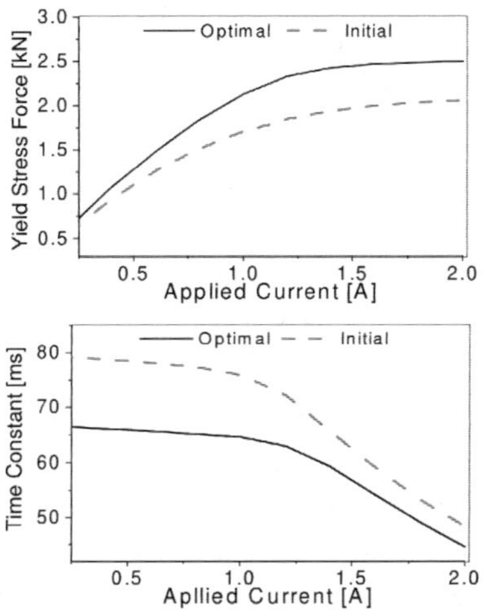

Figure 4. Yield stress force and time constant.

yield stress force, dynamic range, inductive time constant and objective function are 2078*N*, 6.66, 48.5*ms*, and 3, respectively. From the figure, it is observed that the solution converges after 7 iterations at which the minimal value of the objective function is 2.376. At the optimum, the yield stress force, dynamic range, and inductive time constant are 2498*N*, 10.22 and 44.26*ms*, respectively. The optimal values of L_p, w_c, t_d and t_{op} are 21*mm*, 1.92*mm*, 1.1*mm* and 4.45*mm*. The results show that the yield stress force, dynamics range and conductive time constant are significantly improved. The power consumption of the damper is 2.64*W* at the optimum, which is much smaller than that at the initial (8.68*W*). In order to evaluate performance characteristics of the optimized MR damper in the applied current range, the damping force and inductive time constant of initial and optimized dampers at different values of applied current are presented in figure 4. The figure shows that the yield stress force and inductive time constant of the optimized damper are much better than those of the initial one at any value of the current. The figure also shows that, at high applied current value, the yield stress force approaches to saturation while the time constant is reduced in proportion to the applied current.

4. Conclusion

In this study, geometric optimal design of vehicle MR dampers based on FEM solution was performed using ANSYS parametric design language. The optimization problem was to identify geometric dimensions of the valve structure employed in the MR damper that minimizes an objective function. The objective function was proposed considering the damping force, dynamic range and the inductive time constant of the damper and their reference values. The reference values were determined based on the assumption of constant magnetic flux density throughout the magnetic circuit of the damper. It was shown that, by minimizing the objective function, the yield stress force, dynamics range and conductive time constant are significantly improved at any value of applied current. The power consumption of the optimized damper was also significantly reduced.

References

[1] Kim K D and Jeon D Y 1996 *J. Intel. Mat. Syst. Str.* **10** 79
[2] Lee H S and Choi S B 2000 *J. Intel. Mat. Syst. Str.* **11** 80
[3] Sung K G 2007 *PhD. Thesis* Inha University, Incheon, Korea.
[4] Spencer B F, Dyke S J, Sain M K and Carlson J D 1997 *J. Eng. Mech. ASCE* **123** 230
[5] Dominguez A, Sedaghati R and Stiharu I 2006 *Smart Mater. Struct.* **15** 1179
[6] Choi S B and Lee S K 2001 *J. sound vib.* **245** 375
[7] Shen Y, Gonaraghi F and Heppler G R 2006 *J. Vib.Control* **12** 3
[8] Rosenfield N C and Wereley N M 2004 *Smart Mater. Struct.* **13** 1303

Dynamic Response of Magneto-Rheological Fluid Channel Flow with Fluid-Wall Interactions

H. Nishiyama[1], H. Takana[1], K. Mizuki[2], H. Weisbecker[3] and S. Odenbach[3]

[1]Institute of Fluid Science, Tohoku University, 2-1-1, Katahira, Aoba-ku, Sendai, 980-8577, JAPAN

[2]Graduate School of Engineering, Tohoku University, 6-6-04, Aramaki Aza Aoba, Aoba-ku, Sendai, 980-8579, JAPAN

[3]Technische Universitat Dresden, Mollierbau George-Bahr-Str.3 01069 Dresden, GERMANY

E-mail: `nishiyama@ifs.tohoku.ac.jp`

Abstract. The objective of this study is to provide the fundamental data for the novel application of MR fluid plugging(MRF plugging) to a medical or safety devices for flow control. Experimental evaluation on dynamic response of MRF plugging in pressure mode was conducted to clarify the effect of piston velocity, magnetic flux intensity and wall surface roughness to its plugging performance. MRF plugging performance has been discussed through the evaluation of the endurance pressure of MRF plugging and flow rate. It is shown that the endurance pressure of MRF plugging is increased by the piston velocity, the magnetic flux intensity, and the adoption of wall surface structure. Furthermore, the endurance pressure is decreased for the circular and elastic tube, assuming a biological tube.

1. Introduction

MR fluid plugging(MRF plugging) performance by MRF column is recently expected to be one of the most promising applications to medical or safety devices, such as blood flow control, steam shut-down valve or fuel supply[1,2]. Researches of MR fluid have been actively conducted on the evaluation of rheological properties and its industrial applications, due to its attractive characteristics of controllable yield stress and significant viscous force under even small applied magnetic field. Furthermore, the evaluation of flow control ability has been actively investigated in these days, considering the channel permeability or wall configurations[3-6].

In this study, the effect of surface roughness on rheological properties of MR fluid was experimentally clarified. Then, dynamic response of MRF plugging in pressure mode with fluid-wall interactions was investigated, considering the effects of magnetic flux intensity, piston velocity, the channel configurations, its surface structure and tube elasticity. The final objective of this study is to provide the fundamental data for the novel application of MR plugging performance to a medical or safety devices for flow control.

2. Experimental Procedure and Evaluation

In this study, MRF-132DG manufactured by LORD Corporation is used as a test sample.

(1) MR fluid is filled inside the piston cylinder and horizontal flow channel shown in Fig.1. A MR channel is made of acrylics with rectangular cross section of 10 mm × 10 mm. After forming MR fluid column in a channel by applying magnetic field with electromagnet,

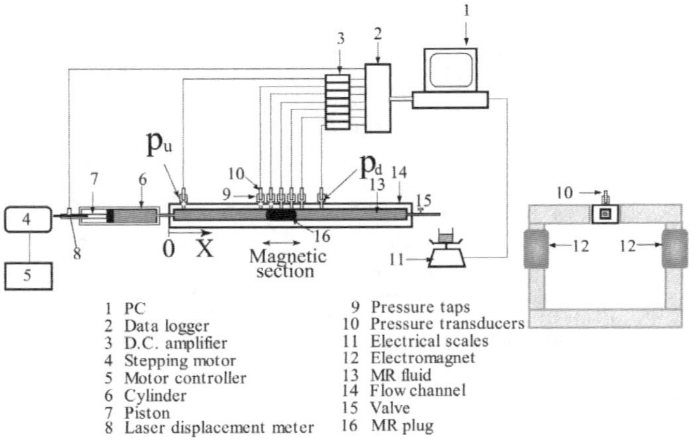

Figure 1. Schematic illustration of experimental setup.

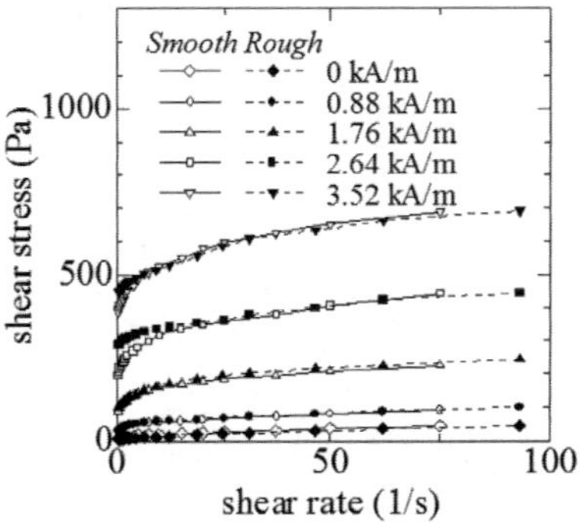

Figure 2. Shear stress as a function of shear rate.

piston is driven at the constant head velocity controlled by a stepping motor to produce the pressure flow mode against the MRF plugging.

(2) Pressures inside the channel are measured by transducer piezoresistive pressure sensors.

(3) Exit of flow channel is opened to atmospheric pressure. The mass of exhaust MR fluid is measured by electrical scale for the evaluation of flow rate.

(4) The same procedures are taken for the different channels: the rectangular channel with the wall surface structure (distorted lattice 1 mm-depth-groove pattern at the magnetic pole side of the internal wall), a semi-rigid circular tube (d_i = 12 mm, Young's modulus of 6.7 kPa), an elastic circular tube (d_i = 12 mm, Young's modulus of 2.5 kPa).

3. Results and Discussion

3.1. Effect of surface roughness on rheological properties in rotating shear mode

Figure 2 shows shear stress measured by cone-plate rheometer with smooth cone ($Rz = 0.5$) and the rough cone ($Rz = 4.5$) in rotating shear mode at various magnetic field intensities. The static yield stress is higher for the rough cone especially under effective magnetic field intensity larger than 2.64 kA/m due to an anchoring effect even in rotating shear mode.

3.2. Effect of piston speed on MRF plugging performance in pressure mode

Figure 3 shows the time evolutions of channel pressures p_{MR}, piston displacement d_{piston} and the exhausted flow rate Q_{out} at various piston head speeds under the applied magnetic flux density of 800 G. Piston starts at $t = 0$, after 3 minutes in resting state under a magnetic field for MR column formation. p_u, p_{d1}, and p_{d2} represent pressures at upstream and downstream of MR column, respectively, as shown in Fig.1. For the references, Fig.3(a) also includes p_u without magnetic field. Under magnetic field, p_u increases without the exhausted flow rate as the piston moves.This shows the MR fluid column 'plugs' the flow. To evaluate its plugging performance, its endurance pressure p_{plug} is defined as p_u at the moment that MR fluid starts to flow detected by Q_{out}. With increasing piston head velocity, which corresponds to the given work per unit volume of the MR fluid column, the endurance pressure and flow rate increase. Furthermore, their time gradients also get steeper and steeper. This difference in the time gradient shows the time evolution of MRF plugging breakdown behavior. The increase in p_{plug} is due to the robust resistance of cluster and a higher local concentration of magnetic particles at the upstream of MRF column with increasing piston speed.

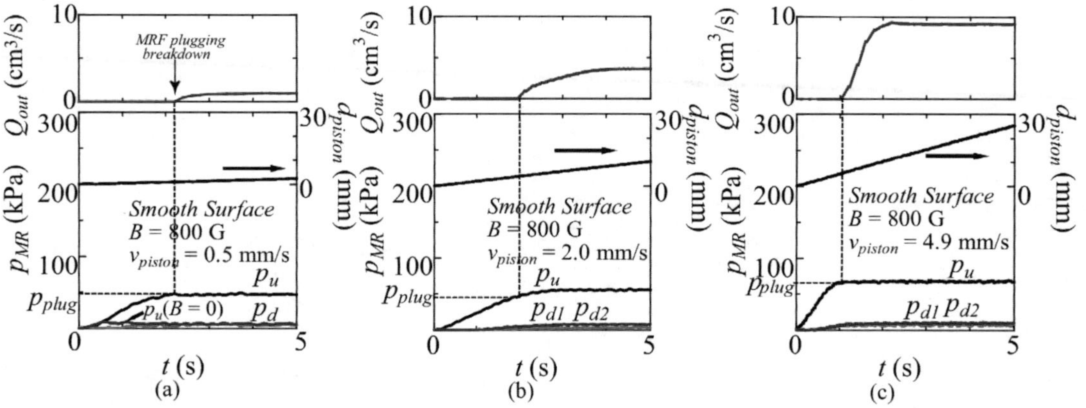

Figure 3. Time evolutions of pressure and flow rate at various piston speeds under $B = 800$ G.

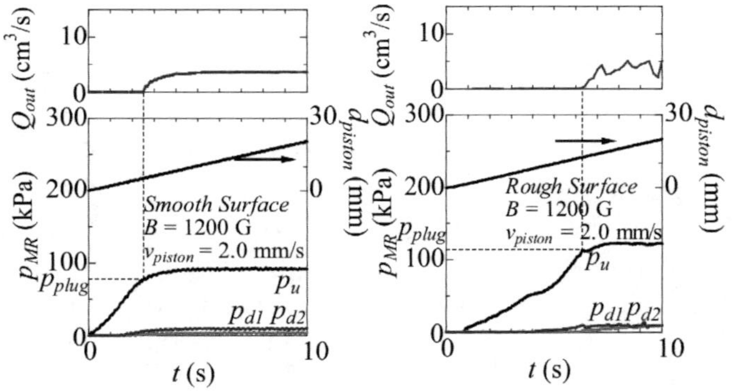

Figure 4. Time evolutions of pressure and flow rate for smooth and rough surface channels under $B = 1200$ G.

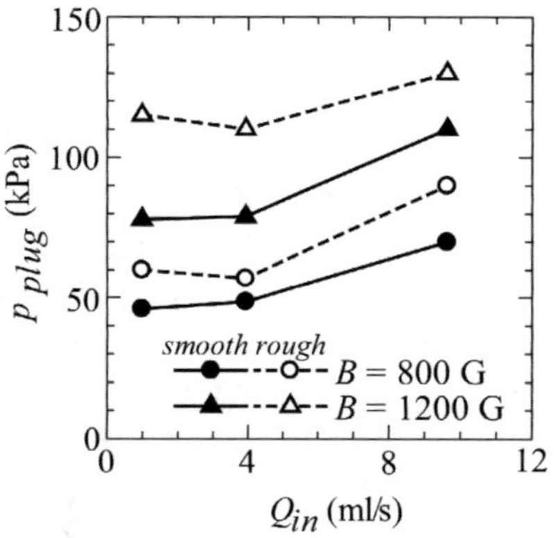

Figure 5. Roughness effect on MRF endurance pressure.

3.3. Effect of wall roughness on MR fluid plugging performance in pressure mode

Figure 4 shows time evolutions of internal pressure, piston displacement and exhausted MR flow rate for smooth and rough surface channels under $B = 1200$ G. The increase of endurance pressure with the increase in applied magnetic field well corresponds to the increase of static yield stress in its rheological property[7]. Comparing with two endurance pressures in Fig.4, p_{plug} for the rough surface channel is 1.5 times larger than that for the smooth one, although there is not much difference in the case of $B = 0$ G. The effect of wall roughness on the endurance pressure is clearly shown in Fig.5. Furthermore, the higher magnetic flux density, the more remarkable effect of surface roughness appears. Referring to the discontinuous pressure response and fluctuating flow rate behavior for the rough surface channel, the MR column seems to resist robustly and yield complexly. These can be due to a strong anchoring between clusters' roots and wall roughness structure. The mechanics of this phenomenon in micro scale and macro scale will be further investigated in the future research.

3.4. Effects of cross section configurations and tube elasticity on MR fluid plugging in pressure mode

Figure 6 shows time evolutions of pressure and flow rate for semi-rigid and elastic circular tube. p_u repeats increasing and decreasing, and Q_{out} goes on-and-off intermittently for high elastic tube with the presence of air bubble content. In this case, p_{plug} is defined as the average of local maximum p_u for endurance pressure evaluation. By evaluating p_{plug} in Fig. 7, it is shown that

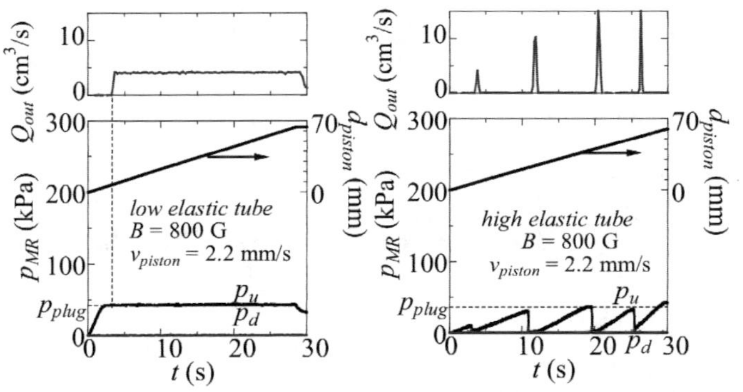

Figure 6. Time evolutions of internal pressure, piston displacement, and flow rate, in case of semi-rigid and elastic circular tube.

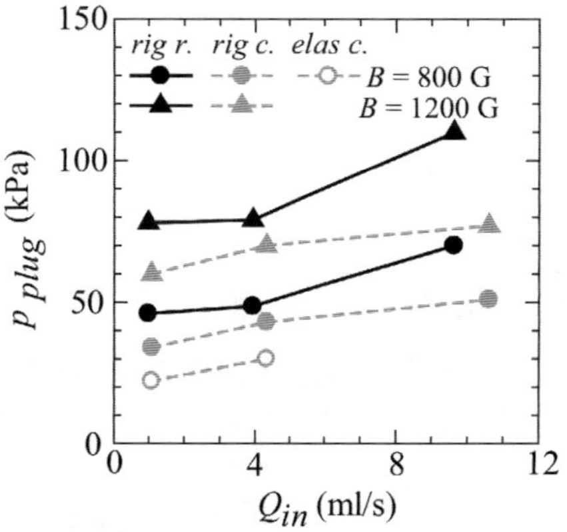

Figure 7. Effects of cross section configurations and tube elasticity on MRF endurance pressure.

the endurance pressure decreases because of circular cross section and high elasticity, assuming a biological tube. The robust cluster region decreases for circular cross section, which leads to the decrease in endurance pressure. Furthermore, the endurance pressure can be decreased for high elasticity tube with its deformation during the cluster formation.

4. Conclusions

The effects of surface roughness on rheological properties and the effect of wall surface structure, channel cross section, and tube elasticity on the dynamic response of MRF plugging in pressure mode are made clear through the experiment analysis as the first step. The obtained results are summarized as follows:

(1) Static yield stress increases for cone with roughness in rotating shear mode.

(2) The endurance pressure of MRF plugging is increased by applied magnetic field intensity, piston velocity and the wall roughness.

(3) The endurance pressure of MRF plugging is decreased by the circular and elastic tube.

5. Acknowledgments

The present study was partly supported by a Grant-in-aid for Exploratory Research from the Japan Society for Promotion of Science(2006, 2007). Authors would like to thank Mr. K. Katagiri, Mr. T. Nakajima, and Mr. J. Jang with Institute of Fluid Science, Tohoku University for their technical supports.

6. References

[1] Sheng R, Flores G A and Liu J 1999 *J. of Magnetism and Magnetic Materials* **194** 167-175

[2] Saito T, Ikeda H and Oguro M 2006 *J. of Intelligent Material Systems and Structures* **17** 341-345

[3] Gordaninejad F, Kavlicoglu B M and Wang X, 2004 *Proc. of the Ninth Int. Conference Electrorheological and Magnetorheological Suspensions* 314-320

[4] Cook E, Hu W and Wereley NM 2007 *J. Intelligent Material Systems and Structures* **18** 1197-1202

[5] Lloyd J R, Hayesmichel M O and Radcliffe C J 2007*J. of Fluids Engineering* **129** 423-428

[6] Lemaire E and Bossis G 1991 *J. Phys. D: Appl. Phys.* **24** 1473-1477

[7] Nisiyama H, Takana H, Mizuki K, Nakajima T and Katagiri K, 2008*AIP Proc. Complex Systems* **982** 592-597

Core-shell typed polymer coated-carbonyl iron suspensions and their magnetorheology

B J Park[1], C W Park[2], S W Yang[2], H M Kim[2] and H J Choi[1]

[1] Department of Polymer Science and Engineering, Inha University, Incheon 402-751, Korea
[2] SK Energy, Institute of Technology, Lubricants Lab, Taejon 305-712, Korea

E-mail: hjchoi@inha.ac.kr

Abstract. In order to resolve aggregation and sedimentation problem of carbonyl iron (CI) for magnetorheological (MR) fluids, polymer coated composite magnetic particles (CIPMMA) were synthesized via a dispersion polymerization method using CI dispersions in methyl methacrylate monomer, and then adopted the product as a dispersed phase of MR fluids. Especially, to improve mechanical property of the coated poly(methyl methacrylate) (PMMA) onto CI microspheres, a cross-linking method was adopted in this study. Flow and viscoelastic properties of the MR fluids were analyzed via a rotational rheometer equipped with a magnetic field supplier using a parallel plate measuring system.

1. Introduction

Magnetorheological (MR) fluid, which is a suspension of magnetically susceptible particles dispersed in a nonmagnetic suspending medium, has been investigated by many research groups [1, 2]. It is regarded as a smart material with controllable properties by an external magnetic field, which is abruptly transformed within milliseconds from a fluid-like to solid-like state under an applied magnetic field, by showing changes of its rheological characteristics such as yield stress and shear viscosity [3-6]. When magnetic field is applied to MR fluids, randomly dispersed magnetic particles are being oriented with a chainlike structure in the direction of magnetic field, which can be deformed and then broken by hydrodynamic drag as they are subjected to the shear force [7, 8]. Therefore, MR fluids have many industrial applications such as dampers, shock absorbers and brakes so as to control resistance due to the tuneable rheological properties and their fast response time [9-11].

For the high performance engineering application, both high yield stress and colloidal dispersion are essential in MR fluids [12]. Among various magnetic materials, pristine carbonyl iron (CI) has been extensively investigated as a MR material regarding MR characteristics due to its high magnetization properties in certain magnetic field strength [13]. However, the CI based MR suspensions are known to possess several drawbacks such as dispersion non-uniformity and sedimentation of the CI particles due to the large density difference. In order to improve this problem, dispersants were added in MR fluids to improve dispersion stability [14-16]. In other ways, magnetic particles were coated with polymer or monodispersed polymer cores were coated with magnetic particle to lower density of the magnetic particles [17-19].

In this study, to improve dispersion stability of the MR fluids, core-shell typed polymer coated-CI composite particles via an encapsulation method with poly(methyl methacrylate) (PMMA) [20] are introduced. The polymer coated composite magnetic particles (CIPMMA) were synthesized via a dispersion polymerization method using CI dispersants in methyl methacrylate monomer, and then adopted as dispersed phase of MR fluids. Especially, to improve mechanical property of the coated PMMA onto CI microspheres, a cross-linking method was adopted using ethylene glycol dimethacrylate (EGDMA) as a crosslinker in this study. MR properties were analyzed via a rotational rheometer equipped with a magnetic field supplier using a measuring system of parallel plate geometry.

2. Experimental

2.1. Preparation

We used carbonyl iron (CI) (standard CD grade, BASF, Germany) particles whose average particle size and density are 2.57 μm and 7.86 g/cm^3, respectively. Initially, the CI surface was modified by methacylic acid (MAA, Junsei, Japan) for 4h to induce easy polymerization reaction on the CI surface. The modified CI particles were then dispersed in methanol containing poly(vinyl pyrrolidone) (PVP) (Aldrich, USA, Mw = 1,300,000 g/mole) as a stabilizer. Methyl methtacrylate (MMA) (Dongyang Chem. Co., Korea) monomer and 2, 2-azobisisobutyronitrile (AIBN) (Junsei, Japan) as a radical initiator were dissolved in the reaction system. The AIBN was used after being recrystallization in acetone. In addition, ethylene glycol dimethacrylate (EGDMA) (Aldrich, USA) was used as a cross-linking agent of PMMA. The mass ratio of CI to MMA was 2.0. The temperature was increased up to 55 °C, and the dispersion polymerization was continued for 24 h under the nitrogen purging. After the polymerization, the crosslinked CIPMMA was separated by magnet and washed with distilled water and methanol to remove PVP, unreacted monomer, and PMMA oligomers. Finally, the CIPMMA composite particles were dried at 60 °C in oven.

2.2. Characterization

External and internal morphology of the produced particle was observed using scanning electron microscope (SEM) (Hitachi S-4300) and transmission electron microscope (TEM). The MR fluids were then prepared by dispersing the crosslinked CIPMMA particles into lubricant oil (Yubase 8, SK Co. Korea). Rheological properties with and without external magnetic field were investigated using a commercial rotational rheometer (MCR 300, Anton-Paar, Germany) with the MR device (MRD 180, Physica, Stuttgart, Germany) which can generate a homogeneous magnetic field. The MR properties were measured in both steady and oscillatory shear modes.

3. Results and discussion

3.1. Morphology of magnetic particles

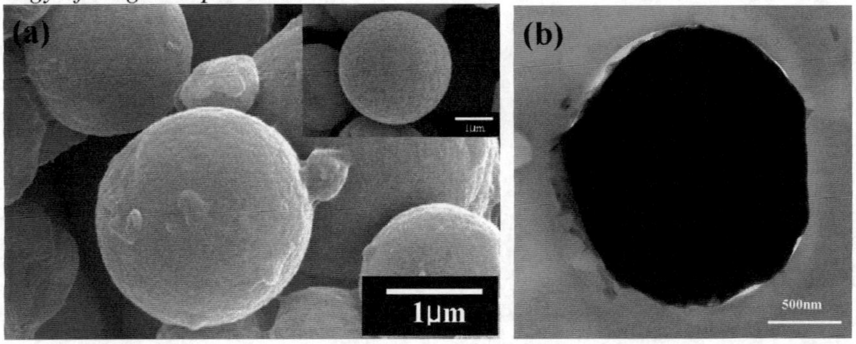

Figure 1. (a) SEM images (inset: pure CI) (b) TEM images of PMMA-coated CI particles

Journal of Physics: Conference Series **149** (2009) 012078 doi:10.1088/1742-6596/149/1/012078

Surface morphology of the cross-linked CIPMMA particles was observed by SEM. Figure 1(a) shows the SEM images of both pure CI and PMMA coated CI particles. The surface of the synthesized cross-linked PMMA-coated CI particles becomes smooth compared to pure CI, indicating that the polymers were encapsulated onto the CI surface as a result of the coating process. It also indicated that the coated particles have spherical shape while showing similar size with pure CI.
The internal structure of the PMMA coated CI particle was shown in Fig. 1(b) which represented TEM image of cross sectioned core-shell particles using an epoxy mold. The CIPMMA particles were found to have about 50nm of PMMA coated layer, of which the PMMA layer was partially sectioned due to difference of hardness between CI and PMMA during microtoming.

3.2. Thermal property of CIPMMA magnetic particles

The encapsulation was also characterized using thermal gravimetry analyzer (TGA). Figure 2 shows thermal degradation of cross-linked CIPMMA magnetic particles with N_2 purging. It displayed typical degradation curve of the dispersion polymerized PMMA, which shows multistep degradation due to solubility difference as PMMA molecular weight, between 150□ and 450□. The final weight loss is estimated to be about 13 wt% which implies the content of PMMA in CIPMMA particles. The density of the CI/PMMA composite at 20□ was measured to be 4.7 g/cm^3 via a gas pycnometer, which is about half of the density of pristine CI.

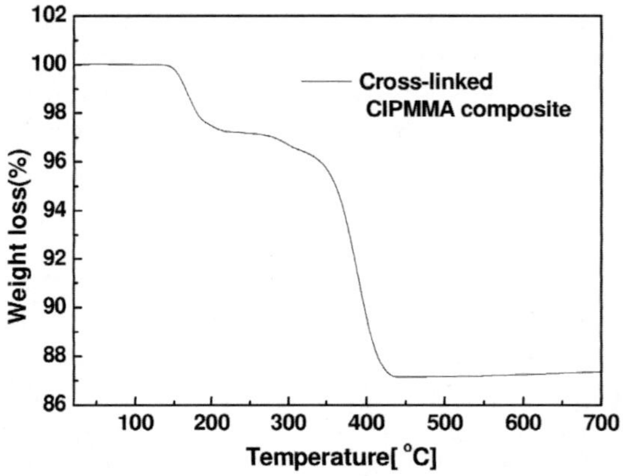

Figure 2. Thermal degradation curve of CIPMMA composite particles

3.3. Magnetorheological properties of MR fluids

The MR properties were characterized via a rotational rheometer equipped with a magnetic field supplier using a measuring system of a parallel plate. Figure 3 shows shear stress as a function of shear rate for the CI and CIPMMA based MR fluids under different external magnetic field strengths ranging from 0 to 342kA/m. Both of pure CI and CIPMMA based MR fluids behave typical Bingham fluids with yield stresses under external magnetic fields, while they behave like Newtonian fluids without a magnetic field, implying the formation of chain structure and the increase of magnetic interaction between the particles. It was also clearly shown that the shear stresses increased for the entire shear rate region with the increase in magnetic field strength, indicating the increase of magnetic dipole moment with increase of magnetic field strength. Although the yield stress of the PMMA encapsulated CI particle based MR fluid was indicated to be smaller than that of the CI based MR fluid because the polymer shell reduces the particle interaction, the CIPMMA based MR fluids exhibit enough yield stresses for their industrial application. Furthermore, the CIPMMA based MR fluids show more stable Newtonian behaviour at 0kA/m magnetic field, indicating that the encapsulation of magnetic particle offers colloidal stability into MR fluids.

Journal of Physics: Conference Series **149** (2009) 012078 doi:10.1088/1742-6596/149/1/012078

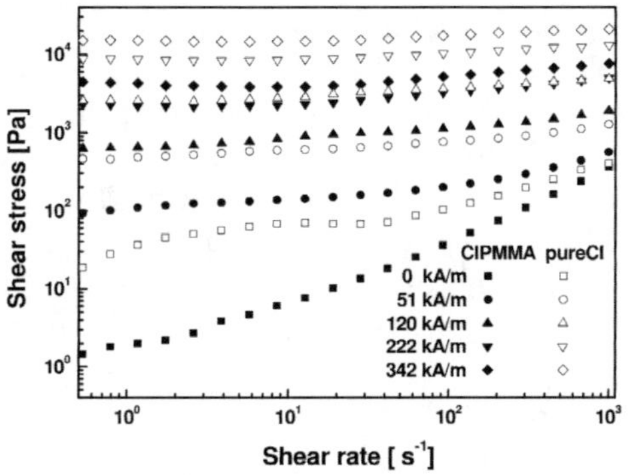

Figure 3. Flow curve of pure CI and CIPMMA composite particle based MR fluids

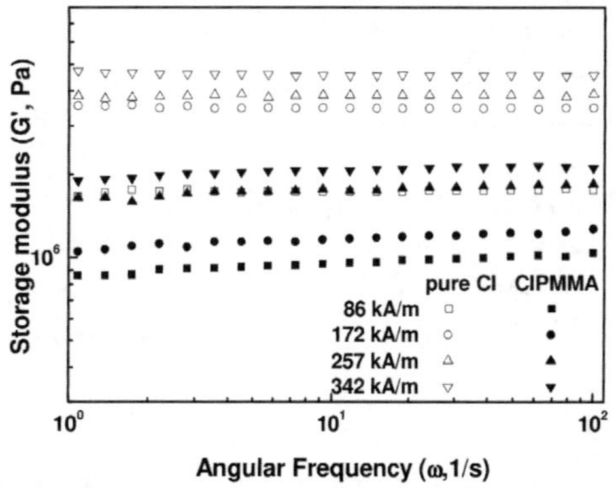

Figure 4. Flow curve of pure CI and CIPMMA composite particle based MR fluids

Viscoelastic behaviour of the MR fluids was also analyzed in the oscillatory shear mode. Figure 4 shows storage modulus of pure CI and CIPMMA based MR fluids as a function of angular frequency. It was found that the storage modulus was frequency independent for both CI and CIPMMA based MR fluids, implying the solid-like properties of MR fluids.

The relationship between shear stress and the applied magnetic field strength (H) is presented in Figure 5 which shows shear stresses of CI and CIPMMA based MR fluid at each magnetic field strength. With the increase in magnetic field strength, the shear stresses were increased. When considering yield stresses of MR fluids at intermediate field strengths, yield stress can be predicted by various factors [21]:

$$\sigma_y = \sqrt{6}\,\phi \mu_0 M_s^{1/2} H^{3/2}$$

where is the volume fraction of the magnetic particles, μ_0 the permeability of free space, M_s magnetic-field saturation, and H is the magnetic field strength. The relationship between yield stress and magnetic field strength exhibits that yield stress increases in proportion to $H^{3/2}$.This prediction of a 3/2 power law for the field-dependence has been confirmed experimentally and we can also confirm in Fig. 5, in which there was a trend for the slope of stress based on the field to start to decline in a high magnetic field strength range.

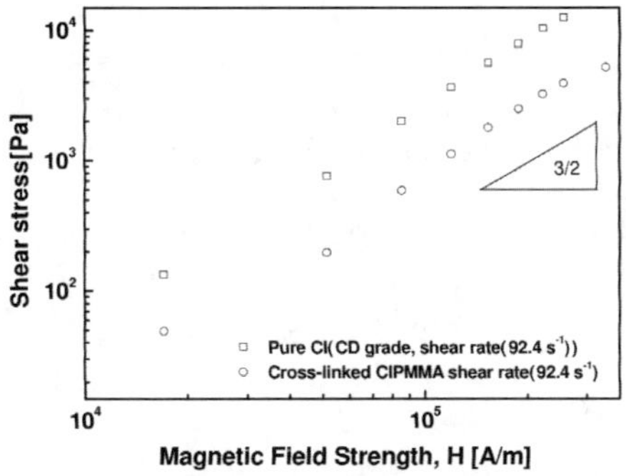

Figure 5. Shear stress curve of MR fluids with different magnetic field strength

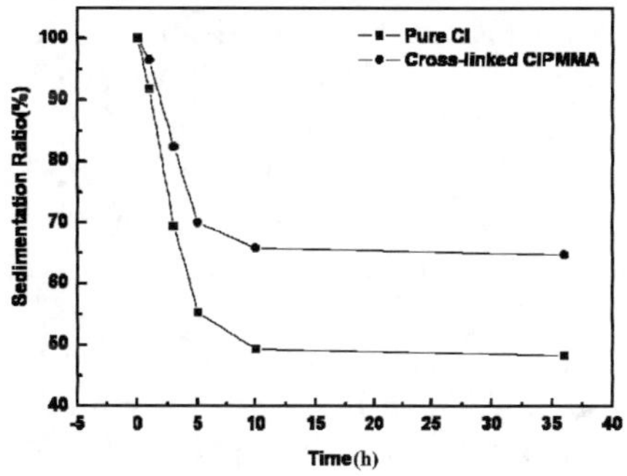

Figure 6. Sedimentation ratio of MR fluids as a function of time

3.4. Sedimentation and re-dispersion of MR fluids

Although the yield stress of MR fluids decreases due to the encapsulation of magnetic particles, the encapsulation has advantages in industrial application because it can enhance the dispersion stability of the suspension. In Fig. 6, the sedimentation ratio of CI and CIPMMA based MR fluid was compared to each other, in which the sedimentation curve of pure CI based MR fluid displayed faster sedimentation than PMMA encapsulated CI based MR fluid due to the density difference between pure CI and CIPMMA particles. After most particle settled down, the pure CI based MR fluid showed a lower level particle height than that of the CIPMMA based MR fluid, indicating that the CIPMMA particle possesses larger effective volume than pure CI in the medium oil. It was also confirmed that the CIPMMA could be easily re-dispersed after removing magnetic fields due to their surface properties.

4. Conclusion

In this study, magnetic CI particles were encapsulated by cross-linked PMMA via in-situ dispersion polymerization. The surface morphologies and cross-sectional images were examined by SEM, TEM. Due to not only decrease of absolute contents of CI with polymer coating but also reduction of the particle interaction because of the polymer shell, the yield stress of CIPMMA composite particle based MR fluid was lower than that of pristine CI based MR fluid. Despite slight lower yield stress and modulus of the MR fluids, it was found that the encapsulation process of magnetic particles improve the sedimentation and dispersion quality.

Acknowledgments

This work was supported by SK Energy, Korea.

References

[1] Bossis G, Khuzir P, Lacis S and Volkova O 2003 *J. Magn. Magn. Mater.* **258-259** 456
[2] Jolly M R, Carlson J D and Munoz B C 1996 *Smart Mater. Struct.* **5** 607
[3] Bica I and Choi H J 2008 *Int. J. Mod. Phys. B* **22** 5041
[4] Ulicny J C and Golden M A 2007 *Int. J. Mod. Phys. B* **21** 4898
[5] See H, Joung C and Ekwebelam C 2007 *Int. J. Mod. Phys. B* **21** 4945
[6] Lim S T, Cho M S, Jang I B and Choi H J 2004 *J. Magn. Magn. Mater.* **282** 170
[7] Lim S T, Cho M S, Choi H J and Jhon M S 2003 *J. Ind. Eng. Chem.* **9** 336
[8] Fang F F, Jang I B and Choi H J 2007 *Diam. Relat. Mat.* **16** 1167
[9] Choi S B, Song H J, Lee H H, Lim S C, Kim J H and Choi H J 2003 *Int. J. Veh. Des.* **33** 2
[10] Saito T, Ikeda H and Oguro M 2006 *J. Intell. Mater. Syst. Struct.* **17** 341
[11] Goncalves F D and Carlson J D 2007 *Int. J. Mod. Phys. B* **21** 4832
[12] Bombard A J F, Joekes I, Alcantara M R and Knobel M 2003 *Adv. Powder Technol.* **416-4** 753
[13] Pan H J, Huang H J, Zhang L Z, Qi J Y and Cao S K 2005 *J. Cent. South Univ. Technol.* **12** 411
[14] Lim S T, Cho M S, Jang I B, Choi H J and Jhon M S 2004 *IEEE Trans. Magn.* **40** 3033
[15] Fang C, Zhao B Y, Chen L, Wu Q, Liu N and Hu K A 2005 *Smart Mater. Struct.* **14** N1
[16] Bell R C, Miller E D, Karli J O, Vavreck A N and Zimmerman D T 2007 *Int. J. Mod. Phys. B* **21** 5018
[17] Jang I B, Kim H B, Lee J Y, You J L, Choi H J and Jhon M S 2005 *J. Appl. Phys.* **97** 10Q912
[18] Li L C, Guan J G, Cheng H B and Tao J Q 2005 *Acta Phys.-Chim. Sin.* **21** 817
[19] Fang F F, Park B J, Choi H J and Jhon M S 2007 *Int. J. Mod. Phys. B* **21** 4981
[20] Cho M S, Lim S T, Jang I B, Choi H J and Jhon M S 2004 *IEEE Trans. Magn.* **40** 3036
[21] Ginder J M, Davis L C and Elie L D 1996 *Int. J. Mod. Phys. B.* **10** 3293

Journal of Physics: Conference Series **149** (2009) 012079 doi:10.1088/1742-6596/149/1/012079

Preview Control of Vehicle Suspension System Featuring MR Shock Absorber

M S Seong[1], S B Choi[1], M W Cho[2] and H G Lee[3]
[1]Smart Structures and Systems Laboratory, Department of Mechanical Engineering, Inha University, Incheon 402-751, Korea
[2]Precision Manufacturing and Inspection Laboratory, Department of Mechanical Engineering, Inha University, Incheon 402-751, Korea
[3]Department of Automotive Engineering, Daeduk College, Daejeon, 305-715, Korea

E-mail: seungbok@inha.ac.kr

Abstract. This paper presents control performance evaluation of optimal preview control algorithm for vehicle suspension featuring MR shock absorber. The optimal preview control algorithm has several advantages such as high control performance over that which is best for a non-preview system. In order to achieve this goal, a commercial MR shock absorber, Delphi Mganeride™, which is applicable to high class passenger vehicle, is adopted and its field-dependent damping force and dynamic responses are experimentally evaluated. Then the governing equation of motion for the full-vehicle model is established and integrated with the MR shock absorber. Subsequently, optimal controller with preview control algorithm is formulated and implemented for vibration suppression of the car body. Control performance of the preview controller is evaluated for the full-vehicle model under random road condition. In addition, the control performances depending on preview distances are evaluated.

1. Introduction

The vehicle dynamic characteristics can be normally improved by suspension systems. So far, a passive oil shock absorber is widely employed for vehicle suspension system. However, its performance is limited by the uncontrollable damping force. Recently, active shock absorber is gradually used to enhance ride comfort as well as steering stability. However, design complication and high cost prevent the popularization of the advanced active suspension systems. Therefore, alternative mechanisms for vehicle suspension systems have been studied to replace the passive and active shock absorbers. For the last decade, the research work on ride comfort and steering stability of a vehicle system using semi-active suspension has been significantly increased. Among several potential candidates, magnetorheological (MR) fluid, whose rheological properties can be rapidly changed by applying a magnetic field to the fluid domain, has been most actively studied for the shock absorber of the semi-active suspension system which can offer a desirable performance without requiring large power consumption and expensive hardware. Carlson et al. [1] proposed a commercially available MR damper which is applicable to on-and-off-highway vehicle suspension system. Spencer Jr. et al. [2] proposed a dynamic model for the prediction of damping force of a MR damper. Choi et al. [3] manufactured an MR damper for a small-sized passenger vehicle and presented control characteristics of the damping force.

In the past, the feedback vibration control algorithms have been widely studied. However, research works on the feedforward control algorithms such as preview control are relatively rare. Preview control of suspension involves the acquisition and use of information concerning the road profile ahead of the vehicle. Hac et al. [4] obtained general analytical solutions including the effects of preview information on ride comfort, road holding and working space of the suspension and power requirements. Thompson et al. [5] proposed spectral decomposition methods to compute accurate RMS values with preview control. Mazbanrad et al. [6] proposed stochastic optimal preview control algorithm of a vehicle suspension on a random road. Gopala et al. [7] proposed semi-active preview control algorithm and half-car model considering the hysteretic nonlinear spring.

In this study, optimal preview control algorithm for the vehicle vibration is constructed by adopting the semi-active MR suspension system. Then preview control performances are evaluated under random road conditions.

2. Vehicle model with MR shock absorber

In this work, damping characteristics of a commercial MR shock absorber, Delphi Corporation's Magneride^{TM}, which is used for high-class passenger vehicle is experimentally evaluated and modelled. Figure 1 presents the measured field-dependent damping force of MR shock absorber with respect to piston velocity. As expected, as the current increases the damping force increases. Figure 2 shows the

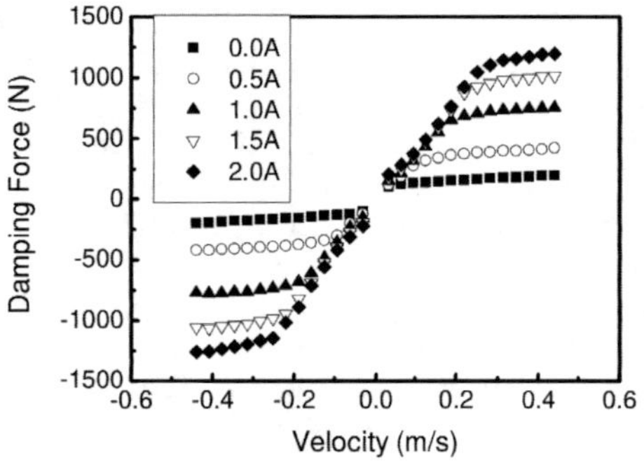

Figure 1. The field-dependent damping force of MR shock absorber.

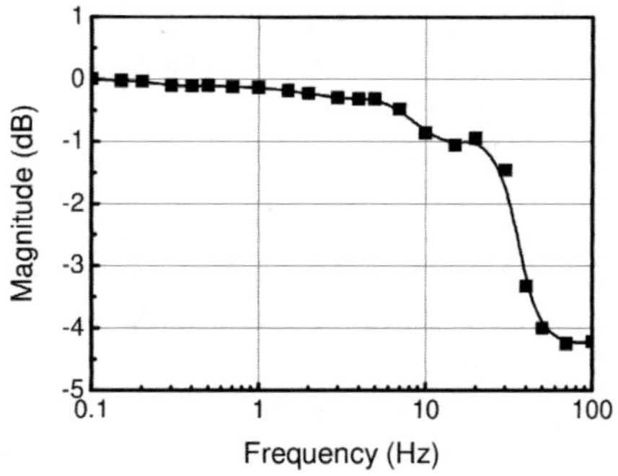

Figure 2. Dynamic bandwidth of MR shock absorber.

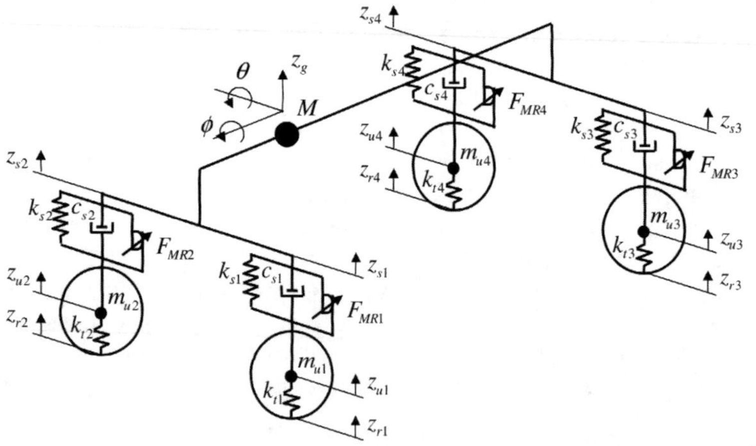

Figure 3. Mechanical model of the full-vehicle MR suspension system.

measured dynamic bandwidth of the damping force in frequency domain. It is identified that the dynamic bandwidth of the proposed MR shock absorber is about *38Hz* at *-3dB*. The MR shock absorber is modelled based on these characteristics. We can construct a mathematical state-space model for a full-vehicle MR suspension system equipped with four MR shock absorbers shown in figure 3:

$$\begin{aligned}&\dot{\mathbf{x}} = \mathbf{Ax} + \mathbf{Bu} + \mathbf{Dw},\\ &\mathbf{u} = [F_{MR1}\ F_{MR2}\ F_{MR3}\ F_{MR4}]^T,\\ &\mathbf{x} = [z_{u1}\ z_{u2}\ z_{u3}\ z_{u4}\ z_g\ \theta\ \phi\ \dot{z}_{u1}\ \dot{z}_{u2}\ \dot{z}_{u3}\ \dot{z}_{u4}\ \dot{z}_g\ \dot{\theta}\ \dot{\phi}]^T,\\ &\mathbf{w} = [z_{r1}\ z_{r2}\ z_{r3}\ z_{r4}]^T\end{aligned} \quad (1)$$

where $\mathbf{x}$ is the state vector, $\mathbf{u}$ is the control input vector, $\mathbf{w}$ is the road input vector. $\mathbf{A}$, $\mathbf{B}$ and $\mathbf{D}$ are the system matrix, input matrix and disturbance matrix, respectively.

3. Preview control

The performance index J is established for ride comfort and vehicle stability by minimizing the acceleration of C.G., suspension travel, tire force and control input [8].

$$J = \lim_{T\to\infty}\frac{1}{2T}\int_0^T \left(\mathbf{x}^T\mathbf{Q}_1\mathbf{x} + 2\mathbf{x}^T\mathbf{N}\mathbf{u} + \mathbf{u}^T\mathbf{R}\mathbf{u} + 2\mathbf{x}^T\mathbf{Q}_{12}\mathbf{w} + \mathbf{w}^T\mathbf{Q}_2\mathbf{w}\right)dt \quad (2)$$

where $\mathbf{Q}_1$, $\mathbf{N}$, $\mathbf{R}$, $\mathbf{Q}_{12}$ and $\mathbf{Q}_2$ are time-invariant weighting matrices. From this performance index, the optimal control inputs are determined as follows:

$$\mathbf{u}(t) = -\mathbf{R}^{-1}\left[\left(\mathbf{N}^T + \mathbf{B}^T\mathbf{P}\right)\mathbf{x}(t) + \mathbf{B}^T\mathbf{r}(t)\right] \quad (3)$$

where $\mathbf{r}(t)$ is the preview function and $\mathbf{P}$ is the positive definite symmetric matrix solution of the algebraic Riccati equation;

$$\begin{aligned}&\mathbf{PA}_n + \mathbf{A}_n{}^T\mathbf{P} - \mathbf{PBR}^{-1}\mathbf{B}^T\mathbf{P} + \mathbf{Q}_n = 0,\\ &\mathbf{A}_n = \mathbf{A} - \mathbf{BR}^{-1}\mathbf{N}^T,\ \mathbf{Q}_n = \mathbf{Q}_1 - \mathbf{NR}^{-1}\mathbf{N}^T\end{aligned} \quad (4)$$

The preview function $\mathbf{r}(t)$ is given by

$$\begin{aligned}&\mathbf{r}(t) = \int_0^{t_p+t_d} e^{\mathbf{A}_c{}^T\sigma}(\mathbf{PD} + \mathbf{Q}_{12})\begin{bmatrix} H(t_p-\sigma)w_1(t+\sigma)\\ w_1(t+\sigma-t_d)\\ H(t_p-\sigma)w_3(t+\sigma)\\ w_3(t+\sigma-t_d)\end{bmatrix}d\sigma,\\ &\mathbf{A}_c = \mathbf{A} - \mathbf{BR}^{-1}(\mathbf{N}^T + \mathbf{B}^T\mathbf{P}),\\ &H(t_p-\sigma) = \begin{cases}1 & for\ t_p - \sigma \ge 0\\ 0 & for\ t_p - \sigma < 0\end{cases}\end{aligned} \quad (5)$$

where t_p is the preview time of front tire, t_d is the preview time of rear tire. In equation (3), control input $\mathbf{u}$ directly represents the damping force of F_{MR}. Therefore, the following semi-active actuating condition is imposed.

$$u_i(t) = \begin{cases}u_i & for\ \ u_i(\dot{z}_{si} - \dot{z}_{ui}) > 0\\ 0 & for\ \ u_i(\dot{z}_{si} - \dot{z}_{ui}) \le 0\end{cases},\quad i = 1, 2, 3, 4 \quad (6)$$

In order to investigate the effect of preview control algorithm, a computer simulation is undertaken under the random road condition using MATLAB. Figure 4 presents vertical displacement of C.G. (center of gravity) point under random road condition. The RMS (root mean square) value of

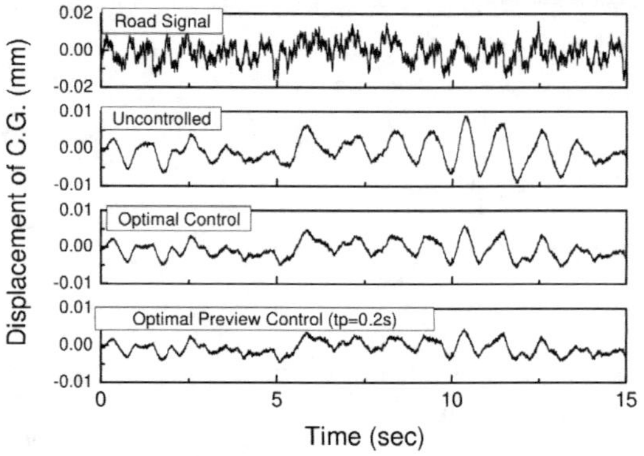

Figure 4. Control performances under random road condition

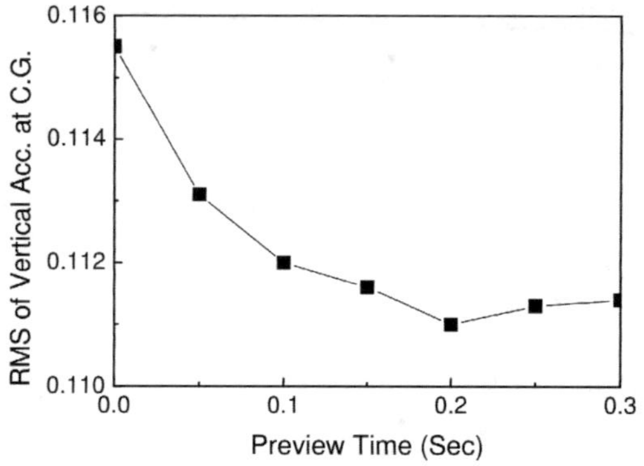

Figure 5. RMS vs. preview time

uncontrolled case is 0.440, optimal control case is 0.314 and optimal preview case is 0.255. From this result, it can be obviously found that unwanted vibrations well suppressed by adopting the optimal preview control algorithm to the MR suspension system. Figure 5 shows the RMS of vertical acceleration at C.G. point vs. preview time. It is observed that the RMS value is sharply decreased over the first 0.1 sec and a little increased when preview time is greater than 0.2 sec.

4. Concluding remarks

In this work, a preview control for vibration control of MR suspension system was proposed and its effectiveness was verified under random road condition. After evaluating damping force characteristics of a commercial MR shock absorber, the full vehicle model with MR shock absorber was constructed. The optimal preview control algorithm was formulated based on MR suspension system. It has been demonstrated that the proposed optimal preview control algorithm with MR suspension system can provide much better vibration control performance compared with the conventional optimal control algorithm. It is finally remarked that the investigation of control performance considering sensor and actuator dynamics will be carried out in near future.

Acknowledgment

This work was supported by Acceleration Research (Center for ER Fluid Technology and Application) of MOST/KOSEF. This financial support is gratefully acknowledged. The authors also thank to GM Korea for providing the MR damper.

References

[1] Carlson J D, Cantanzarite D M and St. Clair K A 1995 *Proc. 5th Int. Conf. on ER Fluid, MR suspensions and Associated Technology* **20** 20
[2] Spencer Jr. B F, Dyke S J, Sain M K and Carlson J D 1997 *J. Eng. Mech.* **123** 230
[3] Choi S B and Sung K G 2008 *Int. J. Vehicle Des.* **46** 94
[4] Hac A 1992 *Vehicle Syst. Dyn.* **21** 167
[5] Thompson A G and Davis B R 2005 *J. Sound Vib.* **285** 571
[6] Marzbanrad J, Ahmadi G, Zohoor H and Hojjat 2004 Y *J. Sound Vib.* **275** 973
[7] Gopala Rao L V V and Narayanan S 2008 *J. Sound Vib.* **310** 352
[8] Marzbanrad J, Ahmadi G, Hojjat Y and Zohoor H 2002 J. Vib. Control **8** 967

Discrete-time Sliding Mode Control for MR Vehicle Suspension System

J W Sohn[1], S B Choi[1] and N M Wereley[2]

[1] Smart Structures and Systems Laboratory, Department of Mechanical Engineering, Inha University, Incheon 402-751, Korea
[2] Smart Structures Laboratory, Department of Aerospace Engineering, University of Maryland, College Park, MD 20742, USA

E-mail: seungbok@inha.ac.kr

Abstract. This paper presents control performance of a full-vehicle suspension system featuring magnetorheological (MR) dampers via a discrete-time sliding mode control algorithm (DSMC). A cylindrical MR damper is designed by incorporating Bingham model of the MR fluid and the field-dependent damping characteristics of the MR damper are evaluated. A full-vehicle suspension model installed with independent four MR dampers is constructed and the governing equations which include vertical, pitch and roll motion are derived. A discrete-time control model is established with considering system uncertainties and a discrete-time sliding mode controller which has inherent robustness to model uncertainty and external disturbance is formulated. Vibration control performances under bump excitation are evaluated and presented.

1. Introduction

In general, vehicle suspension system provides improved passenger comfort as well as improved handling with the effective isolation from road disturbance. Recently, research works on vibration control of the semi-active suspension system using elctrorheological (ER) and magnetorheological (MR) fluids have been significantly increased. The semi-active suspension system can offer a desirable performance generally enhanced in the active mode without requiring large power consumption and expensive hardware compared to passive and active suspension. Petek [1] proposed a electrically controlled semi-active shock absorber using ER fluid and demonstrated the control performance. Carlson et al. [2] proposed a commercially available MR damper which is applicable to on-and-off-highway vehicle suspension system. Most of analytical and experimental researches for the ER and MR suspension system have limited their scope to a quarter car model. Moreover, research work focusing on the reliability of the suspension control system, which is easily subjected to parameter uncertainties and external disturbance in practice, is considerably rare. Choi et al. [3] evaluated control performance of ER full-vehicle suspension system by incorporating continuous-time sliding mode controller (CSMC) into hardware-in-the-loop simulation. In general, the CSMC has been designed based on the continuous-time control model. Thus its implementation by a digital computer requires certain sampling process. However, practical implementation of the CSMC is sometimes very difficult, because the sampling may bring not only degradation of control performance but also instability of the control system with an unnecessary large gain. This leads to the discrete-time sliding

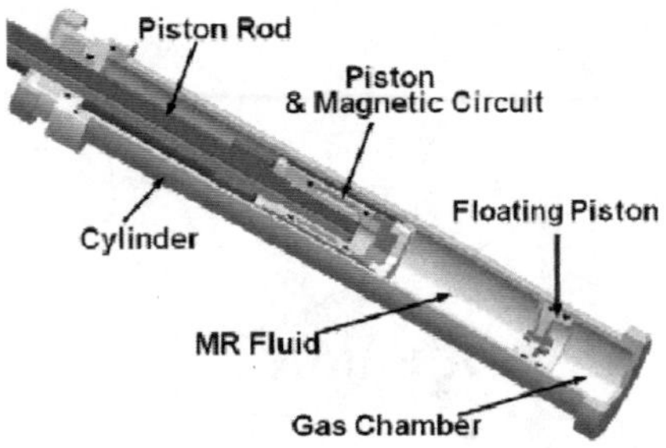

Figure 1. Configuration of the MR damper.

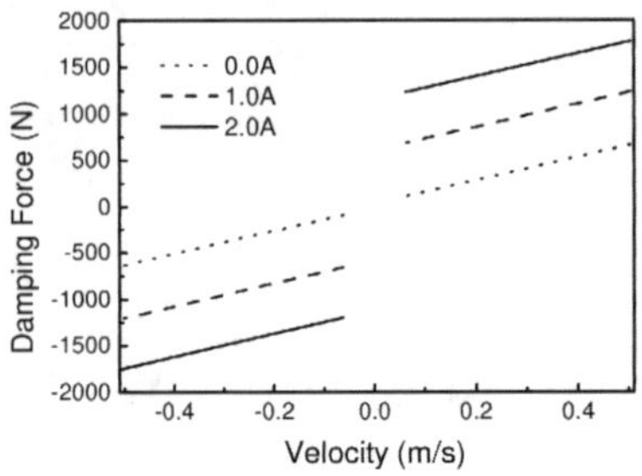

Figure 2. Field-dependent damping force.

mode control (DSMC) system.

Consequently, the main contribution of this study is to evaluate control performance of the discrete-time sliding mode controller (DSMC) for full-vehicle MR suspension system. A cylindrical MR damper is designed based on Bingham characteristics of the MR fluid and the field-dependent damping force is evaluated. A mechanical model of the full-vehicle MR suspension is constructed and equations of motion of the vehicle system are derived by considering vertical, pitch and roll motion. After establishing the discrete-time control model with model uncertainty in the state-space form, a DSMC is designed and implemented to the system. Vibration control performance is evaluated under bump excitation.

2. Modelling of full-vehicle suspension system

The schematic configuration of the proposed MR damper is shown in figure 1. The MR damper consists of cylinder, piston, gas chamber and floating piston which can compensate the volume of rod caused by the movement of piston. It is assumed that the MR fluid is incompressible and that pressure in one chamber is uniformly distributed. Thus, the damping force of the MR damper can be expressed as follows [4];

$$F_{MR} = k_e x_p + c_e \dot{x}_p + (A_p - A_r)(2L/h)\alpha H^{\beta}\,\mathrm{sgn}(\dot{x}_p),\ H = NI/2g \tag{1}$$

where x_p and $\dot{x}_p$ are the piston displacement and velocity, respectively. $\mathrm{sgn}(\cdot)$ is signum function, k_e is the effective stiffness due to gas pressure, c_e is the effective damping due to the fluid viscosity. A_p and A_r represent the piston head and piton rod areas, respectively. L is the length of magnetic pole, h is gap between the magnetic poles, H is applied magnetic field, N is number of coil turns, g is the gap between magnetic poles and I is input current. Here, α and β are intrinsic value of the MR fluid to be experimentally determined. In this study, for the MR fluid, the commercial product (MRF132-LD) from Lord Corporation is used and the experimental constants α and β are identified as 0.083 and 1.25, respectively. Figure 2 presents damping force with respect to the piston velocity at various magnetic fields. It is clearly observed that the damping force is increased as the magnetic field increases. As a specific case, the damping force of 203.1 N at piston velocity of 0.0628 m/s is increased up to 1165.5 N by applying the input current 2.0 A.

Figure 3 presents a mechanical model of the full-vehicle MR suspension system equipped with four MR dampers. The vehicle body itself is assumed to be rigid and has degrees of freedom in the vertical pitch and roll direction. The equations of motion for the full-vehicle model are derived as follows [3];

$$\begin{aligned}
&M\ddot{z}_g = -f_{s1} - f_{s2} - f_{s3} - f_{s4} + F_{MR1} + F_{MR2} + F_{MR3} + F_{MR4} \\
&J_\theta \ddot{\theta} = af_{s1} + af_{s2} - bf_{s3} - bf_{s4} - aF_{MR1} - aF_{MR2} + bF_{MR3} + bF_{MR4} \\
&J_\phi \ddot{\phi} = -cf_{s1} + df_{s2} - cf_{s3} + df_{s4} + cF_{MR1} - dF_{MR2} + cF_{MR3} - dF_{MR4} \\
&m_1\ddot{z}_{u1} = f_{s1} - f_{t1} - F_{MR1},\ m_2\ddot{z}_{u2} = f_{s2} - f_{t2} - F_{MR2}, \\
&m_3\ddot{z}_{u3} = f_{s3} - f_{t3} - F_{MR3},\ m_4\ddot{z}_{u4} = f_{s4} - f_{t4} - F_{MR4}
\end{aligned} \tag{2}$$

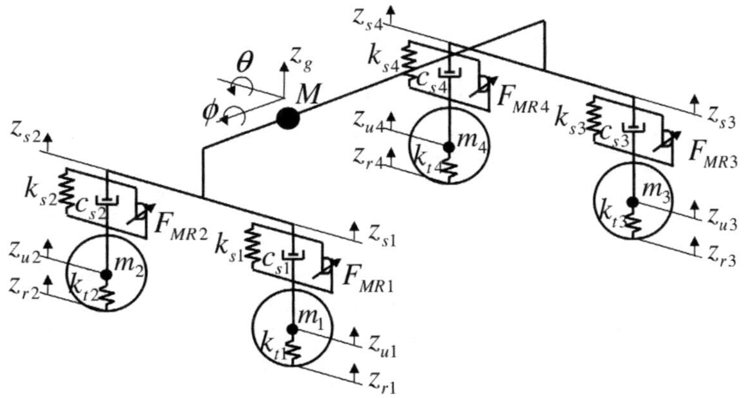

Figure 3. Mechanical model of the full-vehicle MR suspension system.

Here $f_{si} = k_{si}(z_{si} - z_{ui}) + c_{si}(\dot{z}_{si} - \dot{z}_{ui})$ and $f_{ti} = k_{ti}(z_{ui} - z_{ri})$ for $i = 1, 2, 3, 4$. M is the sprung mass and m_i, $i = 1,2,3,4$ is the unspung mass, J_θ is the pitch mass moment of inertia and J_ϕ is the roll mass moment of inertia. k_{si} $(i = 1, 2, 3, 4)$ is the total stiffness coefficient of the suspension spring and k_e. c_{si} $(i = 1, 2, 3, 4)$ is the damping coefficient of the suspension and it is equal to c_e. k_{ti} ($i = 1, 2, 3, 4$) is the stiffness coefficient of tire. Here, z_{si}, z_{ui} and z_{ri} $(i = 1, 2, 3, 4)$ are the vertical displacement of sprung mass, unsprung mass and excitation, respectively. z_g, θ and ϕ are the vertical displacement, pitch angular displacement and roll angular displacement. In equation (2), *a, b, c* and *d* are the distance between the front damper and center of gravity (C. G.) of the sprung mass, the distance between the rear and C.G., the distance between the left and C.G. and the distance between the right and C.G., respectively. By considering system uncertainties such as number of riding person and payload, the state-space control model can be obtained as follows;

$$\dot{\mathbf{x}}(t) = \mathbf{A}(t)\mathbf{x}(t) + \mathbf{B}(t)\mathbf{u}(t) + \mathbf{D}\mathbf{w}(t) = (\mathbf{A}_0 + \Delta\mathbf{A}(t))\mathbf{x}(t) + (\mathbf{B}_0 + \Delta\mathbf{B}(t))\mathbf{u}(t) + \mathbf{D}\mathbf{w}(t) \quad (3)$$

where $\mathbf{x}$, $\mathbf{u}$ and $\mathbf{w}$ are state vector, control input vector and disturbance vector, respectively. $\mathbf{A}_0$, $\Delta\mathbf{A}$, $\mathbf{B}_0$, $\Delta\mathbf{B}$ and $\mathbf{D}$ are the nominal part of the system matrix, uncertain part of the system matrix, nominal part of the control input matrix, uncertain part of the control input matrix and the disturbance matrix, respectively. It is assumed that all elements of the system and input matrices are piecewise smooth functions. By using the zero-order-hold (ZOH) method, the discrete-time control model can be expressed as follows;

$$\dot{\mathbf{x}}(k+1) = (\Phi_0 + \Delta\Phi)\mathbf{x}(k) + (\Gamma_0 + \Delta\Gamma)\mathbf{u}(k) + \mathbf{D}\mathbf{w}(k) \quad (4)$$

where Φ_0 and Γ_0 are the nominal system and nominal input matrix. $\Delta\Phi$ and $\Delta\Gamma$ are corresponding uncertain parts.

3. Controller design and performance evaluation

For the system in equation (4), a sliding surface in the state space is defined as follows;

$$s(k) = \mathbf{C}\mathbf{x}(t) \quad (5)$$

where $\mathbf{C}$ is the surface gradient vector which should be determined so that the sliding surface itself is stable. For the discrete-time system, the controller should be chosen to satisfy the sliding condition and convergence condition as follows;

$$[s(k+1) - s(k)\,\mathrm{sgn}(s(k))] < 0, \ [s(k+1) + s(k)\,\mathrm{sgn}(s(k))] > 0 \quad (6)$$

For the attenuation of vibration chattering in a settled phase and also the enhancement of robustness to the system, equivalent separation method is adopted in this work [5]. At the outside of the sliding region, the equivalent controller and discontinuous controller deliver the state trajectory to the sliding region. At the inside of the sliding region, the β-equivalent controller only delivers the state to the

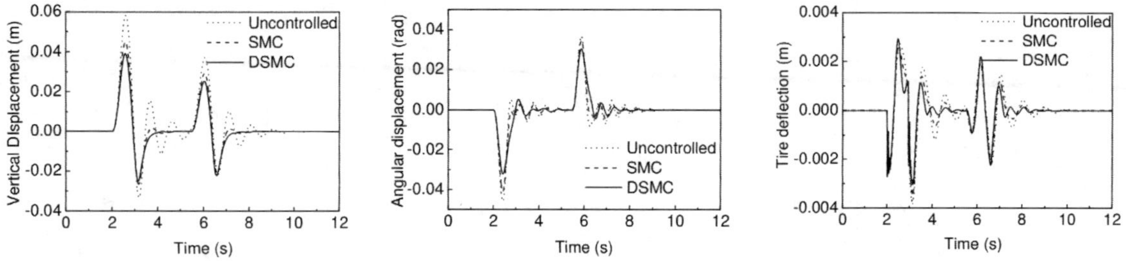

(a) vertical displacement (b) pitch angular displacement (c) tire deflection

Figure 4. Controlled response of the MR suspension system at C.G. for bump excitation.

origin point through the sliding surface. Equivalent controller, β -controller and discontinuous controller are determined as follows;

$$
\begin{aligned}
u_{eq} &= -(\mathbf{C}\Gamma_0)^{-1}(\mathbf{C}\Phi_0 - \mathbf{C})\mathbf{x}(t) \\
u_{eq,\beta} &= -(\mathbf{C}\Gamma_0)^{-1}(\mathbf{C}\Phi_0 - \beta\mathbf{C})\mathbf{x}(t), \;\; 0 \le \beta < 1 \\
u_d &= -h(k)\,\mathrm{sgn}(\mathbf{C}\Gamma_0 s(k))\sum_{i=1}^{4}\left|x_i(k)\right|
\end{aligned}
\tag{7}
$$

where $h(k) = 0$ for the inside of the sliding region and $h_s(k) < h(k) < h_c(k)$ for the outside of the sliding region.

Control performance of the proposed full-vehicle MR suspension system with DSMC is evaluated. Time response of the MR suspension system for the bump excitation with constant velocity of 3.08km/h, which can make first resonant frequency of vehicle suspension system, is presented in figure 4. It is clearly observed that the vertical displacement of sprung mass, pitch angular displacement and tire deflection are reduced by applying the control input determined by the DSMC. This directly implies that ride comport as well as steering stability of a vehicle can be improved by employing the proposed DSMC associated with the semi-active MR suspension system.

4. Conclusion

A DSMC was formulated for vibration control of a full-vehicle MR suspension system. A cylindrical MR damper was designed based on Bingham characteristics of MR fluid and the field-dependent damping force was investigated. After constructing of mechanical model of the full-vehicle MR suspension system, equations of motion of the vehicle system were derived. By considering the system uncertainties, a discrete-time control model of the vehicle system was constructed in the state space. A DSMC was then designed and implemented to the control system. Vibration control performance of the DSMC was evaluated under bump excitation. It has been clearly demonstrated that vibration of the vehicle can be effectively suppressed by applying the proposed control methodology. It is finally remarked that the road test of the proposed control system for MR suspension system will be carried out in the near future.

Acknowledgments

This work was supported by Acceleration Research (Center for ER Fluid Technology and Application) of MOST/KOSEF. This financial support is gratefully acknowledged.

References

[1] Petek N K 1998 *SAE Progress in Technology* **74** 245 (*Technical Paper Series* 920275)
[2] Carlson J D, Cantanzarite D M and St. Clair K A 1995 *Proc. 5th Int. Conf. on ER Fluid, MR Suspensions and Associated Technology* 20
[3] Choi S B, Choi Y T and Park D W 2000 *J. Dyn. Syst. Meas. Control-Tran. ASME* **122** 114
[4] Choi S B, Lee H S and Park Y P 2002 *Veh. Syst. Dyn.* **38** 341
[5] Choi S B, Cheong C C and M S Kim 1997 *J. Intel. Mater. Syst. Struct.* **8** 138

Iron oxidation and its impact on MR behavior

S R Sunkara[1], T W Root[1], J C Ulicny[2] and D J Klingenberg[1]

[1] Chemical and Biological Engineering Dept. and Rheology Research Center, University of Wisconsin, Madison, WI 53706, USA

[2] General Motors R&D and Planning, Warren, MI 48090-9055, USA

E-mail: klingen@engr.wisc.edu

Abstract. The oxidation of particles in MR fluids and its impact on rheology are investigated. The oxidation of iron spheres in an aliphatic oil follows a linear growth law, suggesting that the oxide forms a nonadherent layer. The magnetic field-induced yield stress decreases with increasing extent of oxidation. The rheological behavior is consistent with that predicted using a core-shell model.

1. Introduction

The rheological properties of magnetorheological (MR) fluids change over time. Carlson [1] describes the phenomenon of "in-use-thickening," or IUT, which refers to the increase in off-state viscosity of MR fluids over time during use. IUT can be quite significant, ultimately resulting in device failure, as the MR fluid becomes an unmanageable paste. The cause of this viscosity increase is believed to be spalling of the friable surface layer from the iron particles. This surface layer, composed of iron oxides, is brittle. When subjected to high inter-particle stresses, this surface layer fractures and breaks into small pieces that separate from the primary particle. These small nanometer-sized secondary particles have a large surface area to weight ratio and are less dense than pure iron. This increase in off-state viscosity may therefore be caused by the increase in solid volume, and perhaps colloidal forces acting between small particles.

Ulicny et al. [2] examined the behavior of an MR fan clutch subjected to durability tests performed over extended periods. The field-induced torque capacity, characterized by the maximum output fan speed as the electromagnet current is cycled up and down, decreased slowly with time. Prior to the durability test, the particles were composed of nearly pure iron, containing only 0.4 wt% oxygen. Following the durability tests, the particles were found to contain significantly more oxygen; the particles examined following a 540 hour durability test contained 7 wt% oxygen. The authors also examined the particles using scanning electron microscopy. Prior to the durability test, the particles appeared smooth and spherical, as depicted in Fig. 1. Following the durability test, the particles developed a core-shell structure, with the outer layer containing iron oxide, as illustrated in Fig 2. The authors speculated that the time-dependent decrease in fan speed observed during the durability test may be caused by the oxidation of the iron particles, because the particles' magnetizability decreases with increasing iron oxide content.

In this study, we examine the oxidation of iron particles within MR fluids and its impact on the field-induced rheological properties. We employ experiments in which the particles are oxidized in a controlled environment. The effects of oxidation on rheological properties are examined experimentally and compared with predictions obtained from a model based on dipolar magnetized particles.

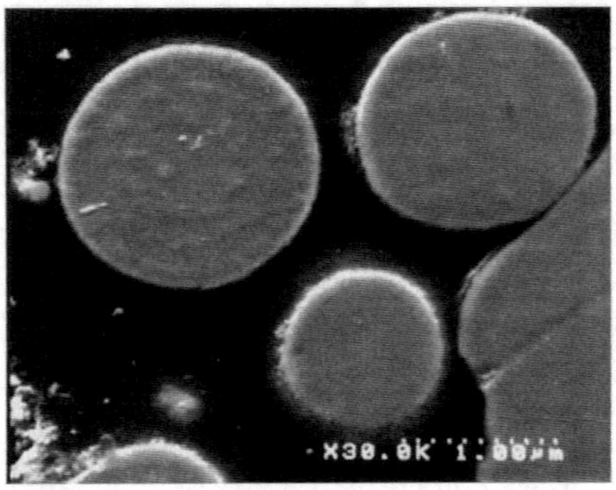

Figure 1. SEM of unused iron particles (from Ref. [2]).

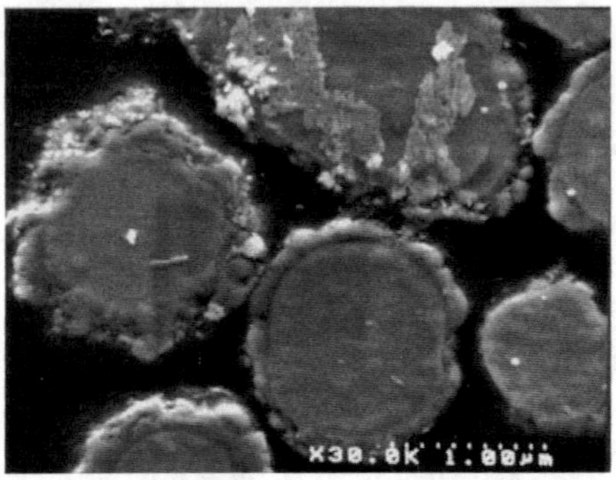

Figure 2. SEM of iron particles after 540 h durability test (from Ref. [2]).

2. Experimental Methods

Iron particles (2 μm diameter; BASF HS) in an aliphatic oil (Spectrasyn 2C; ExxonMobil) were oxidized using a three-neck flask reactor. All oxidation reactions were carried out with 10 wt% suspensions. Air was fed into the flask at a rate of 0.3 liters per minute. A heating mantle was used to maintain the temperature of the reaction mixture at the desired value. A condenser was employed to prevent loss of oil during the reaction. The reaction mixtures were stirred continuously (at 480 rpm) during the reactions.

The extent of oxidation was determined by measuring the decrease in the weight of samples when reduced in hydrogen. The reduction reaction was carried out in a modified Gow-Mac gas chromatograph (GC). Oxidized iron powder was packed in a 0.25 in. I.D. brass tube, and connected to the hydrogen supply inside the GC oven. Glass wool was packed into the tube ends to keep the powder in place. Hydrogen was fed through the bed at a rate of 0.22 liters per minute. Reaction gas products were passed through a Drierite bed. The extent of oxidation determined from the weight loss of the iron particles agreed with that determined from the water weight gain of the Drierite. All reduction reactions were carried out at 400°C, and continued until the sample mass no longer changed with time.

The effect of oxidation on the field-induced rheological properties is determined using a Bohlin VOR rheometer modified for the application of large magnetic fields [3].

3. Results and Discussion

The extent of oxidation of 10 wt% iron suspensions is plotted as a function of oxidation time in Fig. 3 for different oxidation reaction temperatures. The oxygen content grows linearly with time, which suggests that the oxide layer is nonadherent [4]. The activation energy estimated from the results at different temperatures is 92 $\pm$ 36 kJ/mol, which is similar to that obtained for similar systems. Grosvenor et al. [5, 6] obtained an activation energy of 32 $\pm$ 6 kJ/mol for oxidation of pure polycrystalline iron in dry O_2 gas over the temperature range 27–150°C. For oxidation of iron in air, Kubaschewski and Hopkins report an activation energy of 76 kJ/mol over the temperature range 600–1600°C [7], while Paidassi reports an activation energy of 40.5 kJ/mole over the temperature range 400–900°C [8].

Typical results for the effect of particle oxidation on rheological properties are illustrated in Fig. 4, where the yield stress scaled by the yield stress for pure iron particles is plotted as a function of the weight fraction of oxygen within the particles, w_{ox}, for 10 vol% suspensions at an applied flux density of 0.2 T. Also shown is the yield stress measured for a 10 vol% suspension of magnetite particles (10 μm diameter; Cerac Inc.) in Spectrasyn 2C. As expected, the yield stress decreases with increasing oxidation.

The effects of oxidation are also examined via a particle-level model, described in more detail elsewhere [9]. Motivated by the core-shell structures observed by Ulicny et al. [2], the particles are

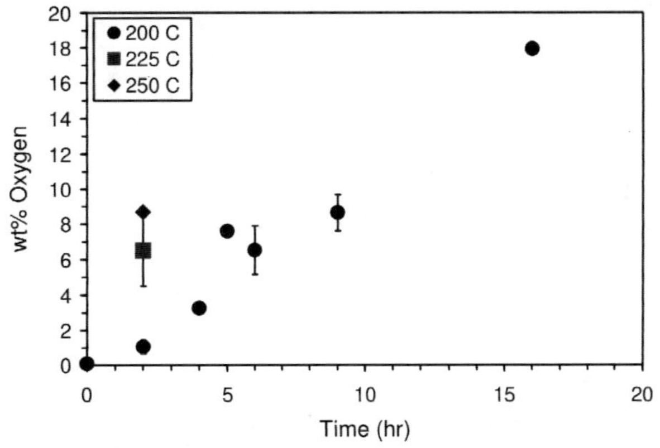

Figure 3. Oxygen content of 10 vol% iron suspensions as a function of oxidation reaction time.

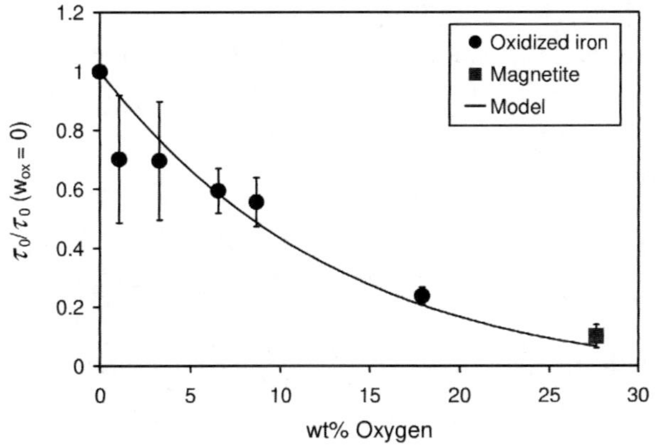

Figure 4. Yield stress (scaled by the value obtained for pure iron particles) as a function of the oxygen weight fraction for 10 vol% suspensions at a flux density of 0.2 T. The solid curve represents the model prediction.

modeled as iron cores (radius R_c) surrounded by an iron oxide (magnetite) spherical shell (outer radius R_s), as illustrated in Fig. 5. The point-dipole approximation to the magnetic field-induced force between two magnetizable particles is [10]

$$\boldsymbol{F}_{ij}^{\text{mag.}} = F_0 \left(\frac{\sigma}{R_{ij}} \right)^4 \left[\left(3\cos^2\theta_{ij} - 1 \right) \boldsymbol{e}_r + \sin 2\theta_{ij} \boldsymbol{e}_\theta \right], \tag{1}$$

where σ is the particle diameter. For core-shell particles such as that depicted in Fig. 5, the force magnitude F_0 in the limit of saturated magnetization is

$$F_0 = \frac{\pi}{12} \mu_0 R_s^2 \left[M_{ss} + \left(\frac{R_c}{R_s} \right)^3 (M_{sc} - M_{ss}) \right]^2, \tag{2}$$

where M_{sc} is the saturation magnetization of the core material, and M_{ss} is the saturation magnetization of the shell material. The rheological properties of suspensions of such particles are determined by exploiting results of dipolar particle-level models. Here, the effect of oxidation appears as a reduction in the core radius and an increase in the shell radius, which results in an overall reduction particle magnetization and thus the field-induced interparticle forces. The shear stress can be expressed

$$\tau = -\frac{1}{V}\sum_{i=1}^{N} z_i F_{x,i} \quad (3)$$

$$= \frac{F_0(f)}{R_s^2(f)}\tau^*(\phi) \quad (4)$$

where f is the degree of oxidation, and $\tau^*(\phi)$ is the dimensionless yield stress determined, for example, from simulations or a chain model. This latter quantity depends only on the particle volume fraction. We note that the particle volume fraction will increase with time in an oxidizing suspension.

Results of this model presented along with the experimental data in Fig. 4. The model agrees well with the experimental data, with the stress decreasing with increasing oxide content (discrepancies are on the order of the experimental uncertainty). This behavior is consistent with results obtained for other systems with iron cores surrounded by nonmagnetizable shells, where the field-induced stress decreases with increasing shell thickness [11, 12, 13].

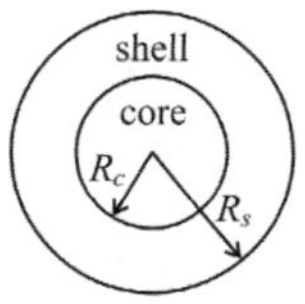

Figure 5. Schematic diagram of the core-shell structure employed in the model.

4. Conclusion

The oxidation of iron spheres in an aliphatic oil follows a linear growth law, suggesting that the oxide forms a nonadherent layer. The magnetic field-induced yield stress decreases with increasing extent of oxidation. The rheological behavior is consistent with that predicted using a core-shell model.

Acknowledgments

This work was supported in part by General Motors. The authors thank Prof. J. A. Dumesic, C. Barrett and E. Kunkes for assistance with the iron oxide reduction experiments.

References

[1] Carlson J D 2001 *Proc. 8th Int. Conf. on Electrorheological Fluids and Magnetorheological Suspensions* (Singapore: World Scientific) p 63
[2] Ulicny J C, Balogh M P, Potter N M and Waldo R A 2007 *Mater. Sci. Eng. A* **443** 16–24
[3] Rankin P J, Horvath A T and Klingenberg D J 1999 *Rheol. Acta* **38** 471–7
[4] Jastrzebski Z D 1987 *The Nature and Properties of Engineering Materials* (New York: Wiley) pp 496-7
[5] Grosvenor A P, Kobe B A and McIntyre N S 2005 *Surf. Sci.* **565** 151-62
[6] Grosvenor A P, Kobe B A and McIntyre N S 2005 *Surf. Sci.* **574** 317-21
[7] Kubaschewski O and Hopkins B E 1962 *Oxidation of Metals and Alloys* (London: Butterworth) p 111
[8] Paidassi J 1958 *Acta Metallurg.* **6** 184-94
[9] Sunkara S R 2008 M.S. Thesis, Univ. of Wisconsin
[10] Kittipoomwong D, Klingenberg D J and Ulicny J C 2005 *J. Rheol.* **49** 1521-38
[11] Cho M S, Lim S T, Jang I B, Choi H J and Jhon M S 2004 *IEEE Trans. Mag.* **40** 3036-8
[12] Choi H J, Park B J, Cho M S and You J L 2007 *J. Magn. Magn. Mat.* **310** 2835-7
[13] Fang F F and Choi H J 2008 *J. Appl. Phys.* **103** 07A301

Journal of Physics: Conference Series **149** (2009) 012082 doi:10.1088/1742-6596/149/1/012082

Multiscale structure-phenomenological study of magnetorheological suspensions in blood

E Yu Taran, V A Gryaznova and O O Melnyk

Faculty of Mechanics and Mathematics, Kyiv Taras Shevchenko National University, 64 Volodymyrska St, 01033 Kyiv, Ukraine.

E-mail: taran@univ.kiev.ua

Abstract. The rheological equations for dilute suspensions of rigid axisymmetric elongated particles possessing a permanent magnetic moment with blood as a carrier fluid are derived within the frame of a structure-phenomenological approach. The Stokes fluid with couple stresses and the uniaxial dumbbell are used, respectively, as the rheological and hydrodynamic models of blood and suspended particles. The influence of an external magnetic field on a rheological behaviour of the suspensions under study is examined.

1. Introduction

Suspensions in blood of magnetically sensitive particles can appear upon an addition to blood of particles of drugs based on magnetic carriers [1, 2]. In particular, such suspensions arise on addition of ferro- or ferrimagnetic micro- and nanoparticles coated with either polysaccharides or proteins designed for diagnostic or hyperthermic treatment of cancer.

While solving medical problems through the use of suspensions in blood, the possible consequences of biomechanical intervention into the human body should be remembered. In order to carry out investigations in this area, a magnetobiorheological model of dilute suspension in blood of axially symmetric elongated microparticles possessing permanent magnetic moment is constructed in our paper owing to the multiscale combination of microfluidic and structure-continual studies of such suspensions within the frames of the structure-phenomenological approach [3].

2. The rheological model of dilute suspension of elongated magnetic particles in blood

For using the structure-phenomenological method [3] in the construction of constitutive equations defining the stressed state in gradient flows of a dilute suspension in blood, we assume that such a suspension satisfy the assumptions of Einsteinian microfluidic study [4]. In particular, the double inequality $l << d << \bar{l}$ is fulfilled; here d is the characteristic size of axially symmetric suspended particles; l is the characteristic size of red blood cells; $\bar{l}$ is the characteristic size of macroflow region of the suspension.

The inequality $l << d$ allows us to consider the interaction of blood with suspended particles as a hydrodynamic interaction. At the first scale level of modeling, we model blood by classical continuum and use, following [5], the Stokes fluid with couple stresses [6]

$$\tau_{(ij)} = -p\delta_{ij} + 2\mu d_{ij}, \tag{2.1}$$

Journal of Physics: Conference Series **149** (2009) 012082 doi:10.1088/1742-6596/149/1/012082

$$\mu_{ij} = 4\eta K_{ij} + 4\eta' K_{ji} \tag{2.2}$$

as a rheological model of blood. The theory of Stokes [6] is the simplest generalization of the Newtonian theory of viscous fluids that takes into account the presence of couple stresses in gradient flows of liquid media. The viscous and couple stresses τ_{ij} and M_{ij} in rheological model of Stokes [6] are completely defined by the field of velocity v_i. In rheological equations (2.1) and (2.2), $\tau_{(ij)}$ is the symmetric part of the stress tensor τ_{ij}; μ_{ij} is the deviatoric part of the couple stress tensor M_{ij}; p is the pressure; μ, η, η' are rheological constants; d_{ij} is the strain rate tensor, $d_{ij} = (1/2)(v_{i,j} + v_{j,i})$; K_{ij} is the gradient tensor of angular microrotational velocities of the fluid particles that is determined as the gradient of the vorticity vector $\omega_i = (1/2)\varepsilon_{irs} v_{s,r}$, $K_{ij} = \omega_{j,i}$; δ_{ij}, ε_{ijk} are the Kronecker and Levi-Civita symbols.

As a hydrodynamic model of suspended particles, a uniaxial dumbbell with axis L is utilized. According to [7], the friction coefficient ξ of dumbbell beads in the Stokes fluid (2.1), (2.2) does not depend on the flow around the beads:

$$\xi = \xi_N (1 + B), \quad B = \frac{2 + \nu}{r_0^2 + (2 + \nu) r_0}. \tag{2.3}$$

In equation (2.3), ξ_N is the friction coefficient of a dumbbell bead of radius r in slow translational motion in the Newtonian fluid with dynamic viscosity μ, $\xi_N = 6\pi\mu r$; $\nu = \eta'/\eta$; $r_0 = r/l_0$, here $l_0 = \eta/\mu$.

In order to model blood by the Stokes fluid (2.1), (2.2), the material characteristic length l_0 is evaluated as function of the haematocrit value of blood C_b on the base of results obtained in [5] and [8] (Table 1).

$l_0 \cdot 10^6$, m	6.666	8.000	14.815
C_b, %	6	13	40

Table 1. The dependence of parameter l_0 on the haematocrit value of blood C_b.

It is assumed in the paper that the suspended particles modeled by uniaxial dumbbells are magnetically sensitive, namely, they possess a permanent magnetic moment $p_i = P n_i$, where P is the value of permanent magnetic moment; n_i is the unit vector characterizing the orientation of an axially symmetric suspended particle as well as the orientation of its dumbbell model. It is also assumed that the suspension is diluted to such an extent that the interaction between the magnetic fields of the suspended particles, as well as the hydrodynamic interaction between them is not taken into account.

Such a modeling allows us to perform the microfluidic study of the considered suspension in blood at a scale of suspended microparticles using the procedures of the Einstein's method [4]. So, we obtain the constitutive equation for vectors n_i and N_i

$$N_i = d_{ik} n_k - d_{km} n_k n_m n_i + \frac{P}{W}(H_i - n_k H_k n_i), \tag{2.4}$$

where N_i is the vector characterising the angular velocity of suspended particles with respect to the carrier fluid, $N_i = \dot{n}_i - \omega_{ij} n_j$; here, the dot over n_i denotes the local time derivation; ω_{ik} is the velocity vortex tensor, $\omega_{ik} = (1/2)(v_{i,k} - v_{k,i})$; W is the rotational friction coefficient of a uniaxial dumbbell in blood as a carrier fluid of the suspension, $W = (1/2)\xi L^2$; H_i is the external magnetic field strength.

In the frame of this microfluidic study we also obtain the rate of mechanical energy dissipation per unit volume of the suspension

$$\Phi = \Phi_0 + n_0 \frac{\xi L^2}{2} \left(\langle N_i N_i \rangle - 2 d_{ij} \langle N_i n_j \rangle + d_{ij} d_{ik} \langle n_j n_k \rangle \right), \tag{2.5}$$

where Φ_0 is the rate of mechanical energy dissipation per unit volume of the carrier fluid of the suspension in the absence of suspended particles; n_0 is the number of suspended particles per unit volume of the suspension; the angular brackets $\langle\ \rangle$ denote averaging with the use of the distribution function F of angular positions of suspended particles, which satisfies the equation

$$\frac{\partial F}{\partial t} + \frac{\partial}{\partial n_i}(F \dot{n}_i) = 0. \tag{2.6}$$

According to the structure-phenomenological approach [3] used in the paper, the inequality $d << \bar{l}$ allows us to model the considered suspension in blood at the scale of the region of suspension's macroflow by microcontinuum with two internal microparameters n_i and N_i defining the orientation and relative angular velocity of suspended microparticles. In this case, the rheological equation defining the stress T_{ij} in the modeled suspension is postulated phenomenologically. According to [3], the most general form of such an equation is

$$\begin{aligned} T_{ij} = t_{ij} + n_0 \big((a_0 + a_1 d_{km} \langle n_k n_m \rangle) \delta_{ij} + a_2 \langle n_i n_j \rangle + a_3 d_{km} \langle n_k n_m n_i n_j \rangle + a_4 d_{ij} + \\ + a_5 d_{ik} \langle n_k n_j \rangle + a_6 d_{jk} \langle n_k n_i \rangle + a_7 \langle n_i N_j \rangle + a_8 \langle n_j N_i \rangle \big), \end{aligned} \tag{2.7}$$

where t_{ij} is the stress tensor in the carrier fluid of the suspension in the absence of the suspended particles; n_0 is the quantity of suspended particles per unit volume of the suspension; a_i $(i = 0, ..., 8)$ are constant phenomenological coefficients.

The phenomenological coefficients a_i $(i = 0, ..., 8)$ in the equation (2.7) are found from comparing the rate of mechanical energy dissipation per unit volume of the suspension defined by the equation (2.5) that was determined in the microfluidic part of the theory, with the rate determined in the same way as in [3] within the framework of the structure-continual approach. As a consequence of this, we obtain the rheological equation of a dilute suspension with blood as the carrier fluid

$$T_{ij} = t_{ij} + \frac{1}{2} n_0 \xi L^2 \left(d_{ik} \langle n_k n_j \rangle - \langle n_j N_i \rangle \right). \tag{2.8}$$

It is demonstrated in [9] that such an equation as (2.4) has the stationary solution $(\dot{n}_i = 0,\ N_i = -\omega_{ik} n_k)$ for steady-state shear flows of a suspension in the presence of the steady-state magnetic field H_i. This means that the suspended dumbbell particles of the considered suspension in blood acquire the same stationary orientation defined by the constitutive equation

$$W \left(v_{i,k} n_k - d_{km} n_k n_m n_i \right) + P (H_i - n_k H_k n_i) = 0 \tag{2.9}$$

under above-mentioned conditions. The rheological equation for stress (2.8) in such an anisotropic suspension takes the form

$$T_{ij} = \tau_{ij} + n_0 W v_{i,k} n_k n_j. \tag{2.10}$$

W is a single rheological parameter in equations (2.9) and (2.10), which characterizes the interaction of suspended particles with blood modeled by the Stokes fluid (2.1), (2.2).

According to equations (2.3), taking into account the couple stresses μ_{ij} arising in blood modeled by the Stokes fluid (2.1), (2.2) leads to an increase in the rotational friction coefficient of the suspended dumbbell particles: $W = W_N(1+B)$, as compared with the value $W_N = (1/2)\xi_N L^2$ in a suspension with the Newtonian model of blood as the suspension carrier fluid.

3. Magnetobiorheological behavior of dilute suspensions in blood of magnetic particles.

The obtained constitutive equations (2.9), (2.10) are used to study the rheological behavior of the considered suspension in the steady simple shear flow $v_x = 0$, $v_y = Kx$, $v_z = 0$ $(K = const)$ in the presence of the cross magnetic field $H_x = H$, $H_y = H_z = 0$ $(H = const)$. It allows to investigate the influence of the shear rate K of the flow, the haematocrit value C_b of blood and the strength H of the external magnetic field on effective viscosity of the considered suspension $\mu_a \equiv (T_{xy} + T_{yx})/(2K)$ and on the non-zero difference of normal stresses $\sigma_1 \equiv T_{yy} - T_{xx}$ arising in it.

The calculations of the numerical values of the characteristic viscosity of the suspension $[\mu_a] \equiv (\mu_a - \mu)/(PH)$ and σ_1^* / K as fuctions of $\alpha = KW_N/(PH)$ (Figures 1 and 2) show that the increase of the shear rate K of the simple shear flow at the fixed strength H of the cross magnetic field leads to the pseudoplastic decrease of the effective viscosity μ_a of the suspension and to the increase of the difference σ_1 (the Weissenberg effect).

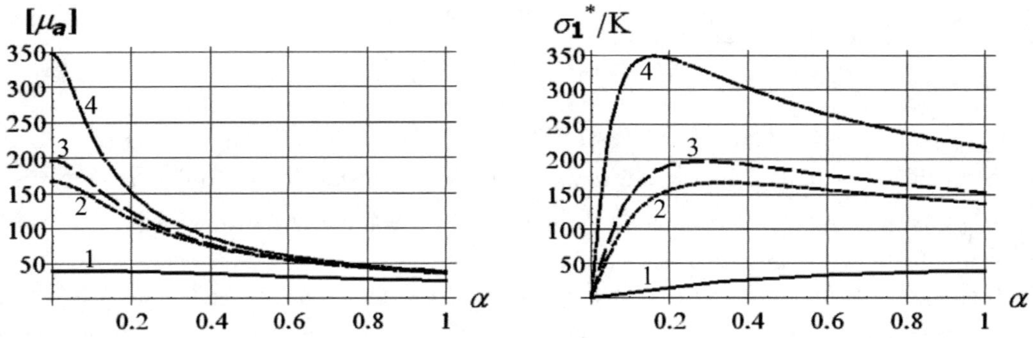

Figure 1: The dependences of $[\mu_a]$ and σ_1^* / K on α at $\bar{p}$ = 10, curves 1 correspond to the suspension with the Newtonian carrier fluid of the viscosity μ; curves 2–4 correspond to the suspension in blood with the haematocrit values C_b = 6%, 13%, 40%, respectively.

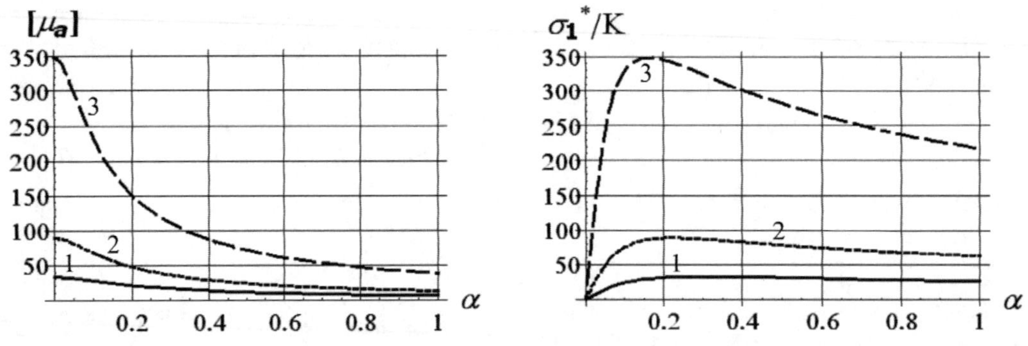

Figure 2: The dependences of $[\mu_a]$ and σ_1^* / K on α at C_b = 40%, curves 1–3 correspond to $\bar{p}$ = 3, 5, 10, respectively.

The following designations are used: $\sigma_1^* = \sigma_1/(\mu V)$; V is the volume concentration of suspended particles, $V = (4/3) n_0 \pi a b^2$, where a and b are semiaxes of ellipsoid of revolution that is equivalent to the dumbbell as a hydrodynamic model of suspended particles in the Newtonian carrier fluid with the viscosity μ; $\overline{p} = a/b$. The calculation was made at the length $L = 4 \cdot 10^{-5}$ m of suspended particles with axial ratios $\overline{p}$ = 3, 5, 10 of them.

The calculations made in the paper show as well that the considered suspension in blood exhibits the magnetorheological properties. We obtain that an increase of the strength H of the magnetic field at the fixed shear rate K of the simple shear flow leads to the increase of the effective viscosity μ_a of the considered suspension in blood and leads in addition to the decrease of the difference σ_1 of normal stresses. The detected variations of rheological characteristics of the suspension are the consequences of variations of orientation of suspended particles under the action of hydrodynamic forces and an external magnetic field.

The constructed rheological model allows us to investigate biorheological properties of the considered suspension. Curves 2–4 on Fig.1 show that the effective viscosity μ_a of the studied suspension in blood and the difference σ_1 of normal stresses in it are augmented with increasing the haematocrit value C_b of blood holding K and H fixed.

4. Conclusions.

The structure-phenomenological method of modeling used in the paper allowed to express explicitly the macrorheological (continual) characteristics of dilute suspension in blood of magnetically sensitive microparticles through the parameters describing microrheological, microstructural and physical properties of blood and microparticles suspended in it. Thanks to such an advantage of the structure-phenomenological method, the biorheological and magnetorheological studies of the considered suspension have been made possible. The results obtained in the paper may be used in medicine. In particular, the change of the strength H of an external magnetic field may be used as a control factor of the rheological properties of a suspension formed on addition of magnetically sensitive elongated particles to blood.

References

[1] Häfeli U, Schütt W, Teller J and Zborowski M, ed 1997 *Scientific and clinical applications of magnetic carriers* (Plenum Press, New York)

[2] Gillies G T, Ritter R C, Broaddus W C, Grady M S, Howard M A III and McNeil R G 1994 Magnetic manipulation instrumentation for medical physics research *Rev Sci Instrum* vol 65 pp 533–562

[3] Taran E Yu 1994 *Structure-phenomenological rheology of dilute suspensions, DSc (Physics and Mathematics) thesis* (Kharkov State University, Kharkov) (in Russian)

[4] Einstein A 1906 Eine neue Bestimmung der Moleküldimensionen *Ann Physik* vol 19 pp 289–306

[5] Valanis K S and Sun C T 1969 Poiseuille flow of a fluid with couple stresses with application to blood flow *Biorheology* vol 65 no 2 pp 85–97

[6] Stokes V K 1966 Couple stresses in fluids *Phys Fluids* vol 9 pp 1709–1715

[7] Stokes V K 1971 Effects of couple stresses in fluids on the creeping flow past a sphere *Phys Fluids* vol. 14 pp 1580–1582

[8] Bugliarello G, Kapur C and Hsiao G 1965 The profile velocity and other characteristics of blood flow in a non-uniform shear field *Proc. of the 4th Int. Congr. on Rheology* (New York: Interscience) pp 351–370

[9] Taran E Yu 1978 Influence of electric field on rheological behaviour of dilute suspension of dipole dumbbells in visco-elastic fluid of Oldroyd *Mekhanika Polimerov* no 3 pp 519–524

Journal of Physics: Conference Series **149** (2009) 012083 doi:10.1088/1742-6596/149/1/012083

Synthesis and magnetorheological effect of Fe3O4-TiO2 nanocomposite

J H Wei[1,2], C J Leng[1] , X Z Zhang[2], W H Li[2], Z Y Liu[1] and J Shi[1]

[1] Key Laboratory of Acoustic and Photonic Materials and Devices of Ministry of Education and Department of Physics, Wuhan University, Wuhan, 430072, P.R. China

[2] School of Mechanical, Materials & Mechatronic Engineering, University of Wollongong, NSW, 2522, Australia

E-mail: jhwei@whu.edu.cn, weihuali@uow.edu.au

Abstract. Aimed to obtain a material with improved rheological property, the Fe_3O_4-TiO_2 nanocomposites were prepared by a modified sol-gel processing. The structure and morphology of the Fe_3O_4-TiO_2 nanocomposites were examined by transmission electron microscopy (TEM), X-ray diffraction (XRD) analysis and Fourier transform infrared spectroscopy (FT-IR) analysis, etc. The magnetic property and the magnetorheological properties of the coated particles were also examined in detail. The experimental results demonstrated that such materials exhibited favorable magnetorheological (MR) effect, and the MR performance of them were dependent on the relative content of TiO_2 in the composite.

1. Introduction

The rheological properties of some fluids can be reversibly modified in the presence of an external field, such as electrorheological (ER) fluids and magnetorheological (MR) fluids. The ER fluids and MR fluids, known as smart liquids for their apparent viscosity capable of experiencing a rapid, reversible change upon application of an external electric field or magnetic field [1-3], have potentially important application in numerous electromechanical devices, such as valves, dampers, and clutches in the automotive and robotics industries, etc. Recently, Wen and other researches [4-6] reported a theoretical research on the electro-magneto-rheological (EMR) effect, the combination of an ER effect and a MR effect, which provides us with a new strategy to control the rheological properties of fluid materials. Although the results for the theoretical research were inspiring, the experimental research on EMR fluids was less reported until now.

In general, the desired particles used in MR or EMR suspensions should have both tiny or no coercive force and large saturated magnetization strength (M_s), such as carbonyl iron, ferrite (including Fe_3O_4) and their composites etc. [7-9] Due to large saturated magnetization strength, Fe_3O_4 nanoparticles have been widely used to form ferrofluids and MR fluids. However, unmodified Fe_3O_4 nanoparticles usually incline to aggregate into the clusters, have poor ability of anti-oxidation and poor thermal stability. To overcome this problem, the coating processing with insulating oxide on the magnetic nanoparticles was usually employed. In this paper, we report a novel composite, a simple chemical synthesis to coat Fe_3O_4 nanoparticles with TiO_2 layer. The TiO_2 dielectric coating was fulfilled by a modified sol-gel process and the thickness of coating layer was governed by solution concentrations and processing durations. The TiO_2 coating on magnetic nanoparticles was beneficial to

Journal of Physics: Conference Series **149** (2009) 012083 doi:10.1088/1742-6596/149/1/012083

minimize aggregation of Fe_3O_4 nanoparticles and therefore, improve the chemical stability and thermal stability, keeping Fe_3O_4 nanoparticles from oxidation. What is more, it also enhance the dielectric response of the nanoparticles. As a result, Fe_3O_4-TiO_2 composites can possess both dielectric and magnetic properties simultaneously with improved stability, which make it an ideal candidate for using in EMR fluid. The Fe_3O_4-TiO_2 nanocoposites were often used as magnetic photocatalyst, but their MR or EMR properties were less reported until now. In the paper, we report the MR effect of Fe_3O_4-TiO_2 nanocomposites based MR fluids. The EMR performance of the materials will be reported later.

2. Experimental

2.1 Preparation of Fe_3O_4-TiO_2 nanocomposites

The nanosized Fe_3O_4 particles were prepared by a chemical coprecipitation method [10]. The Fe_3O_4 nanoparticles were coated with TiO_2 by a modified sol-gel method. In a typical coating procedure, a suitable amount of magnetite particles dispersed in ethanol was ultrasonic for 1h, then, $Ti(OR)_4$, polyethylene glycol, diethylamine and water were slowly added into the mixture. The reaction was allowed at room temperature for 3h under continuous stirring. And then the reaction was continued at 85°C in a rotary evaporation meter to make the solvent fully evaporating. Finally, the resultant powder was heat treated at 100 °C for 8h to remove any trace of water. Here, diethanolamine was used as an additive to prevent the precipitation of titanium butoxide from the alcoholic in the presence of excess water. PEG was added to increase the content of surface hydroxyl in the composite.

2.2 Characterization and measurement

The morphologies of the samples were investigated by transmission electron microscopy (TEM, JEM JEOL-2010). The crystal phase of the prepared products were analyzed by X-ray powder diffraction (XRD, Model Japan Rigaku D/max-γA K_{II} , Cu-target, λ = 0.1541nm). The TEM photographs and XRD patterns reveal that the Fe_3O_4 nanoparticles with diameter about 10nm were completely covered by the amorphous TiO_2 layers. The densities of the particles were measured with a pycnometer by dispersing particles in silicone oil. The densities of the coated particles change from 2.36 to 2.83g/cm^3 with different TiO_2 content, which are far less than the density of Fe_3O_4 4.25g/cm^3. Low particle density is beneficial to the stability of MR fluids.

The chemical structure was determined by a Nicolet Dx-10 Fourier transform infrared spectroscopy (FT-IR) spectrophotometer. The hystersis cycles of magnetite nanoparticles and Fe_3O_4-TiO_2 composites were recorded at room temperature by using a vibrating-sample magnetometer (VSM, EG&G Princeton Applied Research Vibrating Sample Magnetometer, Model 155). An Anton Paar Physica MCR301 rheometer in parallel-plate configuration was used to measure rheological properties of the composite. All the experiments were conducted at the room temperature.

3. Results and Discussion

3.1 Structure and magnetic properties

Figure1 shows the FT-IR spectra of (a) the mixture of Fe_3O_4 and TiO_2 (b) and Fe_3O_4-TiO_2 nanocomposites respectively. The spectrum (a) shows that the broad band around 3413cm^{-1} is the asymmetric and symmetric stretching vibrations of O-H group, whereas the band around 1636cm^{-1} is the H-O-H bending of the coordinated water, the band at 1451 cm^{-1} and 1384 cm^{-1} is the bending vibrations of C-H group, the band at 1087 cm^{-1} and 950cm^{-1} is the charateristic vibrations of Ti-O-C group, and the band at 500-700 cm^{-1} is attributed to the Ti-O-Ti or Fe-O stretching vibrations. In comparing the two spectra in Figure1, it is noted that although the two spectra are similar as a whole, there are some observable differences as marked on spectrum(b), the band at 3413cm^{-1} in spectrum(a) shift to 3385 cm^{-1} in spectrum (b) , and getting into broader, whereas the band at 1636 cm^{-1} in spectrum (a) shift to 1627 cm^{-1} in spectrum (b) and a new peak appeared at around 890 cm^{-1}. So, we infer there must exist some interactions between the Fe-OH and Ti-O-Ti band, the interaction make them uneasy to separate.

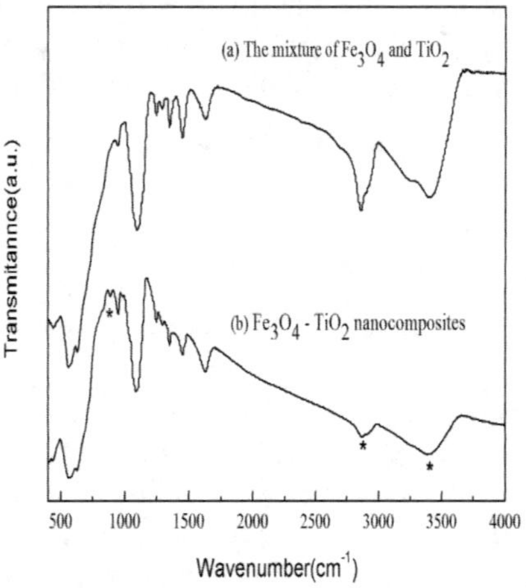

Figure 1. FT-IR spectra for (a) the mixture of Fe_3O_4 and TiO_2 (b) Fe_3O_4-TiO_2 nanocomposites

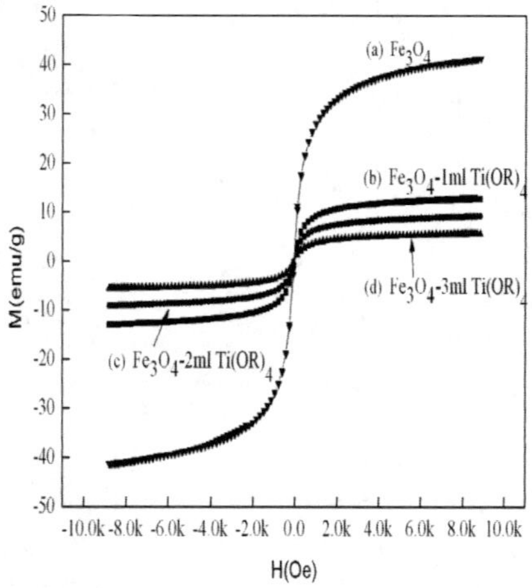

Figure 2. Hysteresis loops of Fe_3O_4 nanoparticles and Fe_3O_4-TiO_2 nanocomposites at room temperature

The hysteresis loops of Fe_3O_4 nanoparticles and Fe_3O_4-TiO_2 nanocomposites are shown in Figure 2. For Fe_3O_4 nanoparticles, saturated magnetization (M_s), remmant magnetization (M_r) and coercive force(H_c) are estimated to be M_s= 41.6emu/g, M_r=0, and H_c=0. No hysteresis loop is observed, indicating a super paramagnetic behavior. For composite structures with 1ml $Ti(OR)_4$ precursor, its saturated magnetization was 12.9emu/g. With the increasing of the TiO_2 content in the composite, the saturation magnetization of the particle decreases. Moreover, these saturation magnetization values are inevitably much lower than that of pure Fe_3O_4 particles, which is consistent with the mass proportion of the magnetite nanoparticles in the coated particles. In addition, Fe_3O_4-TiO_2 nanocomposites also show super paramagnetic character. These particles having no coercivity seem to give a distinct switching response of the MR effect if the field is switched on and off, respectively, which is beneficial to its applications in MR fluids.

3.2 MR effect

MR fluids of pure Fe_3O_4 and Fe_3O_4-TiO_2 nanocomposites at a concentration of 30 wt% in silicone oil were prepared. Rheological behavior of MR suspensions was investigated under steady shear flow and static magnetic field. Figure 3 shows the response of shear stress with shear rate for the Fe_3O_4-TiO_2 nanocomposites based suspension with and without magnetic field. The induced magnetic moment of the particle caused by an applied magnetic field is parallel to the field direction. The MR suspension exhibited significant rheological properties in the range of shear rate from **.1 1/s to .100 1/s**. The flow behavior of suspensions made of Fe_3O_4-TiO_2 nanocomposites only shows a slight departure from Newtonian fluid without the magnetic field. With the application of magnetic field, the MR fluids exhibit the Bingham plastic behavior and the shear stress increases with the increasing of the magnetic field strength.

Figure 4 shows the effects of the TiO_2 contents on MR properties of the Fe_3O_4-TiO_2 nanocomposites based MR fluids. It can be seen that the MR fluid of the composites with 1ml $Ti(OR)_4$ precursor showed the largest shear yield stresses over the other composites based MR fluid due to the higher iron oxide content. However, the composites based MR fluids show significant improved shear stress compared with that of pure magnetite nanoparticles. The possible reason maybe the outer

TiO_2 layer inhibiting the direct contact of Fe_3O_4 particles, which facilitate the particle orientation and the rearrangement upon a magnetic field and make it easy to recover the normal state of dispersion after the removal of the field. In addition, sufficient active groups on the surface of particles promote the surface activity, which further promote the MR performance. However, with the increasing of TiO_2 content in coated particles, the magnetic property decreases, which reduced the MR performance accordingly. So, in this report, the MR fluid of the composites with 1ml $Ti(OR)_4$ precursor showed the optimum shear yield stresses over the other Fe_3O_4-TiO_2 nanocomposites based MR fluid.

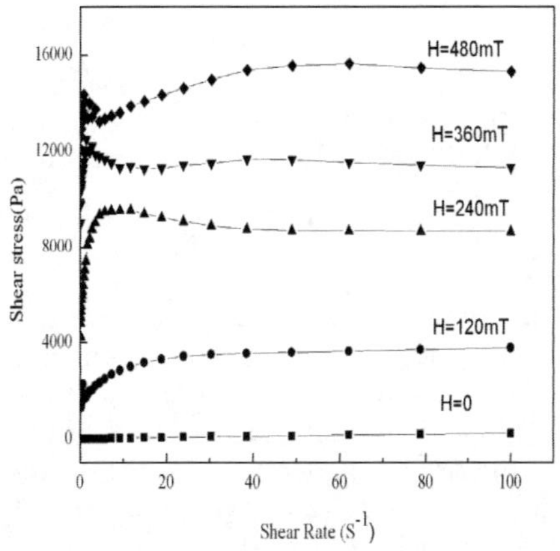

Figure 3. Influence of shear rate on shear stress for Fe_3O_4-TiO_2 nanocomposites based MR fluids under various magnetic fields

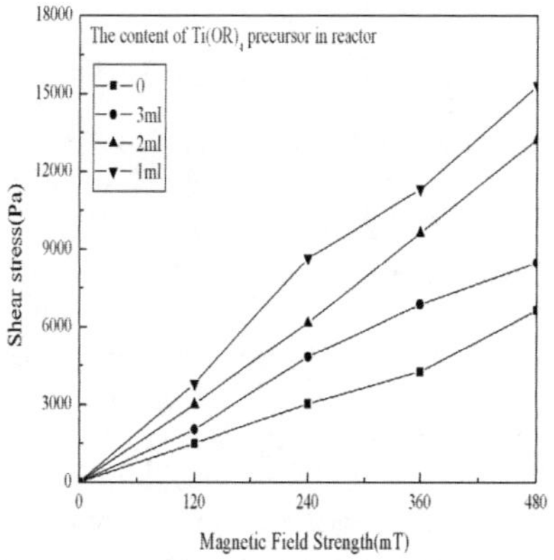

Figure 4. Effects of the TiO_2 contents on MR properties of the Fe_3O_4-TiO_2 nanocomposites based MR fluids.

4. Conclusions

A novel composite, Fe_3O_4-TiO_2 (TiO_2-coated Fe_3O_4) nanocomposites were prepared by modified sol-gel method. The FT-IR spectra infer the existence of interaction between Fe_3O_4 and TiO_2. The Fe_3O_4-TiO_2 nanocomposites show a superparamagnetic, without coercivity or remanence at room temperature, the magnetic property of the particles can be adjusted by the relative content of Fe_3O_4 and TiO_2. The MR fluids containing such materials in silicone oil show significant improved shear stress compared with that of pure magnetite nanoparticles, the shear stress of composite based MR fluids decreases with the increasing of the relative content of TiO_2.

Acknowledgments: We are grateful for the financial support from the National Program on key Basic Research Project (973 Grant No. 2009CB939705) and the National Natural Science Foundation of China (Grant No. 10674105 and 10874131).

References:

[1] Parthasarathya M and Klingenberg D J 1996 *Mater. Sci. Eng. R: Reports*, **17** 57
[2] Ma H R, Wen W J, Tam W Y and Sheng P 2003 *Adv. Phys.***52** 343
[3] Tao R 2001 *J.Phys. Condens. Mater.***13** R979
[4] Wen W J, Wang N, Ma H R, Lin Z F, Chan C T and Sheng P 1999 *Phys. Rev. Lett.* **82** 4248
[5] Zubarev A Y and Iskakova L Y 2003 *Colloid J.* **65** 159
[6] Minagawa K, Watanabe T, Koyama K and Sasaki M. 1994 *Langmuir* **10** 3926
[7] Rwei S P, Lee H Y , Yoo S D, Wang L Y and Lin J G, 2005 *Colloid Polym. Sci.* **283** 1253
[8] Fang F F, Jang I B and Choi H J 2007 *Diamond & Related Mater.***16** 1167
[9] Kim J H, Fang F F and Choi H J and Seo Y 2008 *Mater. Lett.* **62** 2897
[10] Chen C T and Chen Y C 2005 *Anal. Chem.* **77** 5912

Journal of Physics: Conference Series **149** (2009) 012084 doi:10.1088/1742-6596/149/1/012084

A Model of Heat Dissipation for MR based Brake

A Wiehe, V Noack and J Maas

Hochschule Ostwestfalen-Lippe – University of Applied Sciences, Control Engineering and Mechatronics, Liebigstraße 87, D-32657 Lemgo, www.motion-ctrl.de

E-mail: ansgar.wiehe@hs-owl.de

Abstract. In contrast to conventional brakes actuators based on magnetorheological fluids (MRF) offer an advantage in short term, peak load decelerating. The dissipation of a high amount of energy in a short period of time results in a thermal destruction of conventional brakes. Due to the volume based energy dissipation of MR actuators, instead of the surface based energy dissipation of conventional brakes, the rise of temperature and the distribution of energy shows significant advantages. In this paper a design rule for special peak load MR actuators is derived. Furthermore the simplified model, which is the basis of the design rule, is compared to several simulation models, with different levels of detail.

1. Introduction

Electromagnetic brakes and clutches are used for decelerating inertias, controlling force flows and locking positions in applications of industry automation. Functionalities of brakes and clutches can also be realized by smart materials. Especially magneto rheological fluids offer a high potential to substitute conventional braking mechanisms by reducing typical disadvantages like noise and abrasion.

Beside normal operation modes in industrial applications, emergency braking in critical situations is sometimes necessary. Due to the partly high inertias, high braking powers can occur during deceleration. This can result in a thermal overcharging of conventional brakes and thermal destruction of the brake covering. In order to handle this extreme loading on one hand the conventional brakes are oversized and on the other hand an early breakdown is accepted. In the following is shown, that the usage of a MR actuator seems to be an adequate alternative. Therefore a model of energy conversion is presented, which can be used for dimensioning of such MR actuators.

2. Theoretical discussion

To explain the advantages of MR actuators in contrast to conventional brakes the energy balance of both systems is analysed. In braking situation the resulting thermal energy is partly transported to the ambience and partly rising the temperature of braking components. The derivation of the energy balance with respect to the time is given by:

$$\frac{dW}{dt} = P_B = P_A + P_S, \tag{1}$$

where P_B is the braking power, P_A the power dissipated to the environment and P_S the power of the increasing temperature of the reference area. Because conventional and MR brakes have similar materials covering the braking components, power P_A can be assumed as comparable for both devices. Furthermore for short time peak load braking the power P_A can be simplified as zero. The difference between a conventional brake and a MR actuator is caused by the energy assimilation of the braking components, given by

$$P_s = \rho \cdot V \cdot c \cdot \frac{d\Delta \vartheta}{dt}, \tag{2}$$

where ρ is the density of the material, V the volume, c the specific heat and $\Delta\vartheta$ the excess temperature with respect to the ambient temperature. The volume of energy converting parts in conventional brakes is quite small, that peak loads, which can not be dissipated to the ambience immediately, generate unacceptable high temperature rises. The braking material will be thermally destroyed. As described in previous this effect causes a decreasing durability of brakes stressed by emergency stops. In contrast, the energy converting volume of MR actuators can be arbitrary enlarged by choosing appropriated working gaps. Based on common definitions the energy conversion of braking energy is derived by the following approach. Considering an infinitesimal volume of the MR fluid, shown in figure 1, converted braking energy dW_B is given by:

$$dW_B = dM \cdot d\varepsilon = r^2 \cdot \tau \cdot d\varphi \cdot dr \cdot d\varepsilon, \tag{3}$$

where τ is the yield strength and $d\varepsilon$ the relative movement.

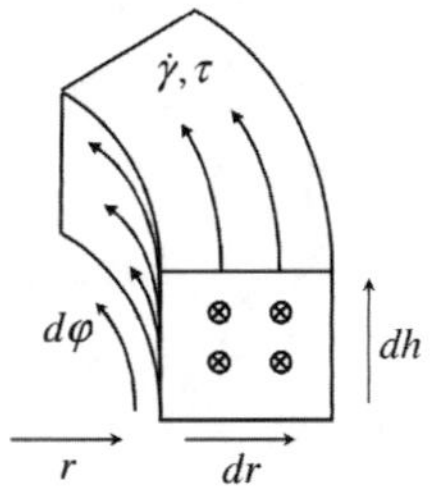

Figure 1. Infinitesimal Volume of MRF.

Substituting the relative movement $d\varepsilon$:

$$d\varepsilon = \dot\gamma \cdot dh \cdot dt, \tag{4}$$

by the local shear rate $\dot\gamma$ gives the dissipated power P_B by:

$$P_B = \int \frac{\partial W_B}{\partial t} = \iiint r^2 \cdot \tau \cdot \dot\gamma \cdot d\varphi \cdot dr \cdot dh. \tag{5}$$

Under consideration of a typical geometry of radial disc MR actuators, some simplifications can be made. By ignoring the viscosity of the MR fluid and assuming an uniformly distributed magnetic field, the yield strength τ is not a function of the geometry $(\tau \neq f(\varphi, r, h))$. In case of discoidial geometries the radius does not depend on angle or height of the working gap $(r \neq f(\varphi, h))$. Assuming a laminar flow, the shear rate can be expressed by [1]:

$$\dot\gamma = \frac{\omega \cdot r}{h_g}, \tag{6}$$

where ω is the angular speed and h_g the high of the working gap. Using these simplifications the braking power of rotary MR actuators can be expressed by:

$$P_B = \frac{2}{3}\pi \cdot \tau \cdot \omega \cdot \left(r_o^3 - r_i^3\right), \tag{7}$$

where r_i describes the inner and r_o the outer radial boundary of the working gap. Combining equation (2) with (7) gives:

$$\frac{3 \cdot \rho \cdot c}{2 \cdot \tau \cdot \omega} \cdot \frac{d\Delta\vartheta}{dt} = \frac{\left(r_o^3 - r_i^3\right)}{h_g \cdot \left(r_o^2 - r_i^2\right)} \tag{8}$$

Equation (8) can be used as a fundamental design rule for MR brakes specially designed for peak loads.

3. Simulation studies

Due to the assumptions made by introducing equation (2), the power P_A, which is dissipated to the environment, was neglected. In a first step the resulting derivation is investigated by comparing simulation results of a model based on equation (2) with more detailed models of the heat transfer.

Based on the modification of a lumped system approach [2] a simple model is proposed. It is assumed that one circular surface of the working gap remains at ambient temperature ϑ_A and the other surfaces are under adiabatic condition. Furthermore assuming the generation of the braking heat in the centre of the working gap the heat flow $\dot{Q}_\lambda$ through the circular surface is given by:

$$\dot{Q}_\lambda = \frac{2 \cdot \pi \cdot \lambda_{MRF} \cdot \left(r_o^2 - r_i^2\right)}{h_g} \cdot \left(\Delta\vartheta_{MRF}\right), \tag{9}$$

where λ_{MRF} is the heat conductance value of the MRF and $\Delta\vartheta_{MRF}$ is the excess temperature in the centre of the working gap. With equation (1) the thermal behavior is given by a first-order differential equation:

$$P_B = \rho_{MRF} \cdot V_{MRF} \cdot c_{MRF} \cdot \Delta\dot{\vartheta}_{MRF} + \dot{Q}_\lambda. \tag{10}$$

A further, more detailed model includes the iron parts around the working gap and the convective heat transfer on the actuators surface. The heat transfer in the real actuator is a multidimensional problem. To simplify the heat transfer to a one dimensional problem, the convective heat transfer of the circular surface is neglected. The way of the mean heat flux through the iron cylinder is assumed at the centre radius in axial direction, on the half height h_{iron} of the iron cylinder the heat flux faces to the outer surface in radial direction. Using the abbreviations:

$$C_{MRF} = \rho_{MRF} \cdot V_{MRF} \cdot c_{MRF},\; C_{iron} = \rho_{iron} \cdot V_{iron} \cdot c_{iron},\; R_{MRF} = \frac{h_g}{2 \cdot \pi \cdot \lambda_{MRF} \cdot \left(r_o^2 - r_i^2\right)}, \tag{11}$$

$$R_{iron,a} = \frac{h_{iron}}{2 \cdot \pi \cdot \lambda_{iron} \cdot \left(r_o^2 - r_i^2\right)},\; R_{iron,r} = \frac{r_o}{\pi \cdot \left(r_o + r_i\right) \cdot \lambda_{iron} \cdot h_{iron}},\; R_A = \frac{1}{2 \cdot \pi \cdot \alpha_{conv} \cdot r_o \cdot h_{iron}}, \tag{12}$$

the thermal behavior can be described by a second degree differential equation system [3]:

$$\begin{bmatrix} \Delta\dot{\vartheta}_{MRF} \\ \Delta\dot{\vartheta}_{Iron} \end{bmatrix} = \begin{bmatrix} -\frac{1}{C_{MRF}\left(R_{MRF} + R_{iron,a}\right)} & \frac{1}{C_{MRF}\left(R_{MRF} + R_{iron,a}\right)} \\ \frac{1}{C_{iron}\left(R_{MRF} + R_{iron,a}\right)} & -\frac{1}{C_{iron}}\left(\frac{1}{R_{MRF} + R_{iron,a}} + \frac{1}{R_{iron,r} + R_A}\right) \end{bmatrix} \cdot \begin{bmatrix} \left(\Delta\vartheta_{MRF}\right) \\ \left(\Delta\vartheta_{Iron}\right) \end{bmatrix} + \begin{bmatrix} 1/C_{MRF} \\ 0 \end{bmatrix} \cdot P_B . \tag{13}$$

Using FE methods instead of the discrete modeling method described above is an alternative way for the simulation of the thermal behavior. To compare adequately the results of the FE method and the discrete modeling method an according FE model is generated. The braking power P_B is in general a function of the radius, given by equation (5). For the simplified FE model the braking power is assumed to be constant relating to the volume. The convective heat transfer of the circular surfaces is neglected. The resulting heat flow of the simplified FE model is shown in figure 2.
To investigate the thermal behavior in a more realistic way a second FE model is generated, where the power dissipation given by equation (5) is taken into consideration. Furthermore the geometry of the working gaps is modified to a double gap geometry, which is typical for radial disk brakes. In figure 4 the transient response of the described models is shown. The braking power is assumed as a falling ramp, which is typical for decelerating inertias to zero velocity. The maximum temperatures of the simulation models are lower than the maximum temperature predicted by the design rule model ensuring safety of operation. In comparison to the simulation models the prediction of the model for the design rule is sufficient accurate to use it of the design procedure.

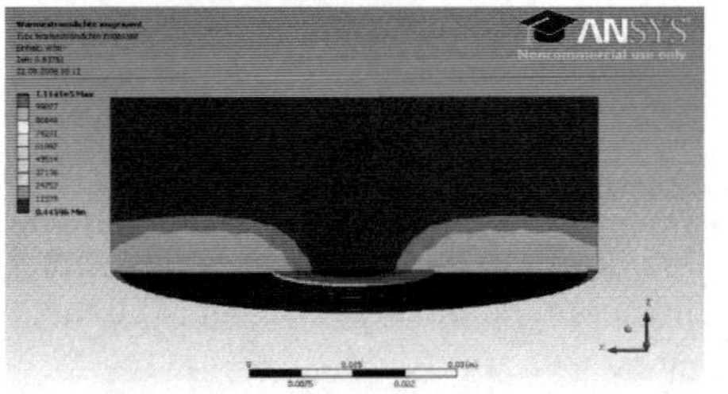

Figure 2. Heat flux of simplified FE model.

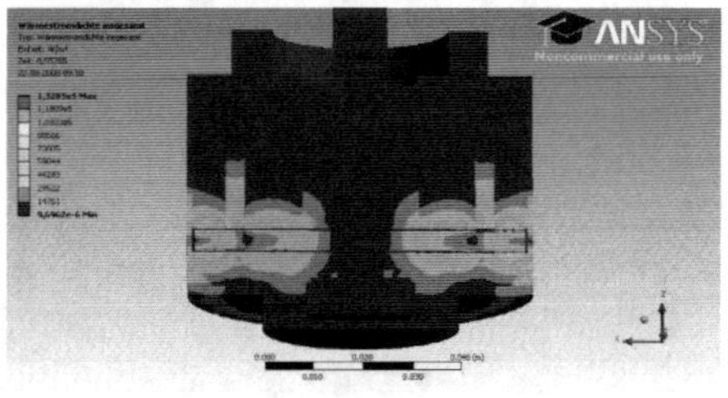

Figure 3. Heat flux of complex FE model.

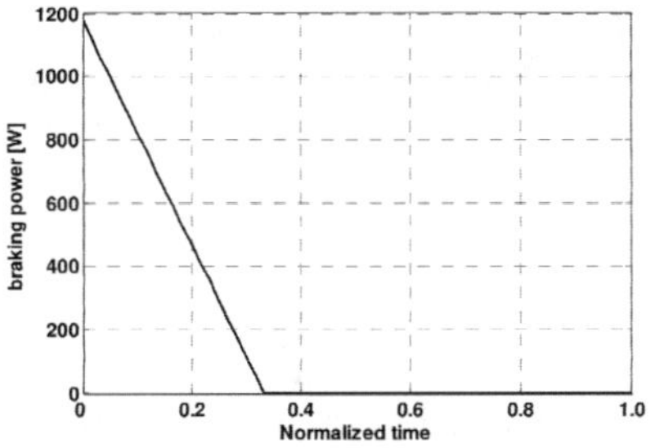

Figure 4. Loading Case for simulation studies.

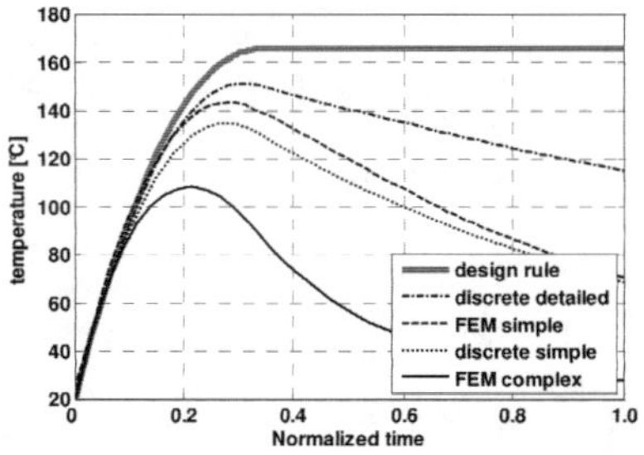

Figure 5. Time response of thermal models.

4. Conclusion

The theoretical discussion of the energy dissipation of conventional and MR based brakes shows an advantage in rise of temperature and distribution of energy. The thermal destruction of braking components in peak load applications can be avoided by the usage of MR brakes. A design rule, derived with a simplified model, can be used for the design of this kind of MR actuators. A first investigation of accuracy of the simplified model gives a sufficient correlation with more complex simulation models. In a further step the simulation models will be validated by measurements.

References

[1] Yamaguchi, H.: *Engineering Fluid Mechanics.* Springer Verlag, 2008.

[2] Kavlicoglu, B.M. et al.: *Heating of a High-torque Magnetorheological Fluid Limited Slip Differential Clutch.* Journal of Intelligent Systems and Structures, Vol. 19 – 2008, pp 235.

[3] Wiehe, A.; Kern, S.; Maas, J.: *Rotatorischer MRF-Aktor für einen Türassistenten.* at - Automatisierungstechnik 56 (2008) 3, S. 155-164, Oldenbourg-Wissenschaftsverlag.

Journal of Physics: Conference Series **149** (2009) 012085 doi:10.1088/1742-6596/149/1/012085

Rotary MR Actuators with Current less Holding Torque for a Vehicle Door Assistant

A Wiehe, K Voth and J Maas

Hochschule Ostwestfalen-Lippe – University of Applied Sciences, Control Engineering and Mechatronics, Liebigstraße 87, D-32657 Lemgo, www.motion-ctrl.de

E-mail: ansgar.wiehe@hs-owl.de

Abstract. Improvement of comfort functionalities is an important distinguishing mark for the competitiveness of automotive manufactures. Therefore the utilization of rotary MR actuators offers several new possibilities like an entry assist functionality for example. Beside requirements like a good controllability and reproducibility of torque the actuator must offer a certain holding torque without electrical feeding, in order to rest (fix) the automotive door in any position for a long period of time. For the realisation of an actuator with a current less torque two approaches are developed. The first one deals with permanent-magnets and offers an almost linear control characteristic. The second approach uses magnetic hysteresis for current less torque generation, which results in a plurivalent control characteristic. The main advantage is a an optimized design with respect to space and weigth. The hysteresis behaviour can be compensated by an adequate control algorithm, resulting in an almost linear torque characteristic.

1. Introduction

Conventional mechanisms provide resting of automotive doors in only two or three positions. Therefor in certain parking situations the passenger is stressed by maintaining the desired position and avoiding a collision with obstacles without any assistance. Today in the luxury segment a full variable detent is implemented by using a special hydraulic damper [1]. Due to the fact that a non-controllable actuator is used, further comfort functionalities cannot be provided. By implementing a controllable semi-active actuator, additional features become possible [2]. To avoid a collision the door will be decelerated after detection of an obstacle. A virtual stop position will be faded in to prevent additional movement of the door in the same direction. Due to this functionality potential damage can be avoided. Getting out is facilitated by enhancement of the door's resting torque. The user can support himself at the fixed door. The impact shock strains of door components can be widely reduced by decelerating the door to a minimal closing speed.

Since the door is hand-guided by the human a semi-active actuator is sufficient for such supporting functionalities. The specifications of an adequate actuator can be summarized to:

- Almost linear and reproducible generation of torque,
- short response time,
- good ratio between outer diameter and maximum torque,
- certain holding torque without electrical feeding (bias torque) and
- noiseless operation.

2. Actuators

Due to their characteristic rotary MR actuators are predestined for this kind of comfort applications. To lock the door over long periods of time, a current less holding torque is strictly required. Since no MR actuators with current less holding torque are established, novel concepts and rules for its design are needed.

For the realization of MR actuators with a current less holding torque two approaches are developed [3]. The first one deals with an arrangement of a permanent magnet and an electromagnet.

As shown in figure 1 this approach uses an arrangement of a permanent magnet and an electromagnet to generate the current less holding torque. In the case of non-fed MR actuator a magnetic field is generated by the permanent magnet (PM), which results in a current less holding torque. The magnetic field in the working gaps can be amplified by feeding the electromagnet (EM) with current. The minimization of the magnetic field can be reached by reversing the current direction. The design procedure is described in [4] in detail.

An alternative approach shown in figure 2 utilizes hysteresis properties of magnetic materials. These materials replace the current less field generation of the PM used in the first approach. The range of semi-hard magnetic materials includes all alloys whose coercivity is between soft and hard magnetic materials. When an external magnetic field, provided in this case by an EM, is applied to the hysteresis material (HM), it becomes magnetized. The grade of magnetization can be controlled by the electric current based on the hysteresis properties of the HM. The current depending torque behavior of the hysteresis based actuator is shown in figure 3 for one closed loop of the hysteresis in addition with the initial magnetization curve (dashed line).

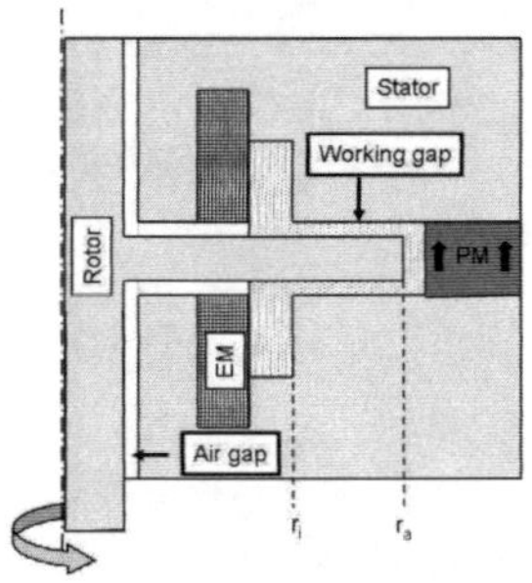

Figure 1. Half-section of the PM based actuator [3].

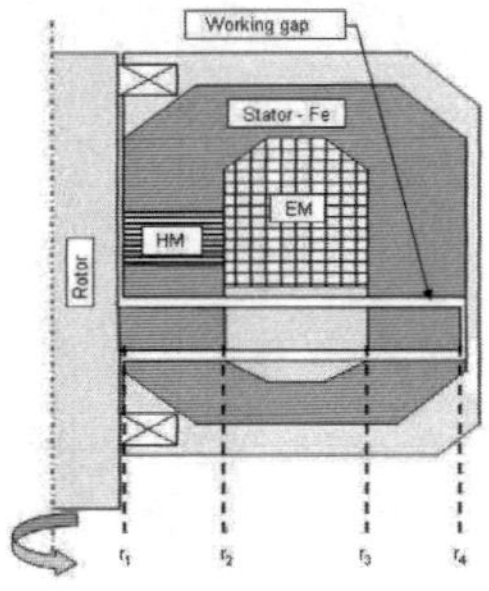

Figure 2. Half-section of the hysteresis based actuator [3].

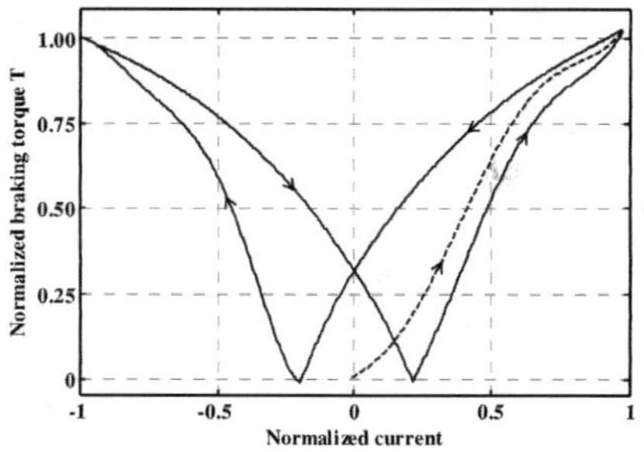

Figure 3. Control characteristics curve for the hysteresis based actuator [3].

3. Hysteresis compensation

Since an accurate control of force is essential when applying the hysteresis based actuator for applications like a vehicle door assistant, the compensation of the hysteresis phenomena is obligatory. In the following two approaches for compensating the hysteresis phenomena are proposed. A fundamental approach to hysteresis control is the inverse compensation as illustrated in figure 4. If an inverse model Γ^{-1} can be constructed of the hysteresis operator Γ, the output B_{wg} (i.e. the magnetic flux density in the working gaps of the hysteresis actuator) will be approximately equal to the reference trajectory B_{des}.

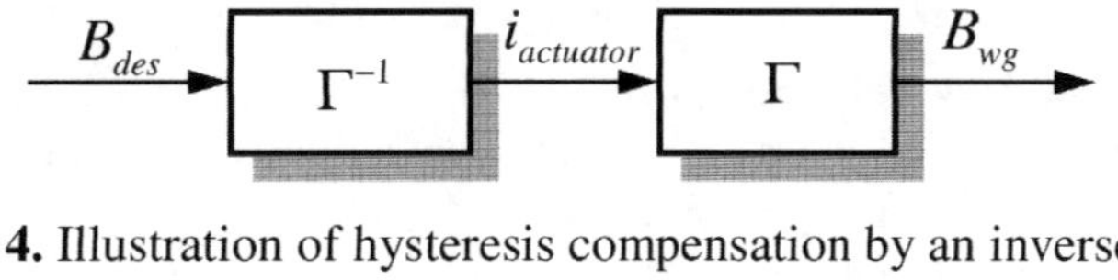

Figure 4. Illustration of hysteresis compensation by an inverse model.

Journal of Physics: Conference Series **149** (2009) 012085 doi:10.1088/1742-6596/149/1/012085

Two static approaches for hysteresis compensation, the Preisach approach and the modified Prandtl-Ishlinskii approach, are investigated.

A detailed description of the Preisach approach can be found in [5]. The basic element of the Preisach model is the hysteresis relay. It is characterized by the thresholds α and β. The output of the hysteresis model is computed as the weighted sum of relay outputs. A disadvantage of the Preisach approach is the high measuring expenditure for the parameter identification. The higher the required precision is the bigger is the measurement effort which is to be perform. Furthermore a comparatively longer computation time is required. An advantage of the Preisach approach is the high precision also in complex hysteresis modeling. The parameter identification can either be accomplished with a generic algorithm or also exclusive by measurements. Both, the major loop and the minor loops of hysteresis are considered for the parameter identification.

The Prandtl-Ishlinksikii approach is defined as the concatenation of a Prandtl-Ishlinskii hysteresis operator H and a Prandtl-Ishlinskii superposition operator S [6].

A large advantage of the Prandtl-Ishlinskii approach is the comparatively low and state-independent computation time. The inner system states are only approximated. Hence the accuracy of modeling and compensation is lower compared to the Preisach approach.

Both compensation algorithms are tested in a simulation study. The results in Figure 5 show the effectiveness of the hysteresis compensation. By using an inverse hysteresis model the behavior of the plant becomes nearly linear. If the hysteresis characteristic is neglected, the error will increase significantly. The measured control characteristics curve shown in figure 6 confirmed the nearly linear characteristic of the hysteresis actuator in combination with the compensation algorithm.

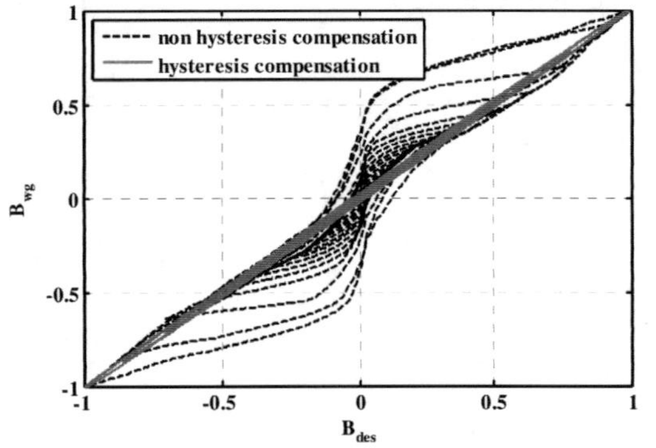

Figure 5. Effectiveness of hysteresis compensation by simulation.

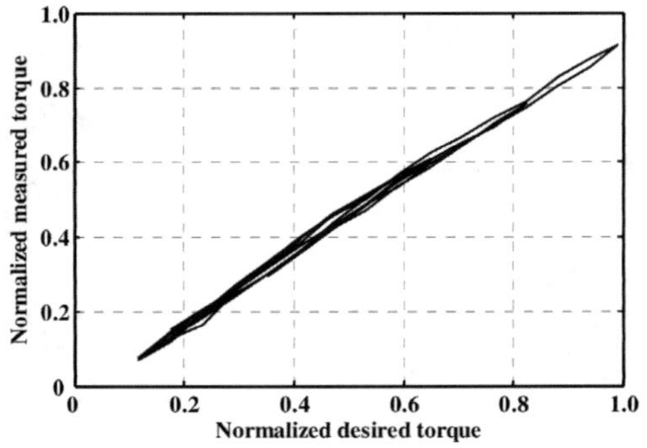

Figure 6. Effectiveness of hysteresis compensation by measurement.

4. Experimental results of the vehicle door assistant

Because of their almost linear control characteristics the PM-based approach as well as the hysteresis based approach in combination with a desired compensation algorithm seems to be applicable for the vehicle door assistant. To evaluate the door's functionality the actuators are tested in a an application environment, which is designed according to dimensions of typical doors of vehicles and which provides the possibility of a reproducible user interaction with the door.

The highest challenge of the vehicle door assistant for the control of the actuator represents the deceleration algorithm. In order to prove the compensation approaches in combination with the hysteresis based actuator the door is accelerated to almost similar angular velocities $\dot{\varphi}_0$ and swinging freely towards an obstacle assumed to be at the position $\varphi_{stop} = 60°$ for example.

With both approaches, the Preisach and the Prandtl-Ishlinskii approach, the door is decelerated precisely towards the desired angle with only a minimum overshot as shown in Figure 5. Additionally, the deceleration trajectory by use of the permanent magnet based actuator is shown. The results of a statistical evaluation of the measurements are given in Table 1.

Journal of Physics: Conference Series **149** (2009) 012085 doi:10.1088/1742-6596/149/1/012085

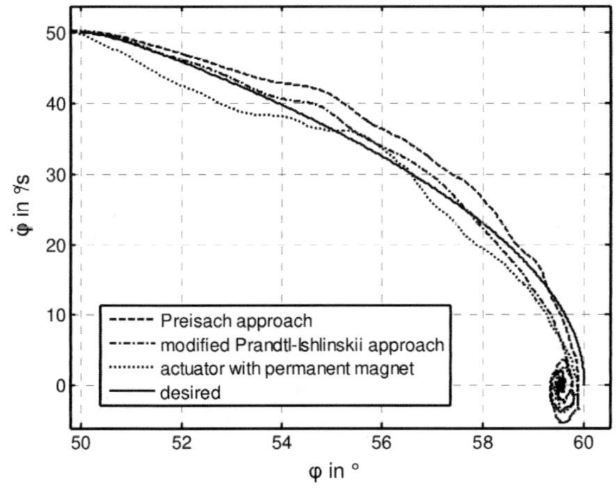

Figure 5. Measured trajectories of deceleration.

Table 1: Statistical data of measurements

	Peisach	Prandtl-Ishlinskii	PM-actuator
number of tests	40	55	52
desired final position φ_{stop}	60 °	60 °	60 °
mean final position φ_{∞}	59.59 °	59.57 °	59.7 °
standard deviation	0.05 °	0.1 °	0.1 °
mean overshot	0.26 °	0.21 °	0.24 °

5. Conclusion

For the generation of a current less holding torque, which is essential for applications like advanced door functionality, two MR actuators are developed. While the first is based on permanent magnets, the second utilizes hysteresis properties of magnetic semi-hard materials. Main advantage of the second approach is the minimized space demand, but the hysteresis behavior has to be compensated. In this paper approaches for hysteresis compensation are introduced. Two concepts the Preisach and the Prandtl-Ishlinskii approach are investigated. The Preisach approach provides better replication accuracy, while the main advantage of the Prandtl-Ishlinskii approach is the comparatively low and state-independent computation time. Since the behavior of the compensated hysteresis based actuator and the permanent magnet based actuator is comparable, both MR actuators are suitable for the vehicle door assistant.

References

[1] ATZ/MTZ, "Türhaltefunktion". ATZ/MTZ extra 10/2005 – Die neue S-Klasse, Oldenbourg Verlag, Munich.

[2] Maas, J., Kern, S.: *Mechatronic Vehicle Door Assistant.* IEEE/ASME International Conference on Advanced Intelligent Mechatronics, Zurich (CH), Nr. 285, 4-7. September 2007.

[3] Wiehe, A., Güth, D., Maas, J.: *Hysteresis based MR Actuator for current less torque generation.* Proceedings of actuator2008 conference, Bremen (DE).

[4] Wiehe, A.; Kern, S.; Maas, J.: Rotatorischer MRF-Aktor für einen Türassistenten. at – Automatisierungstechnik 03/2008 special edition smart actuators, Oldenbourg-Verlag.

[5] Mayergoyz, I. D.: *Mathematical models of hysteresis.* IEEE Transactions on Magnetics 22 (1986) 603-608

[6] Kuhnen, K.: *Inverse Steuerung piezoelektrischer Aktoren mit Hysterese-, Kriech- und Superpositionsoperatoren.* Dissertation, Universität des Saarlandes, 2001

Acknowledgement

This contribution is accomplished within the project "MRF-Bremse - Bremsen und Kupplungen auf Basis magnetorheologischer Flüssigkeiten" (MR brake – MR based brakes and clutches), funded by the Federal Ministry of Education and Research (BMBF) of Germany under grant number 17N0307.

Journal of Physics: Conference Series **149** (2009) 012086 doi:10.1088/1742-6596/149/1/012086

Effect of seven different additives on the properties of MR fluids

J Q Zhang, J Zhang, and Q Jing
Department of Technical Support Engineering, Academy of Armored Force Engineering, Beijing 100072, China

E-mail: zhangjq63@yahoo.com.cn

Abstract. Magnetorheological (MR) fluids have been developed for application in semi-active magnetorheological fluid dampers and other magnetorheological fluid devices. In order to prepare special MR fluids to satisfy the demands of tracked vehicle, two different carrier fluids were chose to prepare MR fluids. Preparation of MR fluids, which are based on carriers such as special shock absorption fluid and 45# transformer oil, was finished. And characteristics of these samples were tested and analyzed. Results indicate, Tween-80 and Span-80 can improve sedimentary stability. Using 45# transformer oil instead of special shock absorption fluid as a carrier, the shear yield stress remains nearly invariable but the viscosity and the sedimentary stability are reduced. MR fluids with diameter of 2.73μm show better sedimentary stability than that of the MR fluids with diameter of 2.3μm, or 4.02μm. Stearic acid obviously improves sedimentary stability and off-state viscosity, but don't perform an obvious function on shear yield stress. In magnetic field of 237KA/m, the shear yield stress of MR fluid based on special shock absorption fluid and 45# transformer oil is 18.34KPa, 14.26KPa, respectively.

1. Introduction

Magnetorheological (MR) fluids can respond to an externally applied magnetic field with a rapid and reversible change in their apparent viscosity. They have attracted much interest in their various engineering applications such as dampers, shock absorbers, isolators and brakes [1, 2, 3, 4], especially in vehicle engineering and civil engineering structures [5, 6, 7].

MR fluids are phase-controllable magnetic suspensions, which contain the magnetizable particles, non-magnetizable suspending fluids and additives. MR fluids behave similarly to Newton fluid without an applied magnetic field; under external magnetic field, the behavior of MR fluids can be described by the Bingham plastic model. When certain magnetic field intensity is imposed, the magnetic particles form chain clusters in the direction of the external field [8].

Though MR fluids do have most of the desirable properties, they also have the following disadvantages [9]: poor sedimentary stability, high viscosity at off-state, and low shear yield stress. Two different carrier fluids are used to prepare MR fluids and use these MR fluids in dampers in semi-active suspension of tracked vehicle to improve performances of vehicle. According to numerous experiments, sedimentary stability of MR fluids was investigated by applying different additives and particle weight fractions. Based on these results, some advanced experiments were made. And off-state viscosity, shear yield stress and sedimentary stability of MR fluids were comprehensively measured and analyzed in order to analyze the effect of different additives on property of MR fluids.

2. Experimental

2.1. The first phase experiment

In order to make clear components and additives, this set of experiments was used to evaluate the sedimentary stability of MR fluid samples. Additives were dispersed uniformly in carrier, and the mixture was kept in 60° for 1 hour, and then carbonyl iron particles were dispersed in the mixture, the MR fluid samples were mixed round in a colloid milling machine for 24 hours. Carbonyl iron particles with diameter of 4.02μm were dispersed in special shock absorption fluid (shown in Table 1) to prepare three kinds of MR fluid, and they were named as MRF11, MRF12, MRF13, their concentrations of carbonyl iron were 80%, and their corresponding additives were Tween-60 and Span-60, OP emulsifier (the structure of OP is shown in Figure 1) and oleic acid, Tween-80 and Span-80 (the components of MR fluids are shown in Table 2). Then sedimentary stability of MR fluid samples was tested and the results were researched.

Table 1. Characteristics of shock absorption fluid.

Component	**Density**(kg/m^3)	**Viscosity(Pa.s)**	
		50°C	**100°C**
70% olefin polymer oil and 30% mineral oil	0.857×10^3	0.0693	0.00857

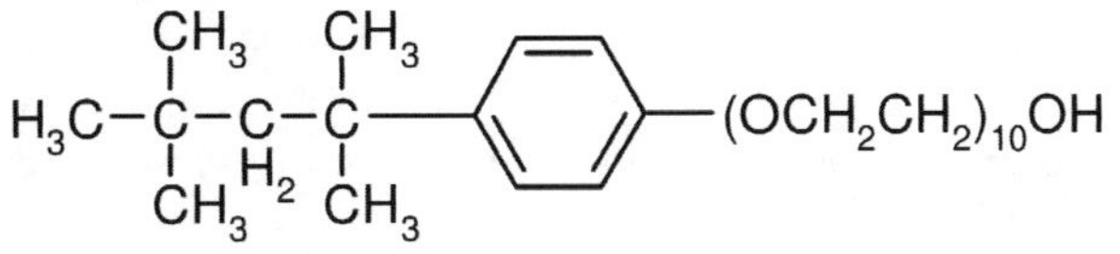

Figure 1. The structure of OP.

Table 2. Components of MR fluids.

	Diameter (μm)	**Concentration (wt.%)**	**Carrier**	**Additives**
MRF11	4.02	70%	shock absorption fluid	Tween-60 and Span-60
MRF12	4.02	70%	shock absorption fluid	OP and oleic acid
MRF13	4.02	70%	shock absorption fluid	Tween-80 and Span-80

2.2. The second phase experiment

To comprehensively investigate and analyze sedimentary stability, off-state viscosity and shear yield stress, the second phase experiment was designed. Carbonyl iron particles with diameter of 2.3μm are produced by Beijing Mountain Technical Development Center for Non-Ferrous Metals and their purity is 98.5%; carbonyl iron particles with other diameters are produced by Beijing Youxinglian Non-Ferrous Metals Co., Ltd and their purity is 99.5%. The components of MR fluids are shown in table 3.

Table 3. Components of MR fluids.

	Diameter (μm)	**Concentration (wt.%)**	**Carrier**	**Additives**
MRF21	2.3	70%	shock absorption fluid	tween-80 span-80
MRF22	2.73	70%	shock absorption fluid	tween-80 span-80
MRF23	4.02	70%	shock absorption fluid	tween-80 span-80
MRF24	2.3	70%	45# transformer oil	tween-80 span-80
MRF25	2.3	70%	shock absorption fluid	tween-80 span-80 Stearic acid

Journal of Physics: Conference Series **149** (2009) 012086 doi:10.1088/1742-6596/149/1/012086

3. Instrumentation

3.1. Sedimentary stability

Since density of the dispersed iron particles was much higher than that of the carrier fluid, sedimentation of the iron particles was generally observed. Placed in the lab for a long time, MR fluids settled and agglomerated so easily that parts of carbonyl iron particles lost their magnetorheological effect. Sedimentary rate (*R*) can be determined by placing MR fluid in a vertical cylindrical container at room temperature [10], *R* can be calculated by:

$$R = \Delta h / h \tag{1}$$

Where Δh is the length of the turbid fluid, and h is the whole length of the fluid.

3.2. Off-state viscosity

In zero magnetic field (off-state), some MR fluids behave similarly to Newton fluid, their off-state viscosities have no relationship with shear strain rate; and some MR fluids at off-state behave characteristics of non-Newton fluid, whose viscosities are inversely proportional to the shear strain rate. Off-state viscosity of MR fluids versus shear strain rate was measured by SNB-1 viscosimeter (Shanghai Exact Science Instruments Co. Ltd).

3.3. Shear yield stress

There are two different yield stress values, dynamic yield stress and static yield stress. The dynamic yield stress of MR fluid is usually described as zero-rate intercept determined through a linear regression curve fit to the measured flow data, while the static yield stress corresponds to the shear stress necessary to initiative flow.

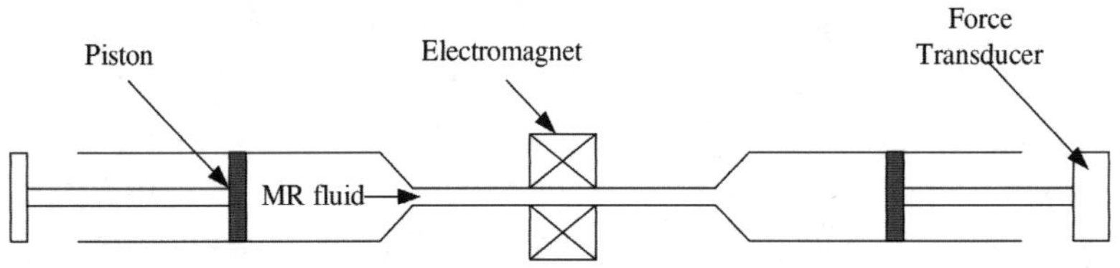

Figure 2. Schematic of flow mode MR rheometer.

Static shear yield stress is one of important parameters in properties of MR fluid. The method of measuring the static shear yield stress is shown in figure 2. By changing current value to alter magnetic flux density, the shear yield stress is evaluated by push force. The shear yield stress [11] can be calculated from equation (2):

$$\tau_y = F_y / \pi r^2 \tag{2}$$

Where, τ_y is shear yield stress, r is radius of the hole, F_y is the press force measured by sensor.

4. Results and discussions

4.1. Sedimentary stability

In the first phase experiment, according to figure 3, sedimentary stability of MR fluid with additives of Tween-80 and Span-80 is better than that of two other MR fluids with additives of OP and oleic acid, or Tween-60 and Span-60. This result indicates that Tween-80 and Span-80 can improve the sedimentary stability of MR fluid.

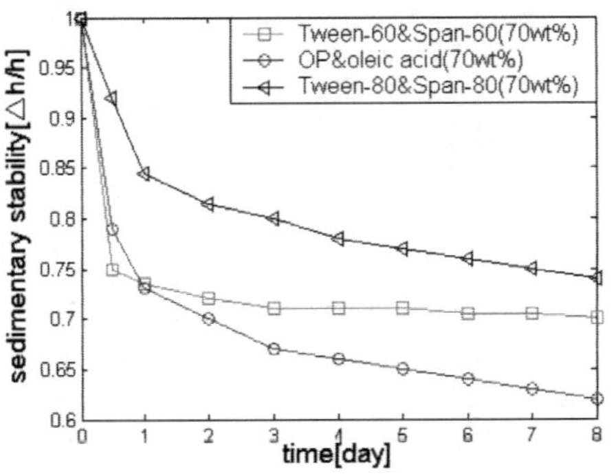

Figure 3. Sedimentary stability versus time.

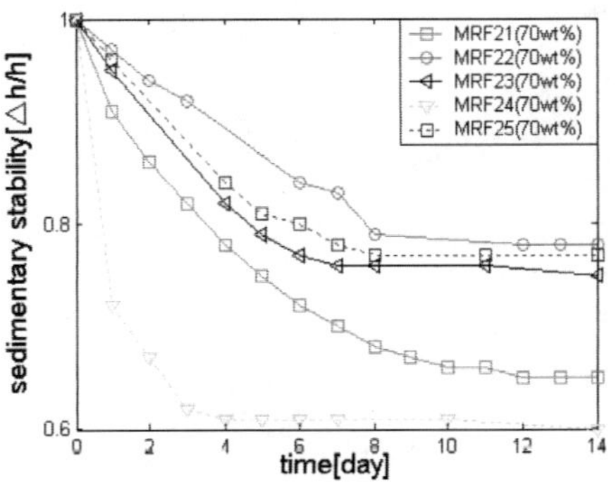

Figure 4. Sedimentary stability versus time.

According to figure 4, the sedimentary stability of MR fluid with diameter of 2.73μm is better than that of MR fluid with diameter of 4.02μm or 2.3μm. This phenomenon can be explained by Stokes equation (3):

$$V = (\rho - \rho_0) g d^2 / 18\eta \tag{3}$$

Where V is the sedimentary velocity of particles in carrier fluid; ρ and d are the density and radius of the particles, respectively; ρ_0 and η are the density and viscosity of the carrier fluid, respectively; g is the gravitational constant.

According to equation (3), the sedimentary velocity of MRF21, MRF22, MRF23 are calculated in table 4, where, gravitational constant g=10(m/s^2), density of the particles $\rho=7.8\times10^3$(kg/m^3), the density of the carrier $\rho_0=0.8\times10^3$(kg/m^3), as shown in table 4.

Table 4. Sedimentary velocity.

	Diameter (m)	**Viscosity** (Pa.s)	**Velocity** (m/s)
MRF21	2.3×10^{-6}	1.309	1.8072×10^{-8}
MRF22	2.73×10^{-6}	3.36	0.8625×10^{-8}
MRF23	4.02×10^{-6}	4.34	1.4480×10^{-8}

Obviously, according to table 4, MRF22 has the slowest sedimentary velocity. This result accords well with the observed phenomenon. Thus, the sedimentary stability of MR fluid with diameter of 2.73μm is better than that of MR fluids with diameter of 4.02μm or 2.3μm.

Additionally, according to figure 4, the sedimentary stability of MRF24 is the worst, mostly because the viscosity of 45# transformer oil is much lower than that of special shock absorption fluid and the viscosity of MR fluid based on 45# transformer oil is lower than that of MR fluid based on special shock absorption fluid. Sedimentary stability of MRF25 is much better than that of MRF21, this indicates that stearic acid can increase the viscosity of MR fluids and improve sedimentary stability of MR fluids.

4.2. Off-state viscosity

Figure 5 represents the viscosities of MR fluids measured in 22.3°C. According to figure 5, viscosities of all MR fluids exhibit shear-thinning behavior. Viscosity of MRF24 based on 45# transformer oil is the lowest, because other MR fluids are based on special shock absorption fluid which has a bigger viscosity than 45# transformer oil. From viscosities of MRF21, MRF22, MRF23, diameter of the carbonyl iron particles has strong influence on the viscosity of MR fluid—means that diameter of the particles is almost in direct proportion to the viscosity of MR fluid. Comparing the viscosity of MRF25 with MRF21, it is easy to observe that the viscosity of MRF25 is higher than that of MRF21, because stearic acid increases the viscosity of MR fluid.

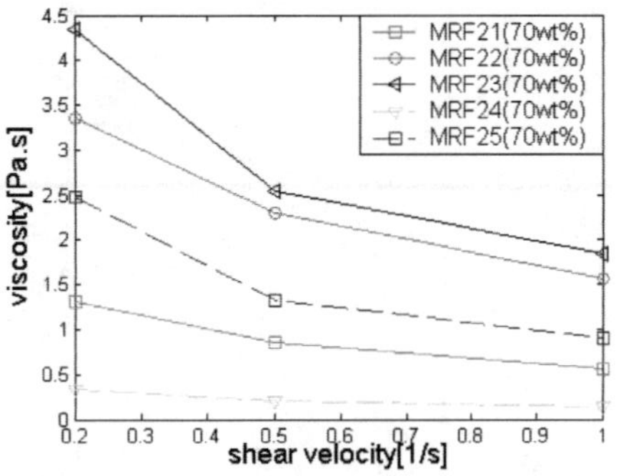

Figure 5. Viscosity versus shear velocity.

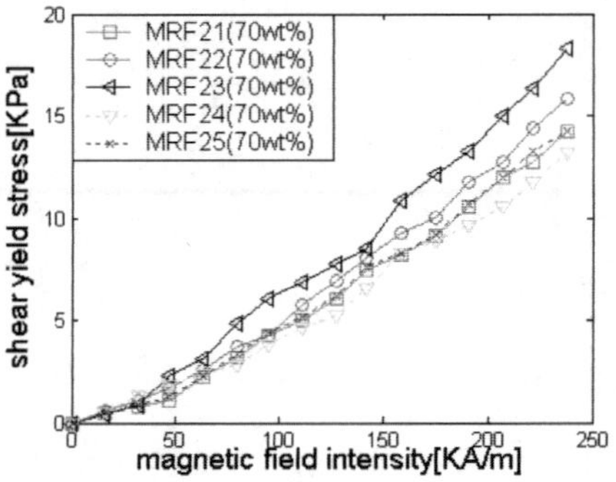

Figure 6. Shear yield stress versus magnetic field intensity.

4.3. Shear yield stress

Figure 6 represents relationship of shear yield stress with different magnetic field intensity. For an increase in magnetic field intensity, a similar increase in the shear yield stress is observed. This is because the stronger magnetic field intensity makes the pull force among the soft magnetic particles bigger, and thus the shear stress becomes bigger to break the column-like structures formed by magnetic particles. According to curves of MRF21, MRF22, MRF23, for an increase in diameter, a similar increase in shear yield stress is observed. From curves of MRF21 and MRF25, the shear yield stress is similar and has not strongly difference, this indicates that stearic acid don't change the shear yield stress of MR fluid. From curves of MRF21 and MRF24, although MR fluids are based on different carriers they have similar shear yield stress but different viscosities. In practice, 45# transformer oil can also be used instead of special shock absorption fluid to prepare MR fluid with low viscosity, to some extent. In magnetic field of 237KA/m, the shear yield stress of MRF23 is 18.34KPa and the shear yield stress of MRF24 is 14.26KPa.

5. Conclusions

Numerous MR fluids based on such carriers as special shock absorption fluid and 45# transformer oil were prepared, and their sedimentary stability, off-state viscosity and shear yield stress were tested and analyzed. The results indicate that tween-80 and Span-80 can improve the sedimentary stability; 45# transformer oil can also be used as a carrier and MR fluids based on this oil can have a low viscosity; sedimentary stability of MR fluids with particle diameter of 2.73μm is the best among the three kinds of diameters referred to this experiment; stearic acid can increase the viscosity and sedimentary stability of MR fluid, but can't improve shear yield stress of MR fluid; in magnetic field of 237KA/m, the shear yield stress of MR fluid based on special shock absorption fluid and 45# transformer oil can reach 18.34KPa, 14.26KPa, respectively.

References

[1] Jolly M R, Bender J W and Carlson J D 1999 *J. Intell. Mater. Syst. Struct.* **10** 5
[2] Carlson J D, Chrzan M J, Raleigh N C and Girard F J 1994 *US Patent:* 5284330
[3] Cao Z, Jiang W, Ye X and Gong X 2008 *J. Magn. Magn. Mater.* **320** 1499
[4] Viota J L, Vicente J D, Duran J D G and Delgado A V 2005 *J. Colloid Interface Sci.* **284**, 527
[5] Ismo V 2002 *Smart Mater. Struct. VTT Research Program 2000-2002*, 51
[6] Gordaninejad F and Kelso S P 2000 *J. Intell. Mater. Syst. Struct.* **11** 395
[7] Yang G Q, Spencer B F and Carlson J D 2002 *Engineering Struct.* **24** 309
[8] Fernando D G and Carlson J D 2006 *Proc. 10th Int. Conf. on ER Fluids, MR Suspensions* 165
[9] Park J H, Chin B D and Park O O 2001 *J. Colloid Interface Sci.* **240** 349
[10] Jun J, Uhm, Ryu J and Suh K 2005 *Colloid Surf. A* **260** 157
[11] Fuchs A et al 2004 *J. Appl. Polymer Sci.* **92** 1176

Journal of Physics: Conference Series **149** (2009) 012087 doi:10.1088/1742-6596/149/1/012087

Optimization Analysis of a New Vane MRF Damper

J Q Zhang, Z Z Feng and Q Jing
Department of Technical Support Engineering, Academy of Armored Force Engineering, Beijing, 100072, China

E-mail: zhangjq63@yahoo.com.cn

Abstract: The primary purpose of this study was to provide the optimization analysis certain characteristics and benefits of a vane MRF damper. Based on the structure of conventional vane hydraulic damper for heavy vehicle, a narrow arc gap between clapboard and rotary vane axle, which one rotates relative to the other, was designed for MRF valve and the mathematical model of damping was deduced. Subsequently, the finite element analysis of electromagnetic circuit was done by ANSYS to perform the optimization process. Some ways were presented to augment the damping adjustable multiple under the condition of keeping initial damping forces and to increase fluid dwell time through the magnetic field. The results show that the method is useful in the design of MR dampers and the damping adjustable range of vane MRF damper can meet the requirement of heavy vehicle semi-active suspension system.

1. Introduction

MR fluids undergo reversible and rapid changes in material characteristics when subjected to a magnetic field. This change is primarily observed as a significant increase of the yield shear stress of MR fluids and can be continuously controlled by tuning the magnetic field intensity [1]. These characteristics have created potential applications in semi-active vibration isolation and motion damping systems in industrial application. In vehicle engineering, energy-dissipating devices of the said systems include mainly piston-cylinder dampers and vane hydraulic dampers. Being compact structure and perfect heat-dissipating ability, the latter are usually adopted in some heavy vehicles. Today, a piston-cylinder MRF damper has been used in the real life applications due to its minimal power dissipation and a board range of capacities [2, 3]. For example, when it is utilized in automotive primary suspension, these automotive primary suspensions can offer combined advantages of both passive and active control systems and show less vehicle motion and superior ride quality in rough road [4, 5]. However, the applications of vane MRF dampers are still few. In this study, we proposed a new vane MRF damper. A phenomenological model that can accurately capture its dynamic behavior is presented. By optimization analysis, dynamic force range of the damper is extended to meet the performance requirement for certain heavy vehicle.

2. Vane MRF damper

Up to now, the structure of the conventional hydraulic damper is already perfect, so it is considered as basic structure of vane MRF damper. The striking dissimilarity between the both is that flow through the orifice valve (i.e. MR valve) of the vane MRF damper is exposed to magnetic field. A partly assembled vane MRF damper is shown in Figure1.

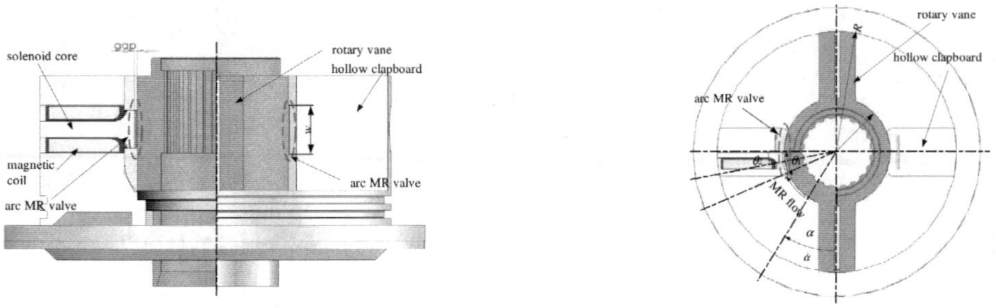

(a) The facade (b) The planform

Figure 1. A cutaway view of a vane MRF damper

Because the yield stress of the MR fluid increases significantly with an increase of magnetic flux density, magnetic field is expected stronger and to plumb up the flow as possibly. So, to expand room for a coil assembly, clapboard is designed to be hollow and the coil with coil core is placed in this hollow. Correspondingly, an arc MR valve is placed between clapboard and rotary vane axle which one rotates relative to the other and they act as two magnetic poles respectively. So the MR valve consists of solenoid core, magnetic coil and rotary vane axle and it controls the flow rate and pressure drop through the gap, and so controls damping level. To evaluate the performance of this MR valve, a generalized MR valve model is developed.

3. Theoretical Modeling

According to the pressure drop equation deduced by Philips and his colleagues (1969) based on parallel-plate model between two paramagnetic poles [6],

$$\Delta P = \frac{12\eta L Q}{wh^3} + \frac{3L\tau_y}{h} \tag{1}$$

Where Q is the volume flow rate due to pressure drop, τ_y is the controllable yield stress of a MR fluid, L is the length of MR valve, w is the width of the plates, η is viscosity of the fluid, and h is the gap between the plates. To apply the above formula for the flow through the MR valve, in this case the following correspondences of symbols are made

$$L \rightarrow \theta \cdot \bar{r}_{MR}$$

Where θ and $\bar{r}_{MR}$ are radian and average radius of MR valve. Then the pressure drop is obtained as,

$$\Delta P = \frac{12\eta Q}{wh^3}\theta_1 \cdot \bar{r}_{MR} + \frac{3\tau_y}{h}\theta_2 \cdot \bar{r}_{MR} = \Delta P_{vis} + \Delta P_{MR} \tag{2}$$

Where $\theta_1 \cdot \bar{r}_{MR}$ is the arc length of the flow channel, $\theta_2 \cdot \bar{r}_{MR}$ is the arc length of the MR valve, ΔP_{vis} is pressure drop due to viscous orifice, and ΔP_{MR} is pressure drop due to MR effect. Suppose that there is no leakage gap and fluid flows only through the MR valve. Conservation of mass for an incompressible fluid yields:

$$\Delta V = \frac{\alpha}{2\pi} \cdot \pi (R^2 - r^2) b \tag{3}$$

$$Q = \frac{dV}{dt} = \frac{b(R^2 - r^2)}{2}\dot{\alpha} \tag{4}$$

where b is the width of the vane, r is inner radius of the vane, R is outer radius of the vane and the rotary angle of the vane around its axis, α, and the rotary velocity , $\dot{\alpha}$,are reconstructed by

$$\alpha = \alpha_0 \sin \omega t \quad \dot{\alpha} = \alpha_0 \omega \cos \omega t \tag{5}$$

where α_0 is the maximum angle of the vane around its axis, and ω is frequency.

To ensure the fluid sufficient amount of dwell time to reach its full yield stress, the high velocity behavior of MR fluid subjected to a magnetic field should be evaluated. Dwell time is defined as the amount of time the fluid spends in the presence of a magnetic field. In other words, dwell time is the time it takes the fluid to flow through the MR valve [7]. So dwell time is calculated from

$$t_{dwell} = \frac{\theta_2 \cdot \bar{r}_{MR}}{v} \tag{6}$$

Where v is the mean fluid velocity thro is the arc length of through the flow channel, as,

$$v = \frac{Q}{wh} \tag{7}$$

4. Optimization of the Structure

In terms of its feasibility for engineering applications, force characteristics of the damper have to be optimized.

4.1. *Dynamic force range Analysis*

A broad dynamic force range is important for force characteristics. Define parameter K [8]

$$K = \frac{\Delta P_{MR}}{\Delta P_{vis}} = \frac{\tau_y \theta_2 w h^2}{2\eta\theta_1 b(R^2 - r^2)\dot{\alpha}} = k_1 k_2 \frac{\tau_y h^2}{2\eta\dot{\alpha}} \tag{8}$$

Where $k_1 = \dfrac{\theta_2}{\theta_1}$ is called magnetic field efficiency coefficient, $k_2 = \dfrac{w}{b(R^2 - r^2)}$ is the geometric parameter of vane MRF damper. It can be seen from equation (8) that the damping adjustable multiple of MRF damper is related to magnetic field efficiency coefficient k_1, geometric parameter k_2, height of gap h and performance parameters of MRF ($\tau_{y\max}$, η).

4.1.1. The relation between K and k_1

K is directly proportional to k_1 which depends on the distribution of magnetic field. In the case of meeting the requirement of damper structure, the proper augmentation of k_1(0□k_1≤1) may enhance K in MRF damper.

4.1.2. The relation between K and k_2

When the structure and the main geometrical parameters (radius of vane R, radius of axis r, width of vane a) of MRF damper are constants, k_2 has a direct ratio relationship with the width w of the MR valve. The initial damping has a ratio relation with $1/w$, K and k_2 present direct ratio. According to the geometrical parameters in vane hydraulic damper, the relation curve of damping adjustable multiple K and w is determined at different dynamic viscosities of MRF damper (Figure 2). Therefore, in the case of meeting the requirement of the initial damping, the proper augmentation of the width w of the MR valve may enhance K. However, w should be restricted by the width of magnetic field.

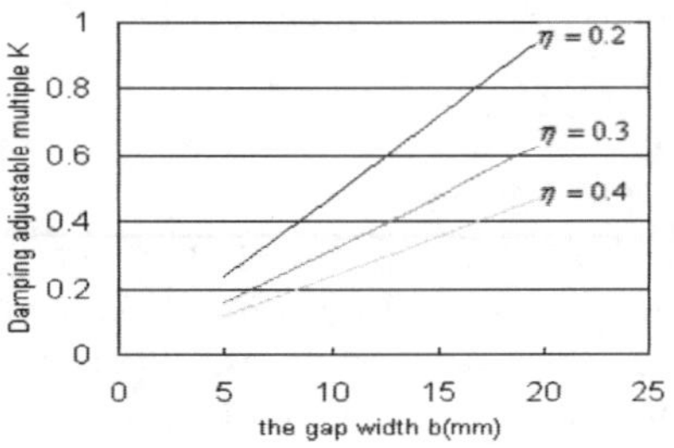

Figure 2. Curves K vs. b

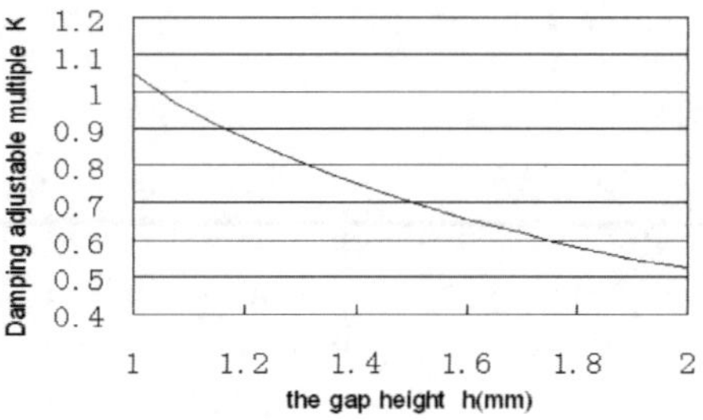

Figure 3. Curve of K vs. h

4.1.3. The relation between K and h

For the vehicle semi-active suspension system, the MRF damper should have the fail-safe capability, which is referred to retain a minimum required damping capacity in the event of a power supply or electronic system failure. So the initial damping of the vane MRF damper should reach or approach to that of vane hydraulic damper. The pressure drop ΔP_{vis} due to viscous orifice has a ratio relationship with $1/h^3$, K has a direct ratio relationship with h^2. From Figure 3, we know that the gap height h cannot be designed too large while keeping the initial damping of the vane MRF damper invariable. Besides, h vs. the magnetic field intensity in the gap should be taken into account.

4.1.4. The relation between K and the performance parameters of MRF

In order to make K as large as possible, the MRF with the higher shear yield stress and the lower initial viscosity should be chosen. Commonly, with the augmentation of the particle volume percent, the shear yield stress and the viscosity are enlarged.

4.2. Magnetic Circuit Analysis

The magnetic field distribution of magnetic circuit is evaluated using ANSYS finite element analysis (FEA) software. A three-dimensional model developed by SolidWorks software, which can be seen in Figure 4, is imported into ANSYS software for specified boundary conditions that include material properties, electromagnetic activation current, and element insulation. A relationship between electromagnet activation current and magnetic field strength is determined using this software. Being an assembly model by SolidWorks software, the sizes of the model are very exact figures. The materials adopted in this model are as followings: coil core-DT4, vane and shell-steel45, constant relative permeability μ_r=700 for steel-45, applying current-2 Amp, turns of coil-1000.

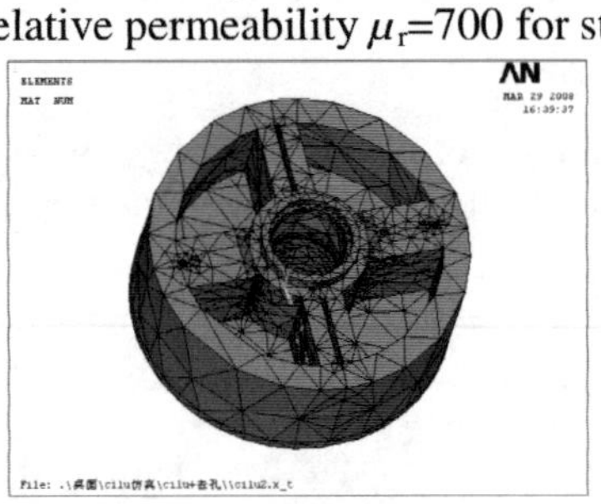

Figure 4. A three-dimensional model developed by SolidWoks software

Figure 5 shows the distribution of magnetic field inside the MR valve. From the results presented in Figure 5(a), the magnetic flux density values of the MRF through MR valve is only 0.128~0.12T, which is too small to MRF, and there is no significant change in the magnetic flux occurs when the applying current is change. Obviously, the cause is that the coil core has become saturated with magnetic flux and magnetic flux of entire circuit is limited to alter further, as presented in Figure 5(b).

The geometrical parameters optimized in the electromagnetic analysis part can be seen in Figure 5. They are the coil turns, coil core radius. Especially, a bell-mouthed cone is designed in coil core. The optimization of each part is complete when no significant change in the magnetic flux occurs when the dimensions of said part are altered further. Furthermore, since the clapboard of the damper is one of the design constraints, coil radius needs to be optimized in this limited space. Because different MR thickness values give different magnetic flux density values inside the MR valve, a thickness between 0.8-1.2mm, is selected as a desired value of MR disk thickness.

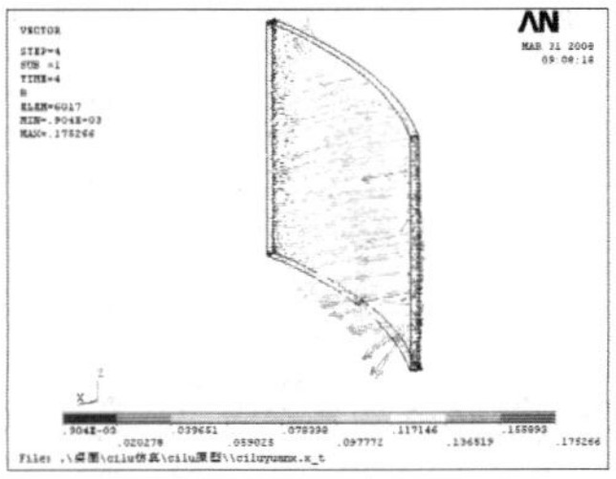

(a) MRF through MR valve

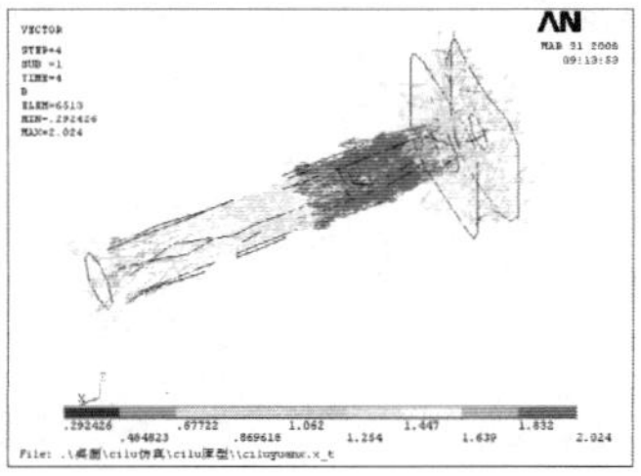

(b) Coil core

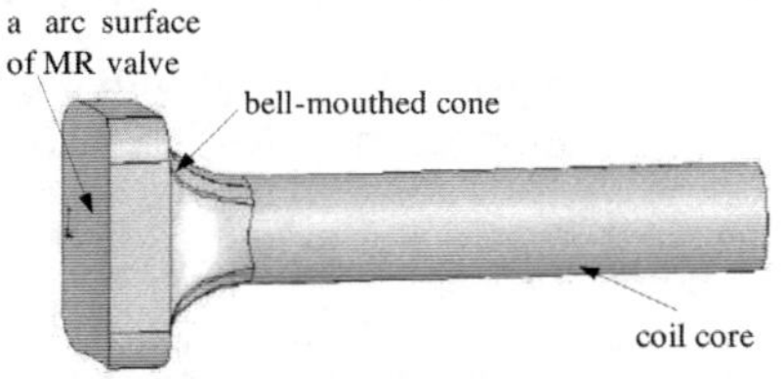

Figure 6. The coil core optimized

Using the final design parameters, electromagnetic analysis is performed for 2A current inputs and 800 turns coil. Figure 6 shows the magnetic flux density values inside magnetic circuit. Obviously, 0.6~0.7T and 1.6~1.7T are achieved inside the MRF through MR valve, which is enough for MR effect, and coil core respectively. In addition to this, the magnetic flux density vectors on the MR surface are uniformly distributed due to the symmetry of the MR valve.

By using the dimensions obtained from the electromagnetic analysis, electromagnet coil resistance is 4 ohm, which is a low resistance. This allows applying currents as high as 2 or 3 Amp and therefore, a low electrical power.

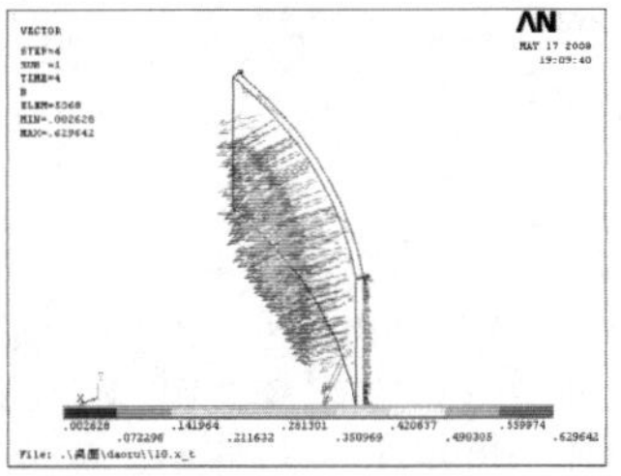

(a) MRF through MR valve

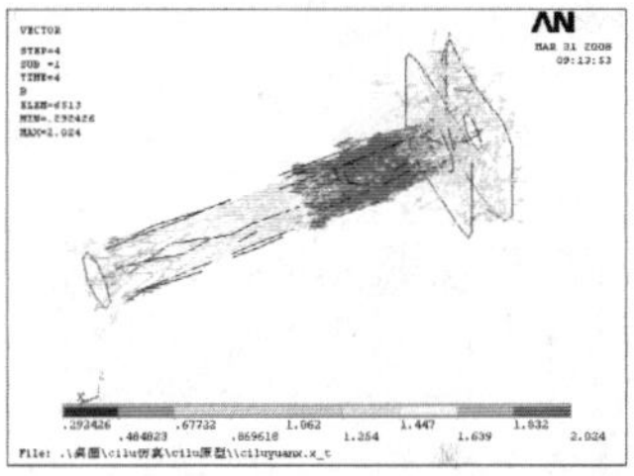

(b) Coil core

5. Conclusions

A new controllable vane MRF damper was developed to determine the viability of implementation of this technology into some heavy vehicles. It has been demonstrated that it is desirable for a semi-active suspension system because of softer damping, fail-safe behaviour, lower dispersion of the damping force, lower heat generation and consuming less power compared to the conventional MR damper. It is believed that improvements of ride quality in rough road should be seen in the positive direction with further optimization of the semi-active algorithm.

References

[1] Ashour A, Rogers C A, Kordonsky W 1996 *J. Intelligent Material Systems and Structures* **7** 123
[2] Ginder J M 1996 *Encyclopedia of Applied Physics* **16** 487.
[3] Carlson J D 2006 *Proceedings 10th Int.Conf.on ER,MR Suspension* 389
[4] Dogruer U, Gordaninejad F and Evrensel C 2004 *Proceedings of SPIE* **5386** 195
[5] Andrea C W and Francis B H 2003 ,Army Tank-automotive Research, Development, and Engineering Center, 1
[6] Guan X C and OuJ P 2000 *Journal of Vibration and Shock* **20** 5
[7] Ahmadian M and Goncalves F 2006 *Proceedings 10th Int.Conf.on ER,MR Suspension* 114
[8] GuanC, Li J H and OuJ P 2005 *Engineering Mechanics* **22** 207

Journal of Physics: Conference Series **149** (2009) 012088 doi:10.1088/1742-6596/149/1/012088

Variable stiffness and damping MR isolator

X Z Zhang, X Y Wang, W H Li and K Kostidis

University of Wollongong, School of Mechanical, Materials & Mechatronic Engineering, NSW 2522, Australia

e-mail: *weihuali@uow.edu.au

Abstract. This paper presents the development of a magnetorheological (MR) fluid-based variable stiffness and damping isolator for vibration suppressions. The MR fluid isolator used a sole MR control unit to achieve the variable stiffness and damping in stepless and relative large scope. A mathematical model of the isolator was derived, and a prototype of the MR fluid isolator was fabricated and its dynamic behavior was measured in vibration under various applied magnetic fields. The parameters of the model under various magnetic fields were identified and the dynamic performances of isolator were evaluated.

1. Introduction

Vibration isolation systems are used to separate the vibration excitation source and the object to be protected in order to reduce the negative effect from the unwanted vibration from the vibration excitation source. Conventional vibration systems are in passive mode, for the intrinsic limitation of passive suspension system, the improvement is effective only in a certain frequency range, because passive design for low frequency is difficult and often represents a compromise between isolation performance and the supported machinery alignment [1].

It is possible to improve the isolation performance over a wide range of frequency by active vibration control. However, active vibration isolation systems are less commonly used than passive systems due to their associated cost, power requirements, complexity and fail-safe problems [2, 3].

Semi-active suspensions can be nearly as effective as fully active suspensions in vibration suppression. They do not require either higher-power actuators or a large power supply. When the control system fails, the semi-active suspension can still work in a passive condition. In early semi-active suspension systems, the adjustable damping force can be achieved by using hydraulic semi-active dampers with electromagnetically controlled valves or Friction damper, of which the damping force is controlled by varying the force normal to a friction interface. More recently, with the introduction of Magnetorheological (MR) fluid, it is easy to achieve the variable damping with an MR damper in isolation system. The effort to improve the performance of the isolation systems by using a variable damping damper based on MR fluid has been done by many researchers [4-5]. However, variable damping and stiffness systems showed significant improvements in vibration reduction application, such as the seismic protection system [6] and the vehicle suspension [7]. However, all the methods to achieve variable stiffness in these devices are usually complicated and low reliability, and it is difficult to accomplish stepless control on the stiffness [8]. Kobori et al. developed a variable stiffness system for protecting the building from the earthquakes [9]. A valve in a cylinder device was controlled in open/close ways to lock and unlock the connection between the beam and the brace and two alternative stiffnesses were produced at each floor. Youn et al. developed a vehicle suspension using air springs, which can be adjusted in three different states: soft, medium and firm, to achieve variable stiffness [10]. Liu et al. proposed a variable damping and stiffness system using two MR dampers and two springs [11]. By controlling the damping of the MR damper, the total stiffness of this system can be varied. However, all the methods to achieve variable stiffness in these devices are usually complicated and low reliability.

In this paper, we proposed a new vibration isolator, which uses a MR valve to connect with two stiffness units. The variable stiffness of the system is accomplished by controlling the damping of the MR valve. The prototype of the vibration isolator was built. The change on the damping and stiffness of system was demonstrated in theoretical and experimental methods.

2. MR fluid isolator

2.1. The Design of the MR Fluid Isolator

The schematic of the MRF isolator is shown in figure 1. The MR isolator consists of an air spring, two connectors, a MR valve and an accumulator. Two connectors link air spring, accumulator and MR valve together with plastic pipes. In each connector, there is a loose separate film, which can move freely, dividing the connector into two chambers. The chambers close to the MR valve are filled with MR fluid. The air spring supports the isolated object. Due to the effect of vibration excitation on the isolated object, the air spring is compressed or tensed by the isolated object in order to generate the deformation of the air spring, which is the cause of the pressure variation in the air spring. Since the change of the pressure in the air spring, the MR fluid is forced to flow from connector 1 to connector 2 through MR valve. When an input current is applied on the coil of the MR valve, a magnetic field is generated in the electromagnet. The shear stress of the MR fluid in MR valve is varied by controlling the magnetic field, which thereby varies the characteristics of the isolator, such as its stiffness and damping.

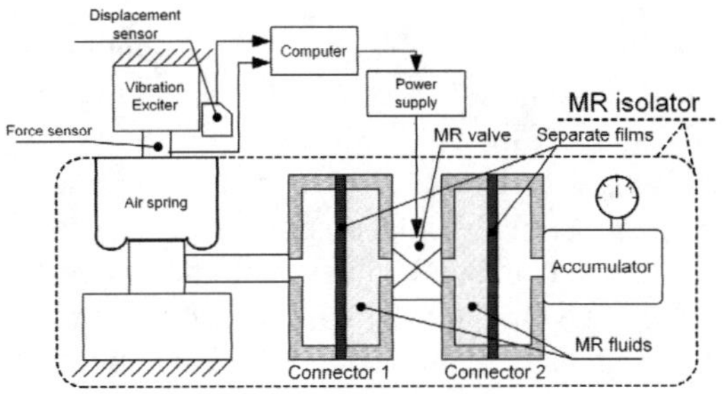

Figure 1 Schematic of the MRF isolator and Experiment Setup

2.2. Mathematical Model

Figure 2a shows the structure of the vibration isolation systems modeled by the linear springs k_1 and k_2, the damping factor of the system c_0, c_1. k_1 and k_2 are the stiffness of the accumulator and the air spring, respectively. The MR damping factor c_1 that can be changed with the magnetic field applied on the MR valve. m_2 is the sprung mass and m_1 is the mass of moving MR fluid. The displacement equation are presented by

$$\begin{aligned} m_2\ddot{x}_2 &= k_2(x_2 - x_1) + c_0(\dot{x}_2 - \dot{x}_0) \\ m_1\ddot{x}_1 &= k_2(x_2 - x_1) - k_1(x_1 - x_0) - c_1(\dot{x}_1 - \dot{x}_0) \end{aligned} \tag{1}$$

where x_0 is the input, x_1, and x_2 are displacements of m_1 and m_2. In this design, m_1 is much less than m_2, so it can be assumed as 0. c_0 is the structure damping of the system.

Figure 2b shows the simplified physical model of MR Fluid isolator. If the damping of the damper is very little, two springs k_1, k_2 can be considered as being in series. The total stiffness of the system is $k_{total} = \dfrac{k_1 \cdot k_2}{k_1 + k_2}$. When the applied current on MR valve is large enough so that the damping force generated by MR valve stop the relative motion between the first layer of the system and foundation, namely $x_0=x_1$, the mass m_2 is only isolated by k_2. The total stiffness of the isolation system is k_2. Therefore, when the input current on the MR valve is adjusted in the range between zero and maximum value, not only the damping of the system is controlled, but also the total stiffness of the system can be changed within the region between $\dfrac{k_1 \cdot k_2}{k_1 + k_2}$ and k_2 as well.

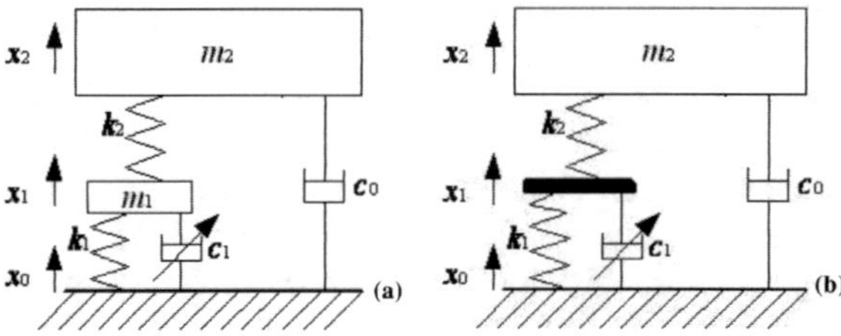

Figure 2 (a) schematic of the MR isolator model; (b) model simplification.

3. Experimental procedures

3.1. Experimental Setup

The schematic of the experimental setup is shown in figure 1. The isolator is connected with a vibration exciter (Type: JZK-5, SINOCERA PIEZOTRONICS, INC. China) which provides the vibration source for experiments. The vibration signal is generated by LabVIEW software in computer and is transmitted to a power amplifier (YE5871-100W) to amplify through the Data Acquisition (DAQ) board (Type: LabVIEW PCI-6221, National Instruments Corporation. U.S.A). After amplifying, the vibration signal is transmitted to vibration exciter. Force sensor (Type: CL-YD-302, SINOCERA PIEZOTRONICS, INC. China) monitors the force F generated by exciter. The displacement sensor (ILD1700-10) measures the displacement x_2 of the compressed surface of the air spring. The signals from force sensor and displacement sensor are amplified by Charge amplifier (YE5851) and processed and transferred by DAQ to the computer. A GW laboratory DC power supply (Type: GPR-3030D, TECPEL CO., LTD. Taiwan) can adjust the input current on the MR valve in order to control the magnetic field intensity of the isolator and change its dynamic performance.

3.2. Testing

The maximum force generated by vibration exciter is 50 N. The initial stiffness of system in absence of any current on the MR valve is 5500N/m, when air spring is in open status. Due to the high initial stiffness of the system, it is difficult to carry out the test with large displacement. The maximum displacement in our tests is 1 mm. For each test, the vibration exciter is programmed to move forward and backward in a sinusoidal wave at certain displacement amplitude and frequency. The resulting force can be measured with force sensor. The MR isolator was initially tested without any applied current. Current was then applied to the coils and the MR valve tested again. The currents varied from 0 to 0.5A, and the experimental data including time, displacement, force, were recorded. All the experiments were carried out at the room temperature of 23°.

4. Experimental and theoretical analysis result

4.1. Experimental Results

The relationships of the MR isolator response damping force versus displacement at various magnetic fields are shown in figures 3a. It can also be seen that the response force of isolator increases with the increasing magnetic field. The maximum force increases from 2.82 N (input current 0A) to 12.37 N (input current 0.4A). The increase on the response force is more the 4 times. Under the uniform displacement input, the rise of the response force indicates that the total stiffness of the isolation system increases, and the increasing extend of the total system stiffness is more 4 times as well. The hysteresis on the response force is usually caused by the increase of damping. Therefore, the experimental result indicates that the damping of system is increasing as the rise of the magnetic field. Furthermore, the response force varies very little when the input current is greater than 0.4A due to the saturating magnetization of MR fluid. Therefore, our experiment was stop when input current was increased to 0.5A. Based on experimental results, the equivalent stiffness and damping capabilities were calculate and shown in figure 3b. These results demonstrate that the developed MR isolator have the capabilities that both its stiffness and damping capacities can be varied by an external field. For the developed MR valve, the flux density is proportional to the coil current, which is given by $B_g = 0.4I \quad (I \leq 0.5A)$.

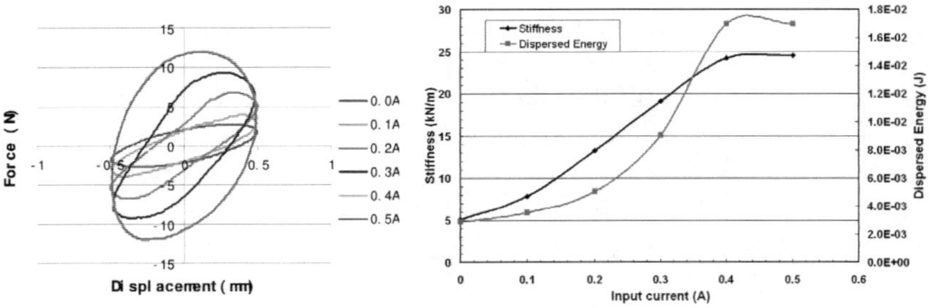

(a) (b)

Figure 3. The relationship of (a) the damping force versus displacement (b) Stiffness and Dispersed Energy at various field strengths (coil currents): $X = 0.5$ mm, $f = 0.5$ Hz, $I = 0$, 0.1, 0.2, 0.3 and 0.5A.

Neglecting the mass of moving MR fluid m_1, as shown in equation (1), the parameters of the system were identified and listed in table 1.

Table 1 Parameters of MR isolator

k_1 (kN/m)	5.5	c_1 I=0.1A (kNs/m)	1.9
k_2 (kN/m)	25	c_1 I=0.2A (kNs/m)	4.9
c_0 (kNs/m)	1	c_1 I=0.3A (kNs/m)	8.9
c_1 I=0.0A (kNs/m)	2E-6	c_1 I=0.4A (kNs/m)	14

With these parameters, a MATLAB Simulink model was built for reconstruction of the data. Fig. 4 shows the reconstructed force versus displacement at various magnetic fields. Obviously, the simulation can present the experimental results well.

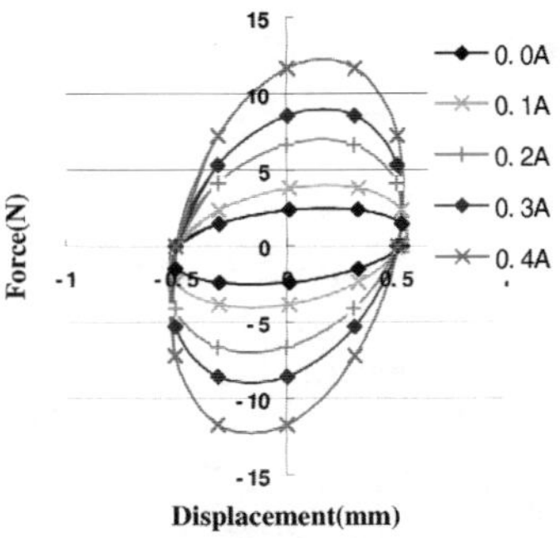

Figure 4. Reconstructed force versus displacement using the identified parameters.

4.2. Discussion

Conventional air spring often needs to control complex and nonlinear air pressure. The control strategy is difficult to be effectively implemented. The developed MRF isolator is easier to be controlled and cheaper, because it does not need air compressor and maintenance system. However, it is relatively expensive than a passive isolator. So it would be more effective to be used for some special applications, which need high suspension effect and low level of maintenance. The applications of MRF damper in buildings have been intensively investigated. Comparing with MRF damper systems, our developed system is more cost-effective because the controlled valve can be separated as a modular and connects with air spring and accumulator via pipes.

5. Conclusion

In this paper, a new semi-active vibration isolation system, which consists of an air spring, a MR valve and an accumulator, is designed and manufactured. A mathematical model of the MR isolator is built for latter theoretical analysis. To describe the practical damping force generated by the MR valve in the isolator, a numerical model for the MR valve is proposed. Then, using experimental method, the parameters of MR isolator are identified. Meanwhile, the dynamic behavior of MR isolator with different current input is measured and compared with that resulting from theoretical analysis. It was shown that the model calculation result was in sufficiently good agreement with the dynamic properties of the isolator measured from the experiment. From the experiment results, it is demonstrated that the stiffness of the MR isolator can be adjusted in a relative large scope, more then 4 times change in the stiffness, using only one MR damping controllable unit.

References

[1] S.Daley, J.Hatonen, D.H.Owens 2006 Active vibration isolation in a smart spring mount1 using a repetitive control approach, *Control Engineering Practice* **14** 991–997

[2] W.H.Zhu, B.Tryggvason, J.C.Piedboeuf 2006 On active acceleration control of vibration isolation systems, *Control Engineering Practice* **14** 863–873

[3] W.K.N.Anakwa, D.R.Thomas, S.C.Jones, J.Bush, D.Green, G.W.Anglin, R.Rio, J.Sheng, S.Garrett, L.Chen 2002 Development and Control of a Prototype Pneumatic Active Suspension System, *IEEE Transactions on Education* **45** 43-49

[4] B.F.Spencer, M.K.Sain, J.D.Carlson 1998 An experimental study of MR dampers for seismic protection, *Smart Materials & Structures* **7** 693-703

[5] X.Wang 2002 Nonlinear behavior of Magnetorheological (MR) fluids and MR dampers for vibration control of structural systems, *PhD thesis, University of Nevada*, Reno

[6] Y.T.Choi, N.M.Wereley, Y.S.Jeon 2005 Semi-Active vibration isolation using Magnetorheological isolators, *Journal of Aircraft*, **42** 1244-1251

[7] T.Kobori, M.Takahashi, T.Nasu, N.Niwa, K.Ogasawara 1993 Seismic Response Controlled Structure with Active Variable Stiffness System, *Earthquake Engineering and Structural Dynamics* **22**(11) 925-941

[8] Y.Liu, H.Matsuhisa, H.Utsuno, J.G.Park 2005 Vibration Control by a Variable Damping and Stiffness System with Magnetorheological Damper, *JSME International Journal, Series C* **48**(2) 305-310

[9] Kobori T., Takahashi M.,Nasu T., Niwa N. and Ogasawara K, "Seismic Response Controlled Structure with Active Variable Stiffness System", Earthquake Engineering and Structural Dynamics, Vol. 22 No.11 925-941 (1993).

[10] Youn I and Hac A. "Semi-Active Suspension with Adaptive Capability", Journal of Sound and Vibration, Vol.180, No.3 475-492 (1995).

[11] Liu Y., Matsuhisa H. Utsuno H. and Park J.G., "Vibration Control by a Variable Damping and Stiffness System with Magnetorheological Damper", JSME International Journal, Series C, Vol. 49, No.2 (006).

Journal of Physics: Conference Series **149** (2009) 012089 doi:10.1088/1742-6596/149/1/012089

Usage of magneto-active elastomers in a bumper of a vehicle for front impact protection

V Bogdanov[1], D Borin[2], G Stepanov[3] and A Andruszkiewicz[2,4]

[1] Technical State University «MAMI», 107023, Moscow, Russia

[2] Technische Universität Dresden, Chair of Magnetofluiddynamics, 01062, Dresden, Germany

[3] State Scientific Research Institute of Chemistry & Technology of Organoelement Compounds, 111123, Moscow, Russia

[4] Wroclaw University of Technology, Institute of Heat Technique and Fluid Mechanics, 50-370, Wroclaw, Poland

E-mail: Dmitry.Borin@tu-dresden.de

Abstract. The control of the rigidity of the bumper can be provided by an application of magneto-active elastomers (MAE), capable to change their properties in a reversible and fast manner under the action of an applied magnetic field. The dependence of the energy taken away by the bumper at the impact on variable weight and rigidity has been found taking into account the presence of controllable MAE elements in the bumper construction. The calculated fields of pressure and deformations are necessary for optimization of a design, form and material of a forward panel of a bumper.

1. Introduction

Notwithstanding the facts of continual improvement of design of vehicles, it is known that the number of automobile accidents has not decreased over the last few years. The bumper of a vehicle is a first element, which perceives the front impact in the most common cases of automobile accidents. Thus, the problem of optimization and improvement of design of a bumper with the purpose of increasing the safety level of the vehicle is an important topic. The dependences allowing to define a zone of optimum increase of mass and rigidity of a bumper have been found earlier, and the design of the bumper's protecting device has been developed [1, 2]. It has been shown that for each value of rigidity of a bumper there is a zone within the limits of which the value of absorbed energy is in direct dependence on rigidity. Thus, carrying out adjustment of rigidity of a bumper it is possible to control the absorbed energy. In this work the process of front impact of the vehicle with a motionless obstacle has been studied with the use of methods of mathematical modeling and the possibility of using of elements with magnetically controllable rigidity for the decrease in a power level of the front impact perceived by the vehicle has been investigated.

2. Mathematical model of impact

To consider dynamical processes at the impact of two objects as *n*-mass mathematical model, one has to use a system of *m* differential equations of order *k* with nonlinear right parts.

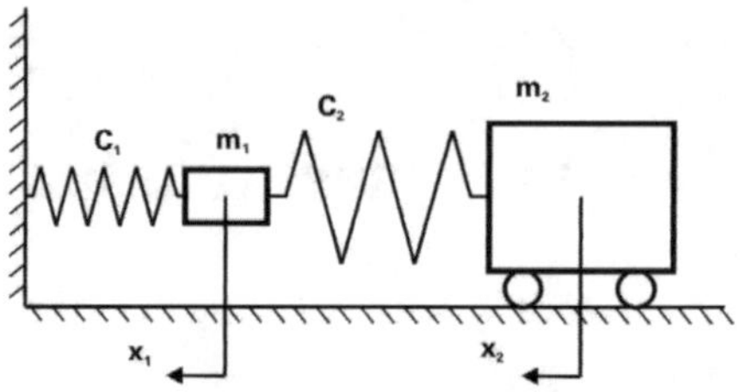

Figure1. Dynamical representing of the front impact by a set of masses and springs: m_1, m_2 – masses of the bumper with fixing elements and vehicle; c_1, c_2 – deformation elements imitating the rigidity of the bumper with longeron fixing elements and rigidity of the vehicle; x_1 , x_2 – coordinates of the centre of mass of the bumper and vehicle.

Movement of this dynamic system is considered in the inertial coordinates system related with a motionless obstacle. The well known second type Lagrange equations are used for the analyses of mathematical model of the dynamic system. The two-mass dynamic system with two degrees of freedom gives a rough but qualitative sufficient evaluation of the potential possibility to decrease the energy of the impact. Figure 1 shows the dynamical analogue of the front impact. The initial phase would be enough to concern the process of a real front impact, since the bumper doesn't play an important role in the process of elastic recovery. In this case the kinetic T and potential energy Π are:

$$\begin{cases} T = \frac{1}{2}(m_1\dot{x}_1^2 + m_2\dot{x}_2^2) \\ \Pi = \frac{1}{2}(c_2(x_2 - x_1)^2 + c_1 x_1^2) \end{cases} \tag{1}$$

Differentiating and considering the Lagrange equation one obtains:

$$\begin{cases} \mathrm{x}_1 = \frac{v_0(\lambda'' - 1)}{p_1(\lambda'' - \lambda')}\sin(\mathrm{p}_1\ \mathrm{t}) + \frac{v_0(1 - \lambda')}{p_2(\lambda'' - \lambda')}\sin(\mathrm{p}_2\ \mathrm{t}) \\ \mathrm{x}_2 = \frac{\lambda' v_0(\lambda'' - 1)}{p_1(\lambda'' - \lambda')}\sin(\mathrm{p}_1\ \mathrm{t}) + \frac{\lambda'' v_0(1 - \lambda')}{p_2(\lambda'' - \lambda')}\sin(\mathrm{p}_2\ \mathrm{t}) \end{cases} \tag{2}$$

It is assumed that the deformation of the bumper take place within 10 ms after impact. For weight and rigidity initial data has been set and the energy absorbed by the bumper E_1 as well as by the vehicle itself E_2 are calculated:

$$E_1 = \frac{c_1 x_1^2}{2} \tag{3}$$

$$E_2 = \frac{c_2(x_2 - x_1)^2}{2} \tag{4}$$

Journal of Physics: Conference Series **149** (2009) 012089 doi:10.1088/1742-6596/149/1/012089

The calculated value of the energy has been compared with the energy absorbed by a vehicle without active protection.

3. Magneto-active elastomers as active elements in construction of a bumper

The control of the rigidity of the bumper can be provided by an application of magneto-active elastomers (MAEs), capable to change their properties in a reversible and fast manner under the action of an applied magnetic field [3-8]. The most important property of a MAE, from the point of view of the considered problem, is the magnetorheological effect [9], i.e. increase of elasticity under the action of an external magnetic field. There are known works on MRE used in some automotive applications, e.g. as suspension bushings etc [3, 10, 11]. In this study elements of a typical anisotropic MAE [12,13] were supplemented in bumper construction to provide a semi-active control of its rigidity.

4. Simulation

Varying weight and rigidity of the bumper and taking into account a controllable MAE based element, it is possible to find dependence of the energy absorbed by the bumper on these parameters and to determine the resulting influence (figure 2). It is assumed in calculations that the deformations of the bumper take place within 10 ms after impact.

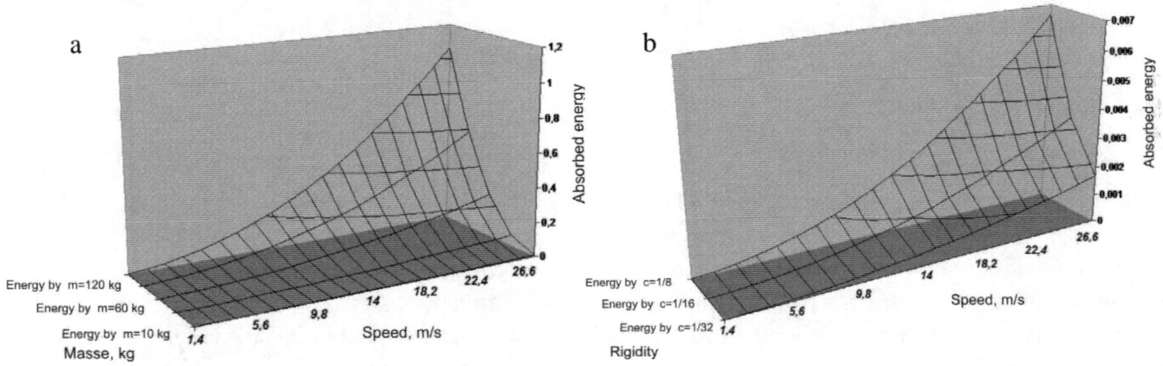

Figure 2. Energy absorbed by the bumper: a) as function of mass and speed of impact at fixed values of the rigidity (1/8), b) as function of rigidity and speed of impact at fixed values of the mass (10 kg).

For each value of rigidity of a bumper (including longeron fixing elements) there is a zone within the limits of which the value of absorbed energy is in direct dependence on rigidity. Out of this zone there is an inverse dependence. Hence one must take into account that changes of rigidity of a bumper can have a negative result, namely, decrease of absorbed energy.

For the calculations of pressure and deformation at a frontal impact a bumpers prototype of a middle class vehicle has been considered. The FEM mesh of the bumper is shown in figure 3.

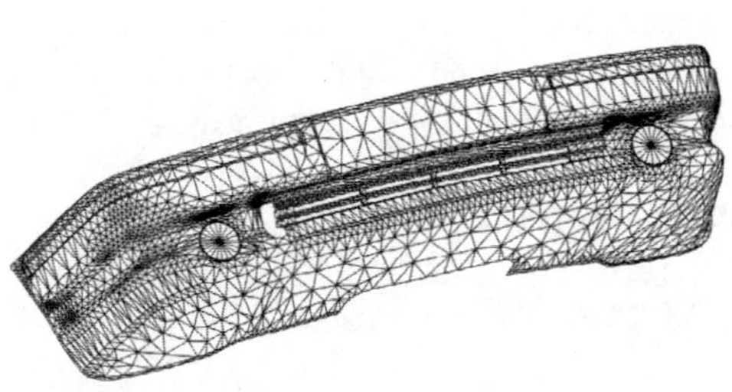

Figure 3. FEM mesh of the bumper

Journal of Physics: Conference Series **149** (2009) 012089 doi:10.1088/1742-6596/149/1/012089

Calculated by means of ANSYS Mechanical software distribution of pressure and deformations (figures 4-5) shows that the load ability of a bumper is defined by the loaded zones which area is in range of 10 – 20 % of the area of a surface of a bumper. The 20-40 % of the area of a surface of a bumper is poorly loaded zones and 40 – 60 % doesn't have considerable pressure. As well as in a case of distribution of deformations, total zones of a field of distribution of pressure will be symmetric concerning vertical axial a bumper facade. Other part of a bumper is under action of a qualitatively lower pressure or even is not loaded, and these zones are supplied with MAEs with adjustable rigidity.

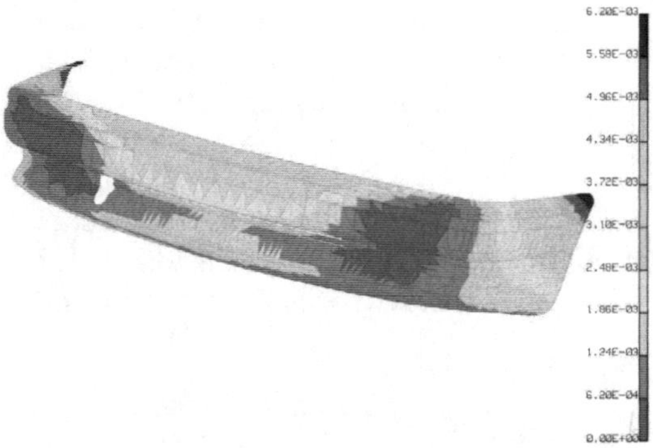

Figure 4. Distribution of deformations

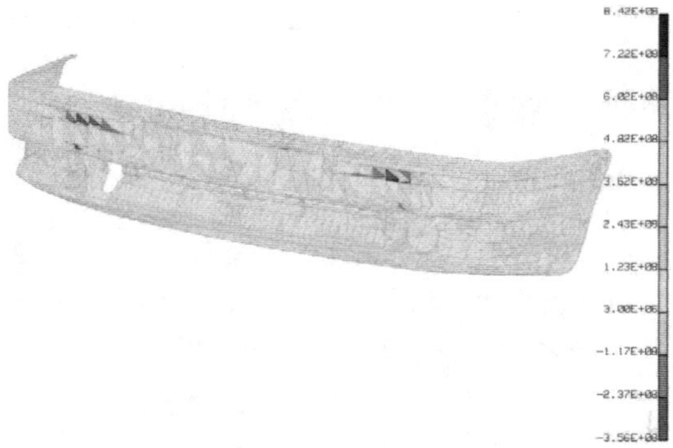

Figure 5. Distribution of pressure

5. Outlooks

Using of a MAE in the bumpers construction is capable not only accumulate and absorb impact influence, but also distribute it on a surface, reduce the effect and consequences of a shock. The calculated fields of pressure and deformations allow to optimize a design and to choose a correct form and material for the forward panel of a bumper. Besides there is a possibility to use a magnetostriction effect [14] as well as memory shape effect of MAE [12, 13] to develop an active surface of a bumper, which can be restored after impact (impacts at slow velocities 1-10 km/h).

References

[1] Bogdanov V, Balabin I and Gladskih A 2003 *Automobile industry* 4 33 (in russian)

[2] Balabin I and Bogdanov V 1999 *Patent of Russian Federation*, 21 2133679

[3] Stewart W, Ginder M, Nichols M and Elie L 1998 US Patent 5816587

[4] Jolly M R, Carlson J D and Munoz B C (1996) *J. Int. Mat. Sys. Struct.* 7 613

[5] Szabo D, Szeghy G and Zrinyi M 1998 Macromolecules 31 6541

[6] Carlson J D and Jolly M R 2000 Mechatronics, 10 (4) 555

[7] Bossis G, Abbo C, Cutillas S, Lacis S and Metayer C 2001 Int. J. Mod. Phys. B 15 564

[8] Zhou G Y and Yiang Z Y 2004 Smart Mater. Stract. 13 309

[9] Nikitin L, Stepanov G and Mironova L (2004) *Polymer Science Ser. A* 46 489

[10] Watson J R 1997 US Patent 5609353

[11] Ginder M, Nichols M, Elie L and Tardiff J (1999) Proc. of SPIE , 3675 131

[12] Nikitin L V, Stepanov G V, Mironova L S and Gorbunov A I 2003 J. Magn. Magn. Mater. 272-276 (III) 2072

[13] Stepanov G V, Abramchuk S S, Grishin D A, Nikitin L V, Kramarenko E Yu and Khokhlov A R 2007 Polymer 48 488

[14] Guana X, Donga X and Ou J (2008) *J. Magn. Magn. Mat.* 320 (3-4) 158

Journal of Physics: Conference Series **149** (2009) 012090 doi:10.1088/1742-6596/149/1/012090

Magnetorheological Elastomers with High Variability of Their Mechanical Properties

H. Böse and R. Röder

Fraunhofer-Institut für Silicatforschung ISC, Neunerplatz 2, D-97082 Würzburg, Germany

E-mail: boese@isc.fraunhofer.de

Abstract. The viscoelastic properties of magnetorheological elastomer (MRE) composites containing different concentrations of iron particles in a silicone matrix and the change of these properties in a magnetic field of variable strength were investigated. MRE samples with small and large iron particles as well as isotropic and anisotropic particle arrangements in the silicone matrix were compared. The results show a strong increase of the storage and loss moduli of the MRE in the magnetic field. A more pronounced influence of the anisotropic configuration was observed for the samples with small particles than for those with large particles. The relative increase factor of the storage modulus in the magnetic field was found to be an order of magnitude higher for the composites with large than with small particles. As a maximum, an enhancement of the storage modulus from less than 30 kPa at zero field to about 2500 kPa at a magnetic flux density of 700 mT, corresponding to an increase factor of 90, could be achieved under the selected experimental conditions.

1. Introduction

Magnetorheological elastomers (MRE) are a relatively new class of smart materials whose mechanical properties can be influenced in a magnetic field. The magnetorheological effect of MRE manifests itself in a considerable increase in the Young's modulus and the shear modulus, which occurs at rising magnetic field strength. In analogy to the shear stress of MR fluids, this effect is reversible.

MRE are soft composite materials which basically consist of magnetically polarizable particles like iron in a deformable polymer matrix. In the magnetic field, the mechanical behavior of the composite is dominated by a superposition of magnetic particles interactions and elastic forces of the elastomer matrix, which causes changes in the elastic as well as in the viscous properties of the MRE.

Strong commercial interest arises from these fascinating properties. The controllable stiffness of the material can possibly be exploited in dynamic vibration absorbers whose frequency of operation may be tuned by the magnetic field strength. Further possible applications of MRE are adaptive vibration dampers, controllable mounts, switchable haptic surfaces as well as controllable seals and valves.

Several papers regarding MRE were published since the 1990s [1-6]. Up to now, most studies were performed on MRE with silicone rubber as the matrix material. Silicone has the advantage that the synthesis starts with liquid precursors, in which the particles can be suspended. Moreover, it could be demonstrated that the mechanical properties of the MRE composite may be influenced by the application of another magnetic field applied already during the curing of the silicone. This effect is ascribed to the formation of particle chains which are "frozen" in the solid state [5].

This paper addresses experimental investigations of the influence of various parameters of the composition of MRE on their mechanical properties and on their capability of the viscoeleastic properties to be tuned in the magnetic field. It was evaluated under which conditions the strongest magnetorheological effects occur. In the investigations silicone rubber was used as matrix material, due to the possibility to prepare very soft composite materials.

2. Experimental

The silicone matrix consisted of polydimethylsiloxane prepared from the reaction of vinyl-terminated polysiloxane chains and polysiloxane with Si-H groups. The basic crosslinking density of the reference elastomer without iron particles was set thus, that an elastic storage modulus of about 50 kPa was achieved. In order to evaluate the influence of the iron particle quantity on the mechanical properties, various silicone-based composites with iron concentrations by volume between 0 and 35 % were prepared. For a part of the samples the silicone matrix was cured only by heat treatment. Another part of the samples was additionally exposed to a magnetic field during the crosslinking process in order to achieve an anisotropic arrangement of the iron particles. Furthermore, two kinds of iron particles with strongly different mean particle sizes of about 5 μm (BASF, Germany) and ca. 40 μm (Höganäs, Sweden), respectively, were used for the preparation of the MRE composites.

The samples were studied by rheological measurements in the oscillation mode in a rheometer Paar-Physica MCR 300 equipped with a magnetorheological measuring unit. For the evaluation of the influence of the magnetic field strength on the viscoelastic properties, the experiments were carried out at a fixed frequency of the oscillation of 10 Hz, an amplitude of 1 % and under the variation of the magnetic flux density.

3. Results and Discussion

3.1. MRE with small iron particles

The measurements yielded the storage modulus which describes the elastic behavior of the sample, as well as the loss modulus which is a measure of the viscous behavior. Both moduli are strongly increased when the magnetic flux density is enhanced. As expected, the base value of the storage modulus without magnetic field rises with increasing concentration of iron particles, indicating the enlarged hardness of the composites with higher iron particle concentrations.

Figure 1 shows the increase of the storage and loss moduli of MRE with various contents of small iron particles in the isotropic distribution with rising magnetic flux density. The higher the iron concentration the larger is the absolute increase of both moduli in the magnetic field, shown here in a logarithmic scale. It becomes apparent, that under all conditions the elastic properties dominate the viscous behavior of the composites.

In Figure 2 the dependence of the storage modulus of the corresponding samples with the same compositions, but cured in the presence of a magnetic field, on the magnetic flux density is revealed. The absolute increase of the storage modulus due to the magnetic field is significantly higher for the anisotropic samples (from 320 kPa at zero field to 2800 kPa at 700 mT) than for the isotropic ones (from 150 kPa at zero field to 700 kPa at 700 mT). This is even true for the relative increase factor, defined as the ratio of the storage modulus at the highest flux density of ca. 700 mT and the storage modulus at zero field, which is about 8 (see Fig. 2) for the anisotropic and less than 5 (not shown, but derivable from Fig. 1) for the isotropic samples as the maximum values. It can be concluded that the magnetic interactions between the iron particles are stronger in the anisotropic than in the isotropic arrangement.

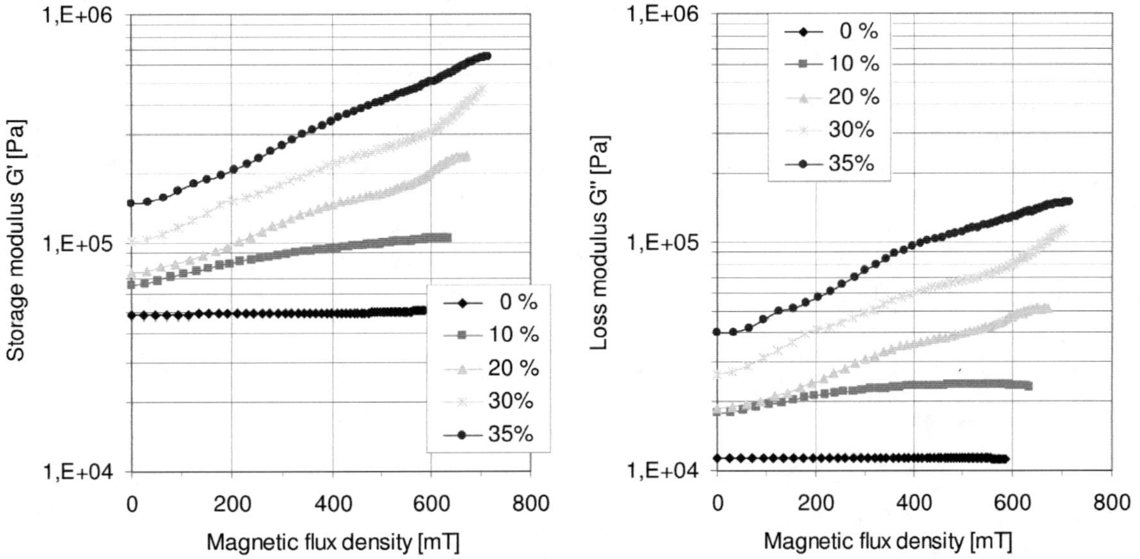

Figure 1. Dependence of storage modulus (left) and loss modulus (right) of isotropic MRE with different concentrations of small iron particles (in vol.%) in silicone on the magnetic flux density

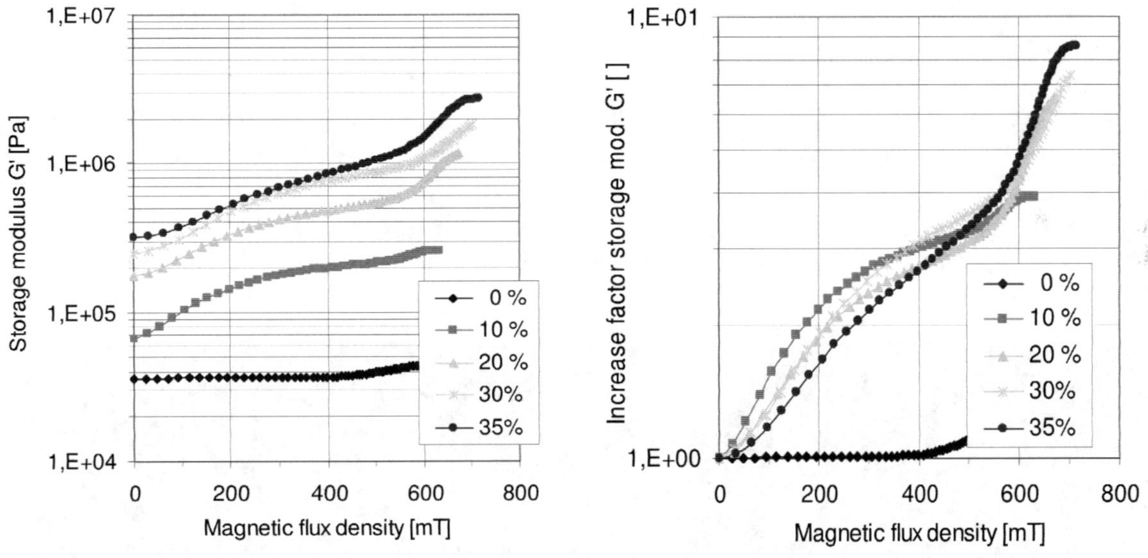

Figure 2. Dependence of storage modulus (left) and its relative enhancement (right) of anisotropic MRE with different contents of small iron particles (in vol.%) in silicone on the magnetic flux density

3.2. MRE with large iron particles

The corresponding results of the MRE samples containing larger iron particles with a mean size of about 40 µm in an isotropic arrangement are depicted in Figure 3. The absolute increase in the storage as well as the loss modulus by the magnetic field is drastically higher than for the composites with small particles. MRE containing 30 or 35 vol.% large iron particles reach storage moduli of about 3

MPa under the chosen experimental conditions at a magnetic flux density of ca. 700 mT. Without the magnetic field, the same samples have storage moduli below 100 kPa, indicating drastic increase factors of the storage modulus due to the magnetic field. Furthermore, in the field-free state very low storage moduli of the samples with small iron contents were observed, which are even lower than the storage modulus of the pure silicone sample without iron. Apparently, the crosslinking of the silicone is affected by the presence of the large iron particles, which results in a reduced elasticity of the composites.

Figure 4 shows scanning electron microscopic photographs of isotropic and anisotropic composites. In contrast to the isotropic material, the formation of fragments of iron particles chains, caused by the application of a magnetic field during the curing process, can be observed in the anisotropic sample.

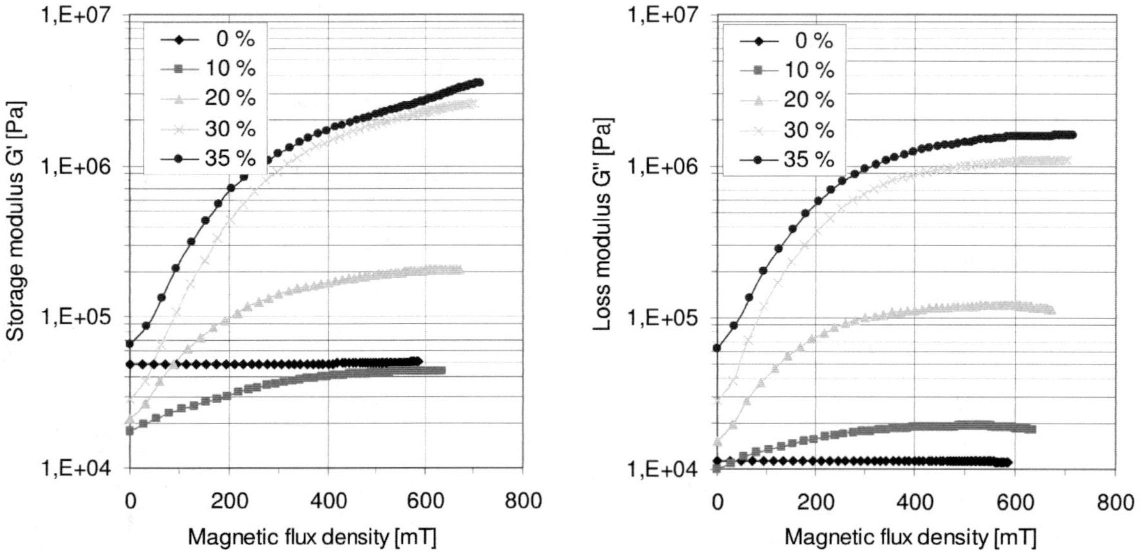

Figure 3. Dependence of storage modulus (left) and loss modulus (right) of isotropic MRE with different concentrations of large iron particles (in vol.%) in silicone on the magnetic flux density

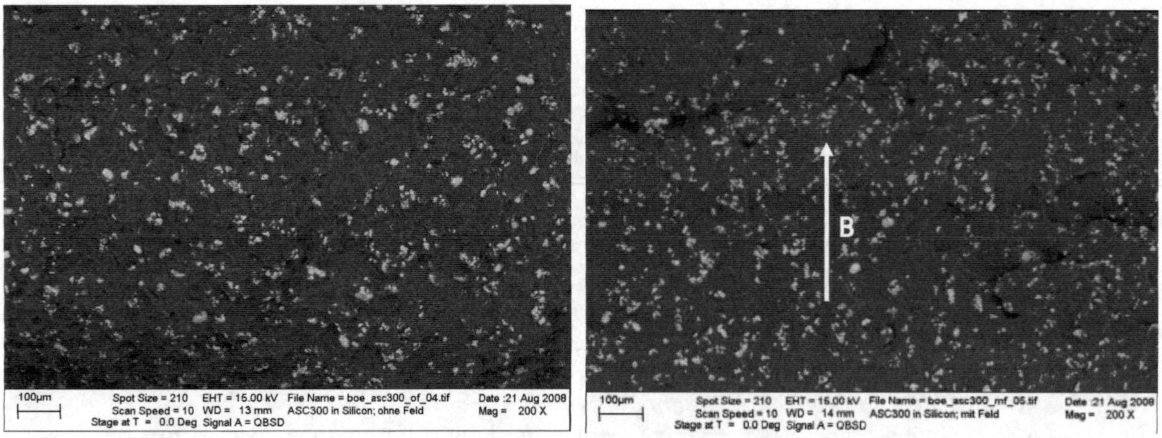

Figure 4. Scanning electron microscopic (SEM) photographs of MRE samples with isotropic (left) and anisotropic arrangement (right, magnetic field direction indicated by the arrow) of large iron particles

Finally, Figure 5 shows the dependence of the storage modulus and its relative increase of the anisotropic composites with large iron particles on the magnetic flux density. In contrast to the samples with small particles, the difference between the maximum storage moduli of isotropic and anisotropic composites is not very pronounced, indicating magnetic interactions of comparable strengths between the iron particles in both configurations. The storage modulus of the isotropic MR elastomer with 30 vol.% iron particles increases from less than 30 kPa at zero field to about 2500 kPa at 700 mT (see Fig. 3). In comparison, for the corresponding anisotropic sample, the storage modulus rises from ca. 50 kPa at zero field to 3000 kPa at 700 mT (see Fig. 4).

The relative increase factor of the anisotropic composite in Figure 5 reaches a maximum of ca. 65. In case of the corresponding isotropic composite, the increase factor reaches even a maximum of about 90 (not shown, but derivable from Fig. 3), due to the very low starting levels of the storage modulus at zero-field. Thus, the maximum increase factors of the MRE composites with large iron particles surmount those of the samples with small particles by an order of magnitude. To the knowledge of the authors, such large increase factors of MRE have not been reported before.

The observed difference between the properties of the isotropic composites with small and large iron particles is referred to the capability of the large particles in the MR elastomer to generate stronger magnetic interactions than the small particles. However, anisotropic MR elastomers with small particles can also generate large storage moduli of nearly 3000 kPa in the magnetic field, but with significantly higher base modulus at zero field than for large particles (see Fig. 2). This behaviour is explained by the chain structure of the iron particles. Due to this ordering, the particles have very low distances from their next neighbours. These low distances enable the small particles to generate strong interparticle interactions in the magnetic field. Furthermore, the larger increase factors of the MRE with large particles may also be caused by the softening influence of the large particles which reduce the hardness of the silicone elastomer.

Another result of the investigations on the MRE with large particles is that the composites with 30 vol.% iron particles yield the highest relative increase of the storage modulus (see Figure 5). This is in agreement with former research, in which an optimum iron concentration in the vicinity of 30 vol.% was reported [4].

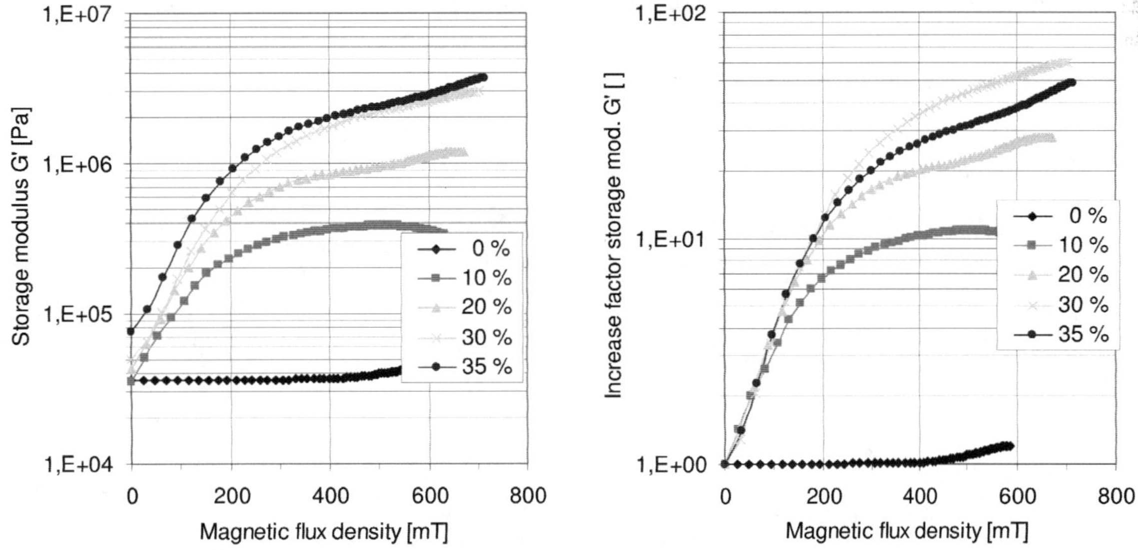

Figure 5. Dependence of storage modulus (left) and its relative enhancement (right) of anisotropic MRE with different contents of large iron particles (in vol.%) in silicone on the magnetic flux density

4. Conclusions

The investigations on various MRE composites containing different concentrations of iron particles in silicone yielded a strong variability of the viscoelastic properties and their change in a magnetic field, caused by the modification of the composition of the MR elastomer in terms of iron particle size and concentration. The enhancement of the storage modulus and the loss modulus is more pronounced for the samples with larger amounts of iron particles than for those with small amounts. Furthermore, composites with large iron particles yielded relative increases of the storage modulus, which are an order of magnitude above those of the corresponding composites containing small iron particles. A maximum factor of increase of the storage modulus of about 90 in the magnetic field could be achieved under the selected experimental conditions, where the absolute storage modulus rises from less than 30 kPa at zero field to about 2500 kPa at 700 mT. These results reveal promising perspectives of this new class of smart materials for various applications like adaptive vibration dampers, switchable haptic surfaces and controllable seals and valves.

References

[1] M. R. Jolly, J. D. Carlson, B. C. Munoz, and T. A. Bullions: The magnetoviscoelastic response of elastomer composites consisting of ferrous particles embedded in a polymer matrix, J. Intelligent Mat. Syst. Struct. 7 (1996) 613-622

[2] L.C.Davis: Model of magnetorheological elastomers, J.Appl.Phys. 85 (1999) 3348-3351

[3] C.Bellan, G.Bossis: Field dependence of viscoelastic properties of MR elastomers, Int. J. Mod. Physics B 16 (2002) 2447-2454

[4] M. Lokander and B. Stenberg: Performance of isotropic magnetorheological rubber materials, Polymer Testing 22 (2003) 245–251

[5] H. Böse: Viscoelastic properties of silicone-based magnetorheological elastomers, Proc. 10th Int. Conf. on Electrorheological Fluids and Magnetorheological Suspensions (F. Gordaninejad, O. A. Graeve, A. Fuchs and D. York, Eds.), World Scientific, Singapore (2007) 51-57

[6] X. L. Gong, L. Chen, J. F. Li: Study of utilizable magnetorheological elastomers, Proc. 10th Int. Conf. on Electrorheological Fluids and Magnetorheological Suspensions (F. Gordaninejad, O. A. Graeve, A. Fuchs and D. York, Eds.), World Scientific, Singapore (2007) 202-209

Journal of Physics: Conference Series **149** (2009) 012091 doi:10.1088/1742-6596/149/1/012091

Isolation and damping properties of magnetorheologic elastomers

C Collette[1], G Kroll[1], G Saive[2], V Guillemier[3], M Avraam[1] and A Preumont[1]

[1] *University of Brussels, 50 av. F.D. Roosevelt, 1050 Brussels, BELGIUM*
[2] *Techspace Aero (SAFRAN Group), 121, route de Liers, 4041 Herstal, BELGIUM*
[3] *MATIS Benelux, 121, route de Liers, 4041 Herstal, BELGIUM*

E-mail: christophe.collette@ulb.ac.be

Abstract. This paper considers two systems based on a magnetorheological elastomer (MRE): a MRE isolator under a frequency varying harmonic excitation and a MRE Dynamic Vibration Absorber (DVA) mounted on a frequency-varying structure under a random excitation. It is shown that the commandability of the elastomer improves the isolation performances in the first case, and decreases the stress level in the structure in the second case.

1. Introduction

Consider a structure fixed on a support through a set of mounts. Often, the mechanical properties of a structure are evolving in time (e.g. changes in mass, temperature, load cycle). If the stiffness and the damping properties of the mounts remain constant, their isolation performances is not always optimal. A technique whose spring constant and damping factor can be changed represents a potentially considerable improvement of the vibration. Along with piezoelectric ceramics and shape memory alloys, the magnetorheological elastomers (MRE) have those capabilities [1, 2].

Measured characteristics from published studies [3] are used to evaluate their isolation and damping performances in two cases. The first is a tunable isolator supporting a structure and subjected to a narrow band excitation. The problem is addressed in section 3, with a single degree of freedom (d.o.f.) MRE isolator. The second is a tunable DVA mounted on a structure whose mass is varying, and that is subjected to a broad band random excitation. The problem is addressed in section 4, with a MRE DVA appended on a flexible structure. In the numerical example, a n-storey like structure has been considered for its simplicity. However, the procedure is applicable to any type of flexible structure. Different kinds of Adaptive DVAs have also been developed ([4, 5] & [6, 7]), but the effect of the MRE DVA on the stress reduction in the structure has not yet been fully addressed. Next section presents the theoretical model of a magnetorheological elastomer, along with its potential applications.

2. Model of a magnetorheological elastomer

Using a simple Kelvin-Voigt model (spring and damper in parallel), the MRE can be characterized by a constant stiffness k and damping coefficient ξ. Experiments [3] have shown that, until a saturation value of the applied magnetic field B_s, the values of k and c increase linearly according to

$$k(B) = a_k + b_k B \quad ; \quad \xi(B) = a_\xi + b_\xi B \tag{1}$$

where B is the magnetic field and a_k, b_k, a_ξ, b_ξ are constant quantities. Note that only the shear mode is considered in this paper, as experimental studies have shown that, in this direction, the relative increase in stiffness is the highest [3]. The following parameters have been used in this study: $a_k = 2000\ N/m$; $b_k = 4.55\ kNT^{-1}m^{-1}$; $a_\xi = 0.15$; $b_\xi = 0.6364\ T^{-1}$; $B_S = 0.11\ T$.

3. MRE isolator

The model used in this section is a single d.o.f. isolator (Fig. 1), subjected to a narrow band harmonic excitation. It consists of a spring with a MRE at one of its end, like in some advanced car suspension systems [8, 9]. The dynamics of the system read

$$m_s\ddot{x}_s + c(\dot{x}_s - \dot{x}_1) + k(x_s - x_1) = 0 \quad ; \quad c(\dot{x}_s - \dot{x}_1) + k(x_s - x_1) = k_1(x_1 - x_0) \tag{2}$$

where m_s is the mass of the oscillator, k_1 is the stiffness of the isolator spring, k and $c = 2\xi\sqrt{km_s}$ the equivalent stiffness and damping coefficients of the elastomer given by Equ.(1). The numerical values used in this study are $m_s = 100\ kg$ and $k_1 = 100\ kN/m$.

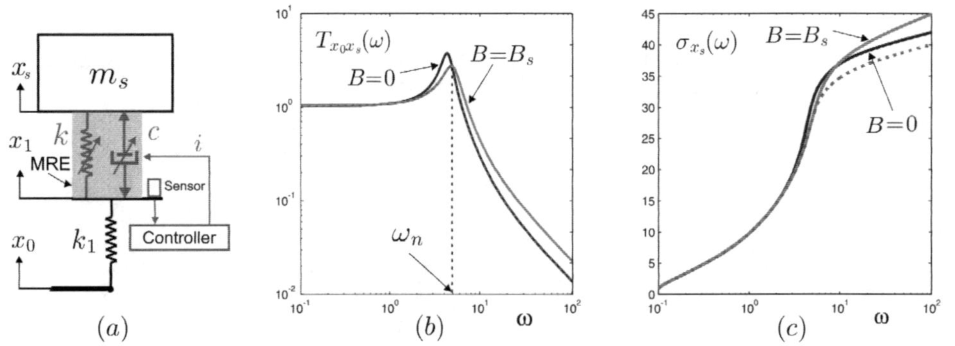

Figure 1. (a) Model of MRE isolator; (b) Transmissibility $T_{x_0x_s}(\omega)$ and (c) RMS value $\sigma_{x_s}(\omega)$ of a MRE isolator for two values of B: $B = 0$ (blue curve) and $B = B_s$ (green curve).

Consider a periodic chirp excitation with a unit amplitude applied in x_0. Figure 1(b) shows the absolute value of the transmissibility $T_{x_0x_s}(\omega)$ of the isolator for two extremal values of the magnetic field : $B = 0$ (blue curve) and $B = B_s$ (green curve). Figure 1(c) shows the cumulative Root Mean Square (RMS) integral of the transmissibilities defined as

$$\sigma_{x_s}(\omega) = \left[\int_0^\omega |T_{x_0x_s}(\nu)|^2 d\nu\right]^{1/2} \tag{3}$$

which describes how the various frequencies contribute to the RMS of the mass displacement. When the frequency of the excitation is lower than ω_n, the transmissibility obtained for $B = B_s$ is the lowest; above this value, $B = 0$ gives the lowest transmissibility. This suggests the following control strategy:

$$\begin{array}{llll} \text{IF} & \omega \le \omega_n & \text{THEN} & B = B_s \\ \text{IF} & \omega > \omega_n & \text{THEN} & B = 0 \end{array}$$

The control strategy is implemented and shown as a dotted red line in Fig. 1. Figure 1(c) shows that the frequency shift can reduce the RMS value of the mass displacement by 10%.

4. MRE Dynamic Vibration Absorber

Consider a n-storey frame subjected to a horizontal base excitation of acceleration $\ddot{x}_0$. The variables of the system are the lateral displacements of each stage $x = (x_1, ..., x_n)^T$. The mass and stiffness matrices are respectively $M = mI_n$ and

$$K = k_S \begin{pmatrix} 2 & -1 & & & \\ -1 & 2 & & & \\ & & \ddots & & \\ & & & 2 & -1 \\ & & & -1 & 1 \end{pmatrix} \tag{4}$$

and I_n is the identity matrix of dimension n.

Consider a DVA attached on floor d and tuned to mode k. According to the equal peak design procedure [10], it can be shown that the optimal values of the stiffness and damping coefficient of the DVA are given by

$$k_a = m_a \left[\omega_k/(1+\lambda)\right]^2 \quad ; \quad \xi_a = \sqrt{3\lambda/[8(1+\lambda)]} \tag{5}$$

where $\lambda = \frac{m_a}{\mu_k}\phi_k^2(d)$, μ_k and ϕ_k respectively the mass and shape of mode k. In such a structure subjected to a seismic excitation, a relevant quantity is the shear force at the base, given by $F_0 = k_s(x_1 - x_0)$. In the numerical example, $m = 10\ kg$, $k_s = 750\ kN/m$, $d = 7$, $m_a = 1\%$ of the total mass ($m_T = 7m$ in this case) and $k = 1$. Using Equ.(5), the optimal values of the stiffness and damping coefficient of the DVA are $k_a = 2620\ N/m$ and $\xi_a = 0.09$. Unfortunately, for an actual material, it is not possible to choose independently the stiffness and damping coefficient. Choosing a stiffness of $k_a = 2620\ N/m$ gives automatically a damping of $\xi_a = 0.23$ (data from [3]). As this value is greater than the optimal value, the DVA is over-damped, providing a lower efficiency than the optimal case. Now consider that the mass of each storey of the structure increases from $m = 10\ kg$ to $m = 13.5\ kg$. Three cases are compared: (i) a DVA with constant properties, optimal for $m = 10\ kg$; (ii) a fictive DVA that would have optimal values for both the stiffness and damping coefficient; (iii) a MRE DVA. Figure 2 (a) to (d) compare respectively the stiffness and damping of the DVA, the reduction ($\Delta\sigma$) of the RMS value of the shear force at the base of the frame

$$\sigma_{F_0} = \left[\int_0^\infty |T_{x_0F_0}(\omega)|^2 d\omega\right]^{1/2} \tag{6}$$

and the reduction (ΔT) of the maximum value of the transmissibility $|T_{x_0F_0}|$ calculated in dB.

As the mass of the structure increases, the DVA with constant properties is not tuned to the first resonance anymore, and ΔT and $\Delta\sigma$ decrease rapidly. The fictive DVA maintains $\Delta\sigma$ above 45% and $\Delta T = 15\ dB$. The MRE DVA, while not optimal for a structure with constant properties, maintains $\Delta\sigma$ at 41 % and $\Delta T = 11\ dB$ throughout the range. Similar results are presented in Fig. 2(a' to d') when the mass of each storey decreases from $m = 13.5\ kg$ to $m = 10\ kg$.

5. Conclusion

In this paper, the isolation and damping properties of the magnetorheological elastomer have been investigated for two cases. Firstly, a single d.o.f. isolator, was subjected to a narrow band excitation. It has been shown that the MRE isolator can decrease the RMS value of the body displacement by 10 %. Secondly, a n-storey mass-varying structure has been considered, on which is appended a MRE DVA. A proper application of the magnetic field to the MRE can maintain a good efficiency compared to a classical DVA. The adaptability is particularly suited for structures whose mass and/or stiffness is varying in a fixed range of values.

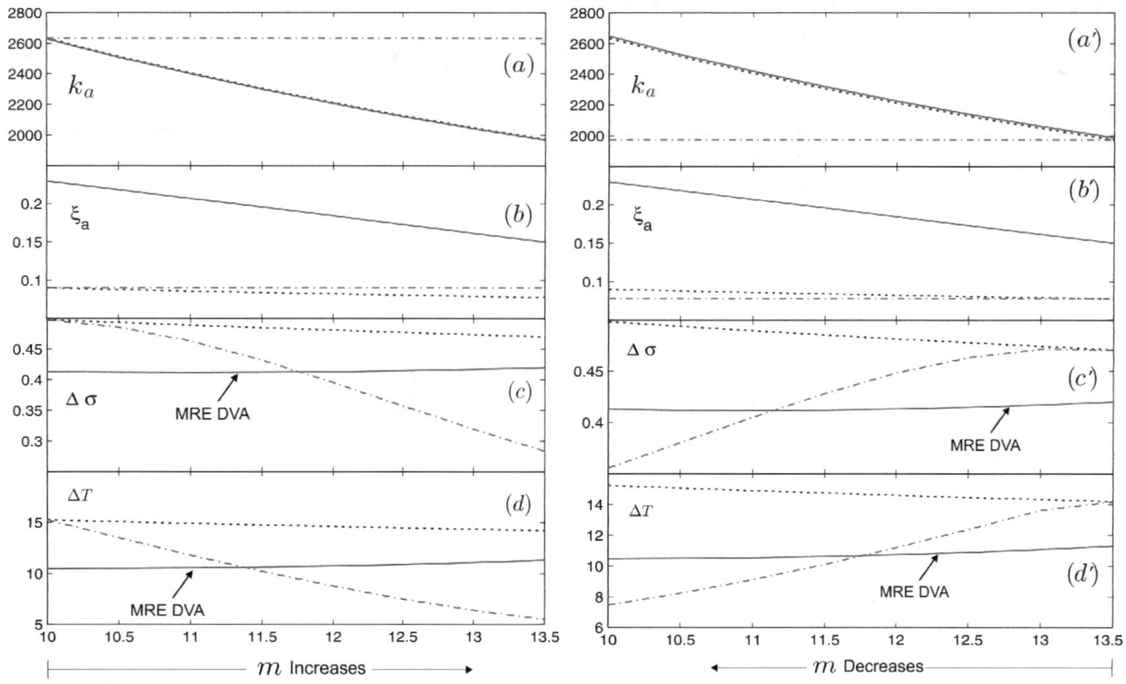

Figure 2. Comparison between three DVAs appended on a mass-varying structure: DVA with constant properties (dashed dotted line), DVA with optimally tunable properties (dotted line), and MRE DVA (solid line). (a, a') Stiffness of the DVA; (b, b') Damping coefficient of the DVA; (c, c') Reduction of the RMS value of the shear force at the base of the structure; (d, d') Reduction of the maximum amplitude of the shear force to excitation transmissibility $|T_{x_0F_0}|$.

6. Acknowledgment

The authors acknowledge the Walloon Government for their financial support (FIRST N 716624). Techspace Aero is also warmly acknowledged for supporting and funding this research.

References

[1] J.D. Carlson and M.R. Jolly. MR fluid, foam and elastomer devices. *Mechatronics*, 10:555–569, 2000.

[2] C. Bellan and G. Bossis. Field dependence of viscoelastic properties of mr elastomer. *International Journal of Modern Physics B*, 16:24472454, 2002.

[3] M. Kallio. *The Elastic and Damping Properties of Magnetorheological Elastomers.* PhD thesis, 2005.

[4] H-x Deng and X-l Gong. Application of magnetorheological elastomer to vibration absorber. *Communications in Nonlinear Science and Numerical Simulation*, 2007.

[5] H-x Deng, L Chen, X-l Gong, W-q Jiang, J-j Yao, and W-h Li. Investigation on magnetorheological elastomers based on natural rubber. *Journal of Material Science*, 42:5483–5489, 2007.

[6] A-M. Lerner and K.A. Cunefare. Adaptable vibration absorber employing a magnetorheological elastomer with variable gap length and methods and systems therefor. *US Patent*, page 2005/0040922, 2005.

[7] A-M. Lerner. The design and implementation of a magnetorheological silicone composite state-switched absorber. Master's thesis, School of Mechanical Engineering, Georgia Institute of Technology, 2005.

[8] J.R. Watson. Method and apparatus for varying the stiffness of a suspension bushing. *US Patent*, page 5609353, 1997.

[9] W.M. Stewart, J.M. Ginder, L.D. Elie, and M.E. Nichols. Method and apparatus for reducing brake shudder. *US Patent*, page 5816587, 1998.

[10] J-P. Den Hartog and J. Ormondroyd. The theory of the damped vibration absorber. *Trans ASME - Journal of Applied Physics*, 50:7, 1928.

Journal of Physics: Conference Series **149** (2009) 012092 doi:10.1088/1742-6596/149/1/012092

Magnetorheological composites as semi-active elements of dampers

Jerzy Kaleta, Daniel Lewandowski, Piotr Zając, Paweł Kustroń

Institute of Materials Science and Applied Mechanics, Faculty of Mechanical Engineering, Wroclaw University of Technology, ul. Mariana Smoluchowskiego 25, 50-370, Wrocław, Poland

E-mail: jerzy.kaleta@pwr.wroc.pl

Abstract. An original magnetorheological composite (MRC) with porous elastomeric matrix and filled with magnetorheological fluid was created at the work. It was used later on to build a damper working in the shearing mode without friction against external surfaces of the so-called skid. This prototype construction was used for damping free vibrations in the beam. An analysis of the effectiveness in the magnetic field function steering the damper was performed. As a result an important relationship between the change of damping in the material under the influence of the magnetic field and the length of time needed for damping the vibrations in the beam was demonstrated.

1. Introduction

At the work characteristics of magnetorheological composites (MRC) were described and the use of them for damping vibration was shown on the simple example. Prototype construction of the damper using material of this type and results of damping vibration of the beam were presented. Described MRC composites appeared as the succession of magnetorheological fluid (MRF). They are built of porous matrix and fluid filling cells up for it. Receiving material with properties similar to the fluid is a result, thanks to which it is possible to shape freely external geometrical dimensions. Most important advantage of this material is capabilities to absorbing energy [1]. In this way the material manufactured enabled us to make the friction damper [2]. The principle of its work consisted in generating the changeable force of friction, dependent on the value of the applied magnetic field, appearing between the moving surface of composite and the casing of the damper.

2. Material test and selection of work mode

Results of examinations met in literature, focused on attempts to use dampers with magnetorheological composites, without analyzing how this composite structure works. It induced authors to examine basic magnetomechanical MRC characteristics. On their own a few MRC types differing in the matrix and the amount and type of ferromagnetic filling up and carrier fluid were produced. The magnetorheological composite was made from two components: an magnetorheological fluid and a porous matrix. MRF was made up according to a recipe similar to first fluid used by J. Rabinow [3] (carbonyl iron powder manufactured by BASF, average particle diameter $\sim 5\ \mu m$ and silicone oil as the carrier fluid - viscosity 100 m^2s^{-1}). The iron powder content in the fluid amounted to 80% by weight. Matrix of the composite was made

Journal of Physics: Conference Series **149** (2009) 012092 doi:10.1088/1742-6596/149/1/012092

from polyurethane sponge characterized by open pores whereby the MRF could freely penetrate inside it. Individual cells size in the matrix was estimated at 300 μm. Tested composites filling with MRF was the same amounting to 70% of the matrix volume. Specimens of material were examined in the arrangement shown in Figure 1.

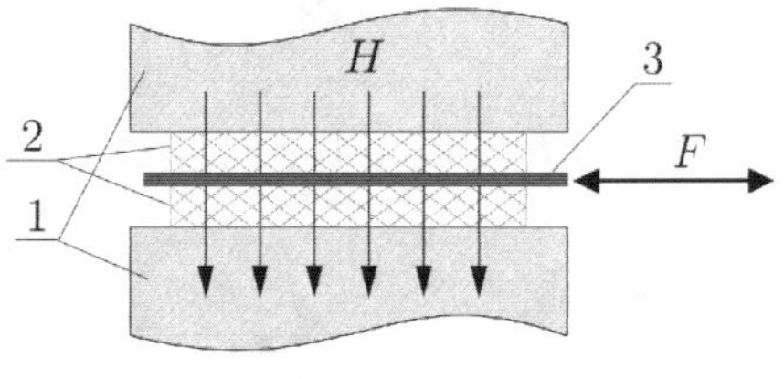

Figure 1. Scheme of the configuration for the examination of magnetorheological composites where: 1 - core of the electromagnet, 2 - MRC, 3 - force F carrying element, H – direction of magnetic field.

The example of hysteresis loops in coordinates $\tau(\gamma)$ for tested type of MRC materials were shown in Figure 2. The analysis of the results of experiments allows us to state, that in deformed composite it is possible to indicate two states. One, when the material is only in shearing mode and second, when it is breaking away from metal surfaces and slides. Total appearing of these two states is creating the very compound mechanism of the MRC work. They recognized, that in some applications with small displacement, the composite can serve as the damping element where only the first state appears (without MRC breaking away from metal surfaces, that is without the skid).

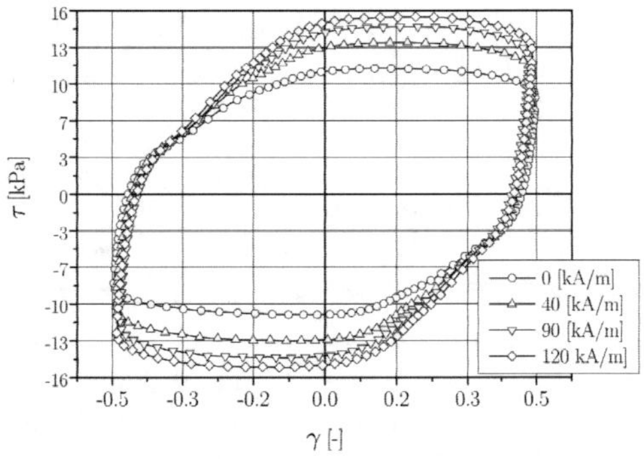

Figure 2. Hysteresis loops $\tau(\gamma)$ for tested magnetorheological composites - with the skid.

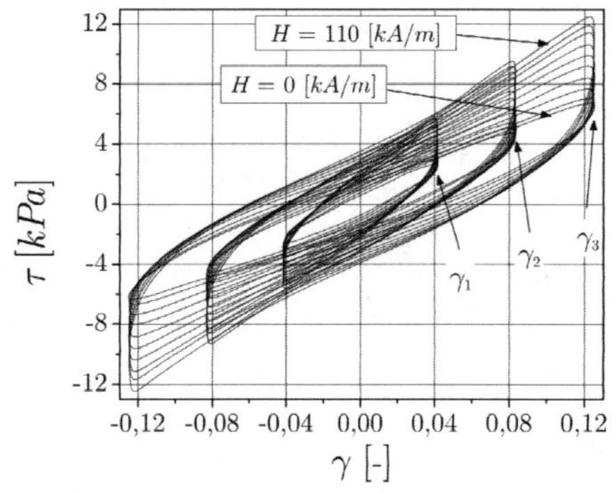

Figure 3. Hysteresis loop $\tau(\gamma)$ for magnetorheological composite - without the skid [4].

Next a MRC group was tested, whose damping properties in the scope of small deformations and no skid were being checked [5]. Results - hysteresis loop under the influence of magnetic field strength for composite were showed on Figure 3.

3. Exemplary application of magnetorheological composites

In the next step prototype damper was built, whose MRC was deformed according to the principle presented in Figure 1 without the skid. Composite material was glued and did not have an opportunity to move with regard to metal elements

The analysis of the experimental researches presented above and in works [5] made it possible to choose parameters of the magnetic circuit and dimensions of the composite in such a way as to attain foreseen range of forces necessary for damping and foreseen displacements. The system of modeling and transmitting the magnetic signal was prepared in such a way that it was possible to limit the waste caused by eddy currents. A gap was made in the core in which we placed two layers of the magnetorheological composite connected to a flat element, by means of which it was possible to transfer the forces from the outside. The scheme of the damper was shown in Figure 4.

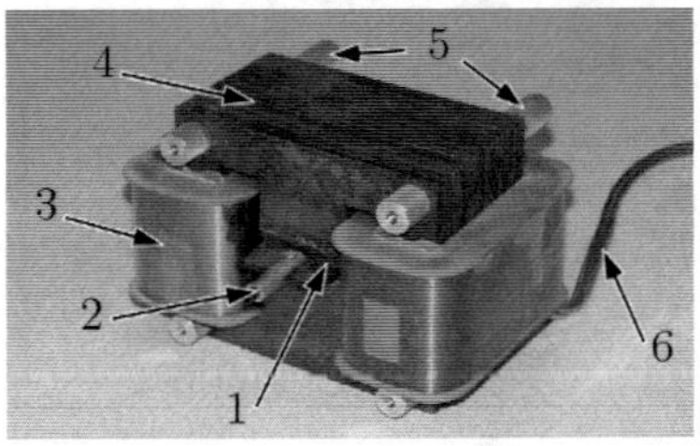

Figure 4. General view of the damper with magnetorheological composite: 1 – MRC, 2 – connecting rod, place of the connection between the damper and vibrating construction, 3 – coils, 4 – magnetic core, 5 – damper mounts, 6 – coils connections to external power supply.

The principal characteristic that was designated while building the damper was the dependence of the forces amplitude upon the current intensity flowing in the coils. For $I = 4\ A$, deformation amplitude $x_a = 1\ mm$ and frequency $f = 5\ Hz$ we attained the maximal force value 40 N (without field 10 N). The existing commercial constructions of dampers with magnetorheological composites (friction damper [2]) feature a similar range of damping forces under the influence of the magnetic field as the prototype construction used here. It is, however difficult to make direct comparisons due to a different mode of work of the material.

3.1. Vibrating object and measurement system

A steel beam on two bearings was used in practical testing of the prototype construction's effectiveness to damp mechanical vibrations. In the relevant literature describing the use of similar dampers with magnetorheological fluids we can find tests in which they are used in discreet vibration systems with one or more degrees of freedom [6]. Here the selected beam should be classed into continuous systems, which are characterized by various types of vibrations. Because of that the problem of damping requires a completely different approach, which must take account of the character of vibrations and the range of frequencies occurring in a given object. The scheme of the measuring stand is presented in Figure 5.

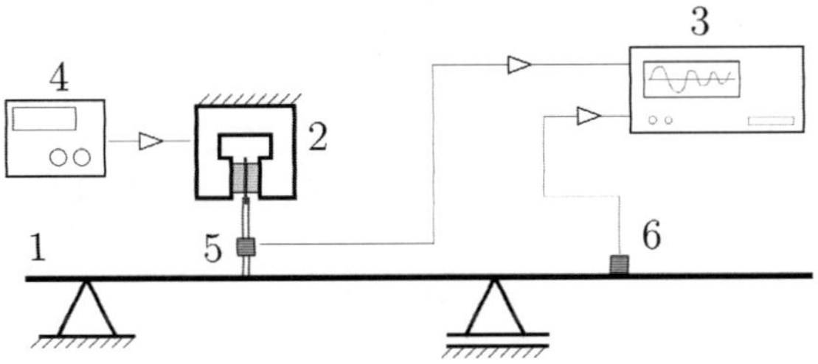

Figure 5. The test stand: 1 – vibrating object, beam on two bearings, 2 – tested damper with magnetorheological composite, 3 – acquisition and processing data system, 4 – adjustable power supply, 5 – force sensor, 6 – accelerometer

3.2. Free vibrations damping

Observations of time-accelerations characteristics were chosen as the tests checking the impact of the damper on the beam vibrations. As a criterion of the efficiency of the damping we took the free vibrations amplitude drop for the lowest resonance frequency at the largest displacements. The beam excitation was realized by the initial displacement for the beam's end without the initial speed. Figure 6 presents time courses of the acceleration amplitude registered for a chosen point (the end of the beam $L = 3\ m$) at two different values of the current intensity. The change in the intensity of the controlled current has, as we see, plays an important role in the beam vibrations. The amplitude drop is the most noticeable effect, that is why we took into account the time (denoted t_w) after which the amplitude is damped up to 20% of its maximal value. For

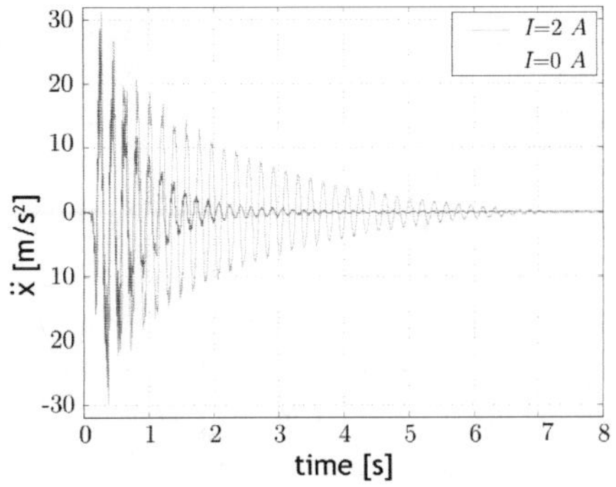

Figure 6. Free vibrations observed at the end of the beam ($L = 3\ m$) for two values of current strength 0 and 2 A

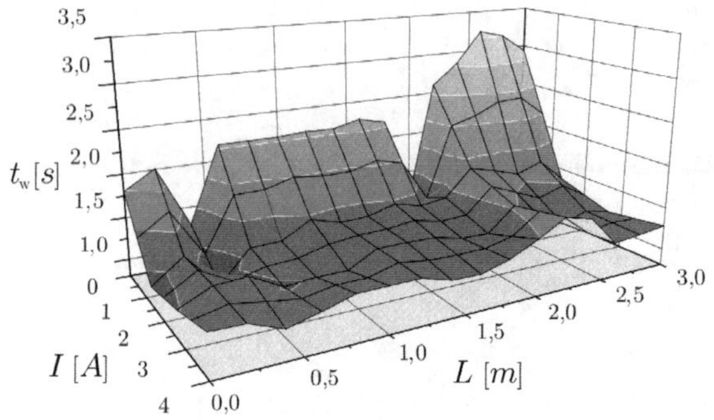

Figure 7. Suppression time t_w of the beam free vibrations in a function of control current strength I for individual points on the beam

the beam as an object simplified to the one-dimension scale there were several measurements made in evenly distributed points in its length. The current intensity value I was changed in the range from 0 to 4 A for each measurement point. The obtained results were presented jointly in the Figure 7. The influence of the current intensity is greatest in the range up to 2 A. Exceeding of this limit proved ineffective only causing unnecessary intake of energy through the coils. The best effects of dumping were attained for points on the free end of the beam, where the time of vibrations extinction underwent the biggest shortening - nearly twofold.

4. Conclusions

- Magnetorheological composite (MRC) was produced with a wide range of damping changes.
- Two state of working magnetorheological composite was tested. For application mode without skid mode was chosen.
- Magnetorheological composite material was used to create construction of the original damper and his usefulness was demonstrated in damping free vibration of the beam.

Acknowledgments

Acknowledgments This research was carried out partially as part of Commissioned Project PBZ-KBN- 115/T08/2004 financed by the Ministry of Education and Science, the State Committee for Scientific Research

References

[1] Deshmukh S S and McKinley G H 2007 *Smart Materials and Structures*
[2] Carlson J D 1999 *Proceedings of the 7th International Conference on Electro-Rheological Fluids and Magneto-Rheological Suspensions* ed Tao R (World Scientific) pp 621–628
[3] Carlson J D 2001 *Proceedings of the Eighth International Conference, Electrorheological Fluids and Magnetorheological Suspensions* ed Bossis G
[4] Lewandowski D 2005 *Damping properties of magnetorheological composites. Tests, models, identification (in Polish)* Ph.D. thesis Wroclaw University of Technology, Institute of Materials Science and Applied Mechanics URL `http://www.immt.pwr.wroc.pl/~lewandowski/praca/phd_dan_lew.pdf`
[5] Kaleta J and Lewandowski D 2007 *Smart Materials and Structures* **16** 1948–1953 URL `http://stacks.iop.org/0964-1726/16/1948`
[6] Müller S 2000 *Zum Einsatz von Semi-Aktiv Aktoren zur Optimale Schwingungsreduktion in Tragwerken* Ph.D. thesis Ruhr–Universität Bochum, Institut für Mechanik

Journal of Physics: Conference Series **149** (2009) 012093 doi:10.1088/1742-6596/149/1/012093

Dynamic Characterization and Modeling of Magneto-Rheological Elastomers Under Compressive Loadings

J H Koo[1], F Khan[1], D D Jang[2] and H J Jung[2]

[1]Department of Mechanical and Manufacturing Engineering, Miami University, Oxford, OH 45056, USA

[2]Department of Civil and Enviornmental Engineering, KAIST, Daejeon, 305-701, Korea

E-mail: koo@muohio.edu

Abstract. The primary goal of this paper is to characterize and model the compression properties of Magneto-Rheological Elastomers (MREs). MRE samples were fabricated by curing a two component elastomer resin with 30 % content of 10 μm sized iron particles by volume. In order to vary the magnetic field during compressive testing, a test fixture was designed and fabricated in which two permanent magnets could be variably positioned on either side of the specimen. By changing the distance between the magnets, the fixture allowed for varying the magnetic field that passes uniformly through the sample. Using this test setup and a dynamic test frame, a series of compression tests of MRE samples was performed by varying the magnetic field and frequency of loading. The results show the MR effect (percent increase in the materials "stiffness") increases as the magnetic field increases and loading frequency increases within the range of the magnetic field and input frequency considered in this study. Furthermore, a phenomenological model was developed to capture the dynamic behaviours of the MREs under compression loadings.

1. Introduction

Magneto-Rheological Elastomers (MREs) are a class of smart materials whose elastic modulus or stiffness can be varied depending on the magnitude of an applied magnetic field. As controllable stiffness elements, MREs can offer innovative engineering solutions to various engineering challenges. For the application of MRE materials, it is essential to understand and model the dynamic characteristics of the materials under various conditions. A number of researchers have investigated the shear properties of MRE materials over the past decade [1]. However, limited research has been performed in characterizing the compression properties of MREs [2]. Compression-tension type combined loadings are realistic loading conditions for various mechanical and structural systems, such as base isolation mounts, and it is imperative that the dynamic behavior of MREs be well understood. Therefore, this study investigates the dynamic characteristics of Magneto-Rheological Elastomers (MREs) under compressive loadings. In particular, it focuses on the stress-strain relationship associated with an MRE's deformation under various input loading conditions. These characteristics can serve as key merits of performance for MRE-based actuators.

2. MRE Sample Preparation

MRE samples were fabricated by curing a two component elastomer (silicone rubber) resin with 30 % content of 10 μm sized iron particles by volume. The shape and dimensions of the samples followed the ASTM standard for compression testing. The cylindrical shape samples are 9.5 mm (3/8 in.) high and 19 mm (¾ in.) in diameter. The silicone rubber contributes elastic properties to the cured sample. The iron particles allow the samples to be influenced by application of a magnetic field. Figure 1 (a) shows an MRE sample with the mold, and an SEM image of the cross section of the sample, which shows the uniformly distributed iron particles in the sample (Figure 1(b)).

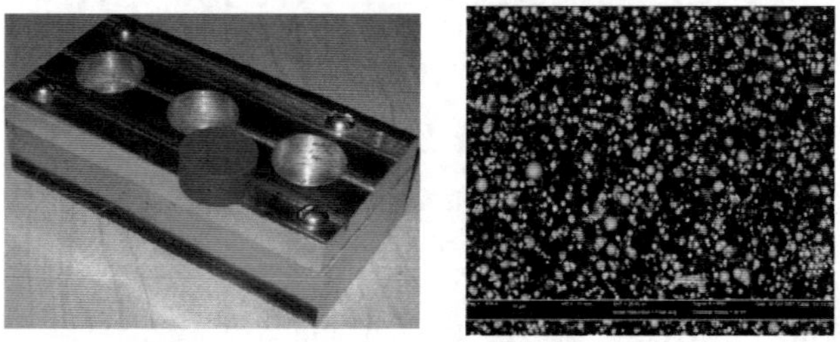

(a) (b)

Figure 1. (a) MRE sample and mold (b) SEM image of an MRE sample

3. Experimental Setup

For dynamic compression testing of MREs, it is tricky to change the magnetic field in the axial direction and measure the mechanical properties (i.e., stress) in the same direction. In this study, in order to vary the magnetic field during compressive testing, a high-stiffness fixture was designed and fabricated in which two permanent high-strength magnets could be variably positioned on either side of the specimen. By changing the distance between the magnets, the fixture allowed for varying the magnetic field that passes uniformly through the sample. Figure 2 (a) shows the test fixture installed in a material testing frame. Using this test setup and a dynamic test frame, a series of compression tests of MRE samples was performed by varying the magnetic field and frequency of loading.

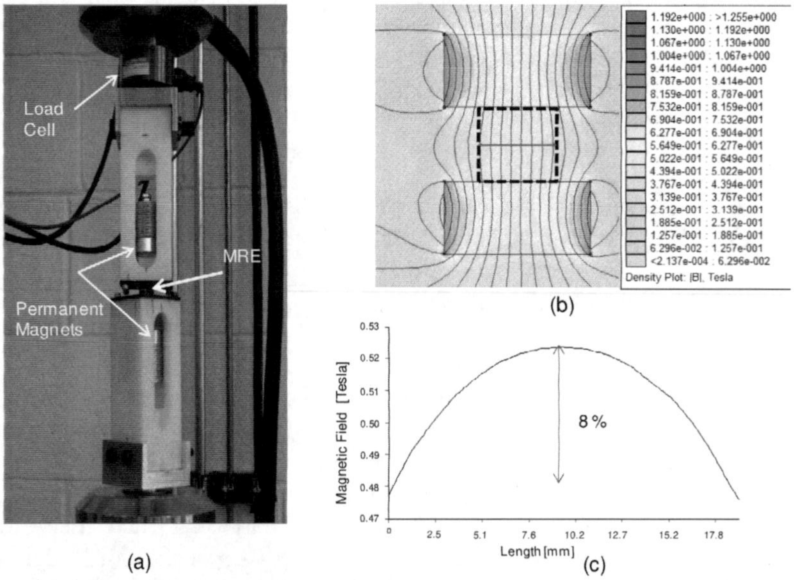

Figure 2. (a) Compression Experimental Setup, (b) Contour Plot of Magnetic Fields (c) Magnetic Field Variation Across the Specimen at the Center (Transverse Direction)

The selection of the magnet was done by using Finite Element Method Magnetics (FEMM). Figure 2 (b) shows the contour plot of magnetic fields when two magnets are the distance of the MRE sample's height away. In order to ensure the uniformity of the magnetic fields across the sample, a magnetic field distribution is measured over the width of the sample, across the center. As shown in Figure 2 (c), the strength of the magnetic field only dropped by about 8% from the center to the edges, which is acceptable amount of variation. Based on the FEMM analysis, a 38.1 mm (1.5 in.) diameter, N52 magnetic was chosen which is rated at 0.6985 Tesla at the surface.

4. Experimental Results

In all compression tests, the MRE samples were axially pre-compressed by 5% and cycled between +/- 5% strain from the pre-compressed state. To investigate the effect of a varying magnetic field, the magnitude of the field was varied from 0 T to 0.6 T at increments of 0.1 T. Moreover, the input frequency was varied between 0.1 Hz and 1 Hz at each value of the magnetic field to study the effects of loading frequency.

Figure 3 (a) shows that the stress-strain variation of an MRE sample with varying magnetic fields at 0.1 Hz harmonic loading frequency. It shows that the stress increases as the magnetic field increases. Moreover, the area of the loop increases as the magnetic field increases, indicating the damping of the material also increases.

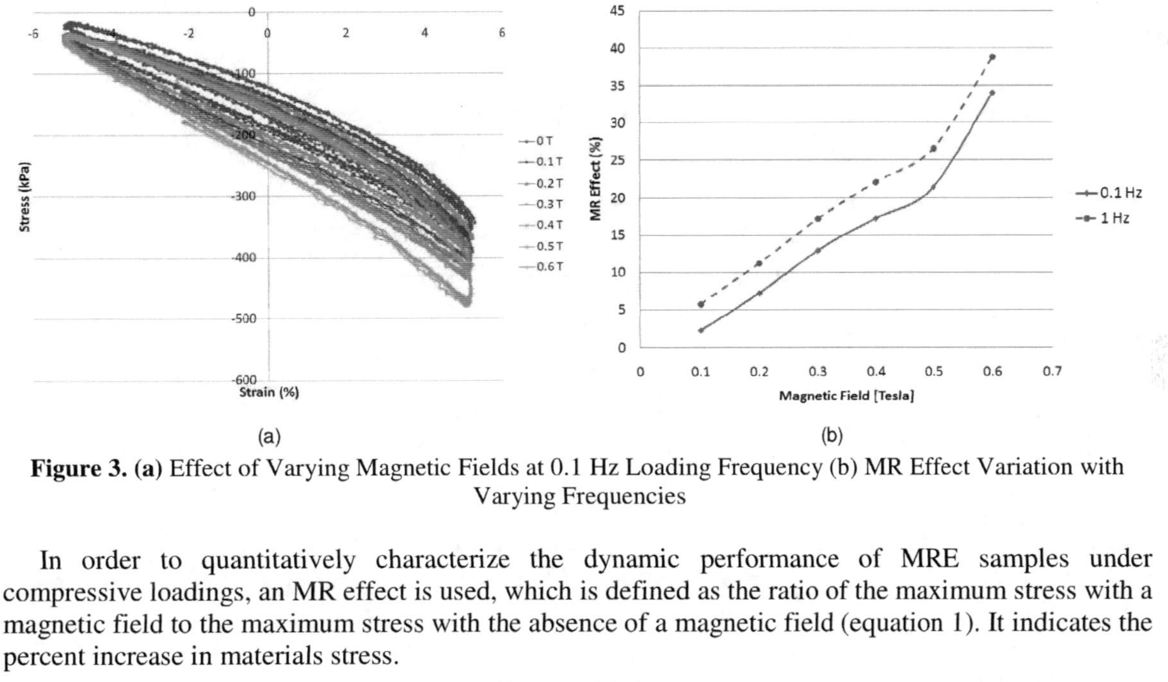

(a) (b)

Figure 3. (a) Effect of Varying Magnetic Fields at 0.1 Hz Loading Frequency (b) MR Effect Variation with Varying Frequencies

In order to quantitatively characterize the dynamic performance of MRE samples under compressive loadings, an MR effect is used, which is defined as the ratio of the maximum stress with a magnetic field to the maximum stress with the absence of a magnetic field (equation 1). It indicates the percent increase in materials stress.

$$MR\ Effect\ (\%) = \frac{\sigma_{\max)B} - \sigma_{\max)B=0}}{\sigma_{\max)B=0}} \times 100 \tag{1}$$

Figure 3 (b) shows the MR effect variation of MRE sample with varying magnetic field at two different loading frequencies (0.1Hz and 1 Hz). Within the loading frequencies considered in this study, the MR effect increases as the loading frequency increases. Moreover, the MR effect increases linearly at low magnetic fields (i.e. less than 0.4 T). However, the MR effects jumped from 0.5 T to 0.6 T. It is anticipated the MR effect would "exponentially" increase at higher magnetic fields before it saturates.

5. Dynamic Modeling of MREs

In order to capture the dynamic behaviour of the MREs under compressive loadings, the MRE element can be modeled using a combination of a linear damping element and a nonlinear spring element (see Figure 4 (a)). In this model, the MRE element's force can be represented as

$$F = c_0 \cdot \dot{x} + f_k(x) \tag{2}$$

where $c_0 = c_{0a} + c_{0b} \cdot B$, $f_k(x) = k_0 \cdot x^2 + k_1 \cdot x$, $k_0 = k_{0a} + k_{0b} \cdot B$, $k_1 = k_{1a} + k_{1b} \cdot B$, and B = magnetic flux density.

To demonstrate an example of modelling, experimental results for the 0.1Hz loading frequency with B = 0.1 T are compared with the results of the numerical modeling. Figure 4 (b) and (c) compares the time history of the stress, and the stress-strain behaviour of the MRE with the simulation models, respectively. The values of the model parameters used in the simulation are summarised in Table 1.

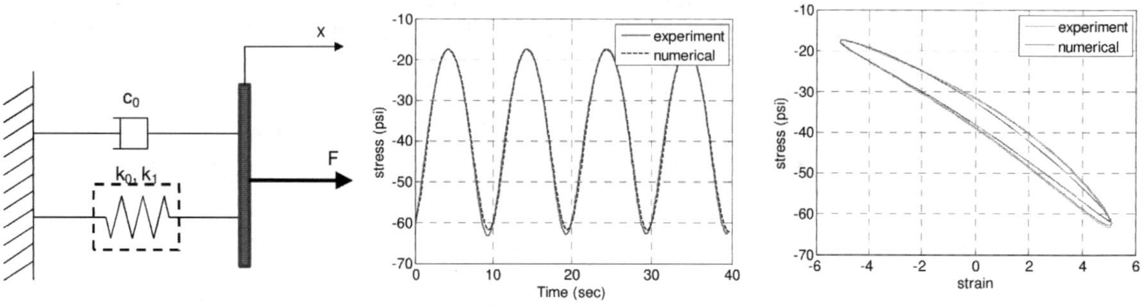

Figure 4. (a) MRE element model (b) Comparison of the Stress Time History and Model Simulation (c) Comparison of the Stress-Strain and Model Simulation

Table 1. MRE Model Parameters

Parameters	c_{0a}	c_{0b}	k_{0a}	k_{0b}	k_{1a}	k_{1b}
Values	0.8342	0.2670	4.1072	1.2435	0.1677	0.0002

6. Conclusions

This study focused on the dynamic characteristics of Magneto-Rheological Elastormers (MREs) under compressive loadings. Cylindrical shape MRE samples were fabricated by curing silicon based rubber resin with the 30 % content of 10 μm iron powers by volume. A test fixture was designed such that two permanent magnets uniformly pass the magnetic fields through the sample. Using this setup, a series of compression tests of MRE samples was performed by varying the magnetic field and frequency of loading. The results show that the mechanical stress of the MRE sample increases nearly 40 % with 0.6 T at 1 Hz loading frequency. Moreover, a phenomenological model that captures their dynamic response characteristics is developed.

Acknowledgments

The authors would like to thank a senior design team (Chris Schafer, Matt Lucchetti, Jeremy Mounce, and Phil Guerrant) at Miami University for their contribution of fabricating the fixture.

References

[1] G. Y. Zhou, "Shear properties of mag-netorheological elastomers", Smart Material. Struct. 12 (2003) 139-46.

[2] M. Kallio, T Lindroos, S Aalto, E Järvinen, T Kärnä, and T Meinander, "Dynamic compression testing of a tunable spring element consisting of a magnetorheological elastomer", Smart Material. Struct. 16 (2007) 506-514.

Journal of Physics: Conference Series **149** (2009) 012094 doi:10.1088/1742-6596/149/1/012094

Field-induced plasticity of soft magnetic elastomers

P V Melenev[1], Yu L Raikher[1], V V Rusakov[1] and G V Stepanov[2]

[1]Institute of Continuous Media Mechanics, Ural Branch of RAS, 614013, Perm, Russia
[2]State Scientific Research Institute of Chemistry & Technology of Organoelement Compounds, Moscow 111123, Russia

E-mail: melenev@icmm.ru

Abstract. A phenomenological model is proposed to describe the plasticity of soft magnetic elastomer composites under an external field. Along with high-elasticity, internal dry friction is introduced whose origin is attributed to the dipole-dipole interaction of the embedded particles. Numerical estimates of the model parameters are obtained from comparison with the measurements performed on iron carbonyl dispersions in soft silicon-rubber matrices.

Introduction

Magnetorheological elastomers are an important class of functional materials. The unique feature ensuring their wide prospects in technical and biomedical applications, is the strong dependence of the mechanical properties on the applied magnetic field [1-9]. Inside this class, soft magnetic elastomers (SMEs) make a separate family distinguished by giant magnetomechanical effects. Typical example is a weakly-linked silicon rubber filled with iron carbonyl grains (2-5 μm) to the volume fraction $\phi \sim$ 30% [2-4,6,7], its Young modulus being ~100 kPa. SMEs are prepared by polymerization of magnetic dispersions carried out either in the absence or in the presence of external field. Accordingly, the resulting SMEs acquire either a homogenous particle distribution or a chain-like texture [7-9].

The mechanics of SMEs depends crucially on the applied field $\boldsymbol{H}$. At $H = 0$ a SME sample responds to elongation up to ~100% in linear elastic (reversible) manner. The same sample under $H \sim$ 100 kA/m, being stretched to the same strain, retains a substantial part of it even after the stress is removed. Moreover, this residual strain is the greater the higher H [6]. As soon as the field is switched off, the sample restores its initial shape and properties [4]. The observed state of *field-induced plasticity* and the effect of *magnetic shape memory* that is based on it, are evidently due to re-grouping of the embedded magnetic particles. Hereby we propose a phenomenological model that ties up the field-induced structure changes in SMEs with their above-mentioned macroscopic behavior.

SME in a strain cycle. Discussion of experiment

In Fig. 1 the points present the stress-strain data obtained in loading cycles of an isotropic, i.e., prepared under zero field, silicon rubber SME with $\phi = 30$ vol.% of iron carbonyl [6]. The sample is a cylinder of aspect ratio ≈ 6 magnetized parallel to its axis. At $\boldsymbol{H} =$ const. the sample is stretched slowly along its axis to a certain length, then the load is reduced to zero with the same rate. When, after the cycle, the field is turned off, the sample practically restores its initial size. Thus, the main facts are:

1) deformation is virtually reversible only under $\boldsymbol{H} = 0$;

2) with the increase of the applied field the initial elastic modulus grows substantially;
3) under deformation in a constant field, the modulus undergoes appreciable changes;
4) being subjected to a loading cycle, a magnetized sample gains a notable residual strain which it retains as long as the field is on.

As the magnetic anisotropy of iron carbonyl is very low, at $\boldsymbol{H} = 0$ the embedded particles are multi-domain, and there is no magnetic interaction between them. When the field is turned on, the induced magnetic moments of the particles enter into dipole-dipole interaction that strives to organize chain-like particle aggregates (clusters). In a soft network the dipole forces, which grow rapidly ($\propto 1/r^4$) with the particles closing in, dominate the elastic ones. However, the network, although distorted, does not break, and remains attached to the particles. Under mechanical loading, the joint reaction of the network and the clusters determines the macroscopic response of the material, Fig. 1 is an example.

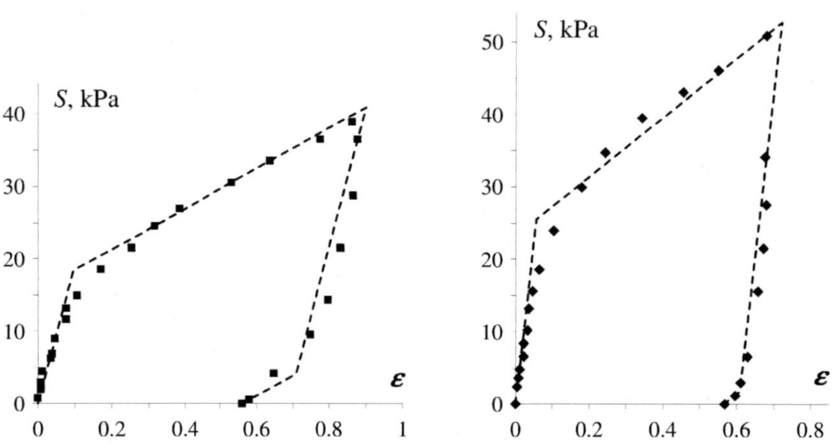

Figure 1. Loading/unloading deformation processes for a SME cylinder under constant fields 96 kA/m (squares) and 207 kA/m (diamonds); in each test the strain is counted from the size that the sample assumes after magnetization in a stress-free state. Dashed lines present a “uniform” fitting with the model; for the numerical values of the parameters see Table 1.

Let a stress-free SME cylinder be magnetized along its axis by a uniform constant field. The occurring cluster formation results in some macroscopic striction of the sample. When a small load is applied, it is counteracted be the polymer network while the existing clusters behave like solid particles. Further loading increases the local stresses in the matrix. As soon as they reach the “magnetic durability” of the clusters, there begins intense destruction of the existing and formation of the new ones. This endows SME with a mechanism of internal dry friction. Indeed, let the load grow a bit. The network transfers this increment to the clusters, and the latter break in their weakest points. Broken clusters do not resist displacements, and locally the material begins to flow. As the filling is sufficiently high, the moving fragments of a given cluster inevitably approach their analogues emerging from the other ones. From this “debris” the magnetodipole attraction builds up new, more dense, clusters capable to stand the enhanced load; thereby the plasticity flow stops. Macroscopically, stretching of a SME in this regime, being a superposition of ductile and elastic reactions, looks like an elastic response with the effective modulus that is appreciably lower than the initial one.

On halting the deformation, the sample would rest in equilibrium with respect to the actual level of external loading and the internal stress distribution. The reverse process, i.e., quasi-static unloading, induces global shrinking of the network and reduction of internal stresses. At the beginning this does not affect the structure of the clusters; with further decrease of the load, the local stresses on the clusters (now their direction is opposite to that while stretching) induce a rearrangement of the same break/rebuild kind (quasi-plasticity). This regime keeps on until the internal stresses fall below the level at which they are able to rebuild the clusters. In result, under complete unloading the sample retains some strain that is maintained by local stresses induced by the set of magnetized clusters whose distribution is different from that existed before the load was applied for the first time.

Turning off the magnetic field radically changes the internal stress balance. As the particles are magnetically soft (no remanence), the magnetodipole forces turn to zero. Accordingly, the particles cease to "sense" each other, and all the clusters fall apart. In result, the local high-elasticity stresses, "frozen" in the presence of the field, get in action. The network (with the particles attached to it) completely restores its configuration, and the SME sample recovers (recalls) its shape.

Model

A simple heuristic scheme (HS) is shown in Fig. 2. Of the two parallel branches, one is a linear elastic spring of rigidity G_1 while another contains a sequence of a spring of rigidity G_2 and a dry-friction element with critical shear S^*. This presents a SME as two interwoven networks which deform affinely. One is purely elastic while another is composed of sequences of elastic and rigid (cluster) segments. The strains of both branches are equal: $\varepsilon = \varepsilon_1 = \varepsilon_2$, the total stress on the HS is the sum of the partial stresses. As follows from Fig. 1, the rheological characteristics of a SME depend on the applied field. Accordingly, we assume that all the parameters of the HS are functions of H.

For small loading (the stress in branch 2 is below S^*) only the springs work, so that the net rigidity of the HS is G_1+G_2. In Fig. 1 this regime corresponds to straight lines departing from zero points. When the stress in branch 2 grows up to S^* (the corresponding strain is $\varepsilon_* = S^*/G_2$), the resistance of the dry-friction element falls down to zero, and the stress in branch 2 becomes strain-independent. In this regime the stress induced by the HS is due only to spring 1. Accordingly, the ascend angles of the lines $S(\varepsilon)$ in Fig. 1 drop down from $\arctan(G_1+G_2)$ to $\arctan(G_1)$. When the strain is halted at some value ε_m, the equilibrium stress equals $S_m = G_1\varepsilon_m + S^*$. Here the dry-friction element stops to slide but the stress S^* in branch 2 is maintained by spring 2 stretched to the strain $\varepsilon_* < \varepsilon_m$, i.e., $S_m = G_1\varepsilon_m + G_2\varepsilon_*$.

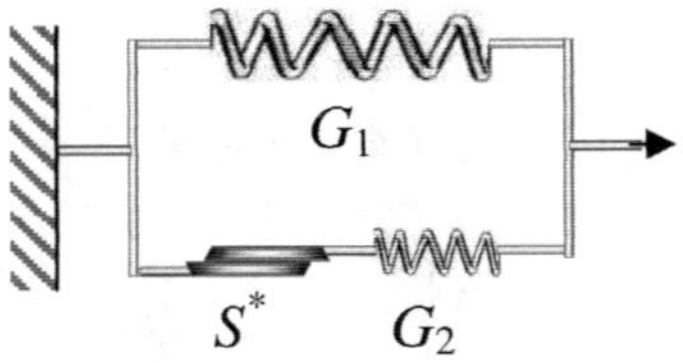

Figure 2. Structure scheme of a SME possessing the field-induced plasticity.

Consider now the diminution of the load. As both springs are stretched, they strive to recover their initial sizes; in the corresponding parts of Fig. 1 the dependencies $S(\varepsilon)$ go down with the tangent $dS/d\varepsilon = -(G_1+G_2)$. When the stain $\varepsilon_m - \varepsilon_*$ is achieved (branch 2 acquires it in result of sliding of the dry-friction element), spring 2 assumes its initial (equilibrium) length while spring 1 is yet stretched. Under further unloading, spring 1 still strives to shrink but to do that it has to compress spring 2. Therefore, the stress in branch 2 becomes opposite to that in the stretching regime. If the elasticity of spring 2 is invariant with respect to elongation/compression, the rigidity of the HS does not change and remains equal G_1+G_2. When and if the stress in branch 2 achieves $-S^*$, the dry-friction element "takes off". In this situation the only source of shrinking of the HS is the excessive energy of spring 1.

Shrinking stops when the stress on the HS turns to zero, that is the partial stresses compensate each other. The residual strain of the HS in this state is ε_r. The value of ε_r depends on the fact whether or not the dry-friction element worked during the unloading process. This, in turn, is determined by the magnitude of the maximum strain ε_m achieved in the cycle. If ε_m exceeds the critical value

$$\varepsilon_{\max} = \varepsilon_*(2 + G_2 / G_1), \quad (1)$$

then the residual strain is $\varepsilon_r = (G_2/G_1)\varepsilon_*$. In the opposite case, it depends linearly on ε_m:

$$\varepsilon_r = G_2\left(\varepsilon_m - \varepsilon_*\right) / (G_1 + G_2). \quad (2)$$

Numerical values of the three parameters of the HS are obtained from comparison of the model loops with the data of Fig. 1. The spring rigidities are determined from the tangents of the $S(\varepsilon)$ curves. The point ε_* at which the tangent falls down corresponds to the onset of the ductile regime. The value $S(\varepsilon_*) = \varepsilon_* G_1 + S^*$ enables one to estimate the dry friction shear S^*. As the experimental data never give any precise tangent breakup points, these numbers contain some uncertainties. At the first sight, the residual strain ε_r, could be determined more accurately. However, to interpret the obtained value it is necessary to find first, whether the maximal achieved strain ε_m exceeds the value $\varepsilon_{\max}$, given by formula (1). For that, one has to employ the estimate of ε_* which accuracy, as mentioned, is not high.

For our HS essential is the quantity ε_*, at which the mechanism of magnetic plasticity (cluster rearrangement) is launched. When choosing the HS parameters to interpret the data of Fig. 1, we try to satisfy two conditions at once: 1) maximal closeness of theoretical and experimental values of the threshold strain ε_*; and 2) maximal geometric resemblance between the model cycle and the experimental dependence $S(\varepsilon)$. The obtained numerical results are given in Table 1.

Table 1. Estimates for the parameters of the model obtained from comparison with the data of Fig. 1 (elongation regime).

H, kA/m	G_1, kPa	G_2, kPa	ε_*	ε_r	ε_m	S^*, kPa
96	28	164	0.096	0.56	0.9	15.7
207	41	413	0.057	0.57	0.72	23.1

According to Fig. 1, at H = 96 kA/m the observed modulus $dS/d\varepsilon$ at the end of unloading falls down significantly: the curve bulges rightward. On the other hand, under the field 207 kA/m the same effect is virtually absent. A probable cause of such a behavior are slow relaxation processes. If so, on diminution of the unloading rate, the bulging of $S(\varepsilon)$ curves should decease.

Conclusions

A remarkable phenomenon of magnetic shape memory displayed by soft magnetic elastomers is based on the field-induced plasticity effect. Although the details of microscopic behavior of real composites are very complicated, a simple phenomenological model quire reasonably accounts for some essential features of magneto-mechanics of SMEs. The presented analysis shows, in particular, that to explain the mechanical response of the material, one has to assume the presence of internal dry friction. A possible origin for such a mechanism in SMEs based on magnetically soft particles is pointed out.

Acknowledgements

The work was supported by the projects: 09-01-99020 from RFBR and 16274QL from ECONET.

References

[1] Zrínyi M, Barsi L, and Büki A 1996 *J. Chem. Phys.* **104** 8750
[2] Nikitin L V, Mironova L S, Stepanov G V and Samus A N 2001 *Polymer Science* A **46** 443
[3] Nikitin L V, Mironova L S, Stepanov G V, Kornev K G and Stepanov G V 2004 *Polymer Science* A **46** 498
[4] Nikitin L V, Mironova L S, Stepanov G V and Gorbunov A I 2004 *J. Magn. Magn. Mater.* **272-276** 2072
[5] Farshad M and Benine A 2004 *Polymer Testing* **23** 347
[6] Stepanov G V 2006 *Proc. Internat. Plyos Conf. on Magnetic Fluids*, Plyos, Russia, 2006, p.41
[7] Abramchuk S, Kramarenko E, Grishin D, Stepanov G, Nikitin L, Filipcsei G, Khokhlov A and Zrínyi M 2007 *Polymers for Advanced Technologies* **18** 513; *ibid* **18** 883
[8] Ginder J M, Clark S M, Schlotter W F and Nichols M E 2002 *Int. J. Mod. Phys. B* **16** 2412
[9] Bellan C and Bossis G 2002 *Int. J. Mod. Phys. B* **16** 2447

Journal of Physics: Conference Series **149** (2009) 012095 doi:10.1088/1742-6596/149/1/012095

MRE properties under shear and squeeze modes and applications

K M Popp[1], X Z Zhang[2] W H Li[2] and P B Kosasih[2]

[1]TU Bergakademie Freiberg, Institute of Machine Elements, Design and Production, Agricolastrasse 1, 09599 Freiberg, Germany

[2]University of Wollongong, School of Mechanical, Materials & Mechatronic Engineering, NSW 2522, Australia

E-mail: Kristin.Popp@imkf.tu-freiberg.de; weihuali@uow.edu.au

Abstract. Magnetorheological elastomers (MREs) belong to the group of so called smart materials. Due to an applied magnetic field the MREs change their material properties like the stiffness. This feature has resulted in a number of novel applications, such as adaptive tuned dynamic vibration absorbers (ATDVA) for suppressing unwanted vibrations over a wide frequency range. MRE based devices operate in different modes, such as shear mode and squeeze mode but the study of mechanical performances of MREs under squeeze mode is very rare. This paper aims to investigate MRE performances under both shear and squeeze modes. Both experimental and simulation studies were conducted to analyze the MR effect in both modes. These studies indicate that MRE working in squeeze mode would result in higher MR effects. As a case study, a MRE based ATDVA was analyzed, which demonstrated good capabilities in reducing vibrations.

1. Introduction

Magnetorheological elastomers (MREs) are composite materials of a rubber like base, micron-sized magnetizable particles and additives [1,2]. Under an applied magnetic field the embedded iron particles form chain-like structures along the magnetic field lines within milliseconds which cause an increase of the stiffness of the material. A number of research groups made use of such materials to develop novel adaptive tuned dynamic vibration absorbers [2-5], where MREs work in shear modes. It is noted that the varying frequency ranges of the vibration absorbers depend much on MR effects and the material properties of the elastomers. However, the shear mode might limit such effects.

This study aims to study the MRE effect under squeeze mode and compare it with that of shear mode. In this study, MRE samples with different compositions were fabricated and tested on a home-made test rig. The mechanical performances under both shear mode and squeeze mode were used to verify simulation analyses based on a one degree of freedom model. This study is expected to provide good guidance to develop high-efficiency MR elastomer based devices.

2. Fabrication of magnetorheological elastomers

For preliminary tests of MREs in shear mode, four different MRE compositions were manufactured, the fabrication approach of which was detailed in [6]. In these tests two elastomers have shown high

MR effects. In shear mode the natural frequency for each of them increases more than twice of their initial state. In this study analyzes of the behavior of one of those MRE compositions under squeeze mode will de discussed in length.

For fabricating the MRE sample three components, including the rubber base, silicon oil and iron powder, were used. The rubber base, a room temperature vulcanizing silicon sealant (PERFIX, Selleys Pty Ltd, Australia), was mixed with a silicon oil DC 200/200cs (Dow Corning Corporation, U.S.A.) to changing its ductility. Iron particles with a 5μm size (SIGMA-ALDRICH CHEMIE GmbH, Germany) were added to the base. The weight ratio of rubber base, silicon oil and iron powder was 1:1:3. The mixture was filled in a mould and placed in a strong magnetic field of 1 Tesla generated by an electromagnetism system (Peking EXCEEDLAN Inc., China). Ten minutes later, the magnetic field was reduced to 0.5 Tesla so that the samples were cured for 7 more hours. The samples will be placed at the room temperature for 24 hours prior to testing.

3. Experiments

The experiments were conducted on a test rig with two different setups, as shown in figure 1(a) and (b), where MREs work in the shear mode and the squeeze mode, respectively. The test rig was built up as a one degree of freedom system with a mass connecting to the base via two pieces of MREs, which were able to move independently to the base. In the squeeze mode, the MREs and mass were directly fixed on the base by a double-sided adhesive tape. In the shear mode the mass was fixed by contact forces of two cylinders which were tightened on the base. On the base two coils were wound to generate a magnetic field. The magnetic field strength up to 150 mT can be obtained when the coil current is 3 A. The base was fixed on a shaker which excited the system with an adjustable frequency range from 20 Hz to 100 Hz. Two acceleration sensors were placed on the upper surface of the mass and the base for measurement of the relative movement of the mass.

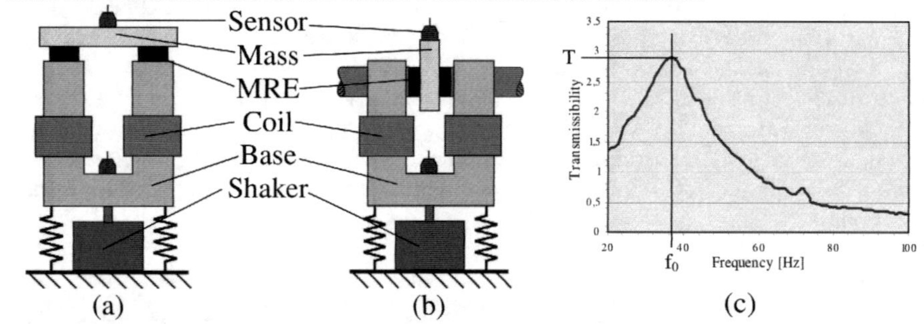

Figure 1. Experimental setup for squeeze mode (a) and shear mode (b) and measured frequency response (c).

The acceleration sensors were used to measure the amplitude and the phase angle between the mass and the base. The natural frequency of the mass was located at the maximal amplitude and the relative MR effect was calculated by the increasing of the frequency. The measured frequency increases of MREs used in shear- and squeeze mode were listed in table 1.

Table 1. Comparison of test data and simulation results for MREs in squeeze- and shear mode.

current intensity [A]	Measurement								Calculation				shear strain [%]
	squeeze mode				shear mode				squeeze mode		shear mode		
	mT	f_0 [Hz]	$T=\left\|\frac{\hat{x}_1}{\hat{x}_e}\right\|$	E[Pa]	mT	f_0 [Hz]	$T=\left\|\frac{\hat{x}_1}{\hat{x}_e}\right\|$	G[Pa]	f_0 [Hz]	$T=\left\|\frac{\hat{x}_1}{\hat{x}_e}\right\|$	f_0 [Hz]	$T=\left\|\frac{\hat{x}_1}{\hat{x}_e}\right\|$	
0	0	48	2.68	182.2E3	0	25	2.53	49.4E3	47	2.13	26	2.31	10
0.76	46.2	53	2.77	222.1E3	32.7	26	2.43	53.5E3	50	2.10	27	2.27	10
1.5	90.1	63	3.26	314.9E3	49.9	35	2.36	96.9E3	60	2.27	34	2.49	5
2.26	136.0	76	3.66	457.8E3	76.4	46	2.52	167.3E3	73	2.50	41	2.71	2
3	155.5	85	3.66	571.4E3	108.2	53	2.69	222.1E3	83	2.52	52	3.04	0.1
increase				**177.1 %**				**212.0 %**		**176.6 %**		**200.0 %**	

From table 1, an obvious MR effect is visible for both modes whereas the relative MR effect is a little higher for the shear mode with 112% (from 25 Hz to 53 Hz) than for the squeeze mode with 77% (48 Hz to 85 Hz). Comparing the absolute frequency values, the frequency change in shear mode of 28 Hz is lower than that in squeeze mode of 37 Hz, which is contrary to the prior statement. It is also shown that in the initial state the natural frequency of the setup with MRE used in squeeze mode is nearly twice as high as for the setup with MRE used in shear mode.

4. Modeling and discussion

The model for describing the elastomers mathematically is a one degree of freedom system as shown in figure 2. Here k* is the complex stiffness of MR elastomers and m is the weight of the mass. The weights of the MREs are neglected.

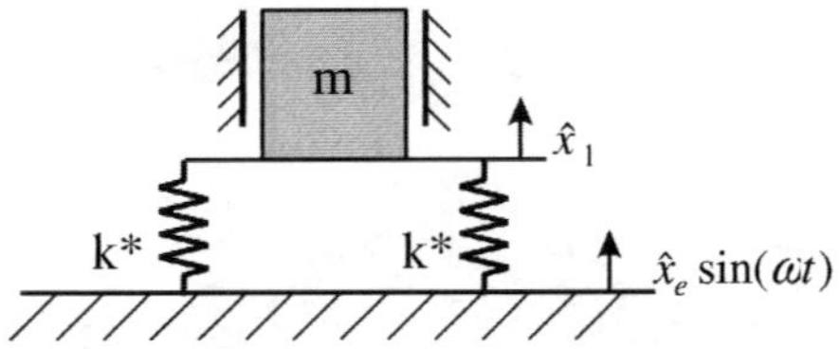

Figure 2. One degree of freedom system for describing the tested system.

For this system the equation of motion is given by

$$m\ddot{x}_1 + k^* x_1 = k^* \hat{x}_e \cdot \sin(\omega t)\,, \tag{1}$$

where the variable excitation of the shaker is $\hat{x}_e \sin(\omega t)$. The stiffness is substituted either by a shear spring or a compression spring. The stiffness of a shear spring depends on the shear modulus while the stiffness of a compression spring is dependent on the Young's modulus. The conversion between these two modulus is given by

$$G = \frac{E}{2\cdot(1+\nu)} \tag{2}$$

where the Poisson ratio ν is 0.5 for rubber like materials. Investigations have shown that the shear modulus for elastomers is complex and has to be expressed by an energy storing part G' and energy dissipating part η, called loss factor [7]. Including these two factors in equation (1), the equation of motion can be written as

$$m\ddot{x}_1 + \frac{G'\cdot A}{h}(1+i\eta)x_1 = \frac{G'\cdot A}{h}(1+i\eta)\hat{x}_e \cdot \sin(\omega t)\,. \tag{3}$$

Both the storage modulus G' and the loss factor η of MREs were measured with a dynamic characterization [7] by using a rheometer (Physika MCR301, Anton Paar GmbH, Germany). Thereby dependencies on the frequency and the shear rates were measured. The calculated storage modulus G' at various strains varying from 10% to 0.1% was shown in Table 1. The weight of the mass, the square face and thickness of the MRE were measured to m = 0.35 kg, A = 480 mm² and h = 5.5 mm respectively. The calculated results of the natural frequencies are listed in table 1. Comparing with experimental values, it is obvious that modeling prediction values agree well with experimental results. Small differences between them may be due to measurement errors of the square face and thickness of the MRE. The transmissibility varies about 0.4, which might be because of the measurement errors of the storage modulus G' and loss factor η.

5. MRE shear mode ATDVA

The fabricated MRE samples were used to develop an adaptive tuned dynamic vibration absorber (ATDVA). The schematic of the MRE ATDVA is shown in figure 3(a). The mechanism of a conventional vibration absorber is to transfer the exciting forces away from the device by an additional oscillating spring mass system. When the natural frequency of the added vibration absorber is equal to the external excitation frequency, the vibration energy of the primary system will be transferred to the ab-

Journal of Physics: Conference Series **149** (2009) 012095 doi:10.1088/1742-6596/149/1/012095

sorber mass, which can significantly reduce unwanted vibrations of the primary system. Conventional vibration absorbers work efficiently only at special frequencies. In the developed MRE ATDVA, the spring element is replaced by the MRE material. The stiffness as well as the working frequency of the absorber can be changed by adjusting the magnetic field, which makes the ATDVA adapts to a wide frequency range. Figure 3(b) shows comparison of the proposed MRE ATDVA with two other typical cases of without absorber and with a passive absorber. Suppose the primary system, as shown in figure 3(a), was excited by a multi disturbance which consisted of three sine sweeping force signals with amplitude 1 and with constant frequencies of 35, 50 and 90 Hz, and two sine sweeping waves with amplitude 1 and varying frequencies from 1 to 200 Hz and 35 to 90 Hz, respectively. The stiffness of the MRE was adjusted by a real time controller which switched the current intensity of the coils between 0A and 2A. The mass ratio between m and m_a was set to 0.01.

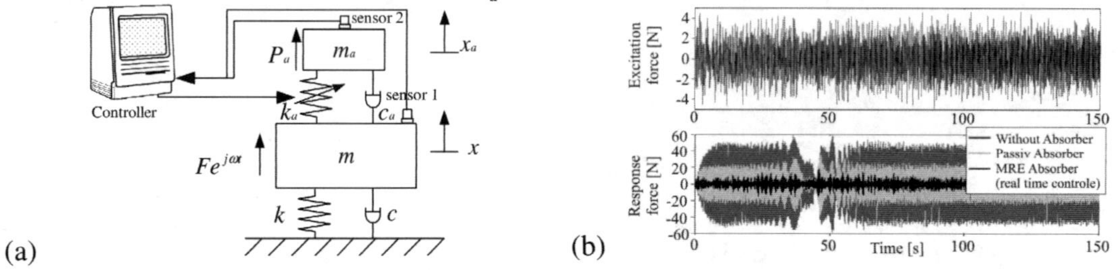

Figure 3. (a) Schematic of an ATDVA and (b) responses for different absorbers.

In figure 3(b) the response signal is shown for the system without any vibration absorber, with a passive tuned dynamic vibration absorber and with an MRE based ATDVA. The simulation results show that the system without control has large force amplitudes because the excitation signals are close to its natural frequency. The passive TDVA can suppress the amplitude of the primary system by half while the MRE based ATDVA using real time control logic can significantly reduce the vibration of the primary system. Thus the MRE ATDVA was effective for even low mass ratios.

6. Conclusion

In this study, the fabrication of MR elastomers and characterization of their mechanical performances under shear and squeeze mode were presented. The results show an increasing of the natural frequency about 28 Hz for squeeze mode and 37 Hz for shear mode. A one degree of freedom model was used to study the MRE performances under both modes. The modeling analysis, in terms of strain and frequency dependence of storage modulus, agreed well with experimental results. The efficiency of the developed MRE ATDVA was simulated and compared with other two cases of without an absorber and with a passive one. The comparison indicates that the MRE ATDVA is super to the other two cases by significantly reducing the responding forces even for low mass ratios.

References

[1] X Z Zhang, W H Li and X L Gong 2008 *Communications in Nonlinear Science and Numerical Simulation* **13** 1910–1916

[2] F Zschunke 2005 *Aktoren auf Basis des magnetorheologischen Effekts* Doktorarbeit, University of Erlangen, Erlangen

[3] M Lita, N C Popa, C Velescu and L N Vékás 2004 *IEEE Tansactions on Magnetics* **40**(2) 496–472

[4] P Pagliarulo, K Kuhnen, C May and H Janocha 2004 *9th International Conference on New Actuators* 698-701

[5] Y Wang, Y Hu, L Chen, X Gong, W Jiang, P Zhang and Z Chen 2006 *Polymer Testing* **25** 262–267

[6] K Popp, P B Kosasih, X Z Zhang and W H Li 2008 *Proceedings of the 15th International Congress on Sound and Vibration*

[7] W H Li, G. Chen, S.H. Yeo 1999 *Smart Material and Structures* **8** 460–468

Journal of Physics: Conference Series **149** (2009) 012096 doi:10.1088/1742-6596/149/1/012096

Characterization of the dynamic mechanical behavior of magneto - elastomers

B Schrittesser[1], Z Major[1,2] and G Filipcsei[3]

[1]Polymer Competence Center Leoben; Leoben, Austria

[2]University of Leoben, Institute of Materials Science and Testing of Plastics, Leoben, Austria

[3]Materials Science and Modeling Research Group of the Hungarian Academy of Sciences, Hungary

E-mail: bernd.schrittesser@pccl.at

Abstract. Due to their magnetic field dependent mechanical response, polymeric materials filled with magnetic particles have become of major practical and theoretical interest. the dynamic-mechanical behavior of polydimethylsiloxane rubber filled with ferro-oxide with and without magnetic field has been investigated in this study. Specimens with different hardness and with two distinct orientations of the fillers (isotropy, anisotropy) were prepared. Dynamic mechanical analysis experiments were performed to determine the storage and loss modulus over a wide amplitude and frequency range with and without external magnetic field. In order to characterize the influence of the ferro-oxide filler particles, a permanent magnetic field to the specimen grips has been attached. The complex-, storage- and loss modulus and the loss factor values were determined. Finally, to compare the isotropic and anisotropic materials a magnetic stiffening factor (MSF) was defined and applied.

1. Introduction and Objectives

The new generation of magnetic gels and elastomers represent novel type of composites, consisting of small (from nano-to-micro sized) magnetic particles dispersed in highly elastic polymeric matrix. To support both material development efforts and engineering design and applications, it is of prime theoretical and practical importance to characterize the complex magneto-dynamic mechanical behavior of elastomeric materials filled by magnetic particles [1,2,5]. Hence, the objectives of this study were: (1) to implement a permanent magnet for an electrodynamic test system and (2) to characterize the complex dynamic mechanical/magnetic behavior of isotropic and anisotropic magnetic particle filled elastomers.

2. Experimental

2.1. Materials and Specimens

Polydimethylsiloxane rubber (PDMS) materials filled with magnetic particles (Fe_3O_4) were prepared in 4 different hardness degrees by varying the curing additive (from 3 to 4.5 wt%) and with two distinct distributions of the filler particles (isotropic and anisotropic) as is shown schematically in figure 1.

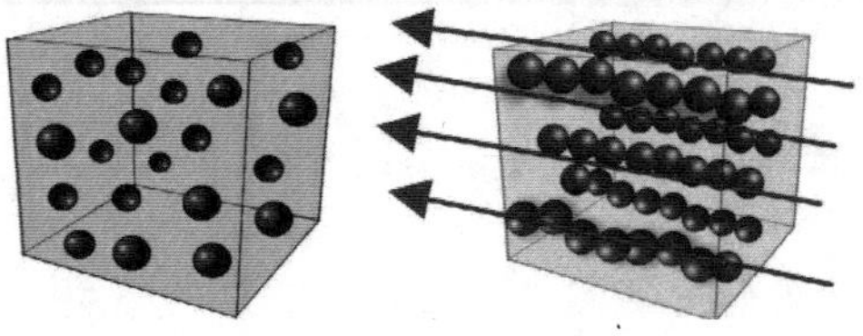

Figure 1. Isotropic and anisotropic distribution of the particles

Figure 2. Fe_3O_4 particles

Fig 2 shows the particle used with an average diameter of approximately 8,4µm. Cylindrical specimens with diameter of 25mm and heights of 4, 6 and 25mm were prepared with and without the application of magnetic field. The cubic specimens with 18 mm edge length were only prepared in magnetic field. The curing process in magnetic field results in a highly anisotropic particle distribution. For more details about the materials and the preparation technique please refer to [3].

2.2. Testing Machines, Test Methods

Dynamic mechanical analysis (DMA) tests were performed using novel electro-dynamical testing machines (BOSE 3100 and 3450 BOSE Corp. MN, USA) in the small-scale deformation regime. All experiments were conducted in compression mode, with and without the application of magnetic field.

2.3. Measurement of the Magnetic Field

To characterize the deformation behavior in magnetic field, a permanent magnet with the nominal magnetic field of about 0,44 Tesla was placed under the specimens. To determine the distance dependence of the magnetic field of the permanent magnet and hence the inhomogenity of the field acting on the particles in the specimen, a calibration curve with a Hall-sensor was first measured and the results are plotted in figure 3. As expected a significant decrease of the magnetic field as function of distance was observed.

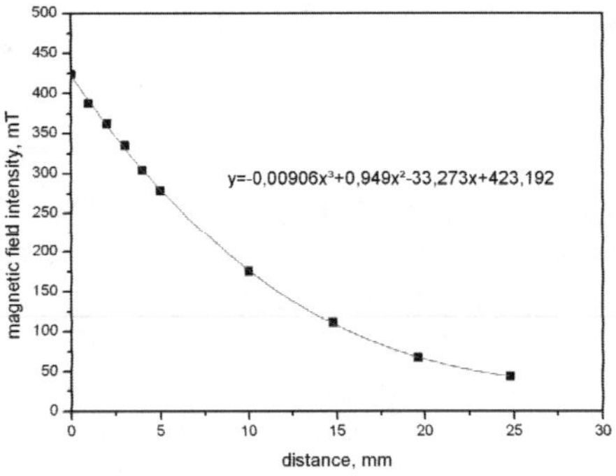

Figure 3. Magnetic field of permanent magnet in dependence of displacement

3. Results

3.1. Amplitude behavior

To characterize the deformation behavior of magneto-elastomers, cyclic tests were performed over a wide deformation amplitude range. All tests were performed at a mean level of 0,5mm in compression at 8 different dynamic amplitudes ranging from 0,02 to 0,8mm and at a constant frequency of 1Hz. The hysteretic deformation behavior of the anisotropic and isotropic specimens with 3 wt% curing additive is shown in figure 4 at the same test amplitude. While a linear behavior of the isotropic specimen with small hysteresis was observed, the anisotropic specimen reveals a highly non-linear behavior and significantly higher hysteresis [4].

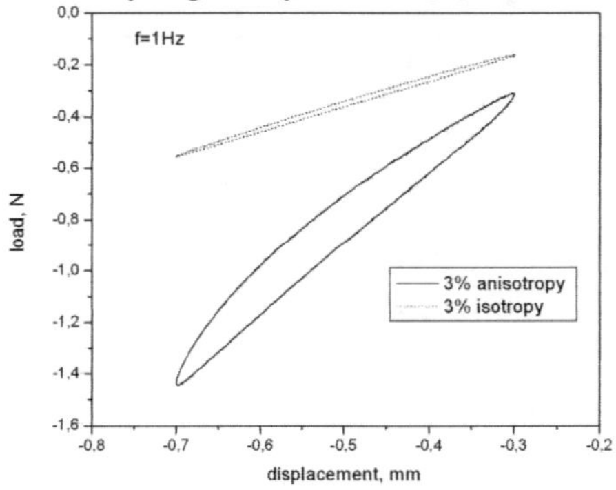

Figure 4. Hysteretic load-displacement curves for isotropic and anisotropic specimens at the same test amplitude.

3.2. Frequency behavior

Dynamic mechanical analysis tests were performed over the frequency range from 0,1 to 100 Hz at a dynamic amplitude of 0,2mm. The dynamic, E*, the storage and loss modulus, E', E'' and tanδ values were determined for all experiments. The frequency dependence of the storage modulus E' for a 3,5wt% cubic specimen with loading perpendicular and normal to the particle orientation is shown in figure 5 without magnetic field. The filled symbols represent the results without magnetic field. A clear increase of the storage modulus was observed for both loading directions.

3.3. Influence of the magnetic field

Furthermore, the frequency dependence of the modulus was also investigated in magnetic field. As it was emphasized before, the permanent magnet produces an inhomogeneous magnetic field, that is, the particles have different magnetic forces as a function of the specimen height in the specimens. In spite of this fact, the increase of the modulus up to about 20% was observed in compare to the test without magnetic field as it is depicted in figure 5. This increase of the modulus was found highly depends on the magnetic particle distribution (iso or anisotropic) and on the hardness (3 to 4.5 wt% curing additive results in about 100% change of the hardness) of the specimens.

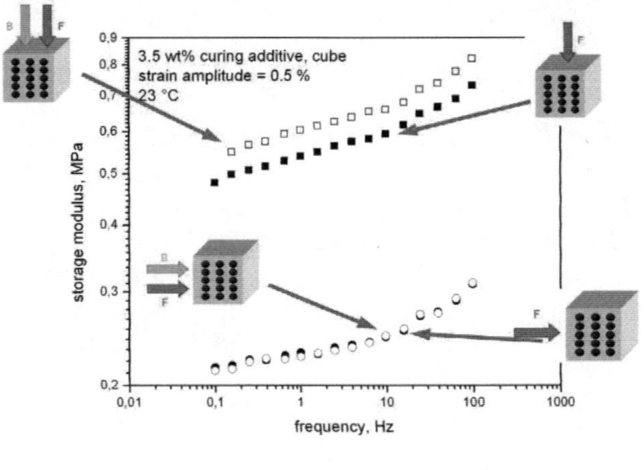

Figure 5. Frequency Scan 3,5% cube, anisotropy and isotropy distribution with magnetic field intensity

3.4. Influence of the load direction combined with magnetic field

While a clear difference between the deformation behavior of the material for various load directions relative to the particle orientation was observed for the anisotropic specimens, the isotropic specimens revealed similar response. A difference in perpendicular loading direction between loading with and without magnetic field intensity could not be observed. Moreover, magnetic field intensity affects on harder material higher than on softer.

3.5. Magnetic Stiffening Factor

The magnetic stiffening factor was defined, which characterizes the ratio between the results with and without of magnetic field.

$$MSF = \frac{E_{MF}}{E_0}$$

The frequency dependence of the magnetic stiffening factor for isotropic particle distribution is shown in figure 6a. As clearly depicted in this figure, isotropic specimens independent from the hardness degree revealed the same MSF values over the frequency range investigated. The frequency dependence of MSF is shown in figure 6b for anisotropic specimens. For anisotropic particle distribution higher magnetic stiffening factor was observed (20 to 35%). Furthermore, no clear tendency was found for the variation of the amount of curing additive

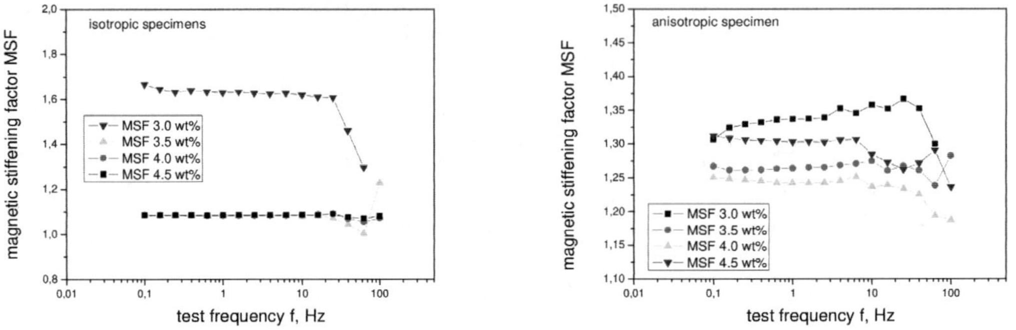

Figure 6: Magnetic stiffening factor for (a) isotropic and (b) anisotropic particle orientation

4. Conclusions

Based on the experimental results described above some general conclusions may be drawn:

- Clear differences of the hysteretic behavior between isotropic and anisotropic specimens were observed.
- Magnetic particle filled elastomers revealed a similar frequency and amplitude dependent behavior with and without magnetic field.
- The combination of the mechanical loading direction, the magnetic field direction and the particle orientation was found clearly influence of the dynamic mechanical behavior of the cubic specimens.

In comparison to the conventional dynamic mechanical analysis, the additional application of magnetic field and the application of anisotropic specimens clearly increase the complexity of the tests and hence the characterization of the material behavior. However, to support practical engineering applications of magneto-elastomers, the determination of the influence of all parameter combinations (mechanical loading, frequency, material anisotropy, magnetic field) on the material behavior is necessary.

References

[1] Ginder J M, Nichols M E, Elie L D and Tardiff J L 1999 *SPIE* **3675** 131–8
[2] Carlson J D and Jolly M R 2000 *Mechatronics* **10** 555–69
[3] Filipcsei G, Csetneki I, Szilágyi A and Zrínyi M 2007 *Oligomers, Polymer Composites, Molecular Imprinting: Magnetic Field-responsive Smart Polymer Composites (Advances in Polymer Science* vol **206***)* (Berlin: Springer) pp 137–89
[4] Zhou G Y and Li J R 2003 *Smart Materials and Structures* **12** 859-72
[5] Stepanov G V, Abramchuk S S, Grishin D A, Nikitin L V, Kramarenko E Yu and Khokhlov A R 2007 *Polymer* **48** 488-95

Journal of Physics: Conference Series **149** (2009) 012097 doi:10.1088/1742-6596/149/1/012097

Field Dependent Response of Magnetorheological Elastomers Utilizing Spherical Fe Particles Versus Fe Nanowires

H J Song[1], N M Wereley[1], R C Bell[2], J L Planinsek[2] and J A Filer II[2]

[1]Department of Aerospace Engineering, University of Maryland, College Park, MD 20742, USA.

[2]Department of Chemistry, Pennsylvania State University, Altoona College, Altoona, PA 16601, USA.

E-mail: Wereley@umd.edu

Abstract. This study compares the dynamic response of nanowire-based magnetorheological elastomers (MREs), to those containing conventional spherical particles. MRE samples were fabricated by curing the iron particle laden elastomeric material in a magnetic field. Material characteristics of the MRE samples were evaluated using a material test machine that was modified to measure static and frequency dependent characteristics of these samples under different magnetic fields. The MRE samples consisted of a silicone rubber matrix containing various weight fractions of iron particles of differing morphology. Nanowires were used to enhance the interaction forces and contact area between particles. The static and dynamic properties of the MREs were evaluated under a compressive load for the various compositions and weight fractions. The stress vs. strain characteristics were measured for each sample. The equivalent damping coefficient of the MRE samples was measured and characterized under magnetic fields of differing intensities. The dynamic characteristic (dynamic stiffness) was measured under sinusoidal excitation in the frequency domain.

1. Introduction

Elastomers are commonly used for passive shock and vibration control. More recently, controllable magnetorheological elastomers (MREs) with varying magnetic fields have attracted great interest [1-3]. MREs are a smart material having static and dynamic properties that can be controlled by varying the intensity of an applied magnetic field. Characteristics of MREs depend on the dispersant volume fraction (or weight fraction or solids loading), dispersant characteristics (such as particle size, composition, and morphology), matrix material, and magnetic field intensity.

Various compositions of MREs have been assessed. For example, iron flakes have been dispersed in a silicone elastomer [1]. MREs have been experimentally shown to have changeable properties with the volume fraction of particles. Magnetostrictive properties of MREs have been presented [2], where irregular carbonyl iron particles with an average size of 5 μm were used. Magnetostriction effects increased with increasing volume fraction. A more recent study examined the properties of MREs consisting of a silicone elastomer matrix containing spherical carbonyl iron (3-5 μm) as the magnetisable filler in which the particles were aligned with an external magnetic field during the curing process [3]. The mechanical properties were examined in a cyclic compressive load with the intensity of magnetic fields in the frequency domain. In this study, the field dependent characteristics of MREs with two different Fe particle morphologies (spheres and nanowires) were evaluated under static and dynamic (harmonic) excitations. Frequency domain characteristics were evaluated under sinusoidal excitation while maintaining a compressive bias load.

2. Composition of MREs and Experimental Setup

Figure 1 shows photographs of MRE samples. The height and diameter of each MRE sample was 25.4 mm and 9.5 mm as shown in Fig. 1(a). The diameter of the spherical particles ranged from 6-10 μm. The nanowires were 15 μm long with a diameter of 298 ± 40 nm. Magnified images of 10 wt% samples are shown in Fig. 1(b). During curing, a magnetic field was applied to the cylindrical samples using permanent magnets. This resulted in the alignment of the particles in the direction of the applied magnetic field (due to the induced magnetic dipoles of the particles). The aligned particles can be seen in Fig. 1(b). For the sphere-based MREs, particle loadings of 10, 30, 50 and 70 wt% were chosen. In the case of nanowires, particle loadings of 10 and 30 wt% were selected.

The experimental setup is presented in Figs. 2 and 3. The internal temperature of the magnetic cell was monitored with a thermocouple and cooled using an external cooling fan. The MRE sample was installed inside of the magnetic cell as shown in Fig. 3. The upper and lower fixtures are connected to the piston rod and the load cell of the Instron test machine. A Hall probe is used to measure the magnetic flux density which is located on top of the lower fixture.

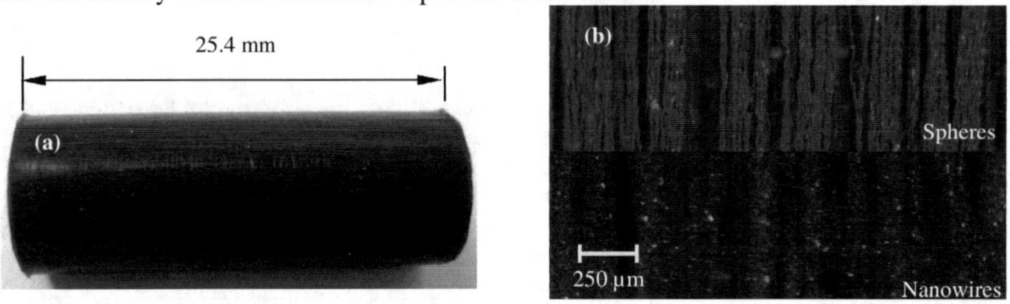

Figure 1. Photograph of an MRE sample (a) with 10 wt% spherical Fe particles and (b) magnified images of MREs containing 10 wt% spherical and nanowire Fe particles.

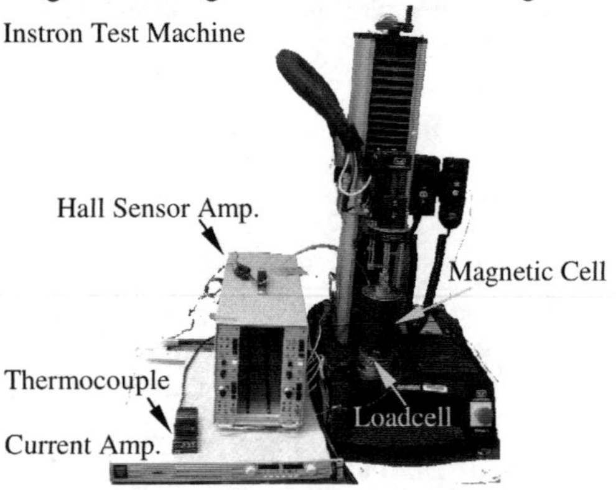

Figure 2. Photograph of the experimental setup for static and dynamic testing.

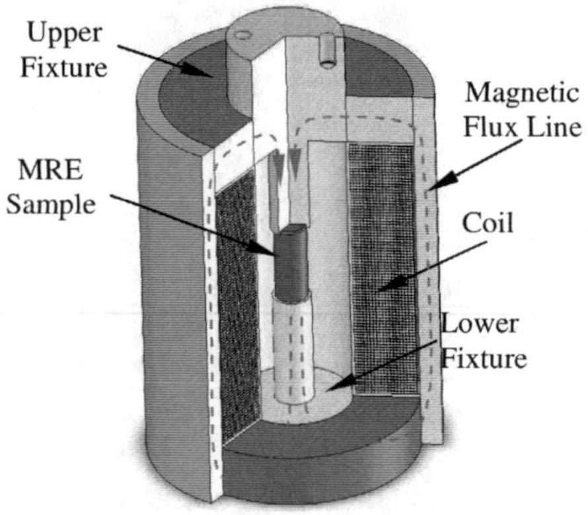

Figure 3. Schematic diagram of the magnetic cell.

Journal of Physics: Conference Series **149** (2009) 012097 doi:10.1088/1742-6596/149/1/012097

3. MRE Characteristics

Characteristics of the MRE samples were evaluated using a modified Instron test machine. All tests were performed under a compressive bias load. The static characteristics of the MRE samples in the absence of magnetic field are shown in Fig. 4. In the case of both particle geometries, the static stiffness increased as the weight fraction increased. However, the increase in stiffness is significantly larger for nanowires than for spherical particles because the contact area of the nanowires with the silicone elastomer is far greater, and the oriented structures within the MRE samples were strengthened as a result. In the case of 30 wt% solids loading, nanowire-based MRE samples had a static stiffness of 55.24 kN/m, which is nominally 2.7 times that of the spherical particle-based MREs.

Figure 5 presents the stress-strain characteristics under a compressive load in the absence of a magnetic field. The maximum applied strain level was set to 1%. As seen from these results, the hysteresis behavior generally increases as the weight fraction is increased. The area of the hysteresis loop is a measure of the dissipated energy (damping).

Stress-displacement data are shown in Figs. 6 and 7 for MREs with a solids loading of 30 wt% for both spherical particles and nanowires. MRE samples were excited sinusoidally under compressive preloads that correspond to 1% of strain. The amplitude of displacement corresponded to a maximum 1% strain in the MRE sample. From the force vs. harmonic displacement data, the damping properties were calculated. The controllable damping effect with the applied magnetic field can be represented using the equivalent viscous damping as a function of excitation frequency [4]. Figs. 8 and 9 show the equivalent damping coefficient in the frequency domain. Here, the equivalent damping coefficient can be obtained from the dissipated energy per cycle. In the case of 10 wt% nanowires, the equivalent

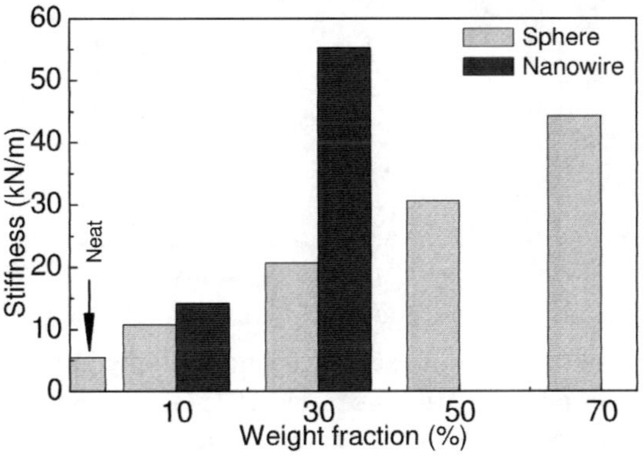

Figure 4. Static compressive stiffness measured in the absence of magnetic field.

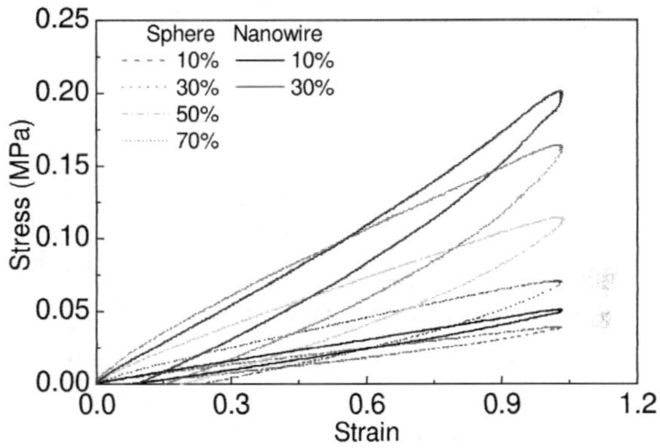

Figure 5. Stress vs. strain under compressive load measured in the absence of magnetic field.

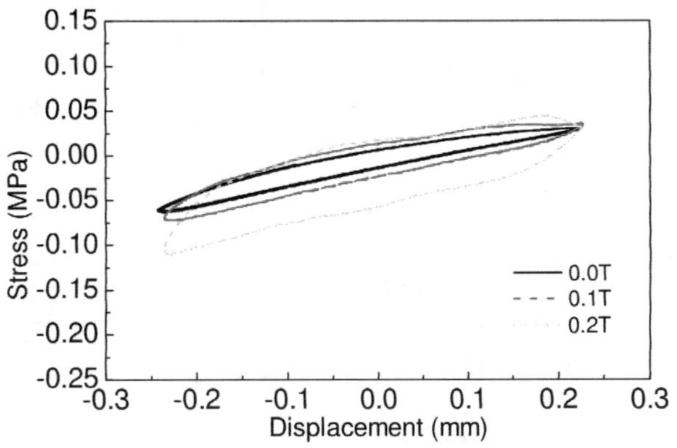

Figure 6. Force vs. displacement with magnetic field (30 wt% spheres, excitation frequency: 1Hz, preload: 5 N).

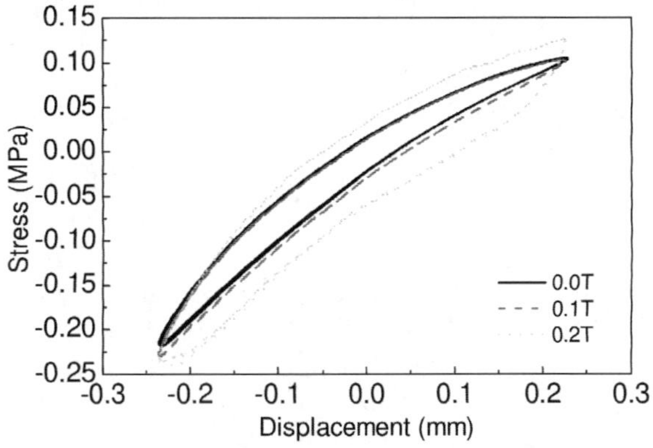

Figure 7. Force vs. displacement with magnetic field (30 wt% nanowires, excitation frequency: 1Hz, preload: 10 N).

Journal of Physics: Conference Series **149** (2009) 012097 doi:10.1088/1742-6596/149/1/012097

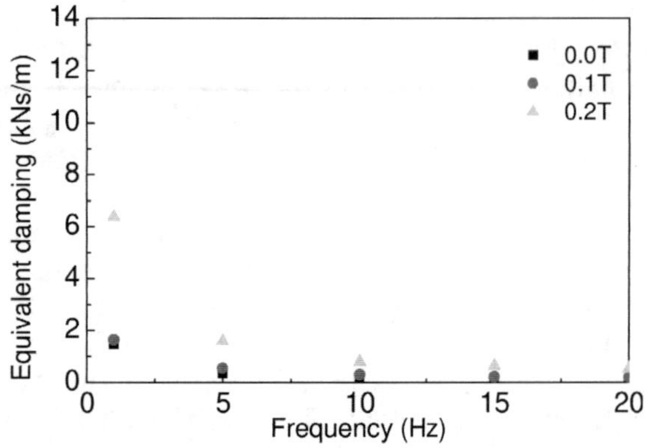

Figure 8. Equivalent damping measured as a function of magnetic field (spheres: 10 wt%, excitation amplitude: 0.254 mm, preload: 5 N).

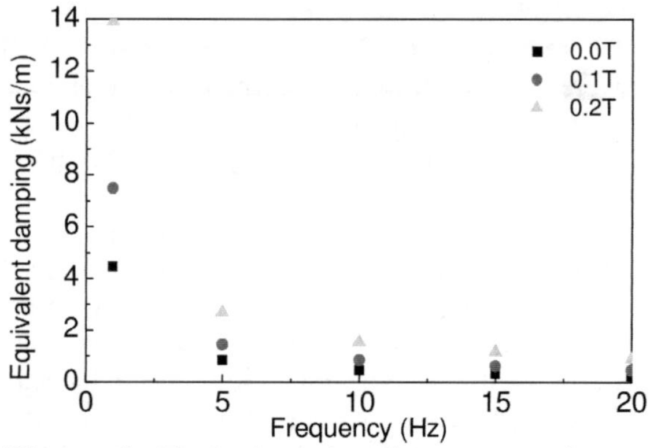

Figure 9. Equivalent damping measured as a function of magnetic field (nanowires: 10 wt%, excitation amplitude: 0.254 mm, preload: 5 N).

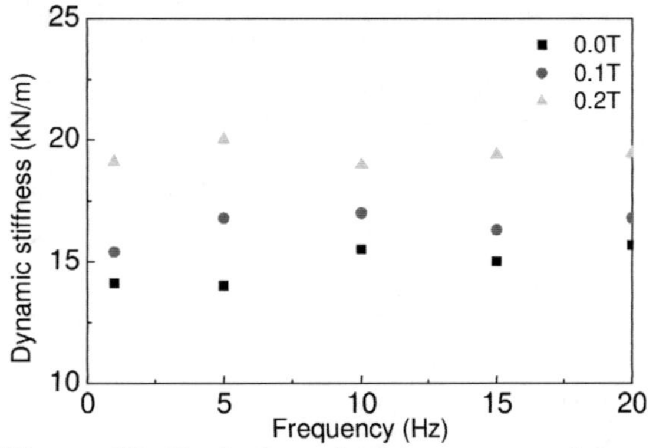

Figure 10. Equivalent damping measured as a function of magnetic field (spheres: 30 wt%, excitation amplitude: 0.254 mm, preload: 5 N).

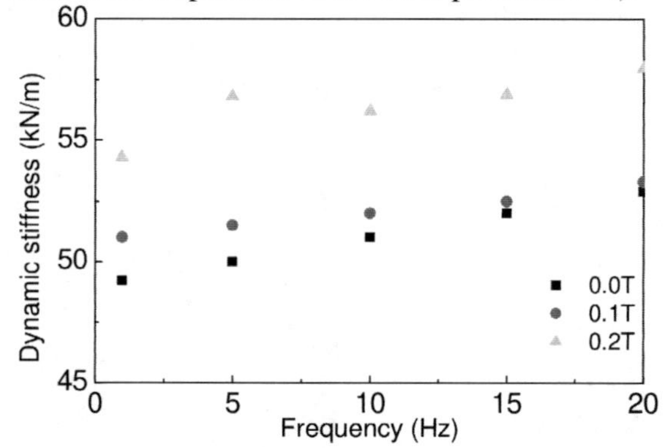

Figure 11. Equivalent damping measured as a function of magnetic field (nanowires: 30 wt%, excitation amplitude: 0.254 mm, preload: 10 N).

damping increased remarkably as the intensity of the magnetic field increased and decreased as the excitation frequency increased as shown in Fig. 9. The dynamic stiffness of the MREs is presented in Figs. 10 and 11. The values increased as the applied magnetic field is increased and also slightly increased as excitation frequency increased. The controllable stiffness range is larger for nanowire-based MREs than for the spherical particles-based MREs.

4. Conclusion

MRE samples were prepared using iron particles of differing morphology (spheres and nanowires) and varying weight fraction. The nanowire-based MREs had static stiffness greater than that of the sphere-based MREs in the absence of a magnetic field at the same weight fraction (10 and 30 wt%). Stiffness and damping were measured for frequencies ranging from 1 – 20 Hz. The 30 wt% nanowire-based MRE had an off-state dynamic stiffness about 3.5 times that of the 30 wt% sphere-based MRE, although both materials exhibited similar increments in dynamic stiffness at 0.2 T. The off-state damping of the 30 wt% nanowire-based MRE was greater than that of the 30 wt% sphere-based MRE. The controllable field-dependent damping was much greater for the nanowire-based MRE - at 1 Hz, the controllable damping was nearly twice as much as that of the sphere-based MRE.

References

[1] Woods B, Wereley N, Hoffmaster R and Nersessian N 2007 *Int. J. Modrn. Phys. B* **21** 5010.
[2] Guan X, Dong X and Ou J 2008 *J. Magnetism Magnetic Mater.* **320** 158.
[3] Kallio M, Lindroos T, Jarvinen E and Meinander T 2007 *Smart Mater. Struct.* **16** 506.
[4] Snyder R, Kamath G, and Wereley N 2001 *AIAA J.* **39**(7) 1240.

Journal of Physics: Conference Series **149** (2009) 012098 doi:10.1088/1742-6596/149/1/012098

Magnetorheological effect of magneto-active elastomers containing large particles

G Stepanov[1], D Borin[2], S Odenbach[2]

[1] State Scientific Research Institute of Chemistry and Technology of Organoelement Compounds, 111123, Moscow, Russia

[2] Technische Universität Dresden, Chair of Magnetofluiddynamics, 01062, Dresden, Germany

E-mail: Dmitry.Borin@tu-dresden.de.

Abstract. The magnetorheological effect of elastomer composites containing a mixture of large (50-80 μm) and small (3-5 μm) particles has been experimentally examined. The data shows that elasticity in the range of small deformations (1%) for a magnetic field strength of 290 mT increased by two order of magnitude. This effect can be explained with the presence of the large particles in the structure of the composite assisting the aggregation effect. Due to the strong increase of the interparticle interaction compared with the restoring elastic forces, the presence of the large particles leads to the observed steep increase of Young's modulus.

1. Introduction

Magnetorheological elastomers (MRE) are in the focus of scientific interests of many research groups since more than decade [1-8]. Because of the variety of magnetically controllable properties (e.g. magnetostriction, shape memory effect, compression in alternating field etc) going for beyond rheological effect it might be preferable to call these smart materials magneto-active elastomers (MAE). The most prominent property of MAE is the so-called magnetorheological (MR) effect, i.e. the increase of elasticity in an external magnetic field. Technical applications using MAE, like variable stiffness devices or active dampers, are based on the MR effect [9-13]. The improvement of the working parameters of such devices requires the development of MAE with a more pronounced MR effect. Typically, the MR effect is a characteristic of anisotropic samples, i.e. elastomers which have been polymerized in a magnetic field in a way that the particles of the disperse phase have formed chain-like aggregates, remaining trapped in the polymeric matrix. A manifestation of the MR effect in isotropic samples, i.e. samples being polymerized in a non-structured state, has not or very small been observed [14-17]. The authors in [18] reported on some MR effect achieved in an isotropic sample filled with irregularly shaped iron particles and in [19] the influence of particle interactions on the MR effect of an isotropic sample has been studied. The goal of the present work was generation of a significant MR effect in isotropic samples, comparable with the effects found in anisotropic MAE.

2. Sample preparation

In a typical MAE [20] consisting of a polymeric silicon matrix (compound SIEL™) and small carbonyl iron particles with a characteristic size of 3-5 µm, large particles with a size of 50-80 µm with identical magnetic properties have been added. The standard compound SIEL™ consists of 2 components, namely of component A which is a mix of low molecular vinyl-containing rubber (V-R) and hydride-containing cross-linking agent (H-R) ($(CH_3)_3SiO\{[(CH_3)_2SiO]a–[CH_3(H)SiO]b\}x–Si(CH_3)_3$ *and* $(CH_2\ CH)_3SiO\ [CH_3SiO]y–Si(CH\ CH_2)_3$) and of component B, which is prepared from V-R and complex platinum catalyst ($(CH_2{=}CH)_3SiO[CH_3SiO]y\text{-}Si(CH{=}CH_2)_3$ + *Pt-catalyst*). The mechanism of the reaction is follow: $\equiv$ Si-CH=CH_2 + HSi $\equiv \rightarrow$ [$\equiv$ $SiCH_2CH_2Si$ $\equiv$]n

Small and large particles have been suspended into the polymeric matrix at identical volume fraction. In this way a maximal homogeneity of the sample can be achieved. One of the special features of the used samples was a highly elastic matrix (modulus of a matrix is 10-20 kPa), as well as a high volume concentration of magnetic particles (36 %). Reduction of sedimentation was achieved by means of high-frequency heating and a regulation of the viscosity of the liquid polymer with plasticizers. In this way samples with a diameter of 4 mm and a length of 10 mm have been prepared.

3. Experiments

3.1. Magnetic field influence on the strain

The MAE has been stretched along the direction of a homogeneous magnetic field. The elasticity modulus of a sample E can be calculated from the stress P – strain L dependence ($E=dP/dL$). Figure 1 shows this dependence for an isotropic sample with an initial elasticity of 40 kPa for various magnetic fields. One can see that the change of the average elasticity of the sample in a magnetic field with an induction of 210 mT exceeds the initial elasticity by more than an order of magnitude.

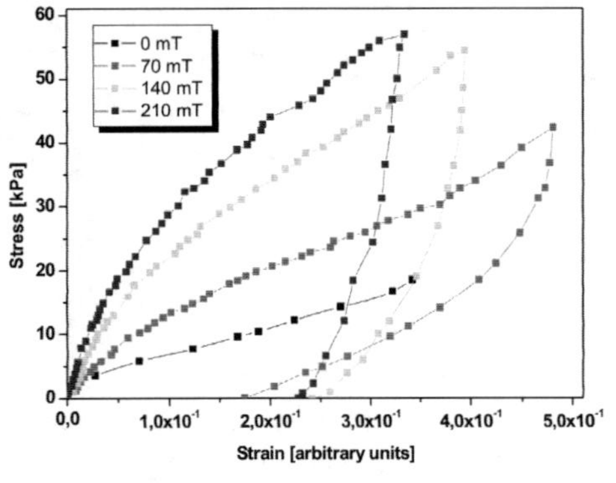

Figure 1. Stress – strain dependence for various magnetic field strength

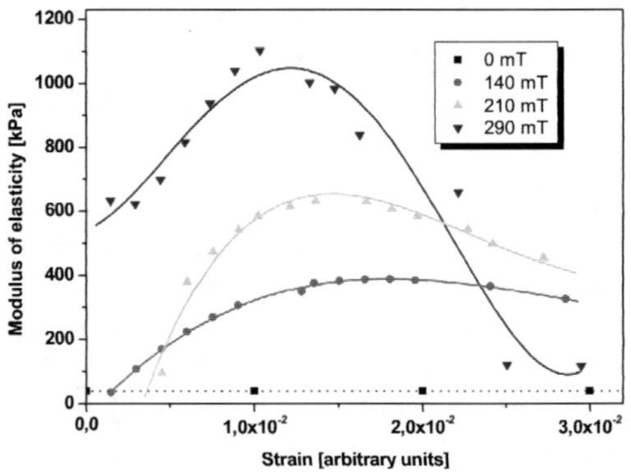

Figure 2. Elasticity modulus versus strain for various magnetic fields strength

The Young's modulus obtained from the initial part of each curves in figure 1 determined in the range of strain from 0-5% is considerably greater than the mean value for the whole curve. Figure 2 shows the dependence of the elasticity modulus on strain for various magnetic fields. An extremum of elasticity occurs for deformations of 1-2 %. With increasing magnetic field strength the extremum shifts towards lower deformations (1 %). A possible explanation for this effect bases on the microstructure of the isotropic sample. Due to the mixture of large and small particles the interparticle distance can be kept as small as in structured MAE. Therefore the particle interaction is comparable and the resulting effects are similar.

Another feature of the developed material is the soft polymeric matrix. The particles in the matrix are able to move reversibly under the action of the magnetic field like it is described in [21]. Large

Journal of Physics: Conference Series **149** (2009) 012098 doi:10.1088/1742-6596/149/1/012098

particles have a higher probability for a displacement from a stationary position under the influence of a magnetic field and this leads to enhancing of the MR effect.

3.2. Magnetic measurements

Magnetic measurements were performed with help of a vibrating sample magnetometer (Lake Shore 7400). Figure 3 shows typical normalized magnetization curves for the MAE samples with different particles mixture. The shape of the curves changes considerably with the variation of the magnetic particles embedded in the matrix. One observes an increase of magnetic susceptibility and hysteresis with increasing particle size. Figure 4 shows in detail the dependence of magnetic susceptibility on magnetic field strength. Lower curve of hysteresis loop in figure 3 corresponds to the increasing magnetic field and upper curve to the decreasing one. The highest peak in magnetic susceptibility in figure 4 belongs to the decreasing part of a hysteresis loop and can be explained with fact that particles have already been aligned in the field direction.

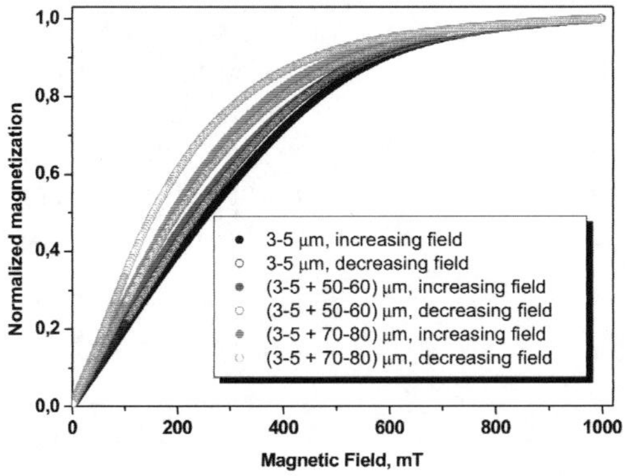

Figure 3. Normalized magnetization curves for MAE samples with different particles mixture.

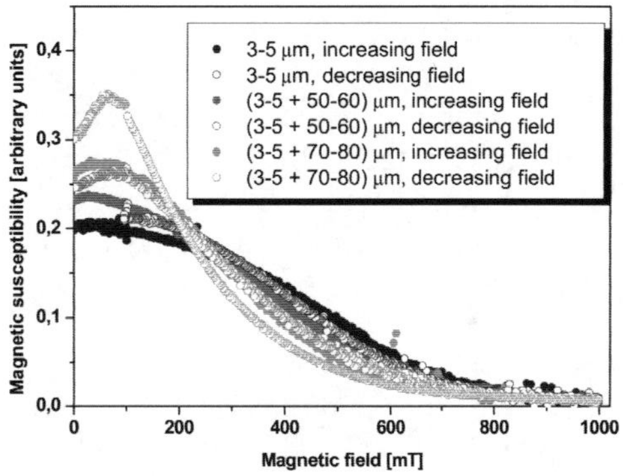

Figure 4. Dependences of a magnetic susceptibility of MAE samples with different particle mixture on a magnetic field strength.

Samples containing small particles only show small hysteresis and the susceptibility increase monotonously over the whole field range. In contrast samples with a mixture of small and large particles have a susceptibility maximum for B~80-100 mT which increases with particle size. Moreover they pronounced hysteresis. These characteristics of the magnetization curves can be explained with structurization processes of the magnetic filler in the polymeric matrix. The formation of structures has a reversible character and depends on magnetic field strength. After remove of the magnetic field the particles return to their initial positions under the influence of the elastic forces. With decreasing magnetic field strength, the structures remaining in samples containing large particles are substantially bigger than in samples containing small particles only. This difference is caused by the significantly larger interparticle interaction of the big particles being able to overcome the restoring elastic forces.

3.3. Optical microscopy

The effect of structurization of small (3-5 μm) and large (70-80 μm) particles in the polymeric matrix has been observed by means of optical microscopy and the results of the observation are presented below in videos recorded for two different magnifications.

With increasing magnetic field strength the particles form chain-like structures aligned in field direction. This structural reorganization is reversible: the particles return to their initial positions with decreasing magnetic field strength.

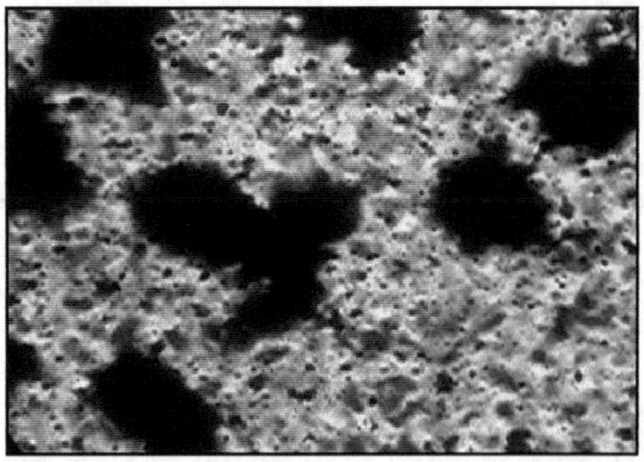

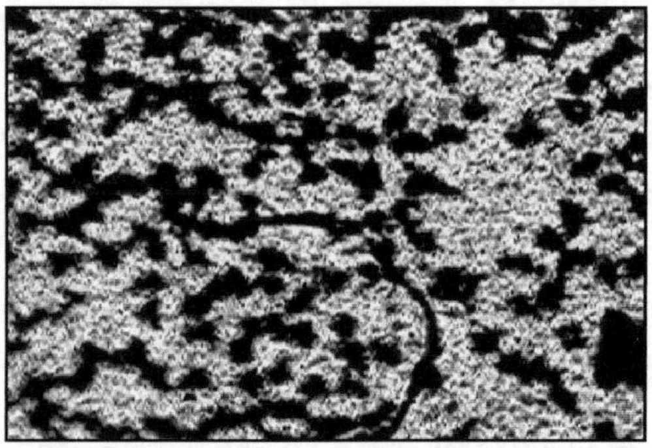

Video. Reversible structure formation in MAE containing small (3-5 μm) and large (70-80 μm) particles under action of applied magnetic field for two different magnifications.

4. Conclusions

The magnetorheological effect in MAE is caused by displacements of the particles inside the matrix. Due to the strong increase of the interparticle interaction with particle size compared to the restoring elastic forces, presence of large particles in MAE leads to the observed steep increase of Young's modulus (increase of the MR effect). In the range of small deformations (1%) the increase of elasticity in a magnetic field strength of 290 mT is about 100 times. This is comparable with the effects obtained for anisotropic MAE. In contrast to MAE containing large particles only or to MR fluids it is still possible to prevent sedimentation during the synthesis of the samples.

References

[1] Zrinyi M, Gacs J and Simon C Patent WO 9702580.
[2] Szabo D, Szeghy G and Zrinyi M 1998 Macromolecules 31 6541
[3] Carlson J D and Jolly M R 2000 Mechatronics, 10 (4) 555
[4] Bossis G, Abbo C, Cutillas S, Lacis S and Metayer C 2001 Int. J. Mod. Phys. B 15 564
[5] Zhou G Y and Yiang Z Y 2004 Smart Mater. Stract. 13 309
[6] Nikitin L V, Stepanov G V, Mironova L S and Gorbunov A I 2004 J. Magn. Magn. Mater. 272-276 (III) 2072
[7] Shen Y and Golnaraghi M F 2005 J. Intell. Mater. Sys. Struct. 15 27
[8] Guan X, Dong X and Qu J 2008 J. Magn. Magn. Mater. 320 3-4 158
[9] Jolly M R, Carlson J D and Munoz B C 1996 J.Intell. Mater. Sys. Struct. 7 613
[10] Stewart W, Ginder M, Nichols M and Elie L US Patent 5,816,587
[11] Watson J R 1997. US Patent 5,609,353
[12] Borin D, Stepanov G, Mikhailov V and Gorbunov A 2007 Magnetohydrodynamics 4 437
[13] Deng H X and Gong X L 2008 Communications in Nonlinear Science and Numerical Simulation 13 9 1938
[14] Bellan C and Bossis G 2002 Int. J. Modern Phys. B 1 6 2447
[15] de Vicente J, Bossis G and Lacis S 2002 J. Magn. Magn. Mater. 251 100
[16] Gong X L, Zhang X Z and Zhang P Q 2005 Polymer Test. 24 5 669
[17] H. Böse: Int. J. Mod. Phys. B 21 (2007) 4790-4797
[18] Lokander M and Stenberg B 2003 Polymer Test. 22 245
[19] Wang Y, Hu Y, Chen L, Gong X, Jiang W, Zhang P and Chen Z 2005 Polymer Test. 25 2 262
[20] Stepanov G V, Abramchuk S S, Grishin D A, Nikitin L V, Kramarenko E Yu and Khokhlov A R 2007 Polymer 48 488
[21] Stepanov G V, Borin D Yu, Raikher Yu L, Melenev P and Perov N 2008 J. Phys. Condens. Matter. 20 2041212008

Journal of Physics: Conference Series **149** (2009) 012099 doi:10.1088/1742-6596/149/1/012099

Numerical Investigation of Smart Base Isolation System Employing MR Elastomer

M Usman[1], S H Sung[1], D D Jang[1], H J Jung[1] and J H Koo[2]

[1]Dept. of Civil & Environmental Engineering, KAIST, 305-701, Guseong-dong, Yuseong-gu, Daejeon 305-701, Korea
[2]Dept. of Mechanical & Manufacturing Engineering, Miami Univ., Oxford, Ohio 45056, USA

E-mail: hjung@kaist.ac.kr

Abstract. This paper evaluates the dynamic performance of a newly proposed smart base isolation system employing Magneto-Rheological Elastomers (MREs). MREs belong to a class of smart materials whose elastic modulus or stiffness can be adjusted by varying the magnitude of the magnetic field. The base isolation systems are considered as one of the most effective devices for vibration reduction of civil engineering structures in the event of earthquakes. The proposed base isolation system strives to enhance the performance of the conventional base-isolation system by using controllable MREs. To validate the effectiveness of the MRE-based isolation system, an extensive simulation study has been performed using a five degree-of-freedom structure under several historical earthquake excitations. The results show that the proposed system outperformed the conventional system in reducing the responses of the structure in all the seismic excitations considered in the study.

1. Introduction

Hybrid-type base-isolation systems have been developed in recent years to improve the performance of passive-type base isolation systems. In the 1980s and early 1990s, considerable attention was paid to an active control device-based hybrid system (e.g., [5],[7]). However, active control-based hybrid systems have many problems, like instability, reliability, power consumption, etc. These issues therefore limited their implementation in large-scale structures. Consequently, later researchers focused on the feasibility of semi-active control devices for seismic response reduction of base-isolated structures which can be both adaptable and stable, while maintaining low external power requirements (e.g., [1]). A smart structural control system employing semi-active control devices (i.e., MR fluid dampers) was proposed and studied by [6] and [8].

Recently, a new type of semi-active base-isolation system has been proposed to incorporate Magneto-Rheological Elastormers (MREs) in base-isolation systems. MREs contain micron-sized magnetisable particles embedded in elastomeric or rubber-like solids (e.g., silicone rubber or natural rubber). The application of a magnetic field to the elastomers increases their spring rate of stiffness [3]. Moreover, MREs may overcome several difficulties that MR fluids have such as deposition, environmental contamination and sealing problems because of their unique feature. For civil engineering applications of MREs, Hwang et al. [4] carried out a conceptual study on the application of MREs to base isolation systems for building structures. However, the feasibility of MREs to base-isolated structures has not been fully explored. As such, this paper intends to investigate the feasibility of the smart base-isolation system employing MREs by conducting numerical simulations. For a dynamic model of MREs, a semi-active stiffness device model proposed by Gandhi et al.[2] is used. In numerical simulations, the six degree-of-freedom structural model coupled with the base-isolation system is considered [6], and three scaled historical earthquakes (El Centro, Northridge, and Hachinohe) are used as input ground motions.

2. Smart Base Isolation System based on MR Elastomers

2.1. System Model

A simple representation of the system is shown in Figure 1. When the isolation layer is added to the structural model, the whole structural model can be treated as a 6-DOF system. In Figure 1, M_b, K_b and C_b represent the mass of the isolation layer, the stiffness and the damping coefficient of the base isolator, respectively. The variable stiffness of the MREs is represented as $K(t)$. The system parameters are: M_b = 6800 kg, M_1=M_2=M_3=M_4=M_5=5897 kg, K_b = 232 kN/m, K_1 = 33732 kN/m, K_2 = 29093 kN/m, K_3 = 28621 kN/m, K_4 = 24954 kN/m, K_5 = 19059 kN/m, C_b = 3.74 kN·s/m, C_1 = 67 kN·s/m, C_2 = 58 kN·s/m, C_3 = 57 kN·s/m, C_4 =50 kN·s/m, C_5 = 38 kN·s/m.

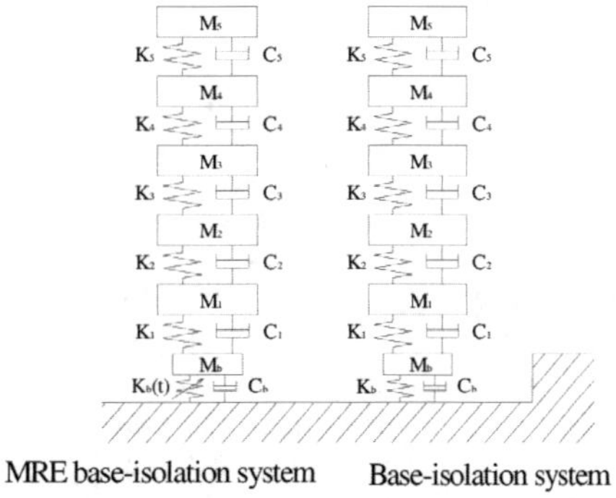

Figure 1. A building structure considered in this study is a 5-DOF system with the fundamental mode [5].

2.2. MRE System Modelling

To describe the dynamic behaviour of MREs, they are considered as a semi-active controllable stiffness device. In other words, their total stiffness, which can vary according to the external input current (magnetic field), is denoted as $K(t)$. It can be expressed as

$$K(t) = K_0 + K_1(u(t)) \tag{1}$$

where K_0 means the baseline stiffness of MREs and $K_1(u(t))$ represents the variation in stiffness due to the command input $u(t)$. The maximum and minimum physically achievable values of the stiffness are K_0 - K_1 and K_0 + K_1, respectively. According to preliminarily experimental tests, the maximum value of K_1 can be 0.7 K_0 (i.e., $K_1 = 0.7K_0$). Therefore, the range of the variation in the stiffness of MREs is

$$0.3K_0 \le K(t) \le 1.7K_0 \tag{2}$$

The range of additional force due to the variable stiffness K_1 from equation (1) is

$$-0.7K_0 x(t) \le f(t) \le 0.7K_0 x(t) \tag{3}$$

where $x(t)$ is the deformation of MREs or the base displacement. The additional force $f(t)$ can be varied with the command input $u(t)$ that is calculated from the semi-active control algorithm. In this study, the LQR (linear quadratic regulator)-based control algorithm is considered and the same weighting matrices for the control algorithm are used as in [6].

2.3. Equations of Motion

Neglecting the nonlinear effects and denoting the base and storey displacements relative to the ground by $x=[x_b\ x_1\ x_2\ x_3\ x_4\ x_5]^T$, the equations of motion of the base-isolated system may be expressed as

$$M\ddot{x} + C\dot{x} + K(t) = -M\Gamma\ddot{x}_g \tag{4}$$

or
$$M\ddot{x} + C\dot{x} + K_0 x = \Lambda f - M\Gamma \ddot{x}_g \tag{5}$$

where M, C and K are the mass, damping, and stiffness matrices are, respectively, f is the supplemental force exerted by the MREs, $\Lambda=[1\ 0\ 0\ 0\ 0\ 0]^T$ the position of the supplemental damper force, Γ vector whose elements are all unity, $\ddot{x}_g$ the ground acceleration. Defining the state vector $z = [x \quad \dot{x}]^T$, the output vector to be regulated $y = [x \quad \dot{x} \quad \ddot{x}]^T$, the state-space form of the equations of motion is given by

$$\dot{z} = Az + Bf + E\ddot{x}_g \tag{6}$$
$$y = Cz + D_y f + F_y \ddot{x}_g \tag{7}$$

3. Numerical Simulation Results

To verify the feasibility of the MRE-based smart base isolation system, numerical simulations are carried out using the simulation model above along with the system parameters provided in section 2. The LGR control was implemented to regulate the stiffness of the MRE. The numerical simulation results of hybrid system employing MREs are compared to those of a passive-type base isolation system using rubber bearings (RB).

Figure 2 shows the storey drifts at the first floor for each system for various earthquake inputs. As shown in the figure, the smart base isolation system employing MREs significantly reduce the response compared with the passive-type base isolation system for various excitations. Figures 3 and 4 compare the peak storey drift at the first floor and the peak acceleration at the top floor for each earthquake input, respectively. The results show that the MRE base-isolation outperforms the conventional base-isolation system.

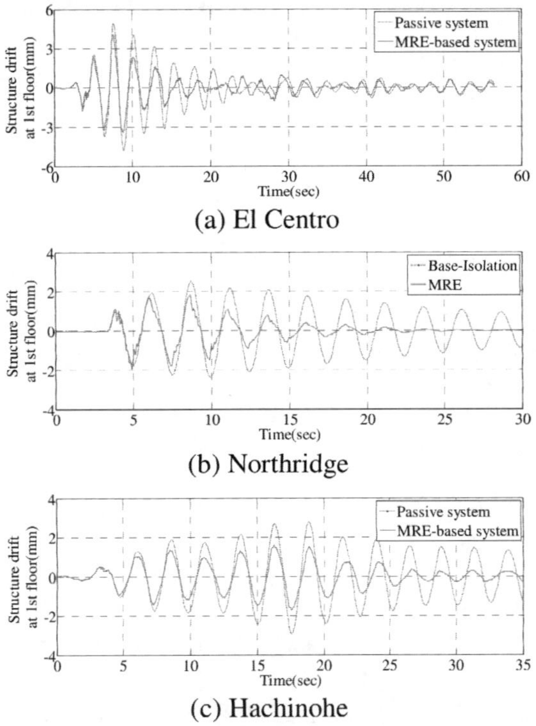

Figure 2. Storey drift at first floor for various earthquake inputs

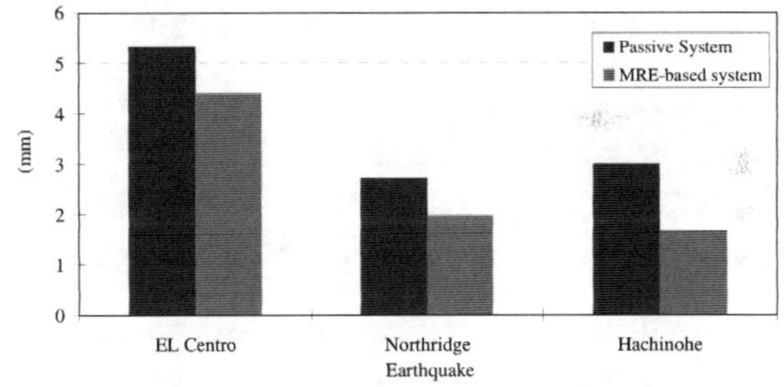

Figure 3. Storey drift at first floor for various earthquakes

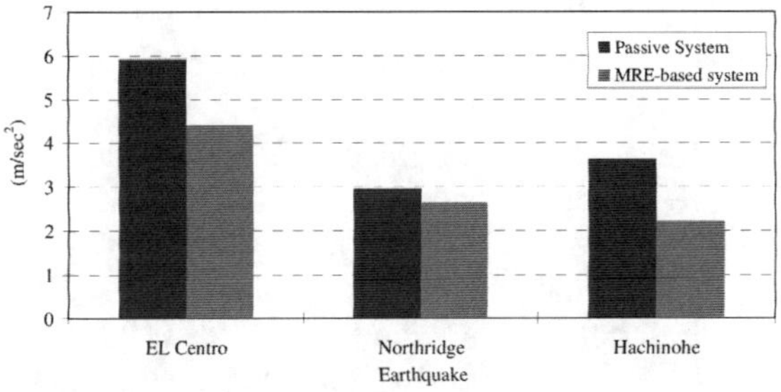

Figure 4. Structural acceleration at top floor for various earthquakes

Table 1 shows the percent response improvement of the base isolation system using MREs as compared with the passive base-isolation system under different earthquake inputs. These results demonstrate that the MRE-based smart base-isolation system can significantly reduce the structural accelerations and storey drifts as well as the base displacements (or isolator deformations).

Table 1. Percent response improvement compared to the conventional base isolation system

Earthquake	Base displacement	Storey drift at first floor	Structural acc. at top floor
El Centro	20.6 %	17.4 %	25.7 %
Northridge	38.5 %	27.3 %	10.3 %
Hachinohe	45.3 %	43.7 %	39.0 %

4. Conclusions

This paper investigated the feasibility of the MRE base-isolation system by evaluating the dynamic performance of the system. In this study, MREs are idealized as a controllable stiffness device. To study the control performance of the MRE-based smart system, numerical simulations were conducted by using a 5-DOF structure model. The performance of the MRE-based system is compared to that of the passive base isolation system, which uses rubber bearings. The numerical simulation results show that all the responses (base drift, structural acceleration and inter-story drift) of the MRE base-isolation system were smaller than those of the passive system, indicating that the MRE system outperformed its passive counterpart. In summary, this simulation study validated the feasibility of the MRE base-isolation system as a new "smart" base-isoation system, and more studies (including experiemnts) should follow to further validte the effectiveness of the MRE system.

Acknowledgements

This research was supported by the Smart Infra-Structure Technology Center (funded by the Korea Science and Engineering Foundation), the Construction Core Technology Program (funded by the Ministry of Land, Transport and Maritime Affairs of Korea), and Higher Education Commission, Government of Pakistan. Their financial supports are gratefully acknowledged.

References

[1] Feng, Q., and Shinozuka, M. (1990), ''Use of a variable damper for hybrid control of bridge response under earthquake,'' Proc., U.S. National Workshop on Structural Control Research, USC Publication No. CE-9013.

[2] Gandhi, F. and Anusonti-Inthra, P. (2003), "Adaptive control of semiactive variable stiffness devices for narrow-band disturbance rejection," Journal of Intelligent Material Systems and Structures, 14, 191-201.

[3] Ginder, J. M., Schlotter, W. F. and Nichols, M. E. (2001), "Magnetorheological elastomers in tunable vibration absorbers" Smart Structures and Materials 2001: Damping and Isolation, 4331, 103-110.

[4] Hwang, I.-H., Lim, J.-H. and Lee, J.-S. (2006)," A study on base isolation performance of magneto-sensitive rubbers," Journal of Earthquake Engineering Society of Korea, 10, 77-84 (in Korean).

[5] Kelly, J. M., Leitmann, G., and Soldatos, A. G. (1987), ''Robust control of base-isolated structures under earthquake excitation,'' J. Optim. Theory Appl., 53, 159–180.

[6] Ramallo, J. C., Johnson, E. A. and Spencer, B. F., Jr. (2002), ''Smart'' base isolation systems," Journal of Engineering Mechanics, ASCE, 128(10), 1088-1099.

[7] Reinhorn, A. M., and Riley, M. (1994), ''Control of bridge vibrations with hybrid devices,'' Proc., 1st World Conf. on Structural Control, Los Angeles, TA2, 50–59.

[8] Yoshioka, H., Ramallo, J. C. and Spencer, B. F., Jr. (2002), "'Smart' base isolation strategies employing magnetorheological damper." *Journal of Structural Engineering*, ASCE, **128**(5), 540-551.

Journal of Physics: Conference Series **149** (2009) 012100 doi:10.1088/1742-6596/149/1/012100

Locomotion Based on Nonlinear Magneto-Elastic Elements

K Zimmermann[1], I Zeidis[1], V Naletova[2,3], V Turkov[3], V Böhm[1] and J Popp[1]

[1]Department of Mechanical Engineering, Technische Universitaet Ilmenau, PF 100565, Ilmenau, Germany
[2]Department of Mechanis and Mathematics, Moscow State University, Moscow 119992, Russia
[3]Institute of Mechanis, Moscow State University, Michurinskii Pr. 1, Moscow 119192, Russia

E-mail: igor.zeidis@tu-ilmenau.de

Abstract. In this paper we discuss an approximately steady motion of two equal mass points, connected by a non linear magneto-elastic element. It is supposed that the system moves along a straight line in the presence of internal excitation and non symmetric Coulomb dry frictional force acting from the surface upon each mass point opposite to the direction of motion. Thereby, the magnitude of this force is also dependent on the direction of motion. Excitation is carried out due to action internal harmonic forces. Such forces arise for a spring made of a magnetizable elastic material by the influence of an external magnetic field.

1. Introduction

The locomotion without the help of legs, i.e. non-pedal locomotion, evoke great interest. Non-pedal forms of locomotion show their advantages in inspection techniques or application in medical technology for diagnostic systems and minimally invasive surgery. Realization of this type of locomotion is possible in some different ways. One of the possibilities is the using of the magnetazable elastic material in an external magnetic field. It is well-known, that by the influence of a magnetic field a ferroelastic element expends. In [1] the motion of a system of two material points, connected by a spring made of magnetizable linear elastic element is considered. In this paper we consider the influence of nonlinearity in magneto-elastic element on the motion.

2. Mathematical model

In [2] a deformation of the elastic composite body, when a magnetic field is applied, is considered. By the action of a magnetic field the sample of the composite with a drop of a magnetic fluid lengthens. In figure 1 the experimental dependencies between the stress p and the strain u without the influence of a magnetic field ($H = 0\ A/m$) and with influence of a field ($H = 1.5 \cdot 10^5\ A/m$) for the sample of a magnetizable material with the length $l_0 = 29mm$, the area of cross section $S_0 = 5.5mm^2$ is shown. The size of ferromagnetic particles is about $110\overset{0}{\mathrm{A}}$. The experiment allows us to define the strain $u_0 = u(0) = 0.071$ (Figure 1).

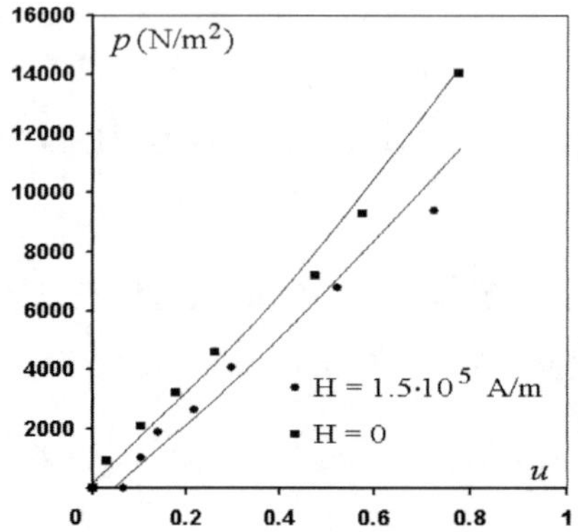

Figure 1. Stress p vs. strain u.

We consider the motion of a system of two material points x_1 and x_2 with the masses m, connected by a spring made of magnetizable elastic element along an axis x (Figure 2).

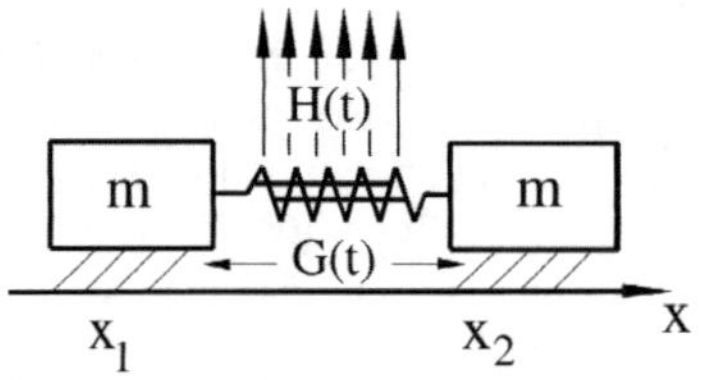

Figure 2. Model of the system.

The spring has a small cubic nonlinearity. Elastic force $f(x)$ in this case is

$$f(x) = -cx - \varepsilon d m x^3, \quad \varepsilon << 1. \tag{1}$$

Here $c > 0$ and $\varepsilon d \geq 0$ are constants of the elastic element, ε is the small parameter. It is supposed that the points are under the action of a small nonsymmetric Coulomb dry friction force $\varepsilon m F(\dot{x})$. The friction force is taken to be different in magnitude depending on the direction of body motion. The function $F(\dot{x})$ can be specified as follows: $F(\dot{x}) = F_-$, *if* $\dot{x} < 0$, $F(\dot{x}) = -F_+$, *if* $\dot{x} > 0$, $F(\dot{x}) = -F_0$, *if* $\dot{x} = 0$. Here $F_- \geq F_+ \geq 0$ are fixed, whereas F_0 may take any value in the interval $(-F_-, F_+)$.
Excitation is carried out due to action of small harmonic internal force by influence of an external magnetic field

$$G(t) = \varepsilon b m \cos\psi, \ \psi = \nu t. \tag{2}$$

Using expressions (1) and (2) and designating $\omega^2 = c/m$, replacing we obtain thefollowing system of equations:

$$\ddot{x}_1 + \omega^2 (x_1 - x_2) = \varepsilon \left(F(\dot{x}_1) - d(x_1 - x_2)^3 - b\cos\psi \right),$$
$$\ddot{x}_2 + \omega^2 (x_2 - x_1) = \varepsilon \left(F(\dot{x}_2) - d(x_2 - x_1)^3 + b\cos\psi \right). \quad (3)$$

To system (3) we apply the procedure of averaging according to [3]. For this purpose we introduce new variables: the velocity of the center of mass V and the deviation of the points relatively to the center of mass z: $V=(\dot{x}_1+\dot{x}_2)/2$, $z=(x_2-x_1)/2$, $\Omega=\sqrt{2}\,\omega$, $z=a\cos\varphi$, $\dot{z}=-a\Omega\sin\varphi$, $\varphi=\Omega t+\vartheta$. Here a is the amplitude, φ is the phase of the oscillation relative the center of mass. We investigate the system in a vicinity of the main resonance $\nu=\Omega+\varepsilon\Delta$. For this purpose we introduce a new slow variable $\xi=\psi-\varphi$ and exclude a fast variable ψ. Details of application of this method are stated in the [4]. After averaging on a fast variable φ we obtain:

$$\dot{V} = \begin{cases} -\varepsilon\left(\dfrac{F_- + F_+}{\pi}\arcsin\dfrac{V}{a\,\Omega} - \dfrac{F_- - F_+}{2}\right), 0 \le V < a\,\Omega, \\ -\varepsilon F_+ \quad \text{if } V \ge a\,\Omega, \end{cases}$$

$$\dot{a} = \begin{cases} -\dfrac{\varepsilon}{\Omega}\left(\dfrac{F_- + F_+}{\pi}\sqrt{1 - \dfrac{V^2}{a^2\Omega^2}} - \dfrac{b}{2}\sin\xi\right), 0 \le V < a\Omega, \\ \varepsilon\dfrac{b}{2\Omega}\sin\xi,\ V \ge a\,\Omega, \end{cases} \quad (4)$$

$$\dot{\xi} = \varepsilon\left(\frac{b}{2a\,\Omega}\cos\xi + \Delta - \frac{3}{\Omega}d\,a^2\right).$$

3. Analysis of the model

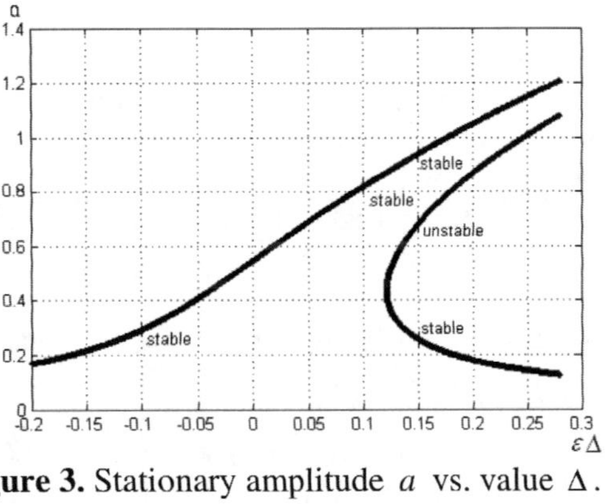

Figure 3. Stationary amplitude a vs. value Δ.

We are interested in an approximately steady motion as a single whole; therefore we seek for the solution $\dot{V}=0$ of the system of equations (4): $V = a\,\Omega\sin\Phi$, $\Phi=\pi(F_- - F)/(2(F_- + F_+))$, $E=(F_- + F_+)/\pi$.

Since $V = const$, it follows from the equations of the system (4) that $a = const$ and $\xi = const$. The equation for stationary amplitude a can be written as follows:

$$a\left|3ad - \Delta\Omega\right| = \sqrt{b^2/4 - E^2\cos^2\Phi}\,. \qquad (5)$$

Figure 3 shows the dependence of the stationary amplitude a on the value of Δ, obtained from equation (5). Since the frequency ν of the driving force is connected with Δ by formula $\nu = \Omega + \varepsilon\Delta$, the curve on the Figure 4 is a resonance curve. The values of the parameters are: $\varepsilon = 0.01$, $\omega = 1.0$, $F_+ = 1.0$, $F_- = 2.0$, $b = 10.0$, $d = 10.0$. For $\Delta < 12$ ($\varepsilon\Delta < 0.12$) there is only one value for the stationary amplitude. This amplitude is stable. For $\Delta > 12$ ($\varepsilon\Delta > 0.12$) there are three values for the stationary amplitude and, accordingly, for the velocity of the system as a single whole, but at most there exists two stable amplitudes.
For the case considered we assume $\Delta = -10$ ($\varepsilon\Delta = -10$) and the equation for stationary amplitudes (5) has only one stable solution, for $a = 0.28$ (Figure 4). The corresponding value for the velocity of the center of mass is $V = 0.2$. Figure 4 shows the velocity of the center mass versus time, obtained from the exact equations of motion (3). The exact velocity converges toward the value $V = 0.2$.

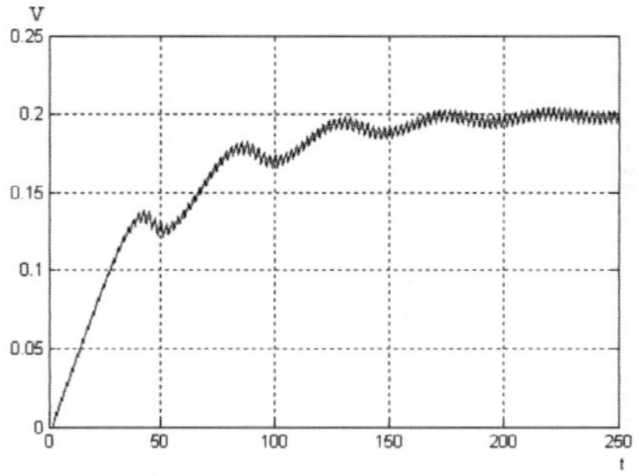

Figure 4. Velocity V vs. time t.

4. Conclusions

We have considered the locomotion of the system with nonlinear magneto-elastic element under the action of an external magnetic field as an actuator. Using the averaging method an expression for the velocity of the system as a whole and an algebraic equation for the corresponding stationary amplitudes can be obtained. Depending on the values of the input parameters the conditions for stability and instability of the stationary amplitudes can be found. At most there exist two stable amplitudes.

Acknowledgments

The work was supported by Deutsche Forschungsgemeinschaft (DFG, Zi 540-11/1).

References

[1] Zimmermann K, Zeidis I, Naletova V A and Turkov V A 2004 Modelling of worm-like motion systems with magneto-elastic elements *Phys. status. Solidi* (c) **1**(12) 3706-3709

[2] Turkov V A 2002 Deformation of an elastic composite involving a magnetic fluid *J. Magn Magnet Mater.* **252** 156-158

[3] Bogoljubow N N and Mitropolski Yu A 1965 *Asymptotische Methoden in der Theorie der nichtlinearen Schwingungen* (Berlin: Akademie-Verlag)

[4] Zimmermann K, Zeidis I and Behn C 2009 *Mechanics of Terrestrial Locomotion* (Heidelberg: Springer-Verlag)

The Cotton-Mouton effect in ferrofluids containing rod-like magnetite particles

V Badescu[1], LE Udrea[1], O Rotariu[1], R Badescu[2] and G Apreotesei[2]

[1] National Inst. of R-D for Technical Physics, 47 Mangeron Blvd., Iasi, 700050, RO
[2] Technical Univ. "Gh. Asachi", Dept. of Physics, 67 Mangeron Blvd., Iasi, 700050, RO

E-mail: bav08@phys-iasi.ro

Abstract. Ferrofluids with rod-like nano-sized magnetite particles were prepared by precipitation from aqueous solutions of ferrous hydroxide and aging them in the presence of a magnetic field. We studied the Couton-Mouton effect measuring the azimuthally distribution of the transmitted light intensity and obtained that its magnetic field dependence differs considerably for ferrofluids containing rod-like and spherical magnetic nanoparticles, respectively. After switching off the magnetizing field, the relaxation of the birefringence was observed, measured and analysed.

1. Introduction

Ferrofluids are very stable colloidal suspensions of ultrafine single domain magnetic particles in a suitable liquid carrier [1, 2]. They have a range of interesting physical properties, from which the magneto-optical properties may provide a basis for numerous applications. The main optical effects studied in ferrofluid films under parallel magnetic fields are linear dichroism and birefringence [3 - 9].

Some recent biomedical applications of ferrofluids require avoiding of field induced chain forming or aggregation of magnetic nanoparticles in suspension. From these we noticed the using of ferrofluids in monitoring of biological binding reactions, particularly by magnetooptical relaxometry [10 - 12].

To enhance the birefringence signal implicated in magnetooptical relaxometry we prepared a ferrofluid containing in suspension rod-like nano-sized magnetite particles with axial ratios >2. This non-typical morphology of particles is the key of our work. For such ferrofluids and at low concentrations the magnetic induced birefringence origin is dominated by form anisotropy of nanocrystallites in suspension. Furthermore, the birefringence relaxation takes place only by rotational diffusion of nanorods about some axis perpendicular to theirs cylindrical axis.

Within the scope of our work were to prepare the ferrofluid containing magnetite particles with above mentioned non-typical morphology, to characterize them and to measure the Cotton-Mouton (C-M) effect both in stationary and transitive regime.

2. Materials and methods

For our studies we prepared ferrofluids having octane as liquid base and oleic acid as stabilizing agent. The magnetic fraction was magnetite, prepared by two synthesis routes introduced by Khalafalla and Reimers [13] and Sugimoto and Matijević [14], respectively. The first route is based on the coprecipitation of $Fe(OH)_2$ and $Fe(OH)_3$ from an aqueous solutions mixture under the action of a strong alkali, followed by the fast aging of those. The obtained nanoparticles were transferred into the octane, resulting a ferrofluid with a magnetite volume content of 0.6%, named further on *sort I* ferrofluid.

The second procedure was analogous to that followed by Zhang et al. [15] and by Vereda et al. [16] with some minor modifications. The process consist in the precipitation of $Fe(OH)_2$ also in aqueous media and the slow aging of that gel in the presence of NO^{3-}. For this aging the reaction system was placed in a hot oil bath preheated to 90^0C and mounted in the cylindrical gap of an electromagnet with closed iron circuit. The solution was slightly stirred for 2 h with the axis of agitator parallel to force lines of the magnetic field. Test probes were extracted from the flask after 0.5 h, 1 h and 1.5h. This stage was repeated at different magnetic field intensities of 80 kA/m, 160 kA/m, 240 kA/m, 320 kA/m and 400 kA/m. The obtained probes of magnetite were cured with oleic acid and dispersed in octane to obtain some ferrofluids with a volume content of magnetite of 0.6%, named further on *sort II* ferrofluids.

The magnetizations of the two sorts of ferrofluids were measured with a vibrating sample magnetometer. In figure 1 are presented the magnetic properties for one *sort II* ferrofluid, the time of aging being 1 h and the intensity of the curing applied magnetic field being 160 kA/m. The measurement showed that the obtained ferrofluid have a superparamagnetic comportment.

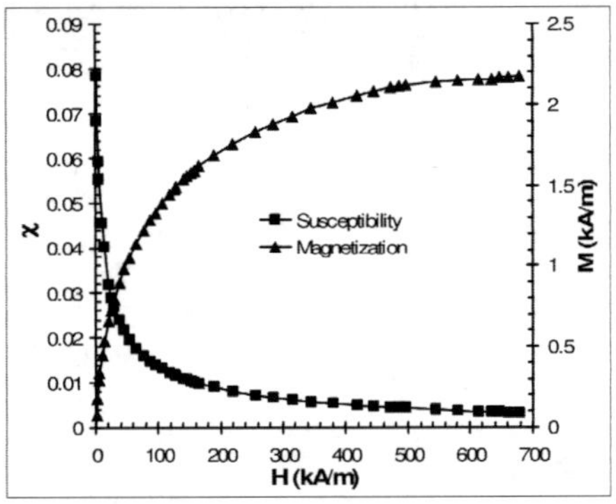

Figure 1. The magnetic properties for a *sort 2* ferrofluid.

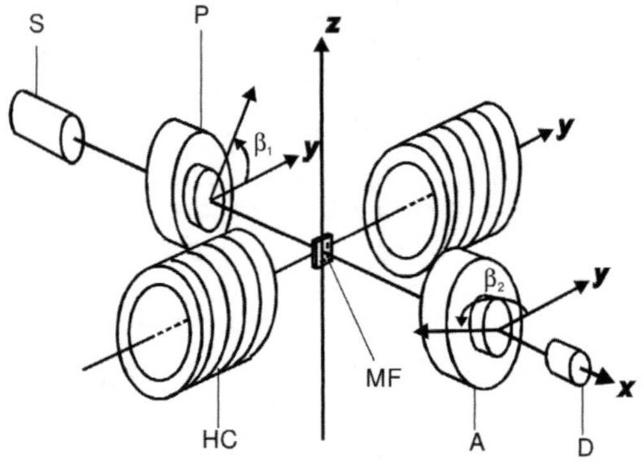

Figure 2. The set-up for the study the azimuthal distribution of the transmitted light intensity.

For a qualitative analysis of the birefringence effect in the prepared ferrofluids we adopted a method relied on the measurement of the azimuthally distribution of the transmitted light intensity through a ferrofluid film magnetized parallel to its plane by an external continuous magnetic field [17, 18]. The experimental set-up is shown in figure 2. A He-Ne laser beam (λ=632 nm) is linearly polarized by the polarizer P and then is incident normally to the ferrofluid film MF. The magnetic field is generated by a pair of Helmholtz coils (HC). The angle β_1 between the transmission axis of the polarizer and the magnetic field is adjustable. The analyzer A is used to examine the polarization state of the transmitted light. The azimuthal angle $\eta = \beta_2 - \beta_1$ is modified rotating the analyzer. Finally, a receiving photodiode (D) is used to detect the intensity I of the transmitted light. In this work the $I-\eta$ curves were measured for the prepared ferrofluids at β_1=45^0 and various H's. Under zero field, the symmetric $I-\eta$ curves exhibits a relative maximum intensity at $\eta_{I=\max}$=0 and zero intensity at $\eta_{I=\min}$=90^0 and - 90^0. The symmetric $I-\eta$ curves under $H \neq 0$ are also obtained with a translational shift $\Delta\eta$ with respect to the zero field curves. This shift implies that the transmission axis of the transmitted light was rotated by an angle $\Delta\eta$ with respect to that of the incident light (i.e. C-M rotation), and it is considered as a qualitative measure of the linear dichroism. On the other hand, the minimum intensity $I_{\min}$ occurred at $\eta_{I=\min}$ is non-zero under $H \neq 0$. This $I_{\min} \neq 0$ implies that the transmitted light is not linearly but elliptically polarized, and it is considered a qualitative measure of birefringence.

Fixing the analyzer at $I_{\min}$ and switching off the magnetizing field, a relaxation of the birefringence can be observed. Assuming that the magnetic nanoparticles in suspension are cylindrical, the birefringence relaxation is determined by a single exponential term

$$\Delta n = \Delta n_0 \exp(-6D_r t) \quad , \tag{1}$$

where Δn_0 is a function of the applied field before its turning off and D_r is the rotational diffusion constant about an axis perpendicular to the cylindrical axis. The D_r can be written as [19]

$$D_r = \frac{3kT}{\pi\eta L^3}\left(\ln\frac{L}{d} - \gamma\right) \quad , \tag{2}$$

where k is Boltzmann's constant, T is temperature, η is the viscosity of the medium, L the length of the cylinder, and d the width. The parameter γ refers to end effects and depends upon the aspect ratio L/d. For sufficiently large L/d, the γ becomes inconsequential; for small L/d, [20]

$$\gamma = 0.662 - 0.92\left(d/L\right) \quad . \tag{3}$$

With known values for the viscosity η of the ferrofluid, the temperature T, Boltzmann's constant k and D_r determined from the magnetobirefringence relaxation measurement it is possible to calculate the aspect ratio L/d of the particles.

3. Results

In figure 3 are shown the $I-\eta$ curves for the *sort 1* ferrofluid. When a magnetic field of H = 10 kA/m is applied, $\eta_{I=\min}$ is moved by an angle $\Delta\eta$ = 1.08^0 and $I_{\min}$ is nonzero. Increasing the field up to 50 kA/m, the $I-\eta$ curves remain approximately in the same position. This behavior can be explained by a low optical anisotropy of the ferrofluid containing nanoparticles with small form anisotropy.

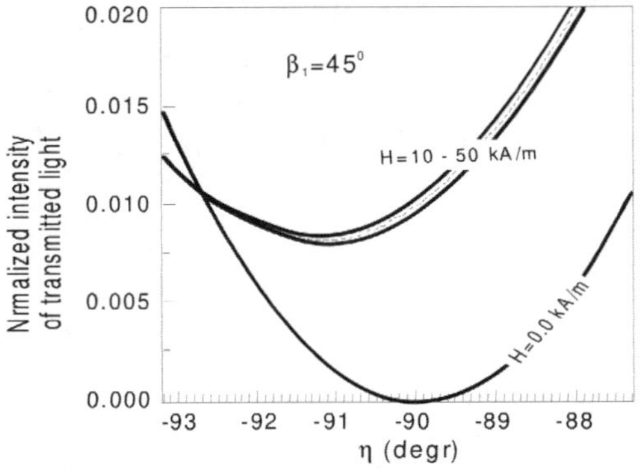

Figure 3. Azimuthally distribution of the transmitted light intensity for the *sort 1* ferrofluid.

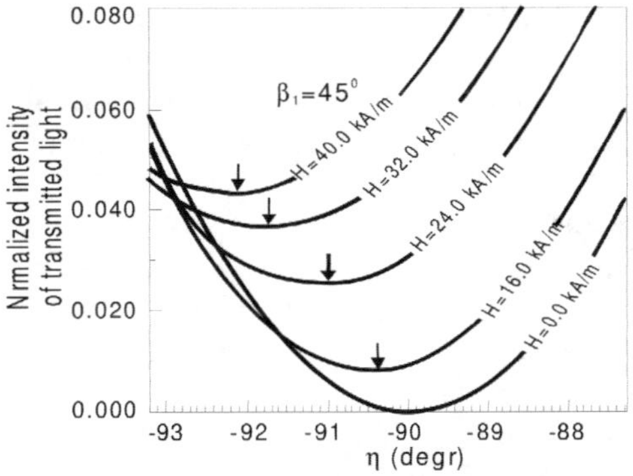

Figure 4. Azimuthally distribution of the transmitted light intensity for the *sort* 2 ferrofluid.

In figure 4 are shown the $I-\eta$ curves for *sort* 2 ferrofluid, magnetically characterized above. The value of $\Delta\eta$'s increased gradually up to 2.14^0 as H increased to 40 kA/m and then remains almost constant as H > 44 kA/m. A similar behavior is observed for $I_{\min}$. The fact that $I_{\min}$ depends on the magnetic field intensity implies that the configuration of the elliptical polarized transmitted light is affected by the applied magnetic field. We thought this effect as a consequence of the induced orientational order of the independent colloidal magnetic nanoparticles with large axial ratios. This result is in accord with those from references [15] and [16].

Figure 5 shows the time evolution of I_{min} for the *sort 1* ferrofluid, after the magnetic field with intensity of 40 kA/m is suddenly cancelled. It is observed a decrease of I_{min} in a very short time (< 0.1 sec) and after that the intensity of the transmitted light remains constant in time.

Figure 6 shows the time evolution of I_{min} for the *sort 2* ferroluid, after the magnetic field with intensity of

(40 kA/m) is suddenly cancelled. In this case two stages are distinguished for the relaxation process. In the first stage I_{min} suddenly decreases with approximately 25 % from the value it had in the presence of magnetic field. After that, I_{min} continuously decreases in a way which cannot be fitted by a simple exponential curve.

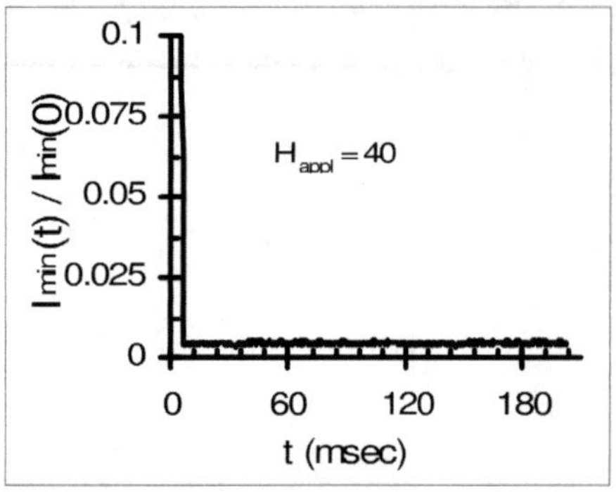

Figure 5. The time evolution of $I_{\min}$ for *sort 1* ferrofluid after switching off the magnetic field.

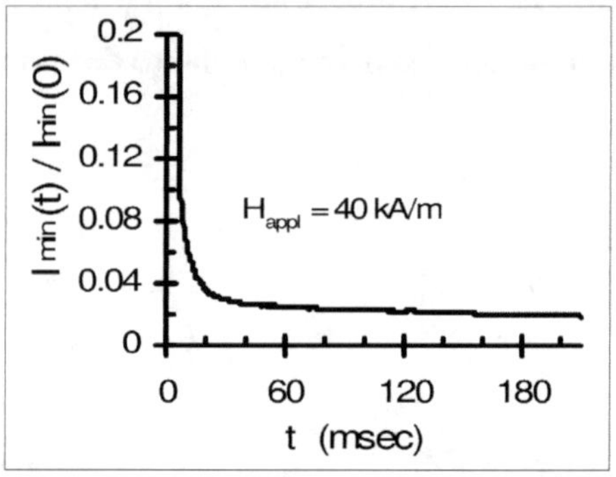

Figure 6. The time evolution of $I_{\min}$ for *sort 2* ferrofluid after switching off the magnetic field.

The fit of the relaxation curve of I_{min} can be approximated with a sum of exponential functions with various amplitudes and time constants. This behavior can be explained by the cumulating of some diffusion processes with different rotation diffusion coefficients, depending on the aspect ratios of nanoparticles in the ferrofluid.

4. Conclusions

For enhancing the birefringence signal implicated in magnetooptical relaxometry we prepared a ferrofluid containing in suspension rod-like nano-sized magnetite particles. At low concentrations these ferrofluids presented a superparamagnetic behaviour. The induced orientational order of the independent nanoparticles with large axial ratios in prepared ferrofluids lead to a clearly C-M effect. Since the time constants for the transitive regime depends on the hydrodynamic size of the particles, we will focus our future work on applying these ferrofluids for the monitoring of binding biological reactions.

References

[1] Rosensweig R E 1997 *Ferrohydrodynamics* (Mineola, New York: Dover)
[2] Odenbach S 2003, *Colloids Surf. A* **217** 171
[3] Martinet A 1974 *Rheol. Acta* **13**, 260
[4] Skibin Yu N and Chekanov V V 1977 *Magnetohydrodynamics* **13** 249
[5] Scholten P C 1980 *IEEE Tran.s Magn.* **MAG-16** 221
[6] Davies H W and Llewellyn J P 1980 *J. Phys. D: Appl. Phys.* **13** 2327
[7] Taketomi S 1983 *Jpn. J. Appl. Phys.* **22** 1137
[8] Xu M and Ridler P J 1997 *J. Appl. Phys.* **82** 326
[9] Hasmonay E, Dubois E, Bacri J –C, Perzynski R et al. 1998 *Eur. Phys. J. B* **5** 859
[10] Romanus E, Groβ C, Kötitz R, Prass S, Lange J et al. 2001 *Magnetohydrodynamics* **37** 328
[11] Wilhelm C, Gazeau F, Roger J, Pons J N et al. 2002 *Phys Rev E* **65** 031404
[12] Glöckl G, Brinkmeier K, Aurich K, Romanus E et al. 2005 *J. Magn. Magn. Mater.* **289** 480
[13] Khalafalla S E and Reimers G W 1980 *IEEE Trans. Magn.* **16** 178
[14] Sugimoto T and Matijević E J 1980 *J. Colloid Interface Sci.* **74** 227
[15] Zhang Y, Shi O H, Xiong H Q and Zhai Y 2005 *Int. J. Mod. Phys.B.* **19** 2757
[16] Vereda F, de Vicente J and Hidalgo-Álvarez R 2007 *Langmuir* **23** 3581
[17] Horng H E, Yang S Y, Lee S L, Wu J M et al. 2000 *Magnetohydrodynamics* **36** 39
[18] Fosa G, Bădescu R and Călugăru Gh 2004 *Rom. J. Phys.* 49 312
[19] Elias J G and Eden D 1981 *Biopolimers* **20** 2369
[20] Tirado J G and Eden D 1980 *J. Chem. Phys.* **73** 1986

Journal of Physics: Conference Series **149** (2009) 012102 doi:10.1088/1742-6596/149/1/012102

Double dipolar Halbach array for rheological measurements on magnetic fluids at variable magnetic flux density B

V C Barroso[1,2,3], **H Raich**[2], **P Blümler**[4] **and M Wilhelm**[1,2,3,5]

[1] Institute for Chemical Technology and Polymer Chemistry, Universität Karlsruhe (TH), Engesserstr. 18, 76128 Karlsruhe, Germany

[2] Max Planck Institute for Polymer Research, Ackermannweg 10, 55128 Mainz, Germany

[3] Faculty for Mechanical Engineering, Technische Universität Darmstadt, Hochschulstr. 1, 64289 Darmstadt, Germany

[4] Research Center Jülich, ICG-3: Phytosphere, Leo-Brandt-Str., 52425 Jülich, Germany

wilhelm@polymer.uni-karlsruhe.de

Abstract. A new experimental setup for measuring rheological properties of magnetic fluids at variable magnetic flux density B with increased homogeneity is described. The proposed setup is mounted on a commercial strain-controlled ARES rotational rheometer. The magnetic flux is generated via two concentric Halbach cylindrical arrays made from permanent NdFeB magnets. The use of permanent magnets overcomes some of the disadvantages of electromagnets (e.g., excessive heating of the coils, formation of large radial stray fields, cost of electricity, cooling, etc). The performance of the new setup is tested for a magnetorheological fluid in both steady and oscillatory shear regimes.

1. Introduction

Magnetic fluids are stable suspensions of ferromagnetic particles in a carrier liquid, divided into two main classes: i) ferrofluids, containing nano-sized particles, and ii) magnetorheological (MR) fluids, with micron-sized particles [1].

The measurement of their rheological properties at different magnetic field strengths has been recently receiving increasing interest [2-6]. A commercial apparatus to perform such measurements is available from Anton Paar GmbH, the MRD-Magneto Rheological Device [5-8]. However, most of the work reported in the literature refers to custom-built devices, either adaptations to commercially available rheometers [3,4] or purpose-built devices [9-11].

Arrays of permanent magnets possessing a one-sided flux were first proposed in 1973 in a theoretical study by Mallinson [12]. Later, Klaus Halbach [13] constructed arrangements of permanent multipole magnets, which are nowadays known as Halbach arrays. These arrays can have a different number of magnetic dipoles, depending on the specific layout of the magnets. The optimized design and construction of dipolar Halbach circular arrays using permanent NdFeB magnets was described by Raich and Blümler [14]. These circular arrays confine the magnetic field mostly to the inner part of the cylinder, with a high homogeneity (typically $\Delta B/B < 1$ %) being achieved.

In this work, a new experimental setup for measuring rheological properties of magnetic fluids at variable magnetic flux density B with increased homogeneity is described. The proposed setup is mounted on a commercial strain-controlled ARES rotational rheometer, and consists of two concentric Halbach cylindrical arrays made from permanent NdFeB magnets. The use of permanent magnets allows overcoming some of the disadvantages of using electromagnets as source for magnetic field (e.g., excessive heating of the coils, formation of large radial stray fields). The performance of the new setup was tested for a MR-fluid in both steady and oscillatory shear regimes.

2. Experimental

2.1. Description of the setup

Two distinct cylinders, each composed of two layers of dipolar Halbach circular arrays, were constructed. A homogeneous magnetic dipole inside each of the cylinders is obtained with this layout. The cylinders are placed inside one another, and can be rotated relatively to each other (Figure 1a). Hence, the resulting magnetic field at the center is the vector sum of the two individual fields and varies with the displacement angle θ.

The magnetic flux density was evaluated at the center of the setup, for different displacement angles θ. The B_y and B_z components of the magnetic flux were scanned in an area comprised of a circle with 28 mm diameter, in a grid with 2 mm spacing between each measurement point. The B_x component of the flux (along the axial direction of the setup) was found to be of the order of 0.1 mT for the maximum possible magnetic field on the setup (i.e., at $\theta = 0°$), and was therefore neglected for the other angles (see also [14]). The total planar magnetic flux density was determined by the vectorial sum of the two measured components, according to Equation 1:

$$\left|\vec{B}\right| = \sqrt{\left|\vec{B}_y\right|^2 + \left|\vec{B}_z\right|^2} \tag{1}$$

The magnetic flux density of this new setup covers a range of nearly zero to almost 0.6 T, with a homogeneity $\Delta B/B$ better than 1.8 % (Figure 1b). Non-uniformities in both the permanent magnets and in the dimensional tolerances of the parts hinder the attainment of the zero-field situation [14].

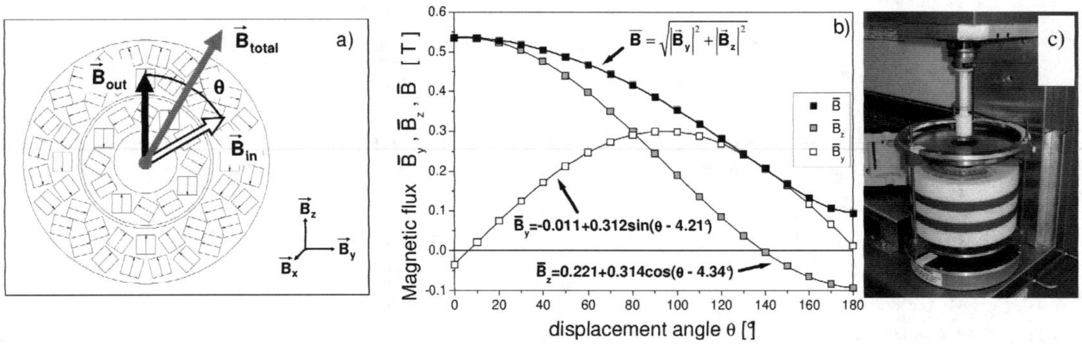

Figure 1. Basic characteristics of the new setup: a) schematics of the arrangement of double dipolar Halbach arrays; b) total magnetic flux density obtained, as a function of the displacement angle θ; c) the setup as mounted on the ARES rheometer.

With this Halbach layout, the magnetic stray field in the direction radial to the cylinders is greatly reduced. In the axial direction, the whole setup is magnetically shielded with metal sheets to avoid any interference coming from the magnetic stray field on the motor and torque transducer of the ARES. The measurement geometries consist of a cup combined with a concentric cylinder (Couette) or a 4-bladed vane. The geometries are made of polyoxymethylene to avoid disturbing the magnetic field during the measurements, as well as the formation of eddy currents (Figure 1c).

A commercial magnetorheological fluid grade MRF-132AD from LORD Corporation (Cary, North Carolina, USA) was used to assess the new setup. According to the manufacturer, the fluid contains 81.64 wt.-% of solids, suspended in a hydrocarbon-based carrier liquid. The mean size of the iron particles is around 6 μm.

2.2. Rheological experiments in steady shear

Stress growth curves at startup of steady shear were obtained for different constant shear rates and magnetic field strengths, until a steady shear regime was observed. For each shear rate, the viscosity was determined by averaging the steady shear zone. A consistent increase of the shear viscosity with increasing magnetic field was found (Figure 2a). This is better observed by plotting the relative viscosity, where an increase of the shear viscosity by a factor of about 30 to 175 was found for $B \neq 0$ (Figure 2b).

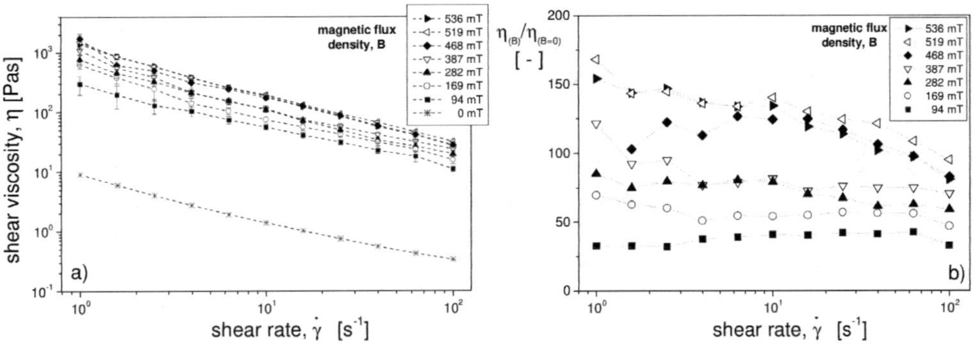

Figure 2. Steady shear viscosity as a function of shear rate, for different applied magnetic flux densities: a) absolute values; b) relative shear viscosity (for $B \neq 0$).

2.3. Rheological experiments in oscillatory shear

Strain sweep tests in oscillatory shear ($\omega_1/2\pi = 1$ Hz) were also performed. In the case of the shear storage modulus, G', an increase of the modulus in the linear regime ($\gamma_0 < 0.1$ %) by a factor of 175 was observed in the presence of the magnetic field. Also, a shift of the linear regime to higher strains was found to occur with increasing magnetic field (Figure 3a). In the presence of the magnetic field, a peak in G'' appears near the transition to the non-linear regime of deformations. This peak is also observed to shift to higher strain amplitudes with increasing magnetic flux density B (Figure 3b).

The rheological properties of the MR fluid used in this work were also measured independently in an external laboratory, using the commercially available Anton Paar MRD flow cell. In the presence of magnetic field, the results shown in the present work (both shear viscosity and dynamic moduli) are approximately one order of magnitude lower than the ones measured in the standard equipment. Nevertheless, both results are consistent and show exactly the same features, being only affected by a shift in the ordinate axis. Further work is on the way to evaluate whether a calibration procedure can be established, that could clarify the present results. Also, the implementation of a Hall sensor on the measuring geometry during the measurements can further help in clarifying this issue.

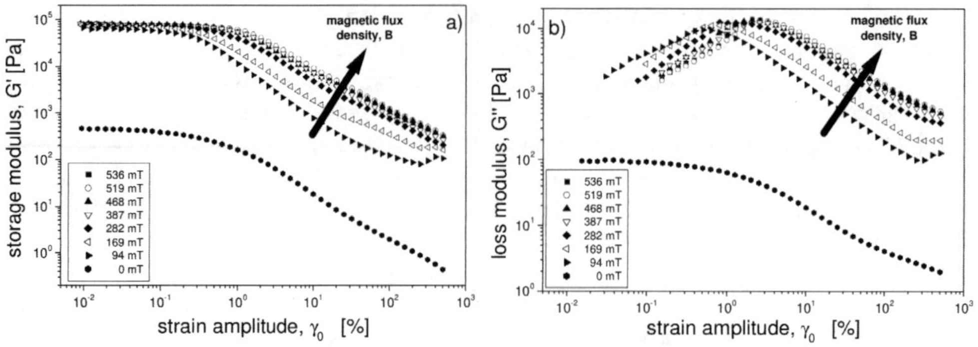

Figure 3. Dependence of the dynamic moduli on the shear strain amplitude, at different *B*.

3. Conclusions

The double dipolar Halbach array allows a varying magnetic field to be obtained, whilst maintaining a high field homogeneity. This setup is mounted on an ARES rheometer, and a commercial MR-fluid was used to evaluate its performance. A dependence on the magnetic field is observed for the dynamic moduli in oscillatory shear and also for the steady shear viscosity. These results are consistent with independent measurements performed in the standard accepted equipment.

Acknowledgements

Manfred Hehn and Frank Keller (MPIP-Mainz), and the staff of the workshops at the MPIP and University Karlsruhe are gratefully acknowledged. Loredana Pop and Jörg Läuger from Anton Paar Germany GmbH are gratefully acknowledged for the comparison measurements. V. C. Barroso acknowledges the financial support from the Technische Universität Darmstadt and the Deutsche Forschungsgemeinschaft.

References

[1] Charles S W 2002 *Ferrofluids: Magnetically Controllable Fluids and Their Applications*, ed Odenbach S (Berlin: Springer Verlag) p 3
[2] Odenbach S 2004 *J. Phys.: Condens. Matter* **16** R1135
[3] Claracq J, Sarrazin J and Montfort J P 2004 *Rheol. Acta* **43** 38
[4] Zschunke F, Rivas R and Brunn P O 2005 *Appl. Rheol.* **15** 116
[5] Laun H M and Gabriel C 2007 *Rheol. Acta* **46** 665
[6] Laun H M, Gabriel C and Schmidt G 2008 *J. Non-Newt. Fluid Mech.* **148** 47
[7] Wollny K, Läuger J and Huck S 2002 *Appl. Rheol.* **12** 25
[8] Läuger J, Wollny K, Stettin H and Huck S 2005 *Int. J. Mod. Phys.* **B19** 1353
[9] Odenbach S and Störk H 1998 *J. Mag. Magnet. Mater.* **183** 188
[10] Odenbach S, Rylewicz T and Heyen M 1999 *J. Mag. Magnet. Mater.* **201** 155
[11] Shorey A B, Kordonski W I, Gorodkin S R, Jacobs S D, Gans R F, Kwong K M and Farny C H 1999 *Rev. Sci. Inst.* **70** 4200
[12] Mallinson J C 1973 *IEEE Trans. Magnet.* **MAG9** 678
[13] Halbach K 1980 *Nucl. Inst. Meth.* **169** 1
[14] Raich H and Blümler P 2004 *Concepts Magn. Reson. Part B (Magn. Reson. Eng.)* **23B** 16

Journal of Physics: Conference Series **149** (2009) 012103 doi:10.1088/1742-6596/149/1/012103

Properties of twisted ferromagnetic filaments

Mihails Belovs[1], **Andrejs Cēbers**[1]

[1] University of Latvia, Zeļļu 8, LV-1002,Latvia

E-mail: aceb@tesla.sal.lv

Abstract. The full set of equations for twisted ferromagnetic filaments is derived. The linear stability analysis of twisted ferromagnetic filament is carried out. Two different types of the buckling instability are found - monotonous and oscillatory. The first in the limit of large twist leads to the shape of filament reminding pearls on the string, the second to spontaneous rotation of the filament, which may constitute the working of chiral microengine.

1. Introduction

Chains of magnetic particles determine the properties of magnetorheological suspensions and under certain conditions may be described by the model of semiflexible filament. Many features of these filaments are determined by their twist. Twist can be easily applied to magnetic filaments due to the torques acting on the magnetic particles. Twist plays important role in the behavior of biopolymers and in living world [1]. For example in cilia twist is induced by the action of molecular motors and leads to their rotation due to the chiral symmetry breaking [2]. Chiral symmetry breaking due to the induced vortical flows is presumably responsible for left-right symmetry breaking at the embryonic stage of development of vertebrates [3]. Here we describe several phenomena caused by the action of twisting torques on the ferromagnetic filaments. We illustrate that similarly to cilia the chiral symmetry breaking takes place for flexible ferromagnetic filaments under the action of twist. Ferromagnetic filaments exist in the living world and are used by magnetotactic bacteria for the orientation in the magnetic field of the Earth [4,5]. They can be made artificially by linking commercially available ferromagnetic particles functionalized by streptavidin with biotinized DNA fragments [6].

2. Model

The theoretical model, which describes flexible ferromagnetic filaments is based on the Kirchhoff model of elastic rod extended to account for interaction energy of filament with external field (l is the contour length of the filament, subscript $_{,...}$ denotes derivative) [7].

$$E = \frac{1}{2}A\int(\Omega_1^2+\Omega_2^2)dl + \frac{1}{2}C\int\Omega_3^2 dl - MH\int\vec{e}_3\cdot\vec{h}dl - \int\Lambda dl \qquad (1)$$

where the orts of material frame $\vec{e}_i$ obey $\vec{e}_{i,l} = \vec{\Omega}\times\vec{e}_i$. A and C are the bend and twist elastic constants, M is the magnetization of the filament per its unit length, H is the applied magnetic field strength, and $\vec{h}$ is the unit vector along the applied magnetic field. Inextensibility of the filament is enforced by the term $-\int\Lambda dl$ in (1). Introducing the complex curvature

$\psi = (-i\Omega_1 + \Omega_2)\exp(i\int^l \Omega_3 dl')$ and $\vec{\epsilon} = (\vec{e}_1 + i\vec{e}_2)\exp(i\int^l \Omega_3 dl')$ the following expressions for the stress $\vec{F}$ and momentum stress $\vec{T}$ are derived

$$\vec{F} = -A\Re(\vec{\epsilon}^{\,*}\psi_{,l}) - \frac{1}{2}A(\Omega_1^2 + \Omega_2^2)\vec{e}_3 + \Re(iC\Omega_3\vec{\epsilon}^{\,*}\psi) - MH(\vec{e}_1\cdot\vec{h}\vec{e}_1 + \vec{e}_2\cdot\vec{h}\vec{e}_2) - \Lambda\vec{e}_3 \tag{2}$$

$$\vec{T} = A\Omega_1\vec{e}_1 + A\Omega_2\vec{e}_2 + C\Omega_3\vec{e}_3 \tag{3}$$

We are considering the Rouse dynamics (ζ is the drag coefficient per unit length)

$$\zeta\vec{v} = \vec{K} = \vec{F}_{,l} \tag{4}$$

here

$$\vec{K} = \left(-\Lambda_{,l} + MH\Re(\vec{\epsilon}^{\,*}\cdot\vec{h}\psi)\right)\vec{e}_3 + \Re(\vec{\epsilon}^{\,*}F_\perp) \tag{5}$$

and

$$F_\perp = -A(\psi_{,ll} + |\psi|^2\psi/2) + iC(\Omega_3\psi)_{,l} - \Lambda\psi + MH\vec{e}_3\cdot\vec{h}\psi \tag{6}$$

$F_\perp$ besides the terms characteristic for nonmagnetic filaments [8,9] contains term $MH\vec{e}_3\cdot\vec{h}\psi$ due to the torque from the magnetic field. Equations for the complex curvature and twist are the following (ζ_r is the rotational drag coefficient per unit length)

$$\zeta\psi_{,t} = F_{\perp,ll} + |\psi|^2F_\perp - \psi_{,l}\Lambda_{,l} + \psi_{,l}MH\Re(\vec{\epsilon}^{\,*}\cdot\vec{h}\psi) + i\psi\int^l F_\perp^*\psi_{,l}dl' \tag{7}$$

$$\zeta_r\Omega_{3,t} = C\Omega_{3,ll} - \frac{\zeta_r}{\zeta}\Im(\psi F^*_{\perp,l}) \tag{8}$$

Inextensibility condition $\vec{e}_3\cdot\vec{K}_{,l} = 0$ gives

$$-\Lambda_{,ll} = -MH\vec{h}\cdot\vec{e}_{3,ll} + \Re(F_\perp\psi^*) \tag{9}$$

3. Linear stability analysis

Here we report the results of the linear stability analysis of straight twisted ferromagnetic filament with the free, unclamped ends and applied torque $\vec{T}(\pm L, t) = C\Omega_3\vec{e}_3$ ($2L$ is the filament length), which is oriented in the direction of the magnetic field (z axis). For small perturbations the relation $\psi = \xi_{,ll}$ ($\xi = x + iy$) is valid. As a result from Eqs.(4),(5) the following equation for the small deformation of the filament is obtained ($\vec{e}_3^{\,0}$ is the tangent vector of straight filament)

$$\zeta\xi_{,t} = -A\xi_{,llll} + iC\Omega_3\xi_{,lll} + MH\vec{e}_3^{\,0}\cdot\vec{h}\xi_{,ll} \tag{10}$$

with the boundary conditions of zero force and torque

$$\xi_{,ll}|_{l=\pm L} = 0; (-A\xi_{,lll} + iC\Omega_3\xi_{,ll} + MH\vec{e}_3^{\,0}\cdot\vec{h}\xi_{,l})|_{l=\pm L} = 0 \tag{11}$$

Dimensionless set of equations is obtained scaling the length with L, time with $\tau = \zeta L^4/A$, introducing the magnetoelastic number $Cm = MH\vec{e}_3^{\,0}\cdot\vec{h}L^2/A$ and $\chi = C\Omega_3L/A$. Looking for the solution of Eq.(10) in form $\xi \to \exp(\lambda t)\xi(l)$ the eigenvalue problem for not self-adjoint operator is obtained:

$$\lambda\xi = -\xi_{,llll} + i\chi\xi_{,lll} + Cm\xi_{,ll} \tag{12}$$

with boundary conditions

$$\xi_{,ll}(\pm 1) = 0; \quad -\xi_{,lll}(\pm 1) + i\chi\xi_{,ll}(\pm 1) + Cm\xi_{,l}(\pm 1) = 0 \tag{13}$$

Journal of Physics: Conference Series **149** (2009) 012103 doi:10.1088/1742-6596/149/1/012103

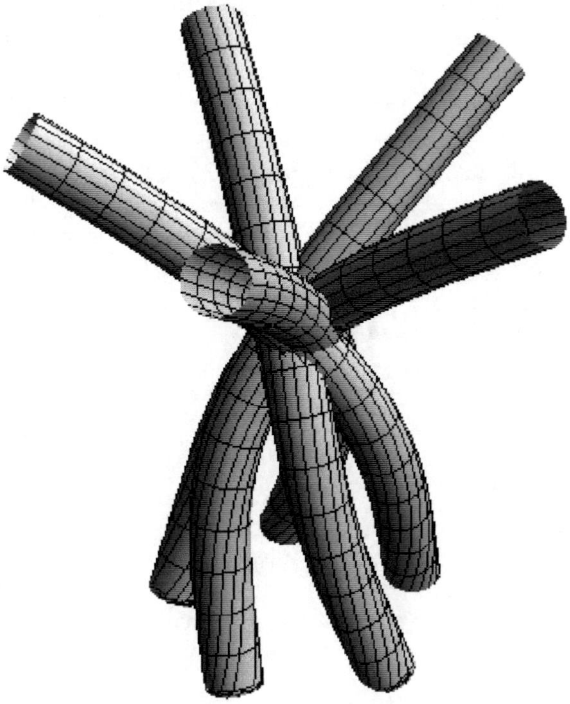

Figure 1. Consecutive shapes of the rotating filament with broken chiral symmetry. $Cm = -1.3; \chi = 3.38$.

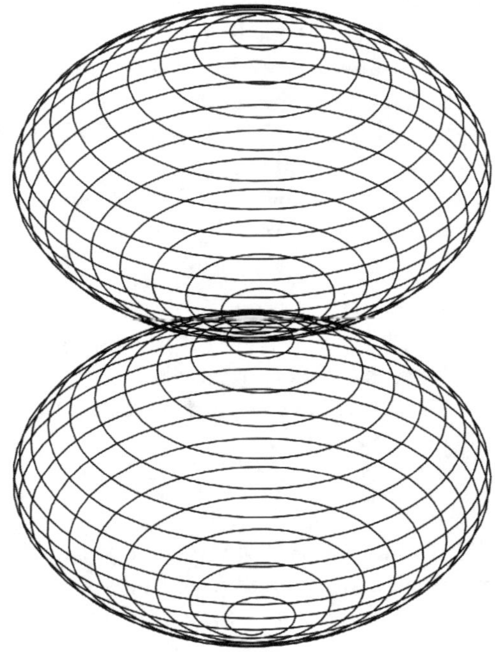

Figure 2. Neutral mode of buckled filament in magnetic field. $n = 2$, $\chi = 2\pi + 250$.

This eigenvalue problem in the absence of the twist is investigated in [10]. It is shown that for all $Cm < 0$ there is positive eigenvalue, which corresponds to odd eigenmode of the filament deformation and for small $|Cm|$ describes the overturning of the magnetic dipole at the inversion of the field. The increment of this mode diminishes with the increase of $|Cm|$ due to the deformation of the filament. At $|Cm|$ larger than the critical value$-\pi^2/4$ the ferromagnetic filament becomes unstable with respect to U-like even deformation mode with increment growing with $|Cm|$. At $|Cm| = -(n\pi)^2/3, (n = 1, 2, ...)$ the eigenvalues of the even and odd modes become equal and degenerate eigenvalue problem arises.

The eigenvalues at small χ may be found by considering twist as perturbation near these degeneracy points. Up to the first order we have $\lambda = \lambda_0 + \lambda_1; \xi = \xi_0 + \xi^1$, ($\lambda_0$ is positive eigenvalue of unperturbed problem [10]). Looking for the solution of the first order equation with $\xi_0 = a\xi^{(e)} + b\xi^{(o)}$, where $\xi^{(e,o)}$ are even and odd eigenfunctions at the degeneracy point, from the solvability condition we obtain

$$\lambda_1^2 = -\chi^2 \frac{\int_{-1}^{1} \xi^{(e)} \xi^{(o)}_{,lll} dl \int_{-1}^{1} \xi^{(o)} \xi^{(e)}_{,lll} dl}{\int_{-1}^{1} \xi^{(e)2} dl \int_{-1}^{1} \xi^{(o)2} dl} \tag{14}$$

Since the numerator in the relation Eq.14 is positive then at the degeneracy point of the spectrum pair of complex conjugated eigenvalues arise. If λ is a complex eigenvalue with corresponding eigenfunction $\xi(l)$ then λ^* also is eigenvalue corresponding to eigenfunction $\xi^*(-l)$. There is not more as one pair of complex eigenvalues. Variable ξ represents vector in plane perpendicular to the axis of the filament and therefore complex eigenvalues λ means that twisted filament rotates with angular velocity $\Im\lambda$. Since the corresponding eigenfunction is the mixture of odd and even eigenfunctions then for a rotating filament the symmetry is broken.

This symmetry breaking is illustrated by Fig. 1 where five consecutive shapes of the filament with broken chiral symmetry are shown at $Cm = -1.3; \chi = 3.38$ for the one period $2\pi/9.1$. As one can see the radiuses of the circles on which move both ends of the filament are different.

This may be confirmed exactly by considering the energy balance equation for the linear problem. Multiplying the real and imaginary parts of Eq.(10) by $x_{,t}$ and $y_{,t}$ correspondingly after adding the results and integration along the filament we obtain in dimensionless units $(\xi(l,t) \to \exp{(i\lambda t)}\xi(l))$

$$|\lambda|^2 \int_{-1}^{+1} |\xi|^2 dl = -2\Re(\lambda)E - \frac{\chi\Im(\lambda)}{2}|\xi_{,l}|^2|_{-1}^{+1} \qquad (15)$$

where E is the sum of bending, magnetic and twist energies:

$$E = \frac{1}{2}\int_{-1}^{1} |\xi_{,ll}|^2 dl + \frac{1}{2}Cm\int_{-1}^{1} |\xi_{,l}|^2 dl - \frac{\chi}{2}\int_{-1}^{1} \Im(\xi_{,l}^*\xi_{,ll})dl$$

From relation (15) follows that in the case of neutral complex eigenvalues ($\Re(\lambda)$=0) the work of twisting torque is dissipated into heat due to the rotation of the filament.

The neutral curves of monotonous perturbations are given by equations (k is natural number)

$$Cm = \frac{\chi^2 - (k\pi)^2}{4} \qquad (16)$$

Shapes of filaments which correspond to neutral modes have $k-1$ nodes, where the curvature of the filament is equal to zero and which coincide with the nodes of buckled filament at $Cm < 0$ in the absence of twist. Its torsion is constant and equals to $\chi/2$ even in the limit of straight filament. An illustration of emerging shape, which reminds the pearls on the string, is shown in Fig. 2 for $k = 2$.

Eigenvalues for arbitrary values of parameters are calculated numerically. Numerical algorithm is based on splines and differentiation matrices with nodes in zeros of classical ortogonal polynoms [11]. Obtained results show that at $\pi < \chi < \chi_c$ application of field parallel to magnetization diminishes the critical twist for buckling instability and antiparallel stabilizes the buckling instability due to the twist (χ_c is critical value of χ when twisted filament buckles in zero field and is given by equation $\tan \chi_c = \chi_c$). For $0 < \chi < \pi$ the situation is opposite to just described.

4. Singular perturbations

Described exchange of stability may be understood by perturbation theory analysis of the eigenvalues of not self-adjoint operator. Eigenvalue $\lambda_0 = 0$ of the problem Eqs.(12) and (13) at $Cm = 0$ has twofold degeneracy. Eigenmodes $\xi_0^{(1)} = 1$ and $\xi_0^{(2)} = l$ correspond to the solid displacement of the filament and its rotation. In the magnetic field degeneracy is broken. The eigenvalue problem in the first order with the respect to Cm reads $(\xi = \xi_0 + \tilde{\xi}; \lambda = \lambda_0 + \lambda_1)$

$$\hat{L}\tilde{\xi} = -\tilde{\xi}_{,llll} + i\chi\tilde{\xi}_{,lll} = \lambda_1\xi_0^{(2)} - Cm\xi_{0,ll}^{(2)} \qquad (17)$$

at boundary conditions

$$\tilde{\xi}_{,ll}(\pm 1) = 0;\ \tilde{\xi}_{,lll}(\pm 1) = Cm\xi_{0,l}^{(2)}(\pm 1) \qquad (18)$$

Adjoint operator $\hat{L}^+$ is determined as

$$\hat{L}^+u = -u_{,llll} + i\chi u_{,lll} \qquad (19)$$

with the different from $\hat{L}$ boundary conditions

$$u_{,lll}(\pm 1) - i\chi u_{,ll}(\pm 1) = 0;\ u_{,ll}(\pm 1) - i\chi u_{,l}(\pm 1) = 0 \tag{20}$$

Operator $\hat{L}^{+}$ has double eigenvalue $\lambda = 0$. The corresponding eigenfunctions are $u_1 = 1$ and $u_2 = \exp(i\chi l)$. Solvability condition of problem Eqs.(17),(18) reads

$$\lambda_1 \int_{-1}^{1} u_2^* \xi_2^{(0)} dl = -Cm u_2^* \xi_{0,l}^{(2)} |_{-1}^{+1}$$

which gives

$$\lambda_1 = \frac{Cm\chi^2 \sin\chi}{\chi\cos\chi - \sin\chi} \tag{21}$$

Relation (21) shows that at $\pi < \chi < \chi_c$ $\lambda_1 > 0$ for $Cm > 0$, which corresponds to the decrease of the critical twist for the buckling of the filament in the field parallel to magnetization of the filament. At $Cm < 0$ and $\pi < \chi < \chi_c$ in agreement with numerical data $\lambda_1 < 0$. At $2\pi > \chi > \chi_c$ application of the field along the magnetization stabilizes the buckling of the filament and applied opposite to the magnetization destabilizes it. From the relation (21) we see that in the point $(Cm = 0, \chi = \chi_c)$ the perturbation theory approach breaks down. This corresponds to the fact that in singular point $(Cm = 0; \chi = \chi_c)$ the algebraic multiplicity of the solution of dispersion equation $\lambda = 0$ is 4 instead of 3 for $0 < \chi < \chi_c$. Analysis of this singular perturbation problem is pending for the future publications.

In conclusion we have demonstrated that twisted ferromagnetic filaments can work as chiral microengines. Competing magnetic torque and twist induced buckling instabilities lead to the rich family of equilibrium shapes, which in the case of stabilizing field direction reminds pearls on the string. There are special points in the parameter plane, where the perturbations are singular.

5. References

[1] Alberts B, Bray D, Lewis J, Raff M, Roberts K, and Watson J D *Molecular Biology of the Cell*, (Third edition,Garland Publishing,Inc., New York, London)
[2] Hilfinger A and Julicher F 2008 *Phys.Biol.* **5** 016003
[3] Bray D *Cell Movements* (Garland Publishing,Inc., Taylor Francis Group)
[4] Scheffel A, Guska M, Faivre D, Linaroudis A, Plitzko J M and Schuler D 2006 *Nature* **440** 110
[5] Ērglis K, Wen Qi, Ose V, Zeltins A, Sharipo A, Janmey P A and Cēbers A 2007 *Biophysical Journal* **93** 1402
[6] Ērglis K, Belovs M and Cēbers A *JMMM to be published*
[7] Cēbers A 2003 *J.Phys.:Condens.Matter* **15** S1335
[8] Goldstein R E, Powers Th R and Wiggins Ch H 1998 *Phys.Rev.Lett.* **80** 5232
[9] Wolgemuth Ch W, Powers Th R and Goldstein R E 2000 *Phys.Rev.Lett.* **84** 1623
[10] Belovs M and Cēbers A 2006 *Phys.Rev.* E **73** 051503
[11] Belovs M and Cīrulis T 2005 *Mathematical Modelling and Analysis* **10** 127

Magnetization dynamics of non-dilute ferrofluids

D Berkov[1], L Iskakova[2] and A Zubarev[2]

[1]Innovent Technology Development, Pruessingstr. 27B. D-07745, Jena, Germany
[2]Ural State University, Lenina Ave 51, 620083,Ekaterinburg, Russia

E-mail: andrey.zubarev@usu.ru

Abstract. The paper is devoted to theoretical studies of the magnetodipolar interparticle interaction effect on magnetization dynamics in moderately concentrated ferrofluids. We consider a homogeneous (without particle aggregates) ferrofluid consisting of identical spherical particles and employ a rigid dipole model, where the particle magnetic moment is fixed with respect to the particle itself. For the magnetization relaxation after an instant change of the external field, we show that the magnetodipolar interaction leads to the increase of the initial magnetization relaxation time. For the complex *ac*-susceptibility χ we find that the this interaction leads to an overall increase of its imaginary part χ'' and shifts the χ''- peak towards lower frequencies. Comparing results obtained with our analytical approach to numerical simulation data, we demonstrate that the theory provides good quantitative agreement with numerical simulations up to the particle volume fraction $\varphi \sim 6$ - 10%

1. Introduction

Experiments demonstrate that magnetodipolar interparticle interaction changes significantly both the equilibrium [1] and dynamical [2] properties of ferrofluids. Theoretical models of dynamical properties of dilute ferrofluids with vanishing interparticle interactions have been proposed in [3]. These models lead to highly accurate results for very dilute ferrofluids but can not explain properties and behavior of ferrofluids where the interparticle interactions are significant. Depending on the energy of the magnetodipolar interaction, it can lead either to appearance of homogeneous long-range interparticle correlations, or to formation of chain-like, drop-like and other heterogeneous internal structures [4]. Appearance of these correlations and, especially, heterogeneous structures can significantly change dynamical properties of ferrofluids [2]. At present there is no general theory which could adequately predict an internal structure of non-dilute ferrofluids at given experimental conditions. Therefore it is reasonable to consider effects of different internal structures on dynamical phenomena in ferrofluids separately. Such idealized models can provide better insights into the influence of various internal structures on the macroscopical ferrofluids properties.

In this paper we study magnetization dynamics of a homogeneous ferrofluid. For maximal simplification of calculations, we consider a monodisperse model, consisting of identical interacting magnetic particles. It is assumed that magnetic moment of each particle has a constant magnitude and is "frozen" into the particle body. Remagnetization dynamics of this model is studied both analytically and using numerical simulations. In order to simplify the mathematical treatment, we consider an idealized ferrofluid consisting of identical particles. We assume that the magnetodipolar interaction energy is less or of the same order of magnitude as the thermal energy kT (otherwise particle aggregates must appear in a ferrofluid; their treatment is beyond the scope of this paper). Next, in order to focus on magnetodipolar interaction effects, we neglect here the hydrodynamical interaction between particles. Effects of this interaction will be considered in a separate publication.

2. Basic equations

We consider a ferrofluid with volume V containing N identical spherical ferromagnetic particles with the diameter d. The magnitude m of the particle magnetic moment $\mathbf{m}$ is constant, the moment is "frozen" into the particle body. We introduce the unit vector $\mathbf{e}_i = \mathbf{m}_i/m$ of the magnetic moment of the i-th particle and denote the radius-vector of this particle by $\mathbf{r}_i$. To calculate the macroscopic characteristics of this system, we must determine the N-particle distribution function $P_N(\mathbf{e}_1, \ldots, \mathbf{e}_N, \mathbf{r}_1, \ldots, \mathbf{r}_N)$. It can be found by solving the appropriate Fokker-Planck equation, where we have to take into account the magnetodipolar interactions between all particles. This equation reads

$$\frac{\partial P_N}{\partial t} = \sum_i \mathbf{I}_i\left(\frac{D_r}{kT}\mathbf{I}_i(UP_N)\right) + \sum_i \nabla_i\left(\left(\frac{D_t}{kT}\nabla_i U\right)P_N\right) + D_r\sum_i \mathbf{I}_i^2 P_N + D_t\sum_i \nabla_i^2 P_N \tag{1}$$

where we have used the standard notation $\mathbf{I}_i = \mathbf{e}_i \times \dfrac{\partial}{\partial \mathbf{e}_i}$, $\nabla_i = \dfrac{\partial}{\partial \mathbf{r}_i}$. The potential energy of the system is $U = -kT\sum_i(\boldsymbol{\kappa}\cdot\mathbf{e}_i) + (1/2)\sum_{i\neq j} w_{ij}$, where the reduced field $\boldsymbol{\kappa} = \mu_0 m\mathbf{H}/kT$ is defined via the vacuum permeability μ_0 and the applied magnetic field $\mathbf{H}$, term w_{ij} is the magnetodipolar interaction energy of particles i and j, D_r and D_t are rotational and translational diffusion coefficients of the particles. The mean magnetization of a ferrofluid can be found as

$$\mathbf{M} = mn\int \mathbf{e}_1 P_N d\mathbf{e}_1 \ldots d\mathbf{e}_N d\mathbf{r}_1 \ldots d\mathbf{r}_N , \tag{2}$$

where n is number of ferrofluid particles per unite volume.

The Fokker-Planck equation (1) can not be solved exactly for two reasons. The first one is the well-known problem of statistical physics – interparticle interaction in a many-particle system does not allow (nearly always) to solve the governing equation for a many-particle distribution function or to calculate the Gibbs statistical integral. The second reason is the purely mathematical difficulty arising by the solution of the Fokker-Planck equation even for a single particle.

In order to take into account interaction between particles as accurately as possible and to avoid unjustified intuitive assumptions, we use the mathematically regular method of virial expansion over the particle concentration. To overcome the second problem, we use the effective-field approach, suggested by M. Shliomis [3], which is a version of the trial function method. According to this approach, we write the function P_N in the form of an equilibrium Gibbs function in some effective magnetic field $\mathbf{H}_e$, which is to be determined, instead of the real field $\mathbf{H}$. To find the components of $\mathbf{H}_e$ we multiply Eq. (1) by the orientation vector $\mathbf{e}$ and integrate the result over all $\mathbf{e}_i$ and $\mathbf{r}_i$. This way we obtain an equation, which includes the first and second statistical moments of the function P_N. By using in these moments the Gibbs function with the effective field $\mathbf{H}_e$, we arrive at ordinary differential equations with respect to the components of this field. These equations can be solved much easier than the basic equation (1). Having found the field $\mathbf{H}_e$ and using the Gibbs function with $\mathbf{H}_e$ in Eq.(2), we determine the macroscopical magnetization **M**. Detailed derivation of the analytical equation for $\mathbf{H}_e$ will be given elsewhere [6].

Numerical simulations are based on the Langevin dynamics formalism, where the equations of motion for the relevant degrees of freedom characterizing the system under study are solved taking into account thermal fluctuations (see [5] for details).

3. Results and discussion

Main results of our analytical and numerical calculations are presented below.

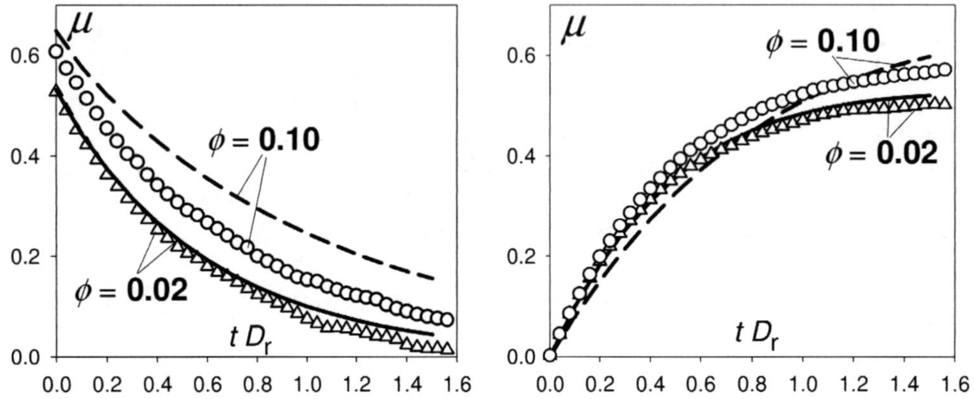

Fig. 1. Time dependence of the reduced magnetization $\mu = M/mn$ after the step-wise change of the external field from 200 Oe to 0 (left), and from 0 to 200 Oe (right). Lines – analytical theory, circles – numerical simulations. Figures near the curves show volume concentrations φ of the magnetic phase.

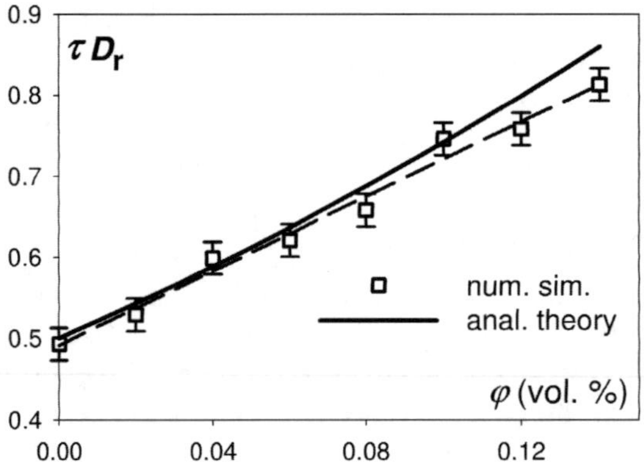

Fig. 2. Relaxation time τ of the magnetization onset after the instant increase of the external field from 0 to 200 Oe vs volume concentration φ.

Fig. 1. shows the effect of the interaction between particles on their mean magnetic moment μ.

Results presented in Fig. 2 demonstrate that the magnetodipolar interaction between particles leads to the increase of the initial rate of the ferrofluid magnetization.

The frequency dependencies of real χ' and imaginary χ'' parts of the reduced complex susceptibility of a ferrofluid $\chi = M_0/(H_0\varphi)$ are presented in Fig.3. Here index 0 marks the amplitudes of the magnetization and applied oscillating field. These results show that interparticle interaction decreases the frequency of χ''-maximum and increases the value of χ'' at this maximum.

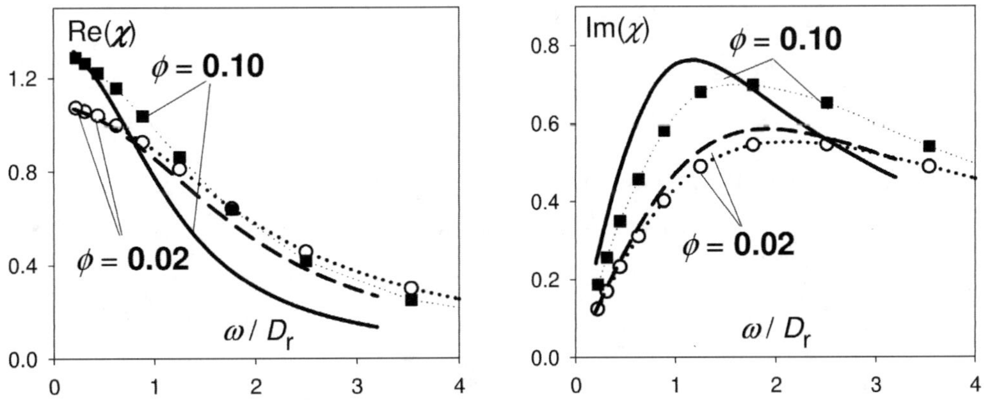

Fig. 3. Real and imaginary parts of the reduced susceptibility χ as functions on the applied field frequency ω. Lines – analytical theory, dots – numerical simulations. Solid lines and squares - $\varphi = 0.1$, dashed lines and circles - $\varphi = 0.02$. Amplitude of the applied field is $H_0 = 200$ Oe.

Conclusion

An analytical model of the magnetization dynamics of idealized monodisperse ferrofluids with interacting particles is presented. The model does not contain any fitting parameters and leads to a good agreement with computer simulations up to high (~10%) volume concentrations of the magnetic phase. Magnetodipolar interaction between particles decreases the rate of the ferrofluid remagnetization, decreases the peak frequency of the imaginary part of the ferrofluid complex susceptibility and increases the height of this peak. Suggested model can be also applied for investigation of other properties (e.g., rheological behavior) of non-dilute ferrofluids and generalized to polydisperse systems.

This work has been supported by RFFI, grants N 06-01-00125, 07-02-00079, 07-01-960769Ural, 08-02-00647, Fund CRDF, PG07-005-02.

References

[1] Pshenichnikov A F 1995 *J. Magn. Magn. Mat.* **145**, 319

[2] Zhang J, Boyd C and Luo W 1996 *Phys. Rev. Lett.* **77** 390; Taketomi S 1998 *Phys. Rev.* **E57**, 3073; Odenbach S 2002 *Magnetoviscous Effects in Ferrofluid* (Berlin, Springer)

[3] Shliomis M I 1974 *Sov. Phys.Usp.* **17**, 153; Martsenyuk M A, Raikher Yu L and Shliomis M I 1974 *J. Exp. Theor. Phys.* **38** 413; Felderhof B U 2000 *Magnetohydrodynamics* **36** 329

[4] Hayers C F 1975 *J. Coll. Int. Sci.*, **52** 239; Peterson E A and Krueger A A 1977 *ibid.* **62** 24; J.C. Bacri J-C and Salin D, 1983 *J. Magn. Magn. Mat.* **9** 48; Pshenichnikov A F 1995 *J. Magn.. Magn. Mat.* **145** 139; Khizgenkov P K, Dorman V L, Bar'jakhtar F G 1989 *Magnetohydrodynamics* **25** 30; Islam M F, Lin K H, Lacoste D, Lubenski T C and Yodh A G 2003 *Phys. Rev.* **E67** 021402

[5] Berkov D V, Gorn N L, Schmitz R, Stock D 2006 *J. Phys.: Cond. Matt.*, **18** S2595

[6] Berkov D V, Iskakova L Y, Zubarev A Y, submitted to *Phys. Rev. E*

Journal of Physics: Conference Series **149** (2009) 012105 doi:10.1088/1742-6596/149/1/012105

Thermomagnetic convection influenced by the magnetoviscous effect

Harald Engler, Dmitry Borin and Stefan Odenbach

Institute of Mechanics, Chair of Magnetofluiddynamics, George-Bähr-Str. 3, 01062 Dresden, Germany

E-mail: Harald.Engler@tu-dresden.de

Abstract. The mechanism of thermomagnetic convection in ferrofluids which is introduced by a magnetic field is well understood. To describe this kind of convection, it is assumed that the material properties, e. g. the viscosity η are independent from the magnetic field. However under certain circumstances ferrofluids show a magnetoviscous effect that means the increase of η in the presence of a magnetic field. Both effects have been well investigated but independently from each other. In this paper it is experimentally shown how the magnetoviscous effect and therefore the increase of η influence the behaviour of the thermomagnetic convection.

1. Introduction

Under certain circumstances thermomagnetic convection arises in a horizontal ferrofluid layer if an external magnetic field is applied and additionally a temperature difference over the fluid gap exists [1, 2]. This situation causes a magnetic body force in the ferrofluid layer which can drive a convective flow [3]. The density of this magnetic force is given by $f_m = \mu_0 \cdot M \cdot \nabla H$. The characteristic of thermomagnetic convection can be described by the dimensionless magnetic Rayleigh number Ra_m which is given by

$$Ra_m = \frac{\mu_0 \cdot \partial M/\partial T\big|_{H_i} \cdot \nabla H_i \cdot \Delta T \cdot d^3}{\kappa \cdot \eta} = \frac{\mu_0 \cdot K\big|_H^2 \cdot \Delta T^2 \cdot d^2}{\kappa \cdot \eta} \qquad \text{eq. 1}$$

where ∇H_i and ΔT denote the inner magnetic field gradient and temperature difference over the fluid gap with thickness d, μ_0 is the vacuum permeability, $K = \partial M/\partial T\big|_{H_i}$ is the pyromagnetic coefficient, and κ and η the temperature conductivity and the dynamic viscosity of the fluid respectively. The critical Rayleigh number $Ra_{m,crit}$ for a horizontal fluid layer which is covered by two rigid boundary plates is 1708 [2]. Below this value the heat is transported by a diffusive process and changes to convective state if $Ra_{m,crit}$ is exceeded. For this case where the convection is described by Ra_m it is assumed that the material properties of the ferrofluid are independent from the applied magnetic field with the exception of the inner magnetic field ∇H_i and K obviously.

A second well known phenomenon appearing in ferrofluids is the magnetoviscous effect [4 - 7] which describes the change of viscosity η under the influence of an applied magnetic field. In many high concentrated ferrofluids a contingent of large particles exists with a magnetic interparticle interaction strong enough to form chainlike structures. This particle formation leads to an increase of η which depends on the magnetic field strength and shear rate. To determine the magnitude of the magnetoviscous effect, an interaction parameter λ has to be introduced defined as the ratio between magnetic interaction energy between two particles which are in contact and their thermal energy. Nevertheless the direct contact of two particles is usually avoided by a surfactant to fight Van der Waalz forces. These forces will otherwise cause agglomeration of the particles leading to sedimentation which destroys the magnetic fluids. The influence of the surfactant layer is considered by a modified interaction parameter λ^* which is given by

$$\lambda^* = \lambda \cdot \left(\frac{d}{d+2\cdot s}\right)^3 = \frac{\mu_0 \cdot M_0^2 \cdot V}{24 \cdot k_B \cdot T} \cdot \left(\frac{d}{d+2\cdot s}\right)^3 \qquad \text{eq. 2}$$

where μ_0 denotes the vacuum permeability, M_0 the spontaneous magnetization of the magnetic material, V the volume of the magnetic particle, k_B the Boltzmann constant, T the absolute temperature, d the mean diameter of the particles and s the thickness of the surfactant. If the value of λ^* is larger than 1 the interparticle interactions are strong enough to build chainlike structures which can affect η. For magnetite particles, chain formation becomes effective for $d \geq 16$ nm.
In this paper it is described how the magnetoviscous effect influences the thermomagnetic convection in ferrofluids. Both phenomena have separately been well investigated in the past but the possibility of a concurrent appearance has never been considered.

2. Experimental data

2.1 Ferrofluid samples

For the investigation of the above described effects, two samples of a commercial ferrofluid called APG 513A from Ferrotec, Germany, are used. The samples, which have in principle the same fluid and magnetic properties, differ in their microstructure. The sample APG 513A_Old which was produced by a former production process has a broad particle size distribution which means that it contains large particles which are able to force the magnetoviscous effect. The second sample APG 513A_NEW, which was produced by a modern production process, has a small particle size distribution, so that most of the particles have a diameter close to the mean diameter which is 10 nm.
The ferrofluid consists of magnetic particles made from magnetite (7,2 vol. %) suspended in synthetic ester. The saturation magnetization M_S is 28,2 kA/m at room temperature of 20 °C and the pyromagnetic coefficient K is 35,3 $Am^{-1}K^{-1}$ for a field strength of 25 kA/m. Further basic fluid properties are the thermal expansion coefficient $\beta_T = 4{,}5 \cdot 10^{-4}$ 1/K, the temperature conductivity $\kappa = 5 \cdot 10^{-8}$ m^2/s and the viscosity $\eta_{(T = 20\,°C)} = 170$ mPas.

2.1. Magnetoviscous effect

The change of η under the influence of an applied magnetic field was investigated with the shear-controlled rheometer described in [8] with some technical modifications. The magnetic field is generated by a pair of Helmholtz coils with a maximum field strength up to 40 kA/m and the field homogeneity in the center of the coils is better than 95 %. The measurement cell of the rheometer consists of a cone-plate geometry and allows shear rate variations in the range from 10^{-4} s^{-1} to 10^3 s^{-1}.
Figure 1 shows the change of viscosity $\Delta\eta$ as a function of the applied magnetic field strength for a shear rate of 0,1 s^{-1}. For the sample APG 513A_NEW only a very weak increase of $\Delta\eta$ can be seen. This change reaches 50 % of the viscosity η at H = 0 kA/m for a field strength of 30 kA/m. The increase of $\Delta\eta$ changes dramatically for the sample APG 513A_OLD. Here a change of viscosity by a factor of 4,4 for a field strength of 30 kA/m can be found. This sample has a contingent of large particles which give rise to a strong viscosity change.

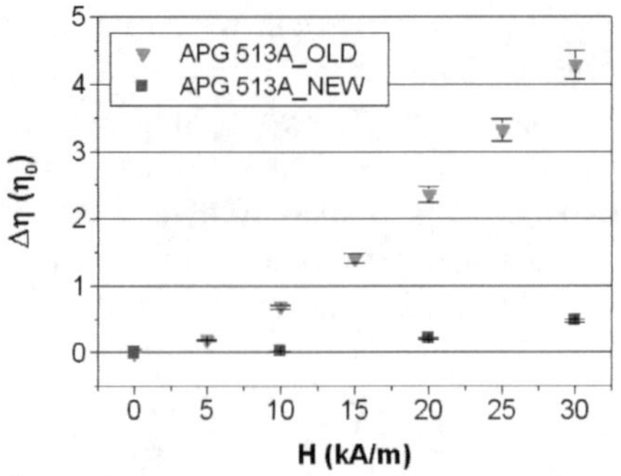

Figure 1. *Increase of viscosity η as a function of the applied magnetic field strength for two samples of ferrofluid.*

Actually a small amount of the particles in this sample - volume concentration Φ = 0,8 % - have a mean diameter of 16 nm, leading to a magnetic interparticle interaction parameter λ^* = 2,87 [9].

2.2. Thermomagnetic convection

The behaviour of thermomagnetic convection under the influence of a magnetic field was investigated with a specialized measuring cell [10] which is modified with a heat flux sensor. The setup provides the possibility to measure the temperature difference ΔT and the heat flux Q over a horizontal fluid gap with a height of 4 mm and a diameter of 88 mm. The mangetic field - maximum possible field strength is 25 kA/m - is generated by two pairs of Helmholtz coils placed in a Fanselau arrangement. The inhomogeneities in the center of the arrangement are smaller than 0,5 %. The temperature is measured by thermistors with an accuracy of 0,01 K. Furthermore the heat flux sensor has a accuracy of 0,1 W/m^2. The temperature difference over the fluid gap is generated by two water chambers above and below the measuring cell with a temperature stability of 0,01 K.

In figure 2 the behaviour of Q as a function of ΔT for a horizotnal fluid layer heated from below is shown for the sample APG 513A_NEW. Without an applied magetic field only the thermal convection which is driven by the bouyancy froce works. The characteristic of the heat flux transfer changes from a diffusive state to a convective state where the slope of Q changes remarkably. At this point the critical Rayleigh number of 1708 and therefore the critical temperature difference ΔT_{crit} of 32 °C are reached. With an applied magnetic field the situation changes because two mechanisms - thermal and thermomagnetic convection - are working simultaneously. The heat flux Q increases in general because of the additional heat flux introduced by the magnetic force. Due to this additional heat flux the critical Rayleigh number is reached for a value of ΔT_{crit} which is smaller by a factor of 2 than the value of ΔT_{crit} for H = 0 kA/m. For this sample it is assumed that the material properties of the ferrofluid sample, e. g. μ_0, κ and η, are independent from the magnetic field. Further it can be seen that for a magnetic field of 25 kA/m the slope of Q does not change abruptly at the transition point, but rather a transition zone exists. This behaviour is caused by the characteristic of the microstructure of ferrofluids, where nano-scaled particles are suspended in an appropriate carrier liquid. This leads to additionally heat and mass transfer phenomena, such as the Soret effect [11].

Figure 3 shows Q as a function of ΔT for the sample APG 513A_OLD. Without an applied magnetic field the typical behaviour of thermal convection with the change from conduction to convection at ΔT_{crit} is shown whereas ΔT_{crit} is again 32 °C. The progress of the heat flux curve is nearly identical to the curve of the sample APG 513A_NEW as shown before which is explained by the fact that the basic fluid properties are similar. The difference of the samples is given by the microstructure of the fluid. For the second fluid an increase of ΔT_{crit} is observed due to the strong field dependent increase of its viscosity.

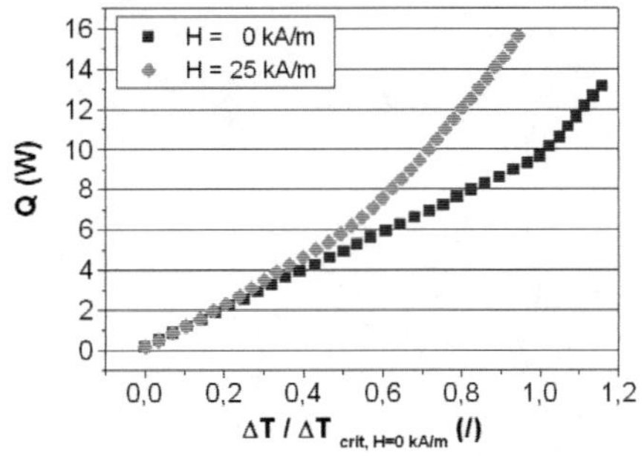

Figure 2. APG 513A_NEW: Heat flux as a function of the temperature difference for field strength of 0 kA/m and 25 kA/m.

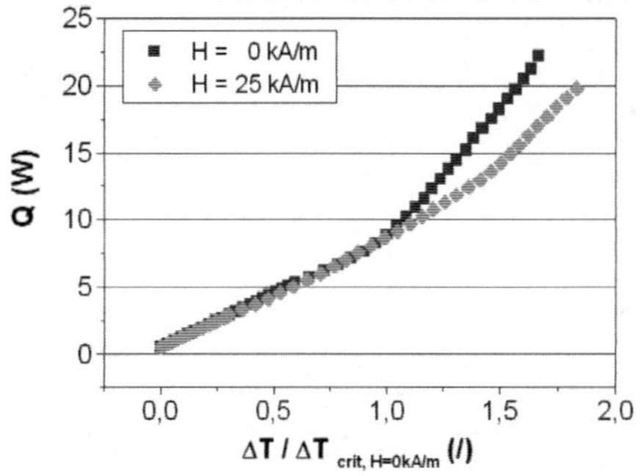

Figure 3. APG 513A_OLD: Sample contains large particles leading to an increase of η. This causes an increase of ΔT_{crit} for an applied magnetic field.

3. Conclusion

The experimental results presented here have shown that under certain circumstances the magnetoviscous effect can influence thermomagnetic convection. By applying a magnetic field the additional heat flux introduced by the magnetic force causes a reduction of $\Delta T_{crit.}$ This situation occurs for the ferrofluid sample called APG 513A_NEW which shows no magnetoviscous effect. In contrast a significant magnetoviscous effect appears for the ferrofluid sample called APG 513A_Old which contains a contingent of large particles. For this sample η increases by a factor of 4,4 for a magnetic field strength of 30 kA/m. This behaviour of η causes a shift of the transition point, where the heat flux characteristic changes from conduction to convection, to higher values of ΔT_{crit}.
For investigations of thermomagnetic convection it should be considered that for certain circumstances the magnetoviscous effect can arise in ferrofluids which will affect the experimental results of the threshold of convection.

4. Acknowledgements

Financial support by Deutsche Forschungsgemeinschaft (DFG) under grant Od18/7-1 and by DLR 50 WM 0639 providing the basis for our investigations is gratefully acknowledged.

References

1. Finlayson B A 1970 *Convective Instability of ferromagnetic fluids* J. Fluid Mech. **40** 753
2. Schwab L 1989 *Konvektion in Ferrofluiden* Dissertation LMU Muenchen
3. Rosensweig R E 1985 *Ferrohydrodynamics* New York Cambridge University Press
4. Odenbach S 2002 *Magnetoviscous Effects in Ferrofluids* LNPm71 Berlin Springer Verlag
5. Hall W F and Busenberg S N 1969 *Viscosity of magnetic suspensions* The Journal of Chemical Physics **51**
6. Shliomis M I 1972 *Effective viscosity of magnetic suspensions* Soviet Phys. JETP **34** (6)
7. McTague J P 1969 *Magnetoviscosity of magnetic colloids* The Journal of Chemical Physics **51**
8. Odenbach S, Rylewicz T and Heyen M 1999 *A rheometer dedicated for the investigation of viscoelastic effects in commercial magnetic fluids* J. Magn. Magn. Mater. **201** 155-8
9. Shahnazian H and Odenbach S 2008 Rheological investigations of ferrofluids with a shear stress controlled rheometer J. Phys.: Condens. Matter **20** 204137 (4pp)
10. Engler H and Odenbach S 2008 *Parametric modulation of thermomagnetic convection in magnetic fluids* J. Phys.: Condens. Matter **20** 204135 (5pp)
11. Odenbach S 1995 *Convection driven by forced diffusion in magnetic fluids under the influence of strong magnetic force* Journal of magnetism and magnetic materials 149

11th Conference on Electrorheological Fluids and Magnetorheological Suspensions IOP Publishing
Journal of Physics: Conference Series **149** (2009) 012106 doi:10.1088/1742-6596/149/1/012106

Thermodiffusive effects in magnetic fluids: application of the ferrofluid dynamics theory

Adrian Lange

Chair Magnetofluiddynamics, Institute of Fluid Mechanics, Technische Universität Dresden, 01062 Dresden, Germany

E-mail: adrian.lange@tu-dresden.de

Abstract. The ferrofluid dynamics theory is applied to thermodiffusive problems in magnetic fluids in the presence of magnetic fields. The most general expression of the mass flux are given. By employing these results to experiments, global Soret coefficients in agreement with measurements are determined. Also an estimate for so far unknown transport coefficient is made.

1. Introduction

The thermodiffusive or Soret effect describes the establishment of concentrations gradients in response to temperature gradients for a two- or multi-component fluid. Since the motion of the ferromagnetic nanoparticles in magnetic fluids (MFs) can be influenced by external magnetic fields, the Soret effect in MFs shows a strong dependence on any nonzero magnetic field strength [1–3]. The changes of the Soret coefficient can be up to six times its zero field value [2] and even a change of its sign was measured [2; 3]. Previous theoretical approaches [4; 5] need as input the magnetophoretic velocity of the nanoparticles. Therefore certain microscopic properties are assumed as e.g. spherical particles of equal size and the applicability of the Stokes hydrodynamic drag. A comparison with the experimental results show great differences: the microscopic approach gives changes of the Soret coefficient which are three orders of magnitude smaller than those in the experiment. That means that concerning thermodiffusive processes in MFs there is a wide gap between experiment and theory. Therefore as an alternative approach, a macroscopic theory, called ferrofluid dynamics (FFD) theory, is tested to describe the measured phenomena.

2. Ferrofluid dynamics theory

The principal structure of the ferrofluid dynamics theory was laid down in [6]. It is "... a general, strictly macroscopic approach relying solely on symmetry considerations, conservation laws, and thermodynamics." [7]. An essential part of the theory is the set of material-dependent parameters like susceptibilities and transport coefficients. They can be determined by well design experiments [8] and used here as fit parameters to describe thermodiffusive processes in magnetic fluids under the influence of magnetic fields.

As usual in thermodynamic theories, one starts with the thermodynamic energy density u. It is taken as a function of the entropy density s, the density $\rho^{(1)}$ of the magnetic part of the fluid, the momentum density $\mathbf{g} = \rho\mathbf{v}$, the total density ρ, and the magnetic induction $\mathbf{B} = \mu_0(\mathbf{M}+\mathbf{H})$

[6],

$$du = Tds + \tilde{\mu}_c d\rho^{(1)} + v_i dg_i + \mu^{(2)} d\rho + H_i dB_i \ , \tag{1}$$

where $\tilde{\mu}_c = \tilde{\mu}^{(1)} - \tilde{\mu}^{(2)}$ is the difference in the chemical potentials of the two constituents and $\rho = \rho^{(1)} + \rho^{(2)} = \phi\rho_{\rm m} + (1-\phi)\rho_{\rm cl}$. The density of the magnetic particles (carrier liquid) is denoted by $\rho_{\rm m}$ ($\rho_{\rm cl}$) and ϕ is the volume fraction of magnetic particles in the fluid. It is assumed that the magnetic part of the chemical potential can be separated, $\tilde{\mu}_c = \mu_c(\rho, \rho^{(1)}, T, \mathbf{v}) + \mu_c^m(\rho, \rho^{(1)}, T, \mathbf{v}, \mathbf{H})$. This assumption guarantees a nonzero chemical potential for zero magnetic field and is confirmed by calculations for MFs, where the magnetic part contributes additively to the total chemical potential [9]. The nonmagnetic part is given in [10]. The dissipative mass flux $\mathbf{j}^D$ is proportional to the gradients of the the temperature and the chemical potential,

$$j_i^D = \xi_1 \nabla_i T + \xi \nabla_i \tilde{\mu}_c + \xi_{\|} M_i M_j \nabla_j \tilde{\mu}_c + \xi_\times \varepsilon_{ijk} M_j \nabla_k \tilde{\mu}_c \ . \tag{2}$$

Whereas the first two terms characterize isotropic mass fluxes, the last two terms describe anisotropic mass fluxes, namely parallel and perpendicular to the direction of $\mathbf{M}$. The first two transport coefficients, ξ_1, ξ, are related to the classical equation of diffusion including the Soret effect (see below). The last two coefficients, $\xi_{\|}$, $\xi_\times$, reflect that the magnetic field may influence the mass flux for any nonzero strength. Starting with these equations, one gets an analytical result for the magnetic part of the chemical potential,

$$\mu_c^m = -\mu_0 \int_0^H \frac{\partial M}{\partial \rho^{(1)}} dH' \ . \tag{3}$$

It allows a direct determination of μ_c^m if the magnetization $M(H, \rho^{(1)}, T)$ is known without any assumption about the properties of the MF in contrast to [9; 11], where the knowledge of the volume concentration of the nanoparticles [11] or the strength of the magnetodipole interaction [9] is needed. One gets also a general expression for the mass flux,

$$\begin{aligned}
\frac{\mathbf{j}^D}{\rho} &= \left(\frac{\xi_1}{\rho} + \frac{\xi}{\rho}\frac{\partial \mu_c}{\partial T}\right)\nabla T + \frac{\xi}{\rho}\frac{\partial \mu_c^m}{\partial T}\nabla T + \frac{\partial \tilde{\mu}_c}{\partial T}\left[\frac{\xi_{\|}}{\rho}\mathbf{M}\left(\mathbf{M}\nabla T\right) + \frac{\xi_\times}{\rho}\left(\mathbf{M}\times\nabla T\right)\right] \\
&+\xi\frac{\partial \mu_c}{\partial \rho^{(1)}}\nabla c_1 + \xi\frac{\partial \mu_c^m}{\partial \rho^{(1)}}\nabla c_1 + \frac{\partial \tilde{\mu}_c}{\partial \rho^{(1)}}\left[\xi_{\|}\mathbf{M}\left(\mathbf{M}\nabla c_1\right) + \xi_\times\left(\mathbf{M}\times\nabla c_1\right)\right] \\
&-\frac{\mu_0}{\rho}\frac{\partial M}{\partial \rho^{(1)}}\left[\xi\nabla H + \xi_{\|}\mathbf{M}\left(\mathbf{M}\nabla H\right) + \xi_\times\left(\mathbf{M}\times\nabla H\right)\right] \ ,
\end{aligned} \tag{4}$$

which contains terms proportional to thermophoresis ($\sim \nabla T$), to diffusiophoresis ($\sim \nabla c_1$), and to magnetophoresis ($\sim \nabla H$). For zero magnetic field, Eq. (4) reduces to the classical result (see Eq. (227), Chapt. XI in [10])

$$\frac{\mathbf{j}^D}{\rho} = \left(\frac{\xi_1}{\rho} + \frac{\xi}{\rho}\frac{\partial \mu_c}{\partial T}\right)\nabla T + \xi\frac{\partial \mu_c}{\partial \rho^{(1)}}\nabla c_1 = c_1 c_2 D_T \nabla T + D_c \nabla c_1 \tag{5}$$

with (D_T) D_c the (thermal) diffusion coefficient and $c_2 = 1 - c_1$. The concentration of the magnetic particles $c_1 = \rho^{(1)}/\rho$ is defined by the mass fraction of the total density ρ.

3. Application to experiments

According to the experiments [3], a horizontal unbounded layer of MF of thickness h sandwiched between two plates is considered, where the lower one is cooled to T_1 and the upper one is heated to $T_2 > T_1$ (Fig. 1). These boundary conditions together with the equation of head conduction yield the temperature profile of the quiescent conductive state, $T(z) = T_0 + (T_2 - T_1)(z/h)$, with $T_0 = (T_1 + T_2)/2$. The diffusion equation, $\partial c_1/\partial t = \text{div}\left(\mathbf{j}^D/\rho\right)$, and the boundary condition for impenetrable plates, $j_z^D(z = \pm h/2) = 0$, in rearranged form,

$$\frac{-h}{c_1 c_2 (T_2 - T_1)} \frac{\partial c_1}{\partial z}\bigg|_{z=\pm h/2} = \frac{D_T}{D_c} = S_T \,, \tag{6}$$

allow the determination of the Soret coefficient in the zero field case. In the same way the global Soret coefficient in the presence of a magnetic field can be determined by using Eq. (4). Depending on the orientation of the magnetic field $\mathbf{H}$ to the temperature gradient ∇T (see Fig. 1), the global Soret coefficients $S_T^{\|}$ and $S_T^{\perp}$ are functions of the material parameters D_c, ξ, $\xi_{\|}$, and $\xi_{\perp}$ if $M(H)$, μ_c, S_T, and $\rho^{(1)}$ is known [12].

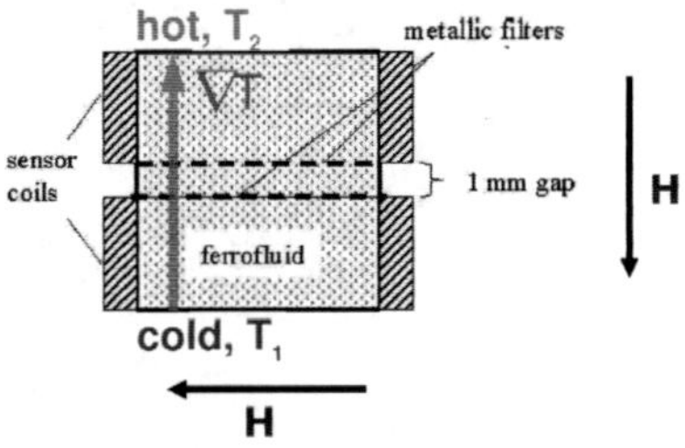

Figure 1. Sketch of the experimental setup with the vertical temperature gradient ∇T inside the ferrofluid layer and the external field $\mathbf{H}$ which is either parallel or perpendicular to ∇T. Courtesy of [3].

To fit $S_T^{\|}$ to the experimental data, the combined quantities ξ/D_c and $\xi_{\|}/D_c$ are used as fit parameters, since $S_T = 0.15\ \text{K}^{-1}$ was measured in the zero field case [3] but not D_c. The blue solid line in Fig. 2 gives the best two parameter fit, yielding $\xi/D_c = 8.2\ \text{kg}\,\text{s}^2/\text{m}^5$ and $\xi_{\|}/D_c = -1.41 \cdot 10^{-7}\ \text{kg}\,\text{s}^2/(\text{m}^3\,\text{A}^2)$. To underline the relevance of $\xi_{\|}/D_c$ even for small fields, the cyan dot-dashed line in Fig. 2 displays $S_T^{\|}$ for $\xi_{\|}/D_c = 0$ and all other parameters as before. Now the theoretical curve misses the measured data ($\square$) clearly. That underlines that the transport coefficients of the anisotropic terms in Eq. (2) are necessary and nonzero for *all* magnetic strengths.

In contrast to the parallel setup, in the perpendicular setup the following coarse approximation is made: $(\partial c_1/\partial y)_{z=\pm h/2}$ shall be a constant C for all H-values tested here. Then the global Soret coefficient $S_T^{\perp}$ depends solely on $F = (\xi_{\perp} C)/D_c$ since ξ/D_c was determined in the parallel setup. The best fit yields $F = 3.75 \cdot 10^{-2}\ \text{kg}\,\text{s}^2/(\text{m}^5\text{A})$. The inferior match with the experimental data (see $\triangle$ and blue dashed line in Fig. 2) in comparison with the parallel setup is due to the approximation that $(\partial c_1/\partial y)_{z=\pm h/2}$ is constant. In the real system it will depend on the magnetic field since the solution for c_1 depends on the magnetic field.

4. Conclusion

The ferrofluid dynamics theory is applied to thermodiffusive problems in magnetic fluids in the presence of magnetic fields. A general expression for the mass flux is given which is independent of the fluid properties, temperature distribution and assumptions about the concentration of the nanoparticles. Applying these results to the experiments [3], it is shown that for thermodiffusive problems in magnetic fluids anisotropic mass fluxes are relevant and no small contributions for any nonzero strengths of the magnetic field. To elucidate this insight, more well designed experiments and further theoretical as well as numerical analyses are needed.

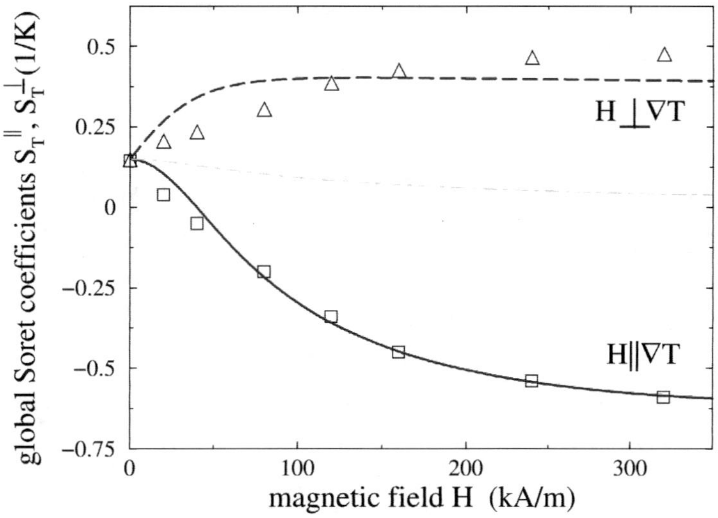

Figure 2. Global Soret coefficients $S_T^{\|}$ and $S_T^{\perp}$ against the magnetic field strength. The blue solid line shows the best fit of $S_T^{\|}$ with $\xi/D_c = 8.2$ kg s^2/m^5 and $\xi_{\|}/D_c = -1.41 \cdot 10^{-7}$ kg s^2/(m^3 A^2) to the experimental data (□). The cyan dot-dashed line displays $S_T^{\|}$ for the same parameters but $\xi_{\|}/D_c = 0$. The blue dashed line indicates the best fit of $S_T^{\perp}$ with $\xi/D_c = 8.2$ kg s^2/m^5 and $F = 3.75 \cdot 10^{-2}$ kg s^2/(m^{5}A) to the experimental data (△) taken from [3].

References

[1] Völker T 2002 *Thermodiffusion in Ferrofluiden: Der Soret-Effekt und sein Beeinflussung durch magnetische Felder* (*Fortschritt-Berichte VDI Reihe 3* no 761) (Düsseldorf: VDI Verlag)

[2] Völker T and Odenbach S 2003 *Phys. Fluids* **15** 2198–2207

[3] Völker T and Odenbach S 2005 *Phys. Fluids* **17** 037104

[4] Blums E, Odenbach S, Mezulis A and Maiorov M 1998 *Phys. Fluids* **10** 2155–2163

[5] Shlioms M I 2002 *Thermal Nonequilibrium Phenomena in Fluid Mixtures* (*Lecture Notes in Physics* no 584) ed Köhler W and Wiegand S (Berlin: Springer Verlag) pp 355–371

[6] Müller H W and Liu M 2001 *Phys. Rev. E* **64** 061405

[7] Müller H W and Liu M 2003 *Phys. Rev. E* **67** 043202

[8] Odenbach S and Müller H W 2002 *Phys. Rev. Lett.* **89** 037202

[9] Zubarev A Y and Iskakova L Y 2002 *Phys. Rev. E* **65** 061406

[10] de Groot S R and Mazur P 1984 *Non-equilibrium thermodynamics* (New York: Dover Publ.)

[11] Bacri J C, Cebers A, Bourdon A, Demouchy G, Heegaard B M and Perzynski R 1995 *Phys. Rev. Lett.* **74** 5032–5035

[12] Lange A 2004 *Phys. Rev. E* **70** 046308

Position feedback control of a nonmagnetic body levitated in magnetic fluid

J H Lee[1], Y J Nam[1], R Yamane[2] and M K Park[1]

[1] Graduate School, Pusan National University, Busan 609-735, Republic of Korea
[2] Kokushikan University, Tokyo 154-8515, Japan

E-mail: magooro@pusan.ac.kr

Abstract. This paper is concerned with the position feedback control of a magnetic fluid actuator which is characterized by the passive levitation of a nonmagnetic body immersed in a magnetic fluid under magnetic fields. First of all, the magnetic fluid actuator is designed based on the ferrohydrostatic relation. After manufacturing the actuator, its static and dynamic characteristics are investigated experimentally. With the aid of the dynamic governing relation obtained experimentally and the proportional-derivative controller, the position tracking control of the actuator is carried out both theoretically and experimentally. As a result, the applicability of the proposed magnetic fluid actuator to various engineering devices is verified.

1. Introduction

Recently, various applications using magnetic fluids has been developed, such as accelerometers, inertial dampers, passive magnetic bearings, pressure rotary seals, loud speaker coolants, and ferrohydrostatic separators [1-2]. Compared with conventional devices, magnetic fluid actuators have many advantages of simplicity in design and construction, high reliability due to the inexistence of any moving mechanical parts, and a large range of electric current for generating a desired magnetic field. Therefore, many investigations have recently been conducted in relation to the development of magnetic fluid actuators [3-7]. However, most of the previous studies have been focused on the basic design of the magnetic fluid actuators or their static characteristics.

In order to improve the dynamic characteristics of these actuators and to verify its availability in various engineering applications, the development of an effective controller and the investigation on its control characteristics are inevitable. Nevertheless, the research data on the practicability of the magnetic fluid actuator are insufficient, and there still remains a wide unexplored domain.

Consequently, the aim of this work is to derive the dynamic governing relation of the magnetic fluid actuator based on the passive levitation and to verify its control performances both theoretically and experimentally.

2. Design of a magnetic fluid actuator and its performances

2.1. Passive levitation condition

A nonmagnetic body of arbitrary shape, volume V, surface area A, and density ρ_b immersed in a stationary magnetic fluid of density ρ_f in the presence of an external magnetic field gradient□H experiences the following forces [8]:

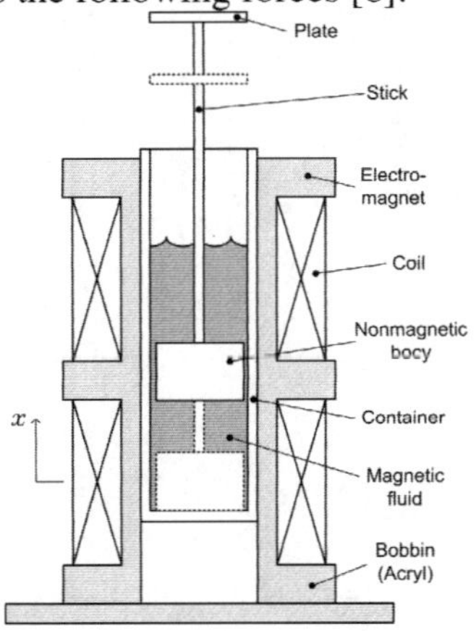

Figure 1. Schematic configuration of the magnetic fluid actuator.

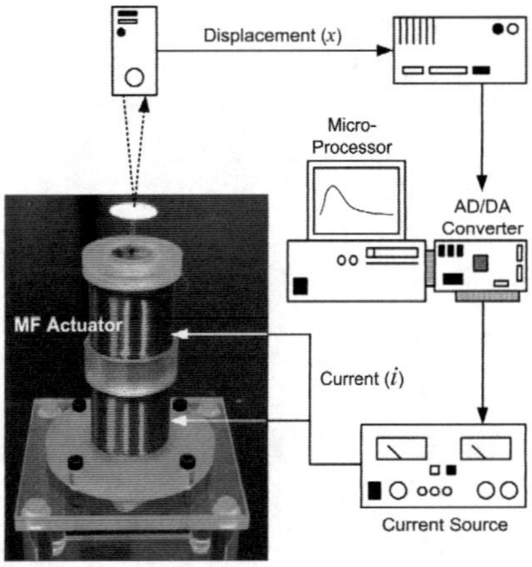

Figure 2. Experimental apparatus for evaluating performances of the actuator.

(i) Gravitational force: $\vec{F}_g = \int_V \rho_b \vec{g} dV = \rho_b \vec{g} V$ (1)

(ii) Buoyancy force: $\vec{F}_b = -\int_V \rho_f \vec{g} dV = -\rho_f \vec{g} V$ (2)

(iii) Magnetic buoyancy force: $\vec{F}_m = -\oint_A \left(\frac{1}{2} \mu_0 M_n^2 + \mu_0 \int_0^H M dH \right) \vec{n} dA$ (3)

where g is the gravitational acceleration, M the magnetization of the fluid, H the magnetic field intensity, and μ_0 the magnetic permeability of vacuum. Also, $\vec{n}$ is the unit vector normal to any point on A. Above equations are written in SI units. The last equalities in (1) and (2) are true only if the densities and the gravitational acceleration are uniform over V. The first term on right of (3) represents distortions of the magnetic field due to the immersed body. Because the error by these distortions is of the order of M/H, they can be reasonably neglected in the common case of $M << H$. By assuming that the magnetic fluid is isothermal material as well as M and ∇H are constant cross V, (3) can be simplified as $\vec{F}_m = -\mu_0 M \nabla H V$. Then, the force acting on the volume unity of the body will be:

$$\vec{F} / V = (\rho_b - \rho_f)\vec{g} - \mu_0 M \nabla H \quad (4)$$

where g is the gravitational acceleration, M the magnetization of the fluid, and μ_0 the magnetic permeability of vacuum. Above equations are written in SI units. As can be seen in above equation, the immersed body sinks when $\rho_b > \rho_f$ in the absence of magnetic field. If the magnetic field gradient has the same orientation to gravitational acceleration, the nonmagnetic body can be levitated at an desired point in the fluid space by a good choice of ∇H; that is, the conditions for floating-up and stable levitation of the body can be expressed by

$$M \nabla H \leq (\rho_b - \rho_f)\vec{g} / \mu_0 \equiv \alpha \quad (5)$$

2.2. *Experimental setup*

Figures 1and 2 show the schematic configuration of a magnetic fluid actuator proposed in this work and the experimental apparatus constructed to investigate its performances. Sample magnetic fluid is a kerosen-based ferricolloid HC-50 (Taiho Industries Co.). Its magnetization curve is shown is shown in Figure 3. For the convenient calculation of $M\nabla H$ in (5), the magnetization equation was obtained sensibly by a curve fitting method. Considering the viscous effect of the magnetic fluid, three cylindrical nonmagnetic bodies with an identical height of 10*mm* but different diameters of 10, 13, and 14mm are prepared. They are made of aluminum (the density ρ_b=2700kg/m^3). In order to measure the

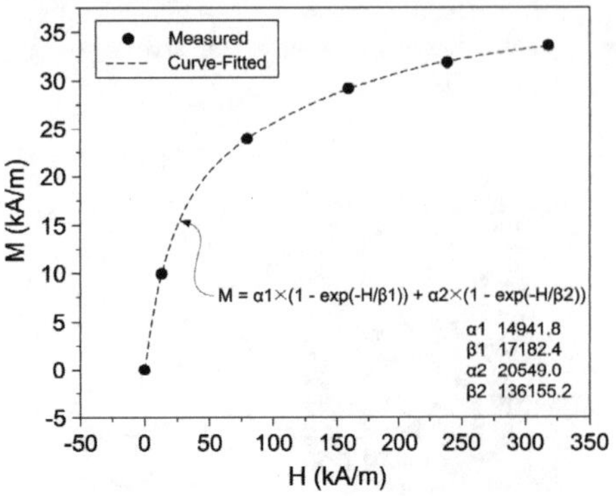

Figure 3. Magnetization curve of HC-50.

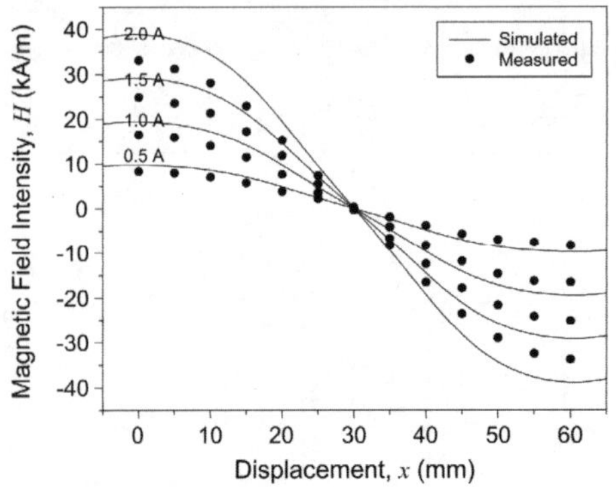

Figure 4. Magnetic field intensity

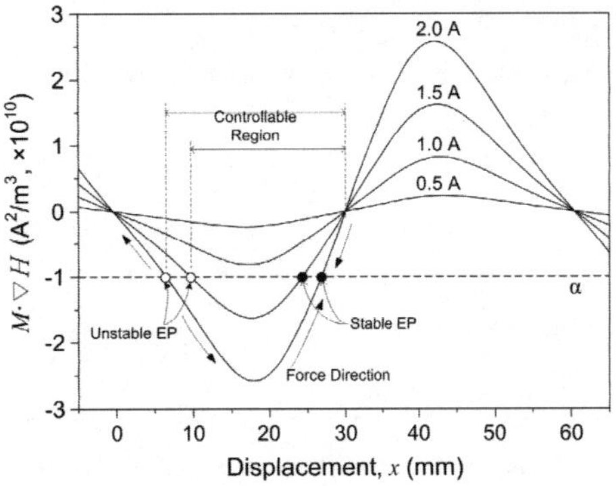

Figure 5. M□H curves.

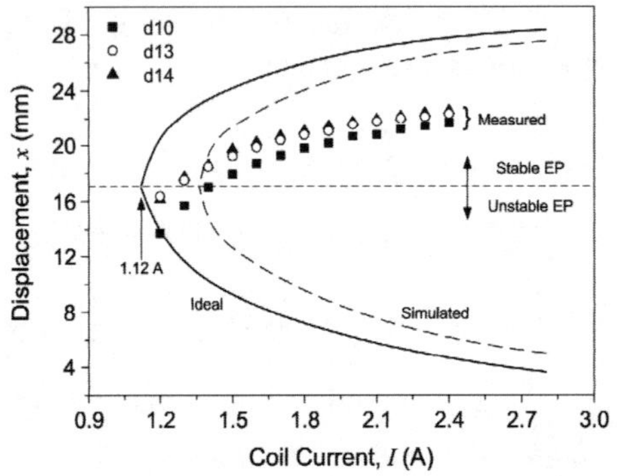

Figure 6. Displacement vs. coil current

displacement x of the body by using an optical displacement transducer, the circular plastic stick with a small plate is attached to the body. A magnetic field gradient suitable for the stable levitation can be obtained by using a pair of hollow solenoids with north poles placed face-to-face with each other. The gap between the magnetic poles is 30mm. Each solenoid is made of 1000 turns of 0.5mm enamel wire (the electric resistance 10.77Ω), and its size is 24mm in inner diameter, 40mm in outer diameter, and 30mm in length. The maximum current available is 2.4A. All experiments were performed at room temperature 20°C. Figure 4 shows the magnetic field intensity obtained both theoretically and experimentally. The coil current was induced to the solenoids in the range of 0.5 to 2.0A by a step of 0.5A, and the magnetic field was measured by a Gauss meter. As predicted, the magnetic field was zero at the center of the gap and gradually stronger as the position was further away from the center. Through the comparison between the experimental results and the simulation analysis, the efficiency of the solenoids was found by approximately 85%.

2.3. Static and Dynamic Performances

Figure 5 shows M□H curves obtained theoretically according to the coil currents. Considering the properties of the materials used in this work, α in (2) was found by $-1.0453\cdot10^{10}A^2/m^3$. From this figure, it can be seen that there are two equilibrium points (EPs) corresponding to the equality of (2) at a certain coil current; one point is unstable and the other is stable. At the unstable EP, the nonmagnetic body tends to sink down or float up in the presence of external mechanical or electromagnetic disturbances. Contrarily, at the stable EP, the body in any displacement is subject to a restoring force returning it to the EP.

Figure 6 shows the displacement of the nonmagnetic body in the magnetic fluid with respect to the coil current. The 'ideal' curve was obtained analytically, and the 'simulated' curve was obtained considering the efficiency of the solenoids. The experiments were performed using the DC current with its range of 1.2 to 2.4A by a step of 0.1A. As shown in Figure 6, the experimental results were not in good accordance with the analytical ones. It is considered that some errors results from the assumptions introduced for the derivation of (1): (i) the magnetic field distortions due to the nonmagnetic body are negligible, (ii) M and ∇H are constant cross the body volume, and (iii) the magnetic fluid is an isothermal material. Other reason for the disagreement may be inhomogeneity of the magnetic field in the radial direction. Nevertheless, it can be clearly seen that the displacement of the immersed body is increased with increasing the coil current. Therefore, it is concluded that the stable levitation of the nonmagnetic body in the magnetic fluid can be achieved only at the stable EP and its levitating position can be controlled actively by coil currents.

Figure 7 shows the step response of the magnetic fluid actuator. Initially, the nonmagnetic body was levitated stably at the position corresponding to the coil current of 1.4A. When the current of 2.2A was suddenly applied after 2sec, the body behaved as a typical step response with the overshoot and viscous damping. In order to investigate the effect of the body diameter, the obtained data were normalized by each minimum and maximum value. Then, the damping coefficients ζ and natural frequencies ω_n for three different bodies were found sensibly through the comparison with simulation results using a second order linear system. From the results, it can be seen that the body with larger diameter has lower natural frequency and larger damping coefficient. This is due to the viscous effect of the fluid at the gap between the body and the container wall. The gap decreased with increasing the body diameter makes the drag force acting on the body strong. Therefore, it is concluded that these design parameters can be optimally determined according to application fields of the proposed actuator.

3. Position Control of the magnetic fluid actuator

3.1. Proportional-Derivative (PD) Controller

The displacement of the body x is measured, and the desired displacement x_d can be determined by substituting the desired current I_d into the relation $x=f(i)$ obtained from curve-fitting results in Figure 5 Also, the proportional-derivative controller is given in the Laplace domain by

$$K(s) = k_p + k_d s \qquad (6)$$

where k_p and k_d are the proportional and derivative gains, respectively. Using second order linear system and (6), the characteristic equation of the closed-loop system is obtained as follows:

$$s^2 + (2\zeta\omega_n + k_d\omega_n^2)s + \omega_n^2(1+k_p) = 0 \qquad (7)$$

Introducing the desired damping coefficient ζ_d and the desired frequency ω_d, so that the closed-loop system lies on a desired dynamic behaviour, the control gains are analytically found by

$$k_p = (\omega^2 - \omega_n^2)/\omega_n^2, \quad k_d = 2(\zeta_c\omega_c - \zeta\omega_n)/\omega_n^2 \qquad (8)$$

In this research, ζ_d =0.7 and ω_d=50*rad/s* are set. Then, the control gains for three different bodies can be determined with the help of ζ and ω_n obtained from the dynamic evaluations. In addition, the applied coil current to the magnetic fluid actuator is obtained by

$$I(t) = I_d + f^{-1}\{k_p e(t) + k_d \dot{e}(t)\} \quad \text{where } e(t)= x_d(t)- x(t) \qquad (9)$$

3.2. Experimental Results and Discussion

Figure 8 shows the results of the position tracking control of the magnetic fluid actuator. In all cases, the results using the controller show improved dynamic characteristics: smaller overshoots, faster rising times, and weaker oscillation behaviours. However, there are some differences between the simulation and experimental results. Although the magnetic body force varies with respect to both the levitating position and applied current, the dynamic model of (2) is obtained neglecting the effect of the levitating position for the convenience. The external disturbances, especially the variation of fluid temperature due to the solenoids, are put in other reason. Note that there are various controllers available in real applications. Therefore, it is expected that the using the detailed model of the actuator as well as the advanced controller leading to better control performances allows further improvement in the dynamic characteristics of the actuator.

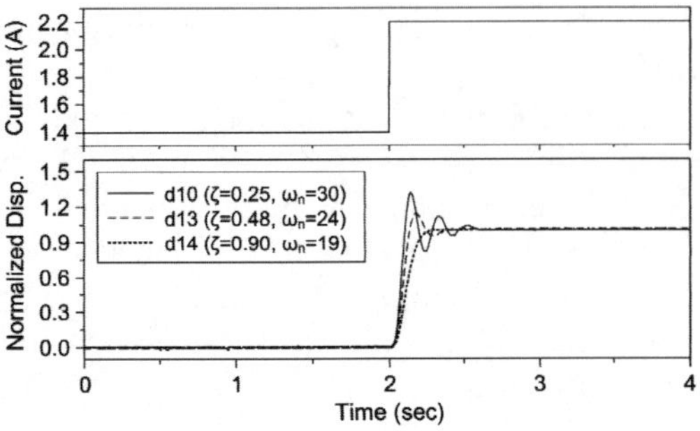

Figure 7. Dynamic characteristics: step responses.

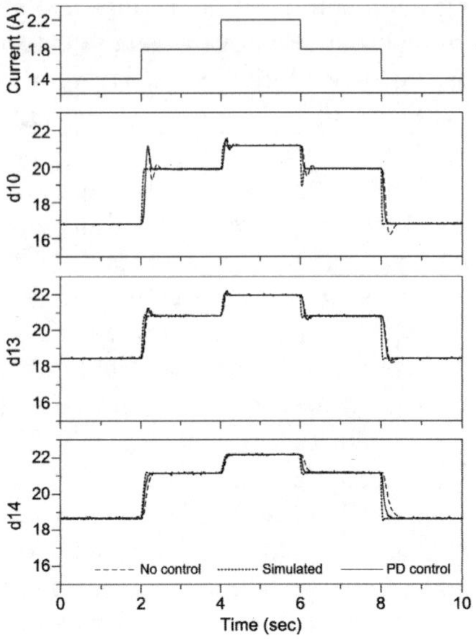

Figure 8. Position tracking control of the magnetic fluid actuator. (unit: mm)

4. Conclusions

The results obtained are summarized as follows; (i) There is one equilibrium point for the stable levitation of any nonmagnetic body immersed in a magnetic fluid in the presence of sufficient magnetic field gradients. In addition, the levitating position can be effectively controlled with the coil currents. (ii) Even with the conventional controller, the dynamic performances of the proposed magnetic fluid actuator can be improved effectively. (iii) Through the comparison study between the simulation analyses and the experimental results, the applicability of the proposed magnetic fluid actuator to various engineering devices was verified.

5. References

[1] Raj K, Moskowitz B and Caciari R 1995 *J. Magn. Magn. Mater.* **149** 174
[2] Popplewell J 1984 *Physics in Technology* **15** 150
[3] Yamane R, Tomita S, Mai J, Park M K and Oshima S 2002 *J. Magn. Magn. Mater.* **252** 268
[4] Sudo S, Ise K and Ikohagi T 2002 *J. Intell. Mater. Syst. Struct.* **13** 539
[5] Oyama T, Ueno K, Katagiri K, Kamiyama S and Nakatani I 2001 *JSME Int. J.* B **44** 1
[6] Kamiyama S, Watanabe J and Sato A 1992 *JSME Int. J.* B **35** 370
[7] Buioca C D, Iusan V, Stanci A and Zoller C 2002 *J. Magn. Magn. Mater.* **252** 318
[8] Rosensweig R E 1985 *Ferrohydrodynamics* Cambridge Cambridge university press

Journal of Physics: Conference Series **149** (2009) 012108 doi:10.1088/1742-6596/149/1/012108

Liquid Gallium based Temperature Sensitive Functional Fluid Dispersing Chemically Synthesized FeMB Nanoparticles

H S Park, L F Cao, G Dodbiba, and T Fujita

Department of Systems Innovation, School of Engineering, The University of Tokyo
7-3-1 Hongo, Bunkyo-ku, Tokyo 113-8656, JAPAN

E-mail: ifpark22@yahoo.co.jp

Abstract. In this work, FeMB (M = Nb, V) nanoparticles were first reported to be synthesized by a chemical method, from reduction of $FeCl_2$, NbF_5 (and NH_4VO_3) using $NaBH_4$ as a reducing agent in aqueous solution. A new temperature sensitive functional fluid was then prepared by dispersing silica coated FeNbVB nanoparticles in liquid gallium. The result shows that the FeNbVB nanoparticles exhibit an oxidation resistance better than that of FeNbB nanoparticles. The FeNbVB nanoparticles were in the size range of 30 - 50 nm and the thickness of silica layer was observed about 10 nm by means of transmission electron microscopy. The magnetization of the synthesized particles and fluid shows a temperature dependency within the testing temperature range of 293 - 353 K, which indicated their application potential in magneto-caloric energy conversion devices.

1. Introduction

A functional fluid is a good choice for the application in magnetocaloric energy conversion and heat exchange devices, where a high thermomagnetic coefficient is required. However, the most widely used functional fluids, prepared by dispersing magnetite nanoparticles in water, can not satisfy these requirements [1]. Therefore, the development of temperature sensitive magnetic nanoparticles for functional fluids is of great interests. Especially, FeMB (M = Zr, Hf, V, Nb, Ti, Ta, Mo and W) nanocrystalline alloy is an excellent type of soft magnetic materials characterized by its high saturation and permeability [2, 3] that, as recently it is reported, show also magneto-caloric effect [4-7]. Nowadays, iron based alloys are mainly fabricated by metallurgical methods, such as arc-melting [8, 9] and mechanical alloying [10, 11], by which nano-sized and homogenously distributed particles are difficult to be obtained. On the other hand, chemically synthesized magnetic materials may have unique magnetic properties, derived from their small particles sizes and uniform size distribution [12, 13]. Thus, developing a chemical synthesis method to prepare temperature sensitive FeMB nanoparticles is of great interests.

On the other hand, when selecting a carrier liquid, it is important to consider the melting point, boiling temperature and evaporation at working temperature. A good carrier for temperature sensitive functional fluids can be a liquid metal, which exhibit a high thermal and electric conductivity as well as fluidity. When ultra fine particles are dispersed in a liquid metal, the functional fluid such as magnetic fluid and magneto-rheological fluid can be prepared. For many years, the synthesis of the mercury based magnetic fluid has been investigated, however, liquid mercury is toxic and difficult to handle. Thus, an environment-friendly alternate, liquid gallium becomes a good choice for the carrier liquid in functional fluid. In particular, gallium has the advantages of remaining in the liquid state throughout a wide temperature range (303 to 2477 K) and has a very low vapour pressure at atmospheric pressure (9.31×10^{-21} Pa at 302.9 K). It has high thermal conductivity which is much higher than that of traditional organic or aqueous carrier liquids [14]. Thus, liquid gallium-based functional fluids can be used in magneto-caloric energy conversion devices or heat exchange devices, where continuous heat diffusion and cooling can be achieved and the elastic properties can be kept at all times [15]. The aim of this work is to prepare the liquid gallium based temperature sensitive functional fluid. Here, we report for the first time the synthesis of FeMB nanoparticles by a chemical method. In order to increase the affinity of the nanoparticles to liquid gallium, silica coating on FeNbVB particle surface was then carried out by the sol-gel method, based on the hydrolysis and condensation of TEOS (tetraethyl-orthosilicate). Our attentions were focused on the preparation and the temperature dependency of magnetization of the synthesized magnetic nanoparticles and functional fluid.

2. Experimental method

2.1 Preparation of FeMB nanoparticles and silica coating procedure

FeNbVB nanoparticles were prepared by a chemical synthesis based on sodium borohydride reduction. A mixed aqueous solution of niobium floride 97 % (NbF_5, Aldrich), ammonium vanadate 99 % (NH_4VO_3, Wako) and iron (II) chloride tetrahydrate 99 % ($FeCl_2{\cdot}4H_2O$, Nacalai,) was precipitated by using 336 mmol of sodium borohydride 98 % ($NaBH_4$, Nacalai) solution. Reagents of NbF_5, $FeCl_2{\cdot}4H_2O$ and $NaBH_4$ were dissolved separately in distilled water. NH_4VO_3 was dissolved in water by adding of 0.5 mol/l sulfuric acid solution. $NaBH_4$ solution was simultaneously added into the mixture containing the Fe, Nb and V ions. The synthesized particles were then washed several times with ethanol. FeNbB nanoparticles were also prepared in the same way, only without the addition of ammonium vanadate. Finally, the synthesized particles were collected by means of filtration, and dried in a desiccator at room temperature. Details of the preparation of the silica coated magnetic nanoparticles were given in a previous publication [16]. Finally, theses core-shell nanoparticles were dispersed in the liquid gallium at the solid weight fraction of 1.67 mass%.

2.2 Composition analysis and characterization

The composition of the sample was determined with inductively coupled plasma optical emission spectrometry (Optima 5300 DV, PerkinElmer) and field emission scanning electron microscope (JSM-7000F, JEOL) equipped with EDS (energy dispersive spectroscopy). The morphology of core-shell type nanoparticles and the thickness of silica layer were observed by using a transmission electron microscope (JEM-2000FX, JEOL) equipped with EDS. Vibrating sample magnetometer (VSM) was also employed to investigate the temperature dependency for the magnetization of samples. The applied magnetic field was varied between 0 and 0.9 T.

3. Results and discussion

3.1 Morphology and composition of FeMB nanoparticles

The content of Fe, Nb, V and B in the samples was qualitatively determined by inductively coupled plasma optical emission spectrometry (ICP-OES), and the results were shown in Table 1. It can be seen that the ratio of iron, niobium and boron is about 78, 7 and 15 mol % for the FeNbB nanoparticles. A drawback is that Nb has lower resistance to oxidation at an elevated temperature. In order to overcome the oxidation problems, 4 mole % of vanadium ion was contained in this work.

Table 1. The normalized element compositions of the samples, determined from ICP-OES analysis.

	Fe	Nb	V	B	Total
FeNbB [mol %]	77.88	7.17	-	14.95	100
FeNbVB [mol %]	80.28	2.84	3.75	13.13	100

The synthesized FeNbVB nanoparticles have a ratio of iron, niobium, vanadium and boron around 80, 3, 4 and 13 at mole %. As shown in Figure 1, the as-synthesized FeNbVB nanoparticles have a smaller particle size in general, and are not aggregated as much as FeNbB particles. The EDS data also showed that the content of oxygen of FeNbVB samples is largely reduced from 46.67 % to 15.25 %.

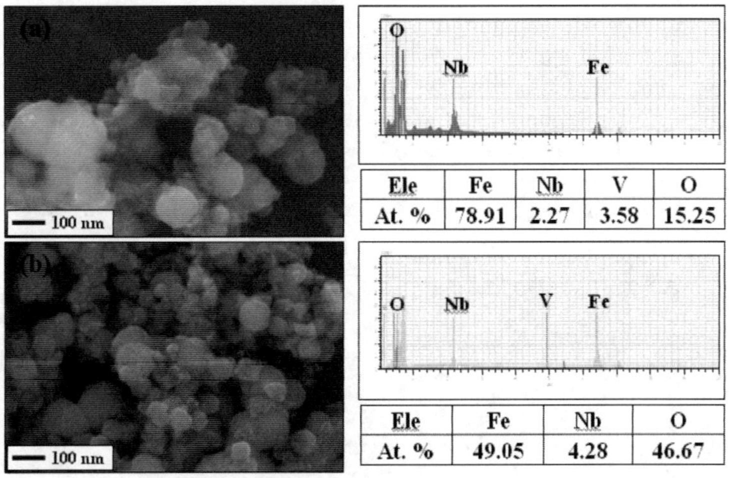

Ele	Fe	Nb	V	O
At. %	78.91	2.27	3.58	15.25

Ele	Fe	Nb	O
At. %	49.05	4.28	46.67

Figure 1. FE-SEM images and energy dispersive spectroscopy (EDS) data of the as-synthesized (a) FeNbB and (b) FeNbVB nanoparticles.

3.2 Morphology and composition of FeNbVB-SiO_2 core-shell nanoparticles

Generally speaking, naked metal powders are not dispersed easily in the liquid gallium directly; on the other hand, refractory metals and ceramics, such as tantalum, tungsten, graphite, stabilized $Zr0_2$, and quartz are most stable in gallium [17]. Therefore, in this work, FeNbVB nanoparticles were coated with silica in order to improve the dispersability of these particles in liquid gallium. Another reason for selecting silica coating, it is possible to prepare the core-shell composite particle consisting of a FeNbVB core and a silica coating layer, with a density similar to that of the liquid gallium [14].

Figure 2 depicts the TEM images and EDS data of the FeNbVB and FeNbVB-SiO_2 core-shell nanoparticles. The FeNbVB particles are mostly in the size range of 30 - 50 nm. By placing the electron beam exactly on these spherical particles, peaks for iron, niobium and vanadium have been observed (Figure 2(a)), and a shell of SiO_2 with a thickness of about 10nm was also be observed on the FeNbVB particles (Figure 2(b)). The formation of silica can be observed from the composition analysis of the coated particles. Also the density of the core-shell particles can be adjusted by controlling the thickness of silica coating layer.

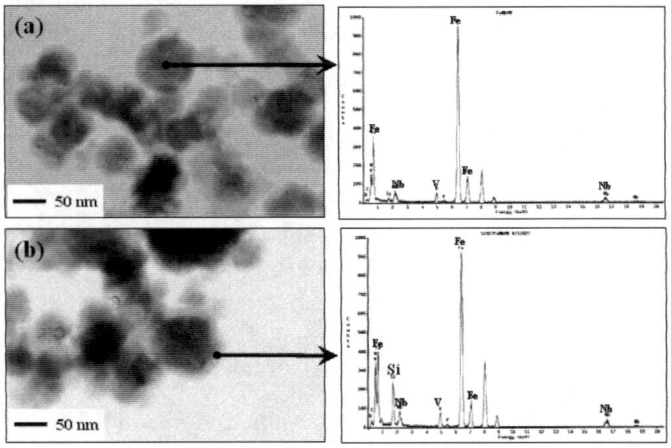

Figure 2. TEM images and energy dispersive spectroscopy data (a) FeNbVB and (b) FeNbVB- SiO_2 nanoparticles.

The macroscopic observation of different particles mixed with liquid gallium is shown in Figure 3. The result clearly shows that the FeNbVB particles without coating can hardly be dispersed into liquid gallium, so that Figure 3(a) shows two separated phase i.e., residual FeNbVB particles and liquid gallium. A small amount of metal powders could be directly dispersed and a small amount of powders tended to remain on the surface of liquid gallium. Figure 3(b) shows that the silica coated FeNbVB particles can be well dispersed into liquid gallium, which keeps its brilliant silvery colour. Moreover, the contact angle at the iron and SiO_2 substrate / liquid gallium interface was 158° and 145°, respectively. Contact angles were measured directly from the image of the drop section by using a digital microscope (Keyence VXH-100), equipped with an 18 million-pixel CCD camera. The measurement of initial contact angel between pure liquid gallium (99.9999 %, 0.1 mL) and substrate samples were conducted at 313 K under atmospheric pressure. This result also shows that silica coated FeNbVB particles can be easily wetted by liquid gallium, relatively when compared with FeNbVB particles without silica coating. Therefore, it is reasonable to conclude that the existence of silica coating improves the dispersion of FeNbVB particles into liquid gallium.

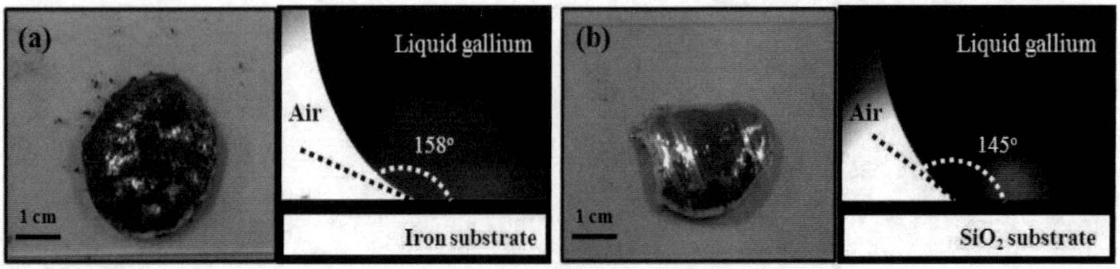

Figure 3. Macroscopic observation of FeNbVB particles mixing with liquid gallium (1.67 mass% of solid fraction, $T = 293$ K) and contact angle of iron and silica for liquid gallium ($T = 313$ K). (a) FeNbVB particles without coating and contact angle of iron for liquid gallium (b) silica coated FeNbVB particles and contact angle of silica for liquid gallium.

3.3 Temperature sensitive magnetic property of as-synthesized samples

Figure 4 shows that the magnetization curve for the samples measured between 293 K and 353 K in 5 K steps and the temperature dependencies of the magnetization for FeNbVB nanoparticles (Figure 4(a)), and silica coated FeNbVB (Figure 4(b)) nanoparticles measured in an applied field 0.9 T. The result shows that the temperature dependency of the magnetization for the silica-coated FeNbVB nanoparticles is almost same as the FeNbVB nanoparticles, though the magnetization values decrease due to the existence of a non-magnetic silica layer (Figure 2(b)). The result shows that the synthesized FeNbVB particles have a little lower saturation magnetization and temperature dependency than FeMB soft magnetic alloy prepared by metallurgical methods [5-7]. However it has relatively high saturation magnetization and temperature dependency when compared with various types of temperature sensitive ferrite such as Ni-Zn and Mn-Zn ferrite etc. [18-20], which are normally applied to magneto caloric energy conversion or heat exchange device. Moreover, chemically synthesized FeNbVB magnetic particles have smaller particles sizes and uniform size distribution than FeMB magnetic alloy obtained from metallurgical methods. Therefore, chemically synthesized FeNbVB magnetic particles more suitable to prepare the liquid gallium based functional fluid. On the other hand, Figure 4(c) shows that the temperature dependency of liquid gallium based functional fluid dispersed silica coated FeNbVB nanoparticles (1.67 mass% of solid fraction). The magnetization value of the synthesized functional fluid decreases, which is caused by the low magnetic property of liquid gallium and low solid fraction of magnetic particles. However, the magnetization value of the synthesized functional fluid still shows the temperature dependence, i.e. decreases with increasing temperature.

4. Conclusion

In this study, the liquid gallium based temperature sensitive functional fluid, dispersing silica coated FeMB nanoparticles was synthesized. The result shows that the FeNbVB nanoparticles exhibited an oxidation resistance higher than that of FeNbB nanoparticles. The FeNbVB nanoparticles were in the size range of 30 - 50 nm and the thickness of silica layer was about 10 nm. Both the synthesized nanoparticles and functional fluid showed a temperature sensitive of magnetization within the testing temperature range of 293 - 353 K. Therefore, the liquid gallium based temperature sensitive functional fluid dispersing silica coated FeNbVB nanoparticles is considered as an attractive working liquid for magneto-caloric energy conversion and heat exchange applications due to a high saturation magnetization and a high temperature dependence even low solid fraction.

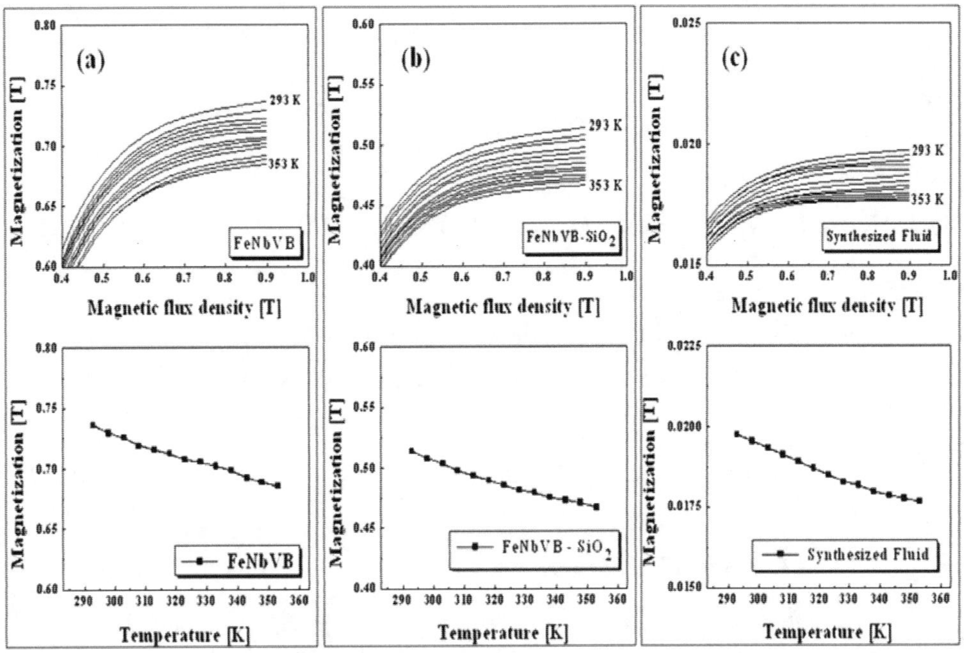

Figure 4. Magnetization curve for the samples measured between 293 K and 353 K in 5 K steps and the temperature dependencies of the magnetization for the samples measured at an applied field of 0.9 T. (a) FeNbVB, (b) FeNbVB-SiO_2 and (c) synthesized liquid gallium based functional fluid.

Acknowledgments
The present research was supported in part through the Global COE Program, "Global Centrer of Excellence for Mechanical Systems Innovation," by the Ministry of Education, Culture, Sports, Science and Technology.

References

[1] Auzans E 1999 Institute of Physics of Latvian University Thesis
[2] Makino A, Bitoha T and Moue A 2003 J. Scripta Mater. **48** 869
[3] Makino A, Inoue A and Masumoto T 1999 J. Nanostructured Materials **12** 825
[4] Chau N, Yhanh P Q, Hoa N Q and The N D 2006 J. M. M. M. **304** 36-40
[5] Škorvánek I and Kováč J 2004 Cze. J. of phys. **54** D 189
[6] Škorvánek I et. al. 2007, Materials Science and Engineering **A 449–451** 460
[7] Chiriac H and Lupu N, 2005, IEEE Transactions on Magnetics **41** (10) 3289
[8] Gschneidner K et al 2003 Nat. Mater. **2** 587
[9] Inoue A, Shen B, Koshiba A, Kato H and Yavari A R 2003 Nat. Mater. **2** 661
[10] Kim J -J, Choi Y, Suresh S and Argon A S 2002 Science **295** 654
[11] Kwon H W and Jeong I C 2004 Phys. Stat. Solidi a **201** 1921
[12] Eckert D, Müller K H and Handstein A 1990 IEEE Trans. Magn. **26** 1834
[13] Sun S, Murray C B, Weller D, Folks L and Moser A 2002 A. Science **287** 1989.
[14] Cao L F , Park H S, Dodbiba G and Fujita T 2008 Magnetohydrodynamics **44**(2) 97
[15] V. Ya. Prokhorenko et. Al., 2000 High Temperature **38** (6) 954
[16] Park H S, Dodbiba G, Cao L F and Fujita T 2008 J. Phys.: Condens. Matter **120** 1
[17] Luebbers P R and chopra O K 1995 16th IEEE/NPSS Symposium on Fusion Engineering, Champaign, Illinois, September 30 – October 4 **1** 232
[18] Fujita T, Mamiya M and Jeyadevan B 1990 J. M. M. M. **85** 203
[19] Poddara P. et. al. 2006 J. M. M. M. **307** 227
[20] Arulmurugan R et. al. 2006 J. M. M. M. 298 83

Journal of Physics: Conference Series **149** (2009) 012109 doi:10.1088/1742-6596/149/1/012109

Flow control of magnetic fluids exposed to magnetic fields

M Reindl[1], A Leschhorn[2], M Lücke[2] and S Odenbach[1]

[1]Institute of Fluid Mechanics, Chair of Magnetofluiddynamics, 01062 Dresden, Germany

[2]Institute of Theoretical Physics, Universität des Saarlandes, 66041 Saarbrücken, Germany

Matthias.Reindl@tu-dresden.de

Abstract. The description of flow in ferrohydrodynamics (Rosensweig, 1985) is based on a combination of equations, namely the continuity equation, the Navier-Stokes equation, the Maxwell equations and particular equations for the magnetization. Since the different models to describe the relaxation of magnetization differ, the adequate one has yet to be identified. By comparing experimental and simulation data of a model system, this goal may get achieved. As a model system, a Taylor-Couette apparatus was chosen. In this paper, experimental results concerning the transition form circular Couette flow to Taylor vortex flow at different field strengths of an axial magnetic field are compared to a linear stability analysis. The relaxation equation established by Shliomis (Shliomis, 1972) and the Debye-Model with a field dependent relaxation time showed to give qualitative accordance with the experimental data.

1. Motivation

The magneto-rotational effect as a cause for an increase of viscosity of ferrofluids exposed to magnetic fields has been the subject of many investigations [1-6]. In a ferrofluid being subject to shear flow, the particles are forced to rotate with their axis of rotation parallel to the orientation of vorticity. When a magnetic field is applied, the suspended particles try to align themselves with their magnetic moment along the orientation of the magnetic field. Assuming magnetically hard particles (i. e. the Brownian relaxation time is shorter than the Néel relaxation time), a (magnetic) torque is exerted on them in case the vorticity of the shear flow and the magnetic field are not collinear. This magnetic torque counteracts the viscous torque exerted by the carrier liquid, causing an increase of the fluid's viscosity. If vorticity and magnetic field are collinear, no increase of the fluid's viscosity will be observable. A flow field, with a not vanishing component of vorticity perpendicular to the magnetic field, turns the magnetic dipoles of magnetically hard particles out of the magnetic field direction. The magnetization of the fluid is now not only dependent on the magnetic field, but also on the flow field. It is thus not describable by the equations for equilibrium magnetization anymore. Models to describe the magnetization outside equilibrium are necessary.

Shliomis [2] was the first to describe the magnetization of a ferrofluid considering a not vanishing component of vorticity. Since then, other equations for the magnetization where developed, all of which are mainly based on the relaxation of magnetization against equilibrium, which is determined by a relaxation time. The existing equations to describe the relaxation of magnetization are different in their physical concepts and the way they predict experimental results. In order to determine the right one, a comparison of analytical and experimental data of a model system is needed. In Section 2 some equations for the relaxation of magnetization are introduced, in section 3 the experimental apparatus is outlined shortly and in section 4 some results are presented.

2. Theoretical

The simplest model to calculate the magnetization outside the equilibrium is the Debye model, denoted here as *Debye*:

$$d_t\mathbf{M} = -\frac{1}{\tau}\left(\mathbf{M}-\mathbf{M}^{eq}\right)+\mathbf{\Omega}\times\mathbf{M} \tag{1}$$

This model implies a relaxation of the magnetization **M** towards the equilibrium magnetization $\mathbf{M}^{eq}$ at a given external magnetic field **H** with a relaxation time τ in a frame rotating with the vorticity $\mathbf{\Omega} = \frac{1}{2}\nabla\times\mathbf{u}$ of the flow **u.** Shliomis [2] developed the relaxation equation (denoted as *S'72*)

$$d_t\mathbf{M} = -\frac{1}{\tau}\left(\mathbf{M}-\mathbf{M}^{eq}\right)+\mathbf{\Omega}\times\mathbf{M}+\kappa(\mathbf{M}\times\mathbf{H})\times\mathbf{M} \tag{2}$$

With an additional constant $\kappa = (6\Phi\tilde{\eta})^{-1}$, where Φ is the volume fraction of the magnetic material and $\tilde{\eta}$ the dynamic viscosity of the carrier liquid. Niklas et al. [4] considered stationary magnetizations ($d_t\mathbf{M} = 0$) near the equilibrium ($\mathbf{M}\text{-}\mathbf{M}^{eq} << 1$), which appear at not too high rotation rates ($|\Omega|\tau << 1$). In this case the magnetization equations presented above can be simplified with (denoted as *Niklas approach*)

$$\mathbf{M}-\mathbf{M}^{eq} = c_N\mathbf{\Omega}\times\mathbf{H} \tag{3}$$

$$c_N = \frac{\chi(H)\tau}{\left(1+\kappa\tau\chi(H)H^2\right)} \tag{4}$$

χ is the susceptibility of the ferrofluid. The relaxation time τ may either be assumed to be constant or dependent on the external magnetic field

$$\tau(H_{ext}) = 2\tau_0\frac{\mathsf{L}(\alpha)}{\alpha-\mathsf{L}(\alpha)} \tag{5}$$

with $\alpha=\alpha_\tau\cdot H_{ext}$ and $\mathsf{L}(x) = \coth(x)-\frac{1}{x}$.

3. Experimental setup

A Taylor-Couette-Setup, consisting of rotary inner and outer cylinders, has been built with a gap height of 20 cm, a radius of the inner cylinder of 1 cm and a gap width of 1 cm, resulting in an aspect ratio of Γ=20 and a radius ratio of η=0.5. Both cylinders may rotate in either direction, whereas they get driven by synchron-servo motors. The magnetic field is generated by four Helmholtz-coils cooled from three sides in Fanselau arrangement. The coils have 1500 windings of copper wire of 1.5 mm in diameter. Powered at around 2350W, a field of about 43kA/m in the homogeneous area (20x20 cm^2) is achieved. The orientation of the magnetic field might get arranged axial or transversal relative to the fluid cell axis. Due to the fact that ferrofluids are opaque, the different flow states and their transitions are measured by ultrasound doppler velocimetry. As a velocimeter, the DOP2125 manufactured by Signal Processing (Lausanne, Switzerland) is used. According to the Doppler-Effect, ultrasound echo reflected from particles immersed in the fluid carries a higher (lower) frequency in case the particles move towards (away from) the transducer. By convention, the velocity of the particle is displayed as positive when the target is moving away from the transducer. Due to geometrical limitations, the ultrasound transducer was placed at the outer cylinder wall. Thereby appearing sinusoidal velocity profiles (**Figure 1**) reveal phase transitions and may be analyzed by spatial and temporal FFT. A 4 MHz ultrasound transducer gave the best results under experimental conditions. Due to the fact that ferrofluid particles are too small to give a usable ultrasound echo, 0.05 vol% tracer particles with 10 μm in diameter consisting of melamin resin had to be added.

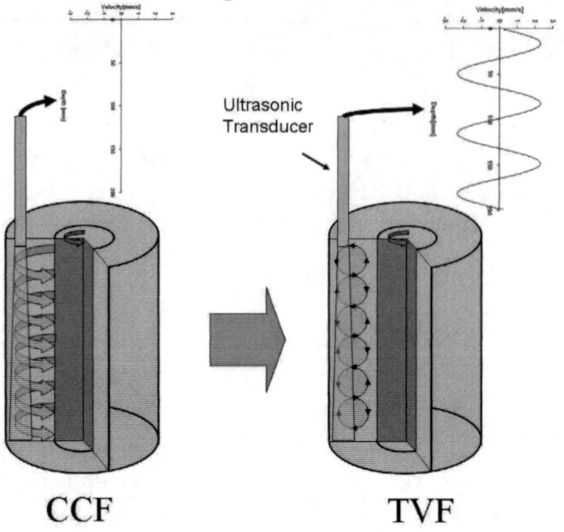

Figure 1. The transition from the ground state (Circular Couette flow, CCF) to Taylor vortex flow (TVF). The ultrasound transducer on top of the fluid cell may evaluate only axial velocity components. Hence, in CCF the velocity signal is zero. In TVF, the velocity signal is sinusoidal, following the taylor vortices streaming upwards and downwards at the outer cylinder wall.

4. Results and discussion

First measurements were carried out with a kerosene based ferrofluid (Table 1) provided by L. Vékás, Timisoara, and concerned the transition from CCF to TVF at different axial magnetic field strengths.

Table 1: Properties of the ferrofluid provided by L. Vékás, Timisoara.

Property	Value
Carrier Liquid	Kerosene
Coating	Oleic acid without excess surfactant
Material	Magnetite
Φ	0.07
M_s	34 kA/m
χ_0	0.93
Viscosity of carrier liquid	0.002 Pas (Lit. datum)
Viscosity	0.0068 Pas

In Figure 2, the experimental data is compared to a linear stability analysis conducted using different equations for the relaxation of magnetization as introduced in section 2. For the given fluid parameters, τ_0 might be calculated to approx. 0.01 ms, assuming a typical mean particle diameter for ferrofluids of 12 nm [1]. However, in order to also consider such crucial parameters like polydispersity and shape anisotropy of the particles [8], τ_0 was chosen to be 0.03 ms. For the analysis, the relaxation time τ_0 was either assumed to be constant or dependent on the external magnetic field H_{ext} whith α_τ = 0.3 m/kA. $\chi(H)$ was determined by magnetization measurements.

The experiments show that Re_{1c} increases strongly until approx. 25 kA/m, when saturation behavior sets in. Theories describing the magnetization outside equilibrium should also reflect this behavior, which is not accomplished by Debye, calculated exactly, or in a Niklas approach with τ_0=*const*. Debye, calculated with a relaxation time depending on the external magnetic field $\tau(H_{ext})$, and S'72 get closer to the experimental observations. However, the significant deviation of the models is a demand for further investigations, including measurements of the relaxation rate and the influence of polydispersity and shape anisotropy effects.

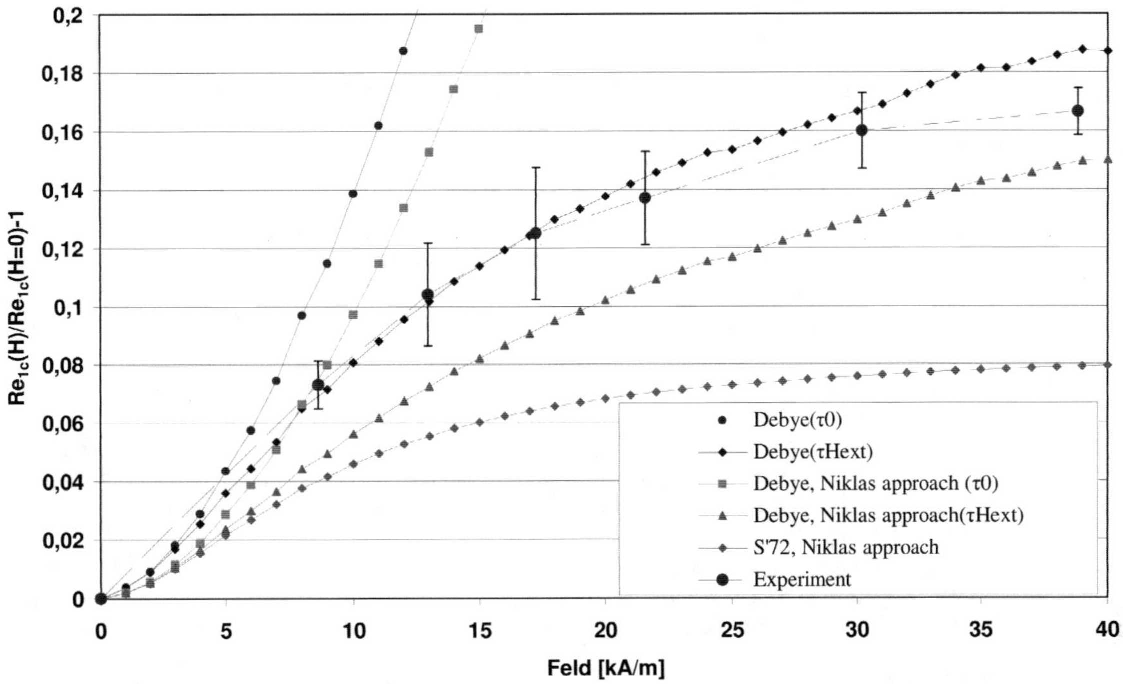

Figure 2. Comparison of experimental data at 20±0.25°C and data obtained by a linear stability analysis using the equations for the relaxation of magnetization as introduced in section 2. The relaxation time is assumed to be τ_0=0.03 ms and the transition is calculated for an axial wave number of k=2.8274 (18 vortices).

References

[1] R. E. Rosensweig: Ferrohydrodynamics, Cambridge University Press, 1985

[2] M. I. Shliomis:Effective viscosity of magnetic suspensions. Sov. Phys. JETP, 34:1291, 1972

[3] A. Leschhorn, M. Luecke: Magnetization of Rotating Ferrofluids: Predictions of Different Theoretical Models, Z. Phys. Chem. 220 1–6, 2006
A. Leschhorn, J. P. Embs and M. Luecke: J. Phys.:Condens. Matter 18, 2633, 2006

[4] M. Niklas: Z. Phys. B 68, 493, 1987
M. Niklas, H. Mueller-Krumbhaar and M. Luecke: J. Magn. Magn. Mater. 81, 29, 1989

[5] M. A. Martsenyuk, Y. L. Raikher and M. I. Shliomis: On the kinetics of magnetization of ferromagnetic particle suspensions. Sov. Phys. JETP, 38:413, 1974

[6] J. T. Waldron, Y. P. Kalmykov and W. T. Coffey: Rotational Brownian motion and dielectric relaxation of polar molecules subjected to a constant bias field: Exact solution. Phys. Rev. E, 49:3976, 1994

[7] S. Odenbach, H. Gilly: Taylor vortex flow of magnetic fluids under the influence of an azimuthal magnetic field. J. Magn. Magn. Mater., 152:123, 1996

[8] S. Odenbach: Magnetoviscous effects in ferrofluids, Lecture Notes in Physics, Springer, 2002

[9] H. Kikura, Y. Takeda, F. Durst: Velocity profile measurement of the Taylor vortex flow of a magnetic fluid using the ultrasonic doppler method. Experiments in Fluids 26, 208-214, 1999

[10] D. Ito, H. Kikura, M. Aritomi, Y. Takeda: Mode bifurcation control of a magnetic fluid on Taylor-Couette vortex flow with small aspect ratio. Journal of Physics: Conference Series 14, 35–41, 2005

[11] G. I. Taylor: Stability of a viscous liquid contained between two rotating cylinders. Phil. Trans. Soc. London A 223: 289-343, 1923

Journal of Physics: Conference Series **149** (2009) 012110 doi:10.1088/1742-6596/149/1/012110

Magnetic capturing and guiding of magnetite-polyvinyl alcohol ferrofluids for targeted drug delivery

O Rotariu[1,3] L E Udrea[1], V Badescu [1], R Badescu[2] and G Apreotesei[2]

[1] National Instit of R- D for Technical Physics, 47 Mangeron Blvd., Iasi, 700050, Romania
[2] Univ "Gh. Asachi", Dept. of Physics, 67 Mangeron Blvd, Iasi, 700050, Romania
[3] Univ. of Aberdeen, School of Biological Sciences, Aberdeen, AB24 3UU, UK.

E-mail: bav08@phys-iasi.ro

Abstract. We presented some results concerning the targeting/capture and guidance of the colloidal magnetic particles (MP) in/through simulated blood vessels/tissues. A bipolar magnetic device was used to investigate the capture of MP in capillary tubes at flow velocities similar to those encountered in the capillary beds of tumors. The capture was influenced by the gradient of the magnetic field, the flow velocity, the concentration and surface properties of the MP. The physical guidance of the MP through simulated tissues was also investigated and it depends on the magnetic field, the MP properties and the interstitial space between the cells.

1. Introduction

The magnetic targeting of MP carrying anticancer agents is efficient for drug delivery, reducing the systemic side effects [1]. Despite real progress obtained with this method [2, 3] some technical and physiological problems hinder the application of the method in practice, so understanding of the physical conditions for the magnetic capture of carrier particles is required. The aim of our study was to evaluate the optimal physical and physiological conditions for the magnetic manipulation of the MP inside the human body, in order to increase the efficiency of the magnetic targeting techniques for tumour treatments. We present an experimental device for studying MP's capture within fluidic systems simulating the flow regime in small blood vessels [4]. It was studied the possibility of MP retention and entrapment inside plastic tubes, within regions that have similar sizes with the cancer tumours. We also present an experimental model for studying the guidance of MP through agar gels (structure and porosity resembles the interstitial space from real tissues) which simulates the magnetic targeting of carrier particles in human tissues [5]. The movement MP through the gels is studied as function of gel porosity, the properties of the magnetic field and the volume of MP colloid.

2. Materials and methods

2.1. Preparation and characterization of ferrofluid

The ferrofluid was prepared by co-precipitation in aqueous media of the iron salts with a concentrated NaOH solution, at 70°C, followed by stabilization with 1% aqueous polyvinyl alcohol solution (MP-PAV). The size of the MP was measured by dynamic light scattering (Nano/Microtrac, Inc.US). The magnetization of the MP was measured using a magnetometer with vibrating probe.

2.2. Capture of magnetic particles

The experimental setup used for the deviation/capturing of MP is presented in Fig. 1. The ferrofluid was pumped through the plastic tube into the bipolar magnetic capture system; the MP were magnetically attracted from the flow field and captured onto the tube's wall. Time necessary for the vessel to be blocked and the length of MP deposit were estimated as a function of the initial fluid velocity, the MP concentration and the position of "blood vessels" within the magnetically active zone.

Journal of Physics: Conference Series **149** (2009) 012110 doi:10.1088/1742-6596/149/1/012110

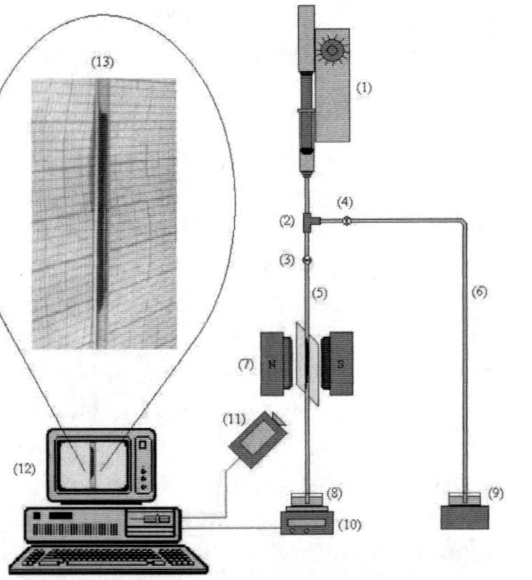

Figure 1. Experimental device: **(1)** injection pump; **(2)** ramification; **(3, 4)** valves; **(5, 6)** capillary plastic tubes with 0.75 mm inner diameter; **(7)** magnetic system – "C form" magnetic circuit with 8.7 cm air gap, made in our laboratories (magnetic field calculated for the symmetry axis of the air gap OX varied between 0.11 to 0.54 T; the gradient of the magnetic field is maximum at the surface of the polar pieces - 0.31 T/cm on the conical polar piece and 0.2 T/cm on the rectangular one and is null at the centre; **(8, 9)** collectors; **(10)** Acculab LA-200 electronic balance with PC interface; **(11)** Handycam camcorder; **(12)** PC; (13) detail of an acquired image.

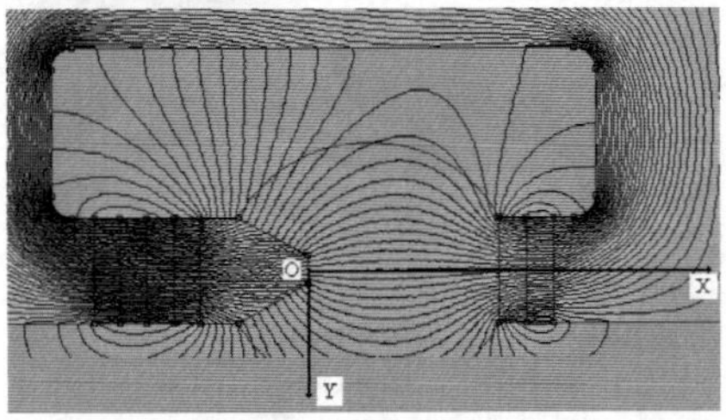

Figure 2. Bipolar "C form" magnetic circuit with active space of 8.7 cm. The yoke of the magnetic circuit was made from massive stainless steel (saturation magnetization ~15,000 Gs) and the circuit was energized by 6 rectangular magnets (Sintered NdFeB, 35 MGOe, 50 × 50 × 12.5 mm, MMG MagDev Ltd, Swindon, UK).

2.3. The guidance of MP through gels

Agar gels with a range of solid concentrations (0.275 to 0.375 %) were prepared by heating agar-deionised water solution at 90° C, followed by cooling at 4° C for 24 hrs. Before cooling, the agar solutions were poured in "Plexigals" rectangular containers (20x20x23 mm) and a 2.5 mm diameter vertical well (13 mm long) was formed by inserting a cylindrical plastic pin in the solution. The obtained well simulates a "reservoir" (e.g. diseased blood vessel) which can accommodate various amounts of MP (0.5 to 3.5 μl). A non-uniform magnetic field generated by a magnetized ferromagnetic rod placed underneath the container and the MP were guided through the gel toward its bottom surface, where the magnetic field is highest. The position of magnetic particles was registered and their velocity through the gel was determined by image analysis.

3. Results and discussion

3.1 The characterization of magnetic particles

The DLS dimensional analysis showed a log-normal distribution, the MP-PAV had sizes between 0.05 to 1.15 μm (mean size = 0.36 μm). We chose such a distribution of the MP because large particles can create a magnetic gradient responsible for the capture of the small particles that otherwise are not captured. The MP-PAV magnetization was 4250 G.

3.2. The capture of MP – PVA

The time interval necessary to block the capillary vessels with MP (blocking time) is an important parameter necessary to establish the time scale for an *in vivo* treatment procedure, giving the approximate time of injecting MPs into the blood stream. Also, the extension of the MP deposits onto the tube's walls (length of the deposit) is important to estimate the volume of tissue which can be treated by the magnetic targeting procedure. The blocking time and the length of particle deposits vary with the process parameters

Journal of Physics: Conference Series **149** (2009) 012110 doi:10.1088/1742-6596/149/1/012110

(tube position in the magnetic field, x, the strength and gradient of the magnetic field, the initial fluid velocity/flow rate, the MPs concentration and their magnetization). The blockage appears earlier in the proximity of the rectangular magnetic poles (t = 677 s at x= 7.7 cm and t = 723s at x =7.2 cm), later in the proximity of conical magnetic poles (t = 1103 s at x = 0.5 cm and t = 1224 s at x = 1.0 cm) and does not appear at x = 1.5 and 6.7 cm. (Fig. 3a). The length of the deposits varies, being higher in the proximity of the rectangular poles, where the gradient of the magnetic field/magnetic force is intermediate (Fig. 3b). However the capture takes place even at larger distances from the poles (x = 1.5), where the gradient is smaller. A maximum deposit length of 7.7 cm was observed, but the deposit was not stable. Moreover, when the tube is placed at x = 6.7 cm the appearance of avalanches is evidenced and the length of deposit fluctuates. The variation of MPs-PAV concentration and the fluid velocity affects the time of blockage and the length of deposit (Fig. 4). At low velocities (v = 0.43 cm/s) that is are similar to that found in diseased capillaries the blockage takes place even at very low concentrations. But, the time necessary to block-up the tube increases and the length of deposit decreases (t = 1074 s, L = 2.79 cm for c = 0.02 g/ml and t = 1954 s, L = 2.23 cm for c = 0.01 g/ml). When the fluid flows at a rate representative of normal arterioles (v = 0.86 cm/s) the deposits of particles are longer (L = 7.7 cm) but unstable and the tubes are not blocked. Using the MPs-PVA and the results for their deposit lengths, and taking into account that the magnetic field has an axial symmetry within the active space of the magnetic circuit, it is possible to appreciate that the maximum volume of diseased tissue hypothetically blocked at the capillary level is ~100 cm3.

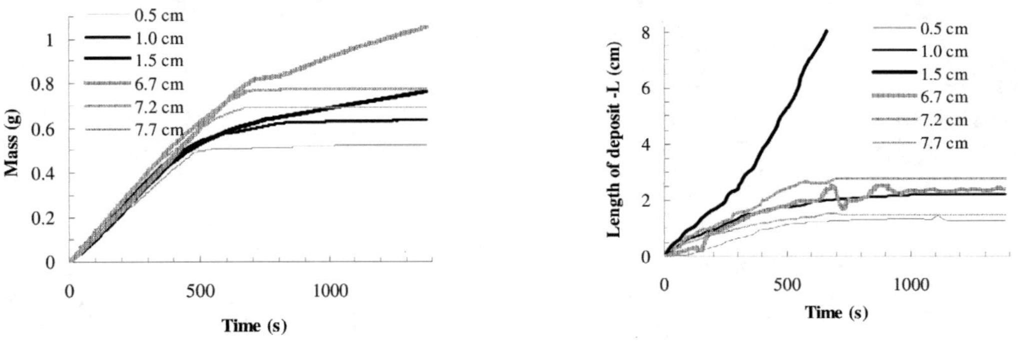

(a) (b)

Figure 3. (a) The mass of fluid collected a function of time. (b) Length of deposit versus time (v = 0.29 cm/s; c = 0.02 g/ml)

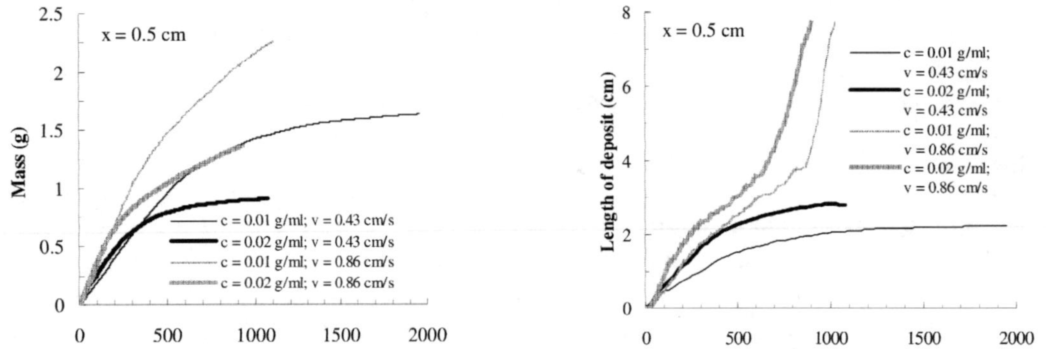

(a) (b)

Figure 4. (a) The mass of fluid collected as a function of time. (b) Length of deposit versus time

3.3 The guiding studies

The velocity of the fastest particles measured by imaging techniques is constant along the MP path through the gel (Fig. 5). Moreover, the MP velocity further depends on the initial amount of colloid from the "reservoir" (Fig.6), the intensity of the background magnetic field (Fig. 7), with agar concentration (Fig. 8). The gel is not permeable for all particles, the larger MP being sieved by the gel and only the smaller ones approaching the bottom of the container. Once the particles arrive onto non-permeable surfaces (e.g membranes surrounding various organs) they spread in shapes which follow the geometry of the magnetic poles. Therefore, it is expected as the pattern of MP distribution inside human tissues will depend on several factors: the characteristics of the tissue, the properties of the magnetic particles, the geometry of magnetic poles and the strength and gradient of the magnetic field.

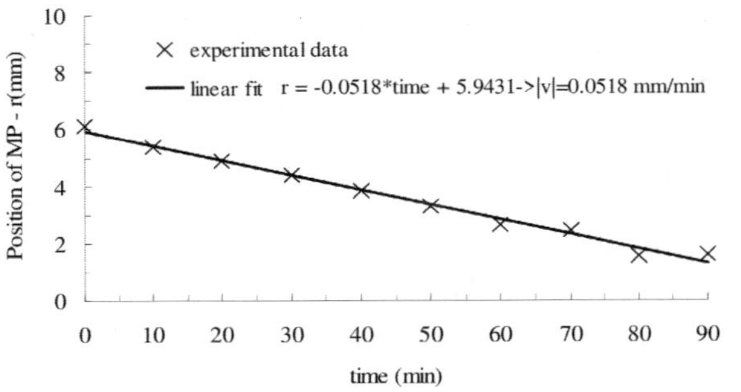

Figure 5. The position of fastest MP

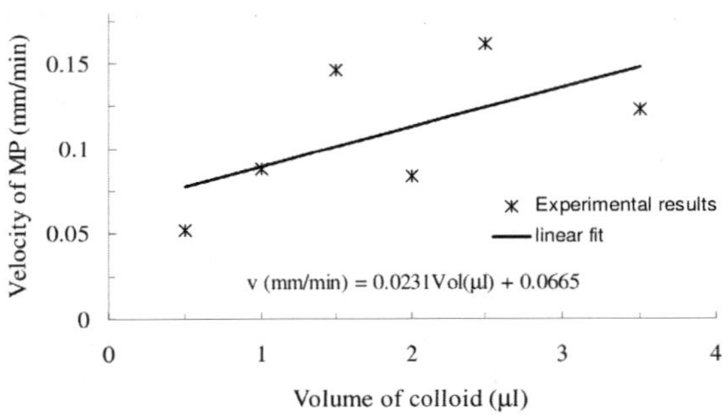

Figure 6. MP velocity / volume of colloid

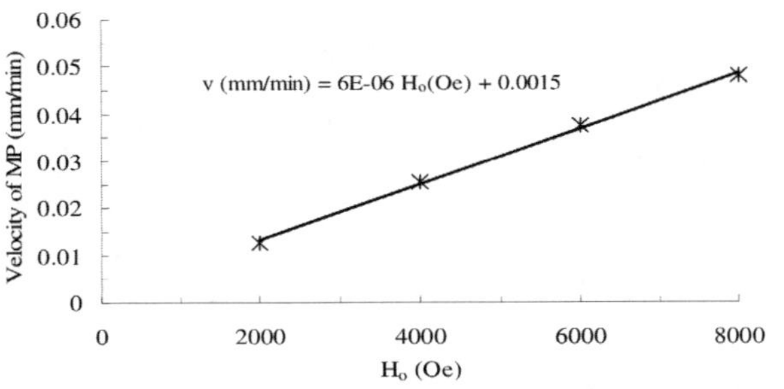

Figure 7. MP velocity function of the intensity of magnetic field, H_o.

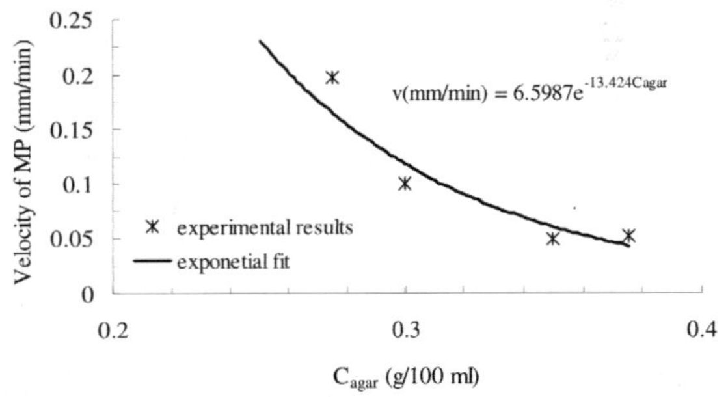

Figure 8. MP velocity versus the agar concentration (gel porosity)

4. Conclusions

The obtained results indicate that the magnetic drug targeting technique using gradient magnetic fields (particularly a "C form" magnetic circuit) is suitable for the treatment of cancers in regions that can be fit to the active space of 8.7 cm (head, neck, breast, hands and legs) and situated at sub-surface of the body because of the values of the magnetic field/gradient, by either concentrating drugs within the tumour or embolization to block the blood supply to the tumour. These results indicate that magnetic techniques using gradient magnetic fields and magnetite colloids have the potential to be practical tools for sub-surface guiding of anti cancerous agents within the human body.

References

[1] Marcucci F and Lefoulon F 2004 *Drug Disc. Today* **9** 219
[2] Alexiou C, Arnold W, Klein R J, Parak F G, Hulin P, Bergemann C, Erhardt W, Wagenpfeil S and Lubbe A S 2000 *Cancer Res.* **60** 6641
[3] Lubbe A S, Bergemann C, Riess H, Schriever F, Reichardt P, Possinger K, Matthias M, Dorken B, Herrmann F, Gurtler R, Hohenberger P, Haas N, Sohr R, Sander B, Lemnke A J, Ohlendorf D, Huhnt W and Huhn D 1996 *Cancer Res.* **56** 4686
[4] Udrea L E, Badescu V, Strachan N J C and Rotariu O 2007 *J. Optoelectr. Adv. Mat.* **9(8)** 2593
[5] Rotariu O, Udrea L E, Strachan N J C and Badescu V 2007 *J. Optoelectr. Adv. Mat.* **9(4)** 942

Investigation of magnetic and hyperthermic effects in ferrofluids with PEG biocompatible surfactant

A. Skumiel, A. Józefczak, T. Hornowski

Department of Molecular Acoustics, Institute of Acoustics, Adam Mickiewicz University, Umultowska 85, 61-614 Poznań, Poland

E-mail: skumiel@amu.edu.pl

Abstract. The aim of this study was to investigate the influence of poly(ethylene glycol) surface-active coating on the magnetic and thermal properties of biocompatible magnetic liquids. The magnetization of the samples was measured using VSM method. From the magnetization data the distribution function of the size of the magnetite particles was determined. Heating ability of the PEG-stabilized magnetic fluids was determined by the calorimetric measurement of the heating rate at 750 kHz frequency and 0.52 kA/m field to evaluate the possibilities of their successful use in hyperthermia treatment. The experimental data of the heating rate were compared with theoretical expression derived by Rosensweig.

1. Introduction

Biocompatible magnetic liquids (MF) are stable colloidal suspensions of nanosized magnetic particles covered with biocompatible molecules and dispersed in a physiological carrier liquid. The biocompatibility of magnetic liquids is determined by both the core ferrite material and the coatings. A good candidate for the core ferrite material is magnetite, Fe_3O_4, which has been proved to be biocompatible [1] and can be conveniently synthesized using chemical coprecipitation reactions [2]. The coating of ferrite particles with long chain polymer molecules serves as protective layer that prevents agglomeration of the particles and minimizes the direct exposure of the ferrite surface to the biological environment. Nanoparticles have a large surface area/volume ratio and they tend to agglomerate and adsorb plasma proteins. When the nanoparticles agglomerate, or are covered with adsorbed plasma proteins, they are quickly cleared by macrophages before they can reach target cells. One possible approach to increase the circulation time of nanoparticles in the blood stream is to modify the particles with poly(ethylene glycol) (PEG) films to disperse them and minimize or eliminate protein adsorption. Surfaces covered with PEG are biocompatible, i.e. nonimmunogenic, nonantigenic and protein-resistant. This is because PEG has uncharged hydrophilic residues, and a very high surface mobility leading to high steric exclusion [3].

The heating effect of a ferrofluid is a result of absorbing energy from the alternating magnetic field and converting it into heat by means of three mechanisms: (1) eddy currents losses; (2) hysteresis losses during reversal of magnetization; and (3) relaxation losses accompanying demagnetization. The eddy currents losses are negligible since a ferrofluid has very low electrical conductivity. Thermal energy from a hysteresis loss depends on the type of the remagnetization process. Over ceratin portions the magnetization curve is irreversible and energy of the magnetic field is dissipated into the medium with each flux-reversal cycle in the form of heat. It is known that hysteresis losses strongly depend on the size of magnetic particles [4]. The third mechanism of the heating effect is associated with a lag between the field and magnetization due to the relaxation nature of the magnetization process in ferrofluid.

The purpose of this study was to assess the thermal characteristics of aqueous-based magnetic liquid with additional PEG coating and to compare them with theoretical predictions.

2. Materials and Methods

The synthesis of our aqueous magnetic fluid was based on coprecipitation of Fe^{2+} and Fe^{3+} salts by NH_4OH at 60 °C. Sodium oleate ($C_{17}H_{33}COONa$) as a first surfactant (10 wt % with respect to Fe_3O_4) was used for the modification of prepared magnetic particles to prevent their agglomeration. To improve stability and increase the circulation half time of the particles, the polyethylene glycol (PEG) as a second surfactant was added to the system magnetite-oleate and stirred over 3 hrs. The sample denoted as MF Oleate+PEG was prepared using PEG from Sigma (1000 g/mol) with the ratio of magnetite to PEG equal to 40 mg/1 mg. The magnetic liquid labled as MF-Oleate contained only oleate surfactant layer and was also studied for comparison. The purpose of this study was to assess the thermal characteristics of PEG-based ferrofluids and to compare them with theoretical predictions.

Magnetic measurements of the samples were carried out with the aid of a vibrating sample magnetometer (VSM) at the temperature of 300K, within the range of variation of external magnetic field induction, $B = 1$ T.

The heating system consisted of sine wave power oscillator, an induction coil (length 78 mm) and an oscilloscope (Philips). A glass tube containing the sample was thermally isolated by a layer of material from the solenoid winding supported on a plastic sleeve. The temperature of the sample was monitored using a thermocouple (LT Lutron TM-917). The measuring setup permitted a continuous recording of the temperature inside the sample upon its heating by the external alternate magnetic field in the range of 500–2000 A/m and frequency of 750 kHz.

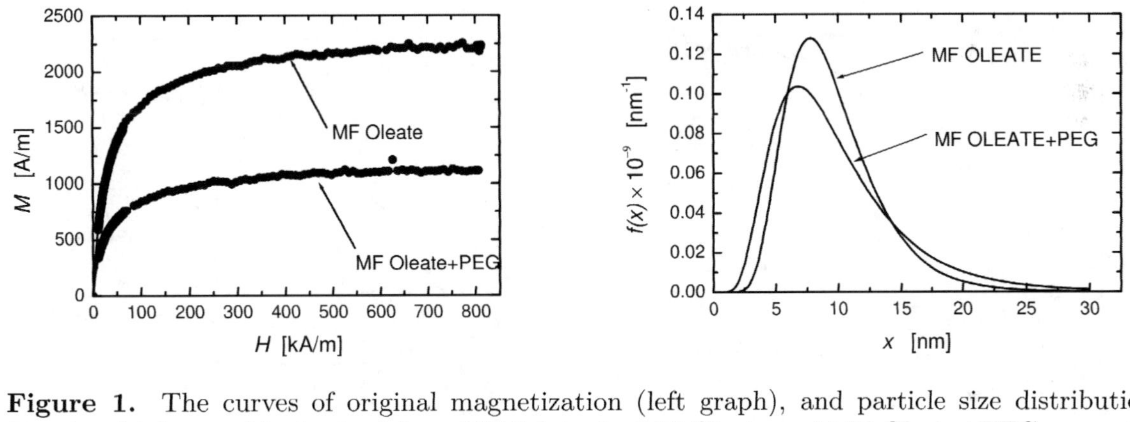

Figure 1. The curves of original magnetization (left graph), and particle size distribution function (right graph) obtained from VSM data for MF Oleate and MF Oleate+PEG magnetic fluids.

3. Results and discussions

The magnetization curves obtained for MF Oleate and MF Oleate+PEG ferrofluids by means of VSM are shown in Figure 1 (left graph). The original magnetization can be expressed as a superposition of the Langevin functions [5]

$$M_L = M_s \int_0^\infty L(\xi) f(x) dx, \tag{1}$$

Table 1. Magnetic parameters of MF-Oleate/PEG samples obtained by means of VSM method from the processing of magnetization curves.

Sample	M_s (A/m)	χ_i	ϕ_V (%)	D_0 (nm)	S	D_m (nm)	σ (nm)
MF Oleate+PEG	1213	0.0598	0.273	8.72	0.499	9.88	5.3
MF Oleate	2331	0.0754	0.522	8.92	0.374	9.57	3.7

where M_s is the saturation magnetization, $L(\xi) = \coth(\xi) - 1/\xi$ is the Langevin function with $\xi = \mu_0 m H / k_B T$, m is the magnetic moment of the particle, μ_0 is the magnetic permeability of vacuum, H is the magnetic field strength, k_B is Boltzmann constant, and T is the absolute temperature.

For the description of the dispersion of magnetite particles the lognormal distribution is usually used [5]

$$f(x) = \frac{1}{xS\sqrt{2\pi}} \exp\left(-\frac{\ln^2(x/D_0)}{2S^2}\right), \tag{2}$$

where x is the magnetic particle diameter, and D_0 and S are the parameters of the distribution function which can be extracted from the magnetization curve. On the basis of these parameters, the mean diameter, D_m, and the standard deviation, σ, of particle size can be determined as well as the saturation magnetization, M_s, and initial susceptibility, χ_i [5]. From the saturation magnetization the volume fraction of magnetite phase can be determined from the expression $\phi_V = M_s/M_d$, where $M_d = 466$ kA m^{-1} is the spontaneous magnetization of magnetite grains.

The particle size functions of the MF Oleate and MF Oleate+PG magnetic liquids extracted from VSM measurements are shown in Figure 1 (right graph). The other magnetic parameters of the studied magnetic fluids studied obtained form the analysis of magnetization curve are presented in Table 1.

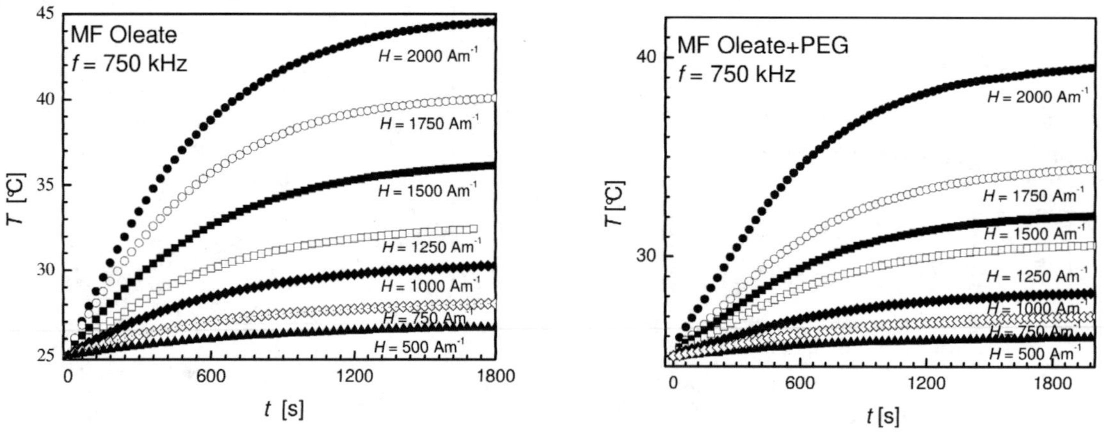

Figure 2. Temperature vs time for the sample MF Oleate (left graph) and MF Oleate+PEG (right graph) for different values of an alternating magnetic field and frequency $f = 750$ kHz.

Figure 2 shows the plot of temperature of the ferrofluids studied as a function of time for different values of the amplitude of the alternate magnetic field and the frequency $f = 750$ kHz. As it is seen from the figure, for small heating times the temperature increase ΔT is roughly linear. From the initial slope the heating rate $(\Delta T/\Delta t)_{t=0}$ can be calculated [6]. Fig. 3 presents the heating rate for the ferrofluids studied as a function of the magnetic field intensity.

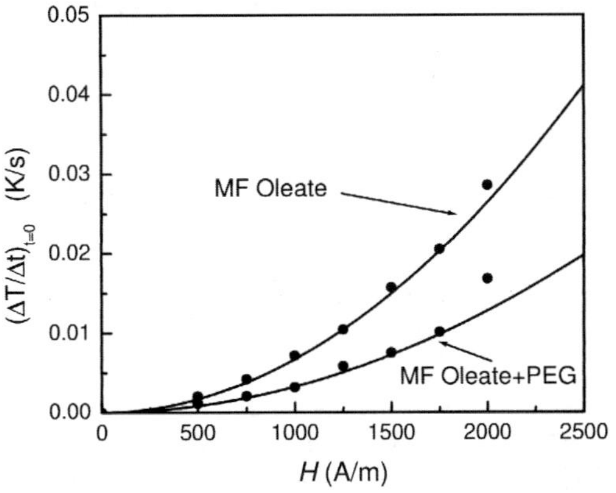

Figure 3. Temperature rise rate against magnetic field strength in polydisperse ferrofluids MF Oleate and MF Oleate+PEG. Solid lines show the theoretical predictions obtained from expression (3) without adjustable parameters.

According to Rosensweig [7] for magnetic fluids with superparamagnetic particles, the temperature rise ΔT of the magnetic fluid during the period of the heating Δt is given by:

$$\frac{\Delta T}{\Delta t} = \frac{1}{C_p \rho} \int_0^{\infty} \pi \mu_0 \chi_0 H_0^2 \frac{2\pi f \tau}{1 + (2\pi f \tau)^2} f(x) dx, \tag{3}$$

where C_p is the sample specific heat capacity, ρ is the mass density of the sample, and $\chi_0 = 3\chi_i L(\xi)/\xi$. There are two distinct mechanisms by which the magnetization of a ferrofluid may relax after removing the applied magnetic field: the Brown and the Néel one, characterized by appropriate relaxation times:

$$\tau_B = \frac{3\eta V_h}{k_B T}, \qquad \tau_N = \tau_0 \exp \frac{K_a V_M}{k_B T}, \qquad \tau = \frac{\tau_B \tau_N}{\tau_B + \tau_N}, \tag{4}$$

where η is the viscosity of the carrier liquid, V_h is the hydrodynamic volume larger than particle volume V_M by the size of the surfactant shell, and K_a is the anisotropy constant of the magnetic material. When both mechanisms act simultaneously the mechanism with the shortest relaxation time is dominant.

Solid lines on Fig 3 show predictions of expression (3). The calculations were done assuming the heat capacity of both samples to be similar to that of water, i.e. 4180 J kg^{-1} K^{-1}. The measured values of mass density and shear viscosity were equal to 1096 kg m^{-3} and 1.91×10^{-3} Pa s and 1015 kg m^{-3} and 2.68×10^{-3} Pa s for MF Oleate and MF Oleate+PEG, respectively. Surfactant layer was set at 2 nm and τ_0 at 10^{-9} s. As it is seen from the figure the theoretical curves, obtained without any adjustable parameters, show excellent agreement with experimental data.

Acknowledgments

The studies were supported by Polish Ministry of Science and Higher Education Grants No. 4T 07B 04130 and N202 097 32/2406.

References

[1] Müller R H, Maaßen S, Weyhers H, Specht F, Lucks J S 1996 *Int. J. Pharm.* **138** 85.
[2] Koneracká Kopčanský M, Antalík M, Timko M, Ramchand C N, Lobo D, Mehta R V, Upadhyay R V 1999 *J. Magn. Magn. Mater.* **201** 427.
[3] Zhang Y, Kohler N, Zhang M 2002 *Biomaterials* **23** 1553.
[4] Kim D -H, Lee S -H, Kim K -N, Kim K -M, Shim I -B, Lee Y -K 2005 *J. Magn. Magn. Mater.* **293** 320.
[5] Raşa M 2000 *Eur. Phys. J.* B **2** 265.
[6] Skumiel A 2006 *J. Magn. Magn. Mater.* **307** 85.
[7] Rosensweig R E 2002 *J. Magn. Magn. Mater.* **252** 370.

Journal of Physics: Conference Series **149** (2009) 012112 doi:10.1088/1742-6596/149/1/012112

Ferrofluid drop rolling on the surface of a liquid

V Sterr[1], K I Morozov[2] and A Engel[1]

[1] Institut für Physik, Carl von Ossietzky Universität Oldenburg, 26111 Oldenburg
[2] Institute of Continuous Media Mechanics, 1 Korolev Street 614013 Perm, Russia

E-mail: sterr@theorie.physik.uni-oldenburg.de,
engel@theorie.physik.uni-oldenburg.de

Abstract. We report on the theoretical analysis of an experiment where a ferrofluid drop swimming on top of a non-magnetic fluid layer is propelled forward by means of a rotating magnetic field. The drop is modelled first as a solid sphere with a Navier slip boundary condition, then as a liquid (half-)sphere with its own inner flow field. In both cases an analytical expression for the drop speed in terms of the experimentally accessible parameters is obtained. While the solution of the Navier slip model contains an unknown parameter, the slip length, the result of the liquid half-sphere model is completely free of fitting parameters and is shown to represent the experimentally measured dependencies very well.

1. Introduction

When a drop of ferrofluid is brought into a rotating magnetic field, the suspended magnetic particles start to move together with this field. Due to the viscous coupling of the particles to their surrounding carrier liquid the angular momentum is transferred to the whole drop and an abundance of phenomena is observed. Drops of ferrofluid which are rotated in the *horizontal* plane while being completely immersed in another (nonmagnetic) liquid of comparable density have been examined both experimentally and theoretically [1, 2]. By contrast, we have investigated a drop of ferrofluid which is swimming *on top* of a layer of nonmagnetic liquid and is rotated in a *vertically* oriented plane. This setup makes it possible to manoeuvre the drop over the fluid surface by tuning of the magnetic field, thereby constituting a pumping device without mechanical parts.

In the experiment, which was carried out at the University of Bayreuth, the constant translation velocity of a mm-sized drop of kerosene-based ferrofluid has been measured in dependence of the drop volume, the magnetic field amplitude and its frequency [3]. Motivated by this experiment, we have developed a simplifying theoretical model which gives an analytical description of a spherical magnetized object that is half-way immersed in a liquid, rotates with constant angular velocity Ω and moves with a constant speed $\boldsymbol{v}_{\text{drop}}$. We have treated two approaches, one in which the drop is approximated by a solid magnetic sphere with a Navier slip boundary condition, and another which describes the full magnetohydrodynamics of a liquid (half-)sphere.

The theoretical description of the 'real' setup poses a very complicated boundary value problem - mainly due to the moving contact line - which would have to be solved by numerical methods. In order to extract the essence of the effect we make some simplifying assumptions

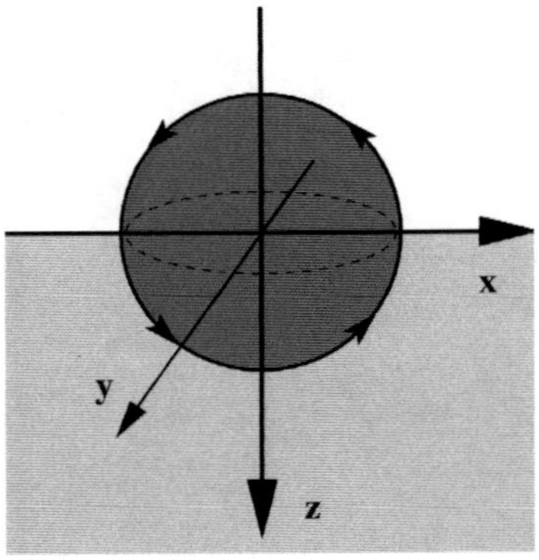

Figure 1. The droplet is considered to be a spherical object half-way immersed into a liquid with an otherwise perfectly flat surface. The liquid below is supposed to cover the infinite half-space. The problem is treated from the refrence frame where the sphere is spinning on the spot. In order to ensure stationarity in this frame, the overall forces and torques acting on the sphere must cancel out. After the velocity field of the surrounding liquid has been determined, its asymptotic value at infinity will give the negative translation velocity of the sphere in the laboratory frame.

which lead to an analytical solution. The basic simplifications and the geometry of the setup are explained in figure 1.

The velocity field of the non-magnetic liquid bearing the sphere (and later also the flow field within the ferrofluid drop) is expanded in vector spherical harmonics according to [4]. Thus the basic differential equations, i.e., the continuity and Stokes equations, $\nabla \cdot \boldsymbol{v} = 0$ and $\nabla^2(\nabla \times \boldsymbol{v}) = \mathbf{0}$, respectively, are transformed into ordinary differential equations which can be solved by a potential ansatz and the appropriate boundary conditions [3].

2. Solid sphere and Navier slip condition

We first treat the case of a solid sphere where the main problem lies in determining the viscous torque acting on it, since it has to compensate for the magnetic torque. In order to avoid a logarithmic divergence due to the large stresses at the moving contact line, a Navier slip condition is employed at the sphere surface [5, 6]. This is a linear relation between the flow field $\boldsymbol{v}$ and its derivative *at* the sphere surface $r = R$ and the velocity $\boldsymbol{U}$ *of* the sphere surface:

$$\left[\partial_r - \frac{1}{R}\right] \boldsymbol{v}(r = R) = \frac{\boldsymbol{v}(r = R) - \boldsymbol{U}}{L_\mathrm{s}} \tag{1}$$

The parameter L_s is called the slip length and denotes the amount of slippage between fluid and solid. This leads to a viscous torque $\boldsymbol{T}_\mathrm{vis} = -\pi\eta\Omega R^3\,\Sigma(L_\mathrm{s})\,\mathbf{e}_y$ on the sphere [7], where η and R denote fluid viscosity and sphere radius, respectively, while Ω is the yet unknown angular velocity of the drop. The abbreviation $\Sigma(L_\mathrm{s})$ includes the infinite series from the expansion in vector spherical harmonics, depends on the slip length and cannot be computed analytically in closed form.

Setting the viscous torque equal to the magnetic torque of the homogeneously magnetized sphere [8, 9]

$$\boldsymbol{T}_\mathrm{mag} = \frac{4\pi}{3} R^3 \frac{\mu_0 H^2 \chi''}{(1+\chi'/3)^2 + (\chi''/3)^2}\,\mathbf{e}_y, \tag{2}$$

the resulting flow field (which is quite a lengthy expression, cf. [3]) at infinity gives the speed

$$\boldsymbol{v}_\mathrm{drop}^\mathrm{sol} = -\frac{2}{3}\frac{\mu_0 H^2 \chi'' R}{\left[(1+\chi'/3)^2 + (\chi''/3)^2\right]\eta\Sigma(L_\mathrm{s})}\,\mathbf{e}_x \tag{3}$$

with which the drop 'rolls' over the surface. Here, H denotes the amplitude of the external magnetic field, and χ' and χ'' are the frequency-dependent real and imaginary parts, respectively, of the ferrofluid's magnetic susceptibility.

Unfortunately, the series in $\Sigma(L_\mathrm{s})$ converges very slowly, making it difficult to determine L_s from experimental data. Therefore, it would be of advantage to have an expression for the drop speed that does not depend on such a parameter.

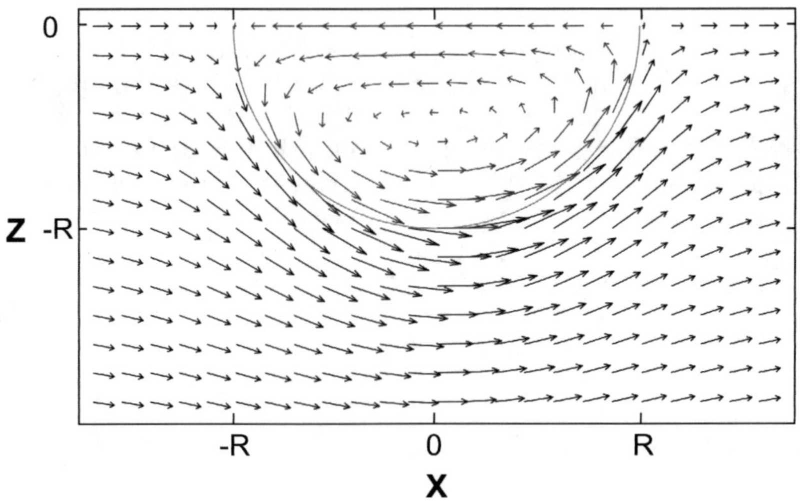

Figure 2. Resulting flow field of the half-drop solution within the x-z-plane.

3. Liquid half-sphere

We now treat the case of a magnetic liquid sphere whose inner flow field is computed along with the flow field of the liquid below. The Navier slip condition and the equality of viscous and magnetic torque are replaced by the continuity of tangential velocities and stresses at the interface of the two fluids.

As a consequence of the requirement of a flat surface it is not possible to obtain a spherical inner velocity field. More precisely, when the vertical velocity component is demanded to vanish at the whole contact line circle, this results in it being zero *throughout the whole drop section* $z = 0$.

However, when the boundary conditions are applied after the *mirror image* of the setup with respect to the fluid surface has been added (which is in order since the equations are linear), a flow field is obtained which proves to be very useful. As the field becomes completely horizontal within the plane of symmetry, it is suggested that only the lower half-sphere is identified with the ferrofluid drop, i.e., after solving the mirror image setup, the whole upper half-space is neglected, resulting in the flow field displayed in figure 2.

The translation velocity of this liquid half-drop now reads [3]

$$\boldsymbol{v}_{\text{drop}}^{\text{liq}} = -\frac{1}{4}\frac{\mu_0 H^2 \chi'' R}{[(1+\chi'/3)^2 + (\chi''/3)^2][2\eta^{(\text{o})} + 3\eta^{(\text{i})}]}\,\mathbf{e}_x \tag{4}$$

with $\eta^{(\text{i})}$ and $\eta^{(\text{o})}$ denoting the viscosities of inner (magnetic) and outer (nonmagnetic) liquid. This expression looks very similar to the previous one (3), but has the desired advantages, i.e., it purely consists of parameters that are experimentally measurable or tunable, and there is no need to calculate numerically an infinite series, so that this result can easily be compared to experimental data.

4. Comparison with the experiment

Figure 3 shows the comparison of the measured values (blue full circles) with the theoretical result (4). The required magnitudes of real part (χ') and imaginary part (χ'') of the ferrofluid's magnetic susceptibility were measured in an independent experiment as functions of the rotation frequency f of the external field H. The data points for the radius dependence were obtained

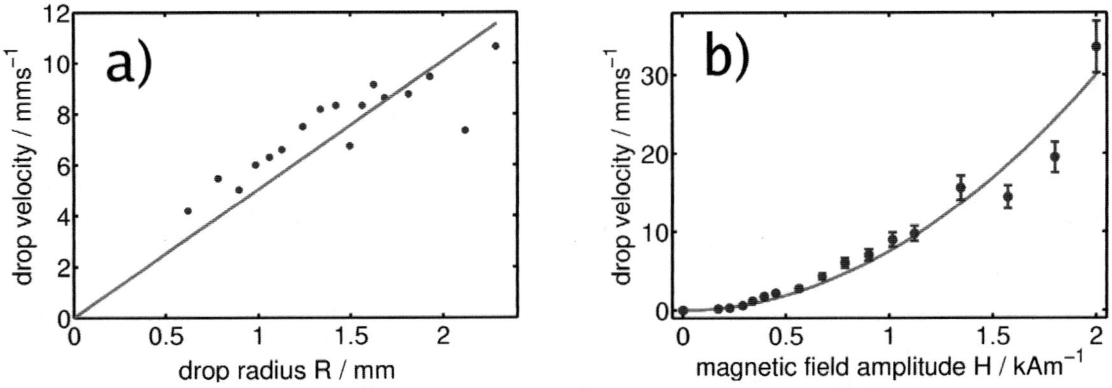

Figure 3. Dependence of the drop speed on **a)** drop radius R for frequency $f = 0.8\,\mathrm{kHz}$ (corresponding to $\chi' = 4.6575$, $\chi'' = 3.2475$) and magnetic field amplitude $H = 0.0844\,\mathrm{kAm}^{-1}$; **b)** magnetic field amplitude H for frequency $f = 0.8\,\mathrm{kHz}$ and drop volume $V = 5\,\mu\mathrm{l}$. Blue dots mark the experimental results, red lines the theoretical outcome (4).

by determining the volume of the drop and assuming a spherical shape. This assumption is well justified for drops up to volumes of about $12\,\mu\mathrm{l}$. Larger drops get considerably deformed by the magnetic field, as can be seen in a movie attached to [3].

Despite this discrepency and the simplifications of our half-sphere model, the magnetic field dependence is in excellent agreement and that of the sphere radius is rather good. For a discussion of the frequency dependence the reader is referred to [3].

5. Conclusions

The experiment has shown that rotating fields can transport ferrofluidic drops. With our analytical modelling we have succeeded in explaining quantitatively the experimental results without any free fitting parameters. Moreover, the theory provides an explicit solution of the flow fields both for a rotating solid magnetic sphere and a spherical ferrofluid drop half-way immersed in a nonmagnetic liquid. The similarity of the final results of both approaches demonstrates the equivalence of Navier slip at a solid surface on the one hand and the continuity of tangential stresses at a fluid-fluid boundary on the other hand.

Acknowledgments

We would like to thank R. Krauß, R. Richter and I. Rehberg for providing us with experimental data as well as fruitful and friendly cooperation.

References

[1] Bacri J C, Cebers A and Perzynski R 1994 *Physical Review Letters* **72** 2705
[2] Lebedev A V, Engel A, Morozov K I and Bauke H 2003 *New Journal of Physics* **5** 57.1
[3] Sterr V, Krauß R, Morozov K I, Rehberg I, Engel A and Richter R 2008 *New Journal of Physics* **10** 063029
[4] Morse P M and Feshbach H 1953 *Methods of theoretical physics* vol 2 (New York: McGraw-Hill book company, Inc.) chapter 13
[5] Navier C L 1823 *Mémoires de l'Académie des Sciences de l'Institut de France* **6** 389
[6] Hocking L M 1977 *Journal of Fluid Mechanics* **79** 209
[7] Sterr V 2007 Diploma thesis `http://www.student.uni-oldenburg.de/verena.sterr/ffdrop.html`
[8] Rosensweig R E 1985 *Ferrohydrodynamics* (New York: Dover Publications, Inc.)
[9] Landau L D and Lifschitz E M 1990 *Lehrbuch der Theoretischen Physik* 5th ed vol 8 (Berlin: Akademie Verlag) Elektrodynamik der Kontinua §8 §29

Journal of Physics: Conference Series **149** (2009) 012113 doi:10.1088/1742-6596/149/1/012113

Hydrodynamic bearing lubricated with magnetic fluids

H Urreta[1], Z Leicht[1], A Sanchez[1], A Agirre[1], P Kuzhir[2] and G Magnac[3]

[1]IDEKO, S. COOP. Arriaga Kalea 2, 20870 Elgoibar, Spain
[2]LPMC, University of Nice – Sophia Antipolis, 06108 Nice Cedex2, France
[3]Cedrat Technologies, 15 Chemin de Malacher - Inovallée - 38246 Meylan, France

E-mail: hurreta@ideko.es

Abstract. This paper summarizes the work carried out in the development of hydrodynamic lubricated journal bearings with magnetic fluids. Two different fluids have been analyzed, one ferrofluid from FERROTEC APG s10n and one magnetorheological fluid from LORD Corp., MRF122-2ED. Theoretical analysis has been carried out with numerical solutions of Reynolds equation, based on apparent viscosity modulation for ferrofluid and Bingham model for MR fluid. To validate this model, one test bench has been designed, manufactured and set up, where preliminary results shown in this paper demonstrate that magnetic fluids can be used to develop active journal bearings.

1. Introduction

Lubricated bearings are well known in industrial applications, and their main advantages compared with rolling and friction bearings are their low friction and very high precision. Lubricated bearings can be externally pressurized, as hydrostatic bearing [1], or they can get the load pressure by relative movement of their pieces, as hydrodynamic bearing [2]. One of the main problems in both types of bearings is the working velocity range. The generated heat into the bearing due to viscous friction can block the mechanism if the dilatation of the shaft is higher than bearing's clearance. To avoid this drawback and with the aim to get high performance smart bearings, lubricated bearings have been developed in different ways. Hydrostatic bearings have implemented semi-active or active valves to work in closed loop [3]. In this field can be classified the work presented by Kuzhir et al. and Leicht et al., in this ERMR08 Conference. In the case of hydrodynamic bearing, the load capacity is in function of the geometry, mainly the gap (technically almost impossible to change in process), the rotational velocity (imposed by operating conditions) and fluid properties. This work is centered in the development of smart and large operating range hydrodynamic bearings lubricated with magnetic fluids [4, 5].

2. Plain journal bearing

A journal bearing consists of two elements, namely the shaft with a radius R_a and the bearing with a radius R_c and width L. Following parameters characterize radial clearance $C = R_c - R_a$, $\overline{O_c O_a}$ distance called eccentricity e, relative eccentricity $\varepsilon = e/C$ and the angle between the load $\overrightarrow{W}$ and the center-line $\overline{O_c O_a}$ called attitude angle ϕ. The plane section trough the bearing can be seen in figure 1.

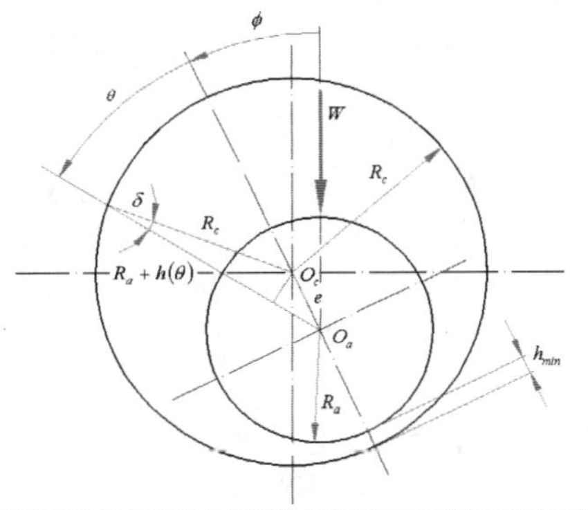

Figure 1. Plane section through the bearing.

If the angular speed is ω, the Reynolds equation [2] for infinitely long journals can be written in polar co-ordinate system shapes as following:

$$\frac{\partial}{\partial \theta}\left(h^3 \frac{\partial p}{\partial \theta}\right) = 6 \mu \rho R^2 \omega \frac{\partial h}{\partial \theta} \tag{1}$$

Thus the pressure field with parabolic distribution in axial axis (Z-axis) is given by:

$$p = \frac{6\mu\,\omega (R/C)^2}{\left(1-\varepsilon^2\right)^{3/2}} \cdot \left(\frac{L^2 - 4z^2}{L^2}\right) \cdot \left[\psi - \varepsilon \cos\psi - \frac{2\psi - 4\varepsilon \sin\psi + \varepsilon^2 \psi + \varepsilon^2 \sin\psi \cos\psi}{2(1-\varepsilon\cos\psi_s)}\right] \tag{2}$$

Where ψ is calculating angle and ψ_s corresponds to the film breaks down location angle. And by integration of the pressure field at boundary conditions, the load and attitude angle are obtained:

$$W = 2 R L \mu \omega \left(\frac{R}{C}\right)^2 \frac{1}{\sqrt{1-\varepsilon^2}\,(1-\varepsilon\cos\psi_s)} \cdot \left[\frac{\varepsilon^2 (1-\cos\psi_s)^4}{1-\varepsilon^2} + 4\left(\sin\psi_s - \psi_s \cos\psi_s\right)^2\right]^{1/2} \tag{3}$$

$$\operatorname{tg}\phi = \frac{2\sqrt{1-\varepsilon^2}\left(\sin\psi_s - \psi_s\cos\psi_s\right)}{\varepsilon\left(1-\cos\psi_s\right)^2} \tag{4}$$

3. Magnetic fluids: Magnetorheological fluids and ferrofluids

Two different magnetic fluids have been used in this work, one ferrofluid from FERROTEC APG s10n and a magnetorheological fluid from LORD Corporation, MRF-122-2ED. The model used for the ferrofluid characterization has been a simple modulation of fluid viscosity:

$$\mu_{app} = k \cdot \mu_{base} \tag{5}$$

Where μ_{app} is the apparent viscosity of the fluid, k is the viscosity change of the fluid in function of magnetic field, and μ_{base} is the viscosity of the fluid without magnetic field. For the case of magnetorheological fluid, Bingham model seems the simplest but the most similar to analyze the MR fluid.

$$\tau = \left(\tau_0(H) + \mu_e \dot{\gamma}\right)\operatorname{sgn}(\dot{\gamma}) \tag{6}$$

The characteristics of this fluid have been taken from manufacturer data sheet, information available in the web: http://www.lordfulfillment.com/upload/DS7027.pdf

4. Journal bearing test bench

The test bench designed and manufactured for this work is composed by a plain journal bearing with two coils to develop the magnetic field into the bearing. The bearing is manufactured in bronze and the pieces around the field in aluminum to focus the field in the desired direction. The bearing is mounted on an aerostatic bearing to avoid any friction. The load is applied with a screw and the relative displacements are measured with two Eddy-current sensors. The load cell (INTERFACE SM 20.000) has ten Newton precision, and displacement sensors (Brüel&Kjaer SD-081) measures well than one micron. The main dimensions of the bearing are: Diameter, 50mm; Length, 50mm; and radial clearance, 100microns. In the figure 2 the pictures of the bearing are shown, with the main elements of the test bench.

- a - - b - - c -

Figure 2. Test bench (a) bearing with two shafts (b) and magnetic pole view (c)

The shaft has been manufactured in two different materials, stainless steel and carbon steel, this way there will be compared magnetic and non-magnetic materials for the shaft. Magnetic poles cover around 73° of the bearing and their coils have 500 wire turns.

5. Theoretical and experimental results

First step in experimental analysis is to determine magnetic field in function of current intensity in the coils. For this task a Gaussimeter MAGNET-PHYSIK FH-54 has been used. The maximum current intensity is three Ampere, and at this current the magnetic field into the bearing is of 40kA/m.

First experiments have been carried out with the ferrofluid, at rotational speed of 100r/min. In the figure 3 can be seen the expected and measured path of the bearing (left side) and the eccentricity of the bearing for applied load (right side), using carbon steel shaft.

The results of the experiments show that the ferrofluid used in this work cannot be used as tunable lubricant. The rheological change of the ferrofluid is very low so its effect in the bearing is negligible.

- a - - b -

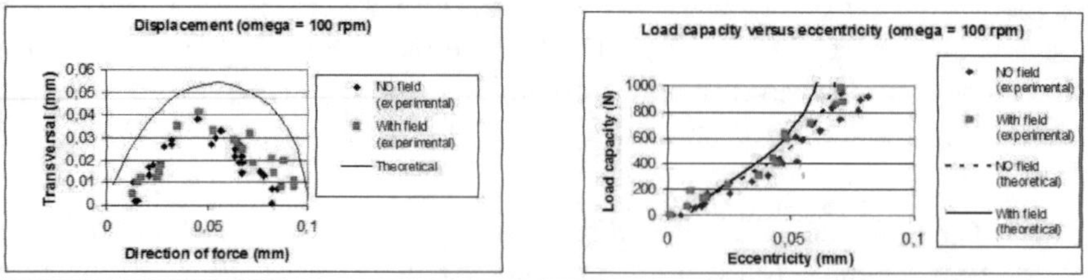

Figure 3. Bearing path (a) and load (b) versus eccentricity at 100r/min, Ferrofluid.

Seeing the results with the ferrofluid, the following experiments have been carried out with magnetorheological fluid, that its shear stress is really high. In the figure 4 are shown the results for stainless steel shaft, and in the figure 5 for carbon steel shaft, in both cases for several speeds (50r/min, 200r/min and 400r/min):

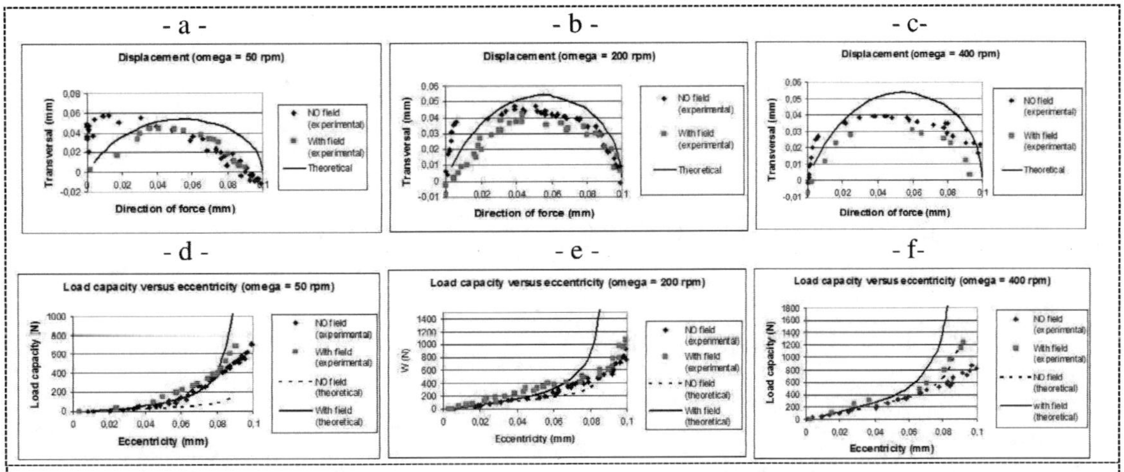

Figure 4. Bearing path (a, b, c) and load (d, e, f) vs. eccentricity, MRF and stainless steel shaft.

Results for stainless steel shaft show that magnetic effect it is only detectable in bearing path (figures 4 a, b, c). Meanwhile the experiments with carbon steel shaft give an increase of bearing's load capacity (figures 5 d, e, f), and its displacement (figures 5 a, b, c) becomes linear and stable.

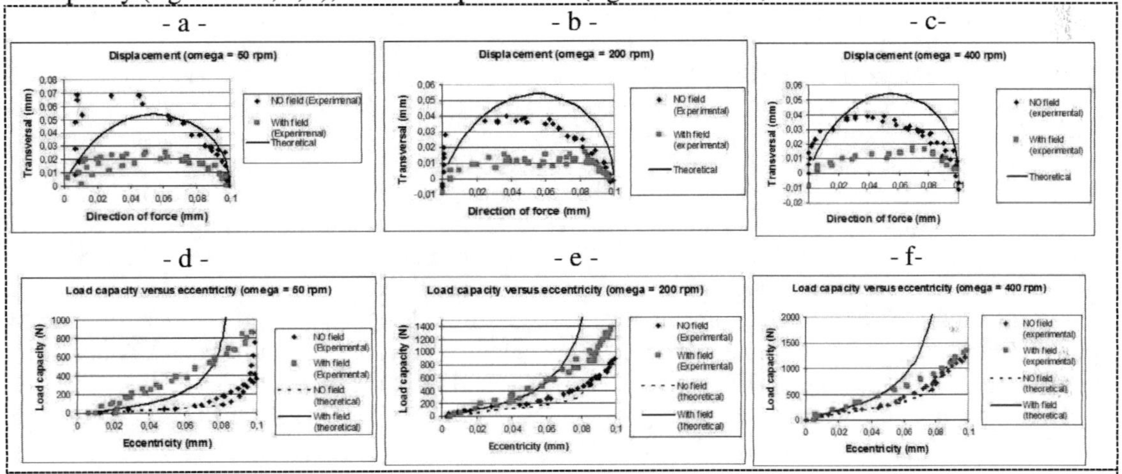

Figure 5. Bearing path (a, b, c) and load (d, e, f) vs. Eccentricity, MRF and carbon steel shaft.

6. Conclusions

It has been set up a test bench for plain journals bearings lubricated with magnetic fluids. Theoretical calculations agree with experimental data for the first half of bearing's clearance. The ferrofluid used in this work cannot be used as active lubricant in journal bearings. Magnetoreological fluids achieve good performance as active fluids in the bearing, this way they could be a good solution to increase the operation range of hydrodynamic bearings.

References

[1] Bassani R., Piccigall B. (1992) Elsevier Tribology Series 22.
[2] Frene J., Nicolas D., Degueurce B., Berthe D., Godet M. (1997) Elsevier Tribology Series 33.
[3] Hesselbach, J.; Abel-Keilhack, C. J. (2003) Ap. Physics, v 93, n10, p 8441-3
[4] Spur G., Patzwald R. (1998) Production Engineering 5 (1) pp. 47-50
[5] Nada G. S., Osman T. A. (2007) Tribology Letter 27: 261 – 268

Journal of Physics: Conference Series **149** (2009) 012114 doi:10.1088/1742-6596/149/1/012114

Magnetic capture of falling magnetic fluid droplet

R Yamane[1], **L J Wang**[2] **and M K Park**[3]

[1]Kokushikan University, 4-28-1 Setagaya, Setagaya-ku, Tokyo 154-8515, Japan

[2]China University of Mining & Technology, Xuzhou, Jiangsu 221116, China

[3]Pusan National University, Changjeon-dong, Kumjeong-ku, Pusan 609-735, Korea

E-mail: yamane@kokushikan.ac.jp

Abstract. As a fundamental research on magnetic drug targeting, trial was made to catch the magnetic fluid droplet falling in the water. The motion of the droplet was experimentally observed and analyzed, and was numerically simulated. Motion of the magnetic droplet is mainly classified into three types according to the strength of the magnetic force. The condition of capture is examined.

1. Introduction

One of the medical applications of the magnetic force is to target the magnetic drug to the diseased part by means of the magnetic force [1-6]. For this purpose the magnetic droplet or the magnetic emulsion should be caught to the wall surface of the vessel in the flow of the blood.

In the present research as a basic approach, the magnetic fluid droplet falling in the water is tried to be captured by the magnetic force. The motion and the trace of the droplet are experimentally analyzed, and are also numerically simulated. The magnetic droplet is directly captured, slides on the wall or falls away according to the strength of the magnetic force. The effects of the size and the concentration of the magnetic fluid on the condition of capture are examined.

2. Experiments

2.1. Experimental arrangements

As shown in figure 1 an experimental setup is consisted of a long reservoir, a nozzle and a permanent magnet. The reservoir is 35 x 35 mm in square cross section and 500 mm in height and is filled with water. The diameter of the nozzles is 1.80, 2.20 and 3.45 mm, through which the magnetic fluid is pushed out to form the droplet by the pressure of the air.

The magnetic fluid is kerosene-base HC-50 manufactured by Taiho Ind. Co. Ltd., which does not blend with the water. The specific gravity is 1.40 and the saturation magnetization is 3.78×10^4 A/m. It is diluted to the concentration of 100% (original), 80 and 60 % in volume. The diameter of the droplet depends on the diameter of the nozzle and the concentration of the fluid, and it ranges from 2.6 to 6.6 mm.

The magnet is a rectangular rare-earth magnet with the size of 50 mm in length, 20 mm in height and 5 mm in width. The distribution of the magnetic flux density is shown in figure 2. s is the distance from the surface of the magnet. The strength of the magnetic force on the droplet is controlled with the distance between the reservoir and the magnet.

The motion and the shape of the droplet are observed with a high-speed CCD camera, and the trace of the droplet is analyzed with the motion analyzing soft ware.

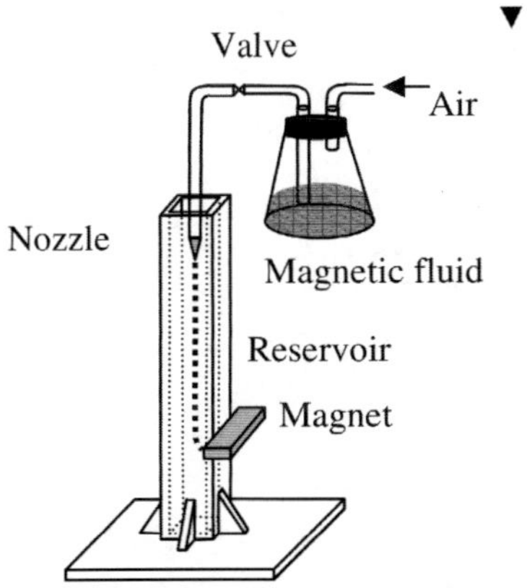

Figure 1.Experimental setup.

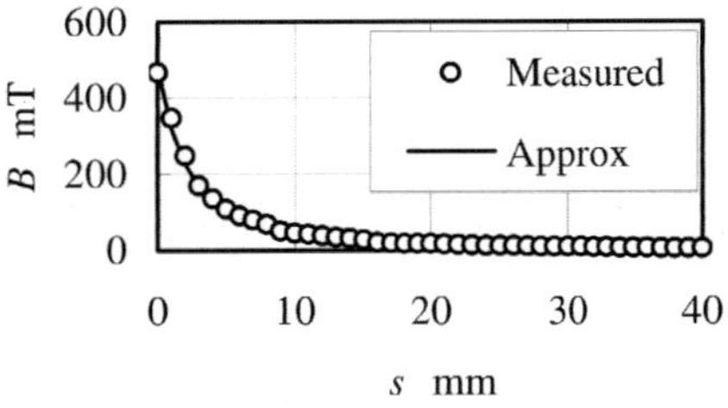

Figure 2.Strength of magnetic field.

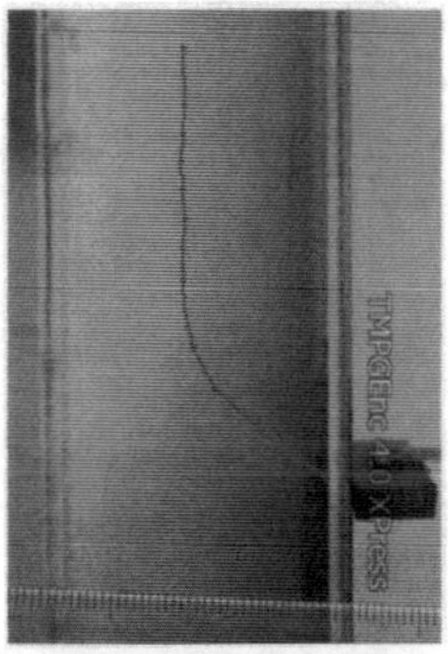
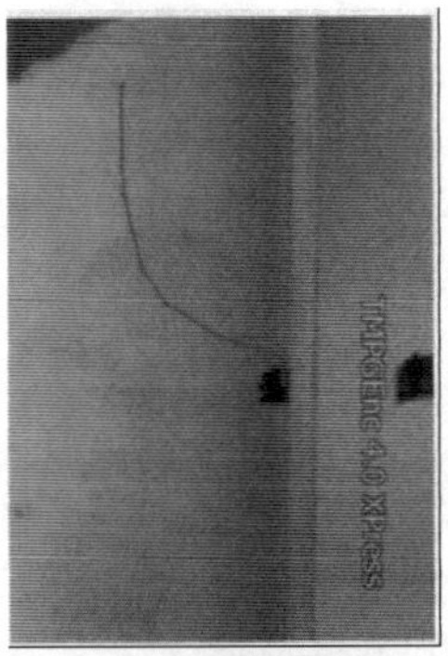

(a) s=5mm (B=14.5mT). (b) s=15mm (B=7mT).

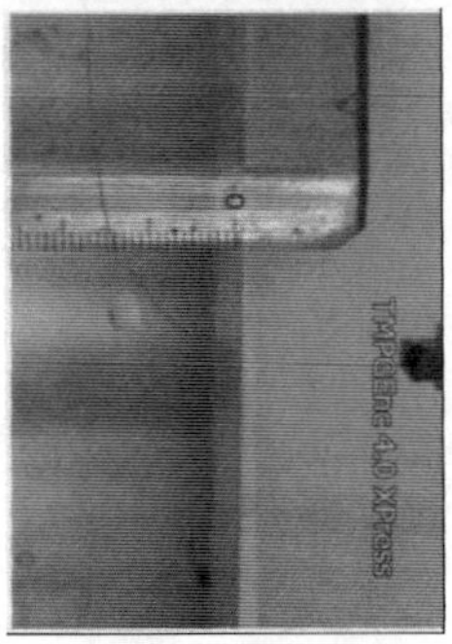
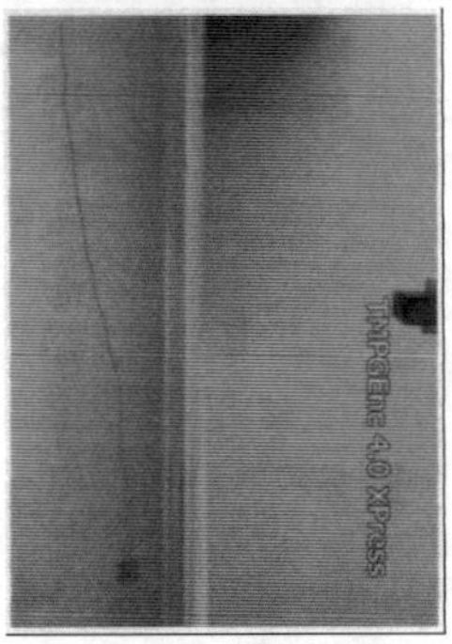

(c) s=25mm (B=4mT). (d) s=35mm (B=1mT).

Figure 3.Observed trace of droplet.

2.2. *Experimental results*

The terminal velocity of the droplet without the magnetic field is 122.5 ~ 171.8 mm/s depending on the diameter of the droplet and the concentration of the magnetic fluid, which is consistent with the velocity of the blood flow in the aorta, 100 ~ 200 mm/s.

The case of the droplet diameter of 2.6 mm and the concentration of 100 % is shown as a typical example: As shown in figure 3, when the magnetic field is strong, the droplet is accelerated toward the wall and is directly captured to the wall surface (figure 3(a)) or captured after sliding on the wall

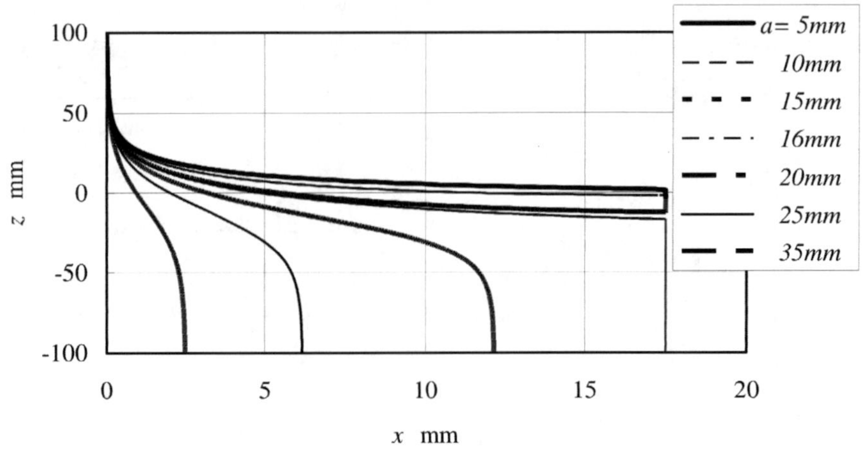

Figure 4.Estimated trace of droplet.

(figure 3(b)). If the magnetic field is weaker, the droplet is drawn to the wall surface but slides away on the wall without being stopped (figure 3(c)). If the magnetic field is weak enough, although the trace of the droplet is curved toward the wall, the droplet continues falling away (figure 3(d)).

3. Numerical simulation

The equation of motion of the droplet is numerically solved, assuming that the droplet is a solid sphere and the magnetic flux density B is curve-fitted as shown in figure 2. The magnetic force F_m which attracts the droplet toward the magnetic pole is represented as

$$F_m = \frac{2\chi B^2 V}{\mu_0 r}$$

where χ is the magnetization, V the volume of the droplet, μ_0 the permeability. r is the distance between the droplet and the magnetic point-pole, which is slightly under the surface of the finite-sized magnet. The depth is assumed to be the thickness of the magnet times sin(π/3). Then r is nearly equal to s in figure 2. The gravity force F_g reduced by the buoyancy is

$$F_g = V(\rho_m - \rho_w)g$$

where ρ_m is the density of the magnetic fluid, ρ_w the density of the water, g the acceleration of gravity.

The trace of the droplet of diameter of 2.6 mm and concentration of 100 % is shown in figure 4. x and z are the horizontal and vertical axes, respectively. The droplet is released at the center of the reservoir cross-section (x=0 mm), and the magnet is placed at z=0 mm. The wall of the reservoir is at x=17.5 mm. When the distance a between the magnetic pole and the wall is less than 15 mm, the droplet is captured to the wall surface. When a is 16-18 mm, the droplet is drawn to the wall surface but slides away on the wall without being stopped. When a is larger, the trace is only curved toward the wall but the droplet continues falling. This is consistent with the experimental results.

4. Condition of capture

Whether the droplet is captured or not depends on the balance between the gravity force F_g and the component of the magnetic force along the wall which is related to the magnetic force F_{mw} on the wall

$$F_{mw} = \frac{2\chi B_w^2 V}{\mu_0 a}$$

where B_w is the magnetic flux density on the wall and a the distance between the wall and the magnet.

Figure 5 show the experimental and simulated limits of capture. Solid symbols show the case that the droplet is captured and open symbols the case not captured in the experiments for each

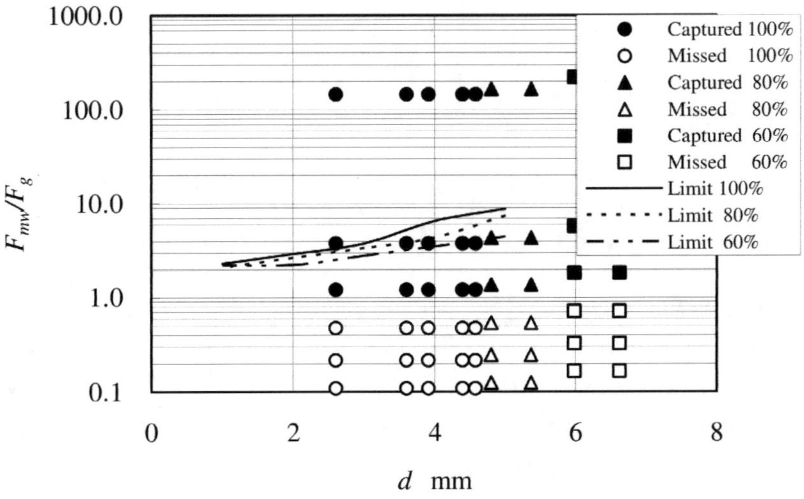

Figure 5.Condition of capture.

concentration. Static balance between the magnetic and gravitational forces shows that the limit of capture is $F_{mw}/F_g > 1.93$, which is consistent with the experimental results. Simulated limit is slightly higher due to neglect of the surface tension and inertia of the droplet. The size of the droplet has almost no effect on the limit, because the volume of the droplet V is contained in both of F_{mw} and F_g, and is cancelled. The concentration of the fluid α has little effect, because the magnetization in F_{mw} and the density of the fluid in F_g are represented as

$$\chi = \alpha \chi_o$$

$$\rho_m = \rho_k + \alpha(\rho_{m0} - \rho_k)$$

where χ_0 is the magnetization of original magnetic fluid, ρ_k the density of kerosene, ρ_{m0} the density of the original magnetic fluid. Then α is almost cancelled in F_{mw}/F_g.

5. Conclusions

(1) Motion of the magnetic droplet is classified according to the strength of the magnetic force into three types: The droplet is captured on the wall, is drawn to the wall but slides away, and the trace is only curved but the droplet continues falling.
(2) The condition of capture is approximately determined by the balance between the magnetic force and the gravity force and is almost not affected by the size and the concentration of the droplet.

Acknowledgement

Authors are deeply indebted to Messrs. T. Ebihara and S. Horiuchi for their contribution in the experiments.

References

[1] Rosensweig R E 1985 *Ferrohydrodynamics* Cambridge Univ. Press 4
[2] Voltairas P A, Fotiadis D I and Michalis L K 2002 *J Biomech.* **35** 813-821
[3] Pankhurst Q A *et al.* 2003 *J Phys D Appl Phys.* **36** R167-R181
[4] Yamane R, Sakiyama N and Park M K 2004 *Proc. 9th Int. Conf. ER Fluids and MR Suspensions* 659-665
[5] Ganguly R et al. 2005 *J. Magnet. Magnet. Mater.* **289** 331-334
[6] Alexiou Ch et al., Odenbach S (Ed.) 2002 *Ferrofluids, Lecture Notes in Physics* Springer 234-251

Measurement of the distribution parameters of size and magnetic properties of magnetic nanoparticles for medical applications

M. Zeisberger, S. Dutz, J. Lehnert, R. Müller
Institute of Photonic Technology, 07745 Jena, Albert-Einstein-Str.9, Germany
E-mail: zeisberger@ipht-jena.de

Abstract. Magnetic nanoparticles usually show a broad size distribution resulting in a corresponding distribution of the magnetic properties. One important goal of particle development is the reduction of the distribution width of the relevant quantities. In this contribution we show that for magnetic nanoparticles with sizes above the superparamagnetic range the measurement of remanence curves provides the required information. From these data we can obtain the remanence ratio and the switching field distribution which can be used as a quality criterion in the development of this type of magnetic nanoparticles.

1. Introduction

Magnetic nanoparticles (MNP) have been proposed for medical applications, e.g. hyperthermia [1] or drug-targeting [2]. As such applications require non-toxic material our work is focussed on the magnetic iron oxides γ-Fe_2O_3 and Fe_3O_4. The most common preparation technique of such particles is the wet-chemical preciptiation which usually provides MNP with a broad distribution of their size resulting in corresponding distributions of the magnetic properties. This is a disadvantage for applications as only a small fraction of the MNP in a given sample shows the optimal properties. The focus of our current work is the preparation of MNP with a sharper distribution of the relevant properties. There are two possible ways to achieve this goal. Firstly, the development of modified preparation methods [3], and secondly, a physical aftertreatment by magnetic separation or centrifugation. The development of such materials requires measurement methods which provide information on the statistical properties of the samples. For small, i.e. superparamagnetic particles this information can be extracted from the Langevin-like magnetization curve using the method of Chantrell [4]. Assuming a lognormal distribution of the MNP size the method provides the mean value and the distribution width parameter. From the size distribution one can calculate the distribution of the Néel relaxation time which determines the heating effect of superparamagnetic particles in an ac magnetic field. Our current work, however, is focussed on larger particles, which show a hysteresis. Here the Chantrell method is not suitable. Moreover, the relevant magnetic parameter is no longer the relaxation time but the anisotropy field H_K and its distribution. In addition, this type of material usually contains a certain fraction of superparamagnetic particles which is indicated by a reduced remanence ratio M_{rs}/M_s (saturation magnetization M_s and maximum remanent

Journal of Physics: Conference Series **149** (2009) 012115 doi:10.1088/1742-6596/149/1/012115

magnetization M_{rs}). In this contribution we demonstrate the characterization of large MNP on the basis of these quantities.

2. Remanence Measurements and Switching Field Distribution

The measurement of the initial remanence curve is performed in the following way. A small field H is applied to a demagnetized sample. Then the field is reduced to zero and the remanent magnetization is measured providing the first point $M_r(H)$ of the initial remanence curve. This procedure is repeated with successively increasing fields until the saturation remanence M_{rs} is reached. In addition, the saturation magnetization M_s is determined from a usual magnetization loop. The remanence curve contains information on the amount of particles which switch their magnetization irreversibly at the field H. Therefore differentiation provides the distribution of this switching field [5]

$$S(H) = \frac{1}{M_{rs}} \frac{\mathrm{d}M_r(H)}{\mathrm{d}H}$$

In the ideal case of monodisperse uniaxial MNP which are aligned parallel to the field $M_r(H)$ is 0 below H_K and equal to M_s above H_K resulting in a remanence ratio $M_{rs}/M_s = 1$ and $S(H) = \delta(H - H_K)$. In general $S(H)$ is broadened by two effects, firstly the statistical orientation of the particles, and secondly, the distribution of H_K. In order to distinguish between these effects $M_r(H)$ and $S(H)$ were simulated for monodisperse (i.e. the same H_K for all particles) statistically oriented samples consisting of particles with uniaxial or cubic anisotropy. For cubic particles two cases, positive or negative anisotropy constant, were regarded. The simulation is based on the minimization of the total energy of an ensemble of isotropically oriented particles. The energy constists of two contributions: the anisotropy energy and the energy of the particles in the external field. Fig. 1 and 2 show the results of the simulation. The switching field distributions were fitted by a lognormal distribution function

$$S(H) = \frac{1}{\sqrt{2\pi}\sigma H} \exp - \frac{\ln^2(H/H_m)}{2\sigma^2}$$

with the mean value H_m and the width σ. Tab. 1 shows the fit parameters. Iron oxide MNP are represented by type N in Tab. 1 if the shape anisotropy is negligable. The lognormal function

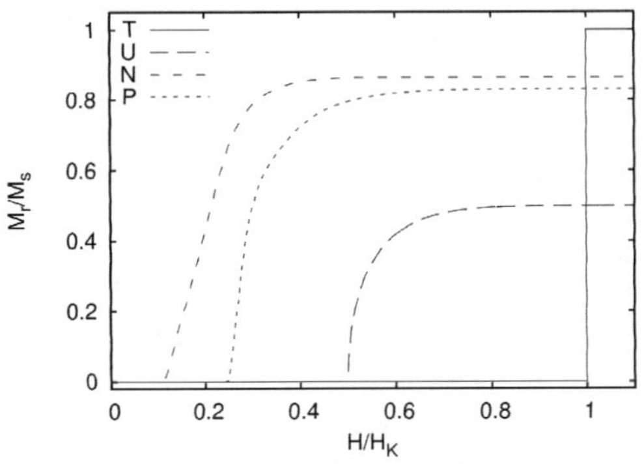

Figure 1. Simulated remanence curves for aligned uniaxial particles (T) and statistically oriented particles of uniaxial (U) or cubic anisotropy with positive (P) or negative (N) anisotropy constant

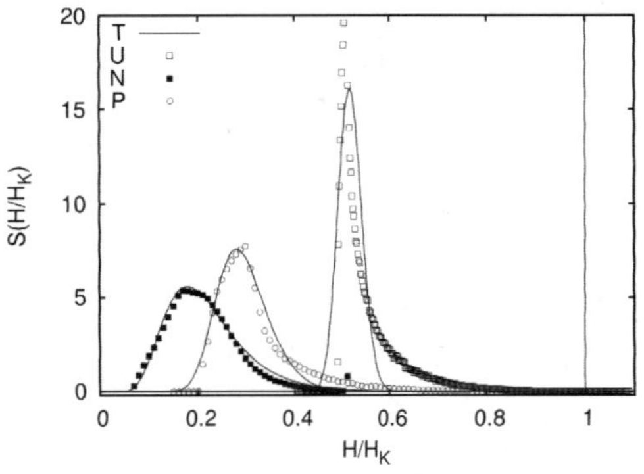

Figure 2. Switching field distributions (symbols) obtained from the simulated data (Fig. 1) and fits using a lognormal distribution function (curves)

Journal of Physics: Conference Series **149** (2009) 012115 doi:10.1088/1742-6596/149/1/012115

Table 1. Parameters of the simulated remanence curves and switching field distributions for aligned uniaxial particles (T) and statistically oriented particles of uniaxial (U) or cubic anisotropy with positive (P) or negative (N) anisotropy constant

	M_r/M_s	H_m/H_K	σ
T	1.00	1.00	0.00
U	0.50	0.52	0.05
P	0.83	0.29	0.18
N	0.86	0.20	0.35

Table 2. Parameters of the measured remanence curves and switching field distributions

sample	M_r/M_s	H_m/Oe	σ
S	0.007	52	0.60
C	0.064	135	0.67
L	0.042	142	0.60
M	0.272	161	0.51

does not fit to the simulated points very well. But we use it here for comparison with the experimental data where the lognormal distribution is suitable.

3. Experimental Investigations

In order to demonstrate the method we present remanence measurements of four iron oxide samples which were prepared by different methods. Three of them were made by alkaline preciptation and coated by carboxymethyl-dextrane. Sample S was prepared by fast precipitation with a strong base and consists mainly of superparamagnetic particles. Sample C was slowly precipitated with a weak base resulting in particle clusters [6]. For sample L a fast precipitation method with a weak base was used which leads to larger single domain particles. For comparison we investigated sample M which consits of magnetosomes [7], i.e. magnetite particles from magnetotactical bacteria. The remanence curves and the resulting switching field distributions as well as their lognormal fits are shown in Fig. 3 and 4. The remanence ratios and the distribution paramters are given in Tab. 2. The most significant difference between the samples is the remanence ratio. Even for the samples L and C which consist of relativly large particles (mean diameter 15 nm or 50 nm) it is one order of magnitude lower than the theoretical value (Tab. 1) indicating a large superparamagnetic fraction. Sample M is much closer to the theoretical values. The distribution width σ is similiar for all samples and differs from the theoretical value by less than a factor of 2. Also the mean values H_m are similiar except for sample S. These results show that a reduction of the superparamagnetic fraction is

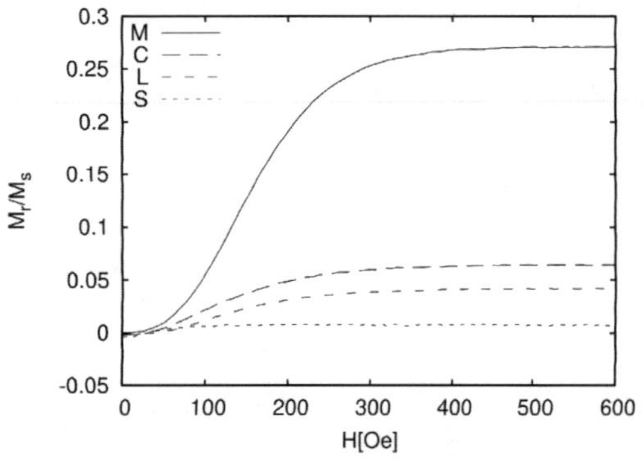

Figure 3. Measured remanence curves of the samples S, C, L, and M

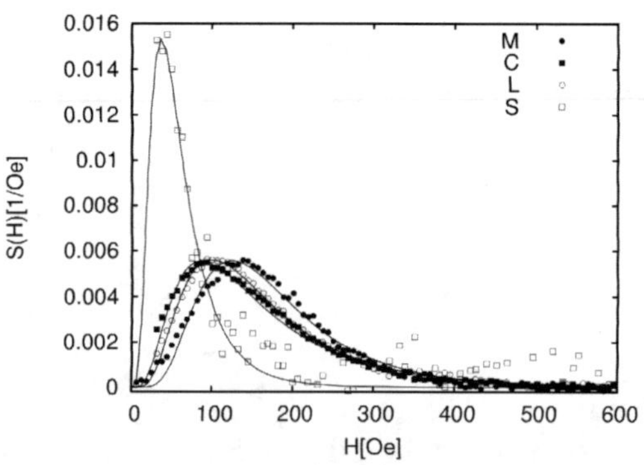

Figure 4. Switching field distributions of the samples S, C, L, and M

Table 3. Fractionation with 10 mm gap: O original sample, A inner fraction, B outer fraction, C final fraction

fraction	M_r/M_s
O	0.0074
A	0.0037
B	0.0041
C	0.0109

Table 4. Fractionation with 17 mm gap: O original sample, A inner fraction, B outer fraction, C final fraction

fraction	H_m/Oe	σ
O	36.6	0.70
A	34.9	0.65
B	33.5	0.69
C	40.4	0.69

the most promising way to improve the material. Nevertheless, the reduction of the distribution width has some potential as well.

4. Characterization of Magnetically Fractionated Samples

Experiments were performed in order to find the optimum parameters (field gradient, flow velocity, particle concentration) for magnetic fractionation of a given particle type. The fractionation set-up consists of a glass tube of 10 mm diameter and 200 mm length which is placed between two oppositely oriented permanent magnets (6 x 10 x 200 mm^3). After flowing through this quadrupol field the fluid is split into two fractions using concentric tubes at the end of the system. The inner part is fraction A, the outer one fraction B. A third fraction (C) is obtained at the end of the process from the last 10 ml of the fluid. The samples obtained by this method were characterized by remanence measurements. Tab. 3 and 4 show the results of two experiments with particles of type S and different gaps between the magnets where an effect of fractionation was observed, i.e. the fractions show increased or decreased values of the remanence ratio or the mean switching field. However, the effects are not very pronounced so far and further optimization of the process is necessary.

5. Conclusions

The measurement of remanence curves and the determination of the switching field distribution is a suitable method to characterize the statistical properties of large MNP samples. The most promising way of improvement of such samples is the reduction of the superparamagnetic fraction. The method was used to characterize samples obtained by magnetic fractionation. In two selected experiments a fractionation effect due to the remanence ratio or the mean switching field was shown.

Acknowledgments

The work was supported by the DFG within the project ZE 825/1-1. The authors thank D. Schüler (LMU München) for providing the magnetosomes.

References

[1] R. K. Gilchrist, R. Medal, W. D. Shorey, R. C. Hanselman, J. C. Parrot, C. B. Taylor, Ann. Surg. 146 (1957) 596
[2] Ch. Alexiou, R. Jurgons, in Magnetism in Medicine, Ed. W. Andrä, H. Nowak, Wiley-VCH, Weinheim 2007
[3] R. Müller, R. Hergt, S. Dutz, M. Zeisberger, W. Gawalek, J. Phys.: Condens. Matter, 18 (2006) S2527-2542
[4] R. W. Chantrell, J. Popplewell, S. W. Charles, IEEE Trans. Magn., 14 no 5 (1978), 975-977
[5] H. Pfeiffer, Phys. Stat. Sol. A 118 (1990) 295
[6] S. Dutz, W. Andrä, R. Hergt, R. Müller, Ch. Oestreich, Ch. Schmidt, J. Töpfer, M. Zeisberger, M.E. Bellemann, J. Mag. Mag. Mat., 311 (2007), 51-54
[7] D. Schüler, Int. Microbiol. 5 (2002), p. 209-214

Journal of Physics: Conference Series **149** (2009) 012116 doi:10.1088/1742-6596/149/1/012116

A contribution about ferrofluid based flow manipulation and locomotion systems

K Zimmermann[1], I Zeidis[1], V Böhm[1], J Popp[1]

TU Ilmenau, Fak. f. Maschinenbau, FG Technische Mechanik, Max-Planck-Ring 12, 98693 Ilmenau, Germany

E-mail: `klaus.zimmermann@tu-ilmenau.de`, `jana.popp@tu-ilmenau.de`

Abstract. With the background of developing apedal bionic inspired locomotion systems for future application fields like autonomous (swarm) robots, medical engineering and inspection systems, this article presents a selection of locomotion systems with bifluidic flow control using ferrofluid. By controlling the change of shape, position and pressure of the ferrofluid in a secondary low viscous fluid by magnetic fields locomotion of objects or the ferrofluid itself can be realised. The locomotion of an object is caused in the first example by a ferrofluid generated flow of the secondary fluid and in the second and third case by the direct alteration of the ferrofluid position.

1. Introduction

With the background of developing apedal bionic inspired locomotion systems [1] for future application fields like autonomous (swarm) robots, medical engineering and inspection systems, this article presents a selection of locomotion systems with bifluidic flow control using ferrofluid. While [2] examines the topic theoretically, the present paper covers in the main applicational matters. Some aspects of the working principles underlying the systems will be discussed, for the aim is extracting the general out of the specific.

2. Classification

According to the definition [3] locomotion is no mere motion but the motion of the centre of mass and of the contact surface to the surrounding medium within a considered period. One has further to differ between passive and active locomotion systems. Whereas the ferrofluid volume in a passive locomotion system is locally fixed and affects the locomotion of a second object or fluid by local (mainly periodic) pressure and shape alteration, the ferrofluid portion of an active locomotion system is part of the locomotive structure and changes shape and position either by a local or global change of pressure. In the following one of the presented systems can be classified as a passive locomotion system. The other two fit in the group of active locomotion systems. All three apparatuses avoid the problem of enclosing the ferrofluid, a highly staining material, by using a secondary fluid as cover and support.

3. Passive locomotion system

The development of ferrofluid based bifluidic pump systems is an actual research field [4][5]. Known approaches provide a spatially changing electromagnetic field, e.g. generated by a temporally shifted excitation of cascaded electromagnets, to cause time-dependent changes of pressure, shape and position of the ferrofluid. Hence, a travelling wave on the ferrofluid surface is generated so that a flow of the surrounding secondary fluid arises. These approaches have the disadvantage of great controlling effort and expenses. The presented system differs in its more simple construction, as the locomotion is generated by a periodically alternating magnetic field of a single electromagnet.

The device (Fig. 1) comprises either one or two separate ferrofluid portions, each locally fixed by a magnetic field of two opposite polarised permanent magnets below the vessel. The ferrofluid volume is set in a channel system filled with an second low viscous fluid.

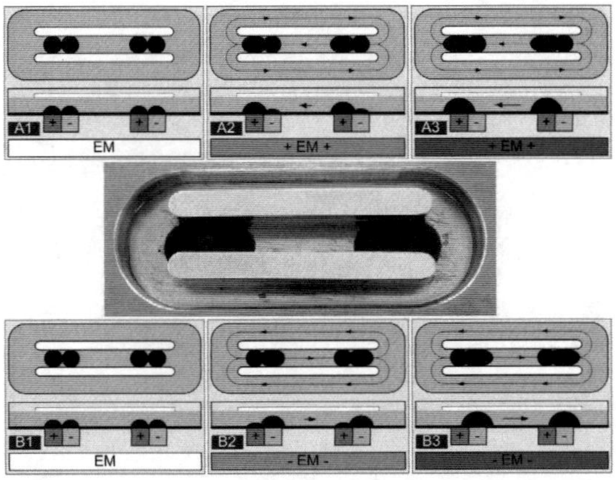

Figure 1. Passive locomotion in the form of flow generation
top (left to right): Increasing global electromagnetic field (+EM+) enhances permanent magnetic (PM(+)) field and suppresses PM(-), ferrofluid shifts accordingly.
bottom: Field direction of (-EM-) field changed polarisation, as (-EM-) increases, PM(-) field is supported and PM(+) diminished, corresponding ferrofluid switches.
Process turns alternatingly. Ferrofluid movement and channel asymmetry induce channel flow of surrounding secondary fluid.

Once exposed to a global, homogeneous, alternating polarised electromagnetic field, the field of the opposite polarised permanent magnets is affected - subdued or supported, respectively - and accordingly the ferrofluid surface changes rocker-like altitude and pressure periodically. A directional channel flow of the secondary fluid is generated by this mechanism, which causes the locomotion of an object on the surface of the secondary fluid. By the use of a single ferrofluid portion a unidirectional flow, in case of two separate ferrofluid portions a bidirectional flow can be accomplished.

The system is experimentally analysed by two methods, particle image velocimetry and recordings by high speed camera.

Flow measurements with particle image velocimetry have been employed to analyse the flow generation and velocity within an excitation period of the electromagnetic field and the subsequent motion of the ferrofluid surface. The flow characteristics are manipulable by excitation frequency, magnetic field strength and the active time of each excitation polarisation within a period. The aim of the measurements is to gain knowledge of the functional parameters firstly for a maximum flow rate and secondly for uniform flow characteristics.

Figure 2 gives an exemplarily screen shot of a vector plot at a fixed time. One can see the secondary fluid flow along the central channel (Fig. 1) in a sideways perspective. Apart from the very left - the flow in direct vicinity of the ferrofluid actuator - a repeatedly arched mode with a globally directed flow can be recognised. In the top layers a middle velocity up to 10 mm/sec can be registered.

Theoretical considerations concerning the interaction between ferrofluid contour and the surrounding fluid dependent of the time within a excitation period will be verified experimentally by high speed recordings.

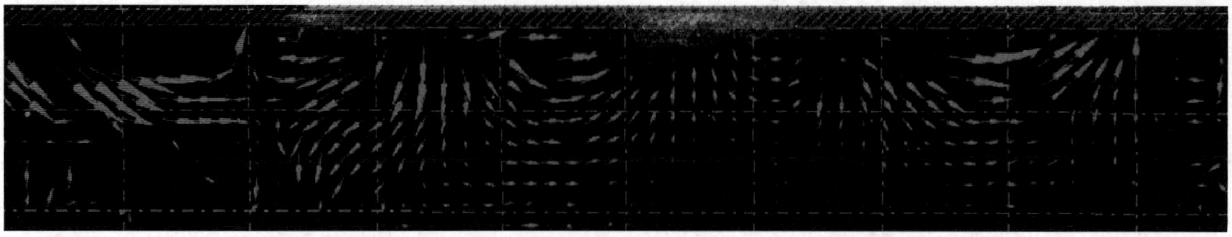

Figure 2. Screen shot of the secondary fluid flow as a vector plot, side wise perspective (at the dark left-side quarter: the ferrofluid portion invisibly)

4. Manipulation system

Figure 3 indicates the general idea of swarm robotics. The behavior of fusing and splitting of ferrofluid portions as locomotive single unit or cluster is to be analysed using an localised, time-dependent electromagnetic excitation array while the ferrofluid portions are suspended in a secondary low viscous fluid.
The flexible shape arrangement of the ferrofluid portions to an unit enables a form- and force-fit connection to one or more in the secondary fluid placed objects of complex contour. The following motion of the ferrofluid due to the change of the magnetic field provides controlled manipulation of the captured objects. The current settings allow to handle small objects up to 1 cm of size with available transporting forces of the range of approximately 0.1 N.

5. Active locomotion system

The last introduced example of a prototype (Fig. 4) refers mostly to bionic models. A single ferrofluid portion (in the range of 1.5 ml) moves uniaxial, bidirectional and actively in a low viscous carrier fluid. The locomotion is forced by a temporary, localised electromagnetic gradient field. A high velocity can be achieved, but in consideration of the controllability and the size of the vessel it should be limited to 10 to 15 millimeter per second.

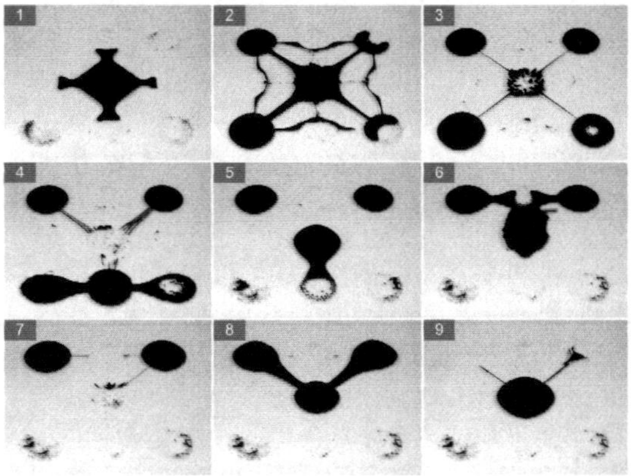

Figure 3. Manipulation of a ferrofluid cluster

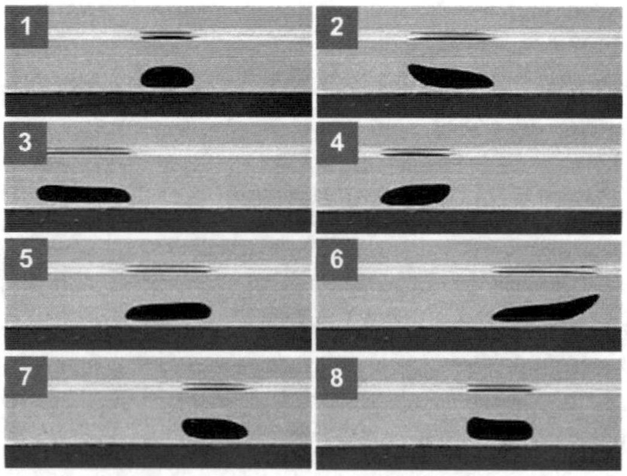

Figure 4. Active, uniaxial locomotion of a ferrofluid portion

6. Conclusion

In the article the aspect of ferrofluid based locomotion systems is given by the presentation of three bifluidic system for flow manipulation and locomotion of objects. While the description comprises the parametric function and its realisation by constructions, which do not afford high expenses and control-effort, the prospect of further experimental or theoretical investigations, respectively, is given in future papers.

Acknowledgments
The work has been supported by the German Science Foundation (DFG) under grant ZI 540/11-1 as well as by the Free State of Thuringia via graduation scholarship.

References
[1] Zimmermann K, Naletova V A, Böhm V, et al. 2007 *Verfahren und Vorrichtung zur Erzeugung einer apedalen translatorischen Bewegung* (Deutsches Patent- & Markenamt) DE 10 2006 059 537 B3

[2] Zimmermann K, Naletova V A, Zeidis I, Turkov V A, Kolev E, Lukashevich M V and Stepanov G V 2007 A deformable magnetizable worm in a magnetic field - a prototype of a mobile crawling robot *Journal of Magnetism and Magnetic Materials.* **311** pp 450-53

[3] Forth E and Schewitzer E 1976 *Meyers Taschenlexikon Bionik* (VEB Bibliographisches Institut Leipzig) pp 229-30

[4] Park G S and Seo K 2004 New design of the magnetic fluid linear pump to reduce the discontinuities of the pumping forces *IEEE Transactions on Magnetics* **40** No.2 pp 916-19

[5] Ando B, Ascia A, Baglio S and Pitrone N 2006 Developement of Novel Ferrofluidic Pumps *Proc. of the 28th IEEE EMBS Annual Int. Conf., New York* pp 2828-31

AUTHOR INDEX

AUTHOR INDEX

AUTHOR INDEX

AUTHOR INDEX

AUTHOR INDEX

Institute of Physics
Dirac House, Temple Back
Bristol BS1 6BE UK